Texts in Applied Mathematics 38

For other volumes published in this series, go to
www.springer.com/series/1214

Jean Gallier

Geometric Methods and Applications

For Computer Science and Engineering

Second Edition

 Springer

Jean Gallier
University of Pennsylvania
Department of Computer and Information Science
3330 Walnut Street
Philadelphia, PA 19104
USA
jean@cis.upenn.edu

Series Editors

S. Antman
Department of Mathematics
and
Institute for Physical Science
and Technology
University of Maryland
College Park, MD 20742-4015
USA
ssa@math.umd.edu

P. Holmes
Department of Mechanical and Aerospace
Engineering
Princeton University
215 Fine Hall
Princeton, NJ 08544
pholmes@math.princeton.edu

L. Sirovich
Laboratory of Applied Mathematics
Department of Biomathematical Sciences
Mount Sinai School of Medicine
New York, NY 10029-6574
lsirovich@rockefeller.edu

K.R. Sreenivasan
Department of Physics
New York University
70 Washington Square South
New York City, NY 10012
katepalli.sreenivasan@nyu.edu

ISSN 0939-2475
ISBN 978-1-4614-2824-4   ISBN 978-1-4419-9961-0 (eBook)
DOI 10.1007/978-1-4419-9961-0
Springer New York Dordrecht Heidelberg London

Mathematics Subject Classification (2010): 51-01, 52-01, 53-01, 57-01, 22-01, 49-01, 15-01

To my wife, Anne, my children, Mia, Philippe, and Sylvie, and my grandchildren, Bahari and Demetrius

Preface

This book is an introduction to fundamental geometric concepts and tools needed for solving problems of a geometric nature with a computer. Our main goal is to present a collection of tools that can be used to solve problems in computer vision, robotics, machine learning, computer graphics, and geometric modeling.

During the ten years following the publication of the first edition of this book, optimization techniques have made a huge comeback, especially in the fields of computer vision and machine learning. In particular, *convex optimization* and its special incarnation, *semidefinite programming (SDP)*, are now widely used techniques in computer vision and machine learning, as one may verify by looking at the proceedings of any conference in these fields. Therefore, we felt that it would be useful to include some material (especially on convex geometry) to prepare the reader for more comprehensive expositions of convex optimization, such as Boyd and Vandenberghe [2], a masterly and encyclopedic account of the subject. In particular, we added Chapter 7, which covers separating and supporting hyperplanes.

We also realized that the importance of the SVD (singular value decomposition) and of the pseudo-inverse had not been sufficiently stressed in the first edition of this book, and we rectified this situation in the second edition. In particular, we added sections on PCA (principal component analysis) and on best affine approximations and showed how they are efficienlty computed using SVD. We also added a section on quadratic optimization and a section on the Schur complement, showing the usefulness of the pseudo-inverse.

In this second edition, many typos and small mistakes have been corrected, some proofs have been shortened, some problems have been added, and some references have been added. Here is a list containing brief descriptions of the chapters that have been modified or added.

- Chapter 3, on the basic properties of convex sets, has been expanded. In particular, we state a version of Carathéodory's theorem for convex cones (Theorem 3.2), a version of Radon's theorem for pointed cones (Theorem 3.6), and Tverberg's theorem (Theorem 3.7), and we define centerpoints and prove their existence (Theorem 3.9).

- Chapter 7 is new. This chapter deals with separating hyperplanes, versions of Farkas's lemma, and supporting hyperplanes. Following Berger [1], various versions of the separation of open or closed convex subsets by hyperplanes are proved as consequences of a geometric version of the Hahn–Banach theorem (Theorem 7.1). We also show how various versions of Farkas's lemma (Lemmas 7.3, 7.4, and 7.5) can be easily deduced from separation results (Corollary 7.4 and Proposition 7.3). Farkas's lemma plays an important result in linear programming. Indeed, it can be used to give a quick proof of so-called strong duality in linear programming. We also prove the existence of supporting hyperplanes for boundary points of closed convex sets (Minkowski's lemma, Proposition 7.4). Unfortunately, lack of space prevents us from discussing polytopes and polyhedra. The reader will find a masterly exposition of these topics in Ziegler [3].

- Chapter 14 is a major revision of Chapter 13 (Applications of Euclidean Geometry to Various Optimization Problems) from the first edition of this book and has been renamed "Applications of SVD and Pseudo-Inverses." Section 14.1, about least squares problems, and the pseudo-inverse has not changed much, but we have added the fact that AA^+ is the orthogonal projection onto the range of A and that A^+A is the orthogonal projection onto $\mathrm{Ker}(A)^\perp$, the orthogonal complement of $\mathrm{Ker}(A)$. We have also added Proposition 14.1, which shows how the pseudo-inverse of a normal matrix A can be obtained from a block diagonalization of A (see Theorem 12.7). Sections 14.2, 14.3, and 14.4 are new.

  In Section 14.2, we define various matrix norms, including operator norms, and we prove Proposition 14.4, showing how a matrix can be best approximated by a rank-k matrix (in the $\| \ \|_2$ norm).

  Section 14.3 is devoted to principal component analysis (PCA). PCA is a very important statistical tool, yet in our experience, most presentations of this concept lack a crisp definition. Most presentations identify the notion of principal components with the result of applying SVD and do not prove why SVD does in fact yield the principal components and directions. To rectify this situation, we give a precise definition of PCAs (Definition 14.3), and we prove rigorously how SVD yields PCA (Theorem 14.3), using the Rayleigh–Ritz ratio (Lemma 14.2).

  In Section 14.4, it is shown how to best approximate a set of data with an affine subspace in the least squares sense. Again, SVD can used to find solutions.

- Chapter 15 is new, except for Section 15.1, which reproduces Section 13.2 from the first edition of this book. We added the definition of the positive semidefinite cone ordering, $\succeq$, on symmetric matrices, since it is extensively used in convex optimization.

  In Section 15.2, we find a necessary and sufficient condition (Proposition 15.2) for the quadratic function $f(x) = \frac{1}{2}x^\top Ax + x^\top b$ to have a minimum in terms of the pseudo-inverse of A (where A is a symmetric matrix). We also show how to accommodate linear constraints of the form $C^\top x = 0$ or affine constraints of the form $C^\top x = t$ (where $t \neq 0$).

  In Section 15.3, we consider the problem of maximizing $f(x) = x^\top Ax$ on the unit sphere $x^\top x = 1$ or, more generally, on the ellipsoid $x^\top Bx = 1$, where A is a symmetric matrix and B is symmetric, positive definite. We show that these problems are completely solved by diagonalizing A with respect to an orthogonal matrix. We also briefly consider the effect of adding linear constraints of the form $C^\top x = 0$ or affine constraints of the form $C^\top x = t$ (where $t \neq 0$).

- Chapter 16 is new. In this chapter, we define the notion of *Schur complement*, and we use it to characterize when a symmetric 2×2 block matrix is either positive semidefinite or positive definite (Proposition 16.1, Proposition 16.2, and Theorem 16.1).

- Chapter 17 is also brand new. In this chapter, we show how a computer vision problem, contour grouping, can be formulated as a quadratic optimization problem involving a Hermitian matrix. Because of the extra dependency on an angle, this optimization problem leads to finding the derivative of eigenvalues and eigenvectors of a normal matrix X. We derive explicit formulas for these derivatives (in the case of eigenvectors, the formula involves the pseudo-inverse of X) and we prove their correctness. It appears to be difficult to find these formulas together with a clean and correct proof in the literature. Our optimization problem leads naturally to the consideration of the *field of values* (or *numerical range*) $F(A)$ of a complex matrix A. A remarkable property of the field of values is that it is a convex subset of the plane, a theorem due to Toeplitz and Hausdorff, for which we give a short proof using a deformation argument (Theorem 17.1). Properties of the fields of values can be exploited to solve our optimization problem. This chapter describes current and exciting research in computer vision.

- Chapter 18 (which used to be Chapter 14 in the first edition) has been slightly expanded and improved. Our experience in teaching the material of this chapter, an introduction to manifolds and Lie groups, is that it is helpful to review carefully the notion of the derivative of a function $f : E \to F$ where E and F are normed vector spaces. Thus we added Section 18.7, which provides such a review. We also state the inverse function theorem and define immersions and submersions. Section 18.8 has also been slightly expanded. We added Proposition 18.6 and Theorem 18.7, which are often useful in proving that various spaces are manifolds; we defined critical and regular values and defined Morse functions; and we made a few cosmetic improvements in the paragraphs following Definition 18.20. A number of new problems on manifolds have been added.

- The only change to Chapter 19 (Chapter 15 in the first edition) is the inclusion of a more complete treatment of the *Frenet frame* for nD curves in Section 19.10.

- Similarly, the only change to Chapter 20 (Chapter 16 in the first edition) is the addition of Section 20.12, on *covariant derivatives* and the *parallel transport*.

Besides adding problems to all the chapters listed above we added one more problem to Chapter 2.

As in the first edition, there is some additional material on the web site `http://www.cis.upenn.edu/~jean/gbooks/geom2.html`

This material has not changed, and the chapter and section numbers are those of the first edition. A graph showing the dependencies of chapters is shown in Figure 0.1.

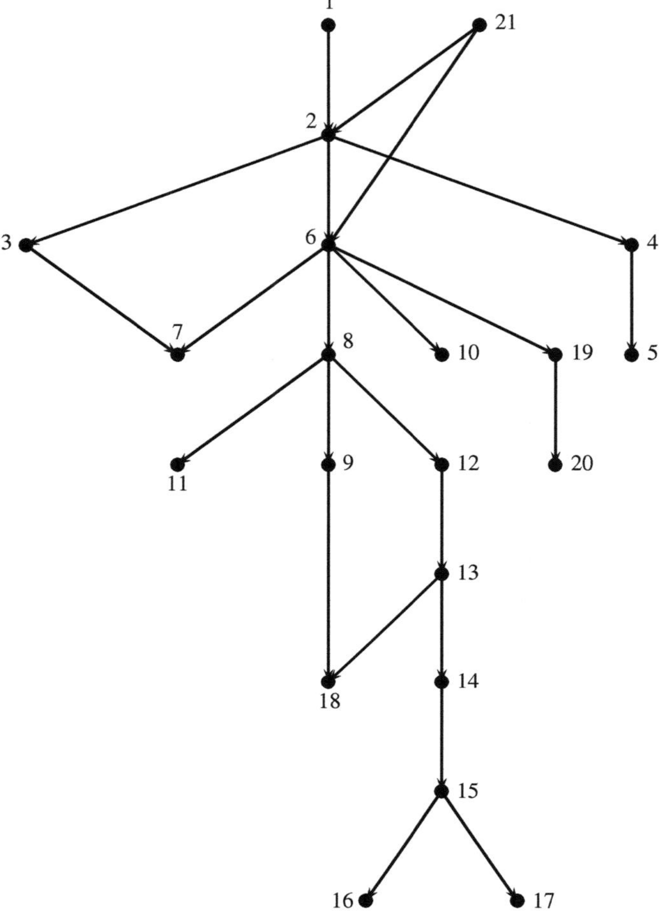

Fig. 0.1 Dependency of chapters.

Acknowledgments

Since the publication of the first edition of this book I have received valuable comments from Kostas Daniilidis, Marcelo Siqueira, Jianbo Shi, Ben Taskar, CJ Taylor, Mickey Brautbar, Katerina Fragiadaki, Ryan Kennedy, Oleg Naroditsky, and Weiyu Zhang. I also want to extend special thanks to David Kramer, who copyedited the first edition of this book over ten years ago, and did a superb job on this second edition.

References

1. Marcel Berger. *Géométrie 2*. Nathan, 1990. English edition: Geometry 2, Universitext, Springer-Verlag.
2. Stephen Boyd and Lieven Vandenberghe. *Convex Optimization*. Cambridge University Press, first edition, 2004.
3. Gunter Ziegler. *Lectures on Polytopes*. GTM No. 152. Springer Verlag, first edition, 1997.

Philadelphia, March 2011 *Jean Gallier*

Preface to the First Edition

Many problems arising in engineering, and notably in computer science and mechanical engineering, require geometric tools and concepts. This is especially true of problems arising in computer graphics, geometric modeling, computer vision, and motion planning, just to mention some key areas. This book is an introduction to fundamental geometric concepts and tools needed for solving problems of a geometric nature with a computer. In a previous text, Gallier [24], we focused mostly on affine geometry and on its applications to the design and representation of polynomial curves and surfaces (and B-splines). The main goal of this book is to provide an introduction to more sophisticated geometric concepts needed in tackling engineering problems of a geometric nature. Many problems in the above areas require some nontrivial geometric knowledge, but in our opinion, books dealing with the relevant geometric material are either too theoretical, or else rather specialized. For example, there are beautiful texts entirely devoted to projective geometry, Euclidean geometry, and differential geometry, but reading each one represents a considerable effort (certainly from a nonmathematician!). Furthermore, these topics are usually treated for their own sake (and glory), with little attention paid to applications.

This book is an attempt to fill this gap. We present a coherent view of geometric methods applicable to many engineering problems at a level that can be understood by a senior undergraduate with a good math background. Thus, this book should be of interest to a wide audience including computer scientists (both students and professionals), mathematicians, and engineers interested in geometric methods (for example, mechanical engineers). In particular, we provide an introduction to affine geometry, projective geometry, Euclidean geometry, basics of differential geometry and Lie groups, and a glimpse of computational geometry (convex sets, Voronoi diagrams, and Delaunay triangulations). This material provides the foundations for the algorithmic treatment of curves and surfaces, some basic tools of geometric modeling. The right dose of projective geometry also leads to a rigorous and yet smooth presentation of rational curves and surfaces. However, to keep the size of this book reasonable, a number of topics could not be included. Nevertheless, they can be found in the additional material on the web site: see `http://www.cis.`

`upenn.edu/~jean/gbooks/geom2.html`. This is the case of the material on rational curves and surfaces.

This book consists of sixteen chapters and an appendix. The additional material on the web site consists of eight chapters and an appendix: see `http://www.cis.upenn.edu/~jean/gbooks/geom2.html`.

- The book starts with a brief introduction (Chapter 1).
- Chapter 2 provides an introduction to affine geometry. This ensures that readers are on firm ground to proceed with the rest of the book, in particular, projective geometry. This is also useful to establish the notation and terminology. Readers proficient in geometry may omit this section, or use it *as needed*. On the other hand, readers totally unfamiliar with this material will probably have a hard time with the rest of the book. These readers are advised do some extra reading in order to assimilate some basic knowledge of geometry. For example, we highly recommend Pedoe [42], Coxeter [9], Snapper and Troyer [52], Berger [2, 3], Fresnel [22], Samuel [51], Hilbert and Cohn–Vossen [31], Boehm and Prautzsch [5], and Tisseron [54].
- Basic properties of convex sets and convex hulls are discussed in Chapter 3. Three major theorems are proved: Carthéodory's theorem, Radon's theorem, and Helly's theorem.
- Chapter 4 presents a construction (the "hat construction") for embedding an affine space into a vector space. An important application of this construction is the projective completion of an affine space, presented in the next chapter. Other applications are treated in Chapter 20 on the web site, see `http://www.cis.upenn.edu/~jean/gbooks/geom2.html`.
- Chapter 5 provides an introduction to projective geometry. Since we are not writing a treatise on projective geometry, we cover only the most fundamental concepts, including projective spaces and subspaces, frames, projective maps, multiprojective maps, the projective completion of an affine space, cross-ratios, duality, and the complexification of a real projective space. This material also provides the foundations for our algorithmic treatment of rational curves and surfaces, to be found on the web site (Chapters 18, 19, 21, 22, 23, 24); see `http://www.cis.upenn.edu/~jean/gbooks/geom2.html`.
- Chapters 6, 8, and 9, provide an introduction to Euclidean geometry, to the groups of isometries $\mathbf{O}(n)$ and $\mathbf{SO}(n)$, the groups of affine rigid motions $\mathbf{Is}(n)$ and $\mathbf{SE}(n)$, and to the quaternions. Several versions of the Cartan–Dieudonné theorem are proved in Chapter 8. The QR-decomposition of matrices is explained geometrically, both in terms of the Gram–Schmidt procedure and in terms of Householder transformations. These chapters are crucial to a firm understanding of the differential geometry of curves and surfaces, and computational geometry.
- Chapter 10 gives a short introduction to some fundamental topics in computational geometry: Voronoi diagrams and Delaunay triangulations.
- Chapter 11 provides an introduction to Hermitian geometry, to the groups of isometries $\mathbf{U}(n)$ and $\mathbf{SU}(n)$, and the groups of affine rigid motions $\mathbf{Is}(n,\mathbb{C})$ and $\mathbf{SE}(n,\mathbb{C})$. The generalization of the Cartan–Dieudonné theorem to Hermitian spaces can be found on the web site, Chapter 25; see `http://www.`

`cis.upenn.edu/˜jean/gbooks/geom2.html`. A short introduction to Hilbert spaces, including the projection theorem, and the isomorphism of every Hilbert space with some space $l^2(K)$, can also be found on the web site in Chapter 26, see `http://www.cis.upenn.edu/˜jean/gbooks/geom2.html`.

- Chapter 12 provides a presentation of the spectral theorems in Euclidean and Hermitian spaces, including normal, self-adjoint, skew self-adjoint, and orthogonal linear maps. Normal form (in terms of block diagonal matrices) for various types of linear maps are presented.
- The singular value decomposition (SVD) and the polar form of linear maps are discussed quite extensively in Chapter 13. The pseudo-inverse of a matrix and its characterization using the Penrose properties are presented.
- Chapter 14 presents some applications of Euclidean geometry to various optimization problems. The method of least squares is presented, as well as the applications of the SVD and QR-decomposition to solve least squares problems. We also describe a method for minimizing positive definite quadratic forms, using Lagrange multipliers.
- Chapter 18 provides an introduction to the linear Lie groups, via a presentation of some of the classical groups and their Lie algebras, using the exponential map. The surjectivity of the exponential map is proved for $\mathbf{SO}(n)$ and $\mathbf{SE}(n)$.
- An introduction to the local differential geometry of curves is given in Chapter 19 (curvature, torsion, the Frenet frame, etc).
- An introduction to the local differential geometry of surfaces based on some lectures by Eugenio Calabi is given in Chapter 20. This chapter is rather unique, as it reflects decades of experience from a very distinguished geometer.
- Chapter 21 is an appendix consisting of short sections consisting of basics of linear algebra and analysis. This chapter has been included to make the material self-contained. Our advice is to use it *as needed*!

A very elegant presentation of rational curves and surfaces can be given using some notions of affine and projective geometry. We push this approach quite far in the material on the web; see `http://www.cis.upenn.edu/˜jean/gbooks/geom2.html`. However, we provide only a cursory coverage of CAGD methods. Luckily, there are excellent texts on CAGD, including Bartels, Beatty, and Barsky [1], Farin [17, 18], Fiorot and Jeannin [20, 21], Riesler [50], Hoschek and Lasser [33], and Piegl and Tiller [43]. Although we cover affine, projective, and Euclidean geometry in some detail, we are far from giving a comprehensive treatment of these topics. For such a treatment, we highly recommend Berger [2, 3], Samuel [51], Pedoe [42], Coxeter [11, 10, 8, 9], Snapper and Troyer [52], Fresnel [22], Tisseron [54], Sidler [45], Dieudonné [13], and Veblen and Young [57, 58], a great classic.

Similarly, although we present some basics of differential geometry and Lie groups, we only scratch the surface. For instance, we refrain from discussing manifolds in full generality. We hope that our presentation is a good preparation for more advanced texts, such as Gray [27], do Carmo [14], Berger and Gostiaux [4], and Lafontaine [36]. The above are still fairly elementary. More advanced texts on differential geometry include do Carmo [15, 16], Guillemin and Pollack [29], Warner

[59], Lang [37], Boothby [6], Lehmann and Sacré [38], Stoker [53], Gallot, Hulin, and Lafontaine [25], Milnor [41], Sharpe [44], Malliavin [39], and Godbillon [26].

It is often possible to reduce interpolation problems involving polynomial curves or surfaces to solving systems of linear equations. Thus, it is very helpful to be aware of efficient methods for numerical matrix analysis. For instance, we present the QR-decomposition of matrices, both in terms of the (modified) Gram–Schmidt method and in terms of Householder transformations, in a novel geometric fashion. For further information on these topics, readers are referred to the excellent texts by Strang [48], Golub and Van Loan [28], Trefethen and Bau [55], Ciarlet [7], and Kincaid and Cheney [34]. Strang's beautiful book on applied mathematics is also highly recommended as a general reference [46]. There are other interesting applications of geometry to computer vision, computer graphics, and solid modeling. Some good references are Trucco and Verri [56], Koenderink [35], and Faugeras [19] for computer vision; Hoffman [32] for solid modeling; and Metaxas [40] for physics-based deformable models.

Novelties

As far as we know, there is no fully developed modern exposition integrating the basic concepts of affine geometry, projective geometry, Euclidean geometry, Hermitian geometry, basics of Hilbert spaces with a touch of Fourier series, basics of Lie groups and Lie algebras, as well as a presentation of curves and surfaces both from the standard differential point of view and from the algorithmic point of view in terms of control points (in the polynomial and rational case).

New Treatment, New Results

This books provides an introduction to affine geometry, projective geometry, Euclidean geometry, Hermitian geometry, Hilbert spaces, a glimpse at Lie groups and Lie algebras, and the basics of local differential geometry of curves and surfaces. We also cover some classics of convex geometry, such as Carathéodory's theorem, Radon's theorem, and Helly's theorem. However, in order to help the reader assimilate all these concepts with the least amount of pain, we begin with some basic notions of affine geometry in Chapter 2. Basic notions of Euclidean geometry come later only in Chapters 6, 8, 9. Generally, noncore material is relegated to appendices or to the web site: see `http://www.cis.upenn.edu/~jean/gbooks/geom2.html`.

We cover the standard local differential properties of curves and surfaces at an elementary level, but also provide an in-depth presentation of polynomial and rational curves and surfaces from an algorithmic point of view. The approach (sometimes called *blossoming*) consists in multilinearizing everything in sight (getting *polar forms*), which leads very naturally to a presentation of polynomial and rational curves and surfaces in terms of control points (Bézier curves and surfaces). We present many algorithms for subdividing and drawing curves and surfaces, all implemented in *Mathematica*. A clean and elegant presentation of control points with weights (and control vectors) is obtained by using a construction for embedding an affine space into a vector space (the so-called "hat construction," originating in Berger [2]). We also give several new methods for drawing efficiently closed rational curves and surfaces, and a method for resolving base points of triangular rational surfaces. We give a quick introduction to the concepts of Voronoi diagrams and Delaunay triangulations, two of the most fundamental concepts in computational geometry. As a general rule, we try to be rigorous, but we always keep the algorithmic nature of the mathematical objects under consideration in the forefront.

Many problems and programming projects are proposed (over 230). Some are routine, some are (very) difficult.

Applications

Although it is core mathematics, geometry has many practical applications. Whenever possible, we point out some of these applications, For example, we mention some (perhaps unexpected) applications of projective geometry to computer vision (camera calibration), efficient communication, error correcting codes, and cryptography (see Section 5.13). As applications of Euclidean geometry, we mention motion interpolation, various normal forms of matrices including QR-decomposition in terms of Householder transformations and SVD, least squares problems (see Section 14.1), and the minimization of quadratic functions using Lagrange multipliers (see Section 15.1). Lie groups and Lie algebras have applications in robot kinematics, motion interpolation, and optimal control. They also have applications in physics. As applications of the differential geometry of curves and surfaces, we mention geometric continuity for splines, and variational curve and surface design (see Section 19.11 and Section 20.13). Finally, as applications of Voronoi diagrams and Delaunay triangulations, we mention the nearest neighbors problem, the largest empty circle problem, the minimum spanning tree problem, and motion planning (see Section 10.5). Of course, rational curves and surfaces have many applications to computer-aided geometric design (CAGD), manufacturing, computer graphics, and robotics.

Many Algorithms and Their Implementation

Although one of our main concerns is to be mathematically rigorous, which implies that we give precise definitions and prove almost all of the results in this book, we are primarily interested in the representation and the implementation of concepts and tools used to solve geometric problems. Thus, we devote a great deal of efforts to the development and implemention of algorithms to manipulate curves, surfaces, triangulations, etc. As a matter of fact, we provide *Mathematica* code for most of the geometric algorithms presented in this book. We also urge the reader to write his own algorithms, and we propose many challenging programming projects.

Open Problems

Not only do we present standard material (although sometimes from a fresh point of view), but whenever possible, we state some open problems, thus taking the reader to the cutting edge of the field. For example, we describe very clearly the problem of resolving base points of rectangular rational surfaces (this material is on the web site, see `http://www.cis.upenn.edu/~jean/gbooks/geom2.html`).

What's Not Covered in This Book

Since this book is already quite long, we have omitted solid modeling techniques, methods for rendering implicit curves and surfaces, the finite elements method, and wavelets. The first two topics are nicely covered in Hoffman [32], and the finite element method is the subject of so many books that we will not attempt to mention any references besides Strang and Fix [47]. As to wavelets, we highly recommend the classics by Daubechies [12], and Strang and Truong [49], among the many texts on this subject. It would also have been nice to include chapters on the algebraic geometry of curves and surfaces. However, this is a very difficult subject that requires a lot of algebraic machinery. Interested readers may consult Fulton [23] or Harris [30].

How to Use This Book for a Course

This books covers three complementary but fairly disjoint topics:

(1) Projective geometry and its applications to rational curves and surfaces (Chapter 5, and on the web page, Chapters 18, 19, 21, 22, 23, 24);
(2) Euclidean geometry, Voronoi diagrams, and Delaunay triangulations, Hermitian geometry, basics of Hilbert spaces, spectral theorems for special kinds of linear maps, SVD, polar form, and basics of Lie groups and Lie algebras (Chapters 6, 8, 9, 10, 11, 12, 13, 14, 18);
(3) Basics of the differential geometry of curves and surfaces (Chapters 19 and 20).

Chapter 21 is an appendix consisting of background material and should be used only *as needed*.

Our experience is that there is too much material to cover in a one–semester course. The ideal situation is to teach the material in the entire book in two semesters. Otherwise, a more algebraically inclined teacher should teach the first or second topic, whereas a more differential-geometrically inclined teacher should teach the third topic. In either case, Chapter 2 on affine geometry should be covered. Chapter 4 is required for the first topic, but not for the second.

Problems are found at the end of each chapter. They range from routine to very difficult. Some programming assignments have been included. They are often quite open-ended, and may require a considerable amount of work. The end of a proof is indicated by a square box ($\square$). The word *iff* is an abbreviation for *if and only if*. References to the web page `http://www.cis.upenn.edu/~jean/gbooks/geom2.html` will be abbreviated as web page.

Hermann Weyl made the following comment in the preface (1938) of his beautiful book [60]:

> The gods have imposed upon my writing the yoke of a foreign tongue that was not sung at my cradle .... Nobody is more aware than myself of the attendant loss in vigor, ease and lucidity of expression.

Being in a similar position, I hope that I was at least successful in conveying my enthusiasm and passion for geometry, and that I have inspired my readers to study some of the books that I respect and admire.

Acknowledgments

This book grew out of lectures notes that I have written as I have been teaching CIS610, *Advanced Geometric Methods in Computer Science*, for the past two years. Many thanks to the copyeditor, David Kramer, who did a superb job. I also wish to thank some students and colleagues for their comments, including Koji Ashida, Doug DeCarlo, Jaydev Desai, Will Dickinson, Charles Erignac, Steve Frye, Edith Haber, Andy Hicks, Paul Hughett, David Jelinek, Marcus Khuri, Hartmut Liefke, Shih-Schon Lin, Ying Liu, Nilesh Mankame, Dimitris Metaxas, Viorel Mihalef, Albert Montillo, Youg-jin Park, Harold Sun, Deepak Tolani, Dianna Xu, and Hui Zhang. Also thanks to Norm Badler for triggering my interest in geometric modeling, and to Marcel Berger, Chris Croke, Ron Donagi, Herman Gluck, David Harbater, Alexandre Kirillov, and Steve Shatz for sharing some of their geometric secrets with me. Finally, many thanks to Eugenio Calabi for teaching me what I know about differential geometry (and much more!). I am very grateful to Professor Calabi for allowing me to write up his lectures on the differential geometry of curves and surfaces given in an undergraduate course in Fall 1994 (as Chapter 20).

References

1. Richard H. Bartels, John C. Beatty, and Brian A. Barsky. *An Introduction to Splines for Use in Computer Graphics and Geometric Modelling*. Morgan Kaufmann, first edition, 1987.
2. Marcel Berger. *Géométrie 1*. Nathan, 1990. English edition: Geometry 1, Universitext, Springer-Verlag.
3. Marcel Berger. *Géométrie 2*. Nathan, 1990. English edition: Geometry 2, Universitext, Springer-Verlag.
4. Marcel Berger and Bernard Gostiaux. *Géométrie différentielle: variétés, courbes et surfaces*. Collection Mathématiques. Puf, second edition, 1992. English edition: Differential geometry, manifolds, curves, and surfaces, GTM No. 115, Springer-Verlag.
5. W. Boehm and H. Prautzsch. *Geometric Concepts for Geometric Design*. AK Peters, first edition, 1994.
6. William M. Boothby. *An Introduction to Differentiable Manifolds and Riemannian Geometry*. Academic Press, second edition, 1986.
7. P.G. Ciarlet. *Introduction to Numerical Matrix Analysis and Optimization*. Cambridge University Press, first edition, 1989. French edition: Masson, 1994.
8. H.S.M. Coxeter. *Non-Euclidean Geometry*. The University of Toronto Press, first edition, 1942.
9. H.S.M. Coxeter. *Introduction to Geometry*. Wiley, second edition, 1989.
10. H.S.M. Coxeter. *The Real Projective Plane*. Springer-Verlag, third edition, 1993.
11. H.S.M. Coxeter. *Projective Geometry*. Springer-Verlag, second edition, 1994.
12. Ingrid Daubechies. *Ten Lectures on Wavelets*. SIAM Publications, first edition, 1992.
13. Jean Dieudonné. *Algèbre Linéaire et Géométrie Elémentaire*. Hermann, second edition, 1965.
14. Manfredo P. do Carmo. *Differential Geometry of Curves and Surfaces*. Prentice-Hall, 1976.
15. Manfredo P. do Carmo. *Riemannian Geometry*. Birkhäuser, second edition, 1992.
16. Manfredo P. do Carmo. *Differential Forms and Applications*. Universitext. Springer-Verlag, first edition, 1994.
17. Gerald Farin. *Curves and Surfaces for CAGD*. Academic Press, fourth edition, 1998.
18. Gerald Farin. *NURB Curves and Surfaces, from Projective Geometry to Practical Use*. AK Peters, first edition, 1995.

19. Olivier Faugeras. *Three-Dimensional Computer Vision, A Geometric Viewpoint*. MIT Press, first edition, 1996.

20. J.-C. Fiorot and P. Jeannin. *Courbes et Surfaces Rationelles*. RMA 12. Masson, first edition, 1989.

21. J.-C. Fiorot and P. Jeannin. *Courbes Splines Rationelles*. RMA 24. Masson, first edition, 1992.

22. Jean Fresnel. *Méthodes Modernes en Géométrie*. Hermann, first edition, 1998.

23. William Fulton. *Algebraic Curves*. Advanced Book Classics. Addison-Wesley, first edition, 1989.

24. Jean H. Gallier. *Curves and Surfaces in Geometric Modeling: Theory and Algorithms*. Morgan Kaufmann, first edition, 1999.

25. S. Gallot, D. Hulin, and J. Lafontaine. *Riemannian Geometry*. Universitext. Springer-Verlag, second edition, 1993.

26. Claude Godbillon. *Géométrie Différentielle et Mécanique Analytique*. Collection Méthodes. Hermann, first edition, 1969.

27. A. Gray. *Modern Differential Geometry of Curves and Surfaces*. CRC Press, second edition, 1997.

28. Gene H. Golub and Charles F. Van Loan. *Matrix Computations*. The Johns Hopkins University Press, third edition, 1996.

29. Victor Guillemin and Alan Pollack. *Differential Topology*. Prentice-Hall, first edition, 1974.

30. Joe Harris. *Algebraic Geometry, A First Course*. GTM No. 133. Springer-Verlag, first edition, 1992.

31. D. Hilbert and S. Cohn-Vossen. *Geometry and the Imagination*. Chelsea Publishing Co., 1952.

32. Christoph M. Hoffman. *Geometric and Solid Modeling*. Morgan Kaufmann, first edition, 1989.

33. J. Hoschek and D. Lasser. *Computer-Aided Geometric Design*. AK Peters, first edition, 1993.

34. D. Kincaid and W. Cheney. *Numerical Analysis*. Brooks/Cole Publishing, second edition, 1996.

35. Jan J. Koenderink. *Solid Shape*. MIT Press, first edition, 1990.

36. Jacques Lafontaine. *Introduction aux Variétés Différentielles*. PUG, first edition, 1996.

37. Serge Lang. *Differential and Riemannian Manifolds*. GTM No. 160. Springer-Verlag, third edition, 1995.

38. Daniel Lehmann and Carlos Sacré. *Géométrie et Topologie des Surfaces*. Puf, first edition, 1982.

39. Paul Malliavin. *Géométrie Différentielle Intrinsèque*. Enseignement des Sciences, No. 14. Hermann, first edition, 1972.

40. Dimitris N. Metaxas. *Physics-Based Deformable Models*. Kluwer Academic Publishers, first edition, 1997.

41. John W. Milnor. *Topology from the Differentiable Viewpoint*. The University Press of Virginia, second edition, 1969.

42. Dan Pedoe. *Geometry, A Comprehensive Course*. Dover, first edition, 1988.

43. Les Piegl and Wayne Tiller. *The NURBS Book*. Monograph in Visual Communications. Springer-Verlag, first edition, 1995.

44. Richard W. Sharpe. *Differential Geometry. Cartan's Generalization of Klein's Erlangen Program*. GTM No. 166. Springer-Verlag, first edition, 1997.

45. J.-C. Sidler. *Géométrie Projective*. InterEditions, first edition, 1993.

46. Gilbert Strang. *Introduction to Applied Mathematics*. Wellesley–Cambridge Press, first edition, 1986.

47. Gilbert Strang and Fix George. *An Analysis of the Finite Element Method*. Wellesley–Cambridge Press, first edition, 1973.

48. Gilbert Strang. *Linear Algebra and Its Applications*. Saunders HBJ, third edition, 1988.

49. Gilbert Strang and Nguyen Truong. *Wavelets and Filter Banks*. Wellesley–Cambridge Press, second edition, 1997.

50. J.-J. Risler. *Mathematical Methods for CAD*. Masson, first edition, 1992.

51. Pierre Samuel. *Projective Geometry*. Undergraduate Texts in Mathematics. Springer-Verlag, first edition, 1988.

52. Ernst Snapper and Troyer Robert J. *Metric Affine Geometry*. Dover, first edition, 1989.

53. J.J. Stoker. *Differential Geometry*. Wiley Classics. Wiley-Interscience, first edition, 1989.

54. Claude Tisseron. *Géométries Affines, Projectives, et Euclidiennes*. Hermann, first edition, 1994.

55. L.N. Trefethen and D. Bau III. *Numerical Linear Algebra*. SIAM Publications, first edition, 1997.

56. Emanuele Trucco and Alessandro Verri. *Introductory Techniques for 3D Computer Vision*. Prentice-Hall, first edition, 1998.

57. O. Veblen and J. W. Young. *Projective Geometry, Vol. 1*. Ginn, second edition, 1938.

58. O. Veblen and J. W. Young. *Projective Geometry, Vol. 2*. Ginn, first edition, 1946.

59. Frank Warner. *Foundations of Differentiable Manifolds and Lie Groups*. GTM No. 94. Springer-Verlag, first edition, 1983.

60. Hermann Weyl. *The Classical Groups. Their Invariants and Representations*. Princeton Mathematical Series, No. 1. Princeton University Press, second edition, 1946.

Philadelphia   *Jean Gallier*

Contents

Chapter 1
Introduction

Je ne crois donc pas avoir fait une œuvre inutile en écrivant le présent Mémoire; je regette seulement qu'il soit trop long; mais quand j'ai voulu me restreindre, je suis tombé dans l'obscurité; j'ai préféré passer pour un peu bavard.
—**Henri Poincaré,** *Analysis Situ,* 1895

1.1 Geometries: Their Origin, Their Uses

What is geometry? According to Veblen and Young [8], geometry deals with the properties of figures in space. Etymologically, geometry means the practical science of measurement. No wonder geometry plays a fundamental role in mathematics, physics, astronomy, and engineering. Historically, as explained in more detail by Coxeter [1], geometry was studied in Egypt about 2000 B.C. Then, it was brought to Greece by Thales (640–456 B.C.). Thales also began the process of abstracting positions and straight edges as points and lines, and studying incidence properties. This line of work was greatly developed by Pythagoras and his disciples, among which we should distinguish Hippocrates. Indeed, Hippocrates attempted a presentation of geometry in terms of logical deductions from a few definitions and assumptions. But it was Euclid (about 300 B.C.) who made fundamental contributions to geometry, recorded in his immortal *Elements*, one of the most widely read books in the world.

Euclid's basic assumptions consist of basic notions concerning magnitudes, and five postulates. Euclid's fifth postulate, sometimes called the "parallel postulate," is historically very significant. It prompted mathematicians to question the traditional foundations of geometry, and led them to realize that there are different kinds of geometries. The fifth postulate can be stated in the following way:

V. *If a straight line meets two other straight lines, so as to make the two interior angles on one side of it together less than two right angles, the other straight lines will meet if produced on that side on which the angles are less than two right angles.*

Euclid's fifth postulate is definitely not self-evident. It is also not simple or natural, and after Euclid, many people tried to deduce it from the other postulates. However, they succeeded only in replacing it by various equivalent assumptions, of which we only mention two:

V′. *Two parallel lines are equidistant.* (Posidonius, first century B.C.).
V″. *The sum of the angles of a triangle is equal to two right angles.* (Legendre, 1752–1833).

According to Euclid, two lines are parallel if they are coplanar without intersecting.

It is remarkable that until the eighteenth century, no serious attempts at proving or disproving Euclid's fifth postulate were made. Saccheri (1667–1733) and Lambert (1728–1777) attempted to prove Euclid's fifth postulate, but of course, this was impossible. This was shown by Lobachevsky (1793–1856) and Bolyai (1802–1860), who proposed some models of non-Euclidean geometries. Actually, Gauss (1777–1855) was the first to consider seriously the possibility that a geometry denying Euclid's fifth postulate was of some interest. However, this was such a preposterous idea in those days that he kept these ideas to himself until others had published them independently.

Thus, circa the 1830s, it was finally realized that there is not just one geometry, but *different kinds of geometries* (spherical, hyperbolic, elliptic). The next big step was taken by Riemann, (1826–1866) who introduced the "infinitesimal approach" to geometry, wherein the differential of distance is expressed as the square root of the sum of the squares of the differentials of the coordinates. Riemann studied spherical spaces of higher dimension, and showed that their geometry is non-Euclidean. Finally, Cayley (1821–1895) and especially Klein (1849–1925) reached a clear understanding of the various geometries and their relationships. Basically, all geometries can be viewed as embedded in a universal geometry, *projective geometry*. Projective geometry itself is non-Euclidean, since two coplanar lines always intersect in a single point.

Projective geometry was developed in the nineteenth century, mostly by Monge, Poncelet, Chasles, Steiner, and Von Staudt (but anticipated by Kepler (1571–1630) and Desargues (1593–1662)). Klein also realized that "a geometry" can be defined by the set of properties invariant under a certain group of transformations. For example, projective properties are invariant under the group of projectivities, affine properties are invariant under the group of affine bijections, and Euclidean properties are invariant under rigid motions. Although it is possible to define these various groups of transformations as certain subgroups of the group of projectivities, such an approach is quite bewildering to a novice. In order to appreciate such acrobatics, one has to already know about projective geometry, affine geometry, and Euclidean geometry.

Since the fifties, geometry has been built on top of linear algebra, as opposed to axiomatically (as in Veblen and Young [8, 9] or Samuel [6]). Even though geometry loses some of its charm presented that way, it has the advantage of receiving a more unified and simpler treatment.

Affine geometry is basically the geometry of linear algebra. Well, not quite, since affine maps are not linear maps. The additional ingredient is that affine geometry is invariant under translations, which are not linear maps! Instead of linear combinations of vectors, we need to consider affine combinations of points, or barycenters (where the scalars add up to 1). Affine maps preserve barycenters. In some sense, affine geometry is the geometry of systems of particles and forces acting on them. Angles and distances are undefined, but parallelism is well defined. The crucial notion is the notion of ratio. Given any two points a, b and any scalar λ, the point $c = (1 - \lambda)a + \lambda b$ is the point on the line (a, b) (assuming $a \neq b$) such that $\overrightarrow{ac} = \lambda \overrightarrow{ab}$, i.e., the point c is "λ of the way between a and b." Even though such a geometry may seem quite restrictive, it allows the handling of polynomial curves and surfaces.

Euclidean geometry is obtained by adding an inner product to affine geometry. This way, angles and distances can be defined. The maps that preserve the inner product are the rigid motions. In Euclidean geometry, orthogonality can be defined. This is a very rich geometry. The structure of rigid motions (rotations and rotations followed by a flip) is well understood, and plays an important role in rigid body mechanics.

Projective geometry is, roughly speaking, linear algebra "up to a scalar." There is no notion of angle or distance, and projective maps are more general than affine maps. What is remarkable is that every affine space can be embedded into a projective space, its projective completion. In such a projective completion, there is a special hyperplane of "points at infinity." Affine maps are the projectivities that preserve (globally) this hyperplane at infinity. Thus, affine geometry can be viewed as a specialization of projective geometry. What is remarkable is that if we consider projective spaces over the complex field, it is possible to introduce the notion of angle in a projective manner (via the cross-ratio). This discovery, due to Poncelet, Laguerre, and Cayley, can be exploited to show that Euclidean geometry is a specialization of projective geometry.

Besides projective geometry and its specializations, there are other important and beautiful facets of geometry, notably differential geometry and algebraic geometry. Nowdays, each one is a major area of mathematics, and it is out of the question to discuss both in any depth. We will present some basics of the differential geometry of curves and surfaces. This topic was studied by many, including Euler and Gauss, who made fundamental contributions. However, we will limit ourselves to the study of local properties and not even attempt to touch manifolds.

These days, projective geometry is rarely taught at any depth in mathematics departments, and similarly for basic differential geometry. Typically, projective spaces are defined at the begining of an algebraic geometry course, but modern algebraic geometry courses deal with much more advanced topics, such as varieties and schemes. Similarly, differential geometry courses proceed quickly to manifolds and Riemannian metrics, but the more elementary "geometry in the small" is cursorily covered, if at all.

Paradoxically, with the advent of faster computers, it was realized by manufacturers (for instance of cars and planes) that it was possible and desirable to use computer-aided methods for their design. Computer vision problems (and some computer graphics problems) can often be formulated in the framework of projective geometry. Thus, there seems to be an interesting turn of events. After being neglected for decades, stimulated by computer science, old-fashioned geometry seems to be making a comeback as a fundamental tool used in manufacturing, computer graphics, computer vision, and motion planning, just to mention some key areas.

We are convinced that geometry will play an important role in computer science and engineering in the years to come. The demand for technology using 3D graphics, virtual reality, animation techniques, etc., is increasing fast, and it is clear that storing and processing complex images and complex geometric models of shapes (face, limbs, organs, etc.) will be required. This book represents an attempt at presenting a coherent view of geometric methods used to tackle problems of a geometric nature with a computer. We believe that this can be a great way of learning some old-fashioned (and some new!) geometry while having fun. Furthermore, there are plenty of opportunities for applying these methods to real-world problems.

While we are interested in the standard (local) differential properties of curves and surfaces (torsion, curvature), we concentrate on methods for discretizing curves and surfaces in order to store them and display them efficiently. However, in order to gain a deeper understanding of this theory of curves and surfaces, we present the underlying geometric concepts in some detail, in particular, affine, projective, and Euclidean geometry.

1.2 Prerequisites and Notation

It is assumed that the reader is familiar with the basics of linear algebra, at the level of Strang [7]. The reader may also consult appropriate chapters on linear algebra in Lang [3]. For the material on the differential geometry of curves and surfaces and Lie groups, familiarity with some basics of analysis are assumed. Lang's text [4] is more than sufficient as background. A general background in classical geometry is helpful, but not mandatory. Two excellent sources are Coxeter [2] and Pedoe [5].

We denote the set $\{0, 1, 2, \ldots\}$ of natural numbers by $\mathbb{N}$, the ring $\{\ldots, -2, -1, 0, 1, 2, \ldots\}$ of integers by $\mathbb{Z}$, the field of rationals by $\mathbb{Q}$, the field of real numbers by $\mathbb{R}$, and the field of complex numbers by $\mathbb{C}$. The multiplicative group $\mathbb{R} - \{0\}$ of reals is denoted by $\mathbb{R}^*$, and similarly, the multiplicative field of complex numbers is denoted by $\mathbb{C}^*$. We let $\mathbb{R}_+ = \{x \in \mathbb{R} \mid x \geq 0\}$ denote the set of nonnegative reals.

The n-dimensional vector space of real n-tuples is denoted by $\mathbb{R}^n$, and the complex n-dimensional vector space of complex n-tuples is denoted by $\mathbb{C}^n$.

Given a vector space E, vectors are usually denoted by lowercase letters from the end of the alphabet, in italic or boldface; for example, $u, v, w, \mathbf{x}, \mathbf{y}, \mathbf{z}$.

The null vector $(0, \ldots, 0)$ is abbreviated as 0 or $\mathbf{0}$. A vector space consisting only of the null vector is called a *trivial vector space*. A trivial vector space $\{0\}$ is sometimes denoted by 0. A vector space $E \neq \{0\}$ is called a *nontrivial vector space*.

When dealing with affine spaces, we will use an arrow in order to distinguish between spaces of points (E, U, etc.) and the corresponding spaces of vectors ($\overrightarrow{E}, \overrightarrow{U}$, etc.).

The dimension of the vector space E is denoted by $\dim(E)$. The direct sum of two vector spaces U, V is denoted by $U \oplus V$. The dual of a vector space E is denoted by E^*. The kernel of a linear map $f \colon E \to F$ is denoted by $\mathrm{Ker}\, f$, and the image by $\mathrm{Im}\, f$. The transpose of a matrix A is denoted by $A^\top$. The identity function is denoted by id, and the $n \times n$-identity matrix is denoted by I_n, or I. The determinant of a matrix A is denoted by $\det(A)$ or $D(A)$.

The cardinality of a set S is denoted by $|S|$. Set difference is denoted by

$$A - B = \{x \mid x \in A \text{ and } x \notin B\}.$$

A list of symbols in their order of appearance in this book is given at the end of the book.

References

1. H.S.M. Coxeter. *Non-Euclidean Geometry*. The University of Toronto Press, first edition, 1942.
2. H.S.M. Coxeter. *Introduction to Geometry*. Wiley, second edition, 1989.
3. Serge Lang. *Algebra*. Addison-Wesley, third edition, 1993.
4. Serge Lang. *Undergraduate Analysis*. UTM. Springer-Verlag, second edition, 1997.
5. Dan Pedoe. *Geometry, A Comprehensive Course*. Dover, first edition, 1988.
6. Pierre Samuel. *Projective Geometry*. Undergraduate Texts in Mathematics. Springer-Verlag, first edition, 1988.
7. Gilbert Strang. *Linear Algebra and Its Applications*. Saunders HBJ, third edition, 1988.
8. O. Veblen and J. W. Young. *Projective Geometry, Vol. 1*. Ginn, second edition, 1938.
9. O. Veblen and J. W. Young. *Projective Geometry, Vol. 2*. Ginn, first edition, 1946.

Chapter 2
Basics of Affine Geometry

L'algèbre n'est qu'une géométrie écrite; la géométrie n'est qu'une algèbre figurée.
—**Sophie Germain**

2.1 Affine Spaces

Geometrically, curves and surfaces are usually considered to be sets of points with some special properties, living in a space consisting of "points." Typically, one is also interested in geometric properties invariant under certain transformations, for example, translations, rotations, projections, etc. One could model the space of points as a vector space, but this is not very satisfactory for a number of reasons. One reason is that the point corresponding to the zero vector (0), called the origin, plays a special role, when there is really no reason to have a privileged origin. Another reason is that certain notions, such as parallelism, are handled in an awkward manner. But the deeper reason is that vector spaces and affine spaces really have different geometries. The geometric properties of a vector space are invariant under the group of bijective linear maps, whereas the geometric properties of an affine space are invariant under the group of bijective affine maps, and these two groups are not isomorphic. Roughly speaking, there are more affine maps than linear maps.

Affine spaces provide a better framework for doing geometry. In particular, it is possible to deal with points, curves, surfaces, etc., in an **intrinsic manner**, that is, independently of any specific choice of a coordinate system. As in physics, this is highly desirable to really understand what is going on. Of course, coordinate systems have to be chosen to finally carry out computations, but one should learn to resist the temptation to resort to coordinate systems until it is really necessary.

Affine spaces are the right framework for dealing with motions, trajectories, and physical forces, among other things. Thus, affine geometry is crucial to a clean presentation of kinematics, dynamics, and other parts of physics (for example, elasticity). After all, a rigid motion is an affine map, but not a linear map in general.

Also, given an $m \times n$ matrix A and a vector $b \in \mathbb{R}^m$, the set $U = \{x \in \mathbb{R}^n \mid Ax = b\}$ of solutions of the system $Ax = b$ is an affine space, but not a vector space (linear space) in general.

Use coordinate systems only when needed!

This chapter proceeds as follows. We take advantage of the fact that almost every affine concept is the counterpart of some concept in linear algebra. We begin by defining affine spaces, stressing the physical interpretation of the definition in terms of points (particles) and vectors (forces). Corresponding to linear combinations of vectors, we define affine combinations of points (barycenters), realizing that we are forced to restrict our attention to families of scalars adding up to 1. Corresponding to linear subspaces, we introduce affine subspaces as subsets closed under affine combinations. Then, we characterize affine subspaces in terms of certain vector spaces called their directions. This allows us to define a clean notion of parallelism. Next, corresponding to linear independence and bases, we define affine independence and affine frames. We also define convexity. Corresponding to linear maps, we define affine maps as maps preserving affine combinations. We show that every affine map is completely defined by the image of one point and a linear map. Then, we investigate briefly some simple affine maps, the translations and the central dilatations. At this point, we give a glimpse of affine geometry. We prove the theorems of Thales, Pappus, and Desargues. After this, the definition of affine hyperplanes in terms of affine forms is reviewed. The section ends with a closer look at the intersection of affine subspaces.

Our presentation of affine geometry is far from being comprehensive, and it is biased toward the algorithmic geometry of curves and surfaces. For more details, the reader is referred to Pedoe [9], Snapper and Troyer [11], Berger [2, 3], Coxeter [4], Samuel [10], Tisseron [13], and Hilbert and Cohn-Vossen [7].

Suppose we have a particle moving in 3D space and that we want to describe the trajectory of this particle. If one looks up a good textbook on dynamics, such as Greenwood [6], one finds out that the particle is modeled as a point, and that the position of this point x is determined with respect to a "frame" in $\mathbb{R}^3$ by a vector. Curiously, the notion of a frame is rarely defined precisely, but it is easy to infer that a frame is a pair $(O, (e_1, e_2, e_3))$ consisting of an origin O (which is a point) together with a basis of three vectors (e_1, e_2, e_3). For example, the standard frame in $\mathbb{R}^3$ has origin $O = (0,0,0)$ and the basis of three vectors $e_1 = (1,0,0)$, $e_2 = (0,1,0)$, and $e_3 = (0,0,1)$. The position of a point x is then defined by the "unique vector" from O to x.

But wait a minute, this definition seems to be defining frames and the position of a point without defining what a point is! Well, let us identify points with elements of $\mathbb{R}^3$. If so, given any two points $a = (a_1, a_2, a_3)$ and $b = (b_1, b_2, b_3)$, there is a unique *free vector*, denoted by $\overrightarrow{ab}$, from a to b, the vector $\overrightarrow{ab} = (b_1 - a_1, b_2 - a_2, b_3 - a_3)$. Note that

$$b = a + \overrightarrow{ab},$$

addition being understood as addition in $\mathbb{R}^3$. Then, in the standard frame, given a point $x = (x_1, x_2, x_3)$, the position of x is the vector $\overrightarrow{Ox} = (x_1, x_2, x_3)$, which coincides with the point itself. In the standard frame, points and vectors are identified. Points and free vectors are illustrated in Figure 2.1.

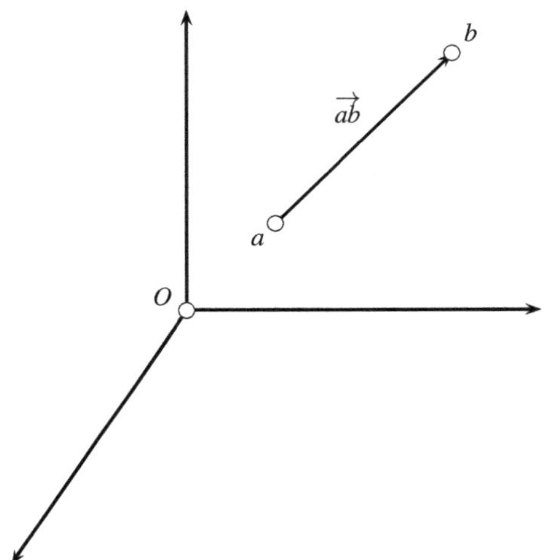

Fig. 2.1 Points and free vectors.

What if we pick a frame with a different origin, say $\Omega = (\omega_1, \omega_2, \omega_3)$, but the same basis vectors (e_1, e_2, e_3)? This time, the point $x = (x_1, x_2, x_3)$ is defined by two position vectors:

$$\overrightarrow{Ox} = (x_1, x_2, x_3)$$

in the frame $(O, (e_1, e_2, e_3))$ and

$$\overrightarrow{\Omega x} = (x_1 - \omega_1, x_2 - \omega_2, x_3 - \omega_3)$$

in the frame $(\Omega, (e_1, e_2, e_3))$.

This is because

$$\overrightarrow{Ox} = \overrightarrow{O\Omega} + \overrightarrow{\Omega x} \quad \text{and} \quad \overrightarrow{O\Omega} = (\omega_1, \omega_2, \omega_3).$$

We note that in the second frame $(\Omega, (e_1, e_2, e_3))$, points and position vectors are no longer identified. This gives us evidence that points are not vectors. It may be computationally convenient to deal with points using position vectors, but such a treatment is not frame invariant, which has undesirable effets.

Inspired by physics, we deem it important to define points and properties of points that are frame invariant. An undesirable side effect of the present approach shows up if we attempt to define linear combinations of points. First, let us review the notion of linear combination of vectors. Given two vectors u and v of coordinates (u_1, u_2, u_3) and (v_1, v_2, v_3) with respect to the basis (e_1, e_2, e_3), for any two scalars λ, μ, we can define the linear combination $\lambda u + \mu v$ as the vector of coordinates

$$(\lambda u_1 + \mu v_1, \lambda u_2 + \mu v_2, \lambda u_3 + \mu v_3).$$

If we choose a different basis (e'_1, e'_2, e'_3) and if the matrix P expressing the vectors (e'_1, e'_2, e'_3) over the basis (e_1, e_2, e_3) is

$$P = \begin{pmatrix} a_1 & b_1 & c_1 \\ a_2 & b_2 & c_2 \\ a_3 & b_3 & c_3 \end{pmatrix},$$

which means that the columns of P are the coordinates of the e'_j over the basis (e_1, e_2, e_3), since

$$u_1 e_1 + u_2 e_2 + u_3 e_3 = u'_1 e'_1 + u'_2 e'_2 + u'_3 e'_3$$

and

$$v_1 e_1 + v_2 e_2 + v_3 e_3 = v'_1 e'_1 + v'_2 e'_2 + v'_3 e'_3,$$

it is easy to see that the coordinates (u_1, u_2, u_3) and (v_1, v_2, v_3) of u and v with respect to the basis (e_1, e_2, e_3) are given in terms of the coordinates (u'_1, u'_2, u'_3) and (v'_1, v'_2, v'_3) of u and v with respect to the basis (e'_1, e'_2, e'_3) by the matrix equations

$$\begin{pmatrix} u_1 \\ u_2 \\ u_3 \end{pmatrix} = P \begin{pmatrix} u'_1 \\ u'_2 \\ u'_3 \end{pmatrix} \quad \text{and} \quad \begin{pmatrix} v_1 \\ v_2 \\ v_3 \end{pmatrix} = P \begin{pmatrix} v'_1 \\ v'_2 \\ v'_3 \end{pmatrix}.$$

From the above, we get

$$\begin{pmatrix} u'_1 \\ u'_2 \\ u'_3 \end{pmatrix} = P^{-1} \begin{pmatrix} u_1 \\ u_2 \\ u_3 \end{pmatrix} \quad \text{and} \quad \begin{pmatrix} v'_1 \\ v'_2 \\ v'_3 \end{pmatrix} = P^{-1} \begin{pmatrix} v_1 \\ v_2 \\ v_3 \end{pmatrix},$$

and by linearity, the coordinates

$$(\lambda u'_1 + \mu v'_1, \lambda u'_2 + \mu v'_2, \lambda u'_3 + \mu v'_3)$$

of $\lambda u + \mu v$ with respect to the basis (e'_1, e'_2, e'_3) are given by

$$\begin{pmatrix} \lambda u'_1 + \mu v'_1 \\ \lambda u'_2 + \mu v'_2 \\ \lambda u'_3 + \mu v'_3 \end{pmatrix} = \lambda P^{-1} \begin{pmatrix} u_1 \\ u_2 \\ u_3 \end{pmatrix} + \mu P^{-1} \begin{pmatrix} v_1 \\ v_2 \\ v_3 \end{pmatrix} = P^{-1} \begin{pmatrix} \lambda u_1 + \mu v_1 \\ \lambda u_2 + \mu v_2 \\ \lambda u_3 + \mu v_3 \end{pmatrix}.$$

Everything worked out because the change of basis does not involve a change of origin. On the other hand, if we consider the change of frame from the frame $(O, (e_1, e_2, e_3))$ to the frame $(\Omega, (e_1, e_2, e_3))$, where $\overrightarrow{O\Omega} = (\omega_1, \omega_2, \omega_3)$, given two points a, b of coordinates (a_1, a_2, a_3) and (b_1, b_2, b_3) with respect to the frame

$(O, (e_1, e_2, e_3))$ and of coordinates (a'_1, a'_2, a'_3) and (b'_1, b'_2, b'_3) with respect to the frame $(\Omega, (e_1, e_2, e_3))$, since

$$(a'_1, a'_2, a'_3) = (a_1 - \omega_1, a_2 - \omega_2, a_3 - \omega_3)$$

and

$$(b'_1, b'_2, b'_3) = (b_1 - \omega_1, b_2 - \omega_2, b_3 - \omega_3),$$

the coordinates of $\lambda a + \mu b$ with respect to the frame $(O, (e_1, e_2, e_3))$ are

$$(\lambda a_1 + \mu b_1, \lambda a_2 + \mu b_2, \lambda a_3 + \mu b_3),$$

but the coordinates

$$(\lambda a'_1 + \mu b'_1, \lambda a'_2 + \mu b'_2, \lambda a'_3 + \mu b'_3)$$

of $\lambda a + \mu b$ with respect to the frame $(\Omega, (e_1, e_2, e_3))$ are

$$(\lambda a_1 + \mu b_1 - (\lambda + \mu)\omega_1, \lambda a_2 + \mu b_2 - (\lambda + \mu)\omega_2, \lambda a_3 + \mu b_3 - (\lambda + \mu)\omega_3),$$

which are different from

$$(\lambda a_1 + \mu b_1 - \omega_1, \lambda a_2 + \mu b_2 - \omega_2, \lambda a_3 + \mu b_3 - \omega_3),$$

unless $\lambda + \mu = 1$.

Thus, we have discovered a major difference between vectors and points: The notion of linear combination of vectors is basis independent, but the notion of linear combination of points is frame dependent. In order to salvage the notion of linear combination of points, some restriction is needed: The scalar coefficients must add up to 1.

A clean way to handle the problem of frame invariance and to deal with points in a more intrinsic manner is to make a clearer distinction between points and vectors. We duplicate $\mathbb{R}^3$ into two copies, the first copy corresponding to points, where we forget the vector space structure, and the second copy corresponding to free vectors, where the vector space structure is important. Furthermore, we make explicit the important fact that the vector space $\mathbb{R}^3$ acts on the set of points $\mathbb{R}^3$: Given any **point** $a = (a_1, a_2, a_3)$ and any **vector** $v = (v_1, v_2, v_3)$, we obtain the **point**

$$a + v = (a_1 + v_1, a_2 + v_2, a_3 + v_3),$$

which can be thought of as the result of translating a to b using the vector v. We can imagine that v is placed such that its origin coincides with a and that its tip coincides with b. This action $+: \mathbb{R}^3 \times \mathbb{R}^3 \to \mathbb{R}^3$ satisfies some crucial properties. For example,

$$a + 0 = a,$$
$$(a + u) + v = a + (u + v),$$

and for any two points a, b, there is a unique free vector $\overrightarrow{ab}$ such that

$$b = a + \overrightarrow{ab}.$$

It turns out that the above properties, although trivial in the case of $\mathbb{R}^3$, are all that is needed to define the abstract notion of affine space (or affine structure). The basic idea is to consider two (distinct) sets E and $\overrightarrow{E}$, where E is a set of points (with no structure) and $\overrightarrow{E}$ is a vector space (of free vectors) acting on the set E.

Did you say "A fine space"?

Intuitively, we can think of the elements of $\overrightarrow{E}$ as forces moving the points in E, considered as physical particles. The effect of applying a force (free vector) $u \in \overrightarrow{E}$ to a point $a \in E$ is a translation. By this, we mean that for every force $u \in \overrightarrow{E}$, the action of the force u is to "move" every point $a \in E$ to the point $a + u \in E$ obtained by the translation corresponding to u viewed as a vector. Since translations can be composed, it is natural that $\overrightarrow{E}$ is a vector space.

For simplicity, it is assumed that all vector spaces under consideration are defined over the field $\mathbb{R}$ of real numbers. Most of the definitions and results also hold for an arbitrary field K, although some care is needed when dealing with fields of characteristic different from zero (see the problems). It is also assumed that all families $(\lambda_i)_{i \in I}$ of scalars have finite support. Recall that a family $(\lambda_i)_{i \in I}$ of scalars has *finite support* if $\lambda_i = 0$ for all $i \in I - J$, where J is a finite subset of I. Obviously, finite families of scalars have finite support, and for simplicity, the reader may assume that all families of scalars are finite. The formal definition of an affine space is as follows.

Definition 2.1. An *affine space* is either the degenerate space reduced to the empty set, or a triple $\langle E, \overrightarrow{E}, + \rangle$ consisting of a nonempty set E (of *points*), a vector space $\overrightarrow{E}$ (of *translations*, or *free vectors*), and an action $+ : E \times \overrightarrow{E} \to E$, satisfying the following conditions.

(A1) $a + 0 = a$, for every $a \in E$.

(A2) $(a + u) + v = a + (u + v)$, for every $a \in E$, and every $u, v \in \overrightarrow{E}$.

(A3) For any two points $a, b \in E$, there is a unique $u \in \overrightarrow{E}$ such that $a + u = b$.

The unique vector $u \in \overrightarrow{E}$ such that $a + u = b$ is denoted by $\overrightarrow{ab}$, or sometimes by **ab**, or even by $b - a$. Thus, we also write

$$b = a + \overrightarrow{ab}$$

(or $b = a + \mathbf{ab}$, or even $b = a + (b - a)$).

The *dimension of the affine space* $\langle E, \overrightarrow{E}, + \rangle$ is the dimension $\dim(\overrightarrow{E})$ of the vector space $\overrightarrow{E}$. For simplicity, it is denoted by $\dim(E)$.

Conditions (A1) and (A2) say that the (abelian) group $\overrightarrow{E}$ acts on E, and condition (A3) says that $\overrightarrow{E}$ acts transitively and faithfully on E. Note that

$$\overrightarrow{a(a+v)} = v$$

for all $a \in E$ and all $v \in \overrightarrow{E}$, since $\overrightarrow{a(a+v)}$ is the unique vector such that $a + v = a + \overrightarrow{a(a+v)}$. Thus, $b = a + v$ is equivalent to $\overrightarrow{ab} = v$. Figure 2.2 gives an intuitive picture of an affine space. It is natural to think of all vectors as having the same origin, the null vector.

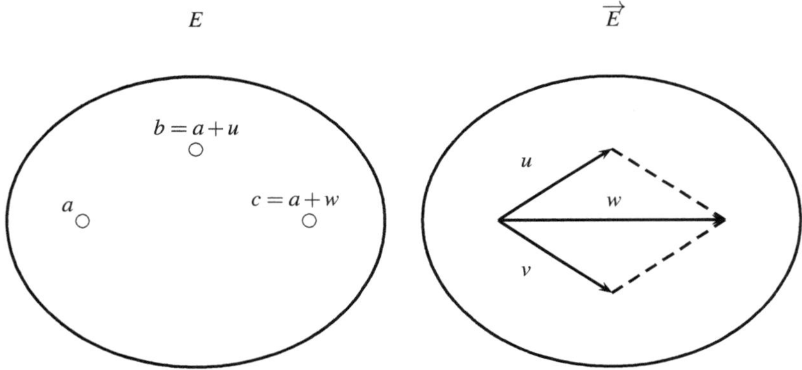

Fig. 2.2 Intuitive picture of an affine space.

The axioms defining an affine space $\langle E, \overrightarrow{E}, + \rangle$ can be interpreted intuitively as saying that E and $\overrightarrow{E}$ are two different ways of looking at the same object, but wearing different sets of glasses, the second set of glasses depending on the choice of an "origin" in E. Indeed, we can choose to look at the points in E, forgetting that every pair (a, b) of points defines a unique vector $\overrightarrow{ab}$ in $\overrightarrow{E}$, or we can choose to look at the vectors u in $\overrightarrow{E}$, forgetting the points in E. Furthermore, if we also pick any point a in E, a point that can be viewed as an *origin* in E, then we can recover all the points in E as the translated points $a + u$ for all $u \in \overrightarrow{E}$. This can be formalized by defining two maps between E and $\overrightarrow{E}$.

For every $a \in E$, consider the mapping from $\overrightarrow{E}$ to E given by

$$u \mapsto a + u,$$

where $u \in \overrightarrow{E}$, and consider the mapping from E to $\overrightarrow{E}$ given by

$$b \mapsto \overrightarrow{ab},$$

where $b \in E$. The composition of the first mapping with the second is

$$u \mapsto a + u \mapsto \overrightarrow{a(a+u)},$$

which, in view of (A3), yields u. The composition of the second with the first mapping is

$$b \mapsto \overrightarrow{ab} \mapsto a + \overrightarrow{ab},$$

which, in view of (A3), yields b. Thus, these compositions are the identity from $\overrightarrow{E}$ to $\overrightarrow{E}$ and the identity from E to E, and the mappings are both bijections.

When we identify E with $\overrightarrow{E}$ via the mapping $b \mapsto \overrightarrow{ab}$, we say that we consider E as the vector space obtained *by taking a as the origin in E*, and we denote it by E_a. Because E_a is a vector space, to be consistent with our notational conventions we should use the notation $\overrightarrow{E_a}$ (using an arrow), instead of E_a. However, for simplicity, we stick to the notation E_a.

Thus, an affine space $\langle E, \overrightarrow{E}, + \rangle$ is a way of defining a vector space structure on a set of points E, without making a commitment to a **fixed** origin in E. Nevertheless, as soon as we commit to an origin a in E, we can view E as the vector space E_a. However, we urge the reader to think of E as a physical set of points and of $\overrightarrow{E}$ as a set of forces acting on E, rather than reducing E to some isomorphic copy of $\mathbb{R}^n$. After all, points are points, and not vectors! For notational simplicity, we will often denote an affine space $\langle E, \overrightarrow{E}, + \rangle$ by $(E, \overrightarrow{E})$, or even by E. The vector space $\overrightarrow{E}$ is called the *vector space associated with E*.

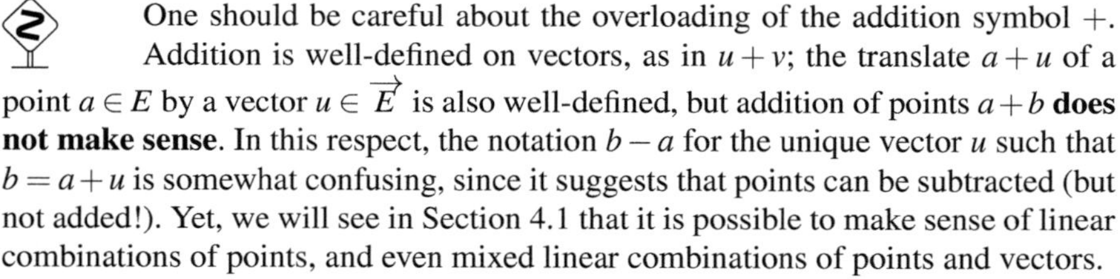

One should be careful about the overloading of the addition symbol $+$. Addition is well-defined on vectors, as in $u + v$; the translate $a + u$ of a point $a \in E$ by a vector $u \in \overrightarrow{E}$ is also well-defined, but addition of points $a + b$ **does not make sense**. In this respect, the notation $b - a$ for the unique vector u such that $b = a + u$ is somewhat confusing, since it suggests that points can be subtracted (but not added!). Yet, we will see in Section 4.1 that it is possible to make sense of linear combinations of points, and even mixed linear combinations of points and vectors.

Any vector space $\overrightarrow{E}$ has an affine space structure specified by choosing $E = \overrightarrow{E}$, and letting $+$ be addition in the vector space $\overrightarrow{E}$. We will refer to the affine structure $\langle \overrightarrow{E}, \overrightarrow{E}, + \rangle$ on a vector space $\overrightarrow{E}$ as the *canonical (or natural) affine structure on* $\overrightarrow{E}$. In particular, the vector space $\mathbb{R}^n$ can be viewed as the affine space $\langle \mathbb{R}^n, \mathbb{R}^n, + \rangle$, denoted by $\mathbb{A}^n$. In general, if K is any field, the affine space $\langle K^n, K^n, + \rangle$ is denoted by $\mathbb{A}_K^n$. In order to distinguish between the double role played by members of $\mathbb{R}^n$, points and vectors, we will denote points by row vectors, and vectors by column vectors. Thus, the action of the vector space $\mathbb{R}^n$ over the set $\mathbb{R}^n$ simply viewed as a set of points is given by

$$(a_1,\ldots,a_n) + \begin{pmatrix} u_1 \\ \vdots \\ u_n \end{pmatrix} = (a_1 + u_1,\ldots,a_n + u_n).$$

We will also use the convention that if $x = (x_1,\ldots,x_n) \in \mathbb{R}^n$, then the column vector associated with x is denoted by $\mathbf{x}$ (in boldface notation). Abusing the notation slightly, if $a \in \mathbb{R}^n$ is a point, we also write $a \in \mathbb{A}^n$. The affine space $\mathbb{A}^n$ is called the *real affine space of dimension n*. In most cases, we will consider $n = 1, 2, 3$.

2.2 Examples of Affine Spaces

Let us now give an example of an affine space that is not given as a vector space (at least, not in an obvious fashion). Consider the subset L of $\mathbb{A}^2$ consisting of all points (x, y) satisfying the equation

$$x + y - 1 = 0.$$

The set L is the line of slope -1 passing through the points $(1, 0)$ and $(0, 1)$ shown in Figure 2.3.

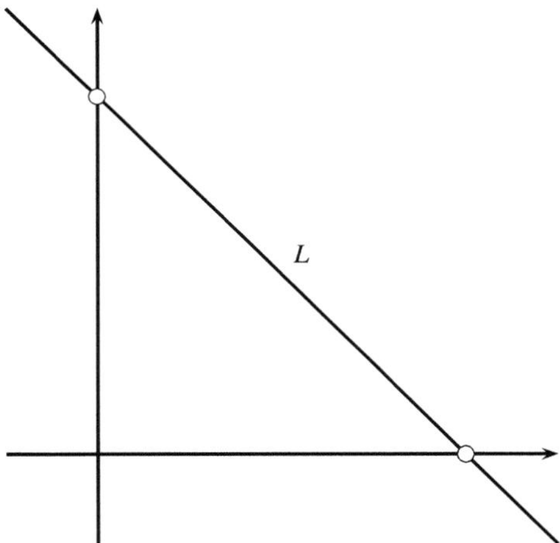

Fig. 2.3 An affine space: the line of equation $x + y - 1 = 0$.

The line L can be made into an official affine space by defining the action $+ : L \times \mathbb{R} \to L$ of $\mathbb{R}$ on L defined such that for every point $(x, 1 - x)$ on L and any $u \in \mathbb{R}$,

$$(x, 1 - x) + u = (x + u, 1 - x - u).$$

It is immediately verified that this action makes L into an affine space. For example, for any two points $a = (a_1, 1 - a_1)$ and $b = (b_1, 1 - b_1)$ on L, the unique (vector) $u \in \mathbb{R}$ such that $b = a + u$ is $u = b_1 - a_1$. Note that the vector space $\mathbb{R}$ is isomorphic to the line of equation $x + y = 0$ passing through the origin.

Similarly, consider the subset H of $\mathbb{A}^3$ consisting of all points (x, y, z) satisfying the equation

$$x + y + z - 1 = 0.$$

The set H is the plane passing through the points $(1, 0, 0)$, $(0, 1, 0)$, and $(0, 0, 1)$. The plane H can be made into an official affine space by defining the action $+: H \times \mathbb{R}^2 \to H$ of $\mathbb{R}^2$ on H defined such that for every point $(x, y, 1 - x - y)$ on H and any $\begin{pmatrix} u \\ v \end{pmatrix} \in \mathbb{R}^2$,

$$(x, y, 1 - x - y) + \begin{pmatrix} u \\ v \end{pmatrix} = (x + u, y + v, 1 - x - u - y - v).$$

For a slightly wilder example, consider the subset P of $\mathbb{A}^3$ consisting of all points (x, y, z) satisfying the equation

$$x^2 + y^2 - z = 0.$$

The set P is a paraboloid of revolution, with axis Oz. The surface P can be made into an official affine space by defining the action $+: P \times \mathbb{R}^2 \to P$ of $\mathbb{R}^2$ on P defined such that for every point $(x, y, x^2 + y^2)$ on P and any $\begin{pmatrix} u \\ v \end{pmatrix} \in \mathbb{R}^2$,

$$(x, y, x^2 + y^2) + \begin{pmatrix} u \\ v \end{pmatrix} = (x + u, y + v, (x + u)^2 + (y + v)^2).$$

This should dispell any idea that affine spaces are dull. Affine spaces not already equipped with an obvious vector space structure arise in projective geometry. Indeed, we will see in Section 5.1 that the complement of a hyperplane in a projective space has an affine structure.

2.3 Chasles's Identity

Given any three points $a, b, c \in E$, since $c = a + \overrightarrow{ac}$, $b = a + \overrightarrow{ab}$, and $c = b + \overrightarrow{bc}$, we get

$$c = b + \overrightarrow{bc} = (a + \overrightarrow{ab}) + \overrightarrow{bc} = a + (\overrightarrow{ab} + \overrightarrow{bc})$$

by (A2), and thus, by (A3),

$$\overrightarrow{ab} + \overrightarrow{bc} = \overrightarrow{ac},$$

which is known as *Chasles's identity*, and illustrated in Figure 2.4.

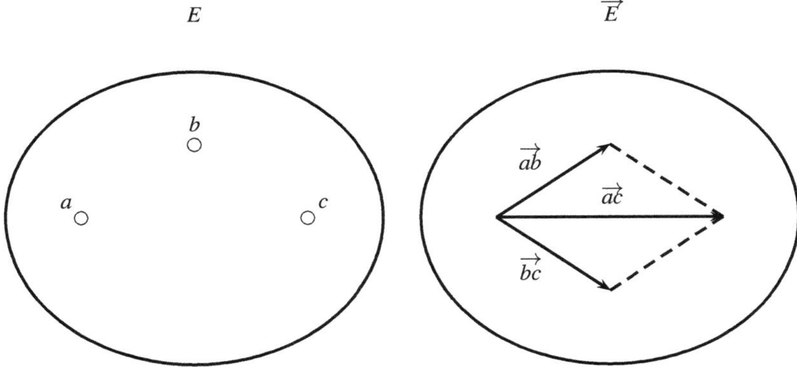

Fig. 2.4 Points and corresponding vectors in affine geometry.

Since $a = a + \overrightarrow{aa}$ and by (A1) $a = a + 0$, by (A3) we get

$$\overrightarrow{aa} = 0.$$

Thus, letting $a = c$ in Chasles's identity, we get

$$\overrightarrow{ba} = -\overrightarrow{ab}.$$

Given any four points $a, b, c, d \in E$, since by Chasles's identity

$$\overrightarrow{ab} + \overrightarrow{bc} = \overrightarrow{ad} + \overrightarrow{dc} = \overrightarrow{ac},$$

we have the *parallelogram law*

$$\overrightarrow{ab} = \overrightarrow{dc} \quad \text{iff} \quad \overrightarrow{bc} = \overrightarrow{ad}.$$

2.4 Affine Combinations, Barycenters

A fundamental concept in linear algebra is that of a linear combination. The corresponding concept in affine geometry is that of an *affine combination*, also called a *barycenter*. However, there is a problem with the naive approach involving a coordinate system, as we saw in Section 2.1. Since this problem is the reason for introducing affine combinations, at the risk of boring certain readers, we give another example showing what goes wrong if we are not careful in defining linear combinations of points.

Consider $\mathbb{R}^2$ as an affine space, under its natural coordinate system with origin $O = (0,0)$ and basis vectors $\begin{pmatrix} 1 \\ 0 \end{pmatrix}$ and $\begin{pmatrix} 0 \\ 1 \end{pmatrix}$. Given any two points $a = (a_1, a_2)$ and

$b = (b_1, b_2)$, it is natural to define the affine combination $\lambda a + \mu b$ as the point of coordinates

$$(\lambda a_1 + \mu b_1, \lambda a_2 + \mu b_2).$$

Thus, when $a = (-1, -1)$ and $b = (2, 2)$, the point $a + b$ is the point $c = (1, 1)$.

Let us now consider the new coordinate system with respect to the origin $c = (1, 1)$ (and the same basis vectors). This time, the coordinates of a are $(-2, -2)$, the coordinates of b are $(1, 1)$, and the point $a + b$ is the point d of coordinates $(-1, -1)$. However, it is clear that the point d is identical to the origin $O = (0, 0)$ of the first coordinate system.

Thus, $a + b$ corresponds to two different points depending on which coordinate system is used for its computation!

This shows that some extra condition is needed in order for affine combinations to make sense. It turns out that if the scalars sum up to 1, the definition is intrinsic, as the following lemma shows.

Lemma 2.1. *Given an affine space E, let $(a_i)_{i \in I}$ be a family of points in E, and let $(\lambda_i)_{i \in I}$ be a family of scalars. For any two points $a, b \in E$, the following properties hold:*

(1) If $\sum_{i \in I} \lambda_i = 1$, then

$$a + \sum_{i \in I} \lambda_i \overrightarrow{ad_i} = b + \sum_{i \in I} \lambda_i \overrightarrow{ba_i}.$$

(2) If $\sum_{i \in I} \lambda_i = 0$, then

$$\sum_{i \in I} \lambda_i \overrightarrow{ad_i} = \sum_{i \in I} \lambda_i \overrightarrow{ba_i}.$$

Proof. (1) By Chasles's identity (see Section 2.3), we have

$$a + \sum_{i \in I} \lambda_i \overrightarrow{ad_i} = a + \sum_{i \in I} \lambda_i (\overrightarrow{ab} + \overrightarrow{ba_i})$$

$$= a + \left(\sum_{i \in I} \lambda_i \right) \overrightarrow{ab} + \sum_{i \in I} \lambda_i \overrightarrow{ba_i}$$

$$= a + \overrightarrow{ab} + \sum_{i \in I} \lambda_i \overrightarrow{ba_i} \qquad \text{since } \sum_{i \in I} \lambda_i = 1$$

$$= b + \sum_{i \in I} \lambda_i \overrightarrow{ba_i} \qquad \text{since } b = a + \overrightarrow{ab}.$$

(2) We also have

$$\sum_{i \in I} \lambda_i \overrightarrow{ad_i} = \sum_{i \in I} \lambda_i (\overrightarrow{ab} + \overrightarrow{ba_i})$$

$$= \left(\sum_{i \in I} \lambda_i \right) \overrightarrow{ab} + \sum_{i \in I} \lambda_i \overrightarrow{ba_i}$$

$$= \sum_{i \in I} \lambda_i \overrightarrow{ba_i},$$

since $\sum_{i \in I} \lambda_i = 0$. $\quad\square$

Thus, by Lemma 2.1, for any family of points $(a_i)_{i \in I}$ in E, for any family $(\lambda_i)_{i \in I}$ of scalars such that $\sum_{i \in I} \lambda_i = 1$, the point

$$x = a + \sum_{i \in I} \lambda_i \overrightarrow{aa_i}$$

is independent of the choice of the origin $a \in E$. This property motivates the following definition.

Definition 2.2. For any family of points $(a_i)_{i \in I}$ in E, for any family $(\lambda_i)_{i \in I}$ of scalars such that $\sum_{i \in I} \lambda_i = 1$, and for any $a \in E$, the point

$$a + \sum_{i \in I} \lambda_i \overrightarrow{aa_i}$$

(which is independent of $a \in E$, by Lemma 2.1) is called the *barycenter (or barycentric combination, or affine combination) of the points a_i assigned the weights λ_i*, and it is denoted by

$$\sum_{i \in I} \lambda_i a_i.$$

In dealing with barycenters, it is convenient to introduce the notion of a *weighted point*, which is just a pair (a, λ), where $a \in E$ is a point, and $\lambda \in \mathbb{R}$ is a scalar. Then, given a family of weighted points $((a_i, \lambda_i))_{i \in I}$, where $\sum_{i \in I} \lambda_i = 1$, we also say that the point $\sum_{i \in I} \lambda_i a_i$ is the *barycenter of the family of weighted points $((a_i, \lambda_i))_{i \in I}$*.

Note that the barycenter x of the family of weighted points $((a_i, \lambda_i))_{i \in I}$ is the unique point such that

$$\overrightarrow{ax} = \sum_{i \in I} \lambda_i \overrightarrow{aa_i} \quad \text{for every } a \in E,$$

and setting $a = x$, the point x is the unique point such that

$$\sum_{i \in I} \lambda_i \overrightarrow{xa_i} = 0.$$

In physical terms, the barycenter is the *center of mass* of the family of weighted points $((a_i, \lambda_i))_{i \in I}$ (where the masses have been normalized, so that $\sum_{i \in I} \lambda_i = 1$, and negative masses are allowed).

Remarks:

(1) Since the barycenter of a family $((a_i, \lambda_i))_{i \in I}$ of weighted points is defined for families $(\lambda_i)_{i \in I}$ of scalars with finite support (and such that $\sum_{i \in I} \lambda_i = 1$), we might as well assume that I is finite. Then, for all $m \geq 2$, it is easy to prove that the barycenter of m weighted points can be obtained by repeated computations of barycenters of two weighted points.

(2) This result still holds, provided that the field K has at least three distinct elements, but the proof is trickier!

(3) When $\sum_{i \in I} \lambda_i = 0$, the vector $\sum_{i \in I} \lambda_i \overrightarrow{ad_i}$ does not depend on the point a, and we may denote it by $\sum_{i \in I} \lambda_i a_i$. This observation will be used in Section 4.1 to define a vector space in which linear combinations of both points and vectors make sense, regardless of the value of $\sum_{i \in I} \lambda_i$.

Figure 2.5 illustrates the geometric construction of the barycenters g_1 and g_2 of the weighted points $\left(a, \frac{1}{4}\right)$, $\left(b, \frac{1}{4}\right)$, and $\left(c, \frac{1}{2}\right)$, and $(a, -1)$, $(b, 1)$, and $(c, 1)$.

The point g_1 can be constructed geometrically as the middle of the segment joining c to the middle $\frac{1}{2}a + \frac{1}{2}b$ of the segment (a, b), since

$$g_1 = \frac{1}{2}\left(\frac{1}{2}a + \frac{1}{2}b\right) + \frac{1}{2}c.$$

The point g_2 can be constructed geometrically as the point such that the middle $\frac{1}{2}b + \frac{1}{2}c$ of the segment (b, c) is the middle of the segment (a, g_2), since

$$g_2 = -a + 2\left(\frac{1}{2}b + \frac{1}{2}c\right).$$

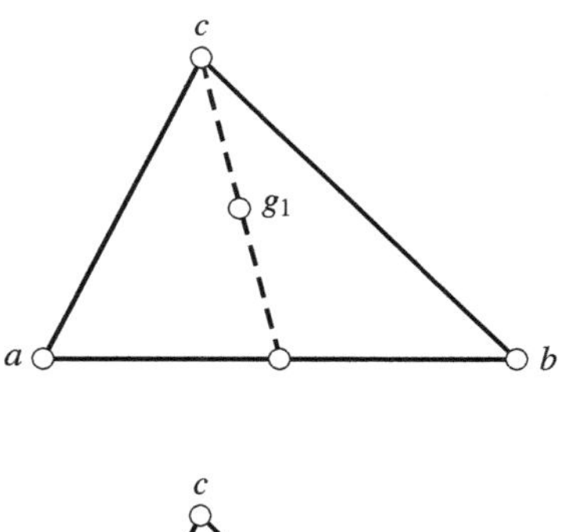

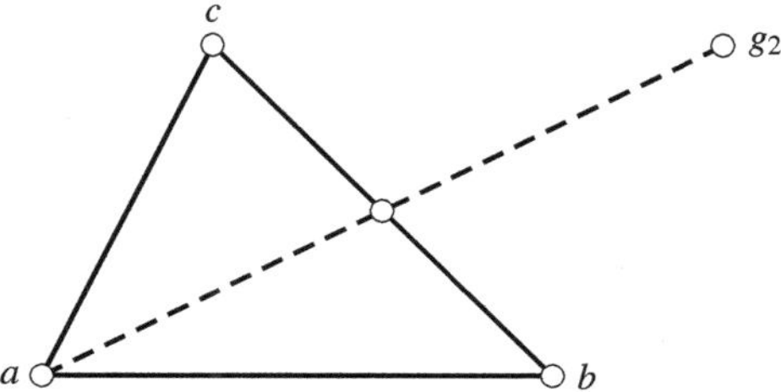

Fig. 2.5 Barycenters, $g_1 = \frac{1}{4}a + \frac{1}{4}b + \frac{1}{2}c$, $g_2 = -a + b + c$.

Later on, we will see that a polynomial curve can be defined as a set of barycenters of a fixed number of points. For example, let (a, b, c, d) be a sequence of points in $\mathbb{A}^2$. Observe that

$$(1-t)^3 + 3t(1-t)^2 + 3t^2(1-t) + t^3 = 1,$$

since the sum on the left-hand side is obtained by expanding $(t+(1-t))^3 = 1$ using the binomial formula. Thus,

$$(1-t)^3 a + 3t(1-t)^2 b + 3t^2(1-t) c + t^3 d$$

is a well-defined affine combination. Then, we can define the curve $F \colon \mathbb{A} \to \mathbb{A}^2$ such that

$$F(t) = (1-t)^3 a + 3t(1-t)^2 b + 3t^2(1-t) c + t^3 d.$$

Such a curve is called a *Bézier curve*, and (a,b,c,d) are called its *control points*. Note that the curve passes through a and d, but generally not through b and c. We show in Chapter 18 (on the web site) how any point $F(t)$ on the curve can be constructed using an algorithm performing affine interpolation steps (the *de Casteljau algorithm*).

2.5 Affine Subspaces

In linear algebra, a (linear) subspace can be characterized as a nonempty subset of a vector space closed under linear combinations. In affine spaces, the notion corresponding to the notion of (linear) subspace is the notion of affine subspace. It is natural to define an affine subspace as a subset of an affine space closed under affine combinations.

Definition 2.3. Given an affine space $\langle E, \overrightarrow{E}, + \rangle$, a subset V of E is an *affine subspace (of $\langle E, \overrightarrow{E}, + \rangle$)* if for every family of weighted points $((a_i, \lambda_i))_{i \in I}$ in V such that $\sum_{i \in I} \lambda_i = 1$, the barycenter $\sum_{i \in I} \lambda_i a_i$ belongs to V.

An affine subspace is also called a *flat* by some authors. According to Definition 2.3, the empty set is trivially an affine subspace, and every intersection of affine subspaces is an affine subspace.

As an example, consider the subset U of $\mathbb{R}^2$ defined by

$$U = \left\{ (x,y) \in \mathbb{R}^2 \mid ax + by = c \right\},$$

i.e., the set of solutions of the equation

$$ax + by = c,$$

where it is assumed that $a \neq 0$ or $b \neq 0$. Given any m points $(x_i, y_i) \in U$ and any m scalars λ_i such that $\lambda_1 + \cdots + \lambda_m = 1$, we claim that

$$\sum_{i=1}^{m} \lambda_i(x_i, y_i) \in U.$$

Indeed, $(x_i, y_i) \in U$ means that

$$ax_i + by_i = c,$$

and if we multiply both sides of this equation by λ_i and add up the resulting m equations, we get

$$\sum_{i=1}^{m} (\lambda_i a x_i + \lambda_i b y_i) = \sum_{i=1}^{m} \lambda_i c,$$

and since $\lambda_1 + \cdots + \lambda_m = 1$, we get

$$a\left(\sum_{i=1}^{m} \lambda_i x_i\right) + b\left(\sum_{i=1}^{m} \lambda_i y_i\right) = \left(\sum_{i=1}^{m} \lambda_i\right) c = c,$$

which shows that

$$\left(\sum_{i=1}^{m} \lambda_i x_i, \sum_{i=1}^{m} \lambda_i y_i\right) = \sum_{i=1}^{m} \lambda_i(x_i, y_i) \in U.$$

Thus, U is an affine subspace of $\mathbb{A}^2$. In fact, it is just a usual line in $\mathbb{A}^2$.

It turns out that U is closely related to the subset of $\mathbb{R}^2$ defined by

$$\overrightarrow{U} = \left\{(x, y) \in \mathbb{R}^2 \mid ax + by = 0\right\},$$

i.e., the set of solutions of the homogeneous equation

$$ax + by = 0$$

obtained by setting the right-hand side of $ax + by = c$ to zero. Indeed, for any m scalars λ_i, the same calculation as above yields that

$$\sum_{i=1}^{m} \lambda_i(x_i, y_i) \in \overrightarrow{U},$$

this time **without any restriction on the** λ_i, since the right-hand side of the equation is null. Thus, $\overrightarrow{U}$ is a subspace of $\mathbb{R}^2$. In fact, $\overrightarrow{U}$ is one-dimensional, and it is just a usual line in $\mathbb{R}^2$. This line can be identified with a line passing through the origin of $\mathbb{A}^2$, a line that is parallel to the line U of equation $ax + by = c$, as illustrated in Figure 2.6.

Now, if (x_0, y_0) is any point in U, we claim that

$$U = (x_0, y_0) + \overrightarrow{U},$$

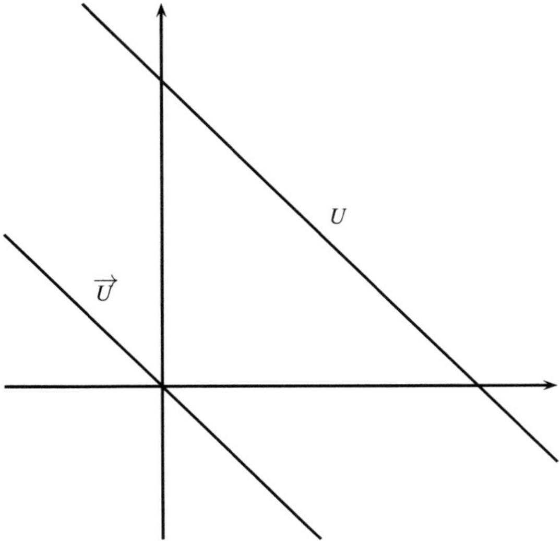

Fig. 2.6 An affine line U and its direction.

where

$$(x_0,y_0) + \overrightarrow{U} = \left\{ (x_0+u_1,y_0+u_2) \mid (u_1,u_2) \in \overrightarrow{U} \right\}.$$

First, $(x_0,y_0) + \overrightarrow{U} \subseteq U$, since $ax_0 + by_0 = c$ and $au_1 + bu_2 = 0$ for all $(u_1,u_2) \in \overrightarrow{U}$. Second, if $(x,y) \in U$, then $ax + by = c$, and since we also have $ax_0 + by_0 = c$, by subtraction, we get

$$a(x - x_0) + b(y - y_0) = 0,$$

which shows that $(x - x_0, y - y_0) \in \overrightarrow{U}$, and thus $(x,y) \in (x_0,y_0) + \overrightarrow{U}$. Hence, we also have $U \subseteq (x_0,y_0) + \overrightarrow{U}$, and $U = (x_0,y_0) + \overrightarrow{U}$.

The above example shows that the affine line U defined by the equation

$$ax + by = c$$

is obtained by "translating" the parallel line $\overrightarrow{U}$ of equation

$$ax + by = 0$$

passing through the origin. In fact, given any point $(x_0,y_0) \in U$,

$$U = (x_0,y_0) + \overrightarrow{U}.$$

More generally, it is easy to prove the following fact. Given any $m \times n$ matrix A and any vector $b \in \mathbb{R}^m$, the subset U of $\mathbb{R}^n$ defined by

$$U = \{x \in \mathbb{R}^n \mid Ax = b\}$$

is an affine subspace of $\mathbb{A}^n$.

Actually, observe that $Ax = b$ should really be written as $Ax^\top = b$, to be consistent with our convention that points are represented by row vectors. We can also use the boldface notation for column vectors, in which case the equation is written as $\mathbf{A}\mathbf{x} = \mathbf{b}$. For the sake of minimizing the amount of notation, we stick to the simpler (yet incorrect) notation $Ax = b$. If we consider the corresponding homogeneous equation $Ax = 0$, the set

$$\overrightarrow{U} = \{x \in \mathbb{R}^n \mid Ax = 0\}$$

is a subspace of $\mathbb{R}^n$, and for any $x_0 \in U$, we have

$$U = x_0 + \overrightarrow{U}.$$

This is a general situation. Affine subspaces can be characterized in terms of subspaces of $\overrightarrow{E}$. Let V be a nonempty subset of E. For every family $(a_1, \ldots, a_n)$ in V, for any family $(\lambda_1, \ldots, \lambda_n)$ of scalars, and for every point $a \in V$, observe that for every $x \in E$,

$$x = a + \sum_{i=1}^{n} \lambda_i \overrightarrow{ad_i}$$

is the barycenter of the family of weighted points

$$\left((a_1, \lambda_1), \ldots, (a_n, \lambda_n), \left(a, 1 - \sum_{i=1}^{n} \lambda_i \right) \right),$$

since

$$\sum_{i=1}^{n} \lambda_i + \left(1 - \sum_{i=1}^{n} \lambda_i \right) = 1.$$

Given any point $a \in E$ and any subset $\overrightarrow{V}$ of $\overrightarrow{E}$, let $a + \overrightarrow{V}$ denote the following subset of E:

$$a + \overrightarrow{V} = \left\{ a + v \mid v \in \overrightarrow{V} \right\}.$$

Lemma 2.2. *Let $\langle E, \overrightarrow{E}, + \rangle$ be an affine space.*

(1) A nonempty subset V of E is an affine subspace iff for every point $a \in V$, the set

$$\overrightarrow{V_a} = \{\overrightarrow{ax} \mid x \in V\}$$

is a subspace of $\overrightarrow{E}$. Consequently, $V = a + \overrightarrow{V_a}$. Furthermore,

$$\overrightarrow{V} = \{\overrightarrow{xy} \mid x, y \in V\}$$

is a subspace of $\overrightarrow{E}$ and $\overrightarrow{V_a} = \overrightarrow{V}$ for all $a \in E$. Thus, $V = a + \overrightarrow{V}$.

(2) For any subspace $\overrightarrow{V}$ of $\overrightarrow{E}$ and for any $a \in E$, the set $V = a + \overrightarrow{V}$ is an affine subspace.

Proof. The proof is straightforward, and is omitted. It is also given in Gallier [5]. $\square$

In particular, when E is the natural affine space associated with a vector space $\overrightarrow{E}$, Lemma 2.2 shows that every affine subspace of E is of the form $u + \overrightarrow{U}$, for a subspace $\overrightarrow{U}$ of $\overrightarrow{E}$. The subspaces of $\overrightarrow{E}$ are the affine subspaces of E that contain 0.

The subspace $\overrightarrow{V}$ associated with an affine subspace V is called the *direction of V*. It is also clear that the map $+ : V \times \overrightarrow{V} \to V$ induced by $+ : E \times \overrightarrow{E} \to E$ confers to $\langle V, \overrightarrow{V}, + \rangle$ an affine structure. Figure 2.7 illustrates the notion of affine subspace.

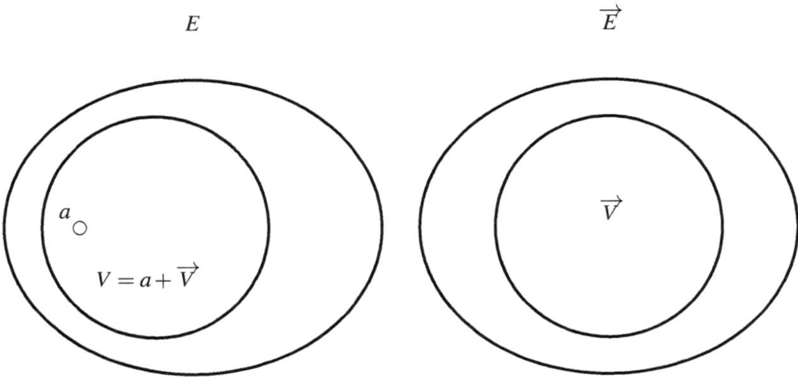

Fig. 2.7 An affine subspace V and its direction $\overrightarrow{V}$.

By the dimension of the subspace V, we mean the dimension of $\overrightarrow{V}$.

An affine subspace of dimension 1 is called a *line*, and an affine subspace of dimension 2 is called a *plane*.

An affine subspace of codimension 1 is called a *hyperplane* (recall that a subspace F of a vector space E has codimension 1 iff there is some subspace G of dimension 1 such that $E = F \oplus G$, the direct sum of F and G, see Strang [12] or Lang [8]).

We say that two affine subspaces U and V are *parallel* if their directions are identical. Equivalently, since $\overrightarrow{U} = \overrightarrow{V}$, we have $U = a + \overrightarrow{U}$ and $V = b + \overrightarrow{U}$ for any $a \in U$ and any $b \in V$, and thus V is obtained from U by the translation $\overrightarrow{ab}$.

In general, when we talk about n points $a_1, \ldots, a_n$, we mean the sequence $(a_1, \ldots, a_n)$, and not the set $\{a_1, \ldots, a_n\}$ (the a_i's need not be distinct).

By Lemma 2.2, a line is specified by a point $a \in E$ and a nonzero vector $v \in \overrightarrow{E}$, i.e., a line is the set of all points of the form $a + \lambda v$, for $\lambda \in \mathbb{R}$.

We say that three points a, b, c are *collinear* if the vectors $\overrightarrow{ab}$ and $\overrightarrow{ac}$ are linearly dependent. If two of the points a, b, c are distinct, say $a \neq b$, then there is a unique $\lambda \in \mathbb{R}$ such that $\overrightarrow{ac} = \lambda \overrightarrow{ab}$, and we define the ratio $\frac{\overrightarrow{ac}}{\overrightarrow{ab}} = \lambda$.

A plane is specified by a point $a \in E$ and two linearly independent vectors $u, v \in \overrightarrow{E}$, i.e., a plane is the set of all points of the form $a + \lambda u + \mu v$, for $\lambda, \mu \in \mathbb{R}$.

We say that four points a, b, c, d are *coplanar* if the vectors $\overrightarrow{ab}, \overrightarrow{ac}$, and $\overrightarrow{ad}$ are linearly dependent. Hyperplanes will be characterized a little later.

Lemma 2.3. *Given an affine space* $\langle E, \overrightarrow{E}, + \rangle$, *for any family* $(a_i)_{i \in I}$ *of points in* E, *the set* V *of barycenters* $\sum_{i \in I} \lambda_i a_i$ *(where* $\sum_{i \in I} \lambda_i = 1$*) is the smallest affine subspace containing* $(a_i)_{i \in I}$.

Proof. If $(a_i)_{i \in I}$ is empty, then $V = \emptyset$, because of the condition $\sum_{i \in I} \lambda_i = 1$. If $(a_i)_{i \in I}$ is nonempty, then the smallest affine subspace containing $(a_i)_{i \in I}$ must contain the set V of barycenters $\sum_{i \in I} \lambda_i a_i$, and thus, it is enough to show that V is closed under affine combinations, which is immediately verified. $\square$

Given a nonempty subset S of E, the smallest affine subspace of E generated by S is often denoted by $\langle S \rangle$. For example, a line specified by two distinct points a and b is denoted by $\langle a, b \rangle$, or even (a, b), and similarly for planes, etc.

Remarks:

(1) Since it can be shown that the barycenter of n weighted points can be obtained by repeated computations of barycenters of two weighted points, a nonempty subset V of E is an affine subspace iff for every two points $a, b \in V$, the set V contains all barycentric combinations of a and b. If V contains at least two points, then V is an affine subspace iff for any two distinct points $a, b \in V$, the set V contains the line determined by a and b, that is, the set of all points $(1 - \lambda)a + \lambda b$, $\lambda \in \mathbb{R}$.

(2) This result still holds if the field K has at least three distinct elements, but the proof is trickier!

2.6 Affine Independence and Affine Frames

Corresponding to the notion of linear independence in vector spaces, we have the notion of affine independence. Given a family $(a_i)_{i \in I}$ of points in an affine space E, we will reduce the notion of (affine) independence of these points to the (linear) independence of the families $(\overrightarrow{a_i a_j})_{j \in (I - \{i\})}$ of vectors obtained by choosing any a_i

as an origin. First, the following lemma shows that it is sufficient to consider only one of these families.

Lemma 2.4. *Given an affine space* $\langle E, \overrightarrow{E}, + \rangle$, *let* $(a_i)_{i \in I}$ *be a family of points in* E. *If the family* $(\overrightarrow{a_i a_j})_{j \in (I - \{i\})}$ *is linearly independent for some* $i \in I$, *then* $(\overrightarrow{a_i a_j})_{j \in (I - \{i\})}$ *is linearly independent for every* $i \in I$.

Proof. Assume that the family $(\overrightarrow{a_i a_j})_{j \in (I - \{i\})}$ is linearly independent for some specific $i \in I$. Let $k \in I$ with $k \neq i$, and assume that there are some scalars $(\lambda_j)_{j \in (I - \{k\})}$ such that

$$\sum_{j \in (I - \{k\})} \lambda_j \overrightarrow{a_k a_j} = 0.$$

Since

$$\overrightarrow{a_k a_j} = \overrightarrow{a_k a_i} + \overrightarrow{a_i a_j},$$

we have

$$\sum_{j \in (I - \{k\})} \lambda_j \overrightarrow{a_k a_j} = \sum_{j \in (I - \{k\})} \lambda_j \overrightarrow{a_k a_i} + \sum_{j \in (I - \{k\})} \lambda_j \overrightarrow{a_i a_j},$$

$$= \sum_{j \in (I - \{k\})} \lambda_j \overrightarrow{a_k a_i} + \sum_{j \in (I - \{i,k\})} \lambda_j \overrightarrow{a_i a_j},$$

$$= \sum_{j \in (I - \{i,k\})} \lambda_j \overrightarrow{a_i a_j} - \left(\sum_{j \in (I - \{k\})} \lambda_j \right) \overrightarrow{a_i a_k},$$

and thus

$$\sum_{j \in (I - \{i,k\})} \lambda_j \overrightarrow{a_i a_j} - \left(\sum_{j \in (I - \{k\})} \lambda_j \right) \overrightarrow{a_i a_k} = 0.$$

Since the family $(\overrightarrow{a_i a_j})_{j \in (I - \{i\})}$ is linearly independent, we must have $\lambda_j = 0$ for all $j \in (I - \{i,k\})$ and $\sum_{j \in (I - \{k\})} \lambda_j = 0$, which implies that $\lambda_j = 0$ for all $j \in (I - \{k\})$. $\square$

We define affine independence as follows.

Definition 2.4. Given an affine space $\langle E, \overrightarrow{E}, + \rangle$, a family $(a_i)_{i \in I}$ of points in E is *affinely independent* if the family $(\overrightarrow{a_i a_j})_{j \in (I - \{i\})}$ is linearly independent for some $i \in I$.

Definition 2.4 is reasonable, since by Lemma 2.4, the independence of the family $(\overrightarrow{a_i a_j})_{j \in (I - \{i\})}$ does not depend on the choice of a_i. A crucial property of linearly independent vectors $(u_1, \ldots, u_m)$ is that if a vector v is a linear combination

$$v = \sum_{i=1}^{m} \lambda_i u_i$$

of the u_i, then the λ_i are unique. A similar result holds for affinely independent points.

Lemma 2.5. *Given an affine space $\langle E, \overrightarrow{E}, + \rangle$, let $(a_0, \ldots, a_m)$ be a family of $m + 1$ points in E. Let $x \in E$, and assume that $x = \sum_{i=0}^{m} \lambda_i a_i$, where $\sum_{i=0}^{m} \lambda_i = 1$. Then, the family $(\lambda_0, \ldots, \lambda_m)$ such that $x = \sum_{i=0}^{m} \lambda_i a_i$ is unique iff the family $(\overrightarrow{a_0 a_1}, \ldots, \overrightarrow{a_0 a_m})$ is linearly independent.*

Proof. The proof is straightforward and is omitted. It is also given in Gallier [5]. $\square$

Lemma 2.5 suggests the notion of affine frame. Affine frames are the affine analogues of bases in vector spaces. Let $\langle E, \overrightarrow{E}, + \rangle$ be a nonempty affine space, and let $(a_0, \ldots, a_m)$ be a family of $m + 1$ points in E. The family $(a_0, \ldots, a_m)$ determines the family of m vectors $(\overrightarrow{a_0 a_1}, \ldots, \overrightarrow{a_0 a_m})$ in $\overrightarrow{E}$. Conversely, given a point a_0 in E and a family of m vectors $(u_1, \ldots, u_m)$ in $\overrightarrow{E}$, we obtain the family of $m + 1$ points $(a_0, \ldots, a_m)$ in E, where $a_i = a_0 + u_i$, $1 \leq i \leq m$.

Thus, for any $m \geq 1$, it is equivalent to consider a family of $m + 1$ points $(a_0, \ldots, a_m)$ in E, and a pair $(a_0, (u_1, \ldots, u_m))$, where the u_i are vectors in $\overrightarrow{E}$. Figure 2.8 illustrates the notion of affine independence.

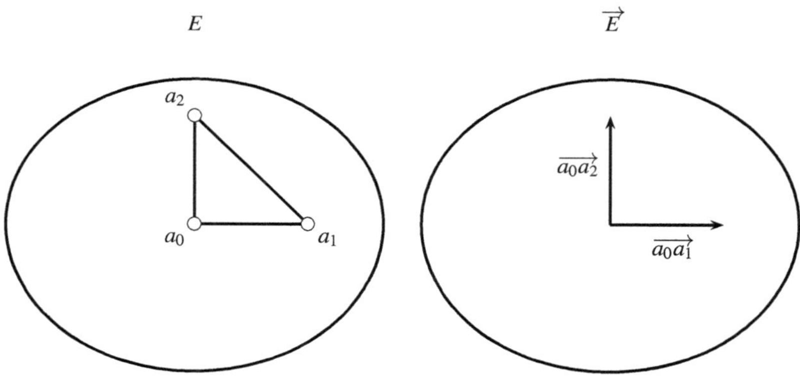

Fig. 2.8 Affine independence and linear independence.

Remark: The above observation also applies to infinite families $(a_i)_{i \in I}$ of points in E and families $(\overrightarrow{u_i})_{i \in I - \{0\}}$ of vectors in $\overrightarrow{E}$, provided that the index set I contains 0.

When $(\overrightarrow{a_0 a_1}, \ldots, \overrightarrow{a_0 a_m})$ is a basis of $\overrightarrow{E}$ then, for every $x \in E$, since $x = a_0 + \overrightarrow{a_0 x}$, there is a unique family $(x_1, \ldots, x_m)$ of scalars such that

$$x = a_0 + x_1 \overrightarrow{a_0 a_1} + \cdots + x_m \overrightarrow{a_0 a_m}.$$

The scalars $(x_1, \ldots, x_m)$ may be considered as coordinates with respect to $(a_0, (\overrightarrow{a_0 a_1}, \ldots, \overrightarrow{a_0 a_m}))$. Since

$$x = a_0 + \sum_{i=1}^{m} x_i \overrightarrow{a_0 a_i} \quad \text{iff} \quad x = \left(1 - \sum_{i=1}^{m} x_i\right) a_0 + \sum_{i=1}^{m} x_i a_i,$$

$x \in E$ can also be expressed uniquely as

$$x = \sum_{i=0}^{m} \lambda_i a_i$$

with $\sum_{i=0}^{m} \lambda_i = 1$, and where $\lambda_0 = 1 - \sum_{i=1}^{m} x_i$, and $\lambda_i = x_i$ for $1 \leq i \leq m$. The scalars $(\lambda_0, \ldots, \lambda_m)$ are also certain kinds of coordinates with respect to $(a_0, \ldots, a_m)$. All this is summarized in the following definition.

Definition 2.5. Given an affine space $\langle E, \overrightarrow{E}, + \rangle$, an *affine frame with origin a_0* is a family $(a_0, \ldots, a_m)$ of $m+1$ points in E such that the list of vectors $(\overrightarrow{a_0 a_1}, \ldots, \overrightarrow{a_0 a_m})$ is a basis of $\overrightarrow{E}$. The pair $(a_0, (\overrightarrow{a_0 a_1}, \ldots, \overrightarrow{a_0 a_m}))$ is also called an *affine frame with origin a_0*. Then, every $x \in E$ can be expressed as

$$x = a_0 + x_1 \overrightarrow{a_0 a_1} + \cdots + x_m \overrightarrow{a_0 a_m}$$

for a unique family $(x_1, \ldots, x_m)$ of scalars, called the *coordinates of x w.r.t. the affine frame* $(a_0, (\overrightarrow{a_0 a_1}, \ldots, \overrightarrow{a_0 a_m}))$. Furthermore, every $x \in E$ can be written as

$$x = \lambda_0 a_0 + \cdots + \lambda_m a_m$$

for some unique family $(\lambda_0, \ldots, \lambda_m)$ of scalars such that $\lambda_0 + \cdots + \lambda_m = 1$ called the *barycentric coordinates of x with respect to the affine frame* $(a_0, \ldots, a_m)$.

The coordinates $(x_1, \ldots, x_m)$ and the barycentric coordinates $(\lambda_0, \ldots, \lambda_m)$ are related by the equations $\lambda_0 = 1 - \sum_{i=1}^{m} x_i$ and $\lambda_i = x_i$, for $1 \leq i \leq m$. An affine frame is called an *affine basis* by some authors. A family $(a_i)_{i \in I}$ of points in E is *affinely dependent* if it is not affinely independent. We can also characterize affinely dependent families as follows.

Lemma 2.6. *Given an affine space* $\langle E, \overrightarrow{E}, + \rangle$, *let* $(a_i)_{i \in I}$ *be a family of points in E. The family* $(a_i)_{i \in I}$ *is affinely dependent iff there is a family* $(\lambda_i)_{i \in I}$ *such that* $\lambda_j \neq 0$ *for some* $j \in I$, $\sum_{i \in I} \lambda_i = 0$, *and* $\sum_{i \in I} \lambda_i \overrightarrow{x a_i} = 0$ *for every* $x \in E$.

Proof. By Lemma 2.5, the family $(a_i)_{i \in I}$ is affinely dependent iff the family of vectors $(\overrightarrow{a_i a_j})_{j \in (I - \{i\})}$ is linearly dependent for some $i \in I$. For any $i \in I$, the family $(\overrightarrow{a_i a_j})_{j \in (I - \{i\})}$ is linearly dependent iff there is a family $(\lambda_j)_{j \in (I - \{i\})}$ such that $\lambda_j \neq 0$ for some j, and such that

$$\sum_{j \in (I - \{i\})} \lambda_j \overrightarrow{a_i a_j} = 0.$$

Then, for any $x \in E$, we have

$$\sum_{j\in(I-\{i\})}\lambda_j\,\overrightarrow{a_ia_j}=\sum_{j\in(I-\{i\})}\lambda_j(\overrightarrow{xa_j}-\overrightarrow{xa_i})$$

$$=\sum_{j\in(I-\{i\})}\lambda_j\,\overrightarrow{xa_j}-\left(\sum_{j\in(I-\{i\})}\lambda_j\right)\overrightarrow{xa_i},$$

and letting $\lambda_i=-\left(\sum_{j\in(I-\{i\})}\lambda_j\right)$, we get $\sum_{i\in I}\lambda_i\overrightarrow{xa_i}=0$, with $\sum_{i\in I}\lambda_i=0$ and $\lambda_j\neq 0$ for some $j\in I$. The converse is obvious by setting $x=a_i$ for some i such that $\lambda_i\neq 0$, since $\sum_{i\in I}\lambda_i=0$ implies that $\lambda_j\neq 0$, for some $j\neq i$. $\square$

Even though Lemma 2.6 is rather dull, it is one of the key ingredients in the proof of beautiful and deep theorems about convex sets, such as Carathéodory's theorem, Radon's theorem, and Helly's theorem (see Section 3.1).

A family of two points (a,b) in E is affinely independent iff $\overrightarrow{ab}\neq 0$, iff $a\neq b$. If $a\neq b$, the affine subspace generated by a and b is the set of all points $(1-\lambda)a+\lambda b$, which is the unique line passing through a and b. A family of three points (a,b,c) in E is affinely independent iff $\overrightarrow{ab}$ and $\overrightarrow{ac}$ are linearly independent, which means that a, b, and c are not on the same line (they are not collinear). In this case, the affine subspace generated by (a,b,c) is the set of all points $(1-\lambda-\mu)a+\lambda b+\mu c$, which is the unique plane containing a, b, and c. A family of four points (a,b,c,d) in E is affinely independent iff $\overrightarrow{ab}$, $\overrightarrow{ac}$, and $\overrightarrow{ad}$ are linearly independent, which means that a, b, c, and d are not in the same plane (they are not coplanar). In this case, a, b, c, and d are the vertices of a tetrahedron. Figure 2.9 shows affine frames and their convex hulls for $|I|=0,1,2,3$.

Given $n+1$ affinely independent points $(a_0,\dots,a_n)$ in E, we can consider the set of points $\lambda_0a_0+\cdots+\lambda_na_n$, where $\lambda_0+\cdots+\lambda_n=1$ and $\lambda_i\geq 0$ $(\lambda_i\in\mathbb{R})$. Such affine combinations are called *convex combinations*. This set is called the *convex hull* of $(a_0,\dots,a_n)$ (or *n-simplex spanned by* $(a_0,\dots,a_n)$). When $n=1$, we get the segment between a_0 and a_1, including a_0 and a_1. When $n=2$, we get the interior of the triangle whose vertices are a_0,a_1,a_2, including boundary points (the edges). When $n=3$, we get the interior of the tetrahedron whose vertices are a_0,a_1,a_2,a_3, including boundary points (faces and edges). The set

$$\{a_0+\lambda_1\overrightarrow{a_0a_1}+\cdots+\lambda_n\overrightarrow{a_0a_n}\mid\text{ where }0\leq\lambda_i\leq 1\ (\lambda_i\in\mathbb{R})\}$$

is called the *parallelotope spanned by* $(a_0,\dots,a_n)$. When E has dimension 2, a parallelotope is also called a *parallelogram*, and when E has dimension 3, a *parallelepiped*.

More generally, we say that a subset V of E is *convex* if for any two points $a,b\in V$, we have $c\in V$ for every point $c=(1-\lambda)a+\lambda b$, with $0\leq\lambda\leq 1$ $(\lambda\in\mathbb{R})$.

Points are not vectors! The following example illustrates why treating points as vectors may cause problems. Let a,b,c be three affinely independent points in $\mathbb{A}^3$. Any point x in the plane (a,b,c) can be expressed as

$$x=\lambda_0a+\lambda_1b+\lambda_2c,$$

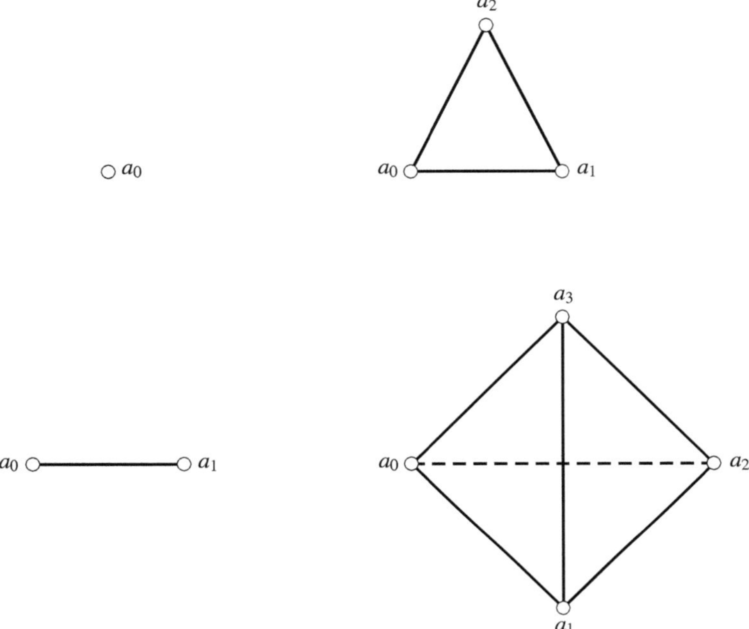

Fig. 2.9 Examples of affine frames and their convex hulls.

where $\lambda_0 + \lambda_1 + \lambda_2 = 1$. How can we compute $\lambda_0, \lambda_1, \lambda_2$? Letting $a = (a_1, a_2, a_3)$, $b = (b_1, b_2, b_3)$, $c = (c_1, c_2, c_3)$, and $x = (x_1, x_2, x_3)$ be the coordinates of a, b, c, x in the standard frame of $\mathbb{A}^3$, it is tempting to solve the system of equations

$$\begin{pmatrix} a_1 & b_1 & c_1 \\ a_2 & b_2 & c_2 \\ a_3 & b_3 & c_3 \end{pmatrix} \begin{pmatrix} \lambda_0 \\ \lambda_1 \\ \lambda_2 \end{pmatrix} = \begin{pmatrix} x_1 \\ x_2 \\ x_3 \end{pmatrix}.$$

However, there is a problem when the origin of the coordinate system belongs to the plane (a, b, c), since in this case, the matrix is not invertible! What we should really be doing is to solve the system

$$\lambda_0 \overrightarrow{Oa} + \lambda_1 \overrightarrow{Ob} + \lambda_2 \overrightarrow{Oc} = \overrightarrow{Ox},$$

where O is any point **not** in the plane (a, b, c). An alternative is to use certain well-chosen cross products.

It can be shown that barycentric coordinates correspond to various ratios of areas and volumes; see the problems.

2.7 Affine Maps

Corresponding to linear maps we have the notion of an affine map. An affine map is defined as a map preserving affine combinations.

Definition 2.6. Given two affine spaces $\langle E, \overrightarrow{E}, + \rangle$ and $\langle E', \overrightarrow{E'}, +' \rangle$, a function $f : E \to E'$ is an *affine map* iff for every family $((a_i, \lambda_i))_{i \in I}$ of weighted points in E such that $\sum_{i \in I} \lambda_i = 1$, we have

$$f\left(\sum_{i \in I} \lambda_i a_i\right) = \sum_{i \in I} \lambda_i f(a_i).$$

In other words, f preserves barycenters.

Affine maps can be obtained from linear maps as follows. For simplicity of notation, the same symbol $+$ is used for both affine spaces (instead of using both $+$ and $+'$).

Given any point $a \in E$, any point $b \in E'$, and any linear map $h \colon \overrightarrow{E} \to \overrightarrow{E'}$, we claim that the map $f : E \to E'$ defined such that

$$f(a + v) = b + h(v)$$

is an affine map. Indeed, for any family $(\lambda_i)_{i \in I}$ of scalars with $\sum_{i \in I} \lambda_i = 1$ and any family $(\overrightarrow{v_i})_{i \in I}$, since

$$\sum_{i \in I} \lambda_i(a + v_i) = a + \sum_{i \in I} \lambda_i \overrightarrow{a(a + v_i)} = a + \sum_{i \in I} \lambda_i v_i$$

and

$$\sum_{i \in I} \lambda_i(b + h(v_i)) = b + \sum_{i \in I} \lambda_i \overrightarrow{b(b + h(v_i))} = b + \sum_{i \in I} \lambda_i h(v_i),$$

we have

$$\begin{aligned}
f\left(\sum_{i \in I} \lambda_i(a + v_i)\right) &= f\left(a + \sum_{i \in I} \lambda_i v_i\right) \\
&= b + h\left(\sum_{i \in I} \lambda_i v_i\right) \\
&= b + \sum_{i \in I} \lambda_i h(v_i) \\
&= \sum_{i \in I} \lambda_i(b + h(v_i)) \\
&= \sum_{i \in I} \lambda_i f(a + v_i).
\end{aligned}$$

$\square$

Note that the condition $\sum_{i \in I} \lambda_i = 1$ was implicitly used (in a hidden call to Lemma 2.1) in deriving that

$$\sum_{i \in I} \lambda_i(a + v_i) = a + \sum_{i \in I} \lambda_i v_i$$

and

$$\sum_{i \in I} \lambda_i(b + h(v_i)) = b + \sum_{i \in I} \lambda_i h(v_i).$$

As a more concrete example, the map

$$\begin{pmatrix} x_1 \\ x_2 \end{pmatrix} \mapsto \begin{pmatrix} 1 & 2 \\ 0 & 1 \end{pmatrix}\begin{pmatrix} x_1 \\ x_2 \end{pmatrix} + \begin{pmatrix} 3 \\ 1 \end{pmatrix}$$

defines an affine map in $\mathbb{A}^2$. It is a "shear" followed by a translation. The effect of this shear on the square (a,b,c,d) is shown in Figure 2.10. The image of the square (a,b,c,d) is the parallelogram (a',b',c',d').

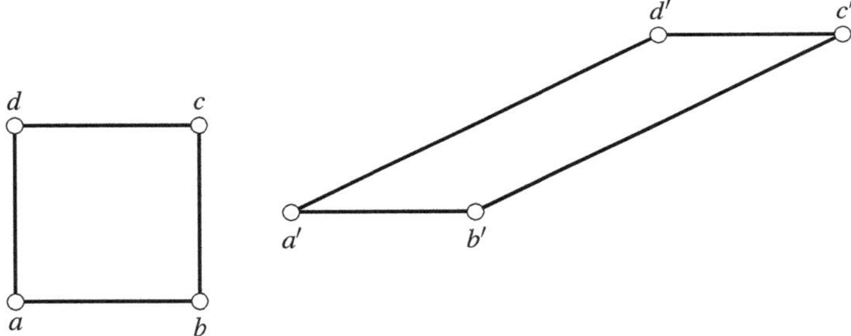

Fig. 2.10 The effect of a shear.

Let us consider one more example. The map

$$\begin{pmatrix} x_1 \\ x_2 \end{pmatrix} \mapsto \begin{pmatrix} 1 & 1 \\ 1 & 3 \end{pmatrix}\begin{pmatrix} x_1 \\ x_2 \end{pmatrix} + \begin{pmatrix} 3 \\ 0 \end{pmatrix}$$

is an affine map. Since we can write

$$\begin{pmatrix} 1 & 1 \\ 1 & 3 \end{pmatrix} = \sqrt{2}\begin{pmatrix} \sqrt{2}/2 & -\sqrt{2}/2 \\ 2/2 & \sqrt{2}/2 \end{pmatrix}\begin{pmatrix} 1 & 2 \\ 0 & 1 \end{pmatrix},$$

this affine map is the composition of a shear, followed by a rotation of angle $\pi/4$, followed by a magnification of ratio $\sqrt{2}$, followed by a translation. The effect of this map on the square (a,b,c,d) is shown in Figure 2.11. The image of the square (a,b,c,d) is the parallelogram (a',b',c',d').

The following lemma shows the converse of what we just showed. Every affine map is determined by the image of any point and a linear map.

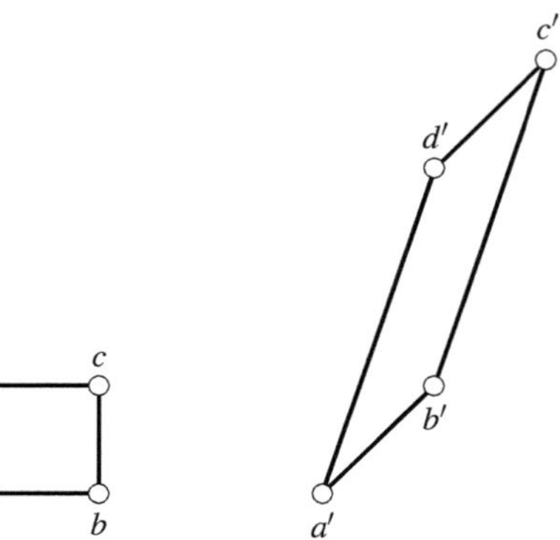

Fig. 2.11 The effect of an affine map.

Lemma 2.7. *Given an affine map* $f : E \to E'$, *there is a unique linear map* $\overrightarrow{f} : \overrightarrow{E} \to \overrightarrow{E'}$ *such that*

$$f(a + v) = f(a) + \overrightarrow{f}(v),$$

for every $a \in E$ *and every* $v \in \overrightarrow{E}$.

Proof. Let $a \in E$ be any point in E. We claim that the map defined such that

$$\overrightarrow{f}(v) = \overrightarrow{f(a)f(a + v)}$$

for every $v \in \overrightarrow{E}$ is a linear map $\overrightarrow{f} : \overrightarrow{E} \to \overrightarrow{E'}$. Indeed, we can write

$$a + \lambda v = \lambda(a + v) + (1 - \lambda)a,$$

since $a + \lambda v = a + \lambda \overrightarrow{a(a + v)} + (1 - \lambda)\overrightarrow{aa}$, and also

$$a + u + v = (a + u) + (a + v) - a,$$

since $a + u + v = a + \overrightarrow{a(a + u)} + \overrightarrow{a(a + v)} - \overrightarrow{aa}$. Since f preserves barycenters, we get

$$f(a + \lambda v) = \lambda f(a + v) + (1 - \lambda)f(a).$$

If we recall that $x = \sum_{i \in I} \lambda_i a_i$ is the barycenter of a family $((a_i, \lambda_i))_{i \in I}$ of weighted points (with $\sum_{i \in I} \lambda_i = 1$) iff

$$\overrightarrow{bx} = \sum_{i \in I} \lambda_i \overrightarrow{ba_i} \quad \text{for every } b \in E,$$

we get

$$\overrightarrow{f(a)f(a + \lambda v)} = \lambda \overrightarrow{f(a)f(a + v)} + (1 - \lambda)\overrightarrow{f(a)f(a)} = \lambda \overrightarrow{f(a)f(a + v)},$$

showing that $\overrightarrow{f}(\lambda v) = \lambda \overrightarrow{f}(v)$. We also have

$$f(a+u+v) = f(a+u) + f(a+v) - f(a),$$

from which we get

$$\overrightarrow{f(a)f(a+u+v)} = \overrightarrow{f(a)f(a+u)} + \overrightarrow{f(a)f(a+v)},$$

showing that $\overrightarrow{f}(u+v) = \overrightarrow{f}(u) + \overrightarrow{f}(v)$. Consequently, $\overrightarrow{f}$ is a linear map. For any other point $b \in E$, since

$$b+v = a+\overrightarrow{ab}+v = a+\overrightarrow{a(a+v)} - \overrightarrow{aa} + \overrightarrow{ab},$$

$b+v = (a+v) - a+b$, and since f preserves barycenters, we get

$$f(b+v) = f(a+v) - f(a) + f(b),$$

which implies that

$$\begin{aligned}
\overrightarrow{f(b)f(b+v)} &= \overrightarrow{f(b)f(a+v)} - \overrightarrow{f(b)f(a)} + \overrightarrow{f(b)f(b)}, \\
&= \overrightarrow{f(a)f(b)} + \overrightarrow{f(b)f(a+v)}, \\
&= \overrightarrow{f(a)f(a+v)}.
\end{aligned}$$

Thus, $\overrightarrow{f(b)f(b+v)} = \overrightarrow{f(a)f(a+v)}$, which shows that the definition of $\overrightarrow{f}$ does not depend on the choice of $a \in E$. The fact that $\overrightarrow{f}$ is unique is obvious: We must have $\overrightarrow{f}(v) = \overrightarrow{f(a)f(a+v)}$. $\square$

The unique linear map $\overrightarrow{f}: \overrightarrow{E} \to \overrightarrow{E'}$ given by Lemma 2.7 is called the *linear map associated with the affine map f*.

Note that the condition

$$f(a+v) = f(a) + \overrightarrow{f}(v),$$

for every $a \in E$ and every $v \in \overrightarrow{E}$, can be stated equivalently as

$$f(x) = f(a) + \overrightarrow{f}(\overrightarrow{ax}), \quad \text{or} \quad \overrightarrow{f(a)f(x)} = \overrightarrow{f}(\overrightarrow{ax}),$$

for all $a, x \in E$. Lemma 2.7 shows that for any affine map $f: E \to E'$, there are points $a \in E$, $b \in E'$, and a unique linear map $\overrightarrow{f}: \overrightarrow{E} \to \overrightarrow{E'}$, such that

$$f(a+v) = b + \overrightarrow{f}(v),$$

for all $v \in \overrightarrow{E}$ (just let $b = f(a)$, for any $a \in E$). Affine maps for which $\overrightarrow{f}$ is the identity map are called *translations*. Indeed, if $\overrightarrow{f} = \mathrm{id}$,

$$f(x) = f(a) + \overrightarrow{f}(\overrightarrow{ax}) = f(a) + \overrightarrow{ax} = x + \overrightarrow{xa} + \overrightarrow{af(a)} + \overrightarrow{ax}$$
$$= x + \overrightarrow{xa} + \overrightarrow{af(a)} - \overrightarrow{xa} = x + \overrightarrow{af(a)},$$

and so

$$\overrightarrow{xf(x)} = \overrightarrow{af(a)},$$

which shows that f is the translation induced by the vector $\overrightarrow{af(a)}$ (which does not depend on a).

Since an affine map preserves barycenters, and since an affine subspace V is closed under barycentric combinations, the image $f(V)$ of V is an affine subspace in E'. So, for example, the image of a line is a point or a line, and the image of a plane is either a point, a line, or a plane.

It is easily verified that the composition of two affine maps is an affine map. Also, given affine maps $f \colon E \to E'$ and $g \colon E' \to E''$, we have

$$g(f(a + v)) = g\left(f(a) + \overrightarrow{f}(v)\right) = g(f(a)) + \overrightarrow{g}\left(\overrightarrow{f}(v)\right),$$

which shows that $\overrightarrow{g \circ f} = \overrightarrow{g} \circ \overrightarrow{f}$. It is easy to show that an affine map $f \colon E \to E'$ is injective iff $\overrightarrow{f} \colon \overrightarrow{E} \to \overrightarrow{E'}$ is injective, and that $f \colon E \to E'$ is surjective iff $\overrightarrow{f} \colon \overrightarrow{E} \to \overrightarrow{E'}$ is surjective. An affine map $f \colon E \to E'$ is constant iff $\overrightarrow{f} \colon \overrightarrow{E} \to \overrightarrow{E'}$ is the null (constant) linear map equal to 0 for all $v \in \overrightarrow{E}$.

If E is an affine space of dimension m and $(a_0, a_1, \ldots, a_m)$ is an affine frame for E, then for any other affine space F and for any sequence $(b_0, b_1, \ldots, b_m)$ of $m + 1$ points in F, there is a unique affine map $f \colon E \to F$ such that $f(a_i) = b_i$, for $0 \leq i \leq m$. Indeed, f must be such that

$$f(\lambda_0 a_0 + \cdots + \lambda_m a_m) = \lambda_0 b_0 + \cdots + \lambda_m b_m,$$

where $\lambda_0 + \cdots + \lambda_m = 1$, and this defines a unique affine map on all of E, since $(a_0, a_1, \ldots, a_m)$ is an affine frame for E.

Using affine frames, affine maps can be represented in terms of matrices. We explain how an affine map $f \colon E \to E$ is represented with respect to a frame $(a_0, \ldots, a_n)$ in E, the more general case where an affine map $f \colon E \to F$ is represented with respect to two affine frames $(a_0, \ldots, a_n)$ in E and $(b_0, \ldots, b_m)$ in F being analogous. Since

$$f(a_0 + x) = f(a_0) + \overrightarrow{f}(x)$$

for all $x \in \overrightarrow{E}$, we have

$$\overrightarrow{a_0 f(a_0 + x)} = \overrightarrow{a_0 f(a_0)} + \overrightarrow{f}(x).$$

Since x, $\overrightarrow{a_0 f(a_0)}$, and $\overrightarrow{a_0 f(a_0 + x)}$, can be expressed as

$$
\begin{aligned}
x &= x_1 \overrightarrow{a_0 a_1} + \cdots + x_n \overrightarrow{a_0 a_n}, \\
\overrightarrow{a_0 f(a_0)} &= b_1 \overrightarrow{a_0 a_1} + \cdots + b_n \overrightarrow{a_0 a_n}, \\
\overrightarrow{a_0 f(a_0 + x)} &= y_1 \overrightarrow{a_0 a_1} + \cdots + y_n \overrightarrow{a_0 a_n},
\end{aligned}
$$

if $A = (a_{ij})$ is the $n \times n$ matrix of the linear map $\overrightarrow{f}$ over the basis $(\overrightarrow{a_0 a_1}, \ldots, \overrightarrow{a_0 a_n})$, letting x, y, and b denote the column vectors of components $(x_1, \ldots, x_n)$, $(y_1, \ldots, y_n)$, and $(b_1, \ldots, b_n)$,

$$
\overrightarrow{a_0 f(a_0 + x)} = \overrightarrow{a_0 f(a_0)} + \overrightarrow{f}(x)
$$

is equivalent to

$$
y = Ax + b.
$$

Note that $b \neq 0$ unless $f(a_0) = a_0$. Thus, f is generally not a linear transformation, unless it has a *fixed point*, i.e., there is a point a_0 such that $f(a_0) = a_0$. The vector b is the "translation part" of the affine map. Affine maps do not always have a fixed point. Obviously, nonnull translations have no fixed point. A less trivial example is given by the affine map

$$
\begin{pmatrix} x_1 \\ x_2 \end{pmatrix} \mapsto \begin{pmatrix} 1 & 0 \\ 0 & -1 \end{pmatrix} \begin{pmatrix} x_1 \\ x_2 \end{pmatrix} + \begin{pmatrix} 1 \\ 0 \end{pmatrix}.
$$

This map is a reflection about the x-axis followed by a translation along the x-axis. The affine map

$$
\begin{pmatrix} x_1 \\ x_2 \end{pmatrix} \mapsto \begin{pmatrix} 1 & -\sqrt{3} \\ \sqrt{3}/4 & 1/4 \end{pmatrix} \begin{pmatrix} x_1 \\ x_2 \end{pmatrix} + \begin{pmatrix} 1 \\ 1 \end{pmatrix}
$$

can also be written as

$$
\begin{pmatrix} x_1 \\ x_2 \end{pmatrix} \mapsto \begin{pmatrix} 2 & 0 \\ 0 & 1/2 \end{pmatrix} \begin{pmatrix} 1/2 & -\sqrt{3}/2 \\ \sqrt{3}/2 & 1/2 \end{pmatrix} \begin{pmatrix} x_1 \\ x_2 \end{pmatrix} + \begin{pmatrix} 1 \\ 1 \end{pmatrix}
$$

which shows that it is the composition of a rotation of angle $\pi/3$, followed by a stretch (by a factor of 2 along the x-axis, and by a factor of $\frac{1}{2}$ along the y-axis), followed by a translation. It is easy to show that this affine map has a unique fixed point. On the other hand, the affine map

$$
\begin{pmatrix} x_1 \\ x_2 \end{pmatrix} \mapsto \begin{pmatrix} 8/5 & -6/5 \\ 3/10 & 2/5 \end{pmatrix} \begin{pmatrix} x_1 \\ x_2 \end{pmatrix} + \begin{pmatrix} 1 \\ 1 \end{pmatrix}
$$

has no fixed point, even though

$$
\begin{pmatrix} 8/5 & -6/5 \\ 3/10 & 2/5 \end{pmatrix} = \begin{pmatrix} 2 & 0 \\ 0 & 1/2 \end{pmatrix} \begin{pmatrix} 4/5 & -3/5 \\ 3/5 & 4/5 \end{pmatrix},
$$

and the second matrix is a rotation of angle θ such that $\cos\theta = \frac{4}{5}$ and $\sin\theta = \frac{3}{5}$. For more on fixed points of affine maps, see the problems.

There is a useful trick to convert the equation $y = Ax + b$ into what looks like a linear equation. The trick is to consider an $(n+1)\times(n+1)$ matrix. We add 1 as the $(n+1)$th component to the vectors x, y, and b, and form the $(n+1)\times(n+1)$ matrix

$$\begin{pmatrix} A & b \\ 0 & 1 \end{pmatrix}$$

so that $y = Ax + b$ is equivalent to

$$\begin{pmatrix} y \\ 1 \end{pmatrix} = \begin{pmatrix} A & b \\ 0 & 1 \end{pmatrix}\begin{pmatrix} x \\ 1 \end{pmatrix}.$$

This trick is very useful in kinematics and dynamics, where A is a rotation matrix. Such affine maps are called *rigid motions*.

If $f\colon E \to E'$ is a bijective affine map, given any three collinear points a, b, c in E, with $a \neq b$, where, say, $c = (1-\lambda)a + \lambda b$, since f preserves barycenters, we have $f(c) = (1-\lambda)f(a) + \lambda f(b)$, which shows that $f(a), f(b), f(c)$ are collinear in E'. There is a converse to this property, which is simpler to state when the ground field is $K = \mathbb{R}$. The converse states that given any bijective function $f\colon E \to E'$ between two real affine spaces of the same dimension $n \geq 2$, if f maps any three collinear points to collinear points, then f is affine. The proof is rather long (see Berger [2] or Samuel [10]).

Given three collinear points a, b, c, where $a \neq c$, we have $b = (1-\beta)a + \beta c$ for some unique β, and we define the *ratio of the sequence $a, b,\ c$*, as

$$\mathrm{ratio}(a,b,c) = \frac{\beta}{(1-\beta)} = \frac{\overrightarrow{ab}}{\overrightarrow{bc}},$$

provided that $\beta \neq 1$, i.e., $b \neq c$. When $b = c$, we agree that $\mathrm{ratio}(a,b,c) = \infty$. We warn our readers that other authors define the ratio of a,b,c as $-\mathrm{ratio}(a,b,c) = \frac{\overrightarrow{ba}}{\overrightarrow{bc}}$. Since affine maps preserve barycenters, it is clear that affine maps preserve the ratio of three points.

2.8 Affine Groups

We now take a quick look at the bijective affine maps. Given an affine space E, the set of affine bijections $f\colon E \to E$ is clearly a group, called the *affine group of E*, and denoted by $\mathbf{GA}(E)$. Recall that the group of bijective linear maps of the vector space $\overrightarrow{E}$ is denoted by $\mathbf{GL}(\overrightarrow{E})$. Then, the map $f \mapsto \overrightarrow{f}$ defines a group homomorphism $L\colon \mathbf{GA}(E) \to \mathbf{GL}(\overrightarrow{E})$. The kernel of this map is the set of translations on E.

The subset of all linear maps of the form $\lambda\,\mathrm{id}_{\overrightarrow{E}}$, where $\lambda \in \mathbb{R} - \{0\}$, is a subgroup of $\mathbf{GL}(\overrightarrow{E})$, and is denoted by $\mathbb{R}^*\mathrm{id}_{\overrightarrow{E}}$ (where $\lambda\,\mathrm{id}_{\overrightarrow{E}}(u) = \lambda u$, and $\mathbb{R}^* = \mathbb{R} - \{0\}$). The subgroup $\mathbf{DIL}(E) = L^{-1}(\mathbb{R}^*\mathrm{id}_{\overrightarrow{E}})$ of $\mathbf{GA}(E)$ is particularly interesting. It turns out that it is the disjoint union of the translations and of the dilatations of ratio $\lambda \neq 1$. The elements of $\mathbf{DIL}(E)$ are called *affine dilatations*.

Given any point $a \in E$, and any scalar $\lambda \in \mathbb{R}$, a *dilatation or central dilatation (or homothety) of center a and ratio* λ is a map $H_{a,\lambda}$ defined such that

$$H_{a,\lambda}(x) = a + \lambda\,\overrightarrow{ax},$$

for every $x \in E$.

Remark: The terminology does not seem to be universally agreed upon. The terms *affine dilatation* and *central dilatation* are used by Pedoe [9]. Snapper and Troyer use the term *dilation* for an affine dilatation and *magnification* for a central dilatation [11]. Samuel uses *homothety* for a central dilatation, a direct translation of the French "homothétie" [10]. Since dilation is shorter than dilatation and somewhat easier to pronounce, perhaps we should use that!

Observe that $H_{a,\lambda}(a) = a$, and when $\lambda \neq 0$ and $x \neq a$, $H_{a,\lambda}(x)$ is on the line defined by a and x, and is obtained by "scaling" $\overrightarrow{ax}$ by λ.

Figure 2.12 shows the effect of a central dilatation of center d. The triangle (a,b,c) is magnified to the triangle (a',b',c'). Note how every line is mapped to a parallel line.

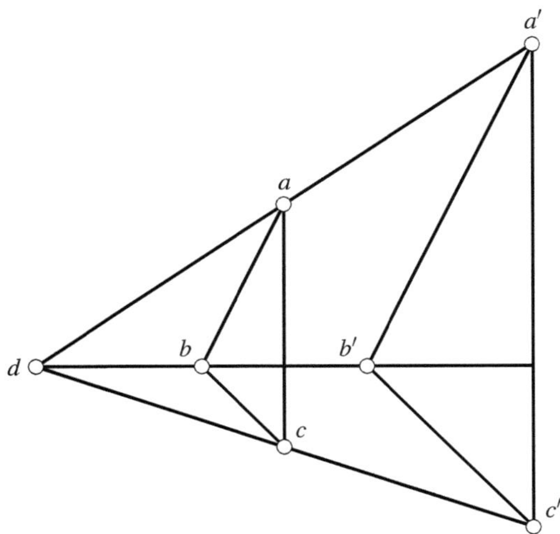

Fig. 2.12 The effect of a central dilatation.

When $\lambda = 1$, $H_{a,1}$ is the identity. Note that $\overrightarrow{H_{a,\lambda}} = \lambda\,\mathrm{id}_{\overrightarrow{E}}$. When $\lambda \neq 0$, it is clear that $H_{a,\lambda}$ is an affine bijection. It is immediately verified that

$$H_{a,\lambda} \circ H_{a,\mu} = H_{a,\lambda\mu}.$$

We have the following useful result.

Lemma 2.8. *Given any affine space E, for any affine bijection $f \in \mathbf{GA}(E)$, if $\overrightarrow{f} = \lambda\,\mathrm{id}_{\overrightarrow{E}}$, for some $\lambda \in \mathbb{R}^*$ with $\lambda \neq 1$, then there is a unique point $c \in E$ such that $f = H_{c,\lambda}$.*

Proof. The proof is straightforward, and is omitted. It is also given in Gallier [5]. $\square$

Clearly, if $\overrightarrow{f} = \mathrm{id}_{\overrightarrow{E}}$, the affine map f is a translation. Thus, the group of affine dilatations $\mathbf{DIL}(E)$ is the disjoint union of the translations and of the dilatations of ratio $\lambda \neq 0, 1$. Affine dilatations can be given a purely geometric characterization.

Another point worth mentioning is that affine bijections preserve the ratio of volumes of parallelotopes. Indeed, given any basis $B = (u_1, \ldots, u_m)$ of the vector space $\overrightarrow{E}$ associated with the affine space E, given any $m + 1$ affinely independent points $(a_0, \ldots, a_m)$, we can compute the determinant $\det_B(\overrightarrow{a_0 a_1}, \ldots, \overrightarrow{a_0 a_m})$ w.r.t. the basis B. For any bijective affine map $f \colon E \to E$, since

$$\det_B\left(\overrightarrow{f}(\overrightarrow{a_0 a_1}), \ldots, \overrightarrow{f}(\overrightarrow{a_0 a_m}) \right) = \det(\overrightarrow{f}) \det_B(\overrightarrow{a_0 a_1}, \ldots, \overrightarrow{a_0 a_m})$$

and the determinant of a linear map is intrinsic (i.e., depends only on $\overrightarrow{f}$, and not on the particular basis B), we conclude that the ratio

$$\frac{\det_B\left(\overrightarrow{f}(\overrightarrow{a_0 a_1}), \ldots, \overrightarrow{f}(\overrightarrow{a_0 a_m}) \right)}{\det_B(\overrightarrow{a_0 a_1}, \ldots, \overrightarrow{a_0 a_m})} = \det(\overrightarrow{f})$$

is independent of the basis B. Since $\det_B(\overrightarrow{a_0 a_1}, \ldots, \overrightarrow{a_0 a_m})$ is the volume of the parallelotope spanned by $(a_0, \ldots, a_m)$, where the parallelotope spanned by any point a and the vectors $(u_1, \ldots, u_m)$ has unit volume (see Berger [2], Section 9.12), we see that affine bijections preserve the ratio of volumes of parallelotopes. In fact, this ratio is independent of the choice of the parallelotopes of unit volume. In particular, the affine bijections $f \in \mathbf{GA}(E)$ such that $\det(\overrightarrow{f}) = 1$ preserve volumes. These affine maps form a subgroup $\mathbf{SA}(E)$ of $\mathbf{GA}(E)$ called the *special affine group of E*. We now take a glimpse at affine geometry.

2.9 Affine Geometry: A Glimpse

In this section we state and prove three fundamental results of affine geometry. Roughly speaking, affine geometry is the study of properties invariant under affine bijections. We now prove one of the oldest and most basic results of affine geometry, the theorem of Thales.

Lemma 2.9. *Given any affine space E, if H_1, H_2, H_3 are any three distinct parallel hyperplanes, and A and B are any two lines not parallel to H_i, letting $a_i = H_i \cap A$ and $b_i = H_i \cap B$, then the following ratios are equal:*

$$\frac{\overrightarrow{a_1 a_3}}{\overrightarrow{a_1 a_2}} = \frac{\overrightarrow{b_1 b_3}}{\overrightarrow{b_1 b_2}} = \rho.$$

Conversely, for any point d on the line A, if $\frac{\overrightarrow{a_1 d}}{\overrightarrow{a_1 a_2}} = \rho$, then $d = a_3$.

Proof. Figure 2.13 illustrates the theorem of Thales. We sketch a proof, leaving the details as an exercise. Since H_1, H_2, H_3 are parallel, they have the same direction $\overrightarrow{H}$, a hyperplane in $\overrightarrow{E}$. Let $u \in \overrightarrow{E} - \overrightarrow{H}$ be any nonnull vector such that $A = a_1 + \mathbb{R}u$. Since A is not parallel to H, we have $\overrightarrow{E} = \overrightarrow{H} \oplus \mathbb{R}u$, and thus we can define the linear map $p \colon \overrightarrow{E} \to \mathbb{R}u$, the projection on $\mathbb{R}u$ parallel to $\overrightarrow{H}$. This linear map induces an affine map $f \colon E \to A$, by defining f such that

$$f(b_1 + w) = a_1 + p(w),$$

for all $w \in \overrightarrow{E}$. Clearly, $f(b_1) = a_1$, and since H_1, H_2, H_3 all have direction $\overrightarrow{H}$, we also have $f(b_2) = a_2$ and $f(b_3) = a_3$. Since f is affine, it preserves ratios, and thus

$$\frac{\overrightarrow{a_1 a_3}}{\overrightarrow{a_1 a_2}} = \frac{\overrightarrow{b_1 b_3}}{\overrightarrow{b_1 b_2}}.$$

The converse is immediate. $\square$

We also have the following simple lemma, whose proof is left as an easy exercise.

Lemma 2.10. *Given any affine space E, given any two distinct points $a, b \in E$, and for any affine dilatation f different from the identity, if $a' = f(a)$, $D = \langle a, b \rangle$ is the line passing through a and b, and D' is the line parallel to D and passing through a', the following are equivalent:*

(i) $b' = f(b)$;

(ii) *If f is a translation, then b' is the intersection of D' with the line parallel to $\langle a, a' \rangle$ passing through b;*

  If f is a dilatation of center c, then $b' = D' \cap \langle c, b \rangle$.

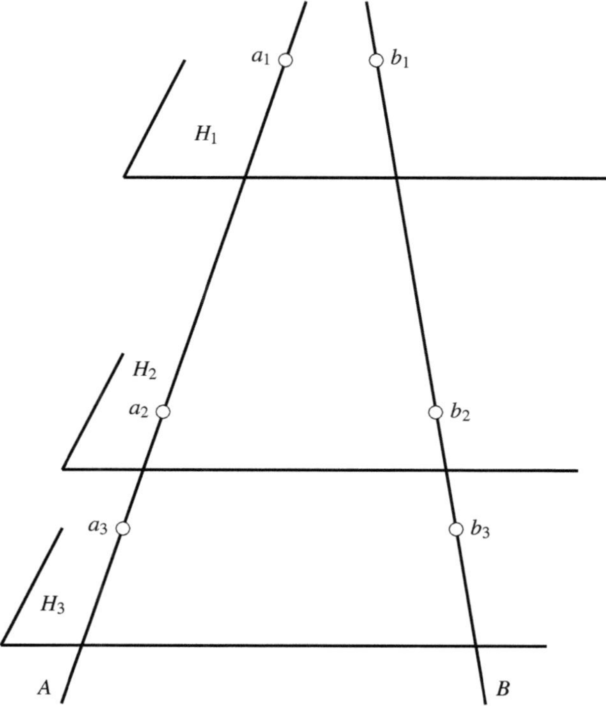

Fig. 2.13 The theorem of Thales.

The first case is the parallelogram law, and the second case follows easily from Thales' theorem.

We are now ready to prove two classical results of affine geometry, Pappus's theorem and Desargues's theorem. Actually, these results are theorems of projective geometry, and we are stating affine versions of these important results. There are stronger versions that are best proved using projective geometry.

Lemma 2.11. *Given any affine plane E, any two distinct lines D and D', then for any distinct points a, b, c on D and a', b', c' on D', if a, b, c, a', b', c' are distinct from the intersection of D and D' (if D and D' intersect) and if the lines $\langle a, b' \rangle$ and $\langle a', b \rangle$ are parallel, and the lines $\langle b, c' \rangle$ and $\langle b', c \rangle$ are parallel, then the lines $\langle a, c' \rangle$ and $\langle a', c \rangle$ are parallel.*

Proof. Pappus's theorem is illustrated in Figure 2.14. If D and D' are not parallel, let d be their intersection. Let f be the dilatation of center d such that $f(a) = b$, and let g be the dilatation of center d such that $g(b) = c$. Since the lines $\langle a, b' \rangle$ and $\langle a', b \rangle$ are parallel, and the lines $\langle b, c' \rangle$ and $\langle b', c \rangle$ are parallel, by Lemma 2.10 we

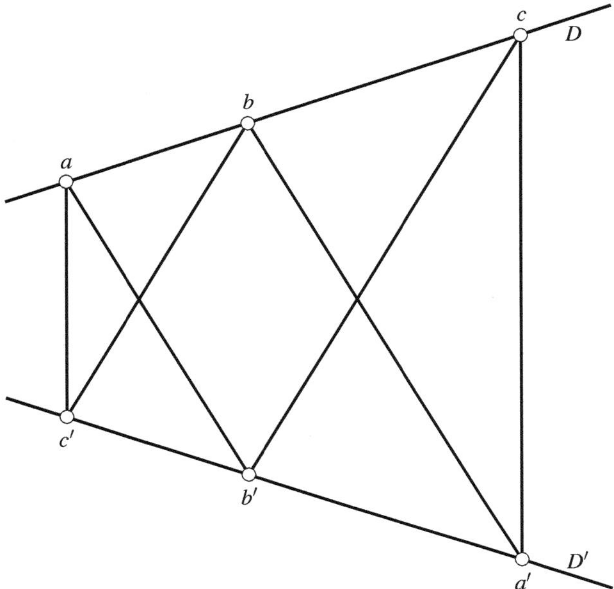

Fig. 2.14 Pappus's theorem (affine version).

have $a' = f(b')$ and $b' = g(c')$. However, we observed that dilatations with the same center commute, and thus $f \circ g = g \circ f$, and thus, letting $h = g \circ f$, we get $c = h(a)$ and $a' = h(c')$. Again, by Lemma 2.10, the lines $\langle a, c' \rangle$ and $\langle a', c \rangle$ are parallel. If D and D' are parallel, we use translations instead of dilatations.   □

There is a converse to Pappus's theorem, which yields a fancier version of Pappus's theorem, but it is easier to prove it using projective geometry. It should be noted that in axiomatic presentations of projective geometry, Pappus's theorem is equivalent to the commutativity of the ground field K (in the present case, $K = \mathbb{R}$). We now prove an affine version of Desargues's theorem.

Lemma 2.12. *Given any affine space E, and given any two triangles (a, b, c) and (a', b', c'), where a, b, c, a', b', c' are all distinct, if $\langle a, b \rangle$ and $\langle a', b' \rangle$ are parallel and $\langle b, c \rangle$ and $\langle b', c' \rangle$ are parallel, then $\langle a, c \rangle$ and $\langle a', c' \rangle$ are parallel iff the lines $\langle a, a' \rangle$, $\langle b, b' \rangle$, and $\langle c, c' \rangle$ are either parallel or concurrent (i.e., intersect in a common point).*

Proof. We prove half of the lemma, the direction in which it is assumed that $\langle a, c \rangle$ and $\langle a', c' \rangle$ are parallel, leaving the converse as an exercise. Since the lines $\langle a, b \rangle$ and $\langle a', b' \rangle$ are parallel, the points a, b, a', b' are coplanar. Thus, either $\langle a, a' \rangle$ and $\langle b, b' \rangle$ are parallel, or they have some intersection d. We consider the second case where they intersect, leaving the other case as an easy exercise. Let f be the dilatation of center d such that $f(a) = a'$. By Lemma 2.10, we get $f(b) = b'$. If $f(c) = c''$, again by Lemma 2.10 twice, the lines $\langle b,c\rangle$ and $\langle b',c''\rangle$ are parallel, and the lines $\langle a,c\rangle$ and $\langle a',c''\rangle$ are parallel. From this it follows that $c'' = c'$. Indeed, recall that $\langle b,c\rangle$ and $\langle b',c'\rangle$ are parallel, and similarly $\langle a,c\rangle$ and $\langle a',c'\rangle$ are parallel. Thus, the lines $\langle b',c''\rangle$ and $\langle b',c'\rangle$ are identical, and similarly the lines $\langle a',c''\rangle$ and $\langle a',c'\rangle$ are identical. Since $\overrightarrow{a'c'}$ and $\overrightarrow{b'c'}$ are linearly independent, these lines have a unique intersection, which must be $c'' = c'$.

The direction where it is assumed that the lines $\langle a,a'\rangle$, $\langle b,b'\rangle$ and $\langle c,c'\rangle$, are either parallel or concurrent is left as an exercise (in fact, the proof is quite similar). $\square$

Desargues's theorem is illustrated in Figure 2.15.

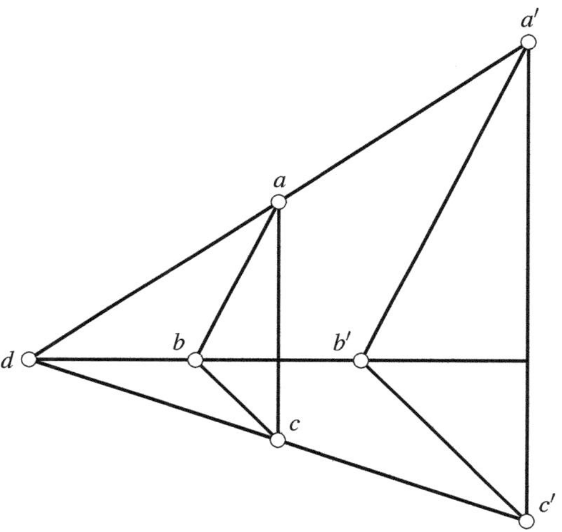

Fig. 2.15 Desargues's theorem (affine version).

There is a fancier version of Desargues's theorem, but it is easier to prove it using projective geometry. It should be noted that in axiomatic presentations of projective geometry, Desargues's theorem is related to the associativity of the ground field K (in the present case, $K = \mathbb{R}$). Also, Desargues's theorem yields a geometric characterization of the affine dilatations. An affine dilatation f on an affine space E is a bijection that maps every line D to a line $f(D)$ parallel to D. We leave the proof as an exercise.

2.10 Affine Hyperplanes

We now consider affine forms and affine hyperplanes. In Section 2.5 we observed that the set L of solutions of an equation

$$ax + by = c$$

is an affine subspace of $\mathbb{A}^2$ of dimension 1, in fact, a line (provided that a and b are not both null). It would be equally easy to show that the set P of solutions of an equation

$$ax + by + cz = d$$

is an affine subspace of $\mathbb{A}^3$ of dimension 2, in fact, a plane (provided that a, b, c are not all null). More generally, the set H of solutions of an equation

$$\lambda_1 x_1 + \cdots + \lambda_m x_m = \mu$$

is an affine subspace of $\mathbb{A}^m$, and if $\lambda_1, \ldots, \lambda_m$ are not all null, it turns out that it is a subspace of dimension $m - 1$ called a *hyperplane*.

We can interpret the equation

$$\lambda_1 x_1 + \cdots + \lambda_m x_m = \mu$$

in terms of the map $f \colon \mathbb{R}^m \to \mathbb{R}$ defined such that

$$f(x_1, \ldots, x_m) = \lambda_1 x_1 + \cdots + \lambda_m x_m - \mu$$

for all $(x_1, \ldots, x_m) \in \mathbb{R}^m$. It is immediately verified that this map is affine, and the set H of solutions of the equation

$$\lambda_1 x_1 + \cdots + \lambda_m x_m = \mu$$

is the *null set*, or *kernel*, of the affine map $f \colon \mathbb{A}^m \to \mathbb{R}$, in the sense that

$$H = f^{-1}(0) = \{x \in \mathbb{A}^m \mid f(x) = 0\},$$

where $x = (x_1, \ldots, x_m)$.

Thus, it is interesting to consider *affine forms*, which are just affine maps $f \colon E \to \mathbb{R}$ from an affine space to $\mathbb{R}$. Unlike linear forms f^*, for which $\operatorname{Ker} f^*$ is never empty (since it always contains the vector 0), it is possible that $f^{-1}(0) = \emptyset$ for an affine form f. Given an affine map $f \colon E \to \mathbb{R}$, we also denote $f^{-1}(0)$ by $\operatorname{Ker} f$, and we call it the *kernel* of f. Recall that an (affine) hyperplane is an affine subspace of codimension 1. The relationship between affine hyperplanes and affine forms is given by the following lemma.

Lemma 2.13. *Let E be an affine space. The following properties hold:*

(a) Given any nonconstant affine form $f: E \to \mathbb{R}$, its kernel $H = \operatorname{Ker} f$ is a hyperplane.

(b) For any hyperplane H in E, there is a nonconstant affine form $f: E \to \mathbb{R}$ such that $H = \operatorname{Ker} f$. For any other affine form $g: E \to \mathbb{R}$ such that $H = \operatorname{Ker} g$, there is some $\lambda \in \mathbb{R}$ such that $g = \lambda f$ (with $\lambda \neq 0$).

(c) Given any hyperplane H in E and any (nonconstant) affine form $f: E \to \mathbb{R}$ such that $H = \operatorname{Ker} f$, every hyperplane H' parallel to H is defined by a nonconstant affine form g such that $g(a) = f(a) - \lambda$, for all $a \in E$ and some $\lambda \in \mathbb{R}$.

Proof. The proof is straightforward, and is omitted. It is also given in Gallier [5]. $\square$

When E is of dimension n, given an affine frame $(a_0, (u_1, \ldots, u_n))$ of E with origin a_0, recall from Definition 2.5 that every point of E can be expressed uniquely as $x = a_0 + x_1 u_1 + \cdots + x_n u_n$, where $(x_1, \ldots, x_n)$ are the *coordinates* of x with respect to the affine frame $(a_0, (u_1, \ldots, u_n))$.

Also recall that every linear form f^* is such that $f^*(x) = \lambda_1 x_1 + \cdots + \lambda_n x_n$, for every $x = x_1 u_1 + \cdots + x_n u_n$ and some $\lambda_1, \ldots, \lambda_n \in \mathbb{R}$. Since an affine form $f: E \to \mathbb{R}$ satisfies the property $f(a_0 + x) = f(a_0) + \overrightarrow{f}(x)$, denoting $f(a_0 + x)$ by $f(x_1, \ldots, x_n)$, we see that we have

$$f(x_1, \ldots, x_n) = \lambda_1 x_1 + \cdots + \lambda_n x_n + \mu,$$

where $\mu = f(a_0) \in \mathbb{R}$ and $\lambda_1, \ldots, \lambda_n \in \mathbb{R}$. Thus, a hyperplane is the set of points whose coordinates $(x_1, \ldots, x_n)$ satisfy the (affine) equation

$$\lambda_1 x_1 + \cdots + \lambda_n x_n + \mu = 0.$$

2.11 Intersection of Affine Spaces

In this section we take a closer look at the intersection of affine subspaces. This subsection can be omitted at first reading.

First, we need a result of linear algebra. Given a vector space E and any two subspaces M and N, there are several interesting linear maps. We have the canonical injections $i: M \to M + N$ and $j: N \to M + N$, the canonical injections $in_1: M \to M \oplus N$ and $in_2: N \to M \oplus N$, and thus, injections $f: M \cap N \to M \oplus N$ and $g: M \cap N \to M \oplus N$, where f is the composition of the inclusion map from $M \cap N$ to M with in_1, and g is the composition of the inclusion map from $M \cap N$ to N with in_2. Then, we have the maps $f + g: M \cap N \to M \oplus N$, and $i - j: M \oplus N \to M + N$.

Lemma 2.14. *Given a vector space E and any two subspaces M and N, with the definitions above,*

$$0 \longrightarrow M \cap N \xrightarrow{f+g} M \oplus N \xrightarrow{i-j} M + N \longrightarrow 0$$

is a short exact sequence, which means that $f+g$ is injective, $i-j$ is surjective, and that $\mathrm{Im}\,(f+g) = \mathrm{Ker}\,(i-j)$. As a consequence, we have the Grassmann relation

$$\dim(M) + \dim(N) = \dim(M+N) + \dim(M \cap N).$$

Proof. It is obvious that $i-j$ is surjective and that $f+g$ is injective. Assume that $(i-j)(u+v) = 0$, where $u \in M$, and $v \in N$. Then, $i(u) = j(v)$, and thus, by definition of i and j, there is some $w \in M \cap N$, such that $i(u) = j(v) = w \in M \cap N$. By definition of f and g, $u = f(w)$ and $v = g(w)$, and thus $\mathrm{Im}\,(f+g) = \mathrm{Ker}\,(i-j)$, as desired. The second part of the lemma follows from standard results of linear algebra (see Artin [1], Strang [12], or Lang [8]). $\square$

We now prove a simple lemma about the intersection of affine subspaces.

Lemma 2.15. *Given any affine space E, for any two nonempty affine subspaces M and N, the following facts hold:*

(1) $M \cap N \neq \emptyset$ iff $\overrightarrow{ab} \in \overrightarrow{M} + \overrightarrow{N}$ for some $a \in M$ and some $b \in N$.

(2) $M \cap N$ consists of a single point iff $\overrightarrow{ab} \in \overrightarrow{M} + \overrightarrow{N}$ for some $a \in M$ and some $b \in N$, and $\overrightarrow{M} \cap \overrightarrow{N} = \{0\}$.

(3) If S is the least affine subspace containing M and N, then $\overrightarrow{S} = \overrightarrow{M} + \overrightarrow{N} + K\overrightarrow{ab}$ (the vector space $\overrightarrow{E}$ is defined over the field K).

Proof. (1) Pick any $a \in M$ and any $b \in N$, which is possible, since M and N are nonempty. Since $\overrightarrow{M} = \{\overrightarrow{ax} \mid x \in M\}$ and $\overrightarrow{N} = \{\overrightarrow{by} \mid y \in N\}$, if $M \cap N \neq \emptyset$, for any $c \in M \cap N$ we have $\overrightarrow{ab} = \overrightarrow{ac} - \overrightarrow{bc}$, with $\overrightarrow{ac} \in \overrightarrow{M}$ and $\overrightarrow{bc} \in \overrightarrow{N}$, and thus, $\overrightarrow{ab} \in \overrightarrow{M} + \overrightarrow{N}$. Conversely, assume that $\overrightarrow{ab} \in \overrightarrow{M} + \overrightarrow{N}$ for some $a \in M$ and some $b \in N$. Then $\overrightarrow{ab} = \overrightarrow{ax} + \overrightarrow{by}$, for some $x \in M$ and some $y \in N$. But we also have

$$\overrightarrow{ab} = \overrightarrow{ax} + \overrightarrow{xy} + \overrightarrow{yb},$$

and thus we get $0 = \overrightarrow{xy} + \overrightarrow{yb} - \overrightarrow{by}$, that is, $\overrightarrow{xy} = 2\overrightarrow{by}$. Thus, b is the middle of the segment $[x,y]$, and since $\overrightarrow{yx} = 2\overrightarrow{yb}$, $x = 2b - y$ is the barycenter of the weighted points $(b,2)$ and $(y,-1)$. Thus x also belongs to N, since N being an affine subspace, it is closed under barycenters. Thus, $x \in M \cap N$, and $M \cap N \neq \emptyset$.

(2) Note that in general, if $M \cap N \neq \emptyset$, then

$$\overrightarrow{M \cap N} = \overrightarrow{M} \cap \overrightarrow{N},$$

because

$$\overrightarrow{M \cap N} = \{\overrightarrow{ab} \mid a,b \in M \cap N\} = \{\overrightarrow{ab} \mid a,b \in M\} \cap \{\overrightarrow{ab} \mid a,b \in N\} = \overrightarrow{M} \cap \overrightarrow{N}.$$

Since $M \cap N = c + \overrightarrow{M \cap N}$ for any $c \in M \cap N$, we have

$$M \cap N = c + \overrightarrow{M} \cap \overrightarrow{N} \quad \text{for any } c \in M \cap N.$$

From this it follows that if $M \cap N \neq \emptyset$, then $M \cap N$ consists of a single point iff $\overrightarrow{M} \cap \overrightarrow{N} = \{0\}$. This fact together with what we proved in (1) proves (2).

(3) This is left as an easy exercise. $\quad \square$

Remarks:

(1) The proof of Lemma 2.15 shows that if $M \cap N \neq \emptyset$, then $\overrightarrow{ab} \in \overrightarrow{M} + \overrightarrow{N}$ for all $a \in M$ and all $b \in N$.

(2) Lemma 2.15 implies that for any two nonempty affine subspaces M and N, if $\overrightarrow{E} = \overrightarrow{M} \oplus \overrightarrow{N}$, then $M \cap N$ consists of a single point. Indeed, if $\overrightarrow{E} = \overrightarrow{M} \oplus \overrightarrow{N}$, then $\overrightarrow{ab} \in \overrightarrow{E}$ for all $a \in M$ and all $b \in N$, and since $\overrightarrow{M} \cap \overrightarrow{N} = \{0\}$, the result follows from part (2) of the lemma.

We can now state the following lemma.

Lemma 2.16. *Given an affine space E and any two nonempty affine subspaces M and N, if S is the least affine subspace containing M and N, then the following properties hold:*

(1) If $M \cap N = \emptyset$, then

$$\dim(M) + \dim(N) < \dim(E) + \dim(\overrightarrow{M} + \overrightarrow{N})$$

and

$$\dim(S) = \dim(M) + \dim(N) + 1 - \dim(\overrightarrow{M} \cap \overrightarrow{N}).$$

(2) If $M \cap N \neq \emptyset$, then

$$\dim(S) = \dim(M) + \dim(N) - \dim(M \cap N).$$

Proof. The proof is not difficult, using Lemma 2.15 and Lemma 2.14, but we leave it as an exercise. $\quad \square$

2.12 Problems

2.1. Given a triangle (a, b, c), give a geometric construction of the barycenter of the weighted points $(a, \frac{1}{4})$, $(b, \frac{1}{4})$, and $(c, \frac{1}{2})$. Give a geometric construction of the barycenter of the weighted points $(a, \frac{3}{2})$, $(b, \frac{3}{2})$, and $(c, -2)$.

2.2. Given a tetrahedron (a, b, c, d) and any two distinct points $x, y \in \{a, b, c, d\}$, let let $m_{x,y}$ be the middle of the edge (x, y). Prove that the barycenter g of the weighted points $(a, \frac{1}{4})$, $(b, \frac{1}{4})$, $(c, \frac{1}{4})$, and $(d, \frac{1}{4})$ is the common intersection of the line segments $(m_{a,b}, m_{c,d})$, $(m_{a,c}, m_{b,d})$, and $(m_{a,d}, m_{b,c})$. Show that if g_d is the barycenter of the weighted points $(a, \frac{1}{3}), (b, \frac{1}{3}), (c, \frac{1}{3})$, then g is the barycenter of $(d, \frac{1}{4})$ and $(g_d, \frac{3}{4})$.

2.3. Let E be a nonempty set, and $\overrightarrow{E}$ a vector space and assume that there is a function $\Phi : E \times E \to \overrightarrow{E}$, such that if we denote $\Phi(a, b)$ by $\overrightarrow{ab}$, the following properties hold:

(1) $\overrightarrow{ab} + \overrightarrow{bc} = \overrightarrow{ac}$, for all $a, b, c \in E$;
(2) For every $a \in E$, the map $\Phi_a : E \to \overrightarrow{E}$ defined such that for every $b \in E$, $\Phi_a(b) = \overrightarrow{ab}$, is a bijection.

Let $\Psi_a : \overrightarrow{E} \to E$ be the inverse of $\Phi_a : E \to \overrightarrow{E}$.

Prove that the function $+ : E \times \overrightarrow{E} \to E$ defined such that

$$a + u = \Psi_a(u)$$

for all $a \in E$ and all $u \in \overrightarrow{E}$ makes $(E, \overrightarrow{E}, +)$ into an affine space.

Note. We showed in the text that an affine space $(E, \overrightarrow{E}, +)$ satisfies the properties stated above. Thus, we obtain an equivalent characterization of affine spaces.

2.4. Given any three points a, b, c in the affine plane $\mathbb{A}^2$, letting (a_1, a_2), (b_1, b_2), and (c_1, c_2) be the coordinates of a, b, c, with respect to the standard affine frame for $\mathbb{A}^2$, prove that a, b, c are collinear iff

$$\begin{vmatrix} a_1 & b_1 & c_1 \\ a_2 & b_2 & c_2 \\ 1 & 1 & 1 \end{vmatrix} = 0,$$

i.e., the determinant is null.

Letting (a_0, a_1, a_2), (b_0, b_1, b_2), and (c_0, c_1, c_2) be the barycentric coordinates of a, b, c with respect to the standard affine frame for $\mathbb{A}^2$, prove that a, b, c are collinear iff

$$\begin{vmatrix} a_0 & b_0 & c_0 \\ a_1 & b_1 & c_1 \\ a_2 & b_2 & c_2 \end{vmatrix} = 0.$$

Given any four points a, b, c, d in the affine space $\mathbb{A}^3$, letting (a_1, a_2, a_3), (b_1, b_2, b_3), (c_1, c_2, c_3), and (d_1, d_2, d_3) be the coordinates of a, b, c, d, with respect to the standard affine frame for $\mathbb{A}^3$, prove that a, b, c, d are coplanar iff

$$\begin{vmatrix} a_1 & b_1 & c_1 & d_1 \\ a_2 & b_2 & c_2 & d_2 \\ a_3 & b_3 & c_3 & d_3 \\ 1 & 1 & 1 & 1 \end{vmatrix} = 0,$$

i.e., the determinant is null.

Letting (a_0,a_1,a_2,a_3), (b_0,b_1,b_2,b_3), (c_0,c_1,c_2,c_3), and (d_0,d_1,d_2,d_3) be the barycentric coordinates of a,b,c,d, with respect to the standard affine frame for $\mathbb{A}^3$, prove that a,b,c,d are coplanar iff

$$\begin{vmatrix} a_0 & b_0 & c_0 & d_0 \\ a_1 & b_1 & c_1 & d_1 \\ a_2 & b_2 & c_2 & d_2 \\ a_3 & b_3 & c_3 & d_3 \end{vmatrix} = 0.$$

2.5. The function $f : \mathbb{A} \to \mathbb{A}^3$ given by

$$t \mapsto (t,t^2,t^3)$$

defines what is called a *twisted cubic* curve. Given any four pairwise distinct values t_1,t_2,t_3,t_4, prove that the points $f(t_1), f(t_2), f(t_3)$, and $f(t_4)$ are not coplanar. *Hint.* Have you heard of the Vandermonde determinant?

2.6. For any two distinct points $a,b \in \mathbb{A}^2$ of barycentric coordinates (a_0,a_1,a_2) and (b_0,b_1,b_2) with respect to any given affine frame (O,i,j), show that the equation of the line $\langle a,b \rangle$ determined by a and b is

$$\begin{vmatrix} a_0 & b_0 & x \\ a_1 & b_1 & y \\ a_2 & b_2 & z \end{vmatrix} = 0,$$

or, equivalently,

$$(a_1b_2 - a_2b_1)x + (a_2b_0 - a_0b_2)y + (a_0b_1 - a_1b_0)z = 0,$$

where (x,y,z) are the barycentric coordinates of the generic point on the line $\langle a,b \rangle$.
  Prove that the equation of a line in barycentric coordinates is of the form

$$ux + vy + wz = 0,$$

where $u \neq v$ or $v \neq w$ or $u \neq w$. Show that two equations

$$ux + vy + wz = 0 \quad \text{and} \quad u'x + v'y + w'z = 0$$

represent the same line in barycentric coordinates iff $(u',v',w') = \lambda(u,v,w)$ for some $\lambda \in \mathbb{R}$ (with $\lambda \neq 0$).
  A triple (u,v,w) where $u \neq v$ or $v \neq w$ or $u \neq w$ is called a system of *tangential coordinates* of the line defined by the equation

$$ux + vy + wz = 0.$$

2.7. Given two lines D and D' in $\mathbb{A}^2$ defined by tangential coordinates (u,v,w) and (u',v',w') (as defined in Problem 2.6), let

$$d = \begin{vmatrix} u & v & w \\ u' & v' & w' \\ 1 & 1 & 1 \end{vmatrix} = vw' - wv' + wu' - uw' + uv' - vu'.$$

(a) Prove that D and D' have a unique intersection point iff $d \neq 0$, and that when it exists, the barycentric coordinates of this intersection point are

$$\frac{1}{d}(vw' - wv',\ wu' - uw',\ uv' - vu').$$

(b) Letting (O, i, j) be any affine frame for $\mathbb{A}^2$, recall that when $x + y + z = 0$, for any point a, the vector

$$x\overrightarrow{aO} + y\overrightarrow{ai} + z\overrightarrow{aj}$$

is independent of a and equal to

$$y\overrightarrow{Oi} + z\overrightarrow{Oj} = (y, z).$$

The triple (x, y, z) such that $x + y + z = 0$ is called the *barycentric coordinates* of the vector $y\overrightarrow{Oi} + z\overrightarrow{Oj}$ w.r.t. the affine frame (O, i, j).

Given any affine frame (O, i, j), prove that for $u \neq v$ or $v \neq w$ or $u \neq w$, the line of equation

$$ux + vy + wz = 0$$

in barycentric coordinates (x, y, z) (where $x + y + z = 1$) has for direction the set of vectors of barycentric coordinates (x, y, z) such that

$$ux + vy + wz = 0$$

(where $x + y + z = 0$).

Prove that D and D' are parallel iff $d = 0$. In this case, if $D \neq D'$, show that the common direction of D and D' is defined by the vector of barycentric coordinates

$$(vw' - wv',\ wu' - uw',\ uv' - vu').$$

(c) Given three lines D, D', and D'', at least two of which are distinct and defined by tangential coordinates (u, v, w), (u', v', w'), and (u'', v'', w''), prove that D, D', and D'' are parallel or have a unique intersection point iff

$$\begin{vmatrix} u & v & w \\ u' & v' & w' \\ u'' & v'' & w'' \end{vmatrix} = 0.$$

2.8. Let (A, B, C) be a triangle in $\mathbb{A}^2$. Let M, N, P be three points respectively on the lines BC, CA, and AB, of barycentric coordinates $(0, m', m'')$, $(n, 0, n'')$, and $(p, p', 0)$, w.r.t. the affine frame (A, B, C).

(a) Assuming that $M \neq C$, $N \neq A$, and $P \neq B$, i.e., $m'n''p \neq 0$, show that

$$\frac{\overrightarrow{MB}}{\overrightarrow{MC}}\,\frac{\overrightarrow{NC}}{\overrightarrow{NA}}\,\frac{\overrightarrow{PA}}{\overrightarrow{PB}} = -\frac{m''np'}{m'n''p}.$$

(b) Prove *Menelaus's theorem*: The points M, N, P are collinear iff

$$m''np' + m'n''p = 0.$$

When $M \neq C$, $N \neq A$, and $P \neq B$, this is equivalent to

$$\frac{\overrightarrow{MB}}{\overrightarrow{MC}}\,\frac{\overrightarrow{NC}}{\overrightarrow{NA}}\,\frac{\overrightarrow{PA}}{\overrightarrow{PB}} = 1.$$

(c) Prove *Ceva's theorem*: The lines AM, BN, CP have a unique intersection point or are parallel iff

$$m''np' - m'n''p = 0.$$

When $M \neq C$, $N \neq A$, and $P \neq B$, this is equivalent to

$$\frac{\overrightarrow{MB}}{\overrightarrow{MC}}\,\frac{\overrightarrow{NC}}{\overrightarrow{NA}}\,\frac{\overrightarrow{PA}}{\overrightarrow{PB}} = -1.$$

2.9. This problem uses notions and results from Problems 2.6 and 2.7. In view of (a) and (b) of Problem 2.7, it is natural to extend the notion of barycentric coordinates of a point in $\mathbb{A}^2$ as follows. Given any affine frame (a, b, c) in $\mathbb{A}^2$, we will say that the barycentric coordinates (x, y, z) of a point M, where $x + y + z = 1$, are the *normalized barycentric coordinates* of M. Then, any triple (x, y, z) such that $x + y + z \neq 0$ is also called a system of barycentric coordinates for the point of normalized barycentric coordinates

$$\frac{1}{x + y + z}\,(x, y, z).$$

With this convention, the intersection of the two lines D and D' is either a point or a vector, in both cases of barycentric coordinates

$$(vw' - wv',\ wu' - uw',\ uv' - vu').$$

When the above is a vector, we can think of it as a point at infinity (in the direction of the line defined by that vector).

Let (D_0, D_0'), (D_1, D_1'), and (D_2, D_2') be three pairs of six distinct lines, such that the four lines belonging to any union of two of the above pairs are neither parallel nor concurrent (have a common intersection point). If D_0 and D_0' have a unique intersection point, let M be this point, and if D_0 and D_0' are parallel, let M denote a nonnull vector defining the common direction of D_0 and D_0'. In either case, let (m, m', m'') be the barycentric coordinates of M, as explained at the beginning of the problem. We call M the *intersection* of D_0 and D_0'. Similarly, define $N = (n, n', n'')$ as the intersection of D_1 and D_1', and $P = (p, p', p'')$ as the intersection of D_2 and D_2'.

Prove that

$$\begin{vmatrix} m & n & p \\ m' & n' & p' \\ m'' & n'' & p'' \end{vmatrix} = 0$$

iff either

(i) (D_0, D_0'), (D_1, D_1'), and (D_2, D_2') are pairs of parallel lines; or
(ii) the lines of some pair (D_i, D_i') are parallel, each pair (D_j, D_j') (with $j \neq i$) has a unique intersection point, and these two intersection points are distinct and determine a line parallel to the lines of the pair (D_i, D_i'); or
(iii) each pair (D_i, D_i') $(i = 0, 1, 2)$ has a unique intersection point, and these points M, N, P are distinct and collinear.

2.10. Prove the following version of *Desargues's theorem*. Let A, B, C, A', B', C' be six distinct points of $\mathbb{A}^2$. If no three of these points are collinear, then the lines AA', BB', and CC' are parallel or collinear iff the intersection points M, N, P (in the sense of Problem 2.7) of the pairs of lines $(BC, B'C')$, $(CA, C'A')$, and $(AB, A'B')$ are collinear in the sense of Problem 2.9.

2.11. Prove the following version of *Pappus's theorem*. Let D and D' be distinct lines, and let A, B, C and A', B', C' be distinct points respectively on D and D'. If these points are all distinct from the intersection of D and D' (if it exists), then the intersection points (in the sense of Problem 2.7) of the pairs of lines (BC', CB'), (CA', AC'), and (AB', BA') are collinear in the sense of Problem 2.9.

2.12. The purpose of this problem is to prove *Pascal's theorem* for the nondegenerate conics. In the affine plane $\mathbb{A}^2$, a *conic* is the set of points of coordinates (x, y) such that

$$\alpha x^2 + \beta y^2 + 2\gamma xy + 2\delta x + 2\lambda y + \mu = 0,$$

where $\alpha \neq 0$ or $\beta \neq 0$ or $\gamma \neq 0$. We can write the equation of the conic as

$$(x, y, 1) \begin{pmatrix} \alpha & \gamma & \delta \\ \gamma & \beta & \lambda \\ \delta & \lambda & \mu \end{pmatrix} \begin{pmatrix} x \\ y \\ 1 \end{pmatrix} = 0.$$

If we now use barycentric coordinates (x, y, z) (where $x + y + z = 1$), we can write

$$\begin{pmatrix} x \\ y \\ 1 \end{pmatrix} = \begin{pmatrix} 1 & 0 & 0 \\ 0 & 1 & 0 \\ 1 & 1 & 1 \end{pmatrix} \begin{pmatrix} x \\ y \\ z \end{pmatrix}.$$

Let

$$B = \begin{pmatrix} \alpha & \gamma & \delta \\ \gamma & \beta & \lambda \\ \delta & \lambda & \mu \end{pmatrix}, \quad C = \begin{pmatrix} 1 & 0 & 0 \\ 0 & 1 & 0 \\ 1 & 1 & 1 \end{pmatrix}, \quad X = \begin{pmatrix} x \\ y \\ z \end{pmatrix}.$$

(a) Letting $A = C^\top B C$, prove that the equation of the conic becomes

$$X^\top AX = 0.$$

Prove that A is symmetric, that $\det(A) = \det(B)$, and that $X^\top AX$ is homogeneous of degree 2. The equation $X^\top AX = 0$ is called the *homogeneous equation* of the conic.

We say that a conic of homogeneous equation $X^\top AX = 0$ is *nondegenerate* if $\det(A) \neq 0$, and *degenerate* if $\det(A) = 0$. Show that this condition does not depend on the choice of the affine frame.

(b) Given an affine frame (A,B,C), prove that any conic passing through A,B,C has an equation of the form

$$ayz + bxz + cxy = 0.$$

Prove that a conic containing more than one point is degenerate iff it contains three distinct collinear points. In this case, the conic is the union of two lines.

(c) Prove *Pascal's theorem*. Given any six distinct points A,B,C,A',B',C', if no three of the above points are collinear, then a nondegenerate conic passes through these six points iff the intersection points M,N,P (in the sense of Problem 2.7) of the pairs of lines (BC',CB'), (CA',AC') and (AB',BA') are collinear in the sense of Problem 2.9.

Hint. Use the affine frame (A,B,C), and let (a,a',a''), (b,b',b''), and (c,c',c'') be the barycentric coordinates of A',B',C' respectively, and show that M,N,P have barycentric coordinates

$$(bc,cb',c''b), \quad (c'a,c'a',c''a'), \quad (ab'',a''b',a''b'').$$

2.13. The *centroid* of a triangle (a,b,c) is the barycenter of $(a,\frac{1}{3})$, $(b,\frac{1}{3})$, $(c,\frac{1}{3})$. If an affine map takes the vertices of triangle $\Delta_1 = \{(0,0),(6,0),(0,9)\}$ to the vertices of triangle $\Delta_2 = \{(1,1),(5,4),(3,1)\}$, does it also take the centroid of Δ_1 to the centroid of Δ_2? Justify your answer.

2.14. Let E be an affine space over $\mathbb{R}$, and let $(a_1,\ldots,a_n)$ be any $n \geq 3$ points in E. Let $(\lambda_1,\ldots,\lambda_n)$ be any n scalars in $\mathbb{R}$, with $\lambda_1 + \cdots + \lambda_n = 1$. Show that there must be some i, $1 \leq i \leq n$, such that $\lambda_i \neq 1$. To simplify the notation, assume that $\lambda_1 \neq 1$. Show that the barycenter $\lambda_1 a_1 + \cdots + \lambda_n a_n$ can be obtained by first determining the barycenter b of the $n-1$ points $a_2,\ldots,a_n$ assigned some appropriate weights, and then the barycenter of a_1 and b assigned the weights λ_1 and $\lambda_2 + \cdots + \lambda_n$. From this, show that the barycenter of any $n \geq 3$ points can be determined by repeated computations of barycenters of two points. Deduce from the above that a nonempty subset V of E is an affine subspace iff whenever V contains any two points $x,y \in V$, then V contains the entire line $(1-\lambda)x + \lambda y$, $\lambda \in \mathbb{R}$.

2.15. Assume that K is a field such that $2 = 1+1 \neq 0$, and let E be an affine space over K. In the case where $\lambda_1 + \cdots + \lambda_n = 1$ and $\lambda_i = 1$, for $1 \leq i \leq n$ and $n \geq 3$, show that the barycenter $a_1 + a_2 + \cdots + a_n$ can still be computed by repeated computations of barycenters of two points.

Finally, assume that the field K contains at least three elements (thus, there is some $\mu \in K$ such that $\mu \neq 0$ and $\mu \neq 1$, but $2 = 1 + 1 = 0$ is possible). Prove that the barycenter of any $n \geq 3$ points can be determined by repeated computations of barycenters of two points. Prove that a nonempty subset V of E is an affine subspace iff whenever V contains any two points $x, y \in V$, then V contains the entire line $(1 - \lambda)x + \lambda y$, $\lambda \in K$.

Hint. When $2 = 0$, $\lambda_1 + \cdots + \lambda_n = 1$ and $\lambda_i = 1$, for $1 \leq i \leq n$, show that n must be odd, and that the problem reduces to computing the barycenter of three points in two steps involving two barycenters. Since there is some $\mu \in K$ such that $\mu \neq 0$ and $\mu \neq 1$, note that μ^{-1} and $(1 - \mu)^{-1}$ both exist, and use the fact that

$$\frac{-\mu}{1 - \mu} + \frac{1}{1 - \mu} = 1.$$

2.16. (i) Let (a, b, c) be three points in $\mathbb{A}^2$, and assume that (a, b, c) are not collinear. For any point $x \in \mathbb{A}^2$, if $x = \lambda_0 a + \lambda_1 b + \lambda_2 c$, where $(\lambda_0, \lambda_1, \lambda_2)$ are the barycentric coordinates of x with respect to (a, b, c), show that

$$\lambda_0 = \frac{\det(\overrightarrow{xb}, \overrightarrow{bc})}{\det(\overrightarrow{ab}, \overrightarrow{ac})}, \qquad \lambda_1 = \frac{\det(\overrightarrow{ax}, \overrightarrow{ac})}{\det(\overrightarrow{ab}, \overrightarrow{ac})}, \qquad \lambda_2 = \frac{\det(\overrightarrow{ab}, \overrightarrow{ax})}{\det(\overrightarrow{ab}, \overrightarrow{ac})}.$$

Conclude that $\lambda_0, \lambda_1, \lambda_2$ are certain signed ratios of the areas of the triangles (a, b, c), (x, a, b), (x, a, c), and (x, b, c).

(ii) Let (a, b, c) be three points in $\mathbb{A}^3$, and assume that (a, b, c) are not collinear. For any point x in the plane determined by (a, b, c), if $x = \lambda_0 a + \lambda_1 b + \lambda_2 c$, where $(\lambda_0, \lambda_1, \lambda_2)$ are the barycentric coordinates of x with respect to (a, b, c), show that

$$\lambda_0 = \frac{\overrightarrow{xb} \times \overrightarrow{bc}}{\overrightarrow{ab} \times \overrightarrow{ac}}, \qquad \lambda_1 = \frac{\overrightarrow{ax} \times \overrightarrow{ac}}{\overrightarrow{ab} \times \overrightarrow{ac}}, \qquad \lambda_2 = \frac{\overrightarrow{ab} \times \overrightarrow{ax}}{\overrightarrow{ab} \times \overrightarrow{ac}}.$$

Given any point O not in the plane of the triangle (a, b, c), prove that

$$\lambda_1 = \frac{\det(\overrightarrow{Oa}, \overrightarrow{Ox}, \overrightarrow{Oc})}{\det(\overrightarrow{Oa}, \overrightarrow{Ob}, \overrightarrow{Oc})}, \qquad \lambda_2 = \frac{\det(\overrightarrow{Oa}, \overrightarrow{Ob}, \overrightarrow{Ox})}{\det(\overrightarrow{Oa}, \overrightarrow{Ob}, \overrightarrow{Oc})},$$

and

$$\lambda_0 = \frac{\det(\overrightarrow{Ox}, \overrightarrow{Ob}, \overrightarrow{Oc})}{\det(\overrightarrow{Oa}, \overrightarrow{Ob}, \overrightarrow{Oc})}.$$

(iii) Let (a, b, c, d) be four points in $\mathbb{A}^3$, and assume that (a, b, c, d) are not coplanar. For any point $x \in \mathbb{A}^3$, if $x = \lambda_0 a + \lambda_1 b + \lambda_2 c + \lambda_3 d$, where $(\lambda_0, \lambda_1, \lambda_2, \lambda_3)$ are the barycentric coordinates of x with respect to (a, b, c, d), show that

$$\lambda_1 = \frac{\det(\overrightarrow{ax}, \overrightarrow{ac}, \overrightarrow{ad})}{\det(\overrightarrow{ab}, \overrightarrow{ac}, \overrightarrow{ad})}, \quad \lambda_2 = \frac{\det(\overrightarrow{ab}, \overrightarrow{ax}, \overrightarrow{ad})}{\det(\overrightarrow{ab}, \overrightarrow{ac}, \overrightarrow{ad})}, \quad \lambda_3 = \frac{\det(\overrightarrow{ab}, \overrightarrow{ac}, \overrightarrow{ax})}{\det(\overrightarrow{ab}, \overrightarrow{ac}, \overrightarrow{ad})},$$

and

$$\lambda_0 = \frac{\det(\overrightarrow{xb},\overrightarrow{bc},\overrightarrow{bd})}{\det(\overrightarrow{ab},\overrightarrow{ac},\overrightarrow{ad})}.$$

Conclude that $\lambda_0, \lambda_1, \lambda_2, \lambda_3$ are certain signed ratios of the volumes of the five tetrahedra (a,b,c,d), (x,a,b,c), (x,a,b,d), (x,a,c,d), and (x,b,c,d).

(iv) Let $(a_0,\ldots,a_m)$ be $m+1$ points in $\mathbb{A}^m$, and assume that they are affinely independent. For any point $x \in \mathbb{A}^m$, if $x = \lambda_0 a_0 + \cdots + \lambda_m a_m$, where $(\lambda_0,\ldots,\lambda_m)$ are the barycentric coordinates of x with respect to $(a_0,\ldots,a_m)$, show that

$$\lambda_i = \frac{\det(\overrightarrow{a_0 a_1},\ldots,\overrightarrow{a_0 a_{i-1}},\overrightarrow{a_0 x},\overrightarrow{a_0 a_{i+1}},\ldots,\overrightarrow{a_0 a_m})}{\det(\overrightarrow{a_0 a_1},\ldots,\overrightarrow{a_0 a_{i-1}},\overrightarrow{a_0 a_i},\overrightarrow{a_0 a_{i+1}},\ldots,\overrightarrow{a_0 a_m})}$$

for every i, $1 \le i \le m$, and

$$\lambda_0 = \frac{\det(\overrightarrow{x a_1},\overrightarrow{a_1 a_2},\ldots,\overrightarrow{a_1 a_m})}{\det(\overrightarrow{a_0 a_1},\ldots,\overrightarrow{a_0 a_i},\ldots,\overrightarrow{a_0 a_m})}.$$

Conclude that λ_i is the signed ratio of the volumes of the simplexes $(a_0,\ldots,x,\ldots a_m)$ and $(a_0,\ldots,a_i,\ldots a_m)$, where $0 \le i \le m$.

2.17. With respect to the standard affine frame for the plane $\mathbb{A}^2$, consider the three geometric transformations f_1, f_2, f_3 defined by

$$x' = -\frac{1}{4}x - \frac{\sqrt{3}}{4}y + \frac{3}{4}, \quad y' = \frac{\sqrt{3}}{4}x - \frac{1}{4}y + \frac{\sqrt{3}}{4},$$

$$x' = -\frac{1}{4}x + \frac{\sqrt{3}}{4}y - \frac{3}{4}, \quad y' = -\frac{\sqrt{3}}{4}x - \frac{1}{4}y + \frac{\sqrt{3}}{4},$$

$$x' = \frac{1}{2}x, \quad y' = \frac{1}{2}y + \frac{\sqrt{3}}{2}.$$

(a) Prove that these maps are affine. Can you describe geometrically what their action is (rotation, translation, scaling)?

(b) Given any polygonal line L, define the following sequence of polygonal lines:

$$S_0 = L,$$
$$S_{n+1} = f_1(S_n) \cup f_2(S_n) \cup f_3(S_n).$$

Construct S_1 starting from the line segment $L = ((-1,0),(1,0))$.

Can you figure out what S_n looks like in general? (You may want to write a computer program.) Do you think that S_n has a limit?

2.18. In the plane $\mathbb{A}^2$, with respect to the standard affine frame, a point of coordinates (x,y) can be represented as the complex number $z = x + iy$. Consider the set of geometric transformations of the form

$$z \mapsto az + b,$$

where a, b are complex numbers such that $a \neq 0$.

(a) Prove that these maps are affine. Describe what these maps do geometrically.

(b) Prove that the above set of maps is a group under composition.

(c) Consider the set of geometric transformations of the form

$$z \mapsto az + b \quad \text{or} \quad z \mapsto a\bar{z} + b,$$

where a, b are complex numbers such that $a \neq 0$, and where $\bar{z} = x - iy$ if $z = x + iy$. Describe what these maps do geometrically. Prove that these maps are affine and that this set of maps is a group under composition.

2.19. Given a group G, a subgroup H of G is called a *normal subgroup* of G iff $xHx^{-1} = H$ for all $x \in G$ (where $xHx^{-1} = \{xhx^{-1} \mid h \in H\}$).

(i) Given any two subgroups H and K of a group G, let

$$HK = \{hk \mid h \in H,\, k \in K\}.$$

Prove that every $x \in HK$ can be written in a unique way as $x = hk$ for $h \in H$ and $k \in K$ iff $H \cap K = \{1\}$, where 1 is the identity element of G.

(ii) If H and K are subgroups of G, and H is a normal subgroup of G, prove that HK is a subgroup of G. Furthermore, if $G = HK$ and $H \cap K = \{1\}$, prove that G is isomorphic to $H \times K$ under the multiplication operation

$$(h_1, k_1) \cdot (h_2, k_2) = (h_1 k_1 h_2 k_1^{-1},\, k_1 k_2).$$

When $G = HK$, where H, K are subgroups of G, H is a normal subgroup of G, and $H \cap K = \{1\}$, we say that G is the *semidirect product of H and K*.

(iii) Let $(E, \overrightarrow{E})$ be an affine space. Recall that the *affine group of E*, denoted by $\mathbf{GA}(E)$, is the set of affine bijections of E, and that the *linear group of $\overrightarrow{E}$*, denoted by $\mathbf{GL}(\overrightarrow{E})$, is the group of bijective linear maps of $\overrightarrow{E}$. The map $f \mapsto \overrightarrow{f}$ defines a group homomorphism $L\colon \mathbf{GA}(E) \to \mathbf{GL}(\overrightarrow{E})$, and the kernel of this map is the set of translations on E, denoted as $T(E)$. Prove that $T(E)$ is a normal subgroup of $\mathbf{GA}(E)$.

(iv) For any $a \in E$, let

$$\mathbf{GA}_a(E) = \{f \in \mathbf{GA}(E) \mid f(a) = a\},$$

the set of affine bijections leaving a fixed. Prove that that $\mathbf{GA}_a(E)$ is a subgroup of $\mathbf{GA}(E)$, and that $\mathbf{GA}_a(E)$ is isomorphic to $\mathbf{GL}(\overrightarrow{E})$. Prove that $\mathbf{GA}(E)$ is isomorphic to the direct product of $T(E)$ and $\mathbf{GA}_a(E)$.

Hint. Note that if $u = \overrightarrow{f(a)a}$ and t_u is the translation associated with the vector u, then $t_u \circ f \in \mathbf{GA}_a(E)$ (where the translation t_u is defined such that $t_u(a) = a + u$ for every $a \in E$).

(v) Given a group G, let $\mathbf{Aut}(G)$ denote the set of homomorphisms $f\colon G \to G$. Prove that the set $\mathbf{Aut}(G)$ is a group under composition (called the *group of automorphisms of G).* Given any two groups H and K and a homomorphism $\theta \colon K \to$ **Aut**(H), we define $H \times_\theta K$ as the set $H \times K$ under the multiplication operation

$$(h_1, k_1) \cdot (h_2, k_2) = (h_1 \,\theta(k_1)(h_2), \, k_1 k_2).$$

Prove that $H \times_\theta K$ is a group.

Hint. The inverse of (h, k) is $(\theta(k^{-1})(h^{-1}), \, k^{-1})$.

Prove that the group $H \times_\theta K$ is the semidirect product of the subgroups $\{(h, 1) \mid h \in H\}$ and $\{(1, k) \mid k \in K\}$. The group $H \times_\theta K$ is also called the *semidirect product of H and K relative to* θ.

Note. It is natural to identify $\{(h, 1) \mid h \in H\}$ with H and $\{(1, k) \mid k \in K\}$ with K.

If G is the semidirect product of two subgroups H and K as defined in (ii), prove that the map $\gamma \colon K \to$ **Aut**(H) defined by conjugation such that

$$\gamma(k)(h) = khk^{-1}$$

is a homomorphism, and that G is isomorphic to $H \times_\gamma K$.

(vi) Define the map $\theta \colon$ **GL**$(\overrightarrow{E}) \to$ **Aut**$(\overrightarrow{E})$ as follows: $\theta(f) = f$, where $f \in$ **GL**$(\overrightarrow{E})$ (note that θ can be viewed as an inclusion map). Prove that **GA**(E) is isomorphic to the semidirect product $\overrightarrow{E} \times_\theta$ **GL**$(\overrightarrow{E})$.

(vii) Let **SL**$(\overrightarrow{E})$ be the subgroup of **GL**$(\overrightarrow{E})$ consisting of the linear maps such that $\det(f) = 1$ (the *special linear group of* $\overrightarrow{E}$), and let **SA**(E) be the subgroup of **GA**(E) (the *special affine group of E*) consisting of the affine maps f such that $\overrightarrow{f} \in$ **SL**$(\overrightarrow{E})$. Prove that **SA**(E) is isomorphic to the semidirect product $\overrightarrow{E} \times_\theta$ **SL**$(\overrightarrow{E})$, where $\theta \colon$ **SL**$(\overrightarrow{E}) \to$ **Aut**$(\overrightarrow{E})$ is defined as in (vi).

(viii) Assume that $(E, \overrightarrow{E})$ is a Euclidean affine space, as defined in Chapter 6. Let **SO**$(\overrightarrow{E})$ be the *special orthogonal group of* $\overrightarrow{E}$, as defined in Definition 6.6 (the isometries with determinant $+1$), and let **SE**(E) be the subgroup of **SA**(E) (the *special Euclidean group of E*) consisting of the affine isometries f such that $\overrightarrow{f} \in$ **SO**$(\overrightarrow{E})$. Prove that **SE**(E) is isomorphic to the semidirect product $\overrightarrow{E} \times_\theta$ **SO**$(\overrightarrow{E})$, where $\theta \colon$ **SO**$(\overrightarrow{E}) \to$ **Aut**$(\overrightarrow{E})$ is defined as in (vi).

2.20. The purpose of this problem is to study certain affine maps of $\mathbb{A}^2$.

(1) Consider affine maps of the form

$$\begin{pmatrix} x_1 \\ x_2 \end{pmatrix} \mapsto \begin{pmatrix} \cos\theta & -\sin\theta \\ \sin\theta & \cos\theta \end{pmatrix} \begin{pmatrix} x_1 \\ x_2 \end{pmatrix} + \begin{pmatrix} b_1 \\ b_2 \end{pmatrix}.$$

Prove that such maps have a unique fixed point c if $\theta \neq 2k\pi$, for all integers k. Show that these are rotations of center c, which means that with respect to a frame with origin c (the unique fixed point), these affine maps are represented by rotation matrices.

(2) Consider affine maps of the form

$$\begin{pmatrix} x_1 \\ x_2 \end{pmatrix} \mapsto \begin{pmatrix} \lambda\cos\theta & -\lambda\sin\theta \\ \mu\sin\theta & \mu\cos\theta \end{pmatrix} \begin{pmatrix} x_1 \\ x_2 \end{pmatrix} + \begin{pmatrix} b_1 \\ b_2 \end{pmatrix}.$$

Prove that such maps have a unique fixed point iff $(\lambda + \mu)\cos\theta \neq 1 + \lambda\mu$. Prove that if $\lambda\mu = 1$ and $\lambda > 0$, there is some angle θ for which either there is no fixed point, or there are infinitely many fixed points.

(3) Prove that the affine map

$$\begin{pmatrix} x_1 \\ x_2 \end{pmatrix} \mapsto \begin{pmatrix} 8/5 & -6/5 \\ 3/10 & 2/5 \end{pmatrix} \begin{pmatrix} x_1 \\ x_2 \end{pmatrix} + \begin{pmatrix} 1 \\ 1 \end{pmatrix}$$

has no fixed point.

(4) Prove that an arbitrary affine map

$$\begin{pmatrix} x_1 \\ x_2 \end{pmatrix} \mapsto \begin{pmatrix} a_1 & a_2 \\ a_3 & a_4 \end{pmatrix} \begin{pmatrix} x_1 \\ x_2 \end{pmatrix} + \begin{pmatrix} b_1 \\ b_2 \end{pmatrix}$$

has a unique fixed point iff the matrix

$$\begin{pmatrix} a_1 - 1 & a_2 \\ a_3 & a_4 - 1 \end{pmatrix}$$

is invertible.

2.21. Let $(E, \overrightarrow{E})$ be any affine space of finite dimension. For every affine map $f \colon E \to E$, let $\mathrm{Fix}(f) = \{a \in E \mid f(a) = a\}$ be the set of fixed points of f.

(i) Prove that if $\mathrm{Fix}(f) \neq \emptyset$, then $\mathrm{Fix}(f)$ is an affine subspace of E such that for every $b \in \mathrm{Fix}(f)$,

$$\mathrm{Fix}(f) = b + \mathrm{Ker}\,(\overrightarrow{f} - \mathrm{id}).$$

(ii) Prove that $\mathrm{Fix}(f)$ contains a unique fixed point iff $\mathrm{Ker}\,(\overrightarrow{f} - \mathrm{id}) = \{0\}$, i.e., $\overrightarrow{f}(u) = u$ iff $u = 0$.

Hint. Show that

$$\overrightarrow{\Omega f(a)} - \overrightarrow{\Omega a} = \overrightarrow{\Omega f(\Omega)} + \overrightarrow{f}(\overrightarrow{\Omega a}) - \overrightarrow{\Omega a},$$

for any two points $\Omega, a \in E$.

2.22. Given two affine spaces $(E, \overrightarrow{E})$ and $(F, \overrightarrow{F})$, let $\mathscr{A}(E, F)$ be the set of all affine maps $f \colon E \to F$.

(i) Prove that the set $\mathscr{A}(E, \overrightarrow{F})$ (viewing $\overrightarrow{F}$ as an affine space) is a vector space under the operations $f + g$ and λf defined such that

$$(f + g)(a) = f(a) + g(a),$$
$$(\lambda f)(a) = \lambda f(a),$$

for all $a \in E$.

(ii) Define an action

$$+\colon \mathscr{A}(E,F) \times \mathscr{A}(E,\overrightarrow{F}) \to \mathscr{A}(E,F)$$

of $\mathscr{A}(E,\overrightarrow{F})$ on $\mathscr{A}(E,F)$ as follows: For every $a \in E$, every $f \in \mathscr{A}(E,F)$, and every $h \in \mathscr{A}(E,\overrightarrow{F})$,

$$(f+h)(a) = f(a)+h(a).$$

Prove that $(\mathscr{A}(E,F), \mathscr{A}(E,\overrightarrow{F}),+)$ is an affine space.

Hint. Show that for any two affine maps $f,g \in \mathscr{A}(E,F)$, the map $\overrightarrow{fg}$ defined such that

$$\overrightarrow{fg}(a) = \overrightarrow{f(a)g(a)}$$

(for every $a \in E$) is affine, and thus $\overrightarrow{fg} \in \mathscr{A}(E,\overrightarrow{F})$. Furthermore, $\overrightarrow{fg}$ is the unique map in $\mathscr{A}(E,\overrightarrow{F})$ such that

$$f+\overrightarrow{fg} = g.$$

(iii) If $\overrightarrow{E}$ has dimension m and $\overrightarrow{F}$ has dimension n, prove that $\mathscr{A}(E,\overrightarrow{F})$ has dimension $n+mn = n(m+1)$.

2.23. Let $(c_1,\ldots,c_n)$ be $n \geq 3$ points in $\mathbb{A}^m$ (where $m \geq 2$). Investigate whether there is a closed polygon with n vertices $(a_1,\ldots,a_n)$ such that c_i is the middle of the edge (a_i, a_{i+1}) for every i with $1 \leq i \leq n-1$, and c_n is the middle of the edge (a_n, a_0). *Hint.* The parity (odd or even) of n plays an important role. When n is odd, there is a unique solution, and when n is even, there are no solutions or infinitely many solutions. Clarify under which conditions there are infinitely many solutions.

2.24. Given an affine space E of dimension n and an affine frame $(a_0,\ldots,a_n)$ for E, let $f\colon E \to E$ and $g\colon E \to E$ be two affine maps represented by the two $(n+1) \times (n+1)$ matrices

$$\begin{pmatrix} A & b \\ 0 & 1 \end{pmatrix} \quad \text{and} \quad \begin{pmatrix} B & c \\ 0 & 1 \end{pmatrix}$$

w.r.t. the frame $(a_0,\ldots,a_n)$. We also say that f and g are represented by (A,b) and (B,c).

(1) Prove that the composition $f \circ g$ is represented by the matrix

$$\begin{pmatrix} AB & Ac+b \\ 0 & 1 \end{pmatrix}.$$

We also say that $f \circ g$ is represented by $(A,b)(B,c) = (AB,Ac+b)$.

(2) Prove that f is invertible iff A is invertible and that the matrix representing f^{-1} is

$$\begin{pmatrix} A^{-1} & -A^{-1}b \\ 0 & 1 \end{pmatrix}.$$

We also say that f^{-1} is represented by $(A,b)^{-1} = (A^{-1}, -A^{-1}b)$. Prove that if A is an orthogonal matrix, the matrix associated with f^{-1} is

$$\begin{pmatrix} A^\top & -A^\top b \\ 0 & 1 \end{pmatrix}.$$

Furthermore, denoting the columns of A by $A_1, \ldots, A_n$, prove that the vector $A^\top b$ is the column vector of components

$$(A_1 \cdot b, \ldots, A_n \cdot b)$$

(where $\cdot$ denotes the standard inner product of vectors).

(3) Given two affine frames $(a_0, \ldots, a_n)$ and $(a_0', \ldots, a_n')$ for E, any affine map $f \colon E \to E$ has a matrix representation (A,b) w.r.t. $(a_0, \ldots, a_n)$ and $(a_0', \ldots, a_n')$ defined such that $b = \overrightarrow{a_0'f(a_0)}$ is expressed over the basis $(\overrightarrow{a_0'a_1'}, \ldots, \overrightarrow{a_0'a_n'})$, and a_{ij} is the ith coefficient of $f(\overrightarrow{a_0a_j})$ over the basis $(\overrightarrow{a_0'a_1'}, \ldots, \overrightarrow{a_0'a_n'})$. Given any three frames $(a_0, \ldots, a_n)$, $(a_0', \ldots, a_n')$, and $(a_0'', \ldots, a_n'')$, for any two affine maps $f \colon E \to E$ and $g \colon E \to E$, if f has the matrix representation (A,b) w.r.t. $(a_0, \ldots, a_n)$ and $(a_0', \ldots, a_n')$ and g has the matrix representation (B,c) w.r.t. $(a_0', \ldots, a_n')$ and $(a_0'', \ldots, a_n'')$, prove that $g \circ f$ has the matrix representation $(B,c)(A,b)$ w.r.t. $(a_0, \ldots, a_n)$ and $(a_0'', \ldots, a_n'')$.

(4) Given two affine frames $(a_0, \ldots, a_n)$ and $(a_0', \ldots, a_n')$ for E, there is a unique affine map $h \colon E \to E$ such that $h(a_i) = a_i'$ for $i = 0, \ldots, n$, and we let (P, ω) be its associated matrix representation with respect to the frame $(a_0, \ldots, a_n)$. Note that $\omega = \overrightarrow{a_0a_0'}$, and that p_{ij} is the ith coefficient of $\overrightarrow{a_0'a_j'}$ over the basis $(\overrightarrow{a_0a_1}, \ldots, \overrightarrow{a_0a_n})$. Observe that (P, ω) is also the matrix representation of id_E w.r.t. the frames $(a_0', \ldots, a_n')$ and $(a_0, \ldots, a_n)$, **in that order**. For any affine map $f \colon E \to E$, if f has the matrix representation (A,b) over the frame $(a_0, \ldots, a_n)$ and the matrix representation (A',b') over the frame $(a_0', \ldots, a_n')$, prove that

$$(A', b') = (P, \omega)^{-1}(A,b)(P, \omega).$$

Given any two affine maps $f \colon E \to E$ and $g \colon E \to E$, where f is invertible, for any affine frame $(a_0, \ldots, a_n)$ for E, if $(a_0', \ldots, a_n')$ is the affine frame image of $(a_0, \ldots, a_n)$ under f (i.e., $f(a_i) = a_i'$ for $i = 0, \ldots, n$), letting (A,b) be the matrix representation of f w.r.t. the frame $(a_0, \ldots, a_n)$ and (B,c) be the matrix representation of g w.r.t. the frame $(a_0', \ldots, a_n')$ (**not** the frame $(a_0, \ldots, a_n)$), prove that $g \circ f$ is represented by the matrix $(A,b)(B,c)$ w.r.t. the frame $(a_0, \ldots, a_n)$.

Remark: Note that this is the **opposite** of what happens if f and g are both represented by matrices w.r.t. the "fixed" frame $(a_0, \ldots, a_n)$, where $g \circ f$ is represented by the matrix $(B,c)(A,b)$. The frame $(a_0', \ldots, a_n')$ can be viewed as a "moving" frame. The above has applications in robotics, for example to rotation matrices expressed in terms of Euler angles, or "roll, pitch, and yaw."

2.25. (a) Let E be a vector space, and let U and V be two subspaces of E such that they form a direct sum $E = U \oplus V$. Recall that this means that every vector $x \in E$ can be written as $x = u + v$, for some unique $u \in U$ and some unique $v \in V$. Define the function $p_U \colon E \to U$ (resp. $p_V \colon E \to V$) so that $p_U(x) = u$ (resp. $p_V(x) = v$), where $x = u + v$, as explained above. Check that that p_U and p_V are linear.

(b) Now assume that E is an affine space (nontrivial), and let U and V be affine subspaces such that $\overrightarrow{E} = \overrightarrow{U} \oplus \overrightarrow{V}$. Pick any $\Omega \in V$, and define $q_U \colon E \to \overrightarrow{U}$ (resp. $q_V \colon E \to \overrightarrow{V}$, with $\Omega \in U$) so that

$$q_U(a) = p_{\overrightarrow{U}}(\overrightarrow{\Omega a}) \quad (\text{resp.} \quad q_V(a) = p_{\overrightarrow{V}}(\overrightarrow{\Omega a})), \quad \text{for every } a \in E.$$

Prove that q_U does not depend on the choice of $\Omega \in V$ (resp. q_V does not depend on the choice of $\Omega \in U$). Define the map $p_U \colon E \to U$ (resp. $p_V \colon E \to V$) so that

$$p_U(a) = a - q_V(a) \quad (\text{resp.} \quad p_V(a) = a - q_U(a)), \quad \text{for every } a \in E.$$

Prove that p_U (resp. p_V) is affine.

The map p_U (resp. p_V) is called the *projection onto U parallel to V (resp. projection onto V parallel to U)*.

(c) Let $(a_0, \ldots, a_n)$ be $n+1$ affinely independent points in $\mathbb{A}^n$ and let $\Delta(a_0, \ldots, a_n)$ denote the convex hull of $(a_0, \ldots, a_n)$ (an n-simplex). Prove that if $f \colon \mathbb{A}^n \to \mathbb{A}^n$ is an affine map sending $\Delta(a_0, \ldots, a_n)$ inside itself, i.e.,

$$f(\Delta(a_0, \ldots, a_n)) \subseteq \Delta(a_0, \ldots, a_n),$$

then f has some fixed point $b \in \Delta(a_0, \ldots, a_n)$, i.e., $f(b) = b$.

Hint: Proceed by induction on n. First, treat the case $n = 1$. The affine map is determined by $f(a_0)$ and $f(a_1)$, which are affine combinations of a_0 and a_1. There is an explicit formula for some fixed point of f. For the induction step, compose f with some suitable projections.

References

1. Michael Artin. *Algebra*. Prentice-Hall, first edition, 1991.
2. Marcel Berger. *Géométrie 1*. Nathan, 1990. English edition: Geometry 1, Universitext, Springer-Verlag.
3. Marcel Berger. *Géométrie 2*. Nathan, 1990. English edition: Geometry 2, Universitext, Springer-Verlag.
4. H.S.M. Coxeter. *Introduction to Geometry*. Wiley, second edition, 1989.
5. Jean H. Gallier. *Curves and Surfaces in Geometric Modeling: Theory and Algorithms*. Morgan Kaufmann, first edition, 1999.
6. Donald T. Greenwood. *Principles of Dynamics*. Prentice-Hall, second edition, 1988.
7. D. Hilbert and S. Cohn-Vossen. *Geometry and the Imagination*. Chelsea Publishing Co., 1952.
8. Serge Lang. *Algebra*. Addison-Wesley, third edition, 1993.
9. Dan Pedoe. *Geometry, A Comprehensive Course*. Dover, first edition, 1988.

10. Pierre Samuel. *Projective Geometry*. Undergraduate Texts in Mathematics. Springer-Verlag, first edition, 1988.
11. Ernst Snapper and Troyer Robert J. *Metric Affine Geometry*. Dover, first edition, 1989.
12. Gilbert Strang. *Linear Algebra and Its Applications*. Saunders HBJ, third edition, 1988.
13. Claude Tisseron. *Géométries Affines, Projectives, et Euclidiennes*. Hermann, first edition, 1994.

Chapter 3
Basic Properties of Convex Sets

3.1 Convex Sets

Convex sets play a very important role in geometry. In this chapter we state and prove some of the "classics" of convex affine geometry: Carathéodory's theorem, Radon's theorem, Helly's theorem, and Krein and Millman's theorem. These theorems share the property that they are easy to state, but they are deep, and their proof, although rather short, requires a lot of creativity. We introduce the notions of separating and supporting hyperplanes, of vertices, and of extreme points. We also define centerpoints and prove their existence.

Given an affine space E, recall that a subset V of E is *convex* if for any two points $a, b \in V$, we have $c \in V$ for every point $c = (1 - \lambda)a + \lambda b$, with $0 \leq \lambda \leq 1$ ($\lambda \in \mathbb{R}$). Given any two points a, b, the notation $[a, b]$ is often used to denote the line segment between a and b, that is,

$$[a, b] = \{c \in E \mid c = (1 - \lambda)a + \lambda b, 0 \leq \lambda \leq 1\},$$

and thus a set V is convex if $[a, b] \subseteq V$ for any two points $a, b \in V$ ($a = b$ is allowed). The empty set is trivially convex, every one-point set $\{a\}$ is convex, and the entire affine space E is, of course, convex.

It is obvious that the intersection of any family (finite or infinite) of convex sets is convex. Then, given any (nonempty) subset S of E, there is a smallest convex set containing S, denoted by $\mathscr{C}(S)$ or $\mathrm{conv}(S)$ and called the *convex hull of S* (namely, the intersection of all convex sets containing S). The *affine hull* of a subset S of E is the smallest affine set containing S, and it will be denoted by $\langle S \rangle$ or $\mathrm{aff}(S)$.

Definition 3.1. Given any affine space E the *dimension* of a nonempty convex subset S of E, denoted by $\dim S$, is the dimension of the smallest affine subset $\mathrm{aff}(S)$ containing S.

A good understanding of what $\mathscr{C}(S)$ is, and good methods for computing it, are essential. First, we have the following simple but crucial lemma:

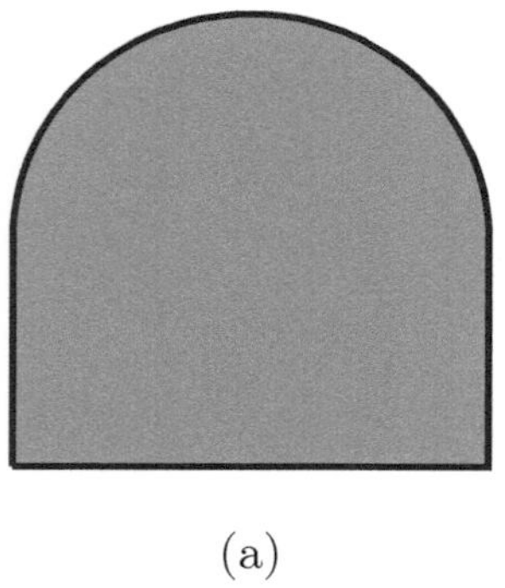

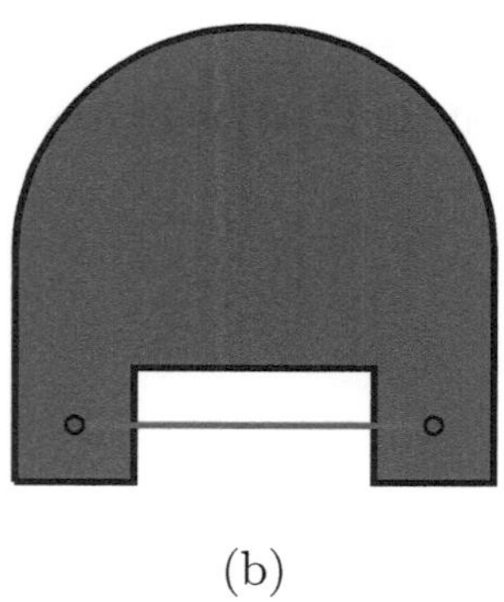

(a)   (b)

Fig. 3.1 (a) A convex set; (b) A nonconvex set.

Lemma 3.1. *Given an affine space* $\langle E, \overrightarrow{E}, + \rangle$, *for any family* $(a_i)_{i \in I}$ *of points in* E, *the set* V *of convex combinations* $\sum_{i \in I} \lambda_i a_i$ *(where* $\sum_{i \in I} \lambda_i = 1$ *and* $\lambda_i \geq 0$*) is the convex hull of* $(a_i)_{i \in I}$.

Proof. If $(a_i)_{i \in I}$ is empty, then $V = \emptyset$, because of the condition $\sum_{i \in I} \lambda_i = 1$. As in the case of affine combinations, it is easily shown by induction that any convex combination can be obtained by computing convex combinations of two points at a time. As a consequence, if $(a_i)_{i \in I}$ is nonempty, then the smallest convex subspace containing $(a_i)_{i \in I}$ must contain the set V of all convex combinations $\sum_{i \in I} \lambda_i a_i$. Thus, it is enough to show that V is closed under convex combinations, which is immediately verified. $\square$

In view of Lemma 3.1, it is obvious that any affine subspace of E is convex. Convex sets also arise in terms of hyperplanes. Given a hyperplane H, if $f \colon E \to \mathbb{R}$ is any nonconstant affine form defining H (i.e., $H = \mathrm{Ker}\, f$), we can define the two subsets

$$H_+(f) = \{a \in E \mid f(a) \geq 0\} \quad \text{and} \quad H_-(f) = \{a \in E \mid f(a) \leq 0\},$$

called *(closed) half-spaces associated with* f.

Observe that if $\lambda > 0$, then $H_+(\lambda f) = H_+(f)$, but if $\lambda < 0$, then $H_+(\lambda f) = H_-(f)$, and similarly for $H_-(\lambda f)$. However, the set

$$\{H_+(f), H_-(f)\}$$

depends only on the hyperplane H, and the choice of a specific f defining H amounts to the choice of one of the two half-spaces. For this reason, we will also say that $H_+(f)$ and $H_-(f)$ are the *closed half-spaces associated with* H. Clearly, $H_+(f) \cup H_-(f) = E$ and $H_+(f) \cap H_-(f) = H$. It is immediately verified that $H_+(f)$ and $H_-(f)$ are convex. Bounded convex sets arising as the intersection of a finite family

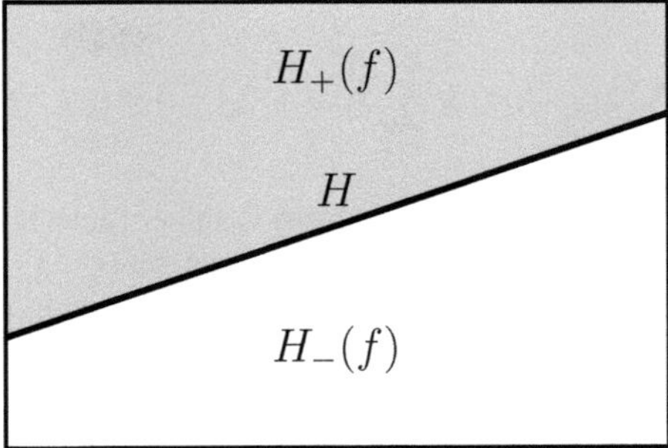

Fig. 3.2 The two half-spaces determined by a hyperplane H.

of half-spaces associated with hyperplanes play a major role in convex geometry and topology (they are called *convex polytopes*).

It is natural to wonder whether Lemma 3.1 can be sharpened in two directions: (1) Is it possible to have a fixed bound on the number of points involved in the convex combinations? (2) Is it necessary to consider convex combinations of all points, or is it possible to consider only a subset with special properties?

The answer is yes in both cases. In case 1, assuming that the affine space E has dimension m, Carathéodory's theorem asserts that it is enough to consider convex combinations of $m + 1$ points. For example, in the plane $\mathbb{A}^2$, the convex hull of a set S of points is the union of all triangles (interior points included) with vertices in S. In case 2, the theorem of Krein and Milman asserts that a convex set that is also compact is the convex hull of its extremal points (given a convex set S, a point $a \in S$ is extremal if $S - \{a\}$ is also convex; see Berger [2] or Lang [4]). Next, we prove Carathéodory's theorem.

3.2 Carathéodory's Theorem

The proof of Carathéodory's theorem is really beautiful. It proceeds by contradiction and uses a minimality argument.

Theorem 3.1. *(Carathéodory, 1907) Given any affine space E of dimension m, for any (nonvoid) family $S = (a_i)_{i \in L}$ in E, the convex hull $\mathscr{C}(S)$ of S is equal to the set of convex combinations of families of $m + 1$ points of S.*

Proof. By Lemma 3.1,

$$\mathscr{C}(S) = \left\{ \sum_{i \in I} \lambda_i a_i \mid a_i \in S, \sum_{i \in I} \lambda_i = 1, \lambda_i \geq 0, I \subseteq L, I \text{ finite} \right\}.$$

We would like to prove that

$$\mathscr{C}(S) = \left\{ \sum_{i \in I} \lambda_i a_i \mid a_i \in S, \sum_{i \in I} \lambda_i = 1, \lambda_i \geq 0, I \subseteq L, |I| = m+1 \right\}.$$

We proceed by contradiction. If the theorem is false, there is some point $b \in \mathscr{C}(S)$ such that b can be expressed as a convex combination $b = \sum_{i \in I} \lambda_i a_i$, where $I \subseteq L$ is a finite set of cardinality $|I| = q$ with $q \geq m+2$, and b cannot be expressed as any convex combination $b = \sum_{j \in J} \mu_j a_j$ of strictly fewer than q points in S, that is, where $|J| < q$. Such a point $b \in \mathscr{C}(S)$ is a convex combination

$$b = \lambda_1 a_1 + \cdots + \lambda_q a_q,$$

where $\lambda_1 + \cdots + \lambda_q = 1$ and $\lambda_i > 0$ $(1 \leq i \leq q)$. We shall prove that b can be written as a convex combination of $q-1$ of the a_i. Pick any origin O in E. Since there are $q > m+1$ points $a_1, \ldots, a_q$, these points are affinely dependent, and by Lemma 2.6, there is a family $(\mu_1, \ldots, \mu_q)$ of scalars not all null, such that $\mu_1 + \cdots + \mu_q = 0$ and

$$\sum_{i=1}^{q} \mu_i \overrightarrow{Oa_i} = 0.$$

Consider the set $T \subseteq \mathbb{R}$ defined by

$$T = \{ t \in \mathbb{R} \mid \lambda_i + t\mu_i \geq 0, \, \mu_i \neq 0, \, 1 \leq i \leq q \}.$$

The set T is nonempty, since it contains 0. Since $\sum_{i=1}^{q} \mu_i = 0$ and the μ_i are not all null, there are some μ_h, μ_k such that $\mu_h < 0$ and $\mu_k > 0$, which implies that $T = [\alpha, \beta]$, where

$$\alpha = \max_{1 \leq i \leq q} \{-\lambda_i/\mu_i \mid \mu_i > 0\} \quad \text{and} \quad \beta = \min_{1 \leq i \leq q} \{-\lambda_i/\mu_i \mid \mu_i < 0\}$$

(T is the intersection of the closed half-spaces $\{t \in \mathbb{R} \mid \lambda_i + t\mu_i \geq 0, \, \mu_i \neq 0\}$). Observe that $\alpha < 0 < \beta$, since $\lambda_i > 0$ for all $i = 1, \ldots, q$.

We claim that there is some j $(1 \leq j \leq q)$ such that

$$\lambda_j + \alpha \mu_j = 0.$$

Indeed, since

$$\alpha = \max_{1 \leq i \leq q} \{-\lambda_i/\mu_i \mid \mu_i > 0\},$$

and since the set on the right-hand side is finite, the maximum is achieved and there is some index j such that $\alpha = -\lambda_j/\mu_j$. If j is some index such that $\lambda_j + \alpha \mu_j = 0$, since $\sum_{i=1}^{q} \mu_i \overrightarrow{Oa_i} = 0$, we have

$$b = \sum_{i=1}^{q} \lambda_i a_i = O + \sum_{i=1}^{q} \lambda_i \overrightarrow{Oa_i} + 0,$$

$$= O + \sum_{i=1}^{q} \lambda_i \overrightarrow{Oa_i} + \alpha \left(\sum_{i=1}^{q} \mu_i \overrightarrow{Oa_i} \right),$$

$$= O + \sum_{i=1}^{q} (\lambda_i + \alpha \mu_i) \overrightarrow{Oa_i},$$

$$= \sum_{i=1}^{q} (\lambda_i + \alpha \mu_i) a_i,$$

$$= \sum_{i=1, i \neq j}^{q} (\lambda_i + \alpha \mu_i) a_i,$$

since $\lambda_j + \alpha \mu_j = 0$. Since $\sum_{i=1}^{q} \mu_i = 0$, $\sum_{i=1}^{q} \lambda_i = 1$, and $\lambda_j + \alpha \mu_j = 0$, we have

$$\sum_{i=1, i \neq j}^{q} \lambda_i + \alpha \mu_i = 1,$$

and since $\lambda_i + \alpha \mu_i \geq 0$ for $i = 1, \ldots, q$, the above shows that b can be expressed as a convex combination of $q - 1$ points from S. However, this contradicts the assumption that b cannot be expressed as a convex combination of strictly fewer than q points from S, and the theorem is proved. $\square$

If S is a finite (of infinite) set of points in the affine plane $\mathbb{A}^2$, Theorem 3.1 confirms our intuition that $\mathscr{C}(S)$ is the union of triangles (including interior points) whose vertices belong to S. Similarly, the convex hull of a set S of points in $\mathbb{A}^3$ is the union of tetrahedra (including interior points) whose vertices belong to S. We get the feeling that triangulations play a crucial role, which is of course true!

An interesting consequence of Carathéodory's theorem is the following result:

Proposition 3.1. *If K is any compact subset of $\mathbb{A}^m$, then the convex hull $\mathrm{conv}(K)$ of K is also compact.*

Proposition 3.1 can be proved by showing that $\mathrm{conv}(K)$ is the image of some compact subset of $\mathbb{R}^{m+1} \times (\mathbb{A}^m)^{m+1}$ under some well-chosen continuous map.

A closer examination of the proof of Theorem 3.1 reveals that the fact that the μ_i's add up to zero is actually not needed in the proof. This fact ensures that T is a closed interval, but all we need is that T be bounded from below, and this requires only that some μ_j be strictly positive. As a consequence, we can prove a version of Theorem 3.1 for convex cones. This is a useful result, since cones play such an important role in convex optimization. Let us recall some basic definitions about cones.

Definition 3.2. Given any vector space E a subset $C \subseteq E$ is a *convex cone* iff C is closed under *positive linear combinations*, that is, linear combinations of the form

$$\sum_{i \in I} \lambda_i v_i, \quad \text{with} \quad v_i \in C \quad \text{and} \quad \lambda_i \geq 0 \quad \text{for all} \quad i \in I,$$

where I has finite support (all $\lambda_i = 0$ except for finitely many $i \in I$). Given any set of vectors S, the *positive hull* of S, or *cone* spanned by S, denoted by cone(S), is the set of all positive linear combinations of vectors in S,

$$\text{cone}(S) = \left\{ \sum_{i \in I} \lambda_i v_i \mid v_i \in S, \lambda_i \geq 0 \right\}.$$

Note that a cone always contains 0. When S consists of a finite number of vectors, the convex cone cone(S) is called a *polyhedral cone*. We have the following version of Carathéodory's theorem for convex cones:

Theorem 3.2. *Given any vector space E of dimension m, for any (nonvoid) family $S = (v_i)_{i \in L}$ of vectors in E, the cone* cone(S) *spanned by S is equal to the set of positive combinations of families of m vectors in S.*

The proof of Theorem 3.2 can be easily adapted from the proof of Theorem 3.1 and is left as an exercise.

There is an interesting generalization of Carathéodory's theorem known as the *colorful Carathéodory theorem*. This theorem, due to Bárány and proved in 1982, can be used to give a fairly short proof of a generalization of Helly's theorem known as Tverberg's theorem (see Section 3.4).

Theorem 3.3. *(Colorful Carathéodory theorem) Let E be any affine space of dimension m. For any point $b \in E$ and for any sequence of $m + 1$ nonempty subsets $(S_1, \ldots, S_{m+1})$ of E, if $b \in$ conv(S_i) for $i = 1, \ldots, m + 1$, then there exists a sequence of $m + 1$ points $(a_1, \ldots, a_{m+1})$ with $a_i \in S_i$, such that $b \in$ conv$(a_1, \ldots, a_{m+1})$, that is, b is a convex combination of the a_i's.*

Although Theorem 3.3 is not hard to prove, we will not prove it here. Instead, we refer the reader to Matousek [6], Chapter 8, Section 8.2. There is also a stronger version of Theorem 3.3, in which it is enough to assume that $b \in$ conv$(S_i \cup S_j)$ for all i, j with $1 \leq i < j \leq m + 1$.

Now that we have given an answer to the first question posed at the end of Section 3.1, we give an answer to the second question.

3.3 Vertices, Extremal Points, and Krein and Milman's Theorem

First, we define the notions of separation and of separating hyperplanes. For this, recall the definition of the closed (or open) half-spaces determined by a hyperplane.

Given a hyperplane H, if $f: E \to \mathbb{R}$ is any nonconstant affine form defining H (i.e., $H = \text{Ker} f$), recall that we define the *closed half-spaces associated with f* by

$$H_+(f) = \{a \in E \mid f(a) \geq 0\},$$
$$H_-(f) = \{a \in E \mid f(a) \leq 0\}.$$

We also define the *open half-spaces associated with* f as the two sets

$$\overset{\circ}{H}_+(f) = \{a \in E \mid f(a) > 0\},$$
$$\overset{\circ}{H}_-(f) = \{a \in E \mid f(a) < 0\}.$$

The set $\{\overset{\circ}{H}_+(f), \overset{\circ}{H}_-(f)\}$ depends only on the hyperplane H. Clearly, we have $\overset{\circ}{H}_+(f) = H_+(f) - H$ and $\overset{\circ}{H}_-(f) = H_-(f) - H$.

Definition 3.3. Given an affine space X and two nonempty subsets A and B of X, we say that a hyperplane H *separates (resp. strictly separates) A and B* if A is in one and B is in the other of the two half-spaces (resp. open half-spaces) determined by H.

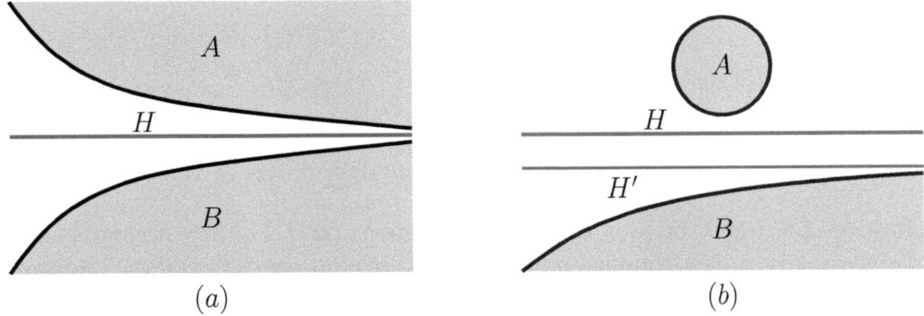

Fig. 3.3 (a) A separating hyperplane H. (b) A strictly separating hyperplane H.

In Figure 3.3 (a), the two closed convex sets A and B are unbounded and both asymptotic to the hyperplane H. The hyperplane H is a separating hyperplane for A and B, but A and B can't be strictly separated. In Figure 3.3 (b), both A and B are convex and closed, B is unbounded and asymptotic to the hyperplane H', but A is bounded. The hyperplane H strictly separates A and B. The hyperplane H' also separates A and B but not strictly.

The special case of separation in which A is convex and $B = \{a\}$ for some point a in A is of particular importance.

Definition 3.4. Let X be an affine space and let A be any nonempty subset of X. A *supporting hyperplane of A* is any hyperplane H containing some point a of A and separating $\{a\}$ and A. We say that H is a *supporting hyperplane of A at a*.

Observe that if H is a supporting hyperplane of A at a, then we must have $a \in \partial A$. Otherwise, there would be some open ball $B(a, \varepsilon)$ of center a contained in A, and so there would be points of A (in $B(a, \varepsilon)$) in both half-spaces determined by H, contradicting the fact that H is a supporting hyperplane of A at a. Furthermore, $H \cap \overset{\circ}{A} = \emptyset$.

One should experiment with various pictures and realize that supporting hyperplanes at a point may not exist (for example, if A is not convex), may not be unique, and may have several distinct supporting points! (See Figure 3.4).

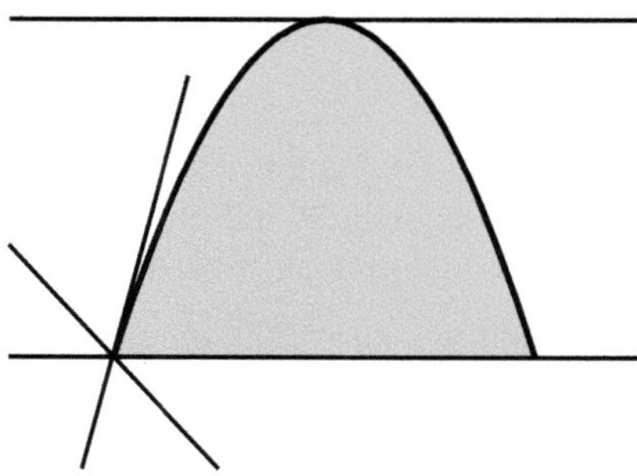

Fig. 3.4 Examples of supporting hyperplanes.

Next, we need to define various types of boundary points of closed convex sets.

Definition 3.5. Let X be an affine space of dimension d. For any nonempty closed and convex subset A of dimension d, a point $a \in \partial A$ has *order* $k(a)$ if the intersection of all the supporting hyperplanes of A at a is an affine subspace of dimension $k(a)$. We say that $a \in \partial A$ is a *vertex* if $k(a) = 0$; we say that a is *smooth* if $k(a) = d - 1$, i.e., if the supporting hyperplane at a is unique.

A vertex is a boundary point a such that there are d independent supporting hyperplanes at a. A d-simplex has boundary points of order $0, 1, \ldots, d - 1$. The following proposition is proved in Berger [2] (Proposition 11.6.2):

Proposition 3.2. *The set of vertices of a closed and convex subset is countable.*

Another important concept is that of an extremal point.

Definition 3.6. Let X be an affine space. For any nonempty convex subset A a point $a \in \partial A$ is *extremal* (or *extreme*) if $A - \{a\}$ is still convex.

It is fairly obvious that a point $a \in \partial A$ is extremal if it does not belong to the interior of any closed nontrivial line segment $[x, y] \subseteq A$ ($x \neq y$, $a \neq x$ and $a \neq y$).

Observe that a vertex is extremal, but the converse is false. For example, in Figure 3.5, all the points on the arc of the parabola, including v_1 and v_2, are extreme points.

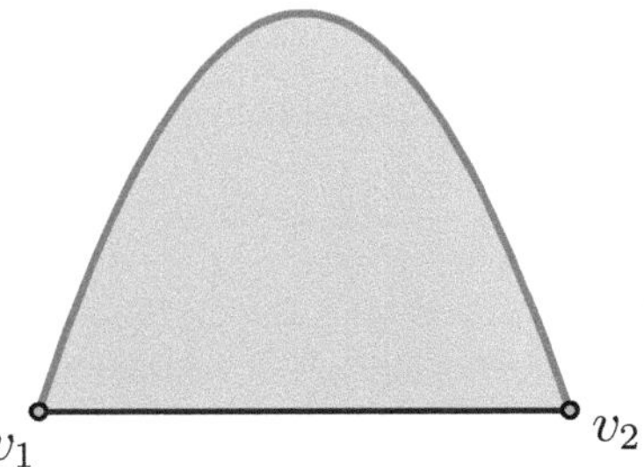

Fig. 3.5 Examples of vertices and extreme points.

However, only v_1 and v_2 are vertices. Also, if $\dim X \geq 3$, the set of extremal points of a compact convex may not be closed.

Actually, it is not at all obvious that a nonempty compact convex set possesses extremal points. In fact, a stronger results holds (Krein and Milman's theorem). In preparation for the proof of this important theorem, observe that any compact (nontrivial) interval of $\mathbb{A}^1$ has two extremal points, its two endpoints. We need the following lemma:

Lemma 3.2. *Let X be an affine space of dimension n, and let A be a nonempty compact and convex set. Then $A = \mathscr{C}(\partial A)$, i.e., A is equal to the convex hull of its boundary.*

Proof. Pick any a in A, and consider any line D through a. Then, $D \cap A$ is closed and convex. However, since A is compact, it follows that $D \cap A$ is a closed interval $[u, v]$ containing a, and $u, v \in \partial A$. Therefore, $a \in \mathscr{C}(\partial A)$, as desired. $\square$

The following important theorem shows that only extremal points matter in determining a compact and convex subset from its boundary. The proof of Theorem 3.4 makes use of a proposition due to Minkowski (Proposition 7.4), which will be proved in Section 7.2.

Theorem 3.4. *(Krein and Milman, 1940) Let X be an affine space of dimension n. Every compact and convex nonempty subset A is equal to the convex hull of its set of extremal points.*

Proof. Denote the set of extremal points of A by $\mathrm{Extrem}(A)$. We proceed by induction on $d = \dim X$. When $d = 1$, the convex and compact subset A must be a closed interval $[u, v]$ or a single point. In either case, the theorem holds trivially. Now assume $d \geq 2$, and assume that the theorem holds for $d - 1$. It is easily verified that

$$\mathrm{Extrem}(A \cap H) = (\mathrm{Extrem}(A)) \cap H,$$

for every supporting hyperplane H of A (such hyperplanes exist, by Minkowski's proposition (Proposition 7.4)). Observe that Lemma 3.2 implies that if we can prove that

$$\partial A \subseteq \mathscr{C}(\mathrm{Extrem}(A)),$$

then, since $A = \mathscr{C}(\partial A)$, we will have established that

$$A = \mathscr{C}(\mathrm{Extrem}(A)).$$

Let $a \in \partial A$, and let H be a supporting hyperplane of A at a (which exists, by Minkowski's proposition). Now A and H are convex, so $A \cap H$ is convex; H is closed and A is compact, so $H \cap A$ is a closed subset of a compact subset A, and thus $A \cap H$ is also compact. Since $A \cap H$ is a compact and convex subset of H and H has dimension $d - 1$, by the induction hypothesis, we have

$$A \cap H = \mathscr{C}(\mathrm{Extrem}(A \cap H)).$$

However,

$$\begin{aligned}
\mathscr{C}(\mathrm{Extrem}(A \cap H)) &= \mathscr{C}((\mathrm{Extrem}(A)) \cap H) \\
&= \mathscr{C}(\mathrm{Extrem}(A)) \cap H \subseteq \mathscr{C}(\mathrm{Extrem}(A)),
\end{aligned}$$

and so $a \in A \cap H \subseteq \mathscr{C}(\mathrm{Extrem}(A))$. Therefore, we have proved that

$$\partial A \subseteq \mathscr{C}(\mathrm{Extrem}(A)),$$

from which we deduce that $A = \mathscr{C}(\mathrm{Extrem}(A))$, as explained earlier. $\square$

Remark: Observe that Krein and Milman's theorem implies that any nonempty compact and convex set has a nonempty subset of extremal points. This is intuitively obvious, but hard to prove! Krein and Milman's theorem also applies to infinite-dimensional affine spaces, provided that they are locally convex; see Valentine [7], Chapter 11, Bourbaki [3], Chapter II, Barvinok [1], Chapter 3, or Lax [5], Chapter 13.

An important consequence of Krein and Milman's theorem is that every convex function on a convex and compact set achieves its maximum at some extremal point.

Definition 3.7. Let A be a nonempty convex subset of $\mathbb{A}^n$. A function $f \colon A \to \mathbb{R}$ is *convex* if

$$f((1 - \lambda)a + \lambda b) \leq (1 - \lambda)f(a) + \lambda f(b)$$

for all $a, b \in A$ and for all $\lambda \in [0, 1]$. The function $f \colon A \to \mathbb{R}$ is *strictly convex* if

$$f((1 - \lambda)a + \lambda b) < (1 - \lambda)f(a) + \lambda f(b)$$

for all $a, b \in A$ with $a \neq b$ and for all λ with $0 < \lambda < 1$. A function $f \colon A \to \mathbb{R}$ is *concave* (resp. *strictly concave*) iff $-f$ is convex (resp. $-f$ is strictly convex).

If f is convex, a simple induction shows that

$$f\left(\sum_{i \in I} \lambda_i a_i\right) \leq \sum_{i \in I} \lambda_i f(a_i)$$

for every finite convex combination in A, i.e., for any finite family $(a_i)_{i \in I}$ of points in A and any family $(\lambda_i)_{i \in I}$ with $\sum_{i \in I} \lambda_i = 1$ and $\lambda_i \geq 0$ for all $i \in I$.

Proposition 3.3. *Let A be a nonempty convex and compact subset of $\mathbb{A}^n$ and let $f \colon A \to \mathbb{R}$ be any function. If f is convex and continuous, then f achieves its maximum at some extreme point of A.*

Proof. Since A is compact and f is continuous, $f(A)$ is a closed interval $[m, M]$ in $\mathbb{R}$, and so f achieves its minimum m and its maximum M. Say $f(c) = M$, for some $c \in A$. By Krein and Milman's theorem, c is some convex combination of extreme points of A,

$$c = \sum_{i=1}^{k} \lambda_i a_i,$$

with $\sum_{i=1}^{k} \lambda_i = 1$, $\lambda_i \geq 0$, and each a_i an extreme point in A. But then, since f is convex,

$$M = f(c) = f\left(\sum_{i=1}^{k} \lambda_i a_i \right) \leq \sum_{i=1}^{k} \lambda_i f(a_i),$$

and if we let

$$f(a_{i_0}) = \max_{1 \leq i \leq k} \{f(a_i)\}$$

for some i_0 such that $1 \leq i_0 \leq k$, then we get

$$M = f(c) \leq \sum_{i=1}^{k} \lambda_i f(a_i) \leq \left(\sum_{i=1}^{k} \lambda_i \right) f(a_{i_0}) = f(a_{i_0}),$$

since $\sum_{i=1}^{k} \lambda_i = 1$. Since M is the maximum value of the function f over A, we have $f(a_{i_0}) \leq M$, and so

$$M = f(a_{i_0}),$$

and f achieves its maximum at the extreme point a_{i_0}, as claimed. $\square$

Proposition 3.3 plays an important role in convex optimization: It guarantees that the maximum value of a convex objective function on a compact and convex set is achieved at some extreme point. Thus, it is enough to look for a maximum at some extreme point of the domain.

Proposition 3.3 fails for minimal values of a convex function. For example, the function $x \mapsto f(x) = x^2$ defined on the compact interval $[-1, 1]$ achieves it minimum at $x = 0$, which is not an extreme point of $[-1, 1]$. However, if f is concave, then f achieves its minimum value at some extreme point of A. In particular, if f is affine, it achieves its minimum and its maximum at some extreme points of A.

We conclude this chapter with three further classics of convex geometry.

3.4 Radon's, Helly's, Tverberg's Theorems and Centerpoints

We begin with *Radon's theorem*.

Theorem 3.5. *(Radon, 1921) Given any affine space E of dimension m, for every subset X of E, if X has at least $m+2$ points, then there is a partition of X into two nonempty disjoint subsets X_1 and X_2 such that the convex hulls of X_1 and X_2 have a nonempty intersection.*

Proof. Pick some origin O in E. Write $X = (x_i)_{i \in L}$ for some index set L (we can let $L = X$). Since by assumption $|X| \geq m+2$, where $m = \dim(E)$, X is affinely dependent, and by Lemma 2.6, there is a family $(\mu_k)_{k \in L}$ (of finite support) of scalars, not all null, such that

$$\sum_{k \in L} \mu_k = 0 \quad \text{and} \quad \sum_{k \in L} \mu_k \overrightarrow{Ox_k} = 0.$$

Since $\sum_{k \in L} \mu_k = 0$, the μ_k are not all null, and $(\mu_k)_{k \in L}$ has finite support, the sets

$$I = \{i \in L \mid \mu_i > 0\} \quad \text{and} \quad J = \{j \in L \mid \mu_j < 0\}$$

are nonempty, finite, and obviously disjoint. Let

$$X_1 = \{x_i \in X \mid \mu_i > 0\} \quad \text{and} \quad X_2 = \{x_i \in X \mid \mu_i \leq 0\}.$$

Again, since the μ_k are not all null and $\sum_{k \in L} \mu_k = 0$, the sets X_1 and X_2 are nonempty, and obviously

$$X_1 \cap X_2 = \emptyset \quad \text{and} \quad X_1 \cup X_2 = X.$$

Furthermore, the definition of I and J implies that $(x_i)_{i \in I} \subseteq X_1$ and $(x_j)_{j \in J} \subseteq X_2$. It remains to prove that $\mathscr{C}(X_1) \cap \mathscr{C}(X_2) \neq \emptyset$. The definition of I and J implies that

$$\sum_{k \in L} \mu_k \overrightarrow{Ox_k} = 0$$

can be written as

$$\sum_{i \in I} \mu_i \overrightarrow{Ox_i} + \sum_{j \in J} \mu_j \overrightarrow{Ox_j} = 0,$$

that is, as

$$\sum_{i \in I} \mu_i \overrightarrow{Ox_i} = \sum_{j \in J} -\mu_j \overrightarrow{Ox_j},$$

where

$$\sum_{i \in I} \mu_i = \sum_{j \in J} -\mu_j = \mu,$$

with $\mu > 0$. Thus, we have

$$\sum_{i \in I} \frac{\mu_i}{\mu} \overrightarrow{Ox_i} = \sum_{j \in J} -\frac{\mu_j}{\mu} \overrightarrow{Ox_j},$$

with

$$\sum_{i\in I}\frac{\mu_i}{\mu} = \sum_{j\in J}-\frac{\mu_j}{\mu} = 1,$$

proving that $\sum_{i\in I}(\mu_i/\mu)x_i \in \mathscr{C}(X_1)$ and $\sum_{j\in J}-(\mu_j/\mu)x_j \in \mathscr{C}(X_2)$ are identical, and thus that $\mathscr{C}(X_1)\cap\mathscr{C}(X_2)\neq\emptyset$. $\square$

A partition (X_1,X_2) of X satisfying the conditions of Theorem 3.5 is sometimes called a *Radon partition* of X, and any point in $\mathrm{conv}(X_1)\cap\mathrm{conv}(X_2)$ is called a *Radon point* of X. Figure 3.6 shows two Radon partitions of five points in the plane.

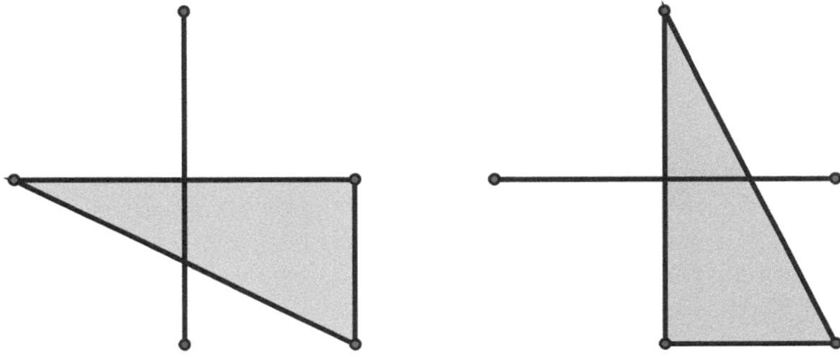

Fig. 3.6 Examples of Radon partitions.

It can be shown that a finite set $X \subseteq E$ has a unique Radon partition iff it has $m+2$ elements and any $m+1$ points of X are affinely independent. For example, there are exactly two possible cases in the plane, as shown in Figure 3.7.

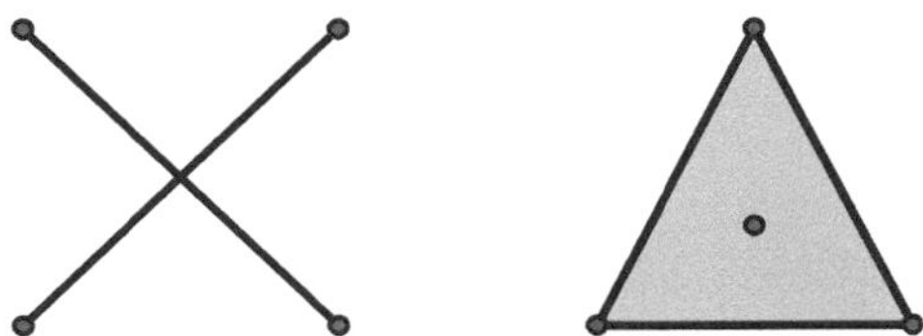

Fig. 3.7 The Radon partitions of four points (in $\mathbb{A}^2$).

There is also a version of Radon's theorem for the class of cones with an apex. Say that a convex cone $C \subseteq E$ has an *apex* (or is a *pointed cone*) iff there is some hyperplane H such that $C \subseteq H_+$ and $H\cap C = \{0\}$. For example, the cone obtained as the intersection of two half-spaces in $\mathbb{R}^3$ is not pointed, since it is a wedge with a line as part of its boundary. Here is the version of Radon's theorem for convex cones:

Theorem 3.6. *Given any vector space E of dimension m, for every subset X of E, if $\mathrm{cone}(X)$ is a pointed cone such that X has at least $m+1$ nonzero vectors, then there is a partition of X into two nonempty disjoint subsets X_1 and X_2 such that the cones $\mathrm{cone}(X_1)$ and $\mathrm{cone}(X_2)$ have a nonempty intersection not reduced to $\{0\}$.*

The proof of Theorem 3.6 is left as an exercise.

There is a beautiful generalization of Radon's theorem known as *Tverberg's theorem*.

Theorem 3.7. *(Tverberg's theorem, 1966) Let E be any affine space of dimension m. For any natural number $r \geq 2$ and every subset X of E, if X has at least $(m+1)(r-1)+1$ points, then there is a partition $(X_1, \ldots, X_r)$ of X into r nonempty pairwise disjoint subsets such that $\bigcap_{i=1}^{r} \mathrm{conv}(X_i) \neq \emptyset$.*

A partition as in Theorem 3.7 is called a *Tverberg partition*, and a point in $\bigcap_{i=1}^{r} \mathrm{conv}(X_i)$ is called a *Tverberg point*. Theorem 3.7 was conjectured by Birch and proved by Tverberg in 1966. Tverberg's original proof was technically quite complicated. Tverberg then gave a simpler proof in 1981, and other simpler proofs were given later, notably by Sarkaria (1992) and Onn (1997), using the colorful Carathéodory theorem. A proof along those lines can be found in Matousek [6], Chapter 8, Section 8.3. A *colored Tverberg theorem* and more can also be found in Matousek [6] (Section 8.3).

Next, we prove a version of *Helly's theorem*.

Theorem 3.8. *(Helly, 1913) Given any affine space E of dimension m, for every family $\{K_1, \ldots, K_n\}$ of n convex subsets of E, if $n \geq m+2$ and the intersection $\bigcap_{i \in I} K_i$ of any $m+1$ of the K_i is nonempty (where $I \subseteq \{1, \ldots, n\}$, $|I| = m+1$), then $\bigcap_{i=1}^{n} K_i$ is nonempty.*

Proof. The proof is by induction on $n \geq m+1$ and uses Radon's theorem in the induction step. For $n = m+1$, the assumption of the theorem is that the intersection of any family of $m+1$ of the K_i's is nonempty, and the theorem holds trivially. Next, let $L = \{1, 2, \ldots, n+1\}$, where $n+1 \geq m+2$. By the induction hypothesis, $C_i = \bigcap_{j \in (L-\{i\})} K_j$ is nonempty for every $i \in L$.

We claim that $C_i \cap C_j \neq \emptyset$ for some $i \neq j$. If so, since $C_i \cap C_j = \bigcap_{k=1}^{n+1} K_k$, we are done. So let us assume that the C_i's are disjoint. Then we can pick a set $X = \{a_1, \ldots, a_{n+1}\}$ such that $a_i \in C_i$, for every $i \in L$. By Radon's theorem, there are two nonempty disjoint sets $X_1, X_2 \subseteq X$ such that $X = X_1 \cup X_2$ and $\mathscr{C}(X_1) \cap \mathscr{C}(X_2) \neq \emptyset$. However, $X_1 \subseteq K_j$ for every j with $a_j \notin X_1$. This is because $a_j \notin K_j$ for every j, and so we get

$$X_1 \subseteq \bigcap_{a_j \notin X_1} K_j.$$

Symmetrically, we also have

$$X_2 \subseteq \bigcap_{a_j \notin X_2} K_j.$$

Since the K_j's are convex and

$$\left(\bigcap_{a_j \notin X_1} K_j \right) \cap \left(\bigcap_{a_j \notin X_2} K_j \right) = \bigcap_{i=1}^{n+1} K_i,$$

it follows that $\mathscr{C}(X_1) \cap \mathscr{C}(X_2) \subseteq \bigcap_{i=1}^{n+1} K_i$, so that $\bigcap_{i=1}^{n+1} K_i$ is nonempty, contradicting the fact that $C_i \cap C_j = \emptyset$ for all $i \neq j$. $\quad\square$

A more general version of Helly's theorem is proved in Berger [2]. An amusing corollary of Helly's theorem is the following result: Consider $n \geq 4$ parallel line segments in the affine plane $\mathbb{A}^2$. If every three of these line segments meet a line, then all of these line segments meet a common line.

We conclude this chapter with a nice application of Helly's theorem to the existence of centerpoints. Centerpoints generalize the notion of median to higher dimensions. Recall that if we have a set of n data points $S = \{a_1, \ldots, a_n\}$ on the real line, a *median* for S is a point x such that both intervals $[x, \infty)$ and $(-\infty, x]$ contain at least $n/2$ of the points in S (by $n/2$, we mean the largest integer greater than or equal to $n/2$).

Given any hyperplane H, recall that the closed half-spaces determined by H are denoted by H_+ and H_- and that $H \subseteq H_+$ and $H \subseteq H_-$. We let $\overset{\circ}{H}_+ = H_+ - H$ and $\overset{\circ}{H}_- = H_- - H$ be the *open half-spaces* determined by H.

Definition 3.8. Let $S = \{a_1, \ldots, a_n\}$ be a set of n points in $\mathbb{A}^d$. A point $c \in \mathbb{A}^d$ is a *centerpoint of S* iff for every hyperplane H, whenever the closed half-space H_+ (resp. H_-) contains c, then H_+ (resp. H_-) contains at least $\frac{n}{d+1}$ points from S (by $\frac{n}{d+1}$, we mean the largest integer greater than or equal to $\frac{n}{d+1}$, namely the ceiling $\lceil \frac{n}{d+1} \rceil$ of $\frac{n}{d+1}$).

So for $d = 2$, for each line D, if the closed half-plane D_+ (resp. D_-) contains c, then D_+ (resp. D_-) contains at least a third of the points from S. For $d = 3$, for each plane H, if the closed half-space H_+ (resp. H_-) contains c, then H_+ (resp. H_-) contains at least a fourth of the points from S, etc. Figure 3.8 shows nine points in the plane and one of their centerpoints (in red). This example shows that the bound $\frac{1}{3}$ is tight.

Observe that a point $c \in \mathbb{A}^d$ is a centerpoint of S iff c belongs to every open half-space $\overset{\circ}{H}_+$ (resp. $\overset{\circ}{H}_-$) containing at least $\frac{dn}{d+1} + 1$ points from S (again, we mean $\lceil \frac{dn}{d+1} \rceil + 1$).

Indeed, if c is a centerpoint of S and H is any hyperplane such that $\overset{\circ}{H}_+$ (resp. $\overset{\circ}{H}_-$) contains at least $\frac{dn}{d+1} + 1$ points from S, then $\overset{\circ}{H}_+$ (resp. $\overset{\circ}{H}_-$) must contain c, since otherwise, the closed half-space H_- (resp. H_+) would contain c and at most $n - \frac{dn}{d+1} - 1 = \frac{n}{d+1} - 1$ points from S, a contradiction. Conversely, assume that c belongs to every open half-space $\overset{\circ}{H}_+$ (resp. $\overset{\circ}{H}_-$) containing at least $\frac{dn}{d+1} + 1$ points from S. Then for any hyperplane H, if $c \in H_+$ (resp. $c \in H_-$) but H_+ contains at

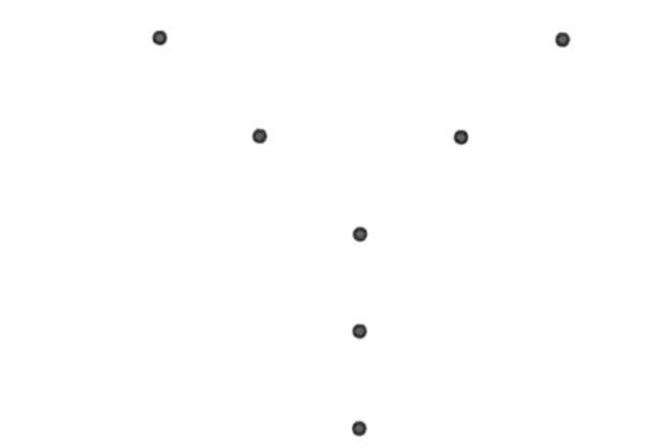

Fig. 3.8 Example of a centerpoint.

most $\frac{n}{d+1} - 1$ points from S, then the open half-space $\mathring{H}_-$ (resp. $\mathring{H}_+$) would contain at least $n - \frac{n}{d+1} + 1 = \frac{dn}{d+1} + 1$ points from S but not c, a contradiction.

We are now ready to prove the existence of centerpoints.

Theorem 3.9. *(Existence of centerpoints) Every finite set* $S = \{a_1, \ldots, a_n\}$ *of* n *points in* $\mathbb{A}^d$ *has some centerpoint.*

Proof. We will use the second characterization of centerpoints involving open half-spaces containing at least $\frac{dn}{d+1} + 1$ points.

Consider the family of sets

$$\mathscr{C} = \left\{ \mathrm{conv}(S \cap \mathring{H}_+) \mid (\exists H)\left(|S \cap \mathring{H}_+| > \frac{dn}{d+1} \right) \right\}$$
$$\cup \left\{ \mathrm{conv}(S \cap \mathring{H}_-) \mid (\exists H)\left(|S \cap \mathring{H}_-| > \frac{dn}{d+1} \right) \right\},$$

where H is a hyperplane.

Since S is finite, $\mathscr{C}$ consists of a finite number of convex sets, say $\{C_1, \ldots, C_m\}$. If we prove that $\bigcap_{i=1}^{m} C_i \neq \emptyset$, we are done, because $\bigcap_{i=1}^{m} C_i$ is the set of centerpoints of S.

First, we prove by induction on k (with $1 \leq k \leq d+1$) that any intersection of k of the C_i's has at least $\frac{(d+1-k)n}{d+1} + k$ elements from S. For $k = 1$, this holds by definition of the C_i's.

Next, consider the intersection of $k+1 \leq d+1$ of the C_i's, say $C_{i_1} \cap \cdots \cap C_{i_k} \cap C_{i_{k+1}}$. Let

$$A = S \cap (C_{i_1} \cap \cdots \cap C_{i_k} \cap C_{i_{k+1}}),$$
$$B = S \cap (C_{i_1} \cap \cdots \cap C_{i_k}),$$
$$C = S \cap C_{i_{k+1}}.$$

Note that $A = B \cap C$. By the induction hypothesis, B contains at least $\frac{(d+1-k)n}{d+1} + k$ elements from S. Since C contains at least $\frac{dn}{d+1} + 1$ points from S, and since

$$|B \cup C| = |B| + |C| - |B \cap C| = |B| + |C| - |A|$$

and $|B \cup C| \le n$, we get $n \ge |B| + |C| - |A|$, that is,

$$|A| \ge |B| + |C| - n.$$

It follows that

$$|A| \ge \frac{(d+1-k)n}{d+1} + k + \frac{dn}{d+1} + 1 - n,$$

that is,

$$|A| \ge \frac{(d+1-k)n + dn - (d+1)n}{d+1} + k + 1 = \frac{(d+1-(k+1))n}{d+1} + k + 1,$$

establishing the induction hypothesis.

Now if $m \le d + 1$, the above claim for $k = m$ shows that $\bigcap_{i=1}^{m} C_i \ne \emptyset$, and we are done. If $m \ge d + 2$, the above claim for $k = d + 1$ shows that any intersection of $d + 1$ of the C_i's is nonempty. Consequently, the conditions for applying Helly's theorem are satisfied, and therefore

$$\bigcap_{i=1}^{m} C_i \ne \emptyset.$$

However, $\bigcap_{i=1}^{m} C_i$ is the set of centerpoints of S, and we are done. $\square$

Remark: The above proof actually shows that the set of centerpoints of S is a convex set. In fact, it is a finite intersection of convex hulls of finitely many points, so it is the convex hull of finitely many points, in other words, a polytope. It should also be noted that Theorem 3.9 can be proved easily using Tverberg's theorem (Theorem 3.7). Indeed, for a judicious choice of r, any Tverberg point is a centerpoint!

Jadhav and Mukhopadhyay have given a linear-time algorithm for computing a centerpoint of a finite set of points in the plane. For $d \ge 3$, it appears that the best that can be done (using linear programming) is $O(n^d)$. However, there are good approximation algorithms (Clarkson, Eppstein, Miller, Sturtivant, and Teng), and in $\mathbb{E}^3$ there is a near-quadratic algorithm (Agarwal, Sharir, and Welzl). Recently, Miller and Sheehy (2009) gave an algorithm for finding an approximate centerpoint in subexponential time together with a polynomial-checkable proof of the approximation guarantee.

3.5 Problems

3.1. Let a, b, c, be any distinct points in $\mathbb{A}^3$, and assume that they are not collinear. Let H be the plane of the equation

$$\alpha x + \beta y + \gamma z + \delta = 0.$$

 (i) What is the intersection of the plane H and the solid triangle determined by a, b, c (the convex hull of a, b, c)?
 (ii) Give an algorithm to find the intersection of the plane H and the triangle determined by a, b, c.
(iii) (**extra credit**) Implement the above algorithm so that the intersection can be visualized (you may use *Maple*, *Mathematica*, *Matlab*, etc.).

3.2. Given any two affine spaces E and F, for any affine map $f \colon E \to F$, any convex set U in E, and any convex set V in F, prove that $f(U)$ is convex and that $f^{-1}(V)$ is convex. Recall that

$$f(U) = \{b \in F \mid \exists a \in U, b = f(a)\}$$

is the *direct image of U under f*, and that

$$f^{-1}(V) = \{a \in E \mid \exists b \in V, b = f(a)\}$$

is the *inverse image of V under f*.

3.3. Consider the subset S of $\mathbb{A}^2$ consisting the points belonging to the right branch of the hyperbola of the equation $x^2 - y^2 = 1$, i.e.,

$$S = \{(x, y) \in \mathbb{R}^2 \mid x^2 - y^2 \geq 1, x \geq 0\}.$$

Prove that S is convex. What is the convex hull of $S \cup \{(0,0)\}$? Is the convex hull of a closed subset of $\mathbb{A}^m$ necessarily a closed set?

3.4. Use the theorem of Carathéodory to prove that if S is a compact subset of $\mathbb{A}^m$, then its convex hull $\mathrm{conv}(S)$ is also compact.

3.5. Let S be any nonempty subset of an affine space E. Given some point $a \in S$, we say that S is *star-shaped with respect to a* if the line segment $[a, x]$ is contained in S for every $x \in S$, i.e., $(1 - \lambda)a + \lambda x \in S$ for all λ such that $0 \leq \lambda \leq 1$. We say that S is *star-shaped* if it is star-shaped w.r.t. to some point $a \in S$.

(1) Prove that every nonempty convex set is star-shaped.
(2) Show that there are star-shaped subsets that are not convex. Show that there are nonempty subsets that are not star-shaped (give an example in $\mathbb{A}^n$, $n = 1, 2, 3$).
(3) Given a star-shaped subset S of E, let $N(S)$ be the set of all points $a \in S$ such that S is star-shaped with respect to a. Prove that $N(S)$ is convex.

3.6. Consider $n \geq 4$ parallel line segments in the affine plane $\mathbb{A}^2$. If every three of these line segments meet a line, then all of these line segments meet a common line. *Hint.* Choose a coordinate system such that the y-axis is parallel to the common direction of the line segments. For any line segment S, let

$$CS = \{(\alpha, \beta) \in \mathbb{R}^2, \text{ the line } y = \alpha x + \beta \text{ meets } S\}.$$

Show that CS is convex and apply Helly's theorem.

3.7. Given any two convex sets S and T in the affine space $\mathbb{A}^m$, and given $\lambda, \mu \in \mathbb{R}$ such that $\lambda + \mu = 1$, the *Minkowski sum* $\lambda S + \mu T$ is the set

$$\lambda S + \mu T = \{\lambda p + \mu q \mid p \in S, q \in T\}.$$

 (i) Prove that $\lambda S + \mu T$ is convex. Draw some Minkowski sums, in particular when S and T are tetrahedra (with T upside down).
(ii) Show that the Minkowski sum does not preserve the center of gravity.

3.8. Prove the version of Carathéodory's theorem for cones (Theorem 3.2), that is: *Given any vector space E of dimension m, for any (nonvoid) family $S = (v_i)_{i \in L}$ of vectors in E, the cone $\mathrm{cone}(S)$ spanned by S is equal to the set of positive combinations of families of m vectors in S.*

3.9. (i) Show that if E is an affine space of dimension m and S is a finite subset of E with n elements, if either $n \geq m + 3$ or $n = m + 2$ and some family of $m + 1$ points of S is affinely dependent, then S has at least two Radon partitions.

(ii) Prove the version of Radon's theorem for cones (Theorem 3.6), namely: *Given any vector space E of dimension m, for every subset X of E, if $\mathrm{cone}(X)$ is a pointed cone such that X has at least $m + 1$ nonzero vectors, then there is a partition of X into two nonempty disjoint subsets X_1 and X_2 such that the cones $\mathrm{cone}(X_1)$ and $\mathrm{cone}(X_2)$ have a nonempty intersection not reduced to $\{0\}$.*

(iii) (**Extra Credit**) Does the converse of (i) hold?

References

1. Alexander Barvinok. *A Course in Convexity*. GSM, Vol. 54. AMS, first edition, 2002.
2. Marcel Berger. *Géométrie 2*. Nathan, 1990. English edition: Geometry 2, Universitext, Springer-Verlag.
3. Nicolas Bourbaki. *Espaces Vectoriels Topologiques*. Eléments de Mathématiques. Hermann, 1981.
4. Serge Lang. *Real and Functional Analysis*. GTM 142. Springer-Verlag, third edition, 1996.
5. Peter D. Lax. *Functional Analysis*. Wiley, first edition, 2002.
6. Jiri Matousek. *Lectures on Discrete Geometry*. GTM No. 212. Springer Verlag, first edition, 2002.
7. Frederick A. Valentine. *Convex Sets*. McGraw–Hill, first edition, 1964.

Chapter 4
Embedding an Affine Space in a Vector Space

4.1 The "Hat Construction," or Homogenizing

For all practical purposes, curves and surfaces live in affine spaces. A disadvantage of the affine world is that points and vectors live in disjoint universes. It is often more convenient, at least mathematically, to deal with linear objects (vector spaces, linear combinations, linear maps), rather than affine objects (affine spaces, affine combinations, affine maps). Actually, it would also be advantageous if we could manipulate points and vectors as if they lived in a common universe, using perhaps an extra bit of information to distinguish between them if necessary.

Such a "homogenization" (or "hat construction") can be achieved. As a matter of fact, such a homogenization of an affine space and its associated vector space will be very useful to define and manipulate rational curves and surfaces. Indeed, the hat construction yields a canonical construction of the projective completion of an affine space. It also leads to a very elegant method for obtaining the various formulae giving the derivatives of a polynomial curve, or the directional derivatives of polynomial surfaces. However, these formulae are not needed in the main text. Thus we omit this topic, referring the readers to Gallier [2].

This chapter proceeds as follows. First, the construction of a vector space $\widehat{E}$ in which both E and $\overrightarrow{E}$ are embedded as (affine) hyperplanes is described. It is shown how affine frames in E become bases in $\widehat{E}$. It turns out that $\widehat{E}$ is characterized by a universality property: Affine maps to vector spaces extend uniquely to linear maps. As a consequence, affine maps between affine spaces E and F extend to linear maps between $\widehat{E}$ and $\widehat{F}$.

Let us first explain how to distinguish between points and vectors practically, using what amounts to a "hacking trick". Then, we will show that such a procedure can be put on firm mathematical grounds.

Assume that we consider the real affine space E of dimension 3, and that we have some affine frame $(a_0, (v_1, v_2, v_2))$. With respect to this affine frame, every point $x \in E$ is represented by its coordinates (x_1, x_2, x_3), where $a = a_0 + x_1 v_1 + x_2 v_2 + x_3 v_3$. A vector $u \in \overrightarrow{E}$ is also represented by its coordinates (u_1, u_2, u_3) over the basis

(v_1, v_2, v_2). One way to distinguish between points and vectors is to add a fourth coordinate, and to agree that points are represented by (row) vectors $(x_1, x_2, x_3, 1)$ whose fourth coordinate is 1, and that vectors are represented by (row) vectors $(v_1, v_2, v_3, 0)$ whose fourth coordinate is 0. This "programming trick" actually works very well. Of course, we are opening the door for strange elements such as $(x_1, x_2, x_3, 5)$, where the fourth coordinate is neither 1 nor 0.

The question is, can we make sense of such elements, and of such a construction? The answer is yes. We will present a construction in which an affine space $\left(E, \overrightarrow{E}\right)$ is embedded in a vector space $\widehat{E}$, in which $\overrightarrow{E}$ is embedded as a hyperplane passing through the origin, and E itself is embedded as an affine hyperplane, defined as $\omega^{-1}(1)$, for some linear form $\omega\colon \widehat{E} \to \mathbb{R}$. In the case of an affine space E of dimension 2, we can think of $\widehat{E}$ as the vector space $\mathbb{R}^3$ of dimension 3 in which $\overrightarrow{E}$ corresponds to the xy-plane, and E corresponds to the plane of equation $z = 1$, parallel to the xy-plane and passing through the point on the z-axis of coordinates $(0, 0, 1)$. The construction of the vector space $\widehat{E}$ is presented in some detail in Berger [1]. Berger explains the construction in terms of vector fields. Ramshaw explains the construction using the symmetric tensor power of an affine space. We prefer a more geometric and simpler description in terms of simple geometric transformations, translations, and dilatations.

Remark: Readers with a good knowledge of geometry will recognize the first step in embedding an affine space into a projective space. We will also show that the homogenization $\widehat{E}$ of an affine space $\left(E, \overrightarrow{E}\right)$, satisfies a universal property with respect to the extension of affine maps to linear maps. As a consequence, the vector space $\widehat{E}$ is unique up to isomorphism, and its actual construction is not so important. However, it is quite useful to visualize the space $\widehat{E}$, in order to understand well rational curves and rational surfaces.

As usual, for simplicity, it is assumed that all vector spaces are defined over the field $\mathbb{R}$ of real numbers, and that all families of scalars (points and vectors) are finite. The extension to arbitrary fields and to families of finite support is immediate. We begin by defining two very simple kinds of geometric (affine) transformations. Given an affine space $\left(E, \overrightarrow{E}\right)$, every $u \in \overrightarrow{E}$ induces a mapping $t_u\colon E \to E$, called a *translation*, and defined such that $t_u(a) = a + u$ for every $a \in E$. Clearly, the set of translations is a vector space isomorphic to $\overrightarrow{E}$. Thus, we will use the same notation u for both the vector u and the translation t_u. Given any point a and any scalar $\lambda \in \mathbb{R}$, we define the mapping $H_{a,\lambda}\colon E \to E$, called *dilatation (or central dilatation, or homothety) of center a and ratio λ*, and defined such that

$$H_{a,\lambda}(x) = a + \lambda \overrightarrow{ax},$$

for every $x \in E$. We have $H_{a,\lambda}(a) = a$, and when $\lambda \neq 0$ and $x \neq a$, $H_{a,\lambda}(x)$ is on the line defined by a and x, and is obtained by "scaling" $\overrightarrow{ax}$ by λ. The effect is a uniform dilatation (or contraction, if $\lambda < 1$). When $\lambda = 0$, $H_{a,0}(x) = a$ for all $x \in E$, and $H_{a,0}$

is the constant affine map sending every point to a. If we assume $\lambda \neq 1$, note that $H_{a,\lambda}$ is never the identity, and since a is a fixed point, $H_{a,\lambda}$ is never a translation.

We now consider the set $\widehat{E}$ of geometric transformations from E to E, consisting of the union of the (disjoint) sets of translations and dilatations of ratio $\lambda \neq 1$. We would like to give this set the structure of a vector space, in such a way that both E and $\overrightarrow{E}$ can be naturally embedded into $\widehat{E}$. In fact, it will turn out that barycenters show up quite naturally too!

In order to "add" two dilatations H_{a_1,λ_1} and H_{a_2,λ_2}, it turns out that it is more convenient to consider dilatations of the form $H_{a,1-\lambda}$, where $\lambda \neq 0$. To see this, let us see the effect of such a dilatation on a point $x \in E$: We have

$$H_{a,1-\lambda}(x) = a + (1-\lambda)\overrightarrow{ax} = a + \overrightarrow{ax} - \lambda\overrightarrow{ax} = x + \lambda\overrightarrow{xa}.$$

For simplicity of notation, let us denote $H_{a,1-\lambda}$ by $\langle a, \lambda \rangle$. Then, we have

$$\langle a, \lambda \rangle(x) = x + \lambda\overrightarrow{xa}.$$

Remarks:

(1) Note that $H_{a,1-\lambda}(x) = H_{x,\lambda}(a)$.

(2) Berger defines a map $h \colon E \to \overrightarrow{E}$ as a *vector field*. Thus, each $\langle a, \lambda \rangle$ can be viewed as the vector field $x \mapsto \lambda\overrightarrow{xa}$. Similarly, a translation u can be viewed as the constant vector field $x \mapsto u$. Thus, we could define $\widehat{E}$ as the (disjoint) union of these two vector fields. We prefer our view in terms of geometric transformations.

Then, since

$$\langle a_1, \lambda_1 \rangle(x) = x + \lambda_1\overrightarrow{xa_1} \quad \text{and} \quad \langle a_2, \lambda_2 \rangle(x) = x + \lambda_2\overrightarrow{xa_2},$$

if we want to define $\langle a_1, \lambda_1 \rangle \mathbin{\widehat{+}} \langle a_2, \lambda_2 \rangle$, we see that we have to distinguish between two cases:

(1) $\lambda_1 + \lambda_2 = 0$. In this case, since

$$\lambda_1\overrightarrow{xa_1} + \lambda_2\overrightarrow{xa_2} = \lambda_1\overrightarrow{xa_1} - \lambda_1\overrightarrow{xa_2} = \lambda_1\overrightarrow{a_2a_1},$$

we let

$$\langle a_1, \lambda_1 \rangle \mathbin{\widehat{+}} \langle a_2, \lambda_2 \rangle = \lambda_1\overrightarrow{a_2a_1},$$

where $\lambda_1\overrightarrow{a_2a_1}$ denotes the translation associated with the vector $\lambda_1\overrightarrow{a_2a_1}$.

(2) $\lambda_1 + \lambda_2 \neq 0$. In this case, the points a_1 and a_2 assigned the weights $\lambda_1/(\lambda_1 + \lambda_2)$ and $\lambda_2/(\lambda_1 + \lambda_2)$ have a barycenter

$$b = \frac{\lambda_1}{\lambda_1 + \lambda_2}a_1 + \frac{\lambda_2}{\lambda_1 + \lambda_2}a_2,$$

such that

$$\overrightarrow{xb} = \frac{\lambda_1}{\lambda_1 + \lambda_2}\overrightarrow{xa_1} + \frac{\lambda_2}{\lambda_1 + \lambda_2}\overrightarrow{xa_2}.$$

Since

$$\lambda_1\overrightarrow{xa_1} + \lambda_2\overrightarrow{xa_2} = (\lambda_1 + \lambda_2)\overrightarrow{xb},$$

we let

$$\langle a_1, \lambda_1 \rangle \,\widehat{+}\, \langle a_2, \lambda_2 \rangle = \left\langle \frac{\lambda_1}{\lambda_1 + \lambda_2}a_1 + \frac{\lambda_2}{\lambda_1 + \lambda_2}a_2, \lambda_1 + \lambda_2 \right\rangle,$$

the dilatation associated with the point b and the scalar $\lambda_1 + \lambda_2$.

Given a translation defined by u and a dilatation $\langle a, \lambda \rangle$, since $\lambda \neq 0$, we have

$$\lambda\overrightarrow{xa} + u = \lambda(\overrightarrow{xa} + \lambda^{-1}u),$$

and so, letting $b = a + \lambda^{-1}u$, since $\overrightarrow{ab} = \lambda^{-1}u$, we have

$$\lambda\overrightarrow{xa} + u = \lambda(\overrightarrow{xa} + \lambda^{-1}u) = \lambda(\overrightarrow{xa} + \overrightarrow{ab}) = \lambda\overrightarrow{xb},$$

and we let

$$\langle a, \lambda \rangle \,\widehat{+}\, u = \langle a + \lambda^{-1}u, \lambda \rangle,$$

the dilatation of center $a + \lambda^{-1}u$ and ratio λ.

The sum of two translations u and v is of course defined as the translation $u + v$. It is also natural to define multiplication by a scalar as follows:

$$\mu \cdot \langle a, \lambda \rangle = \langle a, \lambda\mu \rangle,$$

and

$$\lambda \cdot u = \lambda u,$$

where λu is the product by a scalar in $\overrightarrow{E}$.

We can now use the definition of the above operations to state the following lemma, showing that the "hat construction" described above has allowed us to achieve our goal of embedding both E and $\overrightarrow{E}$ in the vector space $\widehat{E}$.

Lemma 4.1. *The set $\widehat{E}$ consisting of the disjoint union of the translations and the dilatations $H_{a,1-\lambda} = \langle a, \lambda \rangle$, $\lambda \in \mathbb{R}, \lambda \neq 0$, is a vector space under the following operations of addition and multiplication by a scalar: If $\lambda_1 + \lambda_2 = 0$, then*

$$\langle a_1, \lambda_1 \rangle \,\widehat{+}\, \langle a_2, \lambda_2 \rangle = \lambda_1\overrightarrow{a_2a_1};$$

if $\lambda_1 + \lambda_2 \neq 0$, then

$$\langle a_1, \lambda_1 \rangle \,\widehat{+}\, \langle a_2, \lambda_2 \rangle = \left\langle \frac{\lambda_1}{\lambda_1 + \lambda_2}a_1 + \frac{\lambda_2}{\lambda_1 + \lambda_2}a_2, \lambda_1 + \lambda_2 \right\rangle,$$

$$\langle a, \lambda \rangle \,\widehat{+}\, u = u \,\widehat{+}\, \langle a, \lambda \rangle = \langle a + \lambda^{-1}u, \lambda \rangle,$$

$$u \,\widehat{+}\, v = u + v;$$

if $\mu \neq 0$, then

$$\mu \cdot \langle a, \lambda \rangle = \langle a, \lambda \mu \rangle,$$
$$0 \cdot \langle a, \lambda \rangle = 0;$$

and

$$\lambda \cdot u = \lambda u.$$

Furthermore, the map $\omega \colon \widehat{E} \to \mathbb{R}$ defined such that

$$\omega(\langle a, \lambda \rangle) = \lambda,$$
$$\omega(u) = 0,$$

is a linear form, $\omega^{-1}(0)$ is a hyperplane isomorphic to $\overrightarrow{E}$ under the injective linear map $i \colon \overrightarrow{E} \to \widehat{E}$ such that $i(u) = t_u$ (the translation associated with u), and $\omega^{-1}(1)$ is an affine hyperplane isomorphic to E with direction $i(\overrightarrow{E})$, under the injective affine map $j \colon E \to \widehat{E}$, where $j(a) = \langle a, 1 \rangle$ for every $a \in E$. Finally, for every $a \in E$, we have

$$\widehat{E} = i(\overrightarrow{E}) \oplus \mathbb{R} j(a).$$

Proof. The verification that $\widehat{E}$ is a vector space is straightforward. The linear map mapping a vector u to the translation defined by u is clearly an injection $i \colon \overrightarrow{E} \to \widehat{E}$ embedding $\overrightarrow{E}$ as an hyperplane in $\widehat{E}$. It is also clear that ω is a linear form. Note that

$$j(a + u) = \langle a + u, 1 \rangle = \langle a, 1 \rangle \widehat{+} u,$$

where u stands for the translation associated with the vector u, and thus j is an affine injection with associated linear map i. Thus, $\omega^{-1}(1)$ is indeed an affine hyperplane isomorphic to E with direction $i(\overrightarrow{E})$, under the map $j \colon E \to \widehat{E}$. Finally, from the definition of $\widehat{+}$, for every $a \in E$ and every $u \in \overrightarrow{E}$, since

$$i(u) \widehat{+} \lambda \cdot j(a) = u \widehat{+} \langle a, \lambda \rangle = \langle a + \lambda^{-1} u, \lambda \rangle,$$

when $\lambda \neq 0$, we get any arbitrary $v \in \widehat{E}$ by picking $\lambda = 0$ and $u = v$, and we get any arbitrary element $\langle b, \mu \rangle$, $\mu \neq 0$, by picking $\lambda = \mu$ and $u = \mu \overrightarrow{ab}$. Thus,

$$\widehat{E} = i(\overrightarrow{E}) + \mathbb{R} j(a),$$

and since $i(\overrightarrow{E}) \cap \mathbb{R} j(a) = \{0\}$, we have

$$\widehat{E} = i(\overrightarrow{E}) \oplus \mathbb{R} j(a),$$

for every $a \in E$. $\square$

Figure 4.1 illustrates the embedding of the affine space E into the vector space $\widehat{E}$, when E is an affine plane.

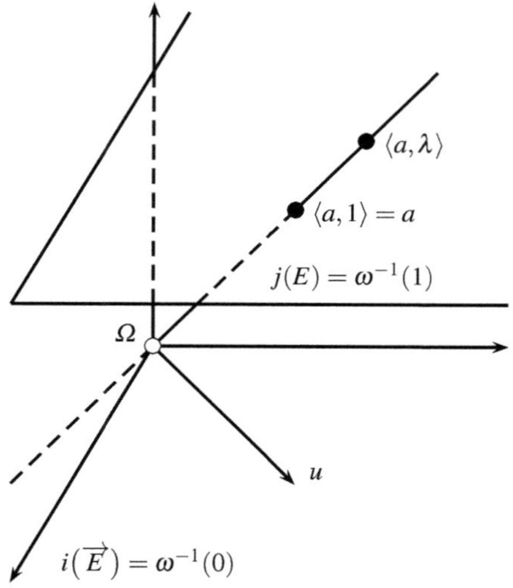

Fig. 4.1 Embedding an affine space $\left(E, \overrightarrow{E}\right)$ into a vector space $\widehat{E}$.

Note that $\widehat{E}$ is isomorphic to $\overrightarrow{E} \cup (E \times \mathbb{R}^*)$. Other authors, such as Ramshaw, use the notation E_* for $\widehat{E}$. Ramshaw calls the linear form $\omega\colon \widehat{E} \to \mathbb{R}$ a *weight (or flavor)*, and he says that an element $z \in \widehat{E}$ such that $\omega(z) = \lambda$ is λ-*heavy (or has flavor λ)* ([3]). The elements of $j(E)$ are 1-heavy and are called *points*, and the elements of $i(\overrightarrow{E})$ are 0-heavy and are called *vectors*. In general, the λ-heavy elements all belong to the hyperplane $\omega^{-1}(\lambda)$ parallel to $i(\overrightarrow{E})$. Thus, intuitively, we can think of $\widehat{E}$ as a stack of parallel hyperplanes, one for each λ, a little bit like an infinite stack of very thin pancakes! There are two privileged pancakes: one corresponding to E, for $\lambda = 1$, and one corresponding to $\overrightarrow{E}$, for $\lambda = 0$.

From now on, we will identify $j(E)$ and E, and $i(\overrightarrow{E})$ and $\overrightarrow{E}$. We will also write λa instead of $\langle a, \lambda \rangle$, which we will call a *weighted point*, and write $1a$ just as a. When we want to be more precise, we may also write $\langle a, 1 \rangle$ as $\overline{a}$ (as Ramshaw does). In particular, when we consider the homogenized version $\widehat{\mathbb{A}}$ of the affine space $\mathbb{A}$ associated with the field $\mathbb{R}$ considered as an affine space, we write $\overline{\lambda}$ for $\langle \lambda, 1 \rangle$, when viewing λ as a point in both $\mathbb{A}$ and $\widehat{\mathbb{A}}$, and simply λ, when viewing λ as a vector in $\mathbb{R}$ and in $\widehat{\mathbb{A}}$. The elements of $\widehat{\mathbb{A}}$ are called *Bézier sites* by Ramshaw. As an example, the expression $2 + 3$ denotes the real number 5, in $\mathbb{A}$, $(\overline{2} + \overline{3})/2$ denotes the midpoint of the segment $[\overline{2}, \overline{3}]$, which can be denoted by $\overline{2.5}$, and $\overline{2} + \overline{3}$ does not make sense in $\mathbb{A}$, since it is not a barycentric combination. However, in $\widehat{\mathbb{A}}$, the expression $\overline{2} + \overline{3}$ makes sense: It is the weighted point $\langle \overline{2.5}, 2 \rangle$.

Then, in view of the fact that

$$\langle a + u, 1 \rangle = \langle a, 1 \rangle \widehat{+} u,$$

and since we are identifying $a + u$ with $\langle a + u, 1 \rangle$ (under the injection j), in the simplified notation the above reads as $a + u = a \widehat{+} u$. Thus, we go one step further, and denote $a \widehat{+} u$ by $a + u$. However, since

$$\langle a, \lambda \rangle \widehat{+} u = \langle a + \lambda^{-1} u, \lambda \rangle,$$

we will refrain from writing $\lambda a \widehat{+} u$ as $\lambda a + u$, because we find it too confusing. From Lemma 4.1, for every $a \in E$, every element of $\widehat{E}$ can be written uniquely as $u \widehat{+} \lambda a$. We also denote

$$\lambda a \widehat{+} (-\mu) b$$

by

$$\lambda a \widehat{-} \mu b.$$

We can now justify rigorously the programming trick of the introduction of an extra coordinate to distinguish between points and vectors. First, we make a few observations. Given any family $(a_i)_{i \in I}$ of points in E, and any family $(\lambda_i)_{i \in I}$ of scalars in $\mathbb{R}$, it is easily shown by induction on the size of I that the following holds:

(1) If $\sum_{i \in I} \lambda_i = 0$, then

$$\sum_{i \in I} \langle a_i, \lambda_i \rangle = \overrightarrow{\sum_{i \in I} \lambda_i a_i},$$

where

$$\overrightarrow{\sum_{i \in I} \lambda_i a_i} = \sum_{i \in I} \lambda_i \overrightarrow{b a_i}$$

for any $b \in E$, which, by Lemma 2.1, is a vector independent of b, or

(2) If $\sum_{i \in I} \lambda_i \neq 0$, then

$$\sum_{i \in I} \langle a_i, \lambda_i \rangle = \left\langle \sum_{i \in I} \frac{\lambda_i}{\sum_{i \in I} \lambda_i} a_i, \sum_{i \in I} \lambda_i \right\rangle.$$

Thus, we see how barycenters reenter the scene quite naturally, and that in $\widehat{E}$, we can make sense of $\sum_{i \in I} \langle a_i, \lambda_i \rangle$, regardless of the value of $\sum_{i \in I} \lambda_i$. When $\sum_{i \in I} \lambda_i = 1$, the element $\sum_{i \in I} \langle a_i, \lambda_i \rangle$ belongs to the hyperplane $\omega^{-1}(1)$, and thus it is a point. When $\sum_{i \in I} \lambda_i = 0$, the linear combination of points $\sum_{i \in I} \lambda_i a_i$ is a vector, and when $I = \{1, \ldots, n\}$, we allow ourselves to write

$$\lambda_1 a_1 \widehat{+} \cdots \widehat{+} \lambda_n a_n,$$

where some of the occurrences of $\widehat{+}$ can be replaced by $\widehat{-}$, as

$$\lambda_1 a_1 + \cdots + \lambda_n a_n,$$

where the occurrences of $\widehat{-}$ (if any) are replaced by $-$.

In fact, we have the following slightly more general property, which is left as an exercise.

Lemma 4.2. *Given any affine space $\left(E, \overrightarrow{E}\right)$, for any family $(a_i)_{i \in I}$ of points in E, any family $(\lambda_i)_{i \in I}$ of scalars in $\mathbb{R}$, and any family $(\overrightarrow{v_j})_{j \in J}$ of vectors in $\overrightarrow{E}$, with $I \cap J = \emptyset$, the following properties hold:*

(1) If $\sum_{i \in I} \lambda_i = 0$, then

$$\sum_{i \in I} \langle a_i, \lambda_i \rangle \,\widehat{+}\, \sum_{j \in J} v_j = \overrightarrow{\sum_{i \in I} \lambda_i a_i} + \sum_{j \in J} v_j,$$

where

$$\overrightarrow{\sum_{i \in I} \lambda_i a_i} = \sum_{i \in I} \lambda_i \overrightarrow{b a_i}$$

for any $b \in E$, which, by Lemma 2.1, is a vector independent of b, or
(2) If $\sum_{i \in I} \lambda_i \neq 0$, then

$$\sum_{i \in I} \langle a_i, \lambda_i \rangle \,\widehat{+}\, \sum_{j \in J} v_j = \left\langle \sum_{i \in I} \frac{\lambda_i}{\sum_{i \in I} \lambda_i} a_i + \sum_{j \in J} \frac{v_j}{\sum_{i \in I} \lambda_i}, \sum_{i \in I} \lambda_i \right\rangle.$$

Proof. By induction on the size of I and the size of J. $\quad\square$

The above formulae show that we have some kind of extended barycentric calculus. Operations on weighted points and vectors were introduced by H. Grassmann, in his book published in 1844! This calculus will be helpful in dealing with rational curves.

4.2 Affine Frames of E and Bases of $\widehat{E}$

There is also a nice relationship between affine frames in $\left(E, \overrightarrow{E}\right)$ and bases of $\widehat{E}$, stated in the following lemma.

Lemma 4.3. *Given any affine space $\left(E, \overrightarrow{E}\right)$, for any affine frame $(a_0, (\overrightarrow{a_0 a_1}, \ldots, \overrightarrow{a_0 a_m}))$ for E, the family $(\overrightarrow{a_0 a_1}, \ldots, \overrightarrow{a_0 a_m}, a_0)$ is a basis for $\widehat{E}$, and for any affine frame $(a_0, \ldots, a_m)$ for E, the family $(a_0, \ldots, a_m)$ is a basis for $\widehat{E}$. Furthermore, given any element $\langle x, \lambda \rangle \in \widehat{E}$, if*

$$x = a_0 + x_1 \overrightarrow{a_0 a_1} + \cdots + x_m \overrightarrow{a_0 a_m}$$

over the affine frame $(a_0, (\overrightarrow{a_0a_1}, \ldots, \overrightarrow{a_0a_m}))$ in E, then the coordinates of $\langle x, \lambda \rangle$ over the basis $(\overrightarrow{a_0a_1}, \ldots, \overrightarrow{a_0a_m}, a_0)$ in $\widehat{E}$ are

$$(\lambda x_1, \ldots, \lambda x_m, \lambda).$$

For any vector $v \in \overrightarrow{E}$, if

$$v = v_1 \overrightarrow{a_0a_1} + \cdots + v_m \overrightarrow{a_0a_m}$$

over the basis $(\overrightarrow{a_0a_1}, \ldots, \overrightarrow{a_0a_m})$ in $\overrightarrow{E}$, then over the basis $(\overrightarrow{a_0a_1}, \ldots, \overrightarrow{a_0a_m}, a_0)$ in $\widehat{E}$, the coordinates of v are

$$(v_1, \ldots, v_m, 0).$$

For any element $\langle a, \lambda \rangle$, where $\lambda \neq 0$, if the barycentric coordinates of a w.r.t. the affine basis $(a_0, \ldots, a_m)$ in E are $(\lambda_0, \ldots, \lambda_m)$ with $\lambda_0 + \cdots + \lambda_m = 1$, then the coordinates of $\langle a, \lambda \rangle$ w.r.t. the basis $(a_0, \ldots, a_m)$ in $\widehat{E}$ are

$$(\lambda \lambda_0, \ldots, \lambda \lambda_m).$$

If a vector $v \in \overrightarrow{E}$ is expressed as

$$v = v_1 \overrightarrow{a_0a_1} + \cdots + v_m \overrightarrow{a_0a_m} = -(v_1 + \cdots + v_m) a_0 + v_1 a_1 + \cdots + v_m a_m,$$

with respect to the affine basis $(a_0, \ldots, a_m)$ in E, then its coordinates w.r.t. the basis $(a_0, \ldots, a_m)$ in $\widehat{E}$ are

$$(-(v_1 + \cdots + v_m), v_1, \ldots, v_m).$$

Proof. We sketch parts of the proof, leaving the details as an exercise. Figure 4.2 shows the basis $(\overrightarrow{a_0a_1}, \overrightarrow{a_0a_2}, a_0)$ corresponding to the affine frame (a_0, a_1, a_2) in E.

If we assume that we have a nontrivial linear combination

$$\lambda_1 \overrightarrow{a_0a_1} \mathbin{\widehat{+}} \cdots \mathbin{\widehat{+}} \lambda_m \overrightarrow{a_0a_m} \mathbin{\widehat{+}} \mu a_0 = 0,$$

if $\mu \neq 0$, then we have

$$\lambda_1 \overrightarrow{a_0a_1} \mathbin{\widehat{+}} \cdots \mathbin{\widehat{+}} \lambda_m \overrightarrow{a_0a_m} \mathbin{\widehat{+}} \mu a_0 = \langle a_0 + \mu^{-1}\lambda_1 \overrightarrow{a_0a_1} + \cdots + \mu^{-1}\lambda_m \overrightarrow{a_0a_m}, \mu \rangle,$$

which is never null, and thus, $\mu = 0$, but since $(\overrightarrow{a_0a_1}, \ldots, \overrightarrow{a_0a_m})$ is a basis of $\overrightarrow{E}$, we must also have $\lambda_i = 0$ for all $i, 1 \leq i \leq m$.

Given any element $\langle x, \lambda \rangle \in \widehat{E}$, if

$$x = a_0 + x_1 \overrightarrow{a_0a_1} + \cdots + x_m \overrightarrow{a_0a_m}$$

over the affine frame $(a_0, (\overrightarrow{a_0a_1}, \ldots, \overrightarrow{a_0a_m}))$ in E, in view of the definition of $\widehat{+}$, we have

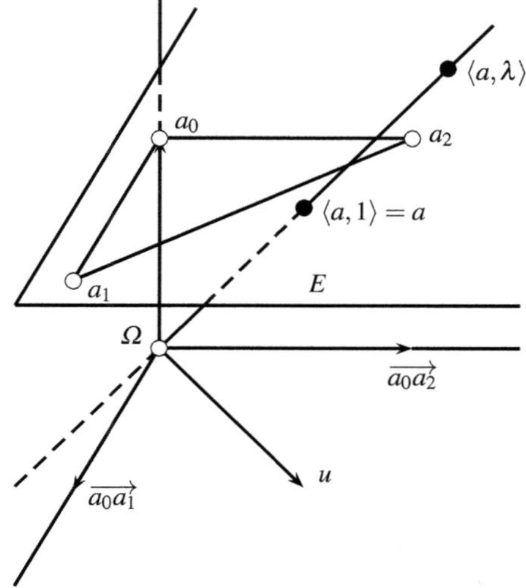

Fig. 4.2 The basis $(\overrightarrow{a_0a_1}, \overrightarrow{a_0a_2}, a_0)$ in $\widehat{E}$.

$$\langle x, \lambda \rangle = \langle a_0 + x_1 \overrightarrow{a_0a_1} + \cdots + x_m \overrightarrow{a_0a_m}, \lambda \rangle$$
$$= \langle a_0, \lambda \rangle \widehat{+} \lambda x_1 \overrightarrow{a_0a_1} \widehat{+} \cdots \widehat{+} \lambda x_m \overrightarrow{a_0a_m},$$

which shows that over the basis $(\overrightarrow{a_0a_1}, \ldots, \overrightarrow{a_0a_m}, a_0)$ in $\widehat{E}$, the coordinates of $\langle x, \lambda \rangle$ are

$$(\lambda x_1, \ldots, \lambda x_m, \lambda).$$

$\square$

If $(x_1, \ldots, x_m)$ are the coordinates of x w.r.t. the affine frame $(a_0, (\overrightarrow{a_0a_1}, \ldots, \overrightarrow{a_0a_m}))$ in E, then $(x_1, \ldots, x_m, 1)$ are the coordinates of x in $\widehat{E}$, i.e., the last coordinate is 1, and if u has coordinates $(u_1, \ldots, u_m)$ with respect to the basis $(\overrightarrow{a_0a_1}, \ldots, \overrightarrow{a_0a_m})$ in $\overrightarrow{E}$, then u has coordinates $(u_1, \ldots, u_m, 0)$ in $\widehat{E}$, i.e., the last coordinate is 0. Figure 4.3 shows the affine frame (a_0, a_1, a_2) in E viewed as a basis in $\widehat{E}$.

Now that we have defined $\widehat{E}$ and investigated the relationship between affine frames in E and bases in $\widehat{E}$, we can give another construction of a vector space $\mathscr{F}$ from E and $\overrightarrow{E}$ that will allow us to "visualize" in a much more intuitive fashion the structure of $\widehat{E}$ and of its operations $\widehat{+}$ and $\cdot$.

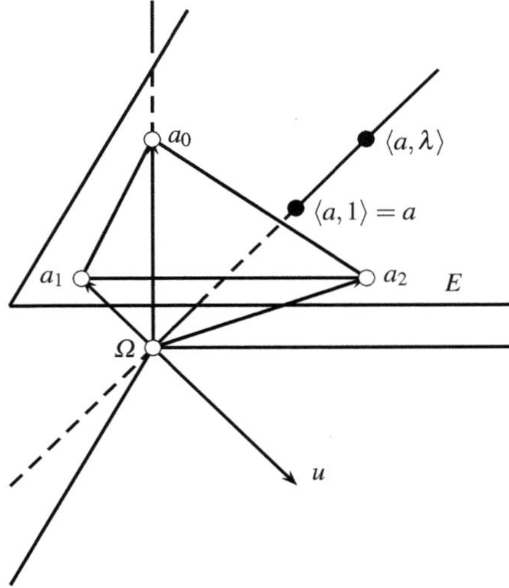

Fig. 4.3 The basis (a_0, a_1, a_2) in $\widehat{E}$.

4.3 Another Construction of $\widehat{E}$

One would probably wish that we could start with this construction of $\mathscr{F}$ first, and then define $\widehat{E}$ using the isomorphism $\widehat{\Omega} : \widehat{E} \to \mathscr{F}$ defined below. Unfortunately, we first need the vector space structure on $\widehat{E}$ to show that $\widehat{\Omega}$ is linear!

Definition 4.1. Given any affine space $\left(E, \overrightarrow{E}\right)$, we define the vector space $\mathscr{F}$ as the direct sum $\overrightarrow{E} \oplus \mathbb{R}$, where $\mathbb{R}$ denotes the field $\mathbb{R}$ considered as a vector space (over itself). Denoting the unit vector in $\mathbb{R}$ by 1, since $\mathscr{F} = \overrightarrow{E} \oplus \mathbb{R}$, every vector $v \in \mathscr{F}$ can be written as $v = u + \lambda 1$, for some unique $u \in \overrightarrow{E}$ and some unique $\lambda \in \mathbb{R}$. Then, for any choice of an origin Ω_1 in E, we define the map $\widehat{\Omega} : \widehat{E} \to \mathscr{F}$, as follows:

$$
\widehat{\Omega}(\theta) = \begin{cases} \lambda(1 + \overrightarrow{\Omega_1 a}) & \text{if } \theta = \langle a, \lambda \rangle, \text{ where } a \in E \text{ and } \lambda \neq 0; \\ u & \text{if } \theta = u, \text{ where } u \in \overrightarrow{E}. \end{cases}
$$

The idea is that, once again, viewing $\mathscr{F}$ as an affine space under its canonical structure, E is embedded in $\mathscr{F}$ as the hyperplane $H = 1 + \overrightarrow{E}$, with direction $\overrightarrow{E}$, the hyperplane $\overrightarrow{E}$ in $\mathscr{F}$. Then, every point $a \in E$ is in bijection with the point $A = 1 + \overrightarrow{\Omega_1 a}$, in the hyperplane H. If we denote the origin 0 of the canonical affine space $\mathscr{F}$ by Ω, the map $\widehat{\Omega}$ maps a point $\langle a, \lambda \rangle \in E$ to a point in $\mathscr{F}$, as follows: $\widehat{\Omega}(\langle a, \lambda \rangle)$ is the point on the line passing through both the origin Ω of $\mathscr{F}$ and the point $A = 1 + \overrightarrow{\Omega_1 a}$ in the hyperplane $H = 1 + \overrightarrow{E}$, such that

$$\widehat{\Omega}(\langle a,\lambda\rangle) = \lambda\overrightarrow{\Omega A} = \lambda(1 + \overrightarrow{\Omega_1 a}).$$

The following lemma shows that $\widehat{\Omega}$ is an isomorphism of vector spaces.

Lemma 4.4. *Given any affine space* $(E, \overrightarrow{E})$, *for any choice* Ω_1 *of an origin in E, the map* $\widehat{\Omega} : \widehat{E} \to \mathscr{F}$ *is a linear isomorphism between* $\widehat{E}$ *and the vector space* $\mathscr{F}$ *of Definition 4.1. The inverse of* $\widehat{\Omega}$ *is given by*

$$\widehat{\Omega}^{-1}(u + \lambda 1) = \begin{cases} \langle\Omega_1 + \lambda^{-1}u, \lambda\rangle) & \text{if } \lambda \neq 0; \\ u & \text{if } \lambda = 0. \end{cases}$$

Proof. It is a straightforward verification. We check that $\widehat{\Omega}$ is invertible, leaving the verification that it is linear as an exercise. We have

$$\langle a,\lambda\rangle \mapsto \lambda 1 + \lambda\overrightarrow{\Omega_1 a} \mapsto \langle\Omega_1 + \overrightarrow{\Omega_1 a}, \lambda\rangle = \langle a,\lambda\rangle$$

and

$$u + \lambda 1 \mapsto \langle\Omega_1 + \lambda^{-1}u, \lambda\rangle \mapsto u + \lambda 1,$$

and since $\widehat{\Omega}$ is the identity on $\overrightarrow{E}$, we have shown that $\widehat{\Omega} \circ \widehat{\Omega}^{-1} = \text{id}$, and $\widehat{\Omega}^{-1} \circ \widehat{\Omega} = \text{id}$. This shows that $\widehat{\Omega}$ is a bijection. $\square$

Figure 4.4 illustrates the embedding of the affine space E into the vector space $\mathscr{F}$, when E is an affine plane.

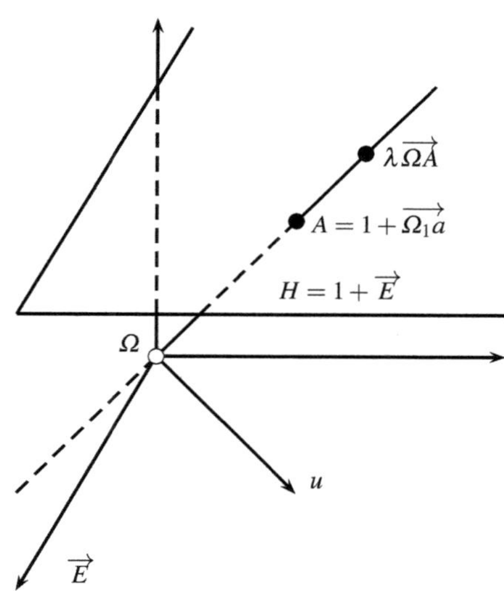

Fig. 4.4 Embedding an affine space $\left(E, \overrightarrow{E}\right)$ into a vector space $\mathscr{F}$.

Lemma 4.4 gives a nice interpretation of the sum operation $\widehat{+}$ of $\widehat{E}$. Given two weighted points $\langle a_1, \lambda_1 \rangle$ and $\langle a_2, \lambda_2 \rangle$, we have

$$\langle a_1, \lambda_1 \rangle \widehat{+} \langle a_2, \lambda_2 \rangle = \widehat{\Omega}^{-1}(\widehat{\Omega}(\langle a_1, \lambda_1 \rangle) + \widehat{\Omega}(\langle a_2, \lambda_2 \rangle)).$$

The operation $\widehat{\Omega}(\langle a_1, \lambda_1 \rangle) + \widehat{\Omega}(\langle a_2, \lambda_2 \rangle)$ has a simple geometric interpretation. If $\lambda_1 + \lambda_2 \neq 0$, then find the points M_1 and M_2 on the lines passing through the origin Ω of $\mathscr{F}$ and the points $A_1 = \widehat{\Omega}(a_1)$ and $A_2 = \widehat{\Omega}(a_2)$ in the hyperplane H, such that $\overrightarrow{\Omega M_1} = \lambda_1 \overrightarrow{\Omega A_1}$ and $\overrightarrow{\Omega M_2} = \lambda_2 \overrightarrow{\Omega A_2}$, add the vectors $\overrightarrow{\Omega M_1}$ and $\overrightarrow{\Omega M_2}$, getting a point N such that $\overrightarrow{\Omega N} = \overrightarrow{\Omega M_1} + \overrightarrow{\Omega M_2}$, and consider the intersection G of the line passing through Ω and N with the hyperplane H. Then, G is the barycenter of A_1 and A_2 assigned the weights $\lambda_1/(\lambda_1 + \lambda_2)$ and $\lambda_2/(\lambda_1 + \lambda_2)$, and if $g = \widehat{\Omega}^{-1}(\overrightarrow{\Omega G})$, then $\widehat{\Omega}^{-1}(\overrightarrow{\Omega N}) = \langle g, \lambda_1 + \lambda_2 \rangle$.

Instead of adding the vectors $\overrightarrow{\Omega M_1}$ and $\overrightarrow{\Omega M_2}$, we can take the middle N' of the segment $M_1 M_2$, and G is the intersection of the line passing through Ω and N' with the hyperplane H.

If $\lambda_1 + \lambda_2 = 0$, then $\langle a_1, \lambda_1 \rangle \widehat{+} \langle a_2, \lambda_2 \rangle$ is a vector determined as follows. Again, find the points M_1 and M_2 on the lines passing through the origin Ω of $\mathscr{F}$ and the points $A_1 = \widehat{\Omega}(a_1)$ and $A_2 = \widehat{\Omega}(a_2)$ in the hyperplane H, such that $\overrightarrow{\Omega M_1} = \lambda_1 \overrightarrow{\Omega A_1}$ and $\overrightarrow{\Omega M_2} = \lambda_2 \overrightarrow{\Omega A_2}$, and add the vectors $\overrightarrow{\Omega M_1}$ and $\overrightarrow{\Omega M_2}$, getting a point N such that $\overrightarrow{\Omega N} = \overrightarrow{\Omega M_1} + \overrightarrow{\Omega M_2}$. The desired vector is $\overrightarrow{\Omega N}$, which is parallel to the line $A_1 A_2$. Equivalently, let N' be the middle of the segment $M_1 M_2$, and the desired vector is $2\overrightarrow{\Omega N'}$.

We can also give a geometric interpretation of $\langle a, \lambda \rangle + u$. Let $A = \widehat{\Omega}(a)$ in the hyperplane H, let D be the line determined by A and u, let M_1 be the point such that $\overrightarrow{\Omega M_1} = \lambda \overrightarrow{\Omega A}$, and let M_2 be the point such that $\overrightarrow{\Omega M_2} = u$, that is, $M_2 = \Omega + u$. By construction, the line D is in the hyperplane H, and it is parallel to $\overrightarrow{\Omega M_2}$, so that D, M_1, and M_2 are coplanar. Then, add the vectors $\overrightarrow{\Omega M_1}$ and $\overrightarrow{\Omega M_2}$, getting a point N such that $\overrightarrow{\Omega N} = \overrightarrow{\Omega M_1} + \overrightarrow{\Omega M_2}$, and let G be the intersection of the line determined by Ω and N with the line D. If $g = \widehat{\Omega}^{-1}(\overrightarrow{\Omega G})$, then, $\widehat{\Omega}^{-1}(\overrightarrow{\Omega N}) = \langle g, \lambda \rangle$. Equivalently, if N' is the middle of the segment $M_1 M_2$, then G is the intersection of the line determined by Ω and N', with the line D.

We now consider the universal property of $\widehat{E}$ mentioned at the beginning of this section.

4.4 Extending Affine Maps to Linear Maps

Roughly, the vector space $\widehat{E}$ has the property that for any vector space $\overrightarrow{F}$ and any affine map $f : E \to \overrightarrow{F}$, there is a unique linear map $\widehat{f} : \widehat{E} \to \overrightarrow{F}$ extending $f : E \to \overrightarrow{F}$. As a consequence, given two affine spaces E and F, every affine map $f : E \to F$

extends uniquely to a linear map $\widehat{f} \colon \widehat{E} \to \widehat{F}$. Other authors, such as Ramshaw, use the notation f_* for $\widehat{f}$. First, we define rigorously the notion of homogenization of an affine space.

Definition 4.2. Given any affine space $\left(E, \overrightarrow{E}\right)$, a *homogenization (or linearization) of* $\left(E, \overrightarrow{E}\right)$ is a triple $\langle \mathscr{E}, j, \omega \rangle$, where $\mathscr{E}$ is a vector space, $j \colon E \to \mathscr{E}$ is an injective affine map with associated injective linear map $i \colon \overrightarrow{E} \to \mathscr{E}$, $\omega \colon \mathscr{E} \to \mathbb{R}$ is a linear form such that $\omega^{-1}(0) = i\left(\overrightarrow{E}\right)$, $\omega^{-1}(1) = j(E)$, and for every vector space $\overrightarrow{F}$ and every affine map $f \colon E \to \overrightarrow{F}$ there is a unique linear map $\widehat{f} \colon \mathscr{E} \to \overrightarrow{F}$ extending f, i.e., $f = \widehat{f} \circ j$, as in the following diagram:

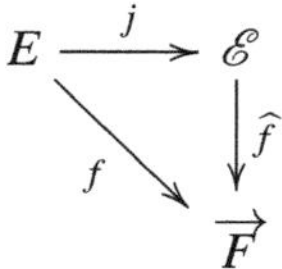

Thus, $j(E) = \omega^{-1}(1)$ is an affine hyperplane with direction $i\left(\overrightarrow{E}\right) = \omega^{-1}(0)$. Note that we could have defined a homogenization of an affine space $\left(E, \overrightarrow{E}\right)$, as a triple $\langle \mathscr{E}, j, H \rangle$, where $\mathscr{E}$ is a vector space, H is an affine hyperplane in $\mathscr{E}$, and $j \colon E \to \mathscr{E}$ is an injective affine map such that $j(E) = H$, and such that the universal property stated above holds. However, Definition 4.2 is more convenient for our purposes, since it makes the notion of weight more evident.

The obvious candidate for $\mathscr{E}$ is the vector space $\widehat{E}$ that we just constructed. The next lemma will show that $\widehat{E}$ indeed has the required extension property. As usual, objects defined by a universal property are unique up to isomorphism. This property is left as an exercise.

Lemma 4.5. *Given any affine space $\left(E, \overrightarrow{E}\right)$ and any vector space $\overrightarrow{F}$, for any affine map $f \colon E \to \overrightarrow{F}$, there is a unique linear map $\widehat{f} \colon \widehat{E} \to \overrightarrow{F}$ extending f such that*

$$\widehat{f}(u \,\widehat{+}\, \lambda a) = \lambda f(a) + \overrightarrow{f}(u)$$

for all $a \in E$, all $u \in \overrightarrow{E}$, and all $\lambda \in \mathbb{R}$, where $\overrightarrow{f}$ is the linear map associated with f. In particular, when $\lambda \neq 0$, we have

$$\widehat{f}(u \,\widehat{+}\, \lambda a) = \lambda f(a + \lambda^{-1} u).$$

Proof. Assuming that $\widehat{f}$ exists, recall that from Lemma 4.1, for every $a \in E$, every element of $\widehat{E}$ can be written uniquely as $u \,\widehat{+}\, \lambda a$. By linearity of $\widehat{f}$ and since $\widehat{f}$ extends f, we have

$$\widehat{f}(u \,\widehat{+}\, \lambda a) = \widehat{f}(u) + \lambda \widehat{f}(a) = \widehat{f}(u) + \lambda f(a) = \lambda f(a) + \widehat{f}(u).$$

If $\lambda = 1$, since $a \,\widehat{+}\, u$ and $a + u$ are identified, and since $\widehat{f}$ extends f, we must have

$$f(a) + \widehat{f}(u) = \widehat{f}(a) + \widehat{f}(u) = \widehat{f}(a \,\widehat{+}\, u) = f(a + u) = f(a) + \overrightarrow{f}(u),$$

and thus $\widehat{f}(u) = \overrightarrow{f}(u)$ for all $u \in \overrightarrow{E}$. Then we have

$$\widehat{f}(u \,\widehat{+}\, \lambda a) = \lambda f(a) + \overrightarrow{f}(u),$$

which proves the uniqueness of $\widehat{f}$. On the other hand, the map $\widehat{f}$ defined as above is clearly a linear map extending f.

When $\lambda \neq 0$, we have

$$\widehat{f}(u \,\widehat{+}\, \lambda a) = \widehat{f}(\lambda(a + \lambda^{-1}u)) = \lambda \widehat{f}(a + \lambda^{-1}u) = \lambda f(a + \lambda^{-1}u).$$

$\square$

Lemma 4.5 shows that $\langle \widehat{E}, j, \omega \rangle$, is a homogenization of $(E, \overrightarrow{E})$. As a corollary, we obtain the following lemma.

Lemma 4.6. *Given two affine spaces E and F and an affine map $f \colon E \to F$, there is a unique linear map $\widehat{f} \colon \widehat{E} \to \widehat{F}$ extending f, as in the diagram below,*

$$
\begin{array}{ccc}
E & \xrightarrow{\ f\ } & F \\
{\scriptstyle j}\big\downarrow & & \big\downarrow{\scriptstyle j} \\
\widehat{E} & \xrightarrow[\widehat{f}]{} & \widehat{F}
\end{array}
$$

such that

$$\widehat{f}(u \,\widehat{+}\, \lambda a) = \overrightarrow{f}(u) \,\widehat{+}\, \lambda f(a),$$

for all $a \in E$, all $u \in \overrightarrow{E}$, and all $\lambda \in \mathbb{R}$, where $\overrightarrow{f}$ is the linear map associated with f. In particular, when $\lambda \neq 0$, we have

$$\widehat{f}(u \,\widehat{+}\, \lambda a) = \lambda f(a + \lambda^{-1}u).$$

Proof. Consider the vector space $\widehat{F}$ and the affine map $j \circ f \colon E \to \widehat{F}$. By Lemma 4.5, there is a unique linear map $\widehat{f} \colon \widehat{E} \to \widehat{F}$ extending $j \circ f$, and thus extending f. $\square$

Note that $\widehat{f} \colon \widehat{E} \to \widehat{F}$ has the property that $\widehat{f}(\overrightarrow{E}) \subseteq \overrightarrow{F}$. More generally, since

$$\widehat{f}(u \,\widehat{+}\, \lambda a) = \overrightarrow{f}(u) \,\widehat{+}\, \lambda f(a),$$

the linear map $\widehat{f}$ is weight-preserving. Also observe that we recover f from $\widehat{f}$, by letting $\lambda = 1$ in $\widehat{f}(u \,\widehat{+}\, \lambda a) = \lambda f(a + \lambda^{-1}u)$, that is, we have

$$f(a+u) = \widehat{f}(u \,\widehat{+}\, a).$$

From a practical point of view, Lemma 4.6 shows us how to homogenize an affine map to turn it into a linear map between the two homogenized spaces. Assume that E and F are of finite dimension, that $(a_0, (u_1, \ldots, u_n))$ is an affine frame of E with origin a_0, and $(b_0, (v_1, \ldots, v_m))$ is an affine frame of F with origin b_0. Then, with respect to the two bases $(u_1, \ldots, u_n, a_0)$ in $\widehat{E}$ and $(v_1, \ldots, v_m, b_0)$ in $\widehat{F}$, a linear map $h: \widehat{E} \to \widehat{F}$ is given by an $(m+1) \times (n+1)$ matrix A. Assume that this linear map h is equal to the homogenized version $\widehat{f}$ of an affine map f. Since

$$\widehat{f}(u \,\widehat{+}\, \lambda a) = \overrightarrow{f}(u) \,\widehat{+}\, \lambda f(a),$$

and since over the basis $(u_1, \ldots, u_n, a_0)$ in $\widehat{E}$, points are represented by vectors whose last coordinate is 1 and vectors are represented by vectors whose last coordinate is 0, the following properties hold.

1. The last row of the matrix $A = M(\widehat{f})$ with respect to the given bases is

$$(0, 0, \ldots, 0, 1)$$

   with m occurrences of 0.
2. The last column of A contains the coordinates

$$(\mu_1, \ldots, \mu_m, 1)$$

   of $f(a_0)$ with respect to the basis $(v_1, \ldots, v_m, b_0)$.
3. The submatrix of A obtained by deleting the last row and the last column is the matrix of the linear map $\overrightarrow{f}$ with respect to the bases $(u_1, \ldots, u_n)$ and $(v_1, \ldots, v_m)$,

Finally, since

$$f(a_0 + u) = \widehat{f}(u \,\widehat{+}\, a_0),$$

given any $x \in E$ and $y \in F$ with coordinates $(x_1, \ldots, x_n, 1)$ and $(y_1, \ldots, y_m, 1)$, for $X = (x_1, \ldots, x_n, 1)^{\top}$ and $Y = (y_1, \ldots, y_m, 1)^{\top}$, we have $y = f(x)$ iff

$$Y = AX.$$

For example, consider the following affine map $f \colon \mathbb{A}^2 \to \mathbb{A}^2$ defined as follows:

$$y_1 = ax_1 + bx_2 + \mu_1,$$
$$y_2 = cx_1 + dx_2 + \mu_2.$$

The matrix of $\widehat{f}$ is

$$\begin{pmatrix} a & b & \mu_1 \\ c & d & \mu_2 \\ 0 & 0 & 1 \end{pmatrix},$$

and we have

$$\begin{pmatrix} y_1 \\ y_2 \\ 1 \end{pmatrix} = \begin{pmatrix} a & b & \mu_1 \\ c & d & \mu_2 \\ 0 & 0 & 1 \end{pmatrix} \begin{pmatrix} x_1 \\ x_2 \\ 1 \end{pmatrix}.$$

In $\widehat{E}$, we have

$$\begin{pmatrix} y_1 \\ y_2 \\ y_3 \end{pmatrix} = \begin{pmatrix} a & b & \mu_1 \\ c & d & \mu_2 \\ 0 & 0 & 1 \end{pmatrix} \begin{pmatrix} x_1 \\ x_2 \\ x_3 \end{pmatrix},$$

which means that the homogeneous map $\widehat{f}$ is is obtained from f by "adding the variable of homogeneity x_3":

$$\begin{aligned} y_1 &= ax_1 + bx_2 + \mu_1 x_3, \\ y_2 &= cx_1 + dx_2 + \mu_2 x_3, \\ y_3 &= x_3. \end{aligned}$$

4.5 Problems

4.1. Prove that $\widehat{E}$ as defined in Lemma 4.1 is indeed a vector space.

4.2. Prove Lemma 4.2.

4.3. Fill in the missing details in the proof of Lemma 4.3.

4.4. Fill in the missing details in the proof of Lemma 4.4.

References

1. Marcel Berger. *Géométrie 1*. Nathan, 1990. English edition: Geometry 1, Universitext, Springer-Verlag.
2. Jean H. Gallier. *Curves and Surfaces in Geometric Modeling: Theory and Algorithms*. Morgan Kaufmann, first edition, 1999.
3. Lyle Ramshaw. Blossoming: A connect-the-dots approach to splines. Technical report, Digital SRC, Palo Alto, CA 94301, 1987. Report No. 19.

Chapter 5
Basics of Projective Geometry

Think geometrically, prove algebraically.
—**John Tate**

5.1 Why Projective Spaces?

For a novice, projective geometry usually appears to be a bit odd, and it is not obvious to motivate why its introduction is inevitable and in fact fruitful. One of the main motivations arises from algebraic geometry.

The main goal of algebraic geometry is to study the properties of geometric objects, such as curves and surfaces, defined implicitly in terms of algebraic equations. For instance, the equation

$$x^2 + y^2 - 1 = 0$$

defines a circle in $\mathbb{R}^2$. More generally, we can consider the curves defined by general equations

$$ax^2 + by^2 + cxy + dx + ey + f = 0$$

of degree 2, known as *conics*. It is then natural to ask whether it is possible to classify these curves according to their generic geometric shape. This is indeed possible. Except for so-called singular cases, we get ellipses, parabolas, and hyperbolas. The same question can be asked for surfaces defined by quadratic equations, known as *quadrics*, and again, a classification is possible. However, these classifications are a bit artificial. For example, an ellipse and a hyperbola differ by the fact that a hyperbola has points at infinity, and yet, their geometric properties are identical, provided that points at infinity are handled properly.

Another important problem is the study of intersection of geometric objects (defined algebraically). For example, given two curves C_1 and C_2 of degree m and n, respectively, what is the number of intersection points of C_1 and C_2? (by degree of the curve we mean the total degree of the defining polynomial).

Well, it depends! Even in the case of lines (when $m = n = 1$), there are three possibilities: either the lines coincide, or they are parallel, or there is a single intersection point. In general, we expect mn intersection points, but some of these points may be missing because they are at infinity, because they coincide, or because they are imaginary.

What begins to transpire is that "points at infinity" cause trouble. They cause exceptions that invalidate geometric theorems (for example, consider the more general versions of the theorems of Pappus and Desargues from Section 2.12), and make it difficult to classify geometric objects. Projective geometry is designed to deal with "points at infinity" and regular points in a uniform way, without making a distinction. Points at infinity are now just ordinary points, and many things become simpler. For example, the classification of conics and quadrics becomes simpler, and intersection theory becomes cleaner (although, to be honest, we need to consider complex projective spaces).

Technically, projective geometry can be defined axiomatically, or by buidling upon linear algebra. Historically, the axiomatic approach came first (see Veblen and Young [28, 29], Emil Artin [1], and Coxeter [7, 8, 5, 6]). Although very beautiful and elegant, we believe that it is a harder approach than the linear algebraic approach. In the linear algebraic approach, all notions are considered up to a scalar. For example, a projective point is really a line through the origin. In terms of coordinates, this corresponds to "homogenizing." For example, the homogeneous equation of a conic is

$$ax^2 + by^2 + cxy + dxz + eyz + fz^2 = 0.$$

Now, regular points are points of coordinates (x, y, z) with $z \neq 0$, and points at infinity are points of coordinates $(x, y, 0)$ (with x, y, z not all null, and up to a scalar). There is a useful model (interpretation) of plane projective geometry in terms of the central projection in $\mathbb{R}^3$ from the origin onto the plane $z = 1$. Another useful model is the spherical (or the half-spherical) model. In the spherical model, a projective point corresponds to a pair of antipodal points on the sphere.

As affine geometry is the study of properties invariant under affine bijections, projective geometry is the study of properties invariant under bijective projective maps. Roughly speaking, projective maps are linear maps up to a scalar. In analogy with our presentation of affine geometry, we will define projective spaces, projective subspaces, projective frames, and projective maps. The analogy will fade away when we define the projective completion of an affine space, and when we define duality.

One of the virtues of projective geometry is that it yields a very clean presentation of rational curves and rational surfaces. The general idea is that a plane rational curve is the projection of a simpler curve in a larger space, a polynomial curve in $\mathbb{R}^3$, onto the plane $z = 1$, as we now explain.

Polynomial curves are curves defined parametrically in terms of polynomials. More specifically, if $\mathscr{E}$ is an affine space of finite dimension $n \geq 2$ and $(a_0, (e_1, \ldots, e_n))$ is an affine frame for $\mathscr{E}$, a polynomial curve of degree m is a map $F : \mathbb{A} \to \mathscr{E}$ such that

$$F(t) = a_0 + F_1(t)e_1 + \cdots + F_n(t)e_n,$$

for all $t \in \mathbb{A}$, where $F_1(t), \ldots, F_n(t)$ are polynomials of degree at most m.

Although many curves can be defined, it is somewhat embarassing that a circle cannot be defined in such a way. In fact, many interesting curves cannot be defined this way, for example, ellipses and hyperbolas. A rather simple way to extend the class of curves defined parametrically is to allow rational functions instead of polynomials. A *parametric rational curve* of degree m is a function $F \colon \mathbb{A} \to \mathscr{E}$ such that

$$F(t) = a_0 + \frac{F_1(t)}{F_{n+1}(t)} e_1 + \cdots + \frac{F_n(t)}{F_{n+1}(t)} e_n,$$

for all $t \in \mathbb{A}$, where $F_1(t), \ldots, F_n(t), F_{n+1}(t)$ are polynomials of degree at most m. For example, a circle in $\mathbb{A}^2$ can be defined by the rational map

$$F(t) = a_0 + \frac{1-t^2}{1+t^2} e_1 + \frac{2t}{1+t^2} e_2.$$

In the above example, the denominator $F_3(t) = 1 + t^2$ never takes the value 0 when t ranges over $\mathbb{A}$, but consider the following curve in $\mathbb{A}^2$:

$$G(t) = a_0 + \frac{t^2}{t} e_1 + \frac{1}{t} e_2.$$

Observe that $G(0)$ is undefined. The curve defined above is a hyperbola, and for t close to 0, the point on the curve goes toward infinity in one of the two asymptotic directions.

A clean way to handle the situation in which the denominator vanishes is to work in a projective space. Intuitively, this means viewing a rational curve in $\mathbb{A}^n$ as some appropriate projection of a polynomial curve in $\mathbb{A}^{n+1}$, back onto $\mathbb{A}^n$.

Given an affine space $\mathscr{E}$, for any hyperplane H in $\mathscr{E}$ and any point a_0 not in H, the *central projection (or conic projection, or perspective projection) of center a_0 onto H*, is the partial map p defined as follows: For every point x not in the hyperplane passing through a_0 and parallel to H, we define $p(x)$ as the intersection of the line defined by a_0 and x with the hyperplane H.

For example, we can view G as a rational curve in $\mathbb{A}^3$ given by

$$G_1(t) = a_0 + t^2 e_1 + e_2 + t e_3.$$

If we project this curve G_1 (in fact, a parabola in $\mathbb{A}^3$) using the central projection (perspective projection) of center a_0 onto the plane of equation $x_3 = 1$, we get the previous hyperbola. For $t = 0$, the point $G_1(0) = a_0 + e_2$ in $\mathbb{A}^3$ is in the plane of equation $x_3 = 0$, and its projection is undefined. We can consider that $G_1(0) = a_0 + e_2$ in $\mathbb{A}^3$ is projected to infinity in the direction of e_2 in the plane $x_3 = 0$. In the setting of projective spaces, this direction corresponds rigorously to a point at infinity.

Let us verify that the central projection used in the previous example has the desired effect. Let us assume that $\mathscr{E}$ has dimension $n + 1$ and that $(a_0, (e_1, \ldots, e_{n+1}))$ is an affine frame for $\mathscr{E}$. We want to determine the coordinates of the central projection $p(x)$ of a point $x \in \mathscr{E}$ onto the hyperplane H of equation $x_{n+1} = 1$ (the center of projection being a_0). If

$$x = a_0 + x_1 e_1 + \cdots + x_n e_n + x_{n+1} e_{n+1},$$

assuming that $x_{n+1} \neq 0$; a point on the line passing through a_0 and x has coordinates of the form $(\lambda x_1, \ldots, \lambda x_{n+1})$; and $p(x)$, the central projection of x onto the hyperplane H of equation $x_{n+1} = 1$, is the intersection of the line from a_0 to x and this hyperplane H. Thus we must have $\lambda x_{n+1} = 1$, and the coordinates of $p(x)$ are

$$\left(\frac{x_1}{x_{n+1}}, \ldots, \frac{x_n}{x_{n+1}}, 1 \right).$$

Note that $p(x)$ is undefined when $x_{n+1} = 0$. In projective spaces, we can make sense of such points.

The above calculation confirms that $G(t)$ is a central projection of $G_1(t)$. Similarly, if we define the curve F_1 in $\mathbb{A}^3$ by

$$F_1(t) = a_0 + (1 - t^2)e_1 + 2t e_2 + (1 + t^2)e_3,$$

the central projection of the polynomial curve F_1 (again, a parabola in $\mathbb{A}^3$) onto the plane of equation $x_3 = 1$ is the circle F.

What we just sketched is a general method to deal with rational curves. We can use our "hat construction" to embed an affine space $\mathscr{E}$ into a vector space $\widehat{\mathscr{E}}$ having one more dimension, then construct the projective space $\mathbf{P}\big(\widehat{\mathscr{E}}\big)$. This turns out to be the "projective completion" of the affine space $\mathscr{E}$. Then we can define a rational curve in $\mathbf{P}\big(\widehat{\mathscr{E}}\big)$, basically as the central projection of a polynomial curve in $\widehat{\mathscr{E}}$ back onto $\mathbf{P}\big(\widehat{\mathscr{E}}\big)$. The same approach can be used to deal with rational surfaces. Due to the lack of space, such a presentation is omitted from the main text. However, it can be found in the additional material on the web site; see `http://www.cis.upenn.edu/~jean/gbooks/geom2.html`.

More generally, the projective completion of an affine space is a very convenient tool to handle "points at infinity" in a clean fashion.

This chapter contains a brief presentation of concepts of projective geometry. The following concepts are presented: projective spaces, projective frames, homogeneous coordinates, projective maps, projective hyperplanes, multiprojective maps, affine patches. The projective completion of an affine space is presented using the "hat construction." The theorems of Pappus and Desargues are proved, using the method in which points are "sent to infinity." We also discuss the cross-ratio and duality. The chapter ends with a very brief explanation of the use of the complexification of a projective space in order to define the notion of angle and orthogonality in a projective setting. We also include a short section on applications of projective geometry, notably to computer vision (camera calibration), efficient communication, and error-correcting codes.

5.2 Projective Spaces

As in the case of affine geometry, our presentation of projective geometry is rather
sketchy and biased toward the algorithmic geometry of curves and surfaces. For a
systematic treatment of projective geometry, we recommend Berger [3, 4], Samuel
[23], Pedoe [21], Coxeter [7, 8, 5, 6], Beutelspacher and Rosenbaum [2], Fresnel [14], Sidler [24], Tisseron [26], Lehmann and Bkouche [20], Vienne [30],
and the classical treatise by Veblen and Young [28, 29], which, although slightly
old-fashioned, is definitely worth reading. Emil Artin's famous book [1] contains,
among other things, an axiomatic presentation of projective geometry, and a wealth
of geometric material presented from an algebraic point of view. Other "oldies but
goodies" include the beautiful books by Darboux [9] and Klein [19]. For a development of projective geometry addressing the delicate problem of orientation, see
Stolfi [25], and for an approach geared towards computer graphics, see Penna and
Patterson [22].

First, we define projective spaces, allowing the field K to be arbitrary (which
does no harm, and is needed to allow finite and complex projective spaces). Roughly
speaking, every projective concept is a linea–algebraic concept "up to a scalar." For
spaces, this is made precise as follows

Definition 5.1. Given a vector space E over a field K, the *projective space* $\mathbf{P}(E)$
induced by E is the set $(E - \{0\})/\sim$ of equivalence classes of nonzero vectors in E
under the equivalence relation $\sim$ defined such that for all $u, v \in E - \{0\}$,

$$u \sim v \quad \text{iff} \quad v = \lambda u, \text{ for some } \lambda \in K - \{0\}.$$

The *canonical projection* $p \colon (E - \{0\}) \to \mathbf{P}(E)$ is the function associating the
equivalence class $[u]_\sim$ modulo $\sim$ to $u \neq 0$. The *dimension* $\dim(\mathbf{P}(E))$ of $\mathbf{P}(E)$ is
defined as follows: If E is of infinite dimension, then $\dim(\mathbf{P}(E)) = \dim(E)$, and if
E has finite dimension, $\dim(E) = n \geq 1$ then $\dim(\mathbf{P}(E)) = n - 1$.

Mathematically, a projective space $\mathbf{P}(E)$ is a set of equivalence classes of vectors
in E. The spirit of projective geometry is to view an equivalence class $p(u) = [u]_\sim$
as an "atomic" object, forgetting the internal structure of the equivalence class. For
this reason, it is customary to call an equivalence class $a = [u]_\sim$ a *point* (the entire
equivalence class $[u]_\sim$ is collapsed into a single object viewed as a point).

Remarks:

(1) If we view E as an affine space, then for any nonnull vector $u \in E$, since

$$[u]_\sim = \{\lambda u \mid \lambda \in K, \lambda \neq 0\},$$

    letting

$$Ku = \{\lambda u \mid \lambda \in K\}$$

    denote the subspace of dimension 1 spanned by u, the map

$$[u]_\sim \mapsto Ku$$

from $\mathbf{P}(E)$ to the set of one-dimensional subspaces of E is clearly a bijection, and since subspaces of dimension 1 correspond to lines through the origin in E, we can view $\mathbf{P}(E)$ as the set of lines in E passing through the origin. So, the projective space $\mathbf{P}(E)$ can be viewed as the set obtained from E when lines through the origin are treated as points.

However, this is a somewhat deceptive view. Indeed, depending on the structure of the vector space E, a line (through the origin) in E may be a fairly complex object, and treating a line just as a point is really a mental game. For example, E may be the vector space of real homogeneous polynomials $P(x,y,z)$ of degree 2 in three variables x,y,z (plus the null polynomial), and a "line" (through the origin) in E corresponds to an algebraic curve of degree 2. Lots of details need to be filled in, but roughly speaking, the curve defined by P is the "zero locus of P," i.e., the set of points $(x,y,z) \in \mathbf{P}(\mathbb{R}^3)$ (or perhaps in $\mathbf{P}(\mathbb{C}^3)$) for which $P(x,y,z) = 0$. We will come back to this point in Section 5.4 after having introduced homogeneous coordinates.

More generally, E may be a vector space of homogeneous polynomials of degree m in 3 or more variables (plus the null polynomial), and the lines in E correspond to such objects as algebraic curves, algebraic surfaces, and algebraic varieties. The point of view where a complex object such as a curve or a surface is treated as a point in a (projective) space is actually very fruitful and is one of the themes of algebraic geometry (see Fulton [15] or Harris [16]).

(2) When $\dim(E) = 1$, we have $\dim(\mathbf{P}(E)) = 0$. When $E = \{0\}$, we have $\mathbf{P}(E) = \emptyset$. By convention, we give it the dimension -1.

We denote the projective space $\mathbf{P}(K^{n+1})$ by $\mathbb{P}^n_K$. When $K = \mathbb{R}$, we also denote $\mathbb{P}^n_\mathbb{R}$ by $\mathbb{RP}^n$, and when $K = \mathbb{C}$, we denote $\mathbb{P}^n_\mathbb{C}$ by $\mathbb{CP}^n$. The projective space $\mathbb{P}^0_K$ is a (projective) point. The projective space $\mathbb{P}^1_K$ is called a *projective line*. The projective space $\mathbb{P}^2_K$ is called a *projective plane*.

The projective space $\mathbf{P}(E)$ can be visualized in the following way. For simplicity, assume that $E = \mathbb{R}^{n+1}$, and thus $\mathbf{P}(E) = \mathbb{RP}^n$ (the same reasoning applies to $E = K^{n+1}$, where K is any field).

Let H be the affine hyperplane consisting of all points $(x_1,\ldots,x_{n+1})$ such that $x_{n+1} = 1$. Every nonzero vector u in E determines a line D passing through the origin, and this line intersects the hyperplane H in a unique point a, unless D is parallel to H. When D is parallel to H, the line corresponding to the equivalence class of u can be thought of as a point at infinity, often denoted by u_∞. Thus, the projective space $\mathbf{P}(E)$ can be viewed as the set of points in the hyperplane H, together with points at infinity associated with lines in the hyperplane H_∞ of equation $x_{n+1} = 0$. We will come back to this point of view when we consider the projective completion of an affine space. Figure 5.1 illustrates the above representation of the projective space when $E = \mathbb{R}^3$.

We refer to the above model of $\mathbf{P}(E)$ as the *hyperplane model*. In this model some hyperplane H_∞ (through the origin) in $\mathbb{R}^{n+1}$ is singled out, and the points of $\mathbf{P}(E)$ arising from the hyperplane H_∞ are declared to be "points at infinity." The purpose

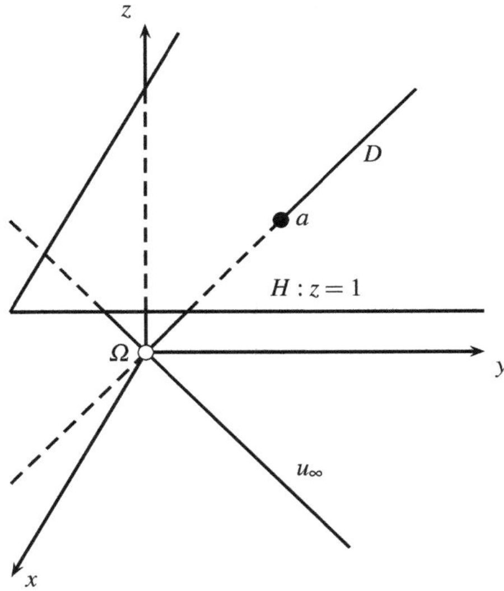

Fig. 5.1 A representation of the projective space $\mathbb{RP}^2$.

of the affine hyperplane H parallel to H_∞ and distinct from H_∞ is to get images for the other points in $\mathbf{P}(E)$ (i.e., those that arise from lines not contained in H_∞). It should be noted that the choice of which points should be considered as infinite is relative to the choice of H_∞. Viewing certain points of $\mathbf{P}(E)$ as points at infinity is convenient for getting a mental picture of $\mathbf{P}(E)$, but there is nothing intrinsic about that. Points of $\mathbf{P}(E)$ are all equal, and unless some additional structure in introduced in $\mathbf{P}(E)$ (such as a hyperplane), a point in $\mathbf{P}(E)$ doesn't know whether it is infinite! The notion of point at infinity is really an affine notion. This point will be made precise in Section 5.6.

Again, for $\mathbb{RP}^n = \mathbf{P}(\mathbb{R}^{n+1})$, instead of considering the hyperplane H, we can consider the n-sphere S^n of center 0 and radius 1, i.e., the set of points $(x_1, \ldots, x_{n+1})$ such that

$$x_1^2 + \cdots + x_n^2 + x_{n+1}^2 = 1.$$

In this case, every line D through the center of the sphere intersects the sphere S^n in two antipodal points a_+ and a_-. The projective space $\mathbb{RP}^n$ is the quotient space obtained from the sphere S^n by identifying antipodal points a_+ and a_-. It is hard to visualize such an object! Nevertheless, some nice projections in $\mathbb{A}^3$ of an embedding of $\mathbb{RP}^2$ into $\mathbb{A}^4$ are given in the surface gallery on the web cite (see `http://www.cis.upenn.edu/~jean/gbooks/geom2.html`, Section 24.7). We call this model of $\mathbf{P}(E)$ the *spherical model*.

A more subtle construction consists in considering the (upper) half-sphere instead of the sphere, where the upper half-sphere S^n_+ is set of points on the sphere S^n such that $x_{n+1} \geq 0$. This time, every line through the center intersects the (upper) half-sphere in a single point, except on the boundary of the half-sphere, where it intersects in two antipodal points a_+ and a_-. Thus, the projective space $\mathbb{RP}^n$ is the quotient space obtained from the (upper) half-sphere S^n_+ by identifying antipodal points a_+ and a_- on the boundary of the half-sphere. We call this model of $\mathbf{P}(E)$ the *half-spherical model*.

When $n = 2$, we get a circle. When $n = 3$, the upper half-sphere is homeomorphic to a closed disk (say, by orthogonal projection onto the xy-plane), and $\mathbb{RP}^2$ is in bijection with a closed disk in which antipodal points on its boundary (a unit circle) have been identified. This is hard to visualize! In this model of the real projective space, projective lines are great semicircles on the upper half-sphere, with antipodal points on the boundary identified. Boundary points correspond to points at infinity. By orthogonal projection, these great semicircles correspond to semiellipses, with antipodal points on the boundary identified. Traveling along such a projective "line," when we reach a boundary point, we "wrap around"! In general, the upper half-sphere S^n_+ is homeomorphic to the closed unit ball in $\mathbb{R}^n$, whose boundary is the $(n-1)$-sphere S^{n-1}. For example, the projective space $\mathbb{RP}^3$ is in bijection with the closed unit ball in $\mathbb{R}^3$, with antipodal points on its boundary (the sphere S^2) identified!

Remarks:

(1) A projective space $\mathbf{P}(E)$ has been defined as a *set* without any topological structure. When the field K is either the field $\mathbb{R}$ of reals or the field $\mathbb{C}$ of complex numbers, the vector space E is a topological space. Thus, the projection map $p\colon (E - \{0\}) \to \mathbf{P}(E)$ induces a topology on the projective space $\mathbf{P}(E)$, namely the quotient topology. This means that a subset V of $\mathbf{P}(E)$ is open iff $p^{-1}(V)$ is an open set in E. Then, for example, it turns out that the real projective space $\mathbb{RP}^n$ is homeomorphic to the space obtained by taking the quotient of the (upper) half-sphere S^n_+, by the equivalence relation identifying antipodal points a_+ and a_- on the boundary of the half-sphere. Another interesting fact is that the complex projective line $\mathbb{CP}^1 = \mathbf{P}(\mathbb{C}^2)$ is homeomorphic to the (real) 2-sphere S^2, and that the real projective space $\mathbb{RP}^3$ is homeomorphic to the group of rotations $\mathbf{SO}(3)$ of $\mathbb{R}^3$.

(2) If H is a hyperplane in E, recall from Lemma 21.1 that there is some nonnull linear form $f \in E^*$ such that $H = \mathrm{Ker}\, f$. Also, given any nonnull linear form $f \in E^*$, its kernel $H = \mathrm{Ker}\, f = f^{-1}(0)$ is a hyperplane, and if $\mathrm{Ker}\, f = \mathrm{Ker}\, g = H$, then $g = \lambda f$ for some $\lambda \neq 0$. These facts can be concisely stated by saying that the map

$$[f]_\sim \mapsto \mathrm{Ker}\, f$$

mapping the equivalence class $[f]_\sim = \{\lambda f \mid \lambda \neq 0\}$ of a nonnull linear form $f \in E^*$ to the hyperplane $H = \mathrm{Ker}\, f$ in E is a bijection between the projective space $\mathbf{P}(E^*)$ and the set of hyperplanes in E. When E is of finite dimension, this bijection yields a useful duality, which will be investigated in Section 5.9.

We now define projective subspaces.

5.3 Projective Subspaces

Projective subspaces of a projective space $\mathbf{P}(E)$ are induced by subspaces of the vector space E.

Definition 5.2. Given a nontrivial vector space E, a *projective subspace (or linear projective variety) of* $\mathbf{P}(E)$ is any subset W of $\mathbf{P}(E)$ such that there is some subspace $V \neq \{0\}$ of E with $W = p(V - \{0\})$. The dimension $\dim(W)$ of W is defined as follows: If V is of infinite dimension, then $\dim(W) = \dim(V)$, and if $\dim(V) = p \geq 1$, then $\dim(W) = p - 1$. We say that a family $(a_i)_{i \in I}$ of points of $\mathbf{P}(E)$ is *projectively independent* if there is a linearly independent family $(\overrightarrow{u_i})_{i \in I}$ in E such that $a_i = p(u_i)$ for every $i \in I$.

Remark: If we allow the empty subset to be a projective subspace, then we have a bijection between the subspaces of E and the projective subspaces of $\mathbf{P}(E)$. In fact, $\mathbf{P}(V)$ is the projective space induced by the vector space V, and we also denote $p(V - \{0\})$ by $\mathbf{P}(V)$, or even by $p(V)$, even though $p(0)$ is undefined.

A projective subspace of dimension 0 is a called a *(projective) point*. A projective subspace of dimension 1 is called a *(projective) line*, and a projective subspace of dimension 2 is called a *(projective) plane*. If H is a hyperplane in E, then $\mathbf{P}(H)$ is called a *projective hyperplane*. It is easily verified that any arbitrary intersection of projective subspaces is a projective subspace. A single point is projectively independent. Two points a, b are projectively independent if $a \neq b$. Two distinct points define a (unique) projective line. Three points a, b, c are projectively independent if they are distinct, and neither belongs to the projective line defined by the other two. Three projectively independent points define a (unique) projective plane.

A closer look at projective subspaces will show some of the advantages of projective geometry: In considering intersection properties, there are no exceptions due to parallelism, as in affine spaces.

Let E be a nontrivial vector space. Given any nontrivial subset S of E, the subset S defines a subset $U = p(S - \{0\})$ of the projective space $\mathbf{P}(E)$, and if $\langle S \rangle$ denotes the subspace of E spanned by S, it is immediately verified that $\mathbf{P}(\langle S \rangle)$ is the intersection of all projective subspaces containing U, and this projective subspace is denoted by $\langle U \rangle$. Given any subspaces M and N of E, recall from Lemma 2.14 that we have the *Grassmann relation*

$$\dim(M) + \dim(N) = \dim(M + N) + \dim(M \cap N).$$

Then the following lemma is easily shown.

Lemma 5.1. *Given a projective space* $\mathbf{P}(E)$, *for any two projective subspaces* U, V *of* $\mathbf{P}(E)$, *we have*

$$\dim(U) + \dim(V) = \dim(\langle U \cup V \rangle) + \dim(U \cap V).$$

Furthermore, if $\dim(U) + \dim(V) \geq \dim(\mathbf{P}(E))$, *then* $U \cap V$ *is nonempty and if* $\dim(\mathbf{P}(E)) = n$, *then:*

(i) The intersection of any n hyperplanes is nonempty.

(ii) For every hyperplane H and every point $a \notin H$, every line D containing a intersects H in a unique point.

(iii) In a projective plane, every two distinct lines intersect in a unique point.

As a corollary, in the projective space $(\dim(\mathbf{P}(E)) = 3)$, for every plane H, every line not contained in H intersects H in a unique point.

It is often useful to deal with projective hyperplanes in terms of nonnull linear forms and equations. Recall that the map

$$[f]_\sim \mapsto \operatorname{Ker} f$$

is a bijection between $\mathbf{P}(E^*)$ and the set of hyperplanes in E, mapping the equivalence class $[f]_\sim = \{\lambda f \mid \lambda \neq 0\}$ of a nonnull linear form $f \in E^*$ to the hyperplane $H = \operatorname{Ker} f$. Furthermore, if $u \sim v$, which means that $u = \lambda v$ for some $\lambda \neq 0$, we have

$$f(u) = 0 \quad \text{iff} \quad f(v) = 0,$$

since $f(v) = \lambda f(u)$ and $\lambda \neq 0$. Thus, there is a bijection

$$\{\lambda f \mid \lambda \neq 0\} \mapsto \mathbf{P}(\operatorname{Ker} f)$$

mapping points in $\mathbf{P}(E^*)$ to hyperplanes in $\mathbf{P}(E)$. Any nonnull linear form f associated with some hyperplane $\mathbf{P}(H)$ in the above bijection (i.e., $H = \operatorname{Ker} f$) is called an *equation of the projective hyperplane* $\mathbf{P}(H)$. We also say that $f = 0$ *is the equation of the hyperplane* $\mathbf{P}(H)$.

Before ending this section, we give an example of a projective space where lines have a nontrivial geometric interpretation, namely as "pencils of lines." If $E = \mathbb{R}^3$, recall that the dual space E^* is the set of all linear maps $f\colon \mathbb{R}^3 \to \mathbb{R}$. As we have just explained, there is a bijection

$$p(f) \mapsto \mathbf{P}(\operatorname{Ker} f)$$

between $\mathbf{P}(E^*)$ and the set of lines in $\mathbf{P}(E)$, mapping every point $a = p(f)$ to the line $D_a = \mathbf{P}(\operatorname{Ker} f)$.

Is there a way to give a geometric interpretation in $\mathbf{P}(E)$ of a line Δ in $\mathbf{P}(E^*)$? Well, a line Δ in $\mathbf{P}(E^*)$ is defined by two distinct points $a = p(f)$ and $b = p(g)$, where $f, g \in E^*$ are two linearly independent linear forms. But f and g define two distinct planes $H_1 = \operatorname{Ker} f$ and $H_2 = \operatorname{Ker} g$ through the origin (in $E = \mathbb{R}^3$), and H_1 and H_2 define two distinct lines $D_1 = p(H_1)$ and $D_2 = p(H_2)$ in $\mathbf{P}(E)$. The line Δ in $\mathbf{P}(E^*)$ is of the form $\Delta = p(V)$, where

$$V = \{\lambda f + \mu g \mid \lambda, \mu \in \mathbb{R}\}$$

is the plane in E^* spanned by f, g. Every nonnull linear form $\lambda f + \mu g \in V$ defines a plane $H = \mathrm{Ker}\,(\lambda f + \mu g)$ in E, and since H_1 and H_2 (in E) are distinct, they intersect in a line L that is also contained in every plane H as above. Thus, the set of planes in E associated with nonnull linear forms in V is just the set of all planes containing the line L. Passing to $\mathbf{P}(E)$ using the projection p, the line L in E corresponds to the point $c = p(L)$ in $\mathbf{P}(E)$, which is just the intersection of the lines D_1 and D_2. Thus, every point of the line Δ in $\mathbf{P}(E^*)$ corresponds to a line in $\mathbf{P}(E)$ passing through c (the intersection of the lines D_1 and D_2), and this correspondence is bijective.

In summary, a line Δ in $\mathbf{P}(E^*)$ corresponds to the set of all lines in $\mathbf{P}(E)$ through some given point. Such sets of lines are called *pencils of lines*.

The above discussion can be generalized to higher dimensions and is discussed quite extensively in Section 5.9. In brief, letting $E = \mathbb{R}^{n+1}$, there is a bijection mapping points in $\mathbf{P}(E^*)$ to hyperplanes in $\mathbf{P}(E)$. A line in $\mathbf{P}(E^*)$ corresponds to a *pencil of hyperplanes* in $\mathbf{P}(E)$, i.e., the set of all hyperplanes containing some given projective subspace $W = p(V)$ of dimension $n - 2$. For $n = 3$, a pencil of planes in $\mathbb{RP}^3 = \mathbf{P}(\mathbb{R}^4)$ is the set of all planes (in $\mathbb{RP}^3$) containing some given line W. Other examples of unusual projective spaces and pencils will be given in Section 5.4.

Next, we define the projective analogues of bases (or frames) and linear maps.

5.4 Projective Frames

As all good notions in projective geometry, the concept of a projective frame turns out to be uniquely defined up to a scalar.

Definition 5.3. Given a nontrivial vector space E of dimension $n + 1$, a family $(a_i)_{1 \leq i \leq n+2}$ of $n + 2$ points of the projective space $\mathbf{P}(E)$ is a *projective frame (or basis) of* $\mathbf{P}(E)$ if there exists some basis $(e_1, \ldots, e_{n+1})$ of E such that $a_i = p(e_i)$ for $1 \leq i \leq n+1$, and $a_{n+2} = p(e_1 + \cdots + e_{n+1})$. Any basis with the above property is said to be *associated with the projective frame* $(a_i)_{1 \leq i \leq n+2}$.

The justification of Definition 5.3 is given by the following lemma.

Lemma 5.2. *If* $(a_i)_{1 \leq i \leq n+2}$ *is a projective frame of* $\mathbf{P}(E)$, *for any two bases* $(u_1, \ldots, u_{n+1})$, $(v_1, \ldots, v_{n+1})$ *of* E *such that* $a_i = p(u_i) = p(v_i)$ *for* $1 \leq i \leq n+1$, *and* $a_{n+2} = p(u_1 + \cdots + u_{n+1}) = p(v_1 + \cdots + v_{n+1})$, *there is a nonzero scalar* $\lambda \in K$ *such that* $v_i = \lambda u_i$, *for all* i, $1 \leq i \leq n+1$.

Proof. Since $p(u_i) = p(v_i)$ for $1 \leq i \leq n+1$, there exist some nonzero scalars $\lambda_i \in K$ such that $v_i = \lambda_i u_i$ for all i, $1 \leq i \leq n+1$. Since we must have

$$p(u_1 + \cdots + u_{n+1}) = p(v_1 + \cdots + v_{n+1}),$$

there is some $\lambda \neq 0$ such that

$$\lambda(u_1 + \cdots + u_{n+1}) = v_1 + \cdots + v_{n+1} = \lambda_1 u_1 + \cdots + \lambda_{n+1} u_{n+1},$$

and thus we have

$$(\lambda - \lambda_1)u_1 + \cdots + (\lambda - \lambda_{n+1})u_{n+1} = 0,$$

and since $(u_1, \ldots, u_{n+1})$ is a basis, we have $\lambda_i = \lambda$ for all i, $1 \leq i \leq n+1$, which implies $\lambda_1 = \cdots = \lambda_{n+1} = \lambda$. $\square$

Lemma 5.2 shows that a projective frame determines a unique basis of E, up to a (nonzero) scalar. This would not necessarily be the case if we did not have a point a_{n+2} such that $a_{n+2} = p(u_1 + \cdots + u_{n+1})$.

When $n = 0$, the projective space consists of a single point a, and there is only one projective frame, the pair (a, a). When $n = 1$, the projective space is a line, and a projective frame consists of any three pairwise distinct points a, b, c on this line. When $n = 2$, the projective space is a plane, and a projective frame consists of any four distinct points a, b, c, d such that a, b, c are the vertices of a nondegenerate triangle and d is not on any of the lines determined by the sides of this triangle. The reader can easily generalize to higher dimensions.

Given a projective frame $(a_i)_{1 \leq i \leq n+2}$ of $\mathbf{P}(E)$, let $(u_1, \ldots, u_{n+1})$ be a basis of E associated with $(a_i)_{1 \leq i \leq n+2}$. For every $a \in \mathbf{P}(E)$, there is some $u \in E - \{0\}$ such that

$$a = [u]_\sim = \{\lambda u \mid \lambda \in K - \{0\}\},$$

the equivalence class of u, and the set

$$\{(x_1, \ldots, x_{n+1}) \in K^{n+1} \mid v = x_1 u_1 + \cdots + x_{n+1} u_{n+1}, \ v \in [u]_\sim = a\}$$

of coordinates of all the vectors in the equivalence class $[u]_\sim$ is called the *set of homogeneous coordinates of a over the basis* $(u_1, \ldots, u_{n+1})$.

Note that for each homogeneous coordinate $(x_1, \ldots, x_{n+1})$ we must have $x_i \neq 0$ for some i, $1 \leq i \leq n+1$, and any two homogeneous coordinates $(x_1, \ldots, x_{n+1})$ and $(y_1, \ldots, y_{n+1})$ for a differ by a nonzero scalar, i.e., there is some $\lambda \neq 0$ such that $y_i = \lambda x_i$, $1 \leq i \leq n+1$. Homogeneous coordinates $(x_1, \ldots, x_{n+1})$ are sometimes denoted by $(x_1 : \cdots : x_{n+1})$, for instance in algebraic geometry.

By Lemma 5.2, any other basis $(v_1, \ldots, v_{n+1})$ associated with the projective frame $(a_i)_{1 \leq i \leq n+2}$ differs from $(u_1, \ldots, u_{n+1})$ by a nonzero scalar, which implies that the set of homogeneous coordinates of $a \in \mathbf{P}(E)$ over the basis $(v_1, \ldots, v_{n+1})$ is identical to the set of homogeneous coordinates of $a \in \mathbf{P}(E)$ over the basis $(u_1, \ldots, u_{n+1})$. Consequently, we can associate a unique set of homogeneous coordinates to every point $a \in \mathbf{P}(E)$ with respect to the projective frame $(a_i)_{1 \leq i \leq n+2}$. With respect to this projective frame, note that a_{n+2} has homogeneous coordinates $(1, \ldots, 1)$, and that a_i has homogeneous coordinates $(0, \ldots, 1, \ldots, 0)$, where the 1 is in the ith position, where $1 \leq i \leq n+1$. We summarize the above discussion in the following definition.

Definition 5.4. Given a nontrivial vector space E of dimension $n + 1$, for any projective frame $(a_i)_{1 \leq i \leq n+2}$ of $\mathbf{P}(E)$ and for any point $a \in \mathbf{P}(E)$, the *set of homogeneous coordinates of a with respect to* $(a_i)_{1 \leq i \leq n+2}$ is the set of $(n+1)$-tuples

$$\{(\lambda x_1,\ldots,\lambda x_{n+1}) \in K^{n+1} \mid x_i \neq 0 \text{ for some } i,\ \lambda \neq 0,$$
$$a = p(x_1 u_1 + \cdots + x_{n+1} u_{n+1})\},$$

where $(u_1,\ldots,u_{n+1})$ is any basis of E associated with $(a_i)_{1 \leq i \leq n+2}$.

Given a projective frame $(a_i)_{1 \leq i \leq n+2}$ for $\mathbf{P}(E)$, if $(x_1,\ldots,x_{n+1})$ are homogeneous coordinates of a point $a \in \mathbf{P}(E)$, we write $a = (x_1,\ldots,x_{n+1})$, and with a slight abuse of language, we may even talk about a point $(x_1,\ldots,x_{n+1})$ in $\mathbf{P}(E)$ and write $(x_1,\ldots,x_{n+1}) \in \mathbf{P}(E)$.

The special case of the projective line $\mathbb{P}^1_K$ is worth examining. The projective line $\mathbb{P}^1_K$ consists of all equivalence classes $[x,y]$ of pairs $(x,y) \in K^2$ such that $(x,y) \neq (0,0)$, under the equivalence relation $\sim$ defined such that

$$(x_1,y_1) \sim (x_2,y_2) \quad \text{iff} \quad x_2 = \lambda x_1 \quad \text{and} \quad y_2 = \lambda y_1,$$

for some $\lambda \in K - \{0\}$. When $y \neq 0$, the equivalence class of (x,y) contains the representative $(xy^{-1},1)$, and when $y = 0$, the equivalence class of $(x,0)$ contains the representative $(1,0)$. Thus, there is a bijection between K and the set of equivalence classes containing some representative of the form $(x,1)$, and we denote the class $[x,1]$ by x. The equivalence class $[1,0]$ is denoted by ∞ and it is called the point at infinity. Thus, the projective line $\mathbb{P}^1_K$ is in bijection with $K \cup \{\infty\}$. The three points $\infty = [1,0]$, $0 = [0,1]$, and $1 = [1,1]$, form a projective frame for $\mathbb{P}^1_K$. The projective frame $(\infty,0,1)$ is often called the *canonical frame of* $\mathbb{P}^1_K$.

Homogeneous coordinates are also very useful to handle hyperplanes in terms of equations. If $(a_i)_{1 \leq i \leq n+2}$ is a projective frame for $\mathbf{P}(E)$ associated with a basis $(u_1,\ldots,u_{n+1})$ for E, a nonnull linear form f is determined by $n+1$ scalars $\alpha_1,\ldots,\alpha_{n+1}$ (not all null), and a point $x \in \mathbf{P}(E)$ of homogeneous coordinates $(x_1,\ldots,x_{n+1})$ belongs to the projective hyperplane $\mathbf{P}(H)$ of equation f iff

$$\alpha_1 x_1 + \cdots + \alpha_{n+1} x_{n+1} = 0.$$

In particular, if $\mathbf{P}(E)$ is a projective plane, a line is defined by an equation of the form $\alpha x + \beta y + \gamma z = 0$. If $\mathbf{P}(E)$ is a projective space, a plane is defined by an equation of the form $\alpha x + \beta y + \gamma z + \delta w = 0$.

We also have the following lemma giving another characterization of projective frames.

Lemma 5.3. *A family* $(a_i)_{1 \leq i \leq n+2}$ *of* $n+2$ *points is a projective frame of* $\mathbf{P}(E)$ *iff for every* i, $1 \leq i \leq n+2$, *the subfamily* $(a_j)_{j \neq i}$ *is projectively independent.*

Proof. We leave as an (easy) exercise the fact that if $(a_i)_{1 \leq i \leq n+2}$ is a projective frame, then each subfamily $(a_j)_{j \neq i}$ is projectively independent. Conversely, pick some $u_i \in E - \{0\}$ such that $a_i = p(u_i)$, $1 \leq i \leq n+2$. Since $(a_j)_{j \neq n+2}$ is projectively independent, $(u_1,\ldots,u_{n+1})$ is a basis of E. Thus, we must have

$$u_{n+2} = \lambda_1 u_1 + \cdots + \lambda_{n+1} u_{n+1},$$

for some $\lambda_i \in K$. However, since for every i, $1 \leq i \leq n+1$, the family $(a_j)_{j \neq i}$ is projectively independent, we must have $\lambda_i \neq 0$, and thus $(\lambda_1 u_1, \ldots, \lambda_{n+1} u_{n+1})$ is also a basis of E, and since

$$u_{n+2} = \lambda_1 u_1 + \cdots + \lambda_{n+1} u_{n+1},$$

it induces the projective frame $(a_i)_{1 \leq i \leq n+2}$. $\square$

Figure 5.2 shows a projective frame (a,b,c,d) in a projective plane. With respect

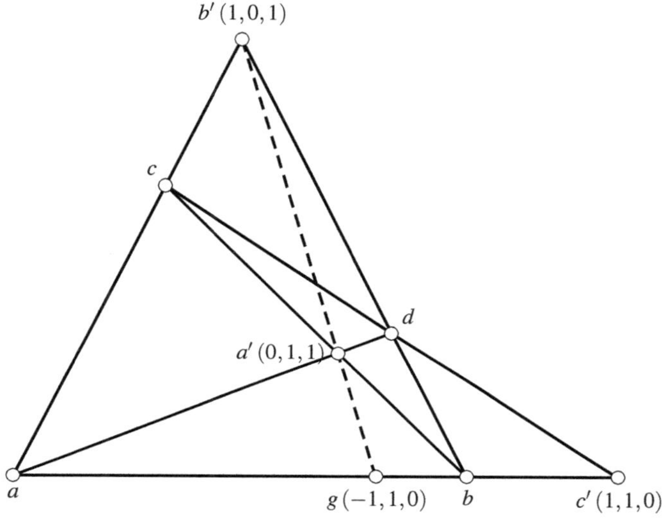

Fig. 5.2 A projective frame (a,b,c,d).

to this projective frame, the points a,b,c,d have homogeneous coordinates $(1,0,0)$, $(0,1,0)$, $(0,0,1)$, and $(1,1,1)$. Let a' be the intersection of $\langle d,a \rangle$ and $\langle b,c \rangle$, b' be the intersection of $\langle d,b \rangle$ and $\langle a,c \rangle$, and c' be the intersection of $\langle d,c \rangle$ and $\langle a,b \rangle$. Then the points a',b',c' have homogeneous coordinates $(0,1,1)$, $(1,0,1)$, and $(1,1,0)$. The diagram formed by the line segments $\langle a,c' \rangle$, $\langle a,b' \rangle$, $\langle b,b' \rangle$, $\langle c,c' \rangle$, $\langle a,d \rangle$, and $\langle b,c \rangle$ is sometimes called a *Möbius net*. It is easily verified that the equations of the lines $\langle a,b \rangle$, $\langle a,c \rangle$, $\langle b,c \rangle$, are $z=0$, $y=0$, and $x=0$, and the equations of the lines $\langle a,d \rangle$, $\langle b,d \rangle$, and $\langle c,d \rangle$, are $y=z$, $x=z$, and $x=y$. If we let e be the intersection of $\langle b,c \rangle$ and $\langle b',c' \rangle$, f be the intersection of $\langle a,c \rangle$ and $\langle a',c' \rangle$, and g be the intersection of $\langle a,b \rangle$ and $\langle a',b' \rangle$, then it easily seen that e,f,g have homogeneous coordinates $(0,-1,1)$, $(1,0,-1)$, and $(-1,1,0)$. These coordinates satisfy the equation $x+y+z=0$, which shows that the points e,f,g are collinear. This is a special case of the projective version of Desargues's theorem. This line is called the *polar line (or fundamental line)* of d with respect to the triangle (a,b,c). The diagram also shows the intersection g of $\langle a,b \rangle$ and $\langle a',b' \rangle$.

The projective space of circles provides a nice illustration of homogeneous coordinates. Let E be the vector space (over $\mathbb{R}$) consisting of all homogeneous polynomials of degree 2 in x, y, z of the form

$$ax^2 + ay^2 + bxz + cyz + dz^2$$

(plus the null polynomial). The projective space $\mathbf{P}(E)$ consists of all equivalence classes

$$[P]_\sim = \{\lambda P \mid \lambda \neq 0\},$$

where $P(x,y,z)$ is a nonnull homogeneous polynomial in E. We want to give a geometric interpretation of the points of the projective space $\mathbf{P}(E)$. In order to do so, pick some projective frame (a_1, a_2, a_3, a_4) for the projective plane $\mathbb{RP}^2$, and associate to every $[P] \in \mathbf{P}(E)$ the subset of $\mathbb{RP}^2$ known as its its *zero locus (or zero set, or variety)* $V([P])$, and defined such that

$$V([P]) = \{a \in \mathbb{RP}^2 \mid P(x,y,z) = 0\},$$

where (x,y,z) are homogeneous coordinates for a.

As explained earlier, we also use the simpler notation

$$V([P]) = \{(x,y,z) \in \mathbb{RP}^2 \mid P(x,y,z) = 0\}.$$

Actually, in order for $V([P])$ to make sense, we have to check that $V([P])$ does not depend on the representative chosen in the equivalence class $[P] = \{\lambda P \mid \lambda \neq 0\}$. This is because

$$P(x,y,z) = 0 \quad \text{iff} \quad \lambda P(x,y,z) = 0 \quad \text{when } \lambda \neq 0.$$

For simplicity of notation, we also denote $V([P])$ by $V(P)$. We also have to check that if $(\lambda x, \lambda y, \lambda z)$ are other homogeneous coordinates for $a \in \mathbb{RP}^2$, where $\lambda \neq 0$, then

$$P(x,y,z) = 0 \quad \text{iff} \quad P(\lambda x, \lambda y, \lambda z) = 0.$$

However, since $P(x,y,z)$ is homogeneous of degree 2, we have

$$P(\lambda x, \lambda y, \lambda z) = \lambda^2 P(x,y,z),$$

and since $\lambda \neq 0$,

$$P(x,y,z) = 0 \quad \text{iff} \quad \lambda^2 P(x,y,z) = 0.$$

The above argument applies to any homogeneous polynomial $P(x_1, \ldots, x_n)$ in n variables of any degree m, since

$$P(\lambda x_1, \ldots, \lambda x_n) = \lambda^m P(x_1, \ldots, x_n).$$

Thus, we can associate to every $[P] \in \mathbf{P}(E)$ the curve $V(P)$ in $\mathbb{RP}^2$. One might wonder why we are considering only homogeneous polynomials of degree 2, and not arbitrary polynomials of degree 2? The first reason is that the polynomials in x, y, z of degree 2 do **not** form a vector space. For example, if $P = x^2 + x$ and $Q = -x^2 + y$, the polynomial $P + Q = x + y$ is not of degree 2. We could consider the set of polynomials of degree ≤ 2, which is a vector space, but now the problem is that $V(P)$ is not necessarily well defined!. For example, if $P(x, y, z) = -x^2 + 1$, we have

$$P(1,0,0) = 0 \quad \text{and} \quad P(2,0,0) = -3,$$

and yet $(2,0,0) = 2(1,0,0)$, so that $P(x, y, z)$ takes different values depending on the representative chosen in the equivalence class $[1,0,0]$. Thus, we are led to restrict ourselves to homogeneous polynomials. Actually, this is usually an advantage more than a disadvantage, because homogeneous polynomials tend to be well behaved. For example, by polarization, they yield multilinear maps.

What are the curves $V(P)$? One way to "see" such curves is to go back to the hyperplane model of $\mathbb{RP}^2$ in terms of the plane H of equation $z = 1$ in $\mathbb{R}^3$. Then the trace of $V(P)$ on H is the circle of equation

$$ax^2 + ay^2 + bx + cy + d = 0.$$

Thus, we may think of $\mathbf{P}(E)$ as a projective space of circles. However, there are some problems. For example, $V(P)$ may be empty! This happens, for instance, for $P(x, y, z) = x^2 + y^2 + z^2$, since the equation

$$x^2 + y^2 + z^2 = 0$$

has only the trivial solution $(0,0,0)$, which does not correspond to any point in $\mathbb{RP}^2$. Indeed, only nonnull vectors in $\mathbb{R}^3$ yield points in $\mathbb{RP}^2$. It is also possible that $V(P)$ is reduced to a single point, for instance when $P(x, y, z) = x^2 + y^2$, since the only homogeneous solution of

$$x^2 + y^2 = 0$$

is $(0,0,1)$. Also, note that the map

$$[P] \mapsto V(P)$$

is not injective. For instance, $P = x^2 + y^2$ and $Q = x^2 + 2y^2$ define the same degenerate circle reduced to the point $(0,0,1)$. We also accept as circles the union of two lines, as in the case

$$(bx + cy + dz)z = 0,$$

where $a = 0$, and even a double line, as in the case

$$z^2 = 0,$$

where $a = b = c = 0$.

A clean way to resolve most of these problems is to switch to homogeneous polynomials over the complex field $\mathbb{C}$ and to consider curves in $\mathbb{CP}^2$. This is what is done in algebraic geometry (see Fulton [15] or Harris [16]). If $P(x, y, z)$ is a homogeneous polynomial over $\mathbb{C}$ of degree 2 (plus the null polynomial), it is easy to show that $V(P)$ is always nonempty, and in fact infinite. It can also be shown that $V(P) = V(Q)$ implies that $Q = \lambda P$ for some $\lambda \in \mathbb{C}$, with $\lambda \neq 0$ (see Samuel [23]). Another advantage of switching to the complex field $\mathbb{C}$ is that the theory of intersection is cleaner. Thus, any two circles that do not contain a common line always intersect in four points, some of which might be multiple points (as in the case of tangent circles). This may seem surprising, since in the real plane, two circles intersect in at most two points. Where are the other two points? They turn out to be the points $(1, i, 0)$ and $(1, -i, 0)$, as one can immediately verify. We can think of them as complex points at infinity! Not only are they at infinity, but they are not real. No wonder we cannot see them! We will come back to these points, called the *circular points*, in Section 5.11.

Going back to the vector space E over $\mathbb{R}$, it is worth saying that it can be shown that if $V(P) = V(Q)$ contains at least two points (in which case, $V(P)$ is actually infinite), then $Q = \lambda P$ for some $\lambda \in \mathbb{R}$ with $\lambda \neq 0$. Thus, even over $\mathbb{R}$, the mapping

$$[P] \mapsto V(P)$$

is injective whenever $V(P)$ is neither empty nor reduced to a single point. Note that the projective space $\mathbf{P}(E)$ of circles has dimension 3. In fact, it is easy to show that three distinct points that are not collinear determine a unique circle (see Samuel [23]).

In a similar vein, we can define the projective space of conics $\mathbf{P}(E)$ where E is the vector space (over $\mathbb{R}$) consisting of all homogeneous polynomials of degree 2 in x, y, z,

$$ax^2 + by^2 + cxy + dxz + eyz + fz^2$$

(plus the null polynomial). The curves $V(P)$ are indeed conics, perhaps degenerate. To see this, we can use the hyperplane model of $\mathbb{RP}^2$. The trace of $V(P)$ on the plane of equation $z = 1$ is the conic of equation

$$ax^2 + by^2 + cxy + dx + ey + f = 0.$$

Another way to see that $V(P)$ is a conic is to observe that in $\mathbb{R}^3$,

$$ax^2 + by^2 + cxy + dxz + eyz + fz^2 = 0$$

defines a cone with vertex $(0, 0, 0)$, and since its section by the plane $z = 1$ is a conic, all of its sections by planes are conics. The mapping

$$[P] \mapsto V(P)$$

is still injective when E is defined over the ground field $\mathbb{C}$, or if $V(P)$ has at least two points when E is defined over $\mathbb{R}$. Note that the projective space $\mathbf{P}(E)$ of conics has dimension 5. In fact, it is easy to show that five distinct points no four of which are not collinear determine a unique conic (see Samuel [23]).

It is also interesting to see what are lines in the space of circles or in the space of conics. In both cases we get pencils (of circles and conics, respectively). For more details, see Samuel [23], Sidler [24], Tisseron [26], Lehmann and Bkouche [20], Pedoe [21], Coxeter [7, 8], and Veblen and Young [28, 29].

We could also investigate algebraic plane curves of any degree m, by letting E be the vector space of homogeneous polynomials of degree m in x, y, z (plus the null polynomial). The zero locus $V(P)$ of P is defined just as before as

$$V(P) = \{(x, y, z) \in \mathbb{RP}^2 \mid P(x, y, z) = 0\}.$$

Observe that when $m = 1$, since homogeneous polynomials of degree 1 are linear forms, we are back to the case where $E = (\mathbb{R}^3)^*$, the dual space of $\mathbb{R}^3$, and $\mathbf{P}(E)$ can be identified with the set of lines in $\mathbb{RP}^2$. But when $m \geq 3$, things are even worse regarding the injectivity of the map $[P] \mapsto V(P)$. For instance, both $P = xy^2$ and $Q = x^2 y$ define the same union of two lines. It is necessary to consider *irreducible* curves, i.e., curves that are defined by irreducible polynomials, and to work over the field $\mathbb{C}$ of complex numbers (recall that a polynomial P is irreducible if it cannot be written as the product $P = Q_1 Q_2$ of two polynomials Q_1, Q_2 of degree ≥ 1).

We can also investigate algebraic surfaces in $\mathbb{RP}^3$ (or $\mathbb{CP}^3$), by letting E be the vector space of homogeneous polynomials of degree m in four variables x, y, z, t (plus the null polynomial). We can also consider the zero locus of a set of equations

$$\mathscr{E} = \{P_1 = 0, P_2 = 0, \ldots, P_n = 0\},$$

where $P_1, \ldots, P_n$ are homogeneous polynomials of degree m in x, y, z, t, defined as

$$V(\mathscr{E}) = \{(x, y, z, t) \in \mathbb{RP}^3 \mid P_i(x, y, z, t) = 0, \ 1 \leq i \leq n\}.$$

This way, we can also deal with space curves.

Finally, we can consider homogeneous polynomials $P(x_1, \ldots, x_{N+1})$ in $N + 1$ variables and of degree m (plus the null polynomial), and study the subsets of $\mathbb{RP}^N$ (or $\mathbb{CP}^N$) defined as the zero locus of a set of equations

$$\mathscr{E} = \{P_1 = 0, P_2 = 0, \ldots, P_n = 0\},$$

where $P_1, \ldots, P_n$ are homogeneous polynomials of degree m in the variables $x_1, \ldots,$ x_{N+1}. For example, it turns out that the set of lines in $\mathbb{RP}^3$ forms a surface of degree 2 in $\mathbb{RP}^5$ (the Klein quadric). However, all this would really take us too far into algebraic geometry, and we simply refer the interested reader to Fulton [15] or Harris [16].

We now consider projective maps.

5.5 Projective Maps

Given two nontrivial vector spaces E and F and a linear map $f \colon E \to F$, observe that for every $u, v \in (E - \operatorname{Ker} f)$, if $v = \lambda u$ for some $\lambda \in K - \{0\}$, then $f(v) = \lambda f(u)$, and thus f restricted to $(E - \operatorname{Ker} f)$ induces a function $\mathbf{P}(f) \colon (\mathbf{P}(E) - \mathbf{P}(\operatorname{Ker} f)) \to \mathbf{P}(F)$ defined such that

$$\mathbf{P}(f)([u]_\sim) = [f(u)]_\sim,$$

as in the following commutative diagram:

$$
\begin{array}{ccc}
E - \operatorname{Ker} f & \xrightarrow{\ f\ } & F - \{0\} \\
\downarrow{\scriptstyle p} & & \downarrow{\scriptstyle p} \\
\mathbf{P}(E) - \mathbf{P}(\operatorname{Ker} f) & \xrightarrow[\mathbf{P}(f)]{} & \mathbf{P}(F)
\end{array}
$$

When f is injective, i.e., when $\operatorname{Ker} f = \{0\}$, then $\mathbf{P}(f) \colon \mathbf{P}(E) \to \mathbf{P}(F)$ is indeed a well-defined function. The above discussion motivates the following definition.

Definition 5.5. Given two nontrivial vector spaces E and F, any linear map $f \colon E \to F$ induces a partial map $\mathbf{P}(f) \colon \mathbf{P}(E) \to \mathbf{P}(F)$ called a *projective map*, such that if $\operatorname{Ker} f = \{u \in E \mid f(u) = 0\}$ is the kernel of f, then $\mathbf{P}(f) \colon (\mathbf{P}(E) - \mathbf{P}(\operatorname{Ker} f)) \to \mathbf{P}(F)$ is a total map defined such that

$$\mathbf{P}(f)([u]_\sim) = [f(u)]_\sim,$$

as in the following commutative diagram:

$$
\begin{array}{ccc}
E - \operatorname{Ker} f & \xrightarrow{\ f\ } & F - \{0\} \\
\downarrow{\scriptstyle p} & & \downarrow{\scriptstyle p} \\
\mathbf{P}(E) - \mathbf{P}(\operatorname{Ker} f) & \xrightarrow[\mathbf{P}(f)]{} & \mathbf{P}(F)
\end{array}
$$

If f is injective, i.e., when $\operatorname{Ker} f = \{0\}$, then $\mathbf{P}(f) \colon \mathbf{P}(E) \to \mathbf{P}(F)$ is a total function called a *projective transformation*, and when f is bijective, we call $\mathbf{P}(f)$ a *projectivity, or projective isomorphism, or homography*. The set of projectivities $\mathbf{P}(f) \colon \mathbf{P}(E) \to \mathbf{P}(E)$ is a group called the *projective (linear) group*, and is denoted by $\mathbf{PGL}(E)$.

One should realize that if a linear map $f \colon E \to F$ is not injective, then the projective map $\mathbf{P}(f) \colon \mathbf{P}(E) \to \mathbf{P}(F)$ is only a *partial map*, i.e., it is undefined on $\mathbf{P}(\operatorname{Ker} f)$. In particular, if $f \colon E \to F$ is the null map (i.e., $\operatorname{Ker} f = E$), the domain of $\mathbf{P}(f)$ is empty and $\mathbf{P}(f)$ is the partial function undefined everywhere. We might want to require in Definition 5.5 that f not be the null map to avoid this degenerate case. Projective maps are often defined only when they are induced by bijective linear maps.

We take a closer look at the projectivities of the projective line $\mathbb{P}^1_K$, since they play a role in the "change of parameters" for projective curves. A projectivity $f\colon \mathbb{P}^1_K \to \mathbb{P}^1_K$ is induced by some bijective linear map $g\colon K^2 \to K^2$ given by some invertible matrix

$$M(g) = \begin{pmatrix} a & b \\ c & d \end{pmatrix}$$

with $ad - bc \neq 0$. Since the projective line $\mathbb{P}^1_K$ is isomorphic to $K \cup \{\infty\}$, it is easily verified that f is defined as follows:

$$c \neq 0 \begin{cases} z \mapsto \dfrac{az+b}{cz+d} & \text{if } z \neq -\dfrac{d}{c}, \\ -\dfrac{d}{c} \mapsto \infty, \\ \infty \mapsto \dfrac{a}{c}; \end{cases} \qquad c = 0 \begin{cases} z \mapsto \dfrac{az+b}{d}, \\ \infty \mapsto \infty. \end{cases}$$

If $K = \mathbb{R}$ or $K = \mathbb{C}$, note that a/c is the limit of $(az+b)/(cz+d)$, as z approaches infinity, and the limit of $(az+b)/(cz+d)$ as z approaches $-d/c$ is ∞ (when $c \neq 0$). Projections between hyperplanes form an important example of projectivities.

Definition 5.6. Given a projective space $\mathbf{P}(E)$, for any two distinct hyperplanes $\mathbf{P}(H)$ and $\mathbf{P}(H')$, for any point $c \in \mathbf{P}(E)$ neither in $\mathbf{P}(H)$ nor in $\mathbf{P}(H')$, the *projection (or perspectivity) of center c between $\mathbf{P}(H)$ and $\mathbf{P}(H')$* is the map $f\colon \mathbf{P}(H) \to \mathbf{P}(H')$ defined such that for every $a \in \mathbf{P}(H)$, the point $f(a)$ is the intersection of the line $\langle c, a \rangle$ through c and a with $\mathbf{P}(H')$.

Let us verify that f is well-defined and a bijective projective transformation. Since the hyperplanes $\mathbf{P}(H)$ and $\mathbf{P}(H')$ are distinct, the hyperplanes H and H' in E are distinct, and since c is neither in $\mathbf{P}(H)$ nor in $\mathbf{P}(H')$, letting $c = p(u)$ for some nonnull vector $u \in E$, then $u \notin H$ and $u \notin H'$, and thus $E = H \oplus Ku = H' \oplus Ku$. If $\pi\colon E \to H'$ is the linear map (projection onto H' parallel to u) defined such that

$$\pi(w + \lambda u) = w,$$

for all $w \in H'$ and all $\lambda \in K$, since $E = H \oplus Ku = H' \oplus Ku$, the restriction $g\colon H \to H'$ of $\pi\colon E \to H'$ to H is a linear bijection between H and H', and clearly $f = \mathbf{P}(g)$, which shows that f is a projectivity.

Remark: Going back to the linear map $\pi\colon E \to H'$ (projection onto H' parallel to u), note that $\mathbf{P}(\pi)\colon \mathbf{P}(E) \to \mathbf{P}(H')$ is also a projective map, but it is not injective, and thus only a partial map. More generally, given a direct sum $E = V \oplus W$, the projection $\pi\colon E \to V$ onto V parallel to W induces a projective map $\mathbf{P}(\pi)\colon \mathbf{P}(E) \to \mathbf{P}(V)$, and given another direct sum $E = U \oplus W$, the restriction of π to U induces a perspectivity f between $\mathbf{P}(U)$ and $\mathbf{P}(V)$. Geometrically, f is defined as follows: Given any point $a \in \mathbf{P}(U)$, if $\langle \mathbf{P}(W), a \rangle$ is the smallest projective subspace containing $\mathbf{P}(W)$ and a, the point $f(a)$ is the intersection of $\langle \mathbf{P}(W), a \rangle$ with $\mathbf{P}(V)$.

Figure 5.3 illustrates a projection f of center c between two projective lines Δ and Δ' (in the real projective plane).

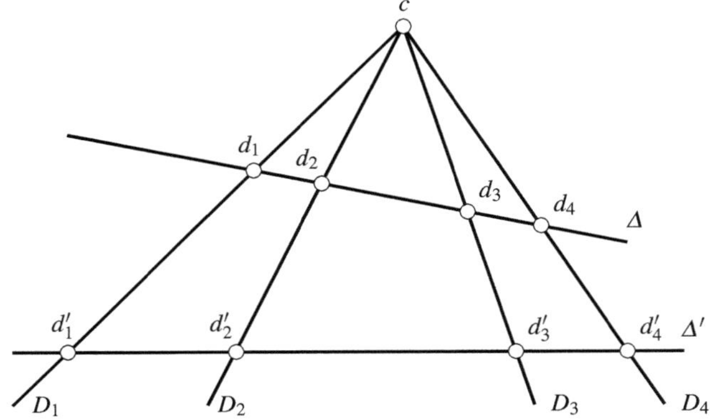

Fig. 5.3 A projection of center c between two lines Δ and Δ'.

If we consider three distinct points d_1, d_2, d_3 on Δ and their images d'_1, d'_2, d'_3 on Δ' under the projection f, then ratios are not preserved, that is,

$$\frac{\overrightarrow{d_3 d_1}}{\overrightarrow{d_3 d_2}} \neq \frac{\overrightarrow{d'_3 d'_1}}{\overrightarrow{d'_3 d'_2}}.$$

However, if we consider four distinct points d_1, d_2, d_3, d_4 on Δ and their images d'_1, d'_2, d'_3, d'_4 on Δ' under the projection f, we will show later that we have the following preservation of the so-called "cross-ratio"

$$\frac{\overrightarrow{d_3 d_1}}{\overrightarrow{d_3 d_2}} \bigg/ \frac{\overrightarrow{d_4 d_1}}{\overrightarrow{d_4 d_2}} = \frac{\overrightarrow{d'_3 d'_1}}{\overrightarrow{d'_3 d'_2}} \bigg/ \frac{\overrightarrow{d'_4 d'_1}}{\overrightarrow{d'_4 d'_2}}.$$

Cross-ratios and projections play an important role in geometry (for some very elegant illustrations of this fact, see Sidler [24]).

We now turn to the issue of determining when two linear maps f, g determine the same projective map, i.e., when $\mathbf{P}(f) = \mathbf{P}(g)$. The following lemma gives us a complete answer.

Lemma 5.4. *Given two nontrivial vector spaces E and F, for any two linear maps $f : E \to F$ and $g : E \to F$, we have $\mathbf{P}(f) = \mathbf{P}(g)$ iff there is some scalar $\lambda \in K - \{0\}$ such that $g = \lambda f$.*

Proof. If $g = \lambda f$, it is clear that $\mathbf{P}(f) = \mathbf{P}(g)$. Conversely, in order to have $\mathbf{P}(f) = \mathbf{P}(g)$, we must have $\operatorname{Ker} f = \operatorname{Ker} g$. If $\operatorname{Ker} f = \operatorname{Ker} g = E$, then f and g are both the null map, and this case is trivial. If $E - \operatorname{Ker} f \neq \emptyset$, by taking a basis of $\operatorname{Im} f$ and some inverse image of this basis, we obtain a basis B of a subspace G of E such that

$E = \operatorname{Ker} f \oplus G$. If $\dim(G) = 1$, the restriction of any linear map $f\colon E \to F$ to G is determined by some nonzero vector $u \in E$ and some scalar $\lambda \in K$, and the lemma is obvious. Thus, assume that $\dim(G) \geq 2$. For any two distinct basis vectors $u, v \in B$, since $\mathbf{P}(f) = \mathbf{P}(g)$, there must be some nonzero scalars $\lambda(u)$, $\lambda(v)$, and $\lambda(u+v)$ such that

$$g(u) = \lambda(u)f(u), \quad g(v) = \lambda(v)f(v), \quad g(u+v) = \lambda(u+v)f(u+v).$$

Since f and g are linear, we get

$$g(u) + g(v) = \lambda(u)f(u) + \lambda(v)f(v) = \lambda(u+v)(f(u)+f(v)),$$

that is,

$$(\lambda(u+v) - \lambda(u))f(u) + (\lambda(u+v) - \lambda(v))f(v) = 0.$$

Since f is injective on G and $u, v \in B \subseteq G$ are linearly independent, $f(u)$ and $f(v)$ are also linearly independent, and thus we have

$$\lambda(u+v) = \lambda(u) = \lambda(v).$$

Now we have shown that $\lambda(u) = \lambda(v)$, for any two distinct basis vectors in B, which proves that $\lambda(u)$ is independent of $u \in G$, and proves that $g = \lambda f$. $\square$

Lemma 5.4 shows that the projective linear group $\mathbf{PGL}(E)$ is isomorphic to the quotient group of the linear group $\mathbf{GL}(E)$ modulo the subgroup $K^* \mathrm{id}_E$ (where $K^* = K - \{0\}$). Using projective frames, we prove the following useful result.

Lemma 5.5. *Given two nontrivial vector spaces E and F of the same dimension $n+1$, for any two projective frames $(a_i)_{1 \leq i \leq n+2}$ for $\mathbf{P}(E)$ and $(b_i)_{1 \leq i \leq n+2}$ for $\mathbf{P}(F)$, there is a unique projectivity $h\colon \mathbf{P}(E) \to \mathbf{P}(F)$ such that $h(a_i) = b_i$ for $1 \leq i \leq n+2$.*

Proof. Let $(u_1, \ldots, u_{n+1})$ be a basis of E associated with the projective frame $(a_i)_{1 \leq i \leq n+2}$, and let $(v_1, \ldots, v_{n+1})$ be a basis of F associated with the projective frame $(b_i)_{1 \leq i \leq n+2}$. Since $(u_1, \ldots, u_{n+1})$ is a basis, there is a unique linear bijection $g\colon E \to F$ such that $g(u_i) = v_i$, for $1 \leq i \leq n+1$. Clearly, $h = \mathbf{P}(g)$ is a projectivity such that $h(a_i) = b_i$, for $1 \leq i \leq n+2$. Let $h'\colon \mathbf{P}(E) \to \mathbf{P}(F)$ be any projectivity such that $h'(a_i) = b_i$, for $1 \leq i \leq n+2$. By definition, there is a linear isomorphism $f\colon E \to F$ such that $h' = \mathbf{P}(f)$. Since $h'(a_i) = b_i$, for $1 \leq i \leq n+2$, we must have $f(u_i) = \lambda_i v_i$, for some $\lambda_i \in K - \{0\}$, where $1 \leq i \leq n+1$, and

$$f(u_1 + \cdots + u_{n+1}) = \lambda(v_1 + \cdots + v_{n+1}),$$

for some $\lambda \in K - \{0\}$. By linearity of f, we have

$$\lambda_1 v_1 + \cdots + \lambda_{n+1} v_{n+1} = \lambda v_1 + \cdots + \lambda v_{n+1},$$

and since $(v_1, \ldots, v_{n+1})$ is a basis of F, we must have

$$\lambda_1 = \cdots = \lambda_{n+1} = \lambda.$$

This shows that $f = \lambda g$, and thus that

$$h' = \mathbf{P}(f) = \mathbf{P}(g) = h,$$

and h is uniquely determined. $\square$

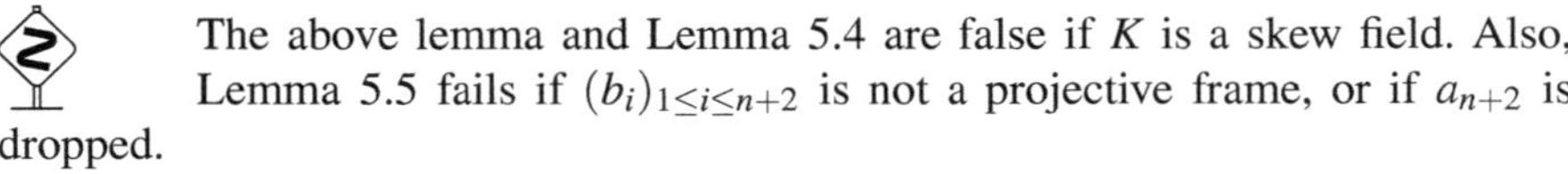 The above lemma and Lemma 5.4 are false if K is a skew field. Also, Lemma 5.5 fails if $(b_i)_{1 \leq i \leq n+2}$ is not a projective frame, or if a_{n+2} is dropped.

As a corollary of Lemma 5.5, given a projective space $\mathbf{P}(E)$, two distinct projective lines D and D' in $\mathbf{P}(E)$, three distinct points a, b, c on D, and any three distinct points a', b', c' on D', there is a unique projectivity from D to D', mapping a to a', b to b', and c to c'. This is because, as we mentioned earlier, any three distinct points on a line form a projective frame.

Remark: As in the affine case, there is "fundamental theorem of projective geometry." For simplicity, we state this theorem assuming that vector spaces are over the field $K = \mathbb{R}$. Given any two projective spaces $\mathbf{P}(E)$ and $\mathbf{P}(F)$ of the same dimension $n \geq 2$, for any bijective function $f\colon \mathbf{P}(E) \to \mathbf{P}(F)$, if f maps any three distinct collinear points a, b, c to collinear points $f(a), f(b), f(c)$, then f is a projectivity. For more general fields, $f = \mathbf{P}(g)$ for some "semilinear" bijection $g\colon E \to F$. A map such as f (preserving collinearity of any three distinct points) is often called a *collineation*. For $K = \mathbb{R}$, collineations and projectivities coincide. For more details, see Samuel [23].

Before closing this section, we illustrate the power of Lemma 5.5 by proving two interesting results. We begin by characterizing perspectivities between lines.

Lemma 5.6. *Given any two distinct lines D and D' in the real projective plane $\mathbb{RP}^2$, a projectivity $f\colon D \to D'$ is a perspectivity iff $f(O) = O$, where O is the intersection of D and D'.*

Proof. If $f\colon D \to D'$ is a perspectivity, then by the very definition of f, we have $f(O) = O$. Conversely, let $f\colon D \to D'$ be a projectivity such that $f(O) = O$. Let a, b be any two distinct points on D also distinct from O, and let $a' = f(a)$ and $b' = f(b)$ on D'. Since f is a bijection and since a, b, O are pairwise distinct, $a' \neq b'$. Let c be the intersection of the lines $\langle a, a' \rangle$ and $\langle b, b' \rangle$, which by the assumptions on a, b, O, cannot be on D or D'. Then we can define the perspectivity $g\colon D \to D'$ of center c, and by the definition of c, we have

$$g(a) = a', \quad g(b) = b', \quad g(O) = O.$$

However, f agrees with g on O, a, b, and since (O, a, b) is a projective frame for D, by Lemma 5.5, we must have $f = g$. $\square$

Using Lemma 5.6, we can give an elegant proof of a version of Desargues's theorem (in the plane).

Lemma 5.7. *Given two triangles* (a,b,c) *and* (a',b',c') *in* $\mathbb{RP}^2$, *where the points* a,b,c,a',b',c' *are pairwise distinct and the lines* $A = \langle b,c\rangle$, $B = \langle a,c\rangle$, $C = \langle a,b\rangle$, $A' = \langle b',c'\rangle$, $B' = \langle a',c'\rangle$, $C' = \langle a',b'\rangle$ *are pairwise distinct, if the lines* $\langle a,a'\rangle$, $\langle b,b'\rangle$, *and* $\langle c,c'\rangle$ *intersect in a common point* d *distinct from* a,b,c, a',b',c', *then the intersection points* $p = \langle b,c\rangle \cap \langle b',c'\rangle$, $q = \langle a,c\rangle \cap \langle a',c'\rangle$, *and* $r = \langle a,b\rangle \cap \langle a',b'\rangle$ *belong to a common line distinct from* A,B,C, A',B',C'.

Proof. In view of the assumptions on a,b,c, a',b',c', and d, the point r is on neither $\langle a,a'\rangle$ nor $\langle b,b'\rangle$, the point p is on neither $\langle b,b'\rangle$ nor $\langle c,c'\rangle$, and the point q is on neither $\langle a,a'\rangle$ nor $\langle c,c'\rangle$. It is also immediately shown that the line $\langle p,q\rangle$ is distinct from the lines A,B,C,A',B',C'. Let $f\colon \langle a,a'\rangle \to \langle b,b'\rangle$ be the perspectivity of center r and $g\colon \langle b,b'\rangle \to \langle c,c'\rangle$ be the perspectivity of center p. Let $h = g \circ f$. Since both $f(d) = d$ and $g(d) = d$, we also have $h(d) = d$. Thus by Lemma 5.6, the projectivity $h\colon \langle a,a'\rangle \to \langle c,c'\rangle$ is a perspectivity. Since

$$h(a) = g(f(a)) = g(b) = c,$$
$$h(a') = g(f(a')) = g(b') = c',$$

the intersection q of $\langle a,c\rangle$ and $\langle a',c'\rangle$ is the center of the perspectivity h. Also note that the point $m = \langle a,a'\rangle \cap \langle p,r\rangle$ and its image $h(m)$ are both on the line $\langle p,r\rangle$, since r is the center of f and p is the center of g. Since h is a perspectivity of center q, the line $\langle m,h(m)\rangle = \langle p,r\rangle$ passes through q, which proves the lemma. $\square$

Desargues's theorem is illustrated in Figure 5.4. It can also be shown that every projectivity between two distinct lines is the composition of two perspectivities (not in a unique way). An elegant proof of Pappus's theorem can also be given using perspectivities. For all this and more, the reader is referred to the problems.

We now consider the projective completion of an affine space.

5.6 Projective Completion of an Affine Space, Affine Patches

Given an affine space E with associated vector space $\overrightarrow{E}$, we can form the vector space $\widehat{E}$, the homogenized version of E, and then, the projective space $\mathbf{P}(\widehat{E})$ induced by $\widehat{E}$. This projective space, also denoted by $\widetilde{E}$, has some very interesting properties. In fact, it satisfies a universal property, but before we can say what it is, we have to take a closer look at $\widetilde{E}$.

Since the vector space $\widehat{E}$ is the disjoint union of elements of the form $\langle a,\lambda\rangle$, where $a \in E$ and $\lambda \in K - \{0\}$, and elements of the form $u \in \overrightarrow{E}$, observe that if $\sim$ is the equivalence relation on $\widehat{E}$ used to define the projective space $\mathbf{P}(\widehat{E})$, then the equivalence class $[\langle a,\lambda\rangle]_\sim$ of a weighted point contains the special representative

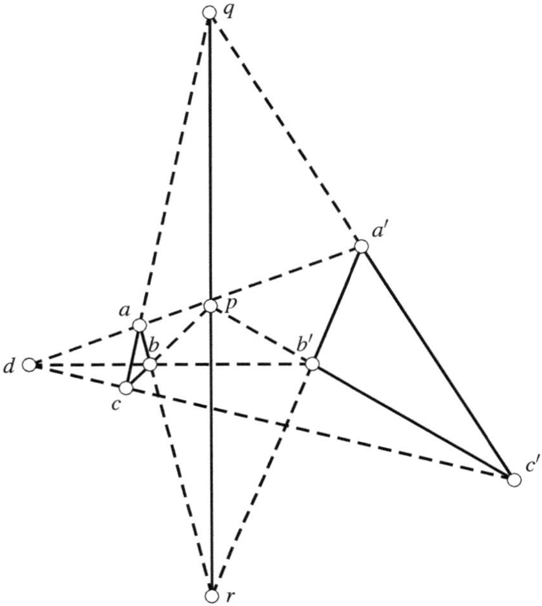

Fig. 5.4 Desargues's theorem (projective version in the plane).

$a = \langle a, 1 \rangle$, and the equivalence class $[u]_\sim$ of a nonzero vector $u \in \overrightarrow{E}$ is just a point of the projective space $\mathbf{P}(\overrightarrow{E})$. Thus, there is a bijection

$$\mathbf{P}(\widehat{E}) \longleftrightarrow E \cup \mathbf{P}(\overrightarrow{E})$$

between $\mathbf{P}(\widehat{E})$ and the disjoint union $E \cup \mathbf{P}(\overrightarrow{E})$, which allows us to view E as being embedded in $\mathbf{P}(\widehat{E})$. The points of $\mathbf{P}(\widehat{E})$ in $\mathbf{P}(\overrightarrow{E})$ will be called *points at infinity*, and the projective hyperplane $\mathbf{P}(\overrightarrow{E})$ is called the *hyperplane at infinity*. We will also denote the point $[u]_\sim$ of $\mathbf{P}(\overrightarrow{E})$ (where $u \neq 0$) by u_∞.

Thus, we can think of $\widetilde{E} = \mathbf{P}(\widehat{E})$ as the projective completion of the affine space E obtained by adding points at infinity forming the hyperplane $\mathbf{P}(\overrightarrow{E})$. As we commented in Section 5.2 when we presented the hyperplane model of $\mathbf{P}(E)$, the notion of point at infinity is really an affine notion. But even if a vector space E doesn't arise from the completion of an affine space, there is an affine structure on the complement of any hyperplane $\mathbf{P}(H)$ in the projective space $\mathbf{P}(E)$. In the case of $\widetilde{E}$, the complement E of the projective hyperplane $\mathbf{P}(\overrightarrow{E})$ is indeed an affine space. This is a general property that is needed in order to figure out the universal property of $\widetilde{E}$.

Lemma 5.8. *Given a vector space E and a hyperplane H in E, the complement $E_H = \mathbf{P}(E) - \mathbf{P}(H)$ of the projective hyperplane $\mathbf{P}(H)$ in the projective space $\mathbf{P}(E)$ can be given an affine structure such that the associated vector space of E_H is H. The affine structure on E_H depends only on H, and under this affine structure, E_H is isomorphic to an affine hyperplane in E.*

Proof. Since H is a hyperplane in E, there is some $w \in E - H$ such that $E = Kw \oplus H$. Thus, every vector u in $E - H$ can be written in a unique way as $\lambda w + h$, where $\lambda \neq 0$ and $h \in H$. As a consequence, for every point $[u]$ in E_H, the equivalence class $[u]$ contains a representative of the form $w + \lambda^{-1}h$, with $\lambda \neq 0$. Then we see that the map $\varphi \colon (w+H) \to E_H$, defined such that

$$\varphi(w+h) = [w+h],$$

is a bijection. In order to define an affine structure on E_H, we define $+ \colon E_H \times H \to E_H$ as follows: For every point $[w+h_1] \in E_H$ and every $h_2 \in H$, we let

$$[w+h_1] + h_2 = [w+h_1+h_2].$$

The axioms of an affine space are immediately verified. Now, $w + H$ is an affine hyperplane is E, and under the affine structure just given to E_H, the map $\varphi \colon (w+H) \to E_H$ is an affine map that is bijective. Thus, E_H is isomorphic to the affine hyperplane $w + H$. If we had chosen a different vector $w' \in E - H$ such that $E = Kw' \oplus H$, then E_H would be isomorphic to the affine hyperplane $w' + H$ parallel to $w + H$. But these two hyperplanes are clearly isomorphic by translation, and thus the affine structure on E_H depends only on H. $\quad\square$

An affine space of the form E_H is called an *affine patch* on $\mathbf{P}(E)$. Lemma 5.8 allows us to view a projective space $\mathbf{P}(E)$ as the result of gluing some affine spaces together, at least when E is of finite dimension. For example, when E is of dimension 2, a hyperplane in E is just a line, and the complement of a point in the projective line $\mathbf{P}(E)$ can be viewed as an affine line. Thus, we can view $\mathbf{P}(E)$ as being covered by two affine lines glued together. When $K = \mathbb{R}$, this shows that topologically, the projective line $\mathbb{RP}^1$ is equivalent to a circle. When E is of dimension 3, a hyperplane in E is just a plane, and the complement of a projective line in the projective plane $\mathbf{P}(E)$ can be viewed as an affine plane. Thus, we can view $\mathbf{P}(E)$ as being covered by three affine planes glued together. However, even when $K = \mathbb{R}$, it is much more difficult to come up with a geometric embedding of the projective plane $\mathbb{RP}^2$ in $\mathbb{A}^3$, and in fact, this is impossible! Nevertheless, there are some fascinating immersions of the projective space $\mathbb{RP}^2$ as $3D$ surfaces with self-intersection, one of which is known as the Boy surface. We urge our readers to consult the remarkable book by Hilbert and Cohn-Vossen [17] for drawings of the Boy surface, and more. Some nice projections in $\mathbb{A}^3$ of an embedding of $\mathbb{RP}^2$ into $\mathbb{A}^4$ are given in the surface gallery on the web page (see `http://www.cis.upenn.edu/~jean/gbooks/geom2.html`, Section 24.7). In fact, we give a control net in $\mathbb{A}^4$ specifying an explicit rational surface homeomorphic to $\mathbb{RP}^2$. One should also consult Fischer's books [12, 11], where many beautiful models of surfaces are displayed, and the commentaries in

Chapter 6 of [11] regarding models of $\mathbb{RP}^2$. More generally, when E is of dimension $n+1$, the projective space $\mathbf{P}(E)$ is covered by $n+1$ affine patches (hyperplanes) glued together. This idea is very fruitful, since it allows the treatment of projective spaces as manifolds, and it is essential in algebraic geometry.

We can now go back to the projective completion $\widetilde{E}$ of an affine space E.

Definition 5.7. Given any affine space E with associated vector space $\overrightarrow{E}$, a *projective completion of the affine space E with hyperplane at infinity $\mathbf{P}(\mathscr{H})$* is a triple $\langle \mathbf{P}(\mathscr{E}), \mathbf{P}(\mathscr{H}), i \rangle$, where $\mathscr{E}$ is a vector space, $\mathscr{H}$ is a hyperplane in $\mathscr{E}$, $i\colon E \to \mathbf{P}(\mathscr{E})$ is an injective map such that $i(E) = \mathscr{E}_{\mathscr{H}}$ and i is affine (where $\mathscr{E}_{\mathscr{H}} = \mathbf{P}(\mathscr{E}) - \mathbf{P}(\mathscr{H})$ is an affine patch), and for every projective space $\mathbf{P}(F)$, every hyperplane H in F, and every map $f\colon E \to \mathbf{P}(F)$ such that $f(E) \subseteq F_H$ and f is affine (where $F_H = \mathbf{P}(F) - \mathbf{P}(H)$ is an affine patch), there is a unique projective map $\widetilde{f}\colon \mathbf{P}(\mathscr{E}) \to \mathbf{P}(F)$ such that

$$f = \widetilde{f} \circ i \quad \text{and} \quad \mathbf{P}(\overrightarrow{f}) = \widetilde{f} \circ \mathbf{P}(i)$$

(where $i\colon \overrightarrow{E} \to \mathscr{H}$ and $\overrightarrow{f}\colon \overrightarrow{E} \to H$ are the linear maps associated with the affine maps $i\colon E \to \mathbf{P}(\mathscr{E})$ and $f\colon E \to \mathbf{P}(F)$), as in the following diagram:

$$
\begin{array}{ccccc}
E & \xrightarrow{\ i\ } & \mathscr{E}_{\mathscr{H}} \subseteq \mathbf{P}(\mathscr{E}) \supseteq \mathbf{P}(\mathscr{H}) & \xleftarrow{\ \mathbf{P}(i)\ } & \mathbf{P}(\overrightarrow{E}) \\
& {}_{f}\searrow & \downarrow{\widetilde{f}} & \swarrow{\mathbf{P}(\overrightarrow{f})} & \\
& & F_H \subseteq \mathbf{P}(F) \supseteq \mathbf{P}(H) & &
\end{array}
$$

The points of $\mathbf{P}(\mathscr{E})$ in $\mathbf{P}(\mathscr{H})$ are called *points at infinity*, and the projective hyperplane $\mathbf{P}(\mathscr{H})$ is called the *hyperplane at infinity*. We will also denote the point $[u]_\sim$ of $\mathbf{P}(\mathscr{H})$ (where $u \neq 0$) by u_∞. As usual, objects defined by a universal property are unique up to isomorphism. We leave the proof as an exercise. The importance of the notion of projective completion stems from the fact that every affine map $f\colon E \to F$ extends *in a unique way* to a projective map $\widetilde{f}\colon \widetilde{E} \to \widetilde{F}$ (provided that the restriction of $\widetilde{f}$ to $\mathbf{P}(\overrightarrow{E})$ agrees with $\mathbf{P}(\overrightarrow{f})$).

We will now show that $\langle \widetilde{E}, \mathbf{P}(\overrightarrow{E}), i \rangle$ is the projective completion of E, where $i\colon E \to \widetilde{E}$ is the injection of E into $\widetilde{E} = E \cup \mathbf{P}(\overrightarrow{E})$. For example, if $E = \mathbb{A}^1_K$ is an affine line, its projective completion $\widetilde{\mathbb{A}^1_K}$ is isomorphic to the projective line $\mathbf{P}(K^2)$, and they both can be identified with $\mathbb{A}^1_K \cup \{\infty\}$, the result of adding a point at infinity (∞) to $\mathbb{A}^1_K$. In general, the projective completion $\widetilde{\mathbb{A}^m_K}$ of the affine space $\mathbb{A}^m_K$ is isomorphic to $\mathbf{P}(K^{m+1})$. Thus, $\widetilde{\mathbb{A}^m}$ is isomorphic to $\mathbb{RP}^m$, and $\widetilde{\mathbb{A}^m_\mathbb{C}}$ is isomorphic to $\mathbb{CP}^m$.

First, let us observe that if E is a vector space and H is a hyperplane in E, then the homogenization $\widehat{E_H}$ of the affine patch E_H (the complement of the projective hyperplane $\mathbf{P}(H)$ in $\mathbf{P}(E)$) is isomorphic to E. The proof is rather simple and uses the fact that there is an affine bijection between E_H and the affine hyperplane $w + H$

in E, where $w \in E - H$ is any fixed vector. Choosing w as an origin in E_H, we know that $\widehat{E_H} = H \,\widehat{+}\, Kw$, and since $E = H \oplus Kw$, it is obvious how to define a linear bijection between $\widehat{E_H} = H \,\widehat{+}\, Kw$ and $E = H \oplus Kw$. As a consequence the projective spaces $\widetilde{E_H}$ and $\mathbf{P}(E)$ are isomorphic, i.e., there is a projectivity between them.

Lemma 5.9. *Given any affine space $(E, \overrightarrow{E})$, for every projective space $\mathbf{P}(F)$, every hyperplane H in F, and every map $f\colon E \to \mathbf{P}(F)$ such that $f(E) \subseteq F_H$ and f is affine (F_H being viewed as an affine patch), there is a unique projective map $\widetilde{f}\colon \widetilde{E} \to \mathbf{P}(F)$ such that*

$$f = \widetilde{f} \circ i \quad and \quad \mathbf{P}(\overrightarrow{f}) = \widetilde{f} \circ \mathbf{P}(i),$$

(where $i\colon \overrightarrow{E} \to \overrightarrow{\widetilde{E}}$ and $\overrightarrow{f}\colon \overrightarrow{E} \to H$ are the linear maps associated with the affine maps $i\colon E \to \widetilde{E}$ and $f\colon E \to \mathbf{P}(F)$), as in the following diagram:

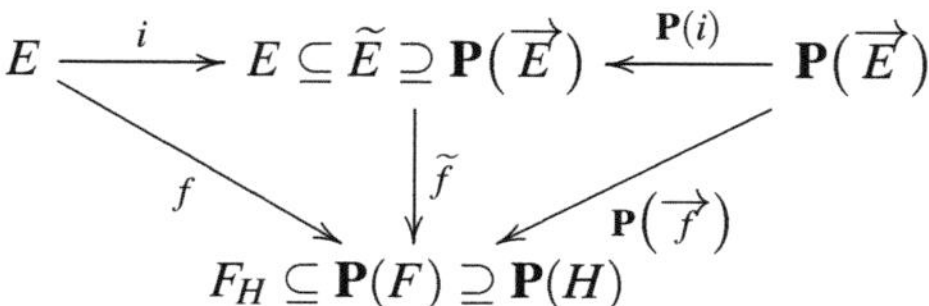

Proof. The existence of $\widetilde{f}$ is a consequence of Lemma 4.5, where we observe that $\widehat{F_H}$ is isomorphic to F. Just take the projective map $\mathbf{P}(\widehat{f})\colon \widetilde{E} \to \mathbf{P}(F)$, where $\widehat{f}\colon \widehat{E} \to F$ is the unique linear map extending f. It remains to prove its uniqueness. Since $f\colon E \to F_H$ is affine, for any $a \in E$ and any $u \in \overrightarrow{E}$, we have

$$f(a + u) = f(a) + \overrightarrow{f}(u),$$

where $\overrightarrow{f}\colon \overrightarrow{E} \to H$ is a linear map. If we fix some $a \in E$, then $f(a) = [w]$, for some $w \in F - H$ and $F = Kw \oplus H$. Assume that $\widetilde{f}\colon \widetilde{E} \to \mathbf{P}(F)$ exists with the desired property. Then there is some linear map $g\colon \widehat{E} \to F$ such that $\widetilde{f} = \mathbf{P}(g)$. Since $f = \widetilde{f} \circ i$, we must have $f(a) = [w] = [g(a)]$, and thus $g(a) = \mu w$, for some $\mu \neq 0$. Also, for every $u \in \overrightarrow{E}$,

$$f(a + u) = [w] + \overrightarrow{f}(u) = [w + \overrightarrow{f}(u)] = [g(a + u)]$$
$$= [g(a) + g(u)] = [\mu w + g(u)],$$

and thus we must have

$$\lambda(u)w + \lambda(u)\overrightarrow{f}(u) = \mu w + g(u),$$

for some $\lambda(u) \neq 0$. If $\mathrm{Ker}\,\overrightarrow{f} = \overrightarrow{E}$, the linear map $\overrightarrow{f}$ is the null map, and since we are requiring that the restriction of $\widetilde{f}$ to $\mathbf{P}(\overrightarrow{E})$ be equal to $\mathbf{P}(\overrightarrow{f})$, the linear map g

must also be the null map on $\overrightarrow{E}$. Thus, $\widetilde{f}$ is unique, and the restriction of $\widetilde{f}$ to $\mathbf{P}(\overrightarrow{E})$ is the partial map undefined everywhere.

If $\overrightarrow{E} - \mathrm{Ker}\,\overrightarrow{f} \neq \emptyset$, by taking a basis of $\mathrm{Im}\,\overrightarrow{f}$ and some inverse image of this basis, we obtain a basis B of a subspace $\overrightarrow{G}$ of $\overrightarrow{E}$ such that $\overrightarrow{E} = \mathrm{Ker}\,\overrightarrow{f} \oplus \overrightarrow{G}$. Since $\overrightarrow{E} = \mathrm{Ker}\,\overrightarrow{f} \oplus \overrightarrow{G}$ where $\dim(\overrightarrow{G}) \geq 1$, for any $x \in \mathrm{Ker}\,\overrightarrow{f}$ and any nonnull vector $y \in \overrightarrow{G}$, we have

$$\lambda(x)w = \mu w + g(x),$$
$$\lambda(y)w + \lambda(y)\overrightarrow{f}(y) = \mu w + g(y),$$

and

$$\lambda(x+y)w + \lambda(x+y)\overrightarrow{f}(x+y) = \mu w + g(x+y),$$

which by linearity yields

$$(\lambda(x+y) - \lambda(x) - \lambda(y) + \mu)w + (\lambda(x+y) - \lambda(y))\overrightarrow{f}(y) = 0.$$

Since $F = Kw \oplus H$ and $\overrightarrow{f}\colon \overrightarrow{E} \to H$, we must have $\lambda(x+y) = \lambda(y)$ and $\lambda(x) = \mu$. Thus, g agrees with $\overrightarrow{f}$ on $\mathrm{Ker}\,\overrightarrow{f}$.

If $\dim(\overrightarrow{G}) = 1$ then for any $y \in \overrightarrow{G}$ we have

$$\lambda(y)w + \lambda(y)\overrightarrow{f}(y) = \mu w + g(y),$$

and for any $\nu \neq 0$ we have

$$\lambda(\nu y)w + \lambda(\nu y)\overrightarrow{f}(\nu y) = \mu w + g(\nu y),$$

which by linearity yields

$$(\lambda(\nu y) - \nu\lambda(y) - \mu + \nu\mu)w + (\nu\lambda(\nu y) - \nu\lambda(y))\overrightarrow{f}(y) = 0.$$

Since $F = Kw \oplus H$, $\overrightarrow{f}\colon \overrightarrow{E} \to H$, and $\nu \neq 0$, we must have $\lambda(\nu y) = \lambda(y)$. Then we must also have $(\lambda(y) - \mu)(1 - \nu) = 0$.

If $K = \{0, 1\}$, since the only nonzero scalar is 1, it is immediate that $g(y) = \overrightarrow{f}(y)$, and we are done. Otherwise, for $\nu \neq 0, 1$, we get $\lambda(y) = \mu$ for all $y \in \overrightarrow{G}$. Then $g = \mu\overrightarrow{f}$ on $\overrightarrow{E}$, and the restriction of $\widetilde{f} = \mathbf{P}(g)$ to $\mathbf{P}(\overrightarrow{E})$ is equal to $\mathbf{P}(\overrightarrow{f})$. But now g is completely determined by

$$g(u \widehat{+} \lambda a) = \lambda g(a) + g(u) = \lambda\mu w + \mu\overrightarrow{f}(u).$$

Thus, we have $g = \lambda\widehat{f}$.

Otherwise, if $\dim\left(\overrightarrow{G}\right) \geq 2$, then for any two distinct basis vectors u and v in B,

$$\lambda(u)w + \lambda(u)\overrightarrow{f}(u) = \mu w + g(u),$$
$$\lambda(v)w + \lambda(v)\overrightarrow{f}(v) = \mu w + g(v),$$

and

$$\lambda(u+v)w + \lambda(u+v)\overrightarrow{f}(u+v) = \mu w + g(u+v),$$

and by linearity, we get

$$(\lambda(u+v) - \lambda(u) - \lambda(v) + \mu)w + (\lambda(u+v) - \lambda(u))\overrightarrow{f}(u)$$
$$+ (\lambda(u+v) - \lambda(v))\overrightarrow{f}(v) = 0.$$

Since $F = Kw \oplus H$, $\overrightarrow{f} \colon \overrightarrow{E} \to H$, and $\overrightarrow{f}(u)$ and $\overrightarrow{f}(v)$ are linearly independent (because $\overrightarrow{f}$ in injective on $\overrightarrow{G}$), we must have

$$\lambda(u+v) = \lambda(u) = \lambda(v) = \mu,$$

which implies that $g = \mu\overrightarrow{f}$ on $\overrightarrow{E}$, and the restriction of $\widetilde{f} = \mathbf{P}(g)$ to $\mathbf{P}\left(\overrightarrow{E}\right)$ is equal to $\mathbf{P}\left(\overrightarrow{f}\right)$. As in the previous case, g is completely determined by

$$g(u \mathbin{\widehat{+}} \lambda a) = \lambda g(a) + g(u) = \lambda\mu w + \mu\overrightarrow{f}(u).$$

Again, we have $g = \lambda\widehat{f}$, and thus $\widetilde{f}$ is unique. $\square$

The requirement that the restriction of $\widetilde{f} = \mathbf{P}(g)$ to $\mathbf{P}\left(\overrightarrow{E}\right)$ be equal to $\mathbf{P}\left(\overrightarrow{f}\right)$ is necessary for the uniqueness of $\widetilde{f}$. The problem comes up when f is a constant map. Indeed, if f is the constant map defined such that $f(a) = [w]$ for some fixed vector $w \in F$, it can be shown that any linear map $g \colon \widehat{E} \to F$ defined such that $g(a) = \mu w$ and $g(u) = \varphi(u)w$ for all $u \in \overrightarrow{E}$, for some $\mu \neq 0$, and some linear form $\varphi \colon \overrightarrow{E} \to F$ satisfies $f = \mathbf{P}(g) \circ i$.

Lemma 5.9 shows that $\left\langle \widetilde{E}, \mathbf{P}\left(\overrightarrow{E}\right), i \right\rangle$ is the projective completion of the affine space E.

The projective completion $\widetilde{E}$ of an affine space E is a very handy place in which to do geometry in, mainly because the following facts can be easily established.

There is a bijection between affine subspaces of E and projective subspaces of $\widetilde{E}$ not contained in $\mathbf{P}\left(\overrightarrow{E}\right)$. Two affine subspaces of E are parallel iff the corresponding projective subspaces of $\widetilde{E}$ have the same intersection with the hyperplane at infinity $\mathbf{P}\left(\overrightarrow{E}\right)$. There is also a bijection between affine maps from E to F and projective maps from $\widetilde{E}$ to $\widetilde{F}$ mapping the hyperplane at infinity $\mathbf{P}\left(\overrightarrow{E}\right)$ into the hyperplane at infinity $\mathbf{P}(\overrightarrow{F})$. In the projective plane, two distinct lines intersect in a single point (possibly at infinity, when the lines are parallel). In the projective space, two distinct planes intersect in a single line (possibly at infinity, when the planes are parallel). In the projective space, a plane and a line not contained in that plane intersect in a single point (possibly at infinity, when the plane and the line are parallel).

5.7 Making Good Use of Hyperplanes at Infinity

Given a vector space E and a hyperplane H in E, we have already observed that the projective spaces $\widetilde{E_H}$ and $\mathbf{P}(E)$ are isomorphic. Thus, $\mathbf{P}(H)$ can be viewed as the hyperplane at infinity in $\mathbf{P}(E)$, and the considerations applying to the projective completion of an affine space apply to the affine patch E_H on $\mathbf{P}(E)$. This fact yields a powerful and elegant method for proving theorems in projective geometry. The general schema is to choose some projective hyperplane $\mathbf{P}(H)$ in $\mathbf{P}(E)$, view it as the "hyperplane at infinity," then prove an affine version of the desired result in the affine patch E_H (the complement of $\mathbf{P}(H)$ in $\mathbf{P}(E)$, which has an affine structure), and then transfer this result back to the projective space $\mathbf{P}(E)$. This technique is often called "sending objects to infinity." We refer the reader to geometry textbooks for a comprehensive development of these ideas (for example, Berger [3, 4], Samuel [23], Sidler [24], Tisseron [26], or Pedoe [21]), but we cannot resist presenting the projective versions of the theorems of Pappus and Desargues. Indeed, the method of sending points to infinity provides some strikingly elegant proofs. We begin with Pappus's theorem, illustrated in Figure 5.5.

Lemma 5.10. *Given any projective plane $\mathbf{P}(E)$ and any two distinct lines D and D', for any distinct points a, b, c, a', b', c', with a, b, c on D and a', b', c' on D', if a, b, c, a', b', c' are distinct from the intersection of D and D', then the intersection points $p = \langle b, c' \rangle \cap \langle b', c \rangle$, $q = \langle a, c' \rangle \cap \langle a', c \rangle$, and $r = \langle a, b' \rangle \cap \langle a', b \rangle$ are collinear.*

Proof. First, since any two lines in a projective plane intersect in a single point, the points p, q, r are well defined. Choose $\Delta = \langle p, r \rangle$ as the line at infinity, and consider the affine plane $X = \mathbf{P}(E) - \Delta$. Since $\langle a, b' \rangle$ and $\langle a', b \rangle$ intersect at a point at infinity r on Δ, $\langle a, b' \rangle$ and $\langle a', b \rangle$ are parallel, and similarly $\langle b, c' \rangle$ and $\langle b', c \rangle$ are parallel. Thus, by the affine version of Pappus's theorem (Lemma 2.11), the lines $\langle a, c' \rangle$ and $\langle a', c \rangle$ are parallel, which means that their intersection q is on the line at infinity $\Delta = \langle p, r \rangle$, which means that p, q, r are collinear. $\quad\square$

By working in the projective completion of an affine plane, we can obtain an improved version of Pappus's theorem for affine planes. The reader will have to figure out how to deal with the special cases where some of p, q, r go to infinity.

Now, we prove a projective version of Desargues's theorem slightly more general than that given in Lemma 5.7. It is interesting that the proof is radically different, depending on the dimension of the projective space $\mathbf{P}(E)$. This is not surprising. In axiomatic presentations of projective plane geometry, Desargues's theorem is independent of the other axioms. Desargues's theorem is illustrated in Figure 5.6.

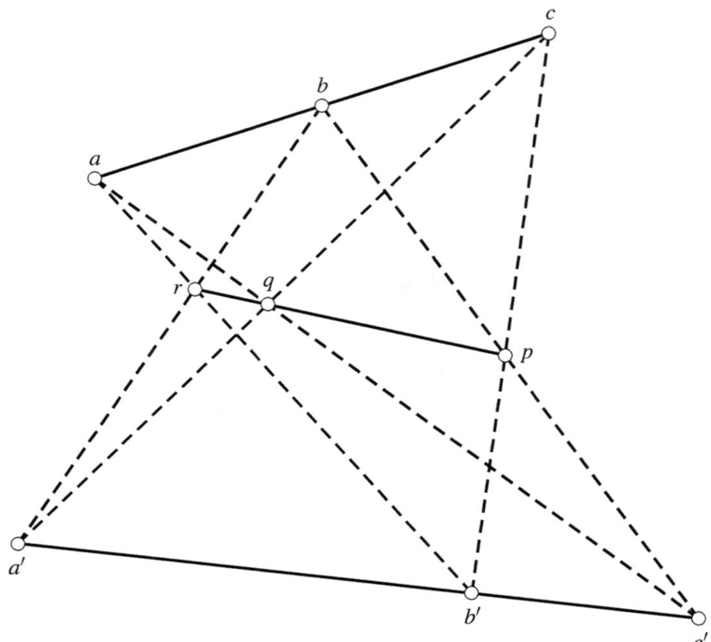

Fig. 5.5 Pappus's theorem (projective version).

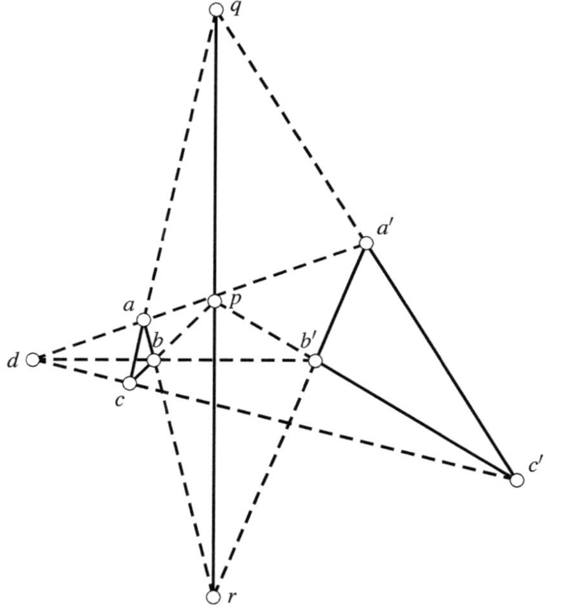

Fig. 5.6 Desargues's theorem (projective version).

Lemma 5.11. *Let* $\mathbf{P}(E)$ *be a projective space. Given two triangles* (a,b,c) *and* (a',b',c'), *where the points* a,b,c,a',b',c' *are pairwise distinct and the lines* $A = \langle b,c \rangle$, $B = \langle a,c \rangle$, $C = \langle a,b \rangle$, $A' = \langle b',c' \rangle$, $B' = \langle a',c' \rangle$, $C' = \langle a',b' \rangle$ *are pairwise distinct, if the lines* $\langle a,a' \rangle$, $\langle b,b' \rangle$, *and* $\langle c,c' \rangle$ *intersect in a common point* d *distinct from* a,b,c, a',b',c', *then the intersection points* $p = \langle b,c \rangle \cap \langle b',c' \rangle$, $q = \langle a,c \rangle \cap \langle a',c' \rangle$, *and* $r = \langle a,b \rangle \cap \langle a',b' \rangle$ *belong to a common line distinct from* A,B,C,A',B',C'.

Proof. First, it is immediately shown that the line $\langle p,q \rangle$ is distinct from the lines A,B,C,A',B',C'. Let us assume that $\mathbf{P}(E)$ has dimension $n \geq 3$. If the seven points d,a,b,c,a',b',c' generate a projective subspace of dimension 3, then by Lemma 5.1, the intersection of the two planes $\langle a,b,c \rangle$ and $\langle a',b',c' \rangle$ is a line, and thus p,q,r are collinear.

If $\mathbf{P}(E)$ has dimension $n = 2$ or the seven points d,a,b,c,a',b',c' generate a projective subspace of dimension 2, we use the following argument. In the projective plane X generated by the seven points d,a,b,c,a',b',c', choose the projective line $\Delta = \langle p,r \rangle$ as the line at infinity. Then in the affine plane $Y = X - \Delta$, the lines $\langle b,c \rangle$ and $\langle b',c' \rangle$ are parallel, and the lines $\langle a,b \rangle$ and $\langle a',b' \rangle$ are parallel, and the lines $\langle a,a' \rangle$, $\langle b,b' \rangle$, and $\langle c,c' \rangle$ are either parallel or concurrent. Then by the converse of the affine version of Desargues's theorem (Lemma 2.12), the lines $\langle a,c \rangle$ and $\langle a',c' \rangle$ are parallel, which means that their intersection q belongs to the line at infinity $\Delta = \langle p,r \rangle$, and thus that p,q,r are collinear. $\square$

The converse of Desargues's theorem also holds (see the problems). Using the projective completion of an affine space, it is easy to state an improved affine version of Desargues's theorem. The reader will have to figure out how to deal with the case where some of the points p,q,r go to infinity. It can also be shown that Pappus's theorem implies Desargues's theorem. Many results of projective or affine geometry can be obtained using the method of "sending points to infinity."

We now discuss briefly the notion of cross-ratio, since it is a major concept of projective geometry.

5.8 The Cross-Ratio

Recall that affine maps preserve the ratio of three collinear points. In general, projective maps do not preserve the ratio of three collinear points. However, bijective projective maps preserve the "ratio of ratios" of any four collinear points (three of which are distinct). Such ratios are called *cross-ratios* (in French, "birapport"). There are several ways of introducing cross-ratios, but since we already have Lemma 5.5 at our disposal, we can circumvent some of the tedious calculations needed if other approaches are chosen.

Given a field K, say $K = \mathbb{R}$, recall that the projective line $\mathbb{P}^1_K$ consists of all equivalence classes $[x,y]$ of pairs $(x,y) \in K^2$ such that $(x,y) \neq (0,0)$, under the equivalence relation $\sim$ defined such that

$$(x_1, y_1) \sim (x_2, y_2) \quad \text{iff} \quad x_2 = \lambda x_1 \quad \text{and} \quad y_2 = \lambda y_1,$$

for some $\lambda \in K - \{0\}$. Letting $\infty = [1, 0]$, the projective line $\mathbb{P}^1_K$ is in bijection with $K \cup \{\infty\}$. Furthermore, letting $0 = [0, 1]$ and $1 = [1, 1]$, the triple $(\infty, 0, 1)$ forms a projective frame for $\mathbb{P}^1_K$. Using this projective frame and Lemma 5.5, we define the cross-ratio of four collinear points as follows.

Definition 5.8. Given a projective line $\Delta = \mathbf{P}(D)$ over a field K, for any sequence (a, b, c, d) of four points in Δ, where a, b, c are distinct (i.e., (a, b, c) is a projective frame), the *cross-ratio* $[a, b, c, d]$ is defined as the element $h(d) \in \mathbb{P}^1_K$, where $h \colon \Delta \to \mathbb{P}^1_K$ is the unique projectivity such that $h(a) = \infty$, $h(b) = 0$, and $h(c) = 1$ (which exists by Lemma 5.5, since (a, b, c) is a projective frame for Δ and $(\infty, 0, 1)$ is a projective frame for $\mathbb{P}^1_K$). For any projective space $\mathbf{P}(E)$ (of dimension ≥ 2) over a field K and any sequence (a, b, c, d) of four collinear points in $\mathbf{P}(E)$, where a, b, c are distinct, the cross-ratio $[a, b, c, d]$ is defined using the projective line Δ that the points a, b, c, d define. For any affine space E and any sequence (a, b, c, d) of four collinear points in E, where a, b, c are distinct, the cross-ratio $[a, b, c, d]$ is defined by considering E as embedded in $\widetilde{E}$.

It should be noted that the definition of the cross-ratio $[a, b, c, d]$ depends on the order of the points. Thus, there could be $24 = 4!$ different possible values depending on the permutation of $\{a, b, c, d\}$. In fact, there are at most 6 distinct values. Also, note that $[a, b, c, d] = \infty$ iff $d = a$, $[a, b, c, d] = 0$ iff $d = b$, and $[a, b, c, d] = 1$ iff $d = c$. Thus, $[a, b, c, d] \in K - \{0, 1\}$ iff $d \notin \{a, b, c\}$.

The following lemma is almost obvious, but very important. It shows that projectivities between projective lines are characterized by the preservation of the cross-ratio of any four points (three of which are distinct).

Lemma 5.12. *Given any two projective lines Δ and Δ', for any sequence (a, b, c, d) of points in Δ and any sequence (a', b', c', d') of points in Δ', if a, b, c are distinct and a', b', c' are distinct, there is a unique projectivity $f \colon \Delta \to \Delta'$ such that $f(a) = a'$, $f(b) = b'$, $f(c) = c'$, and $f(d) = d'$ iff $[a, b, c, d] = [a', b', c', d']$.*

Proof. First, assume that $f \colon \Delta \to \Delta'$ is a projectivity such that $f(a) = a'$, $f(b) = b'$, $f(c) = c'$, and $f(d) = d'$. Let $h \colon \Delta \to \mathbb{P}^1_K$ be the unique projectivity such that $h(a) = \infty$, $h(b) = 0$, and $h(c) = 1$, and let $h' \colon \Delta' \to \mathbb{P}^1_K$ be the unique projectivity such that $h'(a') = \infty$, $h'(b') = 0$, and $h'(c') = 1$. By definition, $[a, b, c, d] = h(d)$ and $[a', b', c', d'] = h'(d')$. However, $h' \circ f \colon \Delta \to \mathbb{P}^1_K$ is a projectivity such that $(h' \circ f)(a) = \infty$, $(h' \circ f)(b) = 0$, and $(h' \circ f)(c) = 1$, and by the uniqueness of h, we get $h = h' \circ f$. But then, $[a, b, c, d] = h(d) = h'(f(d)) = h'(d') = [a', b', c', d']$.

Conversely, assume that $[a, b, c, d] = [a', b', c', d']$. Since (a, b, c) and (a', b', c') are projective frames, by Lemma 5.5, there is a unique projectivity $g \colon \Delta \to \Delta'$ such that $g(a) = a'$, $g(b) = b'$, and $g(c) = c'$. Now, $h' \circ g \colon \Delta \to \mathbb{P}^1_K$ is a projectivity such that $(h' \circ g)(a) = \infty$, $(h' \circ g)(b) = 0$, and $(h' \circ g)(c) = 1$, and thus, $h = h' \circ g$. However, $h'(d') = [a', b', c', d'] = [a, b, c, d] = h(d) = h'(g(d))$, and since h' is injective, we get $d' = g(d)$. $\square$

As a corollary of Lemma 5.12, given any three distinct points a, b, c on a projective line Δ, for every $\lambda \in \mathbb{P}^1_K$ there is a unique point $d \in \Delta$ such that $[a, b, c, d] = \lambda$.

In order to compute explicitly the cross-ratio, we show the following easy lemma.

Lemma 5.13. *Given any projective line $\Delta = \mathbf{P}(D)$, for any three distinct points a, b, c in Δ, if $a = p(u)$, $b = p(v)$, and $c = p(u + v)$, where (u, v) is a basis of D, and for any $[\lambda, \mu]_\sim \in \mathbb{P}^1_K$ and any point $d \in \Delta$, we have*

$$d = p(\lambda u + \mu v) \quad \text{iff} \quad [a, b, c, d] = [\lambda, \mu]_\sim.$$

Proof. If (e_1, e_2) is the basis of K^2 such that $e_1 = (1, 0)$ and $e_2 = (0, 1)$, it is obvious that $p(e_1) = \infty$, $p(e_2) = 0$, and $p(e_1 + e_2) = 1$. Let $f \colon D \to K^2$ be the bijective linear map such that $f(u) = e_1$ and $f(v) = e_2$. Then $f(u + v) = e_1 + e_2$, and thus f induces the unique projectivity $\mathbf{P}(f) \colon \mathbf{P}(D) \to \mathbb{P}^1_K$ such that $\mathbf{P}(f)(a) = \infty$, $\mathbf{P}(f)(b) = 0$, and $\mathbf{P}(f)(c) = 1$. Then

$$\mathbf{P}(f)(p(\lambda u + \mu v)) = [f(\lambda u + \mu v)]_\sim = [\lambda e_1 + \mu e_2]_\sim = [\lambda, \mu]_\sim,$$

that is,

$$d = p(\lambda u + \mu v) \quad \text{iff} \quad [a, b, c, d] = [\lambda, \mu]_\sim.$$

$\square$

We can now compute the cross-ratio explicitly for any given basis (u, v) of D. Assume that a, b, c, d have homogeneous coordinates $[\lambda_1, \mu_1]$, $[\lambda_2, \mu_2]$, $[\lambda_3, \mu_3]$, and $[\lambda_4, \mu_4]$ over the projective frame induced by (u, v). Letting $w_i = \lambda_i u + \mu_i v$, we have $a = p(w_1)$, $b = p(w_2)$, $c = p(w_3)$, and $d = p(w_4)$. Since a and b are distinct, w_1 and w_2 are linearly independent, and we can write $w_3 = \alpha w_1 + \beta w_2$ and $w_4 = \gamma w_1 + \delta w_2$, which can also be written as

$$w_4 = \frac{\gamma}{\alpha} \alpha w_1 + \frac{\delta}{\beta} \beta w_2,$$

and by Lemma 5.13, $[a, b, c, d] = [\gamma/\alpha, \delta/\beta]$. However, since w_1 and w_2 are linearly independent, it is possible to solve for $\alpha, \beta, \gamma, \delta$ in terms of the homogeneous coordinates, obtaining expressions involving determinants:

$$\alpha = \frac{\det(w_3, w_2)}{\det(w_1, w_2)}, \qquad \beta = \frac{\det(w_1, w_3)}{\det(w_1, w_2)},$$

$$\gamma = \frac{\det(w_4, w_2)}{\det(w_1, w_2)}, \qquad \delta = \frac{\det(w_1, w_4)}{\det(w_1, w_2)},$$

and thus, assuming that $d \neq a$, we get

$$[a, b, c, d] = \frac{\begin{vmatrix} \lambda_3 & \lambda_1 \\ \mu_3 & \mu_1 \end{vmatrix}}{\begin{vmatrix} \lambda_3 & \lambda_2 \\ \mu_3 & \mu_2 \end{vmatrix}} \Bigg/ \frac{\begin{vmatrix} \lambda_4 & \lambda_1 \\ \mu_4 & \mu_1 \end{vmatrix}}{\begin{vmatrix} \lambda_4 & \lambda_2 \\ \mu_4 & \mu_2 \end{vmatrix}}.$$

When $d = a$, we have $[a,b,c,d] = \infty$. In particular, if Δ is the projective completion of an affine line D, then $\mu_i = 1$, and we get

$$[a,b,c,d] = \frac{\lambda_3 - \lambda_1}{\lambda_3 - \lambda_2} \Big/ \frac{\lambda_4 - \lambda_1}{\lambda_4 - \lambda_2} = \frac{\overrightarrow{ca}}{\overrightarrow{cb}} \Big/ \frac{\overrightarrow{da}}{\overrightarrow{db}}.$$

When $d = \infty$, we get

$$[a,b,c,\infty] = \frac{\overrightarrow{ca}}{\overrightarrow{cb}},$$

which is just the usual ratio (although we defined it as $-\mathrm{ratio}(a,c,b)$).

We briefly mention some of the properties of the cross-ratio. For example, the cross-ratio $[a,b,c,d]$ is invariant if any two elements and the complementary two elements are transposed, and letting $0^{-1} = \infty$ and $\infty^{-1} = 0$, we have

$$[a,b,c,d] = [b,a,c,d]^{-1} = [a,b,d,c]^{-1}$$

and

$$[a,b,c,d] = 1 - [a,c,b,d].$$

Since the permutations of $\{a,b,c,d\}$ are generated by the above transpositions, the cross-ratio takes at most six values. Letting $\lambda = [a,b,c,d]$, if $\lambda \in \{\infty, 0, 1\}$, then any permutation of $\{a,b,c,d\}$ yields a cross-ratio in $\{\infty, 0, 1\}$, and if $\lambda \notin \{\infty, 0, 1\}$, then there are at most the six values

$$\lambda, \quad \frac{1}{\lambda}, \quad 1 - \lambda, \quad 1 - \frac{1}{\lambda}, \quad \frac{1}{1-\lambda}, \quad \frac{\lambda}{\lambda - 1}.$$

We also define when four points form a harmonic division. For this, we need to assume that K is not of characteristic 2.

Definition 5.9. Given a projective line Δ, we say that a sequence of four collinear points (a,b,c,d) in Δ (where a,b,c are distinct) forms a *harmonic division* if $[a,b,c,d] = -1$. When $[a,b,c,d] = -1$, we also say that c and d are *harmonic conjugates* of a and b.

If a,b,c are distinct collinear points in some affine space, from

$$[a,b,c,\infty] = \frac{\overrightarrow{ca}}{\overrightarrow{cb}},$$

we note that c is the midpoint of (a,b) iff $[a,b,c,\infty] = -1$, that is, if (a,b,c,∞) forms a harmonic division. Figure 5.7 shows a harmonic division (a,b,c,d) on the real line, where the coordinates of (a,b,c,d) are $(-2,2,1,4)$.

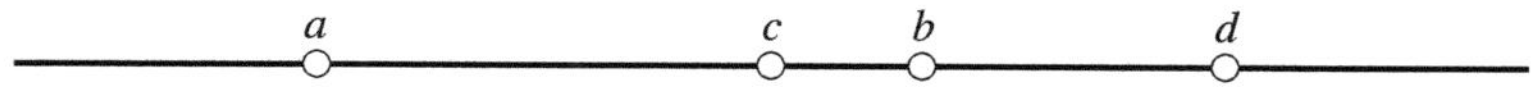

Fig. 5.7 Four points forming a harmonic division.

There is a nice geometric interpretation of harmonic divisions in terms of quadrangles (or complete quadrilaterals). Consider the quadrangle (projective frame) (a,b,c,d) in a projective plane, and let a' be the intersection of $\langle d,a\rangle$ and $\langle b,c\rangle$, b' be the intersection of $\langle d,b\rangle$ and $\langle a,c\rangle$, and c' be the intersection of $\langle d,c\rangle$ and $\langle a,b\rangle$. If we let g be the intersection of $\langle a,b\rangle$ and $\langle a',b'\rangle$, then it is an interesting exercise to show that (a,b,g,c') is a harmonic division.

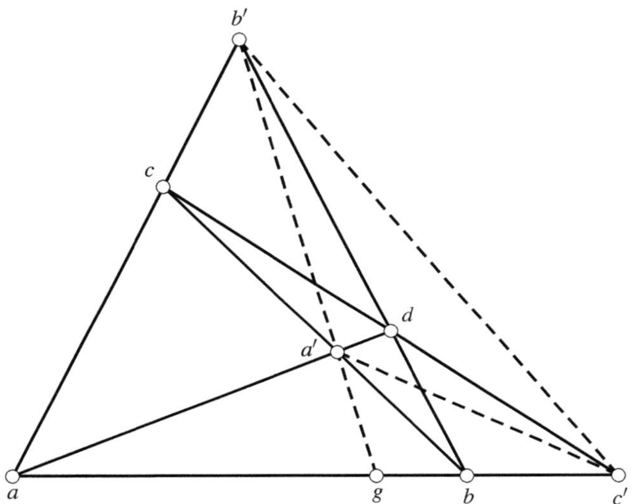

Fig. 5.8 A quadrangle, and harmonic divisions.

In fact, it can be shown that the following quadruples of lines form harmonic divisions: $(\langle c,a\rangle,\langle b',a'\rangle,\langle d,b\rangle,\langle b',c'\rangle)$, $(\langle b,a\rangle,\langle c',a'\rangle,\langle d,c\rangle,\langle c',b'\rangle)$, and $(\langle b,c\rangle,\langle a',c'\rangle,\langle a,d\rangle,\langle a',b'\rangle)$; see Figure 5.8. For more on harmonic divisions, the interested reader should consult any text on projective geometry (for example, Berger [3, 4], Samuel [23], Sidler [24], Tisseron [26], or Pedoe [21]).

Having the notion of cross-ratio at our disposal, we can interpret linear interpolation in the homogenization $\widehat{E}$ of an affine space E as determining a cross-ratio in the projective completion $\widetilde{E}$ of E! This simple fact provides a geometric interpretation of the rational version of the de Casteljau algorithm; see the additional material on the web site (see `http://www.cis.upenn.edu/~jean/gbooks/geom2.html`).

Given any affine space E, let θ_1 and θ_2 be two linearly independent vectors in $\widehat{E}$, and let $t \in K$ be any scalar. Consider

$$\theta_3 = \theta_1 \mathbin{\widehat{+}} \theta_2$$

and

$$\theta_4 = (1-t)\cdot\theta_1 \mathbin{\widehat{+}} t\cdot\theta_2.$$

Observe that the conditions for applying Lemma 5.13 are satisfied, and that the cross-ratio of the points $p(\theta_1)$, $p(\theta_2)$, $p(\theta_3)$, and $p(\theta_4)$ in the projective space $\widetilde{E}$ is given by

$$[p(\theta_1), p(\theta_2), p(\theta_3), p(\theta_4)] = [1-t, t]_{\sim}.$$

Assuming $t \neq 0$ (the case where $\theta_4 \neq \theta_2$), this yields

$$[p(\theta_1), p(\theta_2), p(\theta_3), p(\theta_4)] = \frac{1-t}{t}.$$

Thus, determining θ_4 using the affine interpolation

$$\theta_4 = (1-t) \cdot \theta_1 \,\widehat{+}\, t \cdot \theta_2$$

in $\widehat{E}$ is equivalent to finding the point $p(\theta_4)$ in the projective space $\widetilde{E}$ such that the cross-ratio of the four points $(p(\theta_1), p(\theta_2), p(\theta_3), p(\theta_4))$ is equal to $(1-t)/t$. In the particular case where $\theta_1 = \langle a, \alpha \rangle$ and $\theta_2 = \langle b, \beta \rangle$, where a and b are distinct points of E, if $\alpha + \beta \neq 0$ and $(1-t)\alpha + t\beta \neq 0$, we know that

$$\theta_3 = \left\langle \frac{\alpha}{\alpha+\beta} a + \frac{\beta}{\alpha+\beta} b, \, \alpha+\beta \right\rangle$$

and

$$\theta_4 = \left\langle \frac{(1-t)\alpha}{(1-t)\alpha+t\beta} a + \frac{t\beta}{(1-t)\alpha+t\beta} b, \, (1-t)\alpha+t\beta \right\rangle,$$

and letting

$$c = \frac{\alpha}{\alpha+\beta} a + \frac{\beta}{\alpha+\beta} b$$

and

$$d = \frac{(1-t)\alpha}{(1-t)\alpha+t\beta} a + \frac{t\beta}{(1-t)\alpha+t\beta} b,$$

we also have

$$[a, b, c, d] = \frac{1-t}{t}.$$

Readers may have fun in verifying that when $t = \frac{2}{3}$, the points (a, d, b, c) form a harmonic division!

When $\alpha + \beta = 0$ or $(1-t)\alpha + t\beta = 0$, we have to consider points at infinity, which is better handled in $\widetilde{E}$. In any case, the computation of d can be viewed as determining the unique point d such that $[a, b, c, d] = (1-t)/t$, using

$$c = \frac{\alpha}{\alpha+\beta} a + \frac{\beta}{\alpha+\beta} b.$$

5.9 Duality in Projective Geometry

We now consider duality in projective geometry. Given a vector space E of finite dimension $n+1$, recall that its *dual space* E^* is the vector space of all linear forms $f\colon E \to K$ and that E^* is isomorphic to E. We also have a canonical isomorphism between E and its bidual E^{**}, which allows us to identify E and E^{**}.

Let $\mathscr{H}(E)$ denote the set of hyperplanes in $\mathbf{P}(E)$. In Section 5.3 we observed that the map

$$p(f) \mapsto \mathbf{P}(\operatorname{Ker} f)$$

is a bijection between $\mathbf{P}(E^*)$ and $\mathscr{H}(E)$, in which the equivalence class $p(f) = \{\lambda f \mid \lambda \neq 0\}$ of a nonnull linear form $f \in E^*$ is mapped to the hyperplane $\mathbf{P}(\operatorname{Ker} f)$. Using the above bijection between $\mathbf{P}(E^*)$ and $\mathscr{H}(E)$, a projective subspace $\mathbf{P}(U)$ of $\mathbf{P}(E^*)$ (where U is a subspace of E^*) can be identified with a subset of $\mathscr{H}(E)$, namely the family

$$\{\mathbf{P}(H) \mid H = \operatorname{Ker} f,\, f \in U - \{0\}\}$$

consisting of the projective hyperplanes in $\mathscr{H}(E)$ corresponding to nonnull linear forms in U. Such subsets of $\mathscr{H}(E)$ are called *linear systems (of hyperplanes)*.

The bijection between $\mathbf{P}(E^*)$ and $\mathscr{H}(E)$ allows us to view $\mathscr{H}(E)$ as a projective space, and linear systems as projective subspaces of $\mathscr{H}(E)$. In the projective space $\mathscr{H}(E)$, a point is a hyperplane in $\mathbf{P}(E)$! The duality between subspaces of E and subspaces of E^* (reviewed below) and the fact that there is a bijection between $\mathbf{P}(E^*)$ and $\mathscr{H}(E)$ yields a powerful duality between the set of projective subspaces of $\mathbf{P}(E)$ and the set of linear systems in $\mathscr{H}(E)$ (or equivalently, the set of projective subspaces of $\mathbf{P}(E^*)$).

The idea of duality in projective geometry goes back to Gergonne and Poncelet, in the early nineteenth century. However, Poncelet had a more restricted type of duality in mind (polarity with respect to a conic or a quadric), whereas Gergonne had the more general idea of the duality between points and lines (or points and planes). This more general duality arises from a specific pairing between E and E^* (a nonsingular bilinear form). Here we consider the pairing $\langle -, - \rangle \colon E^* \times E \to K$, defined such that

$$\langle f, v \rangle = f(v),$$

for all $f \in E^*$ and all $v \in E$. Recall that given a subset V of E (respectively a subset U of E^*), the *orthogonal V^0 of V* is the subspace of E^* defined such that

$$V^0 = \{f \in E^* \mid \langle f, v \rangle = 0,\ \text{for every } v \in V\},$$

and that the *orthogonal U^0 of U* is the subspace of E defined such that

$$U^0 = \{v \in E \mid \langle f, v \rangle = 0,\ \text{for every } f \in U\}.$$

Then, by a standard theorem (since E and E^* have the same finite dimension $n+1$), $U = U^{00}$, $V = V^{00}$, and the maps

$$V \mapsto V^0 \quad \text{and} \quad U \mapsto U^0$$

are inverse bijections, where V is a subspace of E, and U is a subspace of E^*.

These maps set up a *duality* between subspaces of E and subspaces of E^*. Furthermore, we know that U has dimension k iff U^0 has dimension $n + 1 - k$, and similarly for V and V^0.

Since a linear system $P = \mathbf{P}(U)$ of hyperplanes in $\mathscr{H}(E)$ corresponds to a subspace U of E^*, and since U^0 is the intersection of all the hyperplanes defined by nonnull linear forms in U, we can view a linear system $P = \mathbf{P}(U)$ in $\mathscr{H}(E)$ as the family of hyperplanes containing $\mathbf{P}(U^0)$.

In view of the identification of $\mathbf{P}(E^*)$ with the set $\mathscr{H}(E)$ of hyperplanes in $\mathbf{P}(E)$, by passing to projective spaces, the above bijection between the set of subspaces of E and the set of subspaces of E^* yields a bijection between the set of projective subspaces of $\mathbf{P}(E)$ and the set of linear systems in $\mathscr{H}(E)$ (or equivalently, the set of projective subspaces of $\mathbf{P}(E^*)$).

More specifically, assuming that E has dimension $n + 1$, so that $\mathbf{P}(E)$ has dimension n, if $Q = \mathbf{P}(V)$ is any projective subspace of $\mathbf{P}(E)$ (where V is any subspace of E) and if $P = \mathbf{P}(U)$ is any linear system in $\mathscr{H}(E)$ (where U is any subspace of E^*), we get a subspace Q^0 of $\mathscr{H}(E)$ defined by

$$Q^0 = \{\mathbf{P}(H) \mid Q \subseteq \mathbf{P}(H), \mathbf{P}(H) \text{ a hyperplane in } \mathscr{H}(E)\},$$

and a subspace P^0 of $\mathbf{P}(E)$ defined by

$$P^0 = \bigcap \{\mathbf{P}(H) \mid \mathbf{P}(H) \in P, \mathbf{P}(H) \text{ a hyperplane in } \mathscr{H}(E)\}.$$

We have $P = P^{00}$ and $Q = Q^{00}$. Since Q^0 is determined by $\mathbf{P}(V^0)$, if $Q = \mathbf{P}(V)$ has dimension k (i.e., if V has dimension $k + 1$), then Q^0 has dimension $n - k - 1$ (since V has dimension $k + 1$ and $\dim(E) = n + 1$, then V^0 has dimension $n + 1 - (k + 1) = n - k$). Thus,

$$\dim(Q) + \dim(Q^0) = n - 1,$$

and similarly, $\dim(P) + \dim(P^0) = n - 1$.

A linear system $P = \mathbf{P}(U)$ of hyperplanes in $\mathscr{H}(E)$ is called a *pencil of hyperplanes* if it corresponds to a projective line in $\mathbf{P}(E^*)$, which means that U is a subspace of dimension 2 of E^*. From $\dim(P) + \dim(P^0) = n - 1$, a pencil of hyperplanes P is the family of hyperplanes in $\mathscr{H}(E)$ containing some projective subspace $\mathbf{P}(V)$ of dimension $n - 2$ (where $\mathbf{P}(V)$ is a projective subspace of $\mathbf{P}(E)$, and $\mathbf{P}(E)$ has dimension n). When $n = 2$, a pencil of hyperplanes in $\mathscr{H}(E)$, also called a *pencil of lines*, is the family of lines passing through a given point. When $n = 3$, a pencil of hyperplanes in $\mathscr{H}(E)$, also called a *pencil of planes*, is the family of planes passing through a given line.

When $n = 2$, the above duality takes a rather simple form. In this case (of a projective plane $\mathbf{P}(E)$), the duality is a bijection between points and lines with the following properties:

- A point a maps to a line D_a (the pencil of lines in $\mathscr{H}(E)$ containing a, also denoted by a^*)
- A line D maps to a point p_D (the line D in $\mathscr{H}(E)$!)
- Two points a, b map to lines D_a, D_b, such that the intersection of D_a and D_b is the point $p_{\langle a,b \rangle}$ corresponding to the line $\langle a,b \rangle$ via duality
- A line D containing two points a, b maps to the intersection p_D of the lines D_a and D_b.
- If $a \in D$, where a is a point and D is a line, then $p_D \in D_a$.

The reader will discover that the dual of Desargues's theorem is its converse. This is a nice way of getting the converse for free! We will not spoil the reader's fun and let him discover the dual of Pappus's theorem.

To conclude our quick tour of projective geometry, we estabish a connection between the cross-ratio of hyperplanes in a pencil of hyperplanes with the cross-ratio of the intersection points of any line not contained in any hyperplane in this pencil with four hyperplanes in this pencil.

5.10 Cross-Ratios of Hyperplanes

Given a pencil $P = \mathbf{P}(U)$ of hyperplanes in $\mathscr{H}(E)$, for any sequence (H_1, H_2, H_3, H_4) of hyperplanes in this pencil, if H_1, H_2, H_3 are distinct, we define the cross-ratio $[H_1, H_2, H_3, H_4]$ as the cross-ratio of the hyperplanes H_i considered as points on the projective line P in $\mathbf{P}(E^*)$. In particular, in a projective plane $\mathbf{P}(E)$, given any four concurrent lines D_1, D_2, D_3, D_4, where D_1, D_2, D_3 are distinct, for any two distinct lines Δ and Δ' not passing through the common intersection c of the lines D_i, letting $d_i = \Delta \cap D_i$, and $d_i' = \Delta' \cap D_i$, note that the projection of center c from Δ to Δ' maps each d_i to d_i'.

Since such a projection is a projectivity, and since projectivities between lines preserve cross-ratios, we have

$$[d_1, d_2, d_3, d_4] = [d_1', d_2', d_3', d_4'],$$

which means that the cross-ratio of the d_i is independent of the line Δ (see Figure 5.9).

In fact, this cross-ratio is equal to $[D_1, D_2, D_3, D_4]$, as shown in the next lemma.

Lemma 5.14. *Let $P = \mathbf{P}(U)$ be a pencil of hyperplanes in $\mathscr{H}(E)$, and let $\Delta = \mathbf{P}(D)$ be any projective line such that $\Delta \not\subseteq H$ for all $H \in P$. The map $h\colon P \to \Delta$ defined such that $h(H) = H \cap \Delta$ for every hyperplane $H \in P$ is a projectivity. Furthermore, for any sequence (H_1, H_2, H_3, H_4) of hyperplanes in the pencil P, if H_1, H_2, H_3 are distinct and $d_i = \Delta \cap H_i$, then $[d_1, d_2, d_3, d_4] = [H_1, H_2, H_3, H_4]$.*

Proof. First, the map $h\colon P \to \Delta$ is well–defined, since in a projective space, every line $\Delta = \mathbf{P}(D)$ not contained in a hyperplane intersects this hyperplane in exactly one point. Since $P = \mathbf{P}(U)$ is a pencil of hyperplanes in $\mathscr{H}(E)$, U has dimension 2,

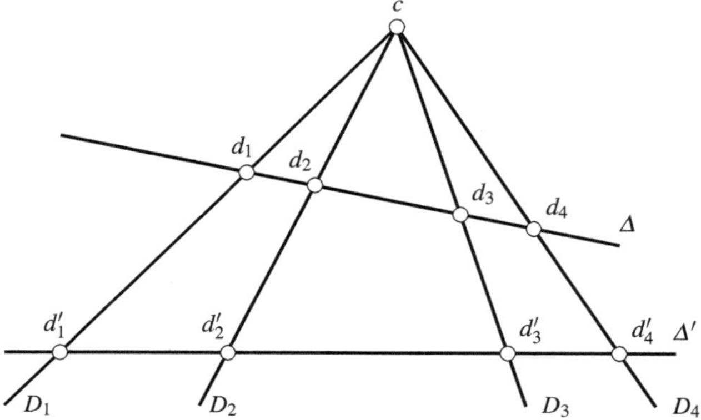

Fig. 5.9 A pencil of lines and its cross-ratio with intersecting lines.

and let φ and ψ be two nonnull linear forms in E^* that constitute a basis of U, and let $F = \varphi^{-1}(0)$ and $G = \psi^{-1}(0)$. Let $a = \mathbf{P}(F) \cap \Delta$ and $b = \mathbf{P}(G) \cap \Delta$. There are some vectors $u, v \in D$ such that $a = p(u)$ and $b = p(v)$, and since φ and ψ are linearly independent, we have $a \neq b$, and we can choose φ and ψ such that $\varphi(v) = -1$ and $\psi(u) = 1$. Also, (u, v) is a basis of D. Then a point $p(\alpha u + \beta v)$ on Δ belongs to the hyperplane $H = p(\gamma \varphi + \delta \psi)$ of the pencil P iff

$$(\gamma \varphi + \delta \psi)(\alpha u + \beta v) = 0,$$

which, since $\varphi(u) = 0$, $\psi(v) = 0$, $\varphi(v) = -1$, and $\psi(u) = 1$, yields $\gamma \beta = \delta \alpha$, which is equivalent to $[\alpha, \beta] = [\gamma, \delta]$ in $\mathbf{P}(K^2)$. But then the map $h \colon P \to \Delta$ is a projectivity. Letting $d_i = \Delta \cap H_i$, since by Lemma 5.12 a projectivity of lines preserves the cross-ratio, we get $[d_1, d_2, d_3, d_4] = [H_1, H_2, H_3, H_4]$. $\square$

5.11 Complexification of a Real Projective Space

Notions such as orthogonality, angles, and distance between points are not projective concepts. In order to define such notions, one needs an inner product on the underlying vector space. We say that such notions belong to *Euclidean geometry*. At first glance, the fact that some important Euclidean concepts are not covered by projective geometry seems a major drawback of projective geometry. Fortunately, geometers of the nineteenth century (including Laguerre, Monge, Poncelet, Chasles, von Staudt, Cayley, and Klein) found an astute way of recovering certain Euclidean notions such as angles and orthogonality (also circles) by embedding real projective spaces into complex projective spaces. In the next two sections we will give a brief account of this method. More details can be found in Berger [3, 4], Pedoe [21], Samuel [23], Coxeter [5, 6], Sidler [24], Tisseron [26], Lehmann and Bkouche [20], and, of course, Volume II of Veblen and Young [29]. Readers may want to consult

Chapter 6, which gives a review of Euclidean geometry, especially Section 8.8, on angles.

The first step is to embed a real vector space E into a complex vector space $E_{\mathbb{C}}$. A quick but somewhat bewildering way to do so is to define the complexification of E as the tensor product $\mathbb{C} \otimes E$. A more tangible way is to define the following structure.

Definition 5.10. Given a real vector space E, let $E_{\mathbb{C}}$ be the structure $E \times E$ under the addition operation

$$(u_1, u_2) + (v_1, v_2) = (u_1 + v_1, u_2 + v_2),$$

and let multiplication by a complex scalar $z = x + iy$ be defined such that

$$(x + iy) \cdot (u, v) = (xu - yv, yu + xv).$$

It is easily shown that the structure $E_{\mathbb{C}}$ is a complex vector space. It is also immediate that

$$(0, v) = i(v, 0),$$

and thus, identifying E with the subspace of $E_{\mathbb{C}}$ consisting of all vectors of the form $(u, 0)$, we can write

$$(u, v) = u + iv.$$

Given a vector $w = u + iv$, its *conjugate* $\overline{w}$ is the vector $\overline{w} = u - iv$. Then conjugation is a map from $E_{\mathbb{C}}$ to itself that is an involution. If $(e_1, \ldots, e_n)$ is any basis of E, then $((e_1, 0), \ldots, (e_n, 0))$ is a basis of $E_{\mathbb{C}}$. We call such a basis a *real basis*.

Given a linear map $f \colon E \to E$, the map f can be extended to a linear map $f_{\mathbb{C}} \colon E_{\mathbb{C}} \to E_{\mathbb{C}}$ defined such that

$$f_{\mathbb{C}}(u + iv) = f(u) + if(v).$$

We define the *complexification* of $\mathbf{P}(E)$ as $\mathbf{P}(E_{\mathbb{C}})$. If $(E, \overrightarrow{E})$ is a real affine space, we define the *complexified projective completion* of $(E, \overrightarrow{E})$ as $\mathbf{P}(\widehat{E_{\mathbb{C}}})$ and denote it by $\widetilde{E}_{\mathbb{C}}$. Then $\widetilde{E}$ is naturally embedded in $\widetilde{E}_{\mathbb{C}}$, and it is called the set of *real points* of $\widetilde{E}_{\mathbb{C}}$.

If E has dimension $n + 1$ and $(e_1, \ldots, e_{n+1})$ is a basis of E, given any homogeneous polynomial $P(x_1, \ldots, x_{n+1})$ over $\mathbb{C}$ of total degree m, because P is homogeneous, it is immediately verified that

$$P(x_1, \ldots, x_{n+1}) = 0$$

iff

$$P(\lambda x_1, \ldots, \lambda x_{n+1}) = 0,$$

for any $\lambda \neq 0$. Thus, we can define the *hypersurface $V(P)$ of equation* $P(x_1, \ldots, x_{n+1}) = 0$ as the subset of $\widetilde{E}_{\mathbb{C}}$ consisting of all points of homogeneous coordinates $(x_1, \ldots, x_{n+1})$ such that $P(x_1, \ldots, x_{n+1}) = 0$. We say that the hypersurface $V(P)$

of equation $P(x_1,\ldots,x_{n+1}) = 0$ is *real* whenever $P(x_1,\ldots,x_{n+1}) = 0$ implies that
$P(\overline{x}_1,\ldots,\overline{x}_{n+1}) = 0$.

 Note that a real hypersurface may have points other than real points, or
no real points at all. For example,

$$x^2 + y^2 - z^2 = 0$$

contains real and complex points such as $(1,i,0)$ and $(1,-i,0)$, and

$$x^2 + y^2 + z^2 = 0$$

contains only complex points. When $m = 2$ (where m is the total degree of P), a
hypersurface is called a *quadric*, and when $m = 2$ and $n = 2$, a *conic*. When $m = 1$,
a hypersurface is just a hyperplane.

Given any homogeneous polynomial $P(x_1,\ldots,x_{n+1})$ over $\mathbb{R}$ of total degree m,
since $\mathbb{R} \subseteq \mathbb{C}$, P viewed as a homogeneous polynomial over $\mathbb{C}$ defines a hypersurface
$V(P)_{\mathbb{C}}$ in $\widetilde{E}_{\mathbb{C}}$, and also a hypersurface $V(P)$ in $\mathbf{P}(E)$. It is clear that $V(P)$ is naturally
embedded in $V(P)_{\mathbb{C}}$, and $V(P)_{\mathbb{C}}$ is called the *complexification of* $V(P)$.

We now show how certain real quadrics without real points can be used to define
orthogonality and angles.

5.12 Similarity Structures on a Projective Space

We begin with a real Euclidean plane $\left(E, \overrightarrow{E}\right)$. We will show that the angle of two
lines D_1 and D_2 can be expressed as a certain cross-ratio involving the lines D_1, D_2
and also two lines D_I and D_J joining the intersection point $D_1 \cap D_2$ of D_1 and D_2 to
two complex points at infinity I and J called the *circular points*. However, there is
a slight problem, which is that we haven't yet defined the angle of two lines! Recall
from Section 8.8 that we define the (oriented) angle $\widehat{u_1 u_2}$ of two unit vectors u_1,
u_2 as the equivalence class of pairs of unit vectors under the equivalence relation
defined such that

$$\langle u_1, u_2 \rangle \equiv \langle u_3, u_4 \rangle$$

iff there is some rotation r such that $r(u_1) = u_3$ and $r(u_2) = u_4$. The set of (oriented)
angles of vectors is a group isomorphic to the group $\mathbf{SO}(2)$ of plane rotations. If the
Euclidean plane is oriented, the measure of the angle of two vectors is defined up
to $2k\pi$ ($k \in \mathbb{Z}$). The angle of two vectors has a measure that is either θ or $2\pi - \theta$,
where $\theta \in [0, 2\pi[$, depending on the orientation of the plane. The problem with
lines is that they are not oriented: A line is defined by a point a and a vector u, but
also by a and $-u$. Given any two lines D_1 and D_2, if r is a rotation of angle θ such
that $r(D_1) = D_2$, note that the rotation $-r$ of angle $\theta + \pi$ also maps D_1 onto D_2.
Thus, in order to define the (oriented) angle $\widehat{D_1 D_2}$ of two lines D_1, D_2, we define an
equivalence relation on pairs of lines as follows:

$$\langle D_1, D_2 \rangle \equiv \langle D_3, D_4 \rangle$$

if there is some rotation r such that $r(D_1) = D_2$ and $r(D_3) = D_4$.

It can be verified that the set of (oriented) angles of lines is a group isomorphic to the quotient group $\mathbf{SO}(2)/\{\mathrm{id}, -\mathrm{id}\}$, also denoted by $\mathbf{PSO}(2)$. In order to define the measure of the angle of two lines, the Euclidean plane E must be oriented. The measure of the angle $\widehat{D_1 D_2}$ of two lines is defined up to $k\pi$ ($k \in \mathbb{Z}$). The angle of two lines has a measure that is either θ or $\pi - \theta$, where $\theta \in [0, \pi[$, depending on the orientation of the plane. We now go back to the circular points.

Let (a_0, a_1, a_2, a_3) be any projective frame for $\widetilde{E}_{\mathbb{C}}$ such that (a_0, a_1) arises from an orthonormal basis (u_1, u_2) of $\overrightarrow{E}$ and the line at infinity H corresponds to $z = 0$ (where (x, y, z) are the homogeneous coordinates of a point w.r.t. (a_0, a_1, a_2, a_3)). Consider the points belonging to the intersection of the real conic Σ of equation

$$x^2 + y^2 - z^2 = 0$$

with the line at infinity $z = 0$. For such points, $x^2 + y^2 = 0$ and $z = 0$, and since

$$x^2 + y^2 = (y - ix)(y + ix),$$

we get exactly two points I and J of homogeneous coordinates $(1, -i, 0)$ and $(1, i, 0)$. The points I and J are called the *circular points*, or the *absolute points*, of $\widetilde{E}_{\mathbb{C}}$. They are complex points at infinity. Any line containing either I or J is called an *isotropic line*.

What is remarkable about I and J is that they allow the definition of the angle of two lines in terms of a certain cross-ratio. Indeed, consider two distinct real lines D_1 and D_2 in E, and let D_I and D_J be the isotropic lines joining $D_1 \cap D_2$ to I and J. We will compute the cross-ratio $[D_1, D_2, D_I, D_J]$. For this, we simply have to compute the cross-ratio of the four points obtained by intersecting D_1, D_2, D_I, D_J with any line not passing through $D_1 \cap D_2$. By changing frame if necessary, so that $D_1 \cap D_2 = a_0$, we can assume that the equations of the lines D_1, D_2, D_I, D_J are of the form

$$y = m_1 x,$$
$$y = m_2 x,$$
$$y = -ix,$$
$$y = ix,$$

leaving the cases $m_1 = \infty$ and $m_2 = \infty$ as a simple exercise. If we choose $z = 1$ as the intersecting line, we need to compute the cross-ratio of the points $(D_1)_\infty = (1, m_1, 0)$, $(D_2)_\infty = (1, m_2, 0)$, $I = (1, -i, 0)$, and $J = (1, i, 0)$, and we get

$$[D_1, D_2, D_I, D_J] = [(D_1)_\infty, (D_2)_\infty, I, J] = \frac{(-i - m_1)}{(i - m_1)} \frac{(i - m_2)}{(-i - m_2)},$$

that is,

$$[D_1, D_2, D_I, D_J] = \frac{m_1 m_2 + 1 + \mathrm{i}(m_2 - m_1)}{m_1 m_2 + 1 - \mathrm{i}(m_2 - m_1)}.$$

However, since m_1 and m_2 are the slopes of the lines D_1 and D_2, it is well known that if θ is the (oriented) angle between D_1 and D_2, then

$$\tan \theta = \frac{m_2 - m_1}{m_1 m_2 + 1}.$$

Thus, we have

$$[D_1, D_2, D_I, D_J] = \frac{m_1 m_2 + 1 + \mathrm{i}(m_2 - m_1)}{m_1 m_2 + 1 - \mathrm{i}(m_2 - m_1)} = \frac{1 + \mathrm{i}\tan \theta}{1 - \mathrm{i}\tan \theta},$$

that is,

$$[D_1, D_2, D_I, D_J] = \cos 2\theta + \mathrm{i}\sin 2\theta = \mathrm{e}^{\mathrm{i}2\theta}.$$

One can check that the formula still holds when $m_1 = \infty$ or $m_2 = \infty$, and also when $D_1 = D_2$. The formula

$$[D_1, D_2, D_I, D_J] = \mathrm{e}^{\mathrm{i}2\theta}$$

is known as *Laguerre's formula*.

If U denotes the group $\{\mathrm{e}^{\mathrm{i}\theta} \mid -\pi \leq \theta \leq \pi\}$ of complex numbers of modulus 1, recall that the map $\Lambda : \mathbb{R} \to U$ defined such that

$$\Lambda(t) = \mathrm{e}^{\mathrm{i}t}$$

is a group homomorphism such that $\Lambda^{-1}(1) = 2k\pi$, where $k \in \mathbb{Z}$. The restriction

$$\Lambda : \,]-\pi, \pi[\, \to (U - \{-1\})$$

of Λ to $\,]-\pi, \pi[\,$ is a bijection, and its inverse will be denoted by

$$\log_U : (U - \{-1\}) \to \,]-\pi, \pi[.$$

For stating Lemma 5.15 more conveniently, we will extend $\log_U$ to U by letting $\log_U(-1) = \pi$, even though the resulting function is not continuous at -1!. Then we can write

$$\theta = \frac{1}{2} \log_U([D_1, D_2, D_I, D_J]).$$

If the orientation of the plane E is reversed, θ becomes $\pi - \theta$, and since

$$\mathrm{e}^{\mathrm{i}2(\pi - \theta)} = \mathrm{e}^{2\mathrm{i}\pi - \mathrm{i}2\theta} = \mathrm{e}^{-\mathrm{i}2\theta},$$

$\log_U(\mathrm{e}^{\mathrm{i}2(\pi-\theta)}) = -\log_U(\mathrm{e}^{\mathrm{i}2\theta})$, and

$$\theta = -\frac{1}{2} \log_U([D_1, D_2, D_I, D_J]).$$

In all cases, we have

$$\theta = \frac{1}{2} |\log_U([D_1, D_2, D_I, D_J])|,$$

a formula due to Cayley. We summarize the above in the following lemma.

Lemma 5.15. *Given any two lines D_1, D_2 in a real Euclidean plane $(E, \overrightarrow{E})$, letting D_I and D_J be the isotropic lines in $\widetilde{E}_{\mathbb{C}}$ joining the intersection point $D_1 \cap D_2$ of D_1 and D_2 to the circular points I and J, if θ is the angle of the two lines D_1, D_2, we have*

$$[D_1, D_2, D_I, D_J] = \mathrm{e}^{i 2\theta},$$

known as Laguerre's formula, and independently of the orientation of the plane, we have

$$\theta = \frac{1}{2} |\log_U([D_1, D_2, D_I, D_J])|,$$

known as Cayley's formula.

In particular, note that $\theta = \pi/2$ iff $[D_1, D_2, D_I, D_J] = -1$, that is, if (D_1, D_2, D_I, D_J) forms a harmonic division. Thus, two lines D_1 and D_2 are orthogonal iff they form a harmonic division with D_I and D_J.

The above considerations show that it is not necessary to assume that $(E, \overrightarrow{E})$ is a real Euclidean plane to define the angle of two lines and orthogonality. Instead, it is enough to assume that two complex conjugate points I, J on the line H at infinity are given. We say that $\langle I, J \rangle$ provides a *similarity structure* on $\widetilde{E}_{\mathbb{C}}$. Note in passing that a circle can be defined as a conic in $\widetilde{E}_{\mathbb{C}}$ that contains the circular points I, J. Indeed, the equation of a conic is of the form

$$ax^2 + by^2 + cxy + dxz + eyz + fz^2 = 0.$$

If this conic contains the circular points $I = (1, -i, 0)$ and $J = (1, i, 0)$, we get the two equations

$$a - b - ic = 0,$$
$$a - b + ic = 0,$$

from which we get $2ic = 0$ and $a = b$, that is, $c = 0$ and $a = b$. The resulting equation

$$ax^2 + ay^2 + dxz + eyz + fz^2 = 0$$

is indeed that of a circle.

Instead of using the function $\log_U : (U - \{-1\}) \to \,]-\pi, \pi[$ as logarithm, one may use the complex logarithm function $\log \colon \mathbb{C}^* \to B$, where $\mathbb{C}^* = \mathbb{C} - \{0\}$ and

$$B = \{x + iy \mid x, y \in \mathbb{R}, \, -\pi < y \le \pi\}.$$

Indeed, the restriction of the complex exponential function $z \mapsto \mathrm{e}^z$ to B is bijective, and thus, log is well-defined on C^* (note that log is a homeomorphism from $\mathbb{C} - \{x \mid$

$x \in \mathbb{R}, x \leq 0\}$ onto $\{x + iy \mid x, y \in \mathbb{R}, -\pi < y < \pi\}$, the interior of B). Then Cayley's formula reads as

$$\theta = \frac{1}{2i} \log([D_1, D_2, D_I, D_J]),$$

with a $\pm$ in front when the plane is nonoriented. Observe that this formula allows the definition of the angle of two complex lines (possibly a complex number) and the notion of orthogonality of complex lines. In this case, note that the isotropic lines are orthogonal to themselves!

The definition of orthogonality of two lines D_1, D_2 in terms of (D_1, D_2, D_I, D_J) forming a harmonic division can be used to give elegant proofs of various results. Cayley's formula can even be used in computer vision to explain modeling and calibrating cameras! (see Faugeras [10]). As an illustration, consider a triangle (a, b, c), and recall that the line a' passing through a and orthogonal to (b, c) is called the *altitude of a*, and similarly for b and c. It is well known that the altitudes a', b', c' intersect in a common point called the *orthocenter* of the triangle (a, b, c). This can be shown in a number of ways using the circular points. Indeed, letting $bc_\infty, ab_\infty,$ $ac_\infty, a'_\infty, b'_\infty,$ and c'_∞ denote the points at infinity of the lines $\langle b, c \rangle, \langle a, b \rangle, \langle a, c \rangle, a', b',$ and c', we have

$$[bc_\infty, a'_\infty, I, J] = -1, \quad [ab_\infty, c'_\infty, I, J] = -1, \quad [ac_\infty, b'_\infty, I, J] = -1,$$

and it is easy to show that there is an involution σ of the line at infinity such that

$$\begin{aligned}
\sigma(I) &= J, \\
\sigma(J) &= I, \\
\sigma(bc_\infty) &= a'_\infty, \\
\sigma(ab_\infty) &= c'_\infty, \\
\sigma(ac_\infty) &= b'_\infty.
\end{aligned}$$

Then, using the result stated in Problem 5.28, the lines a', b', c' are concurrent. For more details and other results, notably on the conics, see Sidler [24], Berger [4], and Samuel [23].

The generalization of what we just did to real Euclidean spaces $(E, \overrightarrow{E})$ of dimension n is simple. Let $(a_0, \ldots, a_{n+1})$ be any projective frame for $\widetilde{E}_{\mathbb{C}}$ such that $(a_0, \ldots, a_{n-1})$ arises from an orthonormal basis $(u_1, \ldots, u_n)$ of $\overrightarrow{E}$ and the hyperplane at infinity H corresponds to $x_{n+1} = 0$ (where $(x_1, \ldots, x_{n+1})$ are the homogeneous coordinates of a point with respect to $(a_0, \ldots, a_{n+1})$). Consider the points belonging to the intersection of the real quadric Σ of equation

$$x_1^2 + \cdots + x_n^2 - x_{n+1}^2 = 0$$

with the hyperplane at infinity $x_{n+1} = 0$. For such points,

$$x_1^2 + \cdots + x_n^2 = 0 \quad \text{and} \quad x_{n+1} = 0.$$

Such points belong to a quadric called the *absolute quadric* of $\widetilde{E}_{\mathbb{C}}$, and denoted by Ω. Any line containing any point on the absolute quadric is called an *isotropic line*. Then, given any two coplanar lines D_1 and D_2 in E, these lines intersect the hyperplane at infinity H in two points $(D_1)_\infty$ and $(D_2)_\infty$, and the line Δ joining $(D_1)_\infty$ and $(D_2)_\infty$ intersects the absolute quadric Ω in two conjugate points I_Δ and J_Δ (also called circular points). It can be shown that the angle θ between D_1 and D_2 is defined by Laguerre's formula:

$$[(D_1)_\infty,(D_2)_\infty,I_\Delta,J_\Delta] = [D_1,D_2,D_{I_\Delta},D_{J_\Delta}] = \mathrm{e}^{\mathrm{i}2\theta},$$

where D_{I_Δ} and D_{J_Δ} are the lines joining the intersection $D_1 \cap D_2$ of D_1 and D_2 to the circular points I_Δ and J_Δ.

As in the case of a plane, the above considerations show that it is not necessary to assume that $(E, \overrightarrow{E})$ is a real Euclidean space to define the angle of two lines and orthogonality. Instead, it is enough to assume that a nondegenerate real quadric Ω in the hyperplane at infinity H and without real points is given. In particular, when $n = 3$, the absolute quadric Ω is a nondegenerate real conic consisting of complex points at infinity. We say that Ω provides a *similarity structure* on $\widetilde{E}_{\mathbb{C}}$.

It is also possible to show that the projectivities of $\widetilde{E}_{\mathbb{C}}$ that leave both the hyperplane H at infinity and the absolute quadric Ω (globally) invariant form a group which is none other than the group of similarities. A *similarity* is a map that is the composition of an isometry (a member of $\mathbf{O}(n)$), a central dilatation, and a translation. For more details on the use of absolute quadrics to obtain some very sophisticated results, the reader should consult Berger [3, 4], Pedoe [21], Samuel [23], Coxeter [5], Sidler [24], Tisseron [26], Lehmann and Bkouche [20], and, of course, Volume II of Veblen and Young [29], which also explains how some non-Euclidean geometries are obtained by chosing the absolute quadric in an appropriate fashion (after Cayley and Klein).

5.13 Some Applications of Projective Geometry

Projective geometry is definitely a jewel of pure mathematics and one of the major mathematical achievements of the nineteenth century. It turns out to be a prerequisite for algebraic geometry, but to our surprise (and pleasure), it also turns out to have applications in engineering. In this short section we summarize some of these applications.

We first discuss applications of projective geometry to camera calibration, a crucial problem in computer vision. Our brief presentation follows quite closely Trucco and Verri [27] (Chapter 2 and Chapter 6). One should also consult Faugeras [10], or Jain, Katsuri, and Schunck [18].

The *pinhole* (or *perspective*) model of a camera is a typical example from computer vision that can be explained very simply in terms of projective transformations. A pinhole camera consists of a point $\mathbf{O}$ called the *center* or *focus of projection*, and a plane π (not containing $\mathbf{O}$) called the *image plane*. The distance f from the image plane π to the center $\mathbf{O}$ is called the *focal length*. The line through $\mathbf{O}$ and perpendicular to π is called the *optical axis*, and the point $\mathbf{o}$, intersection of the optical axis with the image plane is called the *principal point* or *image center*. The way the camera works is that a point P in 3D space is projected onto the image plane (the film) to a point p via the central projection of center $\mathbf{O}$.

It is assumed that an orthonormal frame $\mathscr{F}_c$ is attached to the camera, with its origin at $\mathbf{O}$ and its z-axis parallel to the optical axis. Such a frame is called the *camera reference frame*. With respect to the camera reference frame, it is very easy to write the equations relating the coordinates (x,y) (omitting $z = f$) of the image p (in the image plane π) of a point P of coordinates (X,Y,Z):

$$x = f\frac{X}{Z}, \quad y = f\frac{Y}{Z}.$$

Typically, points in 3D space are defined by their coordinates not with respect to the camera reference frame, but with respect to another frame $\mathscr{F}_w$, called the *world reference frame*. However, for most computer vision algorithms, it is necessary to know the coordinates of a point in 3D space with respect to the camera reference frame. Thus, it is necessary to know the position and orientation of the camera with respect to the frame $\mathscr{F}_w$. The position and orientation of the camera are given by some affine transformation $(R,\mathbf{T})$ mapping the frame $\mathscr{F}_w$ to the frame $\mathscr{F}_c$, where R is a rotation matrix and $\mathbf{T}$ is a translation vector. Furthermore, the coordinates of an image point are typically known in terms of *pixel coordinates*, and it is also necessary to transform the coordinates of an image point with respect to the camera reference frame to pixel coordinates. In summary, it is necessary to know the transformation that maps a point P in world coordinates (w.r.t. $\mathscr{F}_w$) to pixel coordinates.

This transformation of world coordinates to pixel coordinates turns out to be a projective transformation that depends on the extrinsic and the intrinsic parameters of the camera. The *extrinsic parameters* of a camera are the location and orientation of the camera with respect to the world reference frame $\mathscr{F}_w$. It is given by an affine map (in fact, a rigid motion, see Chapter 8, Section 8.4). The *intrinsic parameters* of a camera are the parameters needed to link the pixel coordinates of an image point to the corresponding coordinates in the camera reference frame. If $\mathbf{P_w} = (X_w, Y_w, Z_w)$ and $\mathbf{P_c} = (X_c, Y_c, Z_c)$ are the coordinates of the 3D point P with respect to the frames $\mathscr{F}_w$ and $\mathscr{F}_c$, respectively, we can write

$$\mathbf{P_c} = R(\mathbf{P_w} - \mathbf{T}).$$

Neglecting distorsions possibly introduced by the optics, the correspondence between the coordinates (x,y) of the image point with respect to $\mathscr{F}_c$ and the pixel coordinates $(x_{\text{im}}, y_{\text{im}})$ is given by

$$x = -(x_{\text{im}} - o_x)s_x,$$
$$y = -(y_{\text{im}} - o_y)s_y,$$

where (o_x, o_y) are the pixel coordinates the principal point $\mathbf{o}$ and s_x, s_y are scaling parameters.

After some simple calculations, the upshot of all this is that the transformation between the homogeneous coordinates $(X_w, Y_w, Z_w, 1)$ of a 3D point and its homogeneous pixel coordinates (x_1, x_2, x_3) is given by

$$
\begin{pmatrix} x_1 \\ x_2 \\ x_3 \end{pmatrix} = M \begin{pmatrix} X_w \\ Y_w \\ Z_w \\ 1 \end{pmatrix}
$$

where the matrix M, known as the *projection matrix*, is a 3×4 matrix depending on R, $\mathbf{T}$, o_x, o_y, f (the focal length), and s_x, s_y (for the derivation of this equation, see Trucco and Verri [27], Chapter 2).

The problem of estimating the extrinsic and the instrinsic parameters of a camera is known as the *camera calibration* problem. It is an important problem in computer vision. Now, using the equations

$$
x = -(x_{\text{im}} - o_x)s_x,
$$
$$
y = -(y_{\text{im}} - o_y)s_y,
$$

we get

$$
x_{\text{im}} = -\frac{f}{s_x}\frac{X_c}{Z_c} + o_x,
$$
$$
y_{\text{im}} = -\frac{f}{s_y}\frac{Y_c}{Z_c} + o_y,
$$

relating the coordinates w.r.t. the camera reference frame to the pixel coordinates. This suggests using the parameters $f_x = f/s_x$ and $f_y = f/s_y$ instead of the parameters f, s_x, s_y. In fact, all we need are the parameters $f_x = f/s_x$ and $\alpha = s_y/s_x$, called the *aspect ratio*. Without loss of generality, it can also be assumed that (o_x, o_y) are known. Then we have a total of eight parameters.

One way of solving the calibration problem is to try estimating f_x, α, the rotation matrix R, and the translation vector $\mathbf{T}$ from N image points (x_i, y_i), projections of N suitably chosen world points (X_i, Y_i, Z_i), using the system of equations obtained from the projection matrix. It turns out that if $N \geq 7$ and the points are not coplanar, the rank of the system is 7, and the system has a nontrivial solution (up to a scalar) that can be found using SVD methods (see Chapter 13, Trucco and Verri [27], or Jain, Katsuri, and Schunck [18]).

Another method consists in estimating the whole projection matrix M, which depends on 11 parameters, and then extracting extrinsic and intrinsic parameters. Again, SVD methods are used (see Trucco and Verri [27], and Faugeras [10]).

Cayley's formula can also be used to solve the calibration cameras, as explained in Faugeras [10]. Other problems in computer vision can be reduced to problems in projective geometry (see Faugeras [10]).

In computer graphics, it is also necessary to convert the 3D world coordinates of a point to a two-dimensional representation on a *view plane*. This is achieved by a so-called *viewing system* using a projective transformation. For details on viewing systems see Watt [31] or Foley, van Dam, Feiner, and Hughes [13].

Projective spaces are also the right framework to deal with rational curves and rational surfaces. Indeed, in the projective framework it is easy to deal with vanishing denominators and with "infinite" values of the parameter(s). Such an approach is presented in Chapter 22 for rational curves, and in Chapter 23 and 24 for rational surfaces. In fact, working in a projective framework yields a very simple proof of the method for drawing a rational curve as two Bézier segments (and similarly for surfaces).

It is much less obvious that projective geometry has applications to efficient communication, error-correcting codes, and cryptography, as very nicely explained by Beutelspacher and Rosenbaum [2]. We sketch these applications very briefly, referring our readers to [2] for details. We begin with efficient communication. Suppose that eight students would like to exchange information to do their homework economically. The idea is that each student solves part of the exercises and copies the rest from the others (which we do not recommend, of course!). It is assumed that each student solves his part of the homework at home, and that the solutions are communicated by phone. The problem is to minimize the number of phone calls. An obvious but expensive method is for each student to call each of the other seven students. A much better method is to imagine that the eight students are the vertices of a cube, say with coordinates from $\{0, 1\}^3$. There are three types of edges:

1. Those parallel to the z-axis, called *type* 1;
2. Those parallel to the y-axis, called *type* 2;
3. Those parallel to the x-axis, called *type* 3.

The communication can proceed in three rounds as follows: All nodes connected by type 1 edges exchange solutions; all nodes connected by type 2 edges exchange solutions; and finally all nodes connected by type 3 edges exchange solutions.

It is easy to see that everybody has all the answers at the end of the three rounds. Furthermore, each student is involved only in three calls (making a call or receiving it), and the total number of calls is twelve.

In the general case, N nodes would like to exchange information in such a way that eventually every node has all the information. A good way to to this is to construct certain finite projective spaces, as explained in Beutelspacher and Rosenbaum [2]. We pick q to be an integer (for instance, a prime number) such that there is a finite projective space of any dimension over the finite field of order q. Then, we pick d such that

$$q^{d-1} < N \leq q^d.$$

Since q is prime, there is a projective space $\mathbf{P}(K^{d+1})$ of dimension d over the finite field K of order q, and letting $\mathcal{H}$ be the hyperplane at infinity in $\mathbf{P}(K^{d+1})$, we pick a frame $P_1, \ldots, P_d$ in $\mathcal{H}$. It turns out that the affine space $\mathcal{A} = \mathbf{P}(K^{d+1}) - \mathcal{H}$ has q^d points. Then the communication nodes can be identified with points in the affine space $\mathcal{A}$. Assuming for simplicity that $N = q^d$, the algorithm proceeds in d rounds.

During round i, each node $Q \in \mathscr{A}$ sends the information it has received to all nodes in $\mathscr{A}$ on the line QP_i.

It can be shown that at the end of the d rounds, each node has the total information, and that the total number of transactions is at most

$$(q-1)\log_q(N)N.$$

Other applications of projective spaces to communication systems with switches are described in Chapter 2, Section 8, of Beutelspacher and Rosenbaum [2]. Applications to error-correcting codes are described in Chapter 5 of the same book. Introducing even the most elementary notions of coding theory would take too much space. Let us simply say that the existence of certain types of good codes called *linear $[n, n-r]$-codes with minimum distance d* is equivalent to the existence of certain sets of points called $(n, d-1)$-sets in the finite projective space $\mathbf{P}(\{0, 1\}^r)$. For the sake of completeness, a set of n points in a projective space is an (n, s)-*set* if s is the largest integer such that every subset of s points is projectively independent. For example, an $(n, 3)$-set is a set of n points no three of which are collinear, but at least four of them are coplanar.

Other applications of projective geometry to cryptography are given in Chapter 6 of Beutelspacher and Rosenbaum [2].

5.14 Problems

5.1. (a) Prove that for any field K and any $n \geq 0$, there is a bijection between $\mathbf{P}(K^{n+1})$ and $K^n \cup \mathbf{P}(K^n)$ (which allows us to identify them).

(b) For $K = \mathbb{R}$ or $\mathbb{C}$, prove that $\mathbb{RP}^n$ and $\mathbb{CP}^n$ are connected and compact.
Hint. Recall that $\mathbb{RP}^n = p(\mathbb{R}^{n+1})$ and $\mathbb{CP}^n = p(\mathbb{C}^{n+1})$. If

$$S^n = \{(x_1, \ldots, x_{n+1}) \in K^{n+1} \mid x_1^2 + \cdots + x_{n+1}^2 = 1\},$$

prove that $p(S^n) = p(K^{n+1}) = \mathbf{P}(K^{n+1})$, and recall that S^n is compact for all $n \geq 0$ and connected for $n \geq 1$. For $n = 0$, $\mathbf{P}(K)$ consists of a single point.

5.2. Recall that $\mathbb{R}^2$ and $\mathbb{C}$ can be identified using the bijection $(x, y) \mapsto x + iy$. Also recall that the subset $U(1) \subseteq \mathbb{C}$ consisting of all complex numbers of the form $\cos\theta + i\sin\theta$ is homeomorphic to the circle $S^1 = \{(x, y) \in \mathbb{R}^2 \mid x^2 + y^2 = 1\}$. If $c \colon U(1) \to U(1)$ is the map defined such that

$$c(z) = z^2,$$

prove that $c(z_1) = c(z_2)$ iff either $z_2 = z_1$ or $z_2 = -z_1$, and thus that c induces a bijective map $\widehat{c} \colon \mathbb{RP}^1 \to S^1$. Prove that $\widehat{c}$ is a homeomorphism (remember that $\mathbb{RP}^1$ is compact).

5.3. (i) In $\mathbb{R}^3$, the sphere S^2 is the set of points of coordinates (x,y,z) such that $x^2 + y^2 + z^2 = 1$. The point $N = (0,0,1)$ is called the *north pole*, and the point $S = (0,0,-1)$ is called the *south pole*. The *stereographic projection map* $\sigma_N \colon (S^2 - \{N\}) \to \mathbb{R}^2$ is defined as follows: For every point $M \neq N$ on S^2, the point $\sigma_N(M)$ is the intersection of the line through N and M and the plane of equation $z = 0$. Show that if M has coordinates (x,y,z) (with $x^2 + y^2 + z^2 = 1$), then

$$\sigma_N(M) = \left(\frac{x}{1-z}, \frac{y}{1-z} \right).$$

Prove that σ_N is bijective and that its inverse is given by the map $\tau_N \colon \mathbb{R}^2 \to (S^2 - \{N\})$, with

$$(x,y) \mapsto \left(\frac{2x}{x^2 + y^2 + 1}, \frac{2y}{x^2 + y^2 + 1}, \frac{x^2 + y^2 - 1}{x^2 + y^2 + 1} \right).$$

Similarly, $\sigma_S \colon (S^2 - \{S\}) \to \mathbb{R}^2$ is defined as follows: For every point $M \neq S$ on S^2, the point $\sigma_S(M)$ is the intersection of the line through S and M and the plane of equation $z = 0$. Show that

$$\sigma_S(M) = \left(\frac{x}{1+z}, \frac{y}{1+z} \right).$$

Prove that σ_S is bijective and that its inverse is given by the map $\tau_S \colon \mathbb{R}^2 \to (S^2 - \{S\})$, with

$$(x,y) \mapsto \left(\frac{2x}{x^2 + y^2 + 1}, \frac{2y}{x^2 + y^2 + 1}, \frac{1 - x^2 - y^2}{x^2 + y^2 + 1} \right).$$

Using the complex number $u = x + iy$ to represent the point (x,y), the maps $\tau_N \colon \mathbb{R}^2 \to (S^2 - \{N\})$ and $\sigma_N \colon (S^2 - \{N\}) \to \mathbb{R}^2$ can be viewed as maps from $\mathbb{C}$ to $(S^2 - \{N\})$ and from $(S^2 - \{N\})$ to $\mathbb{C}$, defined such that

$$\tau_N(u) = \left(\frac{2u}{|u|^2 + 1}, \frac{|u|^2 - 1}{|u|^2 + 1} \right)$$

and

$$\sigma_N(u,z) = \frac{u}{1-z},$$

and similarly for τ_S and σ_S. Prove that if we pick two suitable orientations for the xy-plane, we have

$$\sigma_N(M)\sigma_S(M) = 1,$$

for every $M \in S^2 - \{N,S\}$.

(ii) Identifying $\mathbb{C}^2$ and $\mathbb{R}^4$, for $z = x + iy$ and $z' = x' + iy'$, we define

$$\|(z,z')\| = \sqrt{x^2 + y^2 + x'^2 + y'^2}.$$

The sphere S^3 is the subset of $\mathbb{C}^2$ (or $\mathbb{R}^4$) consisting of those points (z,z') such that $\|(z,z')\|^2 = 1$.

Prove that $\mathbf{P}(\mathbb{C}^2) = p(S^3)$, where $p\colon (\mathbb{C}^2 - \{(0,0)\}) \to \mathbf{P}(\mathbb{C}^2)$ is the projection map. If we let $u = z/z'$ (where $z, z' \in \mathbb{C}$) in the map

$$u \mapsto \left(\frac{2u}{|u|^2 + 1}, \frac{|u|^2 - 1}{|u|^2 + 1} \right)$$

and require that $\|(z,z')\|^2 = 1$, show that we get the map $HF\colon S^3 \to S^2$ defined such that

$$HF((z,z')) = (2z\overline{z'},\ |z|^2 - |z'|^2).$$

Prove that $HF\colon S^3 \to S^2$ induces a bijection $\widehat{HF}\colon \mathbf{P}(\mathbb{C}^2) \to S^2$, and thus that $\mathbb{CP}^1 = \mathbf{P}(\mathbb{C}^2)$ is homeomorphic to S^2.

(iii) Prove that the inverse image $HF^{-1}(s)$ of every point $s \in S^2$ is a circle. Thus S^3 can be viewed as a union of disjoint circles. The map HF is called the *Hopf fibration*.

5.4. (i) Prove that the *Veronese map* $V_2\colon \mathbb{R}^3 \to \mathbb{R}^6$ defined such that

$$V_2(x,y,z) = (x^2, y^2, z^2, yz, zx, xy)$$

induces a homeomorphism of $\mathbb{RP}^2$ onto $V_2(S^2)$. Show that $V_2(S^2)$ is a subset of the hyperplane $x_1 + x_2 + x_3 = 1$ in $\mathbb{R}^6$, and thus that $\mathbb{RP}^2$ is homeomorphic to a subset of $\mathbb{R}^5$. Prove that this homeomorphism is smooth.

(ii) Prove that the *Veronese map* $V_3\colon \mathbb{R}^4 \to \mathbb{R}^{10}$ defined such that

$$V_3(x,y,z,t) = (x^2, y^2, z^2, t^2, xy, yz, xz, xt, yt, zt)$$

induces a homeomorphism of $\mathbb{RP}^3$ onto $V_3(S^3)$. Show that $V_3(S^3)$ is a subset of the hyperplane $x_1 + x_2 + x_3 + x_4 = 1$ in $\mathbb{R}^{10}$, and thus that $\mathbb{RP}^3$ is homeomorphic to a subset of $\mathbb{R}^9$. Prove that this homeomorphism is smooth.

5.5. (i) Given a projective plane $\mathbf{P}(E)$ (over any field K) and any projective frame (a,b,c,d) in $\mathbf{P}(E)$, recall that a line is defined by an equation of the form $ux + vy + wz = 0$, where u,v,w are not all zero, and that two lines $ux + vy + wz = 0$ and $u'x + v'y + w'z = 0$ are identical iff $u' = \lambda u$, $v' = \lambda v$, and $w' = \lambda w$, for some $\lambda \neq 0$. Show that any two distinct lines $ux + vy + wz = 0$ and $u'x + v'y + w'z = 0$ intersect in a unique point of homogeneous coordinates

$$(vw' - wv',\ wu' - uw',\ uv' - vu').$$

(ii) Given a projective frame (a,b,c,d), let a' be the intersection of $\langle d,a \rangle$ and $\langle b,c \rangle$, b' be the intersection of $\langle d,b \rangle$ and $\langle a,c \rangle$, and c' be the intersection of $\langle d,c \rangle$ and $\langle a,b \rangle$. Show that the points a',b',c' have homogeneous coordinates $(0,1,1)$, $(1,0,1)$, and $(1,1,0)$. Let e be the intersection of $\langle b,c \rangle$ and $\langle b',c' \rangle$, f be the intersection of $\langle a,c \rangle$ and $\langle a',c' \rangle$, and g be the intersection of $\langle a,b \rangle$ and $\langle a',b' \rangle$. Show that

e, f, g have homogeneous coordinates $(0, -1, 1)$, $(1, 0, -1)$, and $(-1, 1, 0)$, and thus that the points e, f, g are on the line of equation $x + y + z = 0$.

5.6. Prove that if $(a_i)_{1 \leq i \leq n+2}$ is a projective frame, then each subfamily $(a_j)_{j \neq i}$ is projectively independent.

5.7. (i) Given a projective space $\mathbf{P}(E)$ of dimension 3 (over any field K) and any projective frame (A, B, C, D, E) in $\mathbf{P}(E)$, recall that a plane is defined by an equation of the form $ux_0 + vx_1 + wx_2 + tx_3 = 0$ where u, v, w, t are not all zero.

Letting (a_0, a_1, a_2, a_3), (b_0, b_1, b_2, b_3), (c_0, c_1, c_2, c_3), and (d_0, d_1, d_2, d_3) be the homogeneous coordinates of some points a, b, c, d with respect to the projective frame (A, B, C, D, E), prove that a, b, c, d are coplanar iff

$$\begin{vmatrix} a_0 & b_0 & c_0 & d_0 \\ a_1 & b_1 & c_1 & d_1 \\ a_2 & b_2 & c_2 & d_2 \\ a_3 & b_3 & c_3 & d_3 \end{vmatrix} = 0.$$

(ii) Two tetrahedra (A, B, C, D) and (A', B', C', D') are called *Möbius tetrahedra* if A, B, C, D belong respectively to the planes $\langle B', C', D' \rangle$, $\langle C', D', A' \rangle$, $\langle D', A', B' \rangle$, and $\langle A', B', C' \rangle$, and also if A', B', C', D' belong respectively to the planes $\langle B, C, D \rangle$, $\langle C, D, A \rangle$, $\langle D, A, B \rangle$, and $\langle A, B, C \rangle$.

Prove that if A, B, C, D belong respectively to the planes $\langle B', C', D' \rangle$, $\langle C', D', A' \rangle$, $\langle D', A', B' \rangle$, and $\langle A', B', C' \rangle$, and if A', B', C' belong respectively to the planes $\langle B, C, D \rangle$, $\langle C, D, A \rangle$, and $\langle D, A, B \rangle$, then D' belongs to $\langle A, B, C \rangle$. Prove that Möbius tetrahedra exist (Möbius, 1828).

Hint. Let (A, B, C, D, E) be a projective frame based on A, B, C, D. Find the conditions expressing that A', B', C', D' belong respectively to the planes $\langle B, C, D \rangle$, $\langle C, D, A \rangle$, $\langle D, A, B \rangle$, and $\langle A, B, C \rangle$, that A', B', C', D' are not coplanar, and that A, B, C, D belong respectively to the planes $\langle B', C', D' \rangle$, $\langle C', D', A' \rangle$, $\langle D', A', B' \rangle$, and $\langle A', B', C' \rangle$. Show that these conditions are compatible.

5.8. Show that if we relax the hypotheses of Lemma 5.5 to $(a_i)_{1 \leq i \leq n+2}$ being a projective frame in $\mathbf{P}(E)$ and $(b_i)_{1 \leq i \leq n+2}$ being any $n + 2$ points in $\mathbf{P}(F)$, then there may be no projective map $h \colon \mathbf{P}(E) \to \mathbf{P}(F)$ such that $h(a_i) = b_i$ for $1 \leq i \leq n + 2$, or h may not be necessarily unique or bijective.

5.9. For every i, $1 \leq i \leq n + 1$, let U_i be the subset of $\mathbb{RP}^n = \mathbf{P}(\mathbb{R}^{n+1})$ consisting of all points of homogeneous coordinates $(x_1, \ldots, x_i, \ldots, x_{n+1})$ such that $x_i \neq 0$. Show that U_i is an open subset of $\mathbb{RP}^n$. Show that $U_i \cap U_j \neq \emptyset$ for all i, j. Show that there is a bijection between U_i and $\mathbb{A}^n$ defined such that

$$(x_1, \ldots, x_{i-1}, x_i, x_{i+1}, \ldots, x_{n+1}) \mapsto \left(\frac{x_1}{x_i}, \ldots, \frac{x_{i-1}}{x_i}, \frac{x_{i+1}}{x_i}, \ldots, \frac{x_{n+1}}{x_i} \right),$$

whose inverse is the map

$$(x_1, \ldots, x_n) \mapsto (x_1, \ldots, x_{i-1}, 1, x_i, \ldots, x_n).$$

Does the above result extend to $\mathbb{P}^n_K$ where K is any field?

5.10. (i) Given an affine space $(E, \overrightarrow{E})$ (over any field K), prove that there is a bijection between affine subspaces of E and projective subspaces of $\widetilde{E}$ not contained in $\mathbf{P}(\overrightarrow{E})$.

(ii) Prove that two affine subspaces of E are parallel iff the corresponding projective subspaces of $\widetilde{E}$ have the same intersection with the hyperplane at infinity $\mathbf{P}(\overrightarrow{E})$.

(iii) Prove that there is a bijection between affine maps from E to F and projective maps from $\widetilde{E}$ to $\widetilde{F}$ mapping the hyperplane at infinity $\mathbf{P}(\overrightarrow{E})$ into the hyperplane at infinity $\mathbf{P}(\overrightarrow{F})$.

5.11. (i) Consider the map $\varphi \colon \mathbb{RP}^1 \times \mathbb{RP}^1 \to \mathbb{RP}^3$ defined such that

$$\varphi((x_0,x_1), (y_0,y_1)) = (x_0y_0, x_0y_1, x_1y_0, x_1y_1),$$

where (x_0,x_1) and (y_0,y_1) are homogeneous coordinates on $\mathbb{RP}^1$. Prove that φ is well-defined and that $\varphi(\mathbb{RP}^1 \times \mathbb{RP}^1)$ is equal the algebraic subset of $\mathbb{RP}^3$ defined by the homogeneous equation

$$w_{0,0}\, w_{1,1} = w_{0,1}\, w_{1,0},$$

where $(w_{0,0}, w_{0,1}, w_{1,0}, w_{1,1})$ are homogeneous coordinates on $\mathbb{RP}^3$.
Hint. Show that if $w_{0,0}\, w_{1,1} = w_{0,1}\, w_{1,0}$ and for instance $w_{0,0} \neq 0$, then

$$\varphi((w_{0,0},\, w_{1,0}), (w_{0,0},\, w_{0,1})) = w_{0,0}(w_{0,0},\, w_{0,1},\, w_{1,0},\, w_{1,1}),$$

and since $w_{0,0}(w_{0,0},\, w_{0,1},\, w_{1,0},\, w_{1,1})$ and $(w_{0,0},\, w_{0,1},\, w_{1,0},\, w_{1,1})$ are equivalent homogeneous coordinates, the result follows.

Prove that φ is injective.

For $x = (x_0,x_1) \in \mathbb{RP}^1$, show that $\varphi(\{x\} \times \mathbb{RP}^1)$ is a line L^1_x in $\mathbb{RP}^3$, that $L^1_x \cap L^1_{x'} = \emptyset$ whenever $L^1_x \neq L^1_{x'}$, and that the union of all these lines is equal to $\varphi(\mathbb{RP}^1 \times \mathbb{RP}^1)$. Similarly, for $y = (y_0,y_1) \in \mathbb{RP}^1$, show that $\varphi(\mathbb{RP}^1 \times \{y\})$ is a line L^2_y in $\mathbb{RP}^3$, that $L^2_y \cap L^2_{y'} = \emptyset$ whenever $L^2_y \neq L^2_{y'}$, and that the union of all these lines is equal to $\varphi(\mathbb{RP}^1 \times \mathbb{RP}^1)$. Also prove that $L^1_x \cap L^2_y$ consists of a single point.

The embedding φ is called the *Segre embedding*. It shows that $\mathbb{RP}^1 \times \mathbb{RP}^1$ can be embedded as a quadric surface in $\mathbb{RP}^3$. Do the above results extend to $\mathbb{P}^1_K \times \mathbb{P}^1_K$ and $\mathbb{P}^3_K$ where K is any field? Draw as well as possible the affine part of $\varphi(\mathbb{RP}^1 \times \mathbb{RP}^1)$ in $\mathbb{R}^3$ corresponding to $w_{1,1} = 1$.

(ii) Consider the map $\varphi \colon \mathbb{RP}^m \times \mathbb{RP}^n \to \mathbb{RP}^N$ where $N = (m+1)(n+1) - 1$, defined such that

$$\varphi((x_0,\ldots,x_m), (y_0,\ldots,y_n)) = (x_0y_0, \ldots, x_0y_n, x_1y_0, \ldots, x_1y_n, \ldots, x_my_0, \ldots, x_my_n),$$

where $(x_0,\ldots,x_m)$ and $(y_0,\ldots,y_n)$ are homogeneous coordinates on $\mathbb{RP}^m$ and $\mathbb{RP}^n$. Prove that φ is well-defined and that $\varphi(\mathbb{RP}^m \times \mathbb{RP}^n)$ is equal the algebraic subset of $\mathbb{RP}^N$ defined by the set of homogeneous equations

$$\begin{vmatrix} w_{i,j} & w_{i,l} \\ w_{k,j} & w_{k,l} \end{vmatrix} = 0,$$

where $0 \le i,k \le m$ and $0 \le j,l \le n$, and where $(w_{0,0},\ldots,w_{0,m},\ldots,w_{m,0},\ldots,w_{m,n})$ are homogeneous coordinates on $\mathbb{RP}^N$.

Hint. Show that if

$$\begin{vmatrix} w_{i,j} & w_{i,l} \\ w_{k,j} & w_{k,l} \end{vmatrix} = 0,$$

where $0 \le i,k \le m$ and $0 \le j,l \le n$ and for instance $w_{0,0} \ne 0$, then

$$\varphi(x,y) = w_{0,0}(w_{0,0},\ldots,w_{0,m},\ldots,w_{m,0},\ldots,w_{m,n}),$$

where $x = (w_{0,0},\ldots,w_{m,0})$ and $y = (w_{0,0},\ldots,w_{0,n})$.

Prove that φ is injective. The embedding φ is also called the *Segre embedding*. It shows that $\mathbb{RP}^m \times \mathbb{RP}^n$ can be embedded as an algebraic variety in $\mathbb{RP}^N$. Do the above results extend to $\mathbb{P}^m_K \times \mathbb{P}^n_K$ and $\mathbb{P}^N_K$ where K is any field?

5.12. (i) In the projective space $\mathbb{RP}^3$, a line D is determined by two distinct hyperplanes of equations

$$\alpha x + \beta y + \gamma z + \delta t = 0,$$
$$\alpha' x + \beta' y + \gamma' z + \delta' t = 0,$$

where $(\alpha,\beta,\gamma,\delta)$ and $(\alpha',\beta',\gamma',\delta')$ are linearly independent.

Prove that the equations of the two hyperplanes defining D can always be written either as

$$x_1 = ax_3 + a'x_4,$$
$$x_2 = bx_3 + b'x_4,$$

where $\{x_1,x_2,x_3,x_4\} = \{x,y,z,t\}$, $\{x_1,x_2\} \subseteq \{x,y,z\}$, and either $a \ne 0$ or $b \ne 0$, or as

$$t = 0,$$
$$lx + my + nz = 0,$$

where $l \ne 0$, $m \ne 0$, or $n \ne 0$.

In the first case, prove that D is also determined by the intersection of three hyperplanes whose equations are of the form

$$cy - bz = lt,$$
$$az - cx = mt,$$

$$bx - ay = nt,$$

where the equation

$$al + bm + cn = 0$$

holds, and where $a \neq 0$, $b \neq 0$, or $c \neq 0$. We can view (a,b,c,l,m,n) as homogeneous coordinates in $\mathbb{RP}^5$ associated with D. In the case where the equations of D are

$$t = 0,$$
$$lx + my + nz = 0,$$

we let $(0,0,0,l,m,n)$ be the homogeneous coordinates associated with D. Of course, $al + bm + cn = 0$ holds. The homogeneous coordinates (a,b,c,l,m,n) such that $al + bm + cn = 0$ are called the *Plücker coordinates* of D.

(ii) Conversely, given some homogeneous coordinates (a,b,c,l,m,n) in $\mathbb{RP}^5$ satisfying the equation

$$al + bm + cn = 0,$$

show that there is a unique line D with Plücker coordinates (a,b,c,l,m,n).
Hint. If $a = b = c = 0$, the corresponding line has equations

$$t = 0,$$
$$lx + my + nz = 0.$$

Otherwise, the equations

$$cy - bz = lt,$$
$$az - cx = mt,$$
$$bx - ay = nt,$$

are compatible, and they determine a unique line D with Plücker coordinates (a,b,c,l,m,n).

Conclude that the lines in $\mathbb{RP}^3$ can be viewed as the algebraic subset of $\mathbb{RP}^5$ defined by the homogeneous equation

$$x_1 x_3 + x_2 x_5 + x_3 x_6 = 0.$$

This quadric surface in $\mathbb{RP}^5$ is an example of a *Grassmannian variety*. It is often called the *Klein quadric*. Do the above results extend to lines in $\mathbb{P}^3_K$ and $\mathbb{P}^5_K$ where K is any field?

5.13. Given any two distinct point $a, b \in \mathbb{RP}^3$ of homogeneous coordinates (a_1, a_2, a_3, a_4) and (b_1, b_2, b_3, b_4), let $p_{12}, p_{13}, p_{14}, p_{34}, p_{42}, p_{23}$ be the numbers defined as follows:

$$p_{12} = \begin{vmatrix} a_1 & a_2 \\ b_1 & b_2 \end{vmatrix}, \qquad p_{13} = \begin{vmatrix} a_1 & a_3 \\ b_1 & b_3 \end{vmatrix}, \qquad p_{14} = \begin{vmatrix} a_1 & a_4 \\ b_1 & b_4 \end{vmatrix},$$

$$p_{34} = \begin{vmatrix} a_3 & a_4 \\ b_3 & b_4 \end{vmatrix}, \qquad p_{42} = \begin{vmatrix} a_4 & a_2 \\ b_4 & b_2 \end{vmatrix}, \qquad p_{23} = \begin{vmatrix} a_2 & a_3 \\ b_2 & b_3 \end{vmatrix}.$$

(i) Prove that

$$p_{12}p_{34} + p_{13}p_{42} + p_{14}p_{23} = 0.$$

Hint. Expand the determinant

$$\begin{vmatrix} a_1 & b_1 & a_1 & b_1 \\ a_2 & b_2 & a_2 & b_2 \\ a_3 & b_3 & a_3 & b_3 \\ a_4 & b_4 & a_4 & b_4. \end{vmatrix}$$

Conversely, given any six numbers satisfying the equation

$$p_{12}p_{34} + p_{13}p_{42} + p_{14}p_{23} = 0,$$

prove that two points $a = (a_1, a_2, a_3, 0)$ and $b = (b_1, 0, b_3, b_4)$ can be determined such that the p_{ij} are associated with a and b.
Hint. Show that the equations

$$-a_2 b_1 = p_{12},$$
$$a_3 b_4 = p_{34},$$
$$a_1 b_3 - a_3 b_1 = p_{13},$$
$$-a_2 b_4 = p_{42},$$
$$a_1 b_4 = p_{14},$$
$$a_2 b_3 = p_{23},$$

are solvable iff

$$p_{12}p_{34} + p_{13}p_{42} + p_{14}p_{23} = 0.$$

The tuple $(p_{12}, p_{13}, p_{14}, p_{34}, p_{42}, p_{23})$ can be viewed as homogeneous coordinates in $\mathbb{RP}^5$ of the line $\langle a, b \rangle$. They are the *Plücker coordinates* of $\langle a, b \rangle$.

(ii) Prove that two lines of Plücker coordinates $(p_{12}, p_{13}, p_{14}, p_{34}, p_{42}, p_{23})$ and $(p'_{12}, p'_{13}, p'_{14}, p'_{34}, p'_{42}, p'_{23})$ intersect iff

$$p_{12}p'_{34} + p_{13}p'_{42} + p_{14}p'_{23} + p_{34}p'_{12} + p_{42}p'_{13} + p_{23}p'_{14} = 0.$$

Thus, the set of lines that meet a given line in $\mathbb{RP}^3$ correspond to a set of points in $\mathbb{RP}^5$ belonging to a hyperplane, as well as to the Klein quadric. Do the above results extend to lines in $\mathbb{P}_K^3$ and $\mathbb{P}_K^5$ where K is any field?

(iii) Three lines L_1, L_2, L_3 in $\mathbb{RP}^3$ are mutually skew lines iff no pairs of any two of these lines are coplanar. Given any three mutually skew lines L_1, L_2, L_3 and any four lines M_1, M_2, M_3, M_4 in $\mathbb{RP}^3$ such that each line M_i meets every line L_j, show that if any line L meets three of the four lines M_1, M_2, M_3, M_4, then it also meets the fourth. Does the above result extend to $\mathbb{P}_K^3$ where K is any field? Show that the set of lines meeting three given mutually skew lines L_1, L_2, L_3 in $\mathbb{P}_K^3$ is a ruled quadric surface. What do the affine pieces of this quadric look like in $\mathbb{R}^3$?

(iv) Four lines L_1, L_2, L_3, L_4 in $\mathbb{RP}^3$ are mutually skew lines iff no pairs of any two of these lines are coplanar. Given any four mutually skew lines L_1, L_2, L_3, L_4, show that there are at most two lines meeting all four of them. In $\mathbb{CP}^3$, show that there are either two distinct lines or a double line meeting all four of them.

5.14. (i) Prove that the cross-ratio $[a, b, c, d]$ is invariant if any two elements and the complementary two elements are transposed. Prove that

$$[a, b, c, d] = [b, a, c, d]^{-1} = [a, b, d, c]^{-1}$$

and that

$$[a, b, c, d] = 1 - [a, c, b, d].$$

(ii) Letting $\lambda = [a, b, c, d]$, prove that if $\lambda \in \{\infty, 0, 1\}$, then any permutation of $\{a, b, c, d\}$ yields a cross-ratio in $\{\infty, 0, 1\}$, and if $\lambda \notin \{\infty, 0, 1\}$, then there are at most the six values

$$\lambda, \quad \frac{1}{\lambda}, \quad 1 - \lambda, \quad 1 - \frac{1}{\lambda}, \quad \frac{1}{1-\lambda}, \quad \frac{\lambda}{\lambda - 1}.$$

(iii) Prove that the function

$$\lambda \mapsto \frac{(\lambda^2 - \lambda + 1)^3}{\lambda^2 (1 - \lambda)^2}$$

takes a constant value on the six values listed in part (ii).

5.15. Viewing a point (x, y) in $\mathbb{A}^2$ as the complex number $z = x + iy$, prove that four points (a, b, c, d) are cocyclic or collinear iff the cross-ratio $[a, b, c, d]$ is a real number.

5.16. Given any distinct points (x_1, x_2, x_3, x_4) in $\mathbb{RP}^1$, prove that they form a harmonic division, i.e., $[x_1, x_2, x_3, x_4] = -1$ iff

$$2(x_1 x_2 + x_3 x_4) = (x_1 + x_2)(x_3 + x_4).$$

Prove that $[0, x_2, x_3, x_4] = -1$ iff

$$\frac{2}{x_2} = \frac{1}{x_3} + \frac{1}{x_4}.$$

Prove that $[x_1, x_2, x_3, \infty] = -1$ iff

$$2x_3 = x_1 + x_2.$$

Do the above results extend to $\mathbb{P}_K^1$ where K is any field?

5.17. Consider the quadrangle (projective frame) (a,b,c,d) in a projective plane, and let a' be the intersection of $\langle d,a \rangle$ and $\langle b,c \rangle$, b' be the intersection of $\langle d,b \rangle$ and $\langle a,c \rangle$, and c' be the intersection of $\langle d,c \rangle$ and $\langle a,b \rangle$. Show that the following quadruples of lines form harmonic divisions: $(\langle c,a \rangle, \langle b',a' \rangle, \langle d,b \rangle, \langle b',c' \rangle)$, $(\langle b,a \rangle, \langle c',a' \rangle, \langle d,c \rangle, \langle c',b' \rangle)$, and $(\langle b,c \rangle, \langle a',b' \rangle, \langle a,d \rangle, \langle a',c' \rangle)$.

Hint. Send some suitable lines to infinity.

5.18. Let $\mathbf{P}(E)$ be a projective space over any field. For any projective map $\mathbf{P}(f)\colon \mathbf{P}(E) \to \mathbf{P}(E)$, a point $a = p(u)$ is a fixed point of $\mathbf{P}(f)$ iff $\mathbf{P}(f)(a) = a$. Prove that $a = p(u)$ is a fixed point of $\mathbf{P}(f)$ iff u is an eigenvector of the linear map $f\colon E \to E$. Prove that if $E = \mathbb{R}^{2n+1}$, then every projective map $\mathbf{P}(f)\colon \mathbb{RP}^{2n} \to \mathbb{RP}^{2n}$ has a fixed point. Prove that if $E = \mathbb{C}^{n+1}$, then every projective map $\mathbf{P}(f)\colon \mathbb{CP}^n \to \mathbb{CP}^n$ has a fixed point.

5.19. A projectivity $\mathbf{P}(f)\colon \mathbb{RP}^n \to \mathbb{RP}^n$ is an *involution* if $\mathbf{P}(f)$ is not the identity and if $\mathbf{P}(f) \circ \mathbf{P}(f) = \mathrm{id}$. Prove that a projectivity $\mathbf{P}(f)\colon \mathbb{RP}^1 \to \mathbb{RP}^1$ is an involution iff the trace of the matrix of f is null. Does the above result extend to $\mathbb{P}^1_K$ where K is any field?

5.20. Recall Desargues's theorem in the plane: Given any two triangles (a,b,c) and (a',b',c') in $\mathbb{RP}^2$, where the points a,b,c,a',b',c' are distinct and the lines $A = \langle b,c \rangle$, $B = \langle a,c \rangle$, $C = \langle a,b \rangle$, $A' = \langle b',c' \rangle$, $B' = \langle a',c' \rangle$, $C' = \langle a',b' \rangle$ are distinct, if the lines $\langle a,a' \rangle$, $\langle b,b' \rangle$, and $\langle c,c' \rangle$ intersect in a common point d distinct from a,b,c,a',b',c', then the intersection points $p = \langle b,c \rangle \cap \langle b',c' \rangle$, $q = \langle a,c \rangle \cap \langle a',c' \rangle$, and $r = \langle a,b \rangle \cap \langle a',b' \rangle$ belong to a common line distinct from A,B,C,A',B',C'.

Prove that the dual of the above result is its converse. Deduce Desargues's theorem: Given any two triangles (a,b,c) and (a',b',c') in $\mathbb{RP}^2$, where the points a,b,c,a',b',c' are distinct and the lines $A = \langle b,c \rangle$, $B = \langle a,c \rangle$, $C = \langle a,b \rangle$, $A' = \langle b',c' \rangle$, $B' = \langle a',c' \rangle$, $C' = \langle a',b' \rangle$ are distinct, the lines $\langle a,a' \rangle$, $\langle b,b' \rangle$, and $\langle c,c' \rangle$ intersect in a common point d distinct from a,b,c,a',b',c' iff the intersection points $p = \langle b,c \rangle \cap \langle b',c' \rangle$, $q = \langle a,c \rangle \cap \langle a',c' \rangle$, and $r = \langle a,b \rangle \cap \langle a',b' \rangle$ belong to a common line distinct from A,B,C,A',B',C'.

Do the above results extend to $\mathbb{P}^2_K$ where K is any field?

5.21. Let D and D' be any two distinct lines in the real projective plane $\mathbb{RP}^2$, and let $f\colon D \to D'$ be a projectivity. Prove the following facts.

(1) If f is a perspectivity, then for any two distinct points m,n on D, the lines $\langle m,f(n) \rangle$ and $\langle n,f(m) \rangle$ intersect on some fixed line passing through $D \cap D'$.

Hint. Consider any three distinct points a,b,c on D and use Desargues's theorem.

(2) If f is not a perspectivity, then for any two distinct points m,n on D, the lines $\langle m,f(n) \rangle$ and $\langle n,f(m) \rangle$ intersect on the line passing through $f(D \cap D')$ and $f^{-1}(D \cap D')$.

Hint. Use some suitable composition of perspectivities. The line passing through $f(D \cap D')$ and $f^{-1}(D \cap D')$ is called the *axis* of the projectivity.

(iii) Prove that any projectivity $f\colon D \to D'$ between distinct lines is the composition of two perspectivities.

(iv) Use the above facts to give a quick proof of Pappus's theorem: Given any two distinct lines D and D' in a projective plane, for any distinct points a,b,c,a',b',c' with a,b,c on D and a',b',c' on D', if a,b,c,a',b',c' are distinct from the intersection of D and D', then the intersection points $p = \langle b,c'\rangle \cap \langle b',c\rangle$, $q = \langle a,c'\rangle \cap \langle a',c\rangle$, and $r = \langle a,b'\rangle \cap \langle a',b\rangle$ are collinear.

Do the above results extend to $\mathbb{P}_K^2$ where K is any field?

5.22. Recall that in the real projective plane $\mathbb{RP}^2$, by duality, a point a corresponds to the pencil of lines a^* passing through a.

(i) Given any two distinct points a and b in the real projective plane $\mathbb{RP}^2$ and any line L containing neither a nor b, the *perspectivity of axis L between a^* and b^** is the map $f\colon a^* \to b^*$ defined such that for every line $D \in a^*$, the line $f(D)$ is the line through b and the intersection of D and L.

Prove that a projectivity $f\colon a^* \to b^*$ is a perspectivity iff $f(\langle a,b\rangle) = \langle b,a\rangle$.

(ii) Prove that a bijection $f\colon a^* \to b^*$ is a projectivity iff it preserves the cross-ratios of any four distinct lines in the pencil a^*.

(iii) State and prove the dual of Pappus's theorem.

Do the above results extend to $\mathbb{P}_K^2$ where K is any field?

5.23. (i) Prove that every projectivity $f\colon \mathbb{RP}^1 \to \mathbb{RP}^1$ has at most 2 fixed points. A projectivity $f\colon \mathbb{RP}^1 \to \mathbb{RP}^1$ is called *elliptic* if it has no fixed points, *parabolic* if it has a single fixed point, *hyperbolic* if it has two distinct fixed points. Prove that every projectivity $f\colon \mathbb{CP}^1 \to \mathbb{CP}^1$ has 2 distinct fixed points or a double fixed point.

(ii) Recall that a projectivity $f\colon \mathbb{RP}^1 \to \mathbb{RP}^1$ is an *involution* if f is not the identity and if $f \circ f = \mathrm{id}$. Prove that f is an involution iff there is some point $a \in \mathbb{RP}^1$ such that $f(a) \neq a$ and $f(f(a)) = a$.

(iii) Given any two distinct points $a,b \in \mathbb{RP}^1$, prove that there is a unique involution $f\colon \mathbb{RP}^1 \to \mathbb{RP}^1$ having a and b as fixed points. Furthermore, for all $m \neq a,b$, we have

$$[a,b,m,f(m)] = -1.$$

Conversely, the above formula defines an involution with fixed points a and b.

(iv) Prove that every projectivity $f\colon \mathbb{RP}^1 \to \mathbb{RP}^1$ is the composition of at most two involutions.

Do the above results extend to $\mathbb{P}_K^1$ where K is any field?

5.24. Prove that an involution $f\colon \mathbb{RP}^1 \to \mathbb{RP}^1$ has zero or two distinct fixed points. Prove that an involution $f\colon \mathbb{CP}^1 \to \mathbb{CP}^1$ has two distinct fixed points.

5.25. Prove that a bijection $f\colon \mathbb{RP}^1 \to \mathbb{RP}^1$ having two distinct fixed points a and b is a projectivity iff there is some $k \neq 0$ in $\mathbb{R}$ such that for all $m \neq a,b$, we have

$$[a,b,m,f(m)] = k.$$

Does the above result extend to $\mathbb{P}_K^1$ where K is any field?

5.26. Prove that every projectivity $f\colon \mathbb{RP}^1 \to \mathbb{RP}^1$ is the composition of at most three perspectivities.

Hint. Consider some appropriate perspectivities.

Does the above result extend to $\mathbb{P}^1_K$ where K is any field?

5.27. Let (a,b,c,d) be a projective frame in $\mathbb{RP}^2$, and let D be a line not passing through any of a,b,c,d. The line D intersects $\langle a,b \rangle$ and $\langle c,d \rangle$ in p and p', $\langle b,c \rangle$ and $\langle a,d \rangle$ in q and q', and $\langle b,d \rangle$ and $\langle a,c \rangle$ in r and r'. Prove that there is a unique involution mapping p to p', q to q', and r to r'.
Hint. Consider some appropriate perspectivities.

Does the above result extend to $\mathbb{P}^2_K$ where K is any field?

5.28. Let (a,b,c) be a triangle in $\mathbb{RP}^2$, and let D be a line not passing through any of a,b,c, so that D intersects $\langle b,c \rangle$ in p, $\langle c,a \rangle$ in q, and $\langle a,b \rangle$ in r. Let L_a, L_b, L_c be three lines passing through a,b,c, respectively, and intersecting D in p',q',r'. Prove that there is a unique involution mapping p to p', q to q', and r to r' iff the lines L_a, L_b, L_c are concurrent.
Hint. Use Problem 5.27.

Does the above result extend to $\mathbb{P}^2_K$ where K is any field?

5.29. In a projective plane $\mathbf{P}(E)$ where E is a vector space of dimension 3 over any field K, a *conic* is the set of points of homogeneous coordinates (x,y,z) such that

$$\alpha x^2 + \beta y^2 + 2\gamma xy + 2\delta xz + 2\lambda yz + \mu z^2 = 0,$$

where $(\alpha, \beta, \gamma, \delta, \lambda, \mu) \neq (0,0,0,0,0,0)$. We can write the equation of the conic as

$$(x,y,z) \begin{pmatrix} \alpha & \gamma & \delta \\ \gamma & \beta & \lambda \\ \delta & \lambda & \mu \end{pmatrix} \begin{pmatrix} x \\ y \\ z \end{pmatrix} = 0,$$

and letting

$$A = \begin{pmatrix} \alpha & \gamma & \delta \\ \gamma & \beta & \lambda \\ \delta & \lambda & \mu \end{pmatrix}, \qquad X = \begin{pmatrix} x \\ y \\ z \end{pmatrix},$$

the equation of the conic becomes

$$X^\top A X = 0.$$

We say that a conic of equation $X^\top A X = 0$ is *nondegenerate* if $\det(A) \neq 0$ and *degenerate* if $\det(A) = 0$.

(i) For $K = \mathbb{R}$, show that there is only one type of nondegenerate conic, and that there are three kinds of degenerate conics: two distinct lines, a double line, a point, and the empty set. For $K = \mathbb{C}$, show that there is only one type of nondegenerate conic, and that there are two kinds of degenerate conics: two distinct lines or a double line.

(ii) Given any two distinct points a and b in $\mathbb{RP}^2$ and any projectivity $f \colon a^* \to b^*$ that is not a perspectivity, prove that the set of points of the form $L \cap f(L)$ is a nondegenerate conic, where L is any line in the pencil a^*.

What happens when f is a perspectivity? Does the above result hold for any field K?

(iii) Given a nondegenerate conic C, for any point $a \in C$ we can define a bijection $j_a\colon a^* \to C$ as follows: For every line L through a, we define $j_a(L)$ as the other intersection of L and C when L is not the tangent to C at a, and $j_a(L) = a$ otherwise. Given any two distinct points $a, b \in C$, show that the map $f = j_b^{-1} \circ j_a$ is a projectivity $f\colon a^* \to b^*$ that is not a perspectivity. In fact, if O is the intersection of the tangents to C at a and b, show that $f(\langle O, a\rangle) = \langle b, a\rangle$, $f(\langle a, b\rangle) = \langle b, O\rangle$, and for any point $m \neq a, b$ on C, $f(\langle a, m\rangle) = \langle b, m\rangle$. Conclude that C is the set of points of the form $L \cap f(L)$, where L is any line in the pencil a^*.

Hint. In a projective frame where $a = (1, 0, 0)$ and $b = (0, 1, 0)$, the equation of a conic is of the form

$$pz^2 + qxy + ryz + sxz = 0.$$

Remark: The above characterization of the conics is due to Steiner (and Chasles).

(iv) Prove that six points (a, b, c, d, e, f) such that no three of them are collinear belong to a conic iff

$$[\langle a, c\rangle, \langle a, d\rangle, \langle a, e\rangle, \langle a, f\rangle] = [\langle b, c\rangle, \langle b, d\rangle, \langle b, e\rangle, \langle b, f\rangle].$$

5.30. Given a nondegenerate conic C and any six points a, b, c, d, e, f on C such that no three of them are collinear, prove *Pascal's theorem*: The points $z = \langle a, b\rangle \cap \langle d, e\rangle$, $w = \langle b, c\rangle \cap \langle e, f\rangle$, and $t = \langle c, d\rangle \cap \langle f, a\rangle$ are collinear.

Recall that the line $\langle a, a\rangle$ is interpreted as the tangent to C at a.

Hint. By Problem 5.29, for any point m on the conic C, the bijection $j_m\colon m^* \to C$ allows the definition of the cross-ratio of four points a, b, c, d on C as the cross ratio of the lines $\langle m, a\rangle$, $\langle m, b\rangle$, $\langle m, c\rangle$, and $\langle m, d\rangle$ (which does not depend on m). Also recall that the cross-ratio of four lines in the pencil m^* is equal to the cross-ratio of the four intersection points with any line not passing through m. Prove that

$$[z, x, d, e] = [t, c, d, y],$$

and use the perspectivity of center w between $\langle c, y\rangle$ and $\langle e, x\rangle$.

5.31. In a projective plane $\mathbf{P}(E)$ where E is a vector space of dimension 3 over any field K of characteristic different from 2 (say, $K = \mathbb{R}$ or $K = \mathbb{C}$), given a conic C of equation $F(x, y, z) = 0$ where

$$F(x, y, z) = \alpha x^2 + \beta y^2 + 2\gamma xy + 2\delta xz + 2\lambda yz + \mu z^2 = 0$$

(with $(\alpha, \beta, \gamma, \delta, \lambda, \mu) \neq (0, 0, 0, 0, 0, 0)$), using the notation of Problem 5.29 with $X^\top = (x, y, z)$ and $Y^\top = (u, v, w)$, verify that

$$Y^\top A X = \frac{1}{2}\left(u F_x' + v F_y' + w F_z'\right),$$

where F'_x, F'_y, F'_z denote the partial derivatives of $F(x,y,z)$.

If the conic C of equation $X^\top AX = 0$ is nondegenerate, it is well known (and easy to prove) that the tangent line to C at (x_0, y_0, z_0) is given by the equation

$$xF'_{x_0} + yF'_{y_0} + zF'_{z_0} = 0,$$

and thus by the equation $X^\top AX_0 = 0$, with $X^\top = (x,y,z)$ and $X_0^\top = (x_0, y_0, z_0)$. Therefore, the equation of the tangent to C at (x_0, y_0, z_0) is of the form

$$ux + vy + wz = 0,$$

where

$$\begin{pmatrix} u \\ v \\ w \end{pmatrix} = A \begin{pmatrix} x_0 \\ y_0 \\ z_0 \end{pmatrix} \quad \text{and} \quad (x_0, y_0, z_0) A \begin{pmatrix} x_0 \\ y_0 \\ z_0 \end{pmatrix} = 0.$$

(i) If C is a nondegenerate conic of equation $X^\top AX = 0$ in the projective plane $\mathbf{P}(E)$, prove that the set C^* of tangent lines to C is a conic of equation $Y^\top A^{-1} Y = 0$ in the projective plane $\mathbf{P}(E^*)$, where E^* is the dual of the vector space E. Prove that $C^{**} = C$.

Remark: The conic C is sometimes called a *point conic* and the conic C^* a *line conic*. The set of lines defined by the conic C^* is said to be the *envelope* of the conic C.

Conclude that duality transforms the points of a nondegenerate conic into the tangents of the conic, and the tangents of the conic into the points of the conic.

(ii) Given any two distinct lines L and M in $\mathbb{RP}^2$ and any projectivity $f: L \to M$ that is not a perspectivity, prove that the lines of the form $\langle a, f(a) \rangle$ are the tangents enveloping a nondegenerate conic, where a is any point on the line L (use duality).

What happens when f is a perspectivity? Does the above result hold for any field K?

(iii) Given a nondegenerate conic C, for any two distinct tangents L and M to C at a and b, if $O = L \cap M$, show that the map $f: L \to M$ defined such that $f(a) = O$, $f(O) = b$, and $f(L \cap T) = M \cap T$ for any tangent $T \neq L, M$ is a projectivity. Conclude that C is the envelope of the set of lines of the form $\langle m, f(m) \rangle$, where m is any point on L (use duality).

5.32. Given a nondegenerate conic C, prove *Brianchon's theorem*: For any hexagon (a, b, c, d, e, f) circumscribed about C (which means that $\langle a, b \rangle$, $\langle b, c \rangle$, $\langle c, d \rangle$, $\langle d, e \rangle$, $\langle e, f \rangle$, and $\langle f, a \rangle$ are tangent to C), the diagonals $\langle a, d \rangle$, $\langle b, e \rangle$, and $\langle c, f \rangle$ are concurrent.

Hint. Use duality.

5.33. (a) Consider the map $\mathscr{H} : \mathbb{R}^3 \to \mathbb{R}^4$ defined such that

$$(x, y, z) \mapsto (xy, yz, xz, x^2 - y^2).$$

Prove that when it is restricted to the sphere S^2 (in $\mathbb{R}^3$), we have $\mathcal{H}(x,y,z) = \mathcal{H}(x',y',z')$ iff $(x',y',z') = (x,y,z)$ or $(x',y',z') = (-x,-y,-z)$. In other words, the inverse image of every point in $\mathcal{H}(S^2)$ consists of two antipodal points.

Prove that the map $\mathcal{H}$ induces an injective map from the projective plane onto $\mathcal{H}(S^2)$, and that it is a homeomorphism.

(b) The map $\mathcal{H}$ allows us to realize concretely the projective plane in $\mathbb{R}^4$ by choosing any parametrization of the sphere S^2 and applying the map $\mathcal{H}$ to it. Actually, it turns out to be more convenient to use the map $\mathcal{A}$ defined such that

$$(x,y,z) \mapsto (2xy, 2yz, 2xz, x^2 - y^2),$$

because it yields nicer parametrizations. For example, using the stereographic representation where

$$x(u,v) = \frac{2u}{u^2 + v^2 + 1},$$

$$y(u,v) = \frac{2v}{u^2 + v^2 + 1},$$

$$z(u,v) = \frac{u^2 + v^2 - 1}{u^2 + v^2 + 1},$$

show that the following parametrization of the projective plane in $\mathbb{R}^4$ is obtained:

$$x(u,v) = \frac{8uv}{(u^2 + v^2 + 1)^2},$$

$$y(u,v) = \frac{4v(u^2 + v^2 - 1)}{(u^2 + v^2 + 1)^2},$$

$$z(u,v) = \frac{4u(u^2 + v^2 - 1)}{(u^2 + v^2 + 1)^2},$$

$$t(u,v) = \frac{4(u^2 - v^2)}{(u^2 + v^2 + 1)^2}.$$

Investigate the surfaces in $\mathbb{R}^3$ obtained by dropping one of the four coordinates. Show that there are only two of them (up to a rigid motion).

5.34. Give the details of the proof that the altitudes of a triangle are concurrent.

5.35. Let K be the finite field $K = \{0,1\}$. Prove that the projective plane $\mathbf{P}(K^3)$ contains 7 points and 7 lines. Draw the configuration formed by these seven points and lines.

5.36. Prove that if P and Q are two homogeneous polynomials of degree 2 over $\mathbb{R}$ and if $V(P) = V(Q)$ contains at least three elements, then there is some $\lambda \in \mathbb{R}$ such that $Q = \lambda P$, with $\lambda \neq 0$.
Hint. Choose some convenient frame.

5.37. In the Euclidean space $\mathbb{E}^n$ (where $\mathbb{E}^n$ is the affine space $\mathbb{A}^n$ equipped with its usual inner product on $\mathbb{R}^n$), given any $k \in \mathbb{R}$ with $k \neq 0$ and any point a, an *inversion of pole a and power k* is a map $h \colon (\mathbb{E}^n - \{a\}) \to \mathbb{E}^n$ defined such that for every $x \in \mathbb{E}^n - \{a\}$,

$$h(x) = a + k\frac{\overrightarrow{ax}}{\|\overrightarrow{ax}\|^2}.$$

For example, when $n = 2$, choosing any orthonormal frame with origin a, h is defined by the map

$$(x, y) \mapsto \left(\frac{kx}{x^2 + y^2}, \frac{ky}{x^2 + y^2} \right).$$

(a) Assuming for simplicity that $n = 2$, viewing $\mathbb{RP}^2$ as the projective completion of $\mathbb{E}^2$, we can extend h to a partial map $h \colon \mathbb{RP}^2 \to \mathbb{RP}^2$ as follows. Pick any projective frame (a_0, a_1, a_2, a_3) where $a_0 = a + e_1$, $a_1 = a + e_2$, $a_2 = a$, $a_3 = a + e_1 + e_2$, and where (e_1, e_2) is an orthonormal basis for $\mathbb{R}^2$, and define h such that in homogeneous coordinates

$$(x, y, z) \mapsto (kxz, kyz, x^2 + y^2).$$

Show that h is defined on $\mathbb{RP}^2 - \{a\}$. Show that $h \circ h = \mathrm{id}$, except for points on the line at infinity (that are all mapped onto $a = (0, 0, 1)$). Deduce that h is a bijection except for a and the points on the line at infinity. Show that the fixed points of h are on the circle of equation

$$x^2 + y^2 = kz^2.$$

(b) We can also extend h to a partial map $h \colon \mathbb{CP}^2 \to \mathbb{CP}^2$ as in the real case, and define h such that in homogeneous (complex) coordinates

$$(x, y, z) \mapsto (kxz, kyz, x^2 + y^2).$$

Show that h is defined on $\mathbb{CP}^2 - \{a, I, J\}$, where $I = (1, -i, 0)$ and $J = (1, i, 0)$ are the circular points. Show that every point of the line $\langle I, J \rangle$ other than I and J is mapped to A, every point of the line $\langle A, I \rangle$ other than A and I is mapped to I, and every point of the line $\langle A, J \rangle$ other than A and J is mapped to J. Show that $h \circ h = \mathrm{id}$ on the complement of the three lines $\langle I, J \rangle$, $\langle A, I \rangle$, and $\langle A, J \rangle$. Show that the fixed points of h are on the circle of equation

$$x^2 + y^2 = kz^2.$$

Say that a circle of equation

$$ax^2 + ay^2 + bxz + cyz + dz^2 = 0$$

is a *true circle* if $a \neq 0$. We define the *center* of a circle as above (true or not) as the point of homogeneous coordinates $(b, c, -2a)$ and the *radius R* of a true circle is defined as follows: If

$$b^2 + c^2 - 4ad > 0,$$

then $R = \sqrt{b^2 + c^2 - 4ad}/(2a)$; otherwise $R = i\sqrt{4ad - b^2 - c^2}/(2a)$. Note that R can be a complex number. Also, when $a = 0$, we let $R = \infty$.

Verify that in the affine Euclidean plane $\mathbb{E}^2$ (the complement of the line at infinity $z = 0$) the notions of center and radius have the usual meaning (when R is real).

(c) Show that the image of a circle of equation

$$ax^2 + ay^2 + bxz + cyz + dz^2 = 0$$

is the circle of equation

$$dx^2 + dy^2 + kbxz + kcyz + k^2 az^2 = 0.$$

When does a true circle map to a true circle?

(d) Recall the definition of the *stereographic projection map* $\sigma\colon (S^2 - \{N\}) \to \mathbb{R}^2$ from Problem 5.3. Prove that the stereographic projection map is the restriction to S^2 of an inversion of pole N and power $2R^2$ in $\mathbb{E}^3$ (where S^2 a sphere of radius R, N is the north pole of S^2, and the plane of projection is a plane through the center of the sphere).

5.38. As in Problem 5.37, we consider inversions in $\mathbb{RP}^2$ and $\mathbb{CP}^2$, and we assume that some projective frame (a_0, a_1, a_2, a_3) is chosen.

(a) Given two distinct real circles C_1 and C_2 of equations

$$x^2 + y^2 - R^2 z^2 = 0,$$
$$x^2 + y^2 - 2bxz + dz^2 = 0,$$

prove that C_1 and C_2 intersect in two real points iff the line

$$2bx - (d + R^2)z = 0$$

intersects C_1 in two real points iff

$$(R^2 + d - 2bR)(R^2 + d + 2bR) < 0.$$

The line $2bx - (d + R^2)z = 0$ is called the *radical axis* D of the circles C_1 and C_2. If $b = 0$, then C_1 and C_2 have the same center, and the radical axis is the line at infinity. Otherwise, if $b \neq 0$, by chosing a new frame (b_0, b_1, b_2, b_3) such that

$$b_0 = \left(\frac{R^2 + d}{2b} + 1, 0, 0\right), \quad b_1 = \left(\frac{R^2 + d}{2b}, 1, 0\right), \quad b_2 = \left(\frac{R^2 + d}{2b}, 0, 1\right),$$

and

$$b_3 = \left(\frac{R^2 + d}{2b}, 1, 1\right),$$

show that the equations of the circles C_1, C_2 become

$$4b^2(x^2+y^2)+4b(R^2+d)xz+\Delta z^2=0,$$
$$4b^2(x^2+y^2)+4b(R^2+d-2b^2)xz+\Delta z^2=0,$$

where $\Delta=(R^2+d-2bR)(R^2+d+2bR)$.

Letting $C=\Delta/(4b^2)$, the above equations are of the form

$$x^2+y^2-2uxz+Cz^2=0,$$
$$x^2+y^2-2vxz+Cz^2=0,$$

where $u\neq v$.

(b) Consider the pencil of circles defined by C_1 and C_2, i.e., the set of all circles having an equation of the form

$$(\lambda+\mu)(x^2+y^2)-2(\lambda u+\mu v)xz+(\lambda+\mu)Cz^2=0,$$

where $(\lambda,\mu)\neq(0,0)$.

If $C<0$, letting $K^2=-C$ where $K>0$, prove that the circles in the pencil are exactly the circles passing through the points $A=(0,K,1)$ and $B=(0,-K,1)$, called *base points* of the pencil. In this case, prove that the image of all the circles in the pencil by an inversion h of center A is the union of the line at infinity together with the set of all lines through the image $h(B)$ of B under the inversion (pick a convenient frame).

(c) If $C=0$, in which case $A=B=(0,0,1)$, prove that the circles in the pencil are exactly the circles tangent to the radical axis D (at the origin). In this case, prove that the image of all the circles in the pencil by an inversion h of center A is the union of the line at infinity together with the set of all lines parallel to the radical axis D.

(d) If $C>0$, letting $K^2=C$ where $K>0$, prove that there exist two circles in the pencil of radius 0 and of centers $P_1=(K,0,1)$ and $P_2=(-K,0,1)$, called the *Poncelet points* of the pencil. In this case, prove that the image of all the circles in the pencil by an inversion of center P_1 is the set of all circles of center $h(P_2)$ (pick a convenient frame).

Conclude that given any two distinct nonconcentric real circles C_1 and C_2, there is an inversion such that if C_1 and C_2 intersect in two real points, then C_1 and C_2 are mapped to two lines (plus the line at infinity), and if C_1 and C_2 are disjoint (as real circles), then C_1 and C_2 are mapped to two concentric circles.

(e) Given two C^1-curves Γ,Δ in $\mathbb{E}^2$, if Γ and Δ intersect in p, prove that for any inversion h of pole $c\neq p$, h preserves the absolute value of the angle of the tangents to Γ and Δ at p. Conclude that inversions preserve tangency and orthogonality.
Hint. Express Γ,Δ, and h in polar coordinates.

(f) Using (e), prove the following beautiful theorem of Steiner. Let C_1 and C_2 be two disjoint real circles such that C_2 is inside C_1. Construct any sequence $(\Gamma_n)_{n\geq0}$ of circles such that Γ_n is any circle interior to C_1, exterior to C_2, tangent to C_1 and C_2, and furthermore that $\Gamma_{n+1}\neq\Gamma_{n-1}$ and Γ_{n+1} is tangent to Γ_n.

Given a starting circle Γ_0, two cases may arise: Either $\Gamma_n = \Gamma_0$ for some $n \geq 1$, or $\Gamma_n \neq \Gamma_0$ for all $n \geq 1$.

Prove that the outcome is independent of the starting circle Γ_0. In other words, either for every Γ_0 we have $\Gamma_n = \Gamma_0$ for some $n \geq 1$, or for every Γ_0 we have $\Gamma_n \neq \Gamma_0$ for all $n \geq 1$.

5.39. (a) Let $h\colon \mathbb{RP}^2 \to \mathbb{RP}^2$ be the projectivity (w.r.t. any projective frame (a_0, a_1, a_2, a_3)) defined such that

$$(x, y, z) \mapsto (x, y, ax + by + cz),$$

where $c \neq 0$ and h is not the identity.

Prove that the fixed points of h (i.e., those points M such that $h(M) = M$) are the origin $O = a_2 = (0, 0, 1)$ and every point on the line Δ of equation

$$ax + by + (c - 1)z = 0.$$

Prove that every line through the origin is globally invariant under h. Give a geometric construction of $h(M)$ for every point M distinct from O and not on Δ, given any point A distinct from O and not on Δ and its image $A' = h(A)$.

Hint. Consider the intersection P of the line $\langle A, M \rangle$ with the line Δ.

Such a projectivity is called a *homology of center O and of axis Δ* (Poncelet).

Show that in the situation of Desargues's theorem, the triangles (a, b, c) and (a', b', c') are homologous. What is the axis of homology?

(b) Let $h\colon \mathbb{RP}^3 \to \mathbb{RP}^3$ be the projectivity (w.r.t. any projective frame $(a_0, a_1, a_2, a_3, a_4)$) defined such that

$$(x, y, z, t) \mapsto (x, y, z, ax + by + cz + dt),$$

where $d \neq 0$ and h is not the identity.

Prove that the fixed points of h (i.e., those points M such that $h(M) = M$) are the origin $O = a_3 = (0, 0, 0, 1)$ and every point on the plane Π of equation

$$ax + by + cz + (d - 1)t = 0.$$

Prove that every line through the origin is globally invariant under h. Give a geometric construction of $h(M)$ for every point M distinct from O and not on Π, given any point A distinct from O and not on Π and its image $A' = h(A)$.

Hint. Consider the intersection P of the line $\langle A, M \rangle$ with the plane Π.

Such a projectivity is called a *homology of center O and of plane of homology Π* (Poncelet).

(c) Let $h\colon \mathbb{RP}^2 \to \mathbb{RP}^2$ be a projectivity, and assume that h does not preserve (globally) the line at infinity $z = 0$. Prove that there is a rotation R and a point at infinity a_1 such that $h \circ R$ maps all lines through a_1 to lines through a_1.

Chosing a projective frame (a_0, a_1, a_2, a_3) (where a_1 is the point mentioned above), show that $h \circ R$ is defined by a matrix of the form

$$\begin{pmatrix} a & b & c \\ 0 & b' & c' \\ 0 & b'' & c'' \end{pmatrix}$$

where $a \neq 0$ and $b'' \neq 0$. Prove that there exist two translations t_1, t_2 such that $t_2 \circ h \circ R \circ t_1$ is a homology.

If h preserves globally the line at infinity, show that there is a translation t such that $t \circ h$ is defined by a matrix of the form

$$\begin{pmatrix} a & b & c \\ a' & b' & c' \\ 0 & 0 & 1 \end{pmatrix}$$

where $ab' - a'b \neq 0$. Prove that there exist two rotations R_1, R_2 such that $R_2 \circ t \circ h \circ R_1$ has a matrix of the form

$$\begin{pmatrix} A & 0 & 0 \\ 0 & B & 0 \\ 0 & 0 & 1 \end{pmatrix}$$

where $AB = ab' - a'b$. Conclude that $R_2 \circ t \circ h \circ R_1$ is a homology only when $A = B$.

Remark: The above problem is adapted from Darboux.

5.40. Prove that every projectivity $h \colon \mathbb{RP}^2 \to \mathbb{RP}^2$ where $h \neq \mathrm{id}$ and h is not a homology is the composition of two homologies.

5.41. Given any two tetrahedra (a, b, c, d) and (a', b', c', d') in $\mathbb{RP}^3$ where a, b, c, d, a', b', c', d' are pairwise distinct and the lines containing the edges of the two tetrahedra are pairwise distinct, if the lines $\langle a, a' \rangle$, $\langle b, b' \rangle$, $\langle c, c' \rangle$, and $\langle d, d' \rangle$ intersect in a common point O distinct from a, b, c, d, a', b', c', d', prove that the intersection points (of lines) $p = \langle b, c \rangle \cap \langle b', c' \rangle$, $q = \langle a, c \rangle \cap \langle a', c' \rangle$, $r = \langle a, b \rangle \cap \langle a', b' \rangle$, $s = \langle c, d \rangle \cap \langle c', d' \rangle$, $t = \langle b, d \rangle \cap \langle b', d' \rangle$, $u = \langle a, d \rangle \cap \langle a', d' \rangle$, are coplanar.

Prove that the lines of intersection (of planes) $P = \langle b, c, d \rangle \cap \langle b', c', d' \rangle$, $Q = \langle a, c, d \rangle \cap \langle a', c', d' \rangle$, $R = \langle a, b, d \rangle \cap \langle a', b', d' \rangle$, $S = \langle a, b, c \rangle \cap \langle a', b', c' \rangle$, are coplanar.

Hint. Show that there is a homology whose center is O and whose plane of homology is determined by p, q, r, s, t, u.

5.42. Prove that Pappus's theorem implies Desargues's theorem (in the plane).

5.43. If K is a finite field of q elements ($q \geq 2$), prove that the finite projective space $\mathbf{P}(K^{n+1})$ has $q^n + q^{n-1} + \cdots + q + 1$ points and

$$\frac{(q^{n+1} - 1)(q^n - 1)}{(q - 1)^2 (q + 1)}$$

lines.

References

1. Emil Artin. *Geometric Algebra.* Wiley Interscience, first edition, 1957.
2. A. Beutelspacher and U. Rosenbaum. *Projective Geometry.* Cambridge University Press, first edition, 1998.
3. Marcel Berger. *Géométrie 1.* Nathan, 1990. English edition: Geometry 1, Universitext, Springer-Verlag.
4. Marcel Berger. *Géométrie 2.* Nathan, 1990. English edition: Geometry 2, Universitext, Springer-Verlag.
5. H.S.M. Coxeter. *Non-Euclidean Geometry.* The University of Toronto Press, first edition, 1942.
6. H.S.M. Coxeter. *Introduction to Geometry.* Wiley, second edition, 1989.
7. H.S.M. Coxeter. *The Real Projective Plane.* Springer-Verlag, third edition, 1993.
8. H.S.M. Coxeter. *Projective Geometry.* Springer-Verlag, second edition, 1994.
9. Gaston Darboux. *Principes de Géométrie Analytique.* Gauthier-Villars, first edition, 1917.
10. Olivier Faugeras. *Three-Dimensional Computer Vision, A Geometric Viewpoint.* MIT Press, first edition, 1996.
11. Gerd Fischer. *Mathematical Models, Commentary.* Vieweg & Sohn, first edition, 1986.
12. Gerd Fischer. *Mathematische Modelle.* Vieweg & Sohn, first edition, 1986.
13. James Foley, Andries van Dam, Steven Feiner, and John Hughes. *Computer Graphics. Principles and Practice.* Addison-Wesley, second edition, 1993.
14. Jean Fresnel. *Méthodes Modernes en Géométrie.* Hermann, first edition, 1998.
15. William Fulton. *Algebraic Curves.* Advanced Book Classics. Addison-Wesley, first edition, 1989.
16. Joe Harris. *Algebraic Geometry, A First Course.* GTM No. 133. Springer-Verlag, first edition, 1992.
17. D. Hilbert and S. Cohn-Vossen. *Geometry and the Imagination.* Chelsea Publishing Co., 1952.
18. Ramesh Jain, Rangachar Katsuri, and Brian G. Schunck. *Machine Vision.* McGraw-Hill, first edition, 1995.
19. Felix Klein. *Vorlesungen über nicht-Euklidische Geometrie.* AMS Chelsea, first edition, 1927.
20. Daniel Lehmann and Rudolphe Bkouche. *Initiation à la Géométrie.* Puf, first edition, 1988.
21. Dan Pedoe. *Geometry, A Comprehensive Course.* Dover, first edition, 1988.
22. M. Penna and R. Patterson. *Projective Geometry and Its Applications to Computer Graphics.* Prentice-Hall, first edition, 1986.
23. Pierre Samuel. *Projective Geometry.* Undergraduate Texts in Mathematics. Springer-Verlag, first edition, 1988.
24. J.-C. Sidler. *Géométrie Projective.* InterEditions, first edition, 1993.
25. J. Stolfi. *Oriented Projective Geometry.* Academic Press, first edition, 1991.
26. Claude Tisseron. *Géométries Affines, Projectives, et Euclidiennes.* Hermann, first edition, 1994.
27. Emanuele Trucco and Alessandro Verri. *Introductory Techniques for 3D Computer Vision.* Prentice-Hall, first edition, 1998.
28. O. Veblen and J. W. Young. *Projective Geometry, Vol. 1.* Ginn, second edition, 1938.
29. O. Veblen and J. W. Young. *Projective Geometry, Vol. 2.* Ginn, first edition, 1946.
30. Lucas Vienne. *Présentation Algébrique de la Géométrie Classique.* Vuibert, first edition, 1996.
31. Alan Watt. *3D Computer Graphics.* Addison-Wesley, second edition, 1993.

Chapter 6
Basics of Euclidean Geometry

Rien n'est beau que le vrai.
—**Hermann Minkowski**

6.1 Inner Products, Euclidean Spaces

In affine geometry it is possible to deal with ratios of vectors and barycenters of points, but there is no way to express the notion of length of a line segment or to talk about orthogonality of vectors. A Euclidean structure allows us to deal with *metric notions* such as orthogonality and length (or distance).

This chapter covers the bare bones of Euclidean geometry. Deeper aspects of Euclidean geometry are investigated in Chapter 8, in particular the structure of the orthogonal group and the structure of the group of affine rigid motions. One of our main goals is to give the basic properties of the transformations that preserve the Euclidean structure, rotations and reflections, since they play an important role in practice. As affine geometry is the study of properties invariant under bijective affine maps and projective geometry is the study of properties invariant under bijective projective maps, Euclidean geometry is the study of properties invariant under certain affine maps called *rigid motions*. Rigid motions are the maps that preserve the distance between points. Such maps are, in fact, affine and bijective (at least in the finite-dimensional case; see Lemma 8.8). They form a group $\mathbf{Is}(n)$ of affine maps whose corresponding linear maps form the group $\mathbf{O}(n)$ of orthogonal transformations. The subgroup $\mathbf{SE}(n)$ of $\mathbf{Is}(n)$ corresponds to the orientation preserving rigid motions, and there is a corresponding subgroup $\mathbf{SO}(n)$ of $\mathbf{O}(n)$, the group of rotations. These groups play a very important role in geometry, and we will study their structure in some detail in Chapter 8.

Before going any further, a potential point of confusion should be cleared up. Euclidean geometry deals with affine spaces $\left(E, \overrightarrow{E}\right)$, where the associated vector space $\overrightarrow{E}$ is equipped with an inner product. Such spaces are called *Euclidean affine spaces*. However, inner products are defined on vector spaces. Thus, we must first study the properties of vector spaces equipped with an inner product, and the linear maps preserving an inner product (the orthogonal group $\mathbf{SO}(n)$). Such spaces are called *Euclidean spaces* (omitting the word affine). It should be clear from the context whether we are dealing with a Euclidean vector space or a Euclidean affine space, but we will try to be clear about that. For instance, in this chapter, except for Definition 6.3, we are dealing with Euclidean vector spaces and linear maps.

We begin by defining inner products and Euclidean spaces. The Cauchy–Schwarz inequality and the Minkowski inequality are shown. We define orthogonality of vectors and of subspaces, orthogonal bases, and orthonormal bases. We offer a glimpse of Fourier series in terms of the orthogonal families $(\sin px)_{p \geq 1} \cup (\cos qx)_{q \geq 0}$ and $(e^{ikx})_{k \in \mathbb{Z}}$. We prove that every finite-dimensional Euclidean space has orthonormal bases. Orthonormal bases are the Euclidean analogue for affine frames. The first proof uses duality, and the second one the Gram–Schmidt orthogonalization procedure. The *QR*-decomposition for invertible matrices is shown as an application of the Gram–Schmidt procedure. Linear isometries (also called orthogonal transformations) are defined and studied briefly. We conclude with a short section in which some applications of Euclidean geometry are sketched. One of the most important applications, the method of least squares, is discussed in Chapter 14.

For a more detailed treatment of Euclidean geometry, see Berger [2, 3], Snapper and Troyer [22], or any other book on geometry, such as Pedoe [18], Coxeter [6], Fresnel [8], Tisseron [25], or Cagnac, Ramis, and Commeau [4]. Serious readers should consult Emil Artin's famous book [1], which contains an in-depth study of the orthogonal group, as well as other groups arising in geometry. It is still worth consulting some of the older classics, such as Hadamard [10, 11] and Rouché and de Comberousse [19]. The first edition of [10] was published in 1898, and finally reached its thirteenth edition in 1947! In this chapter it is assumed that all vector spaces are defined over the field $\mathbb{R}$ of real numbers unless specified otherwise (in a few cases, over the complex numbers $\mathbb{C}$).

First, we define a Euclidean structure on a vector space. Technically, a Euclidean structure over a vector space E is provided by a symmetric bilinear form on the vector space satisfying some extra properties. Recall that a bilinear form $\varphi : E \times E \to \mathbb{R}$ is *definite* if for every $u \in E$, $u \neq 0$ implies that $\varphi(u,u) \neq 0$, and *positive* if for every $u \in E$, $\varphi(u,u) \geq 0$.

Definition 6.1. A *Euclidean space* is a real vector space E equipped with a symmetric bilinear form $\varphi : E \times E \to \mathbb{R}$ that is *positive definite*. More explicitly, $\varphi : E \times E \to \mathbb{R}$ satisfies the following axioms:

$$\varphi(u_1 + u_2, v) = \varphi(u_1, v) + \varphi(u_2, v),$$
$$\varphi(u, v_1 + v_2) = \varphi(u, v_1) + \varphi(u, v_2),$$
$$\varphi(\lambda u, v) = \lambda \varphi(u, v),$$
$$\varphi(u, \lambda v) = \lambda \varphi(u, v),$$
$$\varphi(u, v) = \varphi(v, u),$$
$$u \neq 0 \text{ implies that } \varphi(u, u) > 0.$$

The real number $\varphi(u, v)$ is also called the *inner product (or scalar product) of u and v*. We also define the *quadratic form associated with φ* as the function $\Phi \colon E \to \mathbb{R}_+$ such that

$$\Phi(u) = \varphi(u, u),$$

for all $u \in E$.

Since φ is bilinear, we have $\varphi(0, 0) = 0$, and since it is positive definite, we have the stronger fact that

$$\varphi(u, u) = 0 \quad \text{iff} \quad u = 0,$$

that is, $\Phi(u) = 0$ iff $u = 0$.

Given an inner product $\varphi \colon E \times E \to \mathbb{R}$ on a vector space E, we also denote $\varphi(u, v)$ by

$$u \cdot v \quad \text{or} \quad \langle u, v \rangle \quad \text{or} \quad (u|v),$$

and $\sqrt{\Phi(u)}$ by $\|u\|$.

Example 6.1. The standard example of a Euclidean space is $\mathbb{R}^n$, under the inner product $\cdot$ defined such that

$$(x_1, \ldots, x_n) \cdot (y_1, \ldots, y_n) = x_1 y_1 + x_2 y_2 + \cdots + x_n y_n.$$

There are other examples.

Example 6.2. For instance, let E be a vector space of dimension 2, and let (e_1, e_2) be a basis of E. If $a > 0$ and $b^2 - ac < 0$, the bilinear form defined such that

$$\varphi(x_1 e_1 + y_1 e_2, x_2 e_1 + y_2 e_2) = a x_1 x_2 + b(x_1 y_2 + x_2 y_1) + c y_1 y_2$$

yields a Euclidean structure on E. In this case,

$$\Phi(x e_1 + y e_2) = a x^2 + 2b x y + c y^2.$$

Example 6.3. Let $\mathscr{C}[a, b]$ denote the set of continuous functions $f \colon [a, b] \to \mathbb{R}$. It is easily checked that $\mathscr{C}[a, b]$ is a vector space of infinite dimension. Given any two functions $f, g \in \mathscr{C}[a, b]$, let

$$\langle f, g \rangle = \int_a^b f(t) g(t) \, dt.$$

We leave as an easy exercise that $\langle -, - \rangle$ is indeed an inner product on $\mathscr{C}[a,b]$. In the case where $a = -\pi$ and $b = \pi$ (or $a = 0$ and $b = 2\pi$, this makes basically no difference), one should compute

$$\langle \sin px, \sin qx \rangle, \quad \langle \sin px, \cos qx \rangle, \quad \text{and} \quad \langle \cos px, \cos qx \rangle,$$

for all natural numbers $p, q \geq 1$. The outcome of these calculations is what makes Fourier analysis possible!

Let us observe that φ can be recovered from Φ. Indeed, by bilinearity and symmetry, we have

$$\begin{aligned}
\Phi(u+v) &= \varphi(u+v, u+v) \\
&= \varphi(u, u+v) + \varphi(v, u+v) \\
&= \varphi(u, u) + 2\varphi(u, v) + \varphi(v, v) \\
&= \Phi(u) + 2\varphi(u, v) + \Phi(v).
\end{aligned}$$

Thus, we have

$$\varphi(u, v) = \frac{1}{2}[\Phi(u+v) - \Phi(u) - \Phi(v)].$$

We also say that φ is the *polar form of* Φ. We will generalize polar forms to polynomials, and we will see that they play a very important role.

One of the very important properties of an inner product φ is that the map $u \mapsto \sqrt{\Phi(u)}$ is a norm.

Lemma 6.1. *Let E be a Euclidean space with inner product φ, and let Φ be the corresponding quadratic form. For all $u, v \in E$, we have the* Cauchy–Schwarz *inequality*

$$\varphi(u,v)^2 \leq \Phi(u)\Phi(v),$$

the equality holding iff u and v are linearly dependent.

We also have the Minkowski *inequality*

$$\sqrt{\Phi(u+v)} \leq \sqrt{\Phi(u)} + \sqrt{\Phi(v)},$$

the equality holding iff u and v are linearly dependent, where in addition if $u \neq 0$ and $v \neq 0$, then $u = \lambda v$ for some $\lambda > 0$.

Proof. For any vectors $u, v \in E$, we define the function $T : \mathbb{R} \to \mathbb{R}$ such that

$$T(\lambda) = \Phi(u + \lambda v),$$

for all $\lambda \in \mathbb{R}$. Using bilinearity and symmetry, we have

$$\begin{aligned}
\Phi(u+\lambda v) &= \varphi(u+\lambda v, u+\lambda v)\\
&= \varphi(u, u+\lambda v) + \lambda\,\varphi(v, u+\lambda v)\\
&= \varphi(u, u) + 2\lambda\,\varphi(u, v) + \lambda^2\,\varphi(v, v)\\
&= \Phi(u) + 2\lambda\,\varphi(u, v) + \lambda^2\,\Phi(v).
\end{aligned}$$

Since φ is positive definite, Φ is nonnegative, and thus $T(\lambda) \geq 0$ for all $\lambda \in \mathbb{R}$. If $\Phi(v) = 0$, then $v = 0$, and we also have $\varphi(u, v) = 0$. In this case, the Cauchy–Schwarz inequality is trivial, and $v = 0$ and u are linearly dependent.

Now, assume $\Phi(v) > 0$. Since $T(\lambda) \geq 0$, the quadratic equation

$$\lambda^2 \Phi(v) + 2\lambda\,\varphi(u, v) + \Phi(u) = 0$$

cannot have distinct real roots, which means that its discriminant

$$\Delta = 4(\varphi(u, v)^2 - \Phi(u)\Phi(v))$$

is null or negative, which is precisely the Cauchy–Schwarz inequality

$$\varphi(u, v)^2 \leq \Phi(u)\Phi(v).$$

If

$$\varphi(u, v)^2 = \Phi(u)\Phi(v),$$

then the above quadratic equation has a double root λ_0, and we have $\Phi(u+\lambda_0 v) = 0$. If $\lambda_0 = 0$, then $\varphi(u, v) = 0$, and since $\Phi(v) > 0$, we must have $\Phi(u) = 0$, and thus $u = 0$. In this case, of course, $u = 0$ and v are linearly dependent. Finally, if $\lambda_0 \neq 0$, since $\Phi(u+\lambda_0 v) = 0$ implies that $u + \lambda_0 v = 0$, then u and v are linearly dependent. Conversely, it is easy to check that we have equality when u and v are linearly dependent.

The Minkowski inequality

$$\sqrt{\Phi(u+v)} \leq \sqrt{\Phi(u)} + \sqrt{\Phi(v)}$$

is equivalent to

$$\Phi(u+v) \leq \Phi(u) + \Phi(v) + 2\sqrt{\Phi(u)\Phi(v)}.$$

However, we have shown that

$$2\varphi(u, v) = \Phi(u+v) - \Phi(u) - \Phi(v),$$

and so the above inequality is equivalent to

$$\varphi(u, v) \leq \sqrt{\Phi(u)\Phi(v)},$$

which is trivial when $\varphi(u, v) \leq 0$, and follows from the Cauchy–Schwarz inequality when $\varphi(u, v) \geq 0$. Thus, the Minkowski inequality holds. Finally, assume that $u \neq 0$ and $v \neq 0$, and that

$$\sqrt{\Phi(u+v)} = \sqrt{\Phi(u)} + \sqrt{\Phi(v)}.$$

When this is the case, we have

$$\varphi(u, v) = \sqrt{\Phi(u)\Phi(v)},$$

and we know from the discussion of the Cauchy–Schwarz inequality that the equality holds iff u and v are linearly dependent. The Minkowski inequality is an equality when u or v is null. Otherwise, if $u \neq 0$ and $v \neq 0$, then $u = \lambda v$ for some $\lambda \neq 0$, and since

$$\varphi(u, v) = \lambda \varphi(v, v) = \sqrt{\Phi(u)\Phi(v)},$$

by positivity, we must have $\lambda > 0$. $\square$

Note that the Cauchy–Schwarz inequality can also be written as

$$|\varphi(u,v)| \leq \sqrt{\Phi(u)}\sqrt{\Phi(v)}.$$

Remark: It is easy to prove that the Cauchy–Schwarz and the Minkowski inequalities still hold for a symmetric bilinear form that is positive, but not necessarily definite (i.e., $\varphi(u,v) \geq 0$ for all $u, v \in E$). However, u and v need not be linearly dependent when the equality holds.

The Minkowski inequality

$$\sqrt{\Phi(u+v)} \leq \sqrt{\Phi(u)} + \sqrt{\Phi(v)}$$

shows that the map $u \mapsto \sqrt{\Phi(u)}$ satisfies the convexity inequality (also known as triangle inequality), condition (N3) of Definition 21.2, and since φ is bilinear and positive definite, it also satisfies conditions (N1) and (N2) of Definition 21.2, and thus it is a *norm* on E. The norm induced by φ is called the *Euclidean norm induced by φ*.

Note that the Cauchy–Schwarz inequality can be written as

$$|u \cdot v| \leq \|u\| \|v\|,$$

and the Minkowski inequality as

$$\|u + v\| \leq \|u\| + \|v\|.$$

Figure 6.1 illustrates the triangle inequality.

We now define orthogonality.

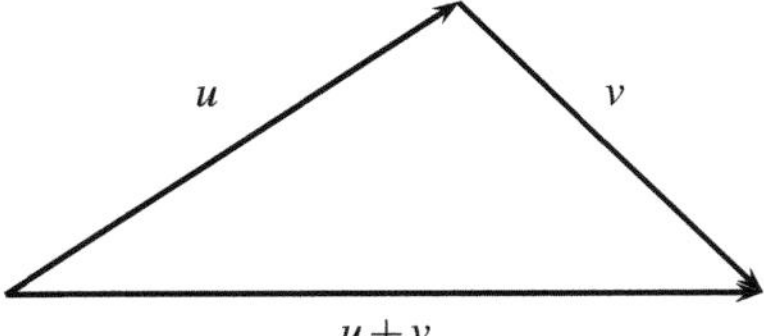

Fig. 6.1 The triangle inequality.

6.2 Orthogonality, Duality, Adjoint of a Linear Map

An inner product on a vector space gives the ability to define the notion of orthogonality. Families of nonnull pairwise orthogonal vectors must be linearly independent. They are called orthogonal families. In a vector space of finite dimension it is always possible to find orthogonal bases. This is very useful theoretically and practically. Indeed, in an orthogonal basis, finding the coordinates of a vector is very cheap: It takes an inner product. Fourier series make crucial use of this fact. When E has finite dimension, we prove that the inner product on E induces a natural isomorphism between E and its dual space E^*. This allows us to define the adjoint of a linear map in an intrinsic fashion (i.e., independently of bases). It is also possible to orthonormalize any basis (certainly when the dimension is finite). We give two proofs, one using duality, the other more constructive using the Gram–Schmidt orthonormalization procedure.

Definition 6.2. Given a Euclidean space E, any two vectors $u, v \in E$ are *orthogonal, or perpendicular,* if $u \cdot v = 0$. Given a family $(u_i)_{i \in I}$ of vectors in E, we say that $(u_i)_{i \in I}$ is *orthogonal* if $u_i \cdot u_j = 0$ for all $i, j \in I$, where $i \neq j$. We say that the family $(u_i)_{i \in I}$ is *orthonormal* if $u_i \cdot u_j = 0$ for all $i, j \in I$, where $i \neq j$, and $\|u_i\| = u_i \cdot u_i = 1$, for all $i \in I$. For any subset F of E, the set

$$F^\perp = \{v \in E \mid u \cdot v = 0, \text{ for all } u \in F\},$$

of all vectors orthogonal to all vectors in F, is called the *orthogonal complement of F*.

Since inner products are positive definite, observe that for any vector $u \in E$, we have

$$u \cdot v = 0 \quad \text{for all } v \in E \quad \text{iff} \quad u = 0.$$

It is immediately verified that the orthogonal complement $F^\perp$ of F is a subspace of E.

Example 6.4. Going back to Example 6.3 and to the inner product

$$\langle f, g \rangle = \int_{-\pi}^{\pi} f(t)g(t)dt$$

on the vector space $\mathscr{C}[-\pi, \pi]$, it is easily checked that

$$\langle \sin px, \sin qx \rangle = \begin{cases} \pi & \text{if } p = q, \, p, q \geq 1, \\ 0 & \text{if } p \neq q, \, p, q \geq 1, \end{cases}$$

$$\langle \cos px, \cos qx \rangle = \begin{cases} \pi & \text{if } p = q, \, p, q \geq 1, \\ 0 & \text{if } p \neq q, \, p, q \geq 0, \end{cases}$$

and

$$\langle \sin px, \cos qx \rangle = 0,$$

for all $p \geq 1$ and $q \geq 0$, and of course, $\langle 1, 1 \rangle = \int_{-\pi}^{\pi} dx = 2\pi$.

As a consequence, the family $(\sin px)_{p \geq 1} \cup (\cos qx)_{q \geq 0}$ is orthogonal. It is not orthonormal, but becomes so if we divide every trigonometric function by $\sqrt{\pi}$, and 1 by $\sqrt{2\pi}$.

Remark: Observe that if we allow complex–valued functions, we obtain simpler proofs. For example, it is immediately checked that

$$\int_{-\pi}^{\pi} e^{ikx} dx = \begin{cases} 2\pi & \text{if } k = 0, \\ 0 & \text{if } k \neq 0, \end{cases}$$

because the derivative of e^{ikx} is ike^{ikx}.

However, beware that something strange is going on. Indeed, unless $k = 0$, we have

$$\langle e^{ikx}, e^{ikx} \rangle = 0,$$

since

$$\langle e^{ikx}, e^{ikx} \rangle = \int_{-\pi}^{\pi} (e^{ikx})^2 dx = \int_{-\pi}^{\pi} e^{i2kx} dx = 0.$$

The inner product $\langle e^{ikx}, e^{ikx} \rangle$ should be strictly positive. What went wrong?

The problem is that we are using the wrong inner product. When we use complex-valued functions, we must use the *Hermitian inner product*

$$\langle f, g \rangle = \int_{-\pi}^{\pi} f(x)\overline{g(x)} dx,$$

where $\overline{g(x)}$ is the *conjugate* of $g(x)$. The Hermitian inner product is not symmetric. Instead,

$$\langle g, f \rangle = \overline{\langle f, g \rangle}.$$

(Recall that if $z = a + ib$, where $a, b \in \mathbb{R}$, then $\bar{z} = a - ib$. Also, $e^{i\theta} = \cos\theta + i\sin\theta$). With the Hermitian inner product, everything works out beautifully! In particular, the family $(e^{ikx})_{k \in \mathbb{Z}}$ is orthogonal. Hermitian spaces and some basics of Fourier series will be discussed more rigorously in Chapter 11.

We leave the following simple two results as exercises.

Lemma 6.2. *Given a Euclidean space E, for any family $(u_i)_{i\in I}$ of nonnull vectors in E, if $(u_i)_{i\in I}$ is orthogonal, then it is linearly independent.*

Lemma 6.3. *Given a Euclidean space E, any two vectors $u, v \in E$ are orthogonal iff*

$$\|u+v\|^2 = \|u\|^2 + \|v\|^2.$$

One of the most useful features of orthonormal bases is that they afford a very simple method for computing the coordinates of a vector over any basis vector. Indeed, assume that $(e_1, \ldots, e_m)$ is an orthonormal basis. For any vector

$$x = x_1 e_1 + \cdots + x_m e_m,$$

if we compute the inner product $x \cdot e_i$, we get

$$x \cdot e_i = x_1 e_1 \cdot e_i + \cdots + x_i e_i \cdot e_i + \cdots + x_m e_m \cdot e_i = x_i,$$

since

$$e_i \cdot e_j = \begin{cases} 1 & \text{if } i = j, \\ 0 & \text{if } i \neq j \end{cases}$$

is the property characterizing an orthonormal family. Thus,

$$x_i = x \cdot e_i,$$

which means that $x_i e_i = (x \cdot e_i) e_i$ is the orthogonal projection of x onto the subspace generated by the basis vector e_i. If the basis is orthogonal but not necessarily orthonormal, then

$$x_i = \frac{x \cdot e_i}{e_i \cdot e_i} = \frac{x \cdot e_i}{\|e_i\|^2}.$$

All this is true even for an infinite orthonormal (or orthogonal) basis $(e_i)_{i\in I}$.

 However, remember that every vector x is expressed as a linear combination

$$x = \sum_{i \in I} x_i e_i$$

where the family of scalars $(x_i)_{i\in I}$ has **finite support**, which means that $x_i = 0$ for all $i \in I - J$, where J is a finite set. Thus, even though the family $(\sin px)_{p\geq 1} \cup (\cos qx)_{q\geq 0}$ is orthogonal (it is not orthonormal, but becomes so if we divide every trigonometric function by $\sqrt{\pi}$, and 1 by $\sqrt{2\pi}$; we won't because it looks messy!), the fact that a function $f \in \mathscr{C}^0[-\pi, \pi]$ can be written as a Fourier series as

$$f(x) = a_0 + \sum_{k=1}^{\infty} (a_k \cos kx + b_k \sin kx)$$

does not mean that $(\sin px)_{p\geq 1} \cup (\cos qx)_{q\geq 0}$ is a basis of this vector space of functions, because in general, the families (a_k) and (b_k) **do not** have finite support! In order for this infinite linear combination to make sense, it is necessary to prove that the partial sums

$$a_0 + \sum_{k=1}^{n} (a_k \cos kx + b_k \sin kx)$$

of the series converge to a limit when n goes to infinity. This requires a topology on the space.

Still, a small miracle happens. If $f \in \mathscr{C}[-\pi, \pi]$ can indeed be expressed as a Fourier series

$$f(x) = a_0 + \sum_{k=1}^{\infty} (a_k \cos kx + b_k \sin kx),$$

the coefficients a_0 and $a_k, b_k, k \geq 1$, can be computed by projecting f over the basis functions, i.e., by taking inner products with the basis functions in $(\sin px)_{p \geq 1} \cup (\cos qx)_{q \geq 0}$. Indeed, for all $k \geq 1$, we have

$$a_0 = \frac{\langle f, 1 \rangle}{\|1\|^2},$$

and

$$a_k = \frac{\langle f, \cos kx \rangle}{\|\cos kx\|^2}, \qquad b_k = \frac{\langle f, \sin kx \rangle}{\|\sin kx\|^2},$$

that is,

$$a_0 = \frac{1}{2\pi} \int_{-\pi}^{\pi} f(x)dx,$$

and

$$a_k = \frac{1}{\pi} \int_{-\pi}^{\pi} f(x) \cos kx\, dx, \qquad b_k = \frac{1}{\pi} \int_{-\pi}^{\pi} f(x) \sin kx\, dx.$$

If we allow f to be complex-valued and use the family $(e^{ikx})_{k \in \mathbb{Z}}$, which is is indeed orthogonal w.r.t. the Hermitian inner product

$$\langle f, g \rangle = \int_{-\pi}^{\pi} f(x)\overline{g(x)}dx,$$

we consider functions $f \in \mathscr{C}[-\pi, \pi]$ that can be expressed as the sum of a series

$$f(x) = \sum_{k \in \mathbb{Z}} c_k e^{ikx}.$$

Note that the index k is allowed to be a negative integer. Then, the formula giving the c_k is very nice:

$$c_k = \frac{\langle f, e^{ikx} \rangle}{\|e^{ikx}\|^2},$$

that is,

$$c_k = \frac{1}{2\pi} \int_{-\pi}^{\pi} f(x)e^{-ikx}dx.$$

Note the presence of the negative sign in e^{-ikx}, which is due to the fact that the inner product is Hermitian. Of course, the real case can be recovered from the complex case. If f is a real-valued function, then we must have

$$a_k = c_k + c_{-k} \quad \text{and} \quad b_k = i(c_k - c_{-k}).$$

Also note that

$$\frac{1}{2\pi} \int_{-\pi}^{\pi} f(x) e^{-ikx} dx$$

is defined not only for all discrete values $k \in \mathbb{Z}$, but for all $k \in \mathbb{R}$, and that if f is continuous over $\mathbb{R}$, the integral makes sense. This suggests defining

$$\widehat{f}(k) = \int_{-\infty}^{\infty} f(x) e^{-ikx} dx,$$

called the *Fourier transform* of f. The Fourier transform analyzes the function f in the "frequency domain" in terms of its spectrum of harmonics. Amazingly, there is an inverse Fourier transform (change e^{-ikx} to e^{+ikx} and divide by the scale factor 2π) that reconstructs f (under certain assumptions on f).

Some basics of Fourier series will be discussed more rigorously in Chapter 11. For more on Fourier analysis, we highly recommend Strang [23] for a lucid introduction with lots of practical examples, and then move on to a good real analysis text, for instance Lang [15, 16], or [20].

A very important property of Euclidean spaces of finite dimension is that the inner product induces a canonical bijection (i.e., independent of the choice of bases) between the vector space E and its dual E^*.

Given a Euclidean space E, for any vector $u \in E$, let $\varphi_u : E \to \mathbb{R}$ be the map defined such that

$$\varphi_u(v) = u \cdot v,$$

for all $v \in E$.

Since the inner product is bilinear, the map φ_u is a linear form in E^*. Thus, we have a map $\flat : E \to E^*$, defined such that

$$\flat(u) = \varphi_u.$$

Lemma 6.4. *Given a Euclidean space E, the map $\flat : E \to E^*$ defined such that*

$$\flat(u) = \varphi_u$$

is linear and injective. When E is also of finite dimension, the map $\flat : E \to E^*$ is a canonical isomorphism.

Proof. That $\flat : E \to E^*$ is a linear map follows immediately from the fact that the inner product is bilinear. If $\varphi_u = \varphi_v$, then $\varphi_u(w) = \varphi_v(w)$ for all $w \in E$, which by definition of φ_u means that

$$u \cdot w = v \cdot w$$

for all $w \in E$, which by bilinearity is equivalent to

$$(v - u) \cdot w = 0$$

for all $w \in E$, which implies that $u = v$, since the inner product is positive definite. Thus, $\flat \colon E \to E^*$ is injective. Finally, when E is of finite dimension n, we know that E^* is also of dimension n, and then $\flat \colon E \to E^*$ is bijective. $\square$

The inverse of the isomorphism $\flat \colon E \to E^*$ is denoted by $\sharp \colon E^* \to E$.

As a consequence of Lemma 6.4, if E is a Euclidean space of finite dimension, every linear form $f \in E^*$ corresponds to a unique $u \in E$ such that

$$f(v) = u \cdot v,$$

for every $v \in E$. In particular, if f is not the null form, the kernel of f, which is a hyperplane H, is precisely the set of vectors that are orthogonal to u.

Remarks:

(1) The "musical map" $\flat \colon E \to E^*$ is not surjective when E has infinite dimension. The result can be salvaged by restricting our attention to continuous linear maps, and by assuming that the vector space E is a *Hilbert space* (i.e., E is a complete normed vector space w.r.t. the Euclidean norm). This is the famous "little" Riesz theorem (or Riesz representation theorem).

(2) Lemma 6.4 still holds if the inner product on E is replaced by a nondegenerate symmetric bilinear form φ. We say that a symmetric bilinear form $\varphi \colon E \times E \to \mathbb{R}$ is *nondegenerate* if for every $u \in E$,

$$\text{if} \quad \varphi(u, v) = 0 \quad \text{for all } v \in E, \quad \text{then} \quad u = 0.$$

For example, the symmetric bilinear form on $\mathbb{R}^4$ defined such that

$$\varphi((x_1, x_2, x_3, x_4), (y_1, y_2, y_3, y_4)) = x_1 y_1 + x_2 y_2 + x_3 y_3 - x_4 y_4$$

is nondegenerate. However, there are nonnull vectors $u \in \mathbb{R}^4$ such that $\varphi(u, u) = 0$, which is impossible in a Euclidean space. Such vectors are called *isotropic*.

The existence of the isomorphism $\flat \colon E \to E^*$ is crucial to the existence of adjoint maps. The importance of adjoint maps stems from the fact that the linear maps arising in physical problems are often self-adjoint, which means that $f = f^*$. Moreover, self-adjoint maps can be diagonalized over orthonormal bases of eigenvectors. This is the key to the solution of many problems in mechanics, and engineering in general (see Strang [23]).

Let E be a Euclidean space of finite dimension n, and let $f \colon E \to E$ be a linear map. For every $u \in E$, the map

$$v \mapsto u \cdot f(v)$$

is clearly a linear form in E^*, and by Lemma 6.4, there is a unique vector in E denoted by $f^*(u)$ such that

$$f^*(u) \cdot v = u \cdot f(v),$$

for every $v \in E$. The following simple lemma shows that the map f^* is linear.

Lemma 6.5. *Given a Euclidean space E of finite dimension, for every linear map $f \colon E \to E$, there is a unique linear map $f^* \colon E \to E$ such that*

$$f^*(u) \cdot v = u \cdot f(v),$$

for all $u, v \in E$. The map f^* is called the adjoint of f (w.r.t. to the inner product).

Proof. Given $u_1, u_2 \in E$, since the inner product is bilinear, we have

$$(u_1 + u_2) \cdot f(v) = u_1 \cdot f(v) + u_2 \cdot f(v),$$

for all $v \in E$, and

$$(f^*(u_1) + f^*(u_2)) \cdot v = f^*(u_1) \cdot v + f^*(u_2) \cdot v,$$

for all $v \in E$, and since by assumption,

$$f^*(u_1) \cdot v = u_1 \cdot f(v)$$

and

$$f^*(u_2) \cdot v = u_2 \cdot f(v),$$

for all $v \in E$, we get

$$(f^*(u_1) + f^*(u_2)) \cdot v = (u_1 + u_2) \cdot f(v),$$

for all $v \in E$. Since $\flat$ is bijective, this implies that

$$f^*(u_1 + u_2) = f^*(u_1) + f^*(u_2).$$

Similarly,

$$(\lambda u) \cdot f(v) = \lambda (u \cdot f(v)),$$

for all $v \in E$, and

$$(\lambda f^*(u)) \cdot v = \lambda (f^*(u) \cdot v),$$

for all $v \in E$, and since by assumption,

$$f^*(u) \cdot v = u \cdot f(v),$$

for all $v \in E$, we get

$$(\lambda f^*(u)) \cdot v = (\lambda u) \cdot f(v),$$

for all $v \in E$. Since $\flat$ is bijective, this implies that

$$f^*(\lambda u) = \lambda f^*(u).$$

Thus, f^* is indeed a linear map, and it is unique, since $\flat$ is a bijection. $\square$

Linear maps $f \colon E \to E$ such that $f = f^*$ are called *self-adjoint* maps. They play a very important role because they have real eigenvalues, and because orthonormal bases arise from their eigenvectors. Furthermore, many physical problems lead to self-adjoint linear maps (in the form of symmetric matrices).

Remark: Lemma 6.5 still holds if the inner product on E is replaced by a nondegenerate symmetric bilinear form φ.

Linear maps such that $f^{-1} = f^*$, or equivalently

$$f^* \circ f = f \circ f^* = \mathrm{id},$$

also play an important role. They are *linear isometries*, or *isometries*. Rotations are special kinds of isometries. Another important class of linear maps are the linear maps satisfying the property

$$f^* \circ f = f \circ f^*,$$

called *normal linear maps*. We will see later on that normal maps can always be diagonalized over orthonormal bases of eigenvectors, but this will require using a Hermitian inner product (over $\mathbb{C}$).

Given two Euclidean spaces E and F, where the inner product on E is denoted by $\langle -, - \rangle_1$ and the inner product on F is denoted by $\langle -, - \rangle_2$, given any linear map $f \colon E \to F$, it is immediately verified that the proof of Lemma 6.5 can be adapted to show that there is a unique linear map $f^* \colon F \to E$ such that

$$\langle f(u), v \rangle_2 = \langle u, f^*(v) \rangle_1$$

for all $u \in E$ and all $v \in F$. The linear map f^* is also called the *adjoint of f*.

Remark: Given any basis for E and any basis for F, it is possible to characterize the matrix of the adjoint f^* of f in terms of the matrix of f, and the symmetric matrices defining the inner products. We will do so with respect to orthonormal bases. Also, since inner products are symmetric, the adjoint f^* of f is also characterized by

$$f(u) \cdot v = u \cdot f^*(v),$$

for all $u, v \in E$.

We can also use Lemma 6.4 to show that any Euclidean space of finite dimension has an orthonormal basis.

Lemma 6.6. *Given any nontrivial Euclidean space E of finite dimension $n \geq 1$, there is an orthonormal basis $(u_1, \ldots, u_n)$ for E.*

Proof. We proceed by induction on n. When $n = 1$, take any nonnull vector $v \in E$, which exists, since we assumed E nontrivial, and let

$$u = \frac{v}{\|v\|}.$$

If $n \geq 2$, again take any nonnull vector $v \in E$, and let

$$u_1 = \frac{v}{\|v\|}.$$

Consider the linear form φ_{u_1} associated with u_1. Since $u_1 \neq 0$, by Lemma 6.4, the linear form φ_{u_1} is nonnull, and its kernel is a hyperplane H. Since $\varphi_{u_1}(w) = 0$ iff $u_1 \cdot w = 0$, the hyperplane H is the orthogonal complement of $\{u_1\}$. Furthermore, since $u_1 \neq 0$ and the inner product is positive definite, $u_1 \cdot u_1 \neq 0$, and thus, $u_1 \notin H$, which implies that $E = H \oplus \mathbb{R}u_1$. However, since E is of finite dimension n, the hyperplane H has dimension $n - 1$, and by the induction hypothesis, we can find an orthonormal basis $(u_2, \ldots, u_n)$ for H. Now, because H and the one dimensional space $\mathbb{R}u_1$ are orthogonal and $E = H \oplus \mathbb{R}u_1$, it is clear that $(u_1, \ldots, u_n)$ is an orthonormal basis for E. $\square$

There is a more constructive way of proving Lemma 6.6, using a procedure known as the *Gram–Schmidt orthonormalization procedure*. Among other things, the Gram–Schmidt orthonormalization procedure yields the *QR-decomposition for matrices*, an important tool in numerical methods.

Lemma 6.7. *Given any nontrivial Euclidean space E of finite dimension $n \geq 1$, from any basis $(e_1, \ldots, e_n)$ for E we can construct an orthonormal basis $(u_1, \ldots, u_n)$ for E, with the property that for every k, $1 \leq k \leq n$, the families $(e_1, \ldots, e_k)$ and $(u_1, \ldots, u_k)$ generate the same subspace.*

Proof. We proceed by induction on n. For $n = 1$, let

$$u_1 = \frac{e_1}{\|e_1\|}.$$

For $n \geq 2$, we also let

$$u_1 = \frac{e_1}{\|e_1\|},$$

and assuming that $(u_1, \ldots, u_k)$ is an orthonormal system that generates the same subspace as $(e_1, \ldots, e_k)$, for every k with $1 \leq k < n$, we note that the vector

$$u'_{k+1} = e_{k+1} - \sum_{i=1}^{k} (e_{k+1} \cdot u_i)\, u_i$$

is nonnull, since otherwise, because $(u_1, \ldots, u_k)$ and $(e_1, \ldots, e_k)$ generate the same subspace, $(e_1, \ldots, e_{k+1})$ would be linearly dependent, which is absurd, since $(e_1, \ldots, e_n)$ is a basis. Thus, the norm of the vector u'_{k+1} being nonzero, we use the following construction of the vectors u_k and u'_k:

$$u'_1 = e_1, \qquad u_1 = \frac{u'_1}{\|u'_1\|},$$

and for the inductive step

$$u'_{k+1} = e_{k+1} - \sum_{i=1}^{k}(e_{k+1} \cdot u_i)\, u_i, \qquad u_{k+1} = \frac{u'_{k+1}}{\|u'_{k+1}\|},$$

where $1 \leq k \leq n-1$. It is clear that $\|u_{k+1}\| = 1$, and since $(u_1, \ldots, u_k)$ is an orthonormal system, we have

$$u'_{k+1} \cdot u_i = e_{k+1} \cdot u_i - (e_{k+1} \cdot u_i)u_i \cdot u_i = e_{k+1} \cdot u_i - e_{k+1} \cdot u_i = 0,$$

for all i with $1 \leq i \leq k$. This shows that the family $(u_1, \ldots, u_{k+1})$ is orthonormal, and since $(u_1, \ldots, u_k)$ and $(e_1, \ldots, e_k)$ generates the same subspace, it is clear from the definition of u_{k+1} that $(u_1, \ldots, u_{k+1})$ and $(e_1, \ldots, e_{k+1})$ generate the same subspace. This completes the induction step and the proof of the lemma. $\square$

Note that u'_{k+1} is obtained by subtracting from e_{k+1} the projection of e_{k+1} itself onto the orthonormal vectors $u_1, \ldots, u_k$ that have already been computed. Then, u'_{k+1} is normalized. The Gram–Schmidt orthonormalization procedure is illustrated in Figure 6.2.

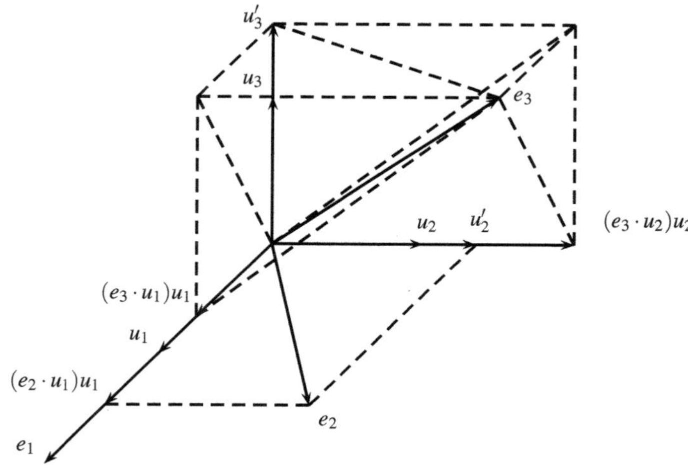

Fig. 6.2 The Gram–Schmidt orthonormalization procedure.

Remarks:

(1) The QR-decomposition can now be obtained very easily, but we postpone this until Section 6.4.

(2) We could compute u'_{k+1} using the formula

$$u'_{k+1} = e_{k+1} - \sum_{i=1}^{k} \left(\frac{e_{k+1} \cdot u'_i}{\|u'_i\|^2} \right) u'_i,$$

and normalize the vectors u'_k at the end. This time, we are subtracting from e_{k+1} the projection of e_{k+1} itself onto the orthogonal vectors $u'_1, \ldots, u'_k$. This might be preferable when writing a computer program.

(3) The proof of Lemma 6.7 also works for a countably infinite basis for E, producing a countably infinite orthonormal basis.

Example 6.5. If we consider polynomials and the inner product

$$\langle f, g \rangle = \int_{-1}^{1} f(t)g(t)dt,$$

applying the Gram–Schmidt orthonormalization procedure to the polynomials

$$1, x, x^2, \ldots, x^n, \ldots,$$

which form a basis of the polynomials in one variable with real coefficients, we get a family of orthonormal polynomials $Q_n(x)$ related to the *Legendre polynomials*.

The Legendre polynomials $P_n(x)$ have many nice properties. They are orthogonal, but their norm is not always 1. The Legendre polynomials $P_n(x)$ can be defined as follows. Letting f_n be the function

$$f_n(x) = (x^2 - 1)^n,$$

we define $P_n(x)$ as follows:

$$P_0(x) = 1, \quad \text{and} \quad P_n(x) = \frac{1}{2^n n!} f_n^{(n)}(x),$$

where $f_n^{(n)}$ is the nth derivative of f_n.

They can also be defined inductively as follows:

$$P_0(x) = 1,$$
$$P_1(x) = x,$$
$$P_{n+1}(x) = \frac{2n+1}{n+1} x P_n(x) - \frac{n}{n+1} P_{n-1}(x).$$

It turns out that the polynomials Q_n are related to the Legendre polynomials P_n as follows:

$$Q_n(x) = \frac{2^n (n!)^2}{(2n)!} P_n(x).$$

As a consequence of Lemma 6.6 (or Lemma 6.7), given any Euclidean space of finite dimension n, if $(e_1, \ldots, e_n)$ is an orthonormal basis for E, then for any two vectors $u = u_1 e_1 + \cdots + u_n e_n$ and $v = v_1 e_1 + \cdots + v_n e_n$, the inner product $u \cdot v$ is expressed as

$$u \cdot v = (u_1 e_1 + \cdots + u_n e_n) \cdot (v_1 e_1 + \cdots + v_n e_n) = \sum_{i=1}^{n} u_i v_i,$$

and the norm $\|u\|$ as

$$\|u\| = \|u_1 e_1 + \cdots + u_n e_n\| = \sqrt{\sum_{i=1}^{n} u_i^2}.$$

We can also prove the following lemma regarding orthogonal spaces.

Lemma 6.8. *Given any nontrivial Euclidean space E of finite dimension $n \geq 1$, for any subspace F of dimension k, the orthogonal complement $F^\perp$ of F has dimension $n - k$, and $E = F \oplus F^\perp$. Furthermore, we have $F^{\perp\perp} = F$.*

Proof. From Lemma 6.6, the subspace F has some orthonormal basis $(u_1, \ldots, u_k)$. This linearly independent family $(u_1, \ldots, u_k)$ can be extended to a basis $(u_1, \ldots, u_k, v_{k+1}, \ldots, v_n)$, and by Lemma 6.7, it can be converted to an orthonormal basis $(u_1, \ldots, u_n)$, which contains $(u_1, \ldots, u_k)$ as an orthonormal basis of F. Now, any vector $w = w_1 u_1 + \cdots + w_n u_n \in E$ is orthogonal to F iff $w \cdot u_i = 0$, for every i, where $1 \leq i \leq k$, iff $w_i = 0$ for every i, where $1 \leq i \leq k$. Clearly, this shows that $(u_{k+1}, \ldots, u_n)$ is a basis of $F^\perp$, and thus $E = F \oplus F^\perp$, and $F^\perp$ has dimension $n - k$. Similarly, any vector $w = w_1 u_1 + \cdots + w_n u_n \in E$ is orthogonal to $F^\perp$ iff $w \cdot u_i = 0$, for every i, where $k + 1 \leq i \leq n$, iff $w_i = 0$ for every i, where $k + 1 \leq i \leq n$. Thus, $(u_1, \ldots, u_k)$ is a basis of $F^{\perp\perp}$, and $F^{\perp\perp} = F$. $\square$

We now define Euclidean affine spaces.

Definition 6.3. An affine space $\left(E, \overrightarrow{E} \right)$ is a *Euclidean affine space* if its underlying vector space $\overrightarrow{E}$ is a Euclidean vector space. Given any two points $a, b \in E$, we define the *distance between a and b, or length of the segment (a, b)*, as $\|\overrightarrow{ab}\|$, the Euclidean norm of $\overrightarrow{ab}$. Given any two pairs of points (a, b) and (c, d), we define their inner product as $\overrightarrow{ab} \cdot \overrightarrow{cd}$. We say that (a, b) and (c, d) are *orthogonal, or perpendicular*, if $\overrightarrow{ab} \cdot \overrightarrow{cd} = 0$. We say that two affine subspaces F_1 and F_2 of E are *orthogonal* if their directions F_1 and F_2 are orthogonal.

The verification that the distance defined in Definition 6.3 satisfies the axioms of Definition 21.1 is immediate. Note that a Euclidean affine space is a normed affine space, in the sense of Definition 21.3. We denote by $\mathbb{E}^m$ the Euclidean affine space obtained from the affine space $\mathbb{A}^m$ by defining on the vector space $\mathbb{R}^m$ the standard inner product

$$(x_1, \ldots, x_m) \cdot (y_1, \ldots, y_m) = x_1 y_1 + \cdots + x_m y_m.$$

The corresponding Euclidean norm is

$$\|(x_1,\ldots,x_m)\| = \sqrt{x_1^2 + \cdots + x_m^2}.$$

6.3 Linear Isometries (Orthogonal Transformations)

In this section we consider linear maps between Euclidean spaces that preserve the Euclidean norm. These transformations, sometimes called *rigid motions*, play an important role in geometry.

Definition 6.4. Given any two nontrivial Euclidean spaces E and F of the same finite dimension n, a function $f : E \to F$ is *an orthogonal transformation, or a linear isometry*, if it is linear and

$$\|f(u)\| = \|u\|,$$

for all $u \in E$.

Remarks:

(1) A linear isometry is often defined as a linear map such that

$$\|f(v) - f(u)\| = \|v - u\|,$$

for all $u, v \in E$. Since the map f is linear, the two definitions are equivalent. The second definition just focuses on preserving the distance between vectors.

(2) Sometimes, a linear map satisfying the condition of Definition 6.4 is called a *metric map*, and a linear isometry is defined as a *bijective* metric map.

An isometry (without the word linear) is sometimes defined as a function $f : E \to F$ (not necessarily linear) such that

$$\|f(v) - f(u)\| = \|v - u\|,$$

for all $u, v \in E$, i.e., as a function that preserves the distance. This requirement turns out to be very strong. Indeed, the next lemma shows that all these definitions are equivalent when E and F are of finite dimension, and for functions such that $f(0) = 0$.

Lemma 6.9. *Given any two nontrivial Euclidean spaces E and F of the same finite dimension n, for every function $f : E \to F$, the following properties are equivalent:*

(1) f is a linear map and $\|f(u)\| = \|u\|$, for all $u \in E$;
(2) $\|f(v) - f(u)\| = \|v - u\|$, for all $u, v \in E$, and $f(0) = 0$;
(3) $f(u) \cdot f(v) = u \cdot v$, for all $u, v \in E$.

Furthermore, such a map is bijective.

Proof. Clearly, (1) implies (2), since in (1) it is assumed that f is linear.

Assume that (2) holds. In fact, we shall prove a slightly stronger result. We prove that if

$$\|f(v) - f(u)\| = \|v - u\|$$

for all $u, v \in E$, then for any vector $\tau \in E$, the function $g \colon E \to F$ defined such that

$$g(u) = f(\tau + u) - f(\tau)$$

for all $u \in E$ is a linear map such that $g(0) = 0$ and (3) holds. Clearly, $g(0) = f(\tau) - f(\tau) = 0$.

Note that from the hypothesis

$$\|f(v) - f(u)\| = \|v - u\|$$

for all $u, v \in E$, we conclude that

$$\begin{aligned}
\|g(v) - g(u)\| &= \|f(\tau + v) - f(\tau) - (f(\tau + u) - f(\tau))\|, \\
&= \|f(\tau + v) - f(\tau + u)\|, \\
&= \|\tau + v - (\tau + u)\|, \\
&= \|v - u\|,
\end{aligned}$$

for all $u, v \in E$. Since $g(0) = 0$, by setting $u = 0$ in

$$\|g(v) - g(u)\| = \|v - u\|,$$

we get

$$\|g(v)\| = \|v\|$$

for all $v \in E$. In other words, g preserves both the distance and the norm.

To prove that g preserves the inner product, we use the simple fact that

$$2u \cdot v = \|u\|^2 + \|v\|^2 - \|u - v\|^2$$

for all $u, v \in E$. Then, since g preserves distance and norm, we have

$$\begin{aligned}
2g(u) \cdot g(v) &= \|g(u)\|^2 + \|g(v)\|^2 - \|g(u) - g(v)\|^2 \\
&= \|u\|^2 + \|v\|^2 - \|u - v\|^2 \\
&= 2u \cdot v,
\end{aligned}$$

and thus $g(u) \cdot g(v) = u \cdot v$, for all $u, v \in E$, which is (3).

In particular, if $f(0) = 0$, by letting $\tau = 0$, we have $g = f$, and f preserves the scalar product, i.e., (3) holds.

Now assume that (3) holds. Since E is of finite dimension, we can pick an orthonormal basis $(e_1, \ldots, e_n)$ for E. Since f preserves inner products, $(f(e_1), \ldots, f(e_n))$ is also orthonormal, and since F also has dimension n, it is a basis of F. Then note that for any $u = u_1 e_1 + \cdots + u_n e_n$, we have

$$u_i = u \cdot e_i,$$

for all i, $1 \leq i \leq n$. Thus, we have

$$f(u) = \sum_{i=1}^{n} (f(u) \cdot f(e_i)) f(e_i),$$

and since f preserves inner products, this shows that

$$f(u) = \sum_{i=1}^{n} (u \cdot e_i) f(e_i) = \sum_{i=1}^{n} u_i f(e_i),$$

which shows that f is linear. Obviously, f preserves the Euclidean norm, and (3) implies (1).

Finally, if $f(u) = f(v)$, then by linearity $f(v - u) = 0$, so that $\|f(v - u)\| = 0$, and since f preserves norms, we must have $\|v - u\| = 0$, and thus $u = v$. Thus, f is injective, and since E and F have the same finite dimension, f is bijective. $\square$

Remarks:

(i) The dimension assumption is needed only to prove that (3) implies (1) when f is not known to be linear, and to prove that f is surjective, but the proof shows that (1) implies that f is injective.

(ii) In (2), when f does not satisfy the condition $f(0) = 0$, the proof shows that f is an affine map. Indeed, taking any vector τ as an origin, the map g is linear, and

$$f(\tau + u) = f(\tau) + g(u)$$

for all $u \in E$, proving that f is affine with associated linear map g.

(iii) Paul Hughett showed me a nice proof of the following interesting fact: The implication that (3) implies (1) holds if we also assume that f is surjective, even if E has infinite dimension. Indeed, observe that

$$\begin{aligned}
(f(\lambda u + \mu v) - \lambda f(u) - \mu f(v)) \cdot f(w) \\
= f(\lambda u + \mu v) \cdot f(w) - \lambda f(u) \cdot f(w) - \mu f(v) \cdot f(w) \\
= (\lambda u + \mu v) \cdot w - \lambda u \cdot w - \mu v \cdot w = 0,
\end{aligned}$$

since f preserves the inner product. However, if f is surjective, every $z \in E$ is of the form $z = f(w)$ for some $w \in E$, and the above equation implies that

$$(f(\lambda u + \mu v) - \lambda f(u) - \mu f(v)) \cdot z = 0$$

for all $z \in E$, which implies that

$$f(\lambda u + \mu v) - \lambda f(u) - \mu f(v) = 0,$$

proving that f is linear.

In view of Lemma 6.9, we will drop the word "linear" in "linear isometry," unless we wish to emphasize that we are dealing with a map between vector spaces.

We are now going to take a closer look at the isometries $f : E \to E$ of a Euclidean space of finite dimension.

6.4 The Orthogonal Group, Orthogonal Matrices

In this section we explore some of the basic properties of the orthogonal group and of orthogonal matrices.

Lemma 6.10. *Let E be any Euclidean space of finite dimension n, and let $f : E \to E$ be any linear map. The following properties hold:*

(1) The linear map $f : E \to E$ is an isometry iff

$$f \circ f^* = f^* \circ f = \mathrm{id}.$$

(2) For every orthonormal basis $(e_1, \ldots, e_n)$ of E, if the matrix of f is A, then the matrix of f^* is the transpose $A^\top$ of A, and f is an isometry iff A satisfies the identities

$$AA^\top = A^\top A = I_n,$$

where I_n denotes the identity matrix of order n, iff the columns of A form an orthonormal basis of E, iff the rows of A form an orthonormal basis of E.

Proof. (1) The linear map $f : E \to E$ is an isometry iff

$$f(u) \cdot f(v) = u \cdot v,$$

for all $u, v \in E$, iff

$$f^*(f(u)) \cdot v = f(u) \cdot f(v) = u \cdot v$$

for all $u, v \in E$, which implies

$$(f^*(f(u)) - u) \cdot v = 0$$

for all $u, v \in E$. Since the inner product is positive definite, we must have

$$f^*(f(u)) - u = 0$$

for all $u \in E$, that is,

$$f^* \circ f = f \circ f^* = \mathrm{id}.$$

(2) If $(e_1, \ldots, e_n)$ is an orthonormal basis for E, let $A = (a_{i,j})$ be the matrix of f, and let $B = (b_{i,j})$ be the matrix of f^*. Since f^* is characterized by

$$f^*(u) \cdot v = u \cdot f(v)$$

for all $u, v \in E$, using the fact that if $w = w_1 e_1 + \cdots + w_n e_n$ we have $w_k = w \cdot e_k$ for all k, $1 \le k \le n$, letting $u = e_i$ and $v = e_j$, we get

$$b_{j,i} = f^*(e_i) \cdot e_j = e_i \cdot f(e_j) = a_{i,j},$$

for all i, j, $1 \le i, j \le n$. Thus, $B = A^\top$. Now, if X and Y are arbitrary matrices over the basis $(e_1, \ldots, e_n)$, denoting as usual the jth column of X by X_j, and similarly for Y, a simple calculation shows that

$$X^\top Y = (X_i \cdot Y_j)_{1 \le i, j \le n}.$$

Then it is immediately verified that if $X = Y = A$, then

$$A^\top A = A A^\top = I_n$$

iff the column vectors $(A_1, \ldots, A_n)$ form an orthonormal basis. Thus, from (1), we see that (2) is clear (also because the rows of A are the columns of $A^\top$). $\quad\square$

Lemma 6.10 shows that the inverse of an isometry f is its adjoint f^*. Lemma 6.10 also motivates the following definition. The set of all real $n \times n$ matrices is denoted by $\mathrm{M}_n(\mathbb{R})$.

Definition 6.5. A real $n \times n$ matrix is an *orthogonal matrix* if

$$A A^\top = A^\top A = I_n.$$

Remark: It is easy to show that the conditions $A A^\top = I_n$, $A^\top A = I_n$, and $A^{-1} = A^\top$, are equivalent. Given any two orthonormal bases $(u_1, \ldots, u_n)$ and $(v_1, \ldots, v_n)$, if P is the change of basis matrix from $(u_1, \ldots, u_n)$ to $(v_1, \ldots, v_n)$ (i.e., the columns of P are the coordinates of the v_j w.r.t. $(u_1, \ldots, u_n)$), since the columns of P are the coordinates of the vectors v_j with respect to the basis $(u_1, \ldots, u_n)$, and since $(v_1, \ldots, v_n)$ is orthonormal, the columns of P are orthonormal, and by Lemma 6.10 (2), the matrix P is orthogonal.

The proof of Lemma 6.9 (3) also shows that if f is an isometry, then the image of an orthonormal basis $(u_1, \ldots, u_n)$ is an orthonormal basis. Students often ask why orthogonal matrices are not called orthonormal matrices, since their columns (and rows) are orthonormal bases! I have no good answer, but isometries do preserve orthogonality, and orthogonal matrices correspond to isometries.

Recall that the determinant $\det(f)$ of a linear map $f : E \to E$ is independent of the choice of a basis in E. Also, for every matrix $A \in \mathrm{M}_n(\mathbb{R})$, we have $\det(A) = \det(A^\top)$, and for any two $n \times n$ matrices A and B, we have $\det(AB) = \det(A) \det(B)$ (for all these basic results, see Lang [14]). Then, if f is an isometry, and A is its matrix with respect to any orthonormal basis, $A A^\top = A^\top A = I_n$ implies that $\det(A)^2 = 1$, that is, either $\det(A) = 1$, or $\det(A) = -1$. It is also clear that the isometries of a Euclidean space of dimension n form a group, and that the isometries of determinant $+1$ form a subgroup. This leads to the following definition.

Definition 6.6. Given a Euclidean space E of dimension n, the set of isometries $f\colon E \to E$ forms a subgroup of $\mathbf{GL}(E)$ denoted by $\mathbf{O}(E)$, or $\mathbf{O}(n)$ when $E = \mathbb{R}^n$, called the *orthogonal group (of E)*. For every isometry f, we have $\det(f) = \pm 1$, where $\det(f)$ denotes the determinant of f. The isometries such that $\det(f) = 1$ are called *rotations, or proper isometries, or proper orthogonal transformations*, and they form a subgroup of the special linear group $\mathbf{SL}(E)$ (and of $\mathbf{O}(E)$), denoted by $\mathbf{SO}(E)$, or $\mathbf{SO}(n)$ when $E = \mathbb{R}^n$, called the *special orthogonal group (of E)*. The isometries such that $\det(f) = -1$ are called *improper isometries, or improper orthogonal transformations, or flip transformations*.

As an immediate corollary of the Gram–Schmidt orthonormalization procedure, we obtain the QR-decomposition for invertible matrices.

6.5 QR-Decomposition for Invertible Matrices

Now that we have the definition of an orthogonal matrix, we can explain how the Gram–Schmidt orthonormalization procedure immediately yields the QR-decomposition for matrices.

Lemma 6.11. *Given any real $n \times n$ matrix A, if A is invertible, then there is an orthogonal matrix Q and an upper triangular matrix R with positive diagonal entries such that $A = QR$.*

Proof. We can view the columns of A as vectors $A_1, \dots, A_n$ in $\mathbb{E}^n$. If A is invertible, then they are linearly independent, and we can apply Lemma 6.7 to produce an orthonormal basis using the Gram–Schmidt orthonormalization procedure. Recall that we construct vectors Q_k and Q'_k as follows:

$$Q'_1 = A_1, \qquad Q_1 = \frac{Q'_1}{\|Q'_1\|},$$

and for the inductive step

$$Q'_{k+1} = A_{k+1} - \sum_{i=1}^{k} (A_{k+1} \cdot Q_i)\, Q_i, \qquad Q_{k+1} = \frac{Q'_{k+1}}{\|Q'_{k+1}\|},$$

where $1 \le k \le n-1$. If we express the vectors A_k in terms of the Q_i and Q'_i, we get the triangular system

$$A_1 = \|Q_1'\|Q_1,$$

$$\vdots$$

$$A_j = (A_j \cdot Q_1)\,Q_1 + \cdots + (A_j \cdot Q_i)\,Q_i + \cdots + \|Q_j'\|Q_j,$$

$$\vdots$$

$$A_n = (A_n \cdot Q_1)\,Q_1 + \cdots + (A_n \cdot Q_{n-1})\,Q_{n-1} + \|Q_n'\|Q_n.$$

Letting $r_{k,k} = \|Q_k'\|$, and $r_{i,j} = A_j \cdot Q_i$ (the reversal of i and j on the right-hand side **is** intentional!), where $1 \le k \le n$, $2 \le j \le n$, and $1 \le i \le j-1$, and letting $q_{i,j}$ be the ith component of Q_j, we note that $a_{i,j}$, the ith component of A_j, is given by

$$a_{i,j} = r_{1,j}q_{i,1} + \cdots + r_{i,j}q_{i,i} + \cdots + r_{j,j}q_{i,j} = q_{i,1}r_{1,j} + \cdots + q_{i,i}r_{i,j} + \cdots + q_{i,j}r_{j,j}.$$

If we let $Q = (q_{i,j})$, the matrix whose columns are the components of the Q_j, and $R = (r_{i,j})$, the above equations show that $A = QR$, where R is upper triangular The diagonal entries $r_{k,k} = \|Q_k'\| = A_k \cdot Q_k$ are indeed positive. $\square$

The reader should try the above procedure on some concrete examples for 2×2 and 3×3 matrices.

Remarks:

(1) Because the diagonal entries of R are positive, it can be shown that Q and R are unique.
(2) The *QR*-decomposition holds even when A is not invertible. In this case, R has some zero on the diagonal. However, a different proof is needed. We will give a nice proof using Householder matrices (see Lemma 8.6, and also Strang [23, 24], Golub and Van Loan [9], Trefethen and Bau [26], Demmel [7], Kincaid and Cheney [13], or Ciarlet [5]).

Example 6.6. Consider the matrix

$$A = \begin{pmatrix} 0 & 0 & 5 \\ 0 & 4 & 1 \\ 1 & 1 & 1 \end{pmatrix}.$$

We leave as an exercise to show that $A = QR$, with

$$Q = \begin{pmatrix} 0 & 0 & 1 \\ 0 & 1 & 0 \\ 1 & 0 & 0 \end{pmatrix} \quad \text{and} \quad R = \begin{pmatrix} 1 & 1 & 1 \\ 0 & 4 & 1 \\ 0 & 0 & 5 \end{pmatrix}.$$

Example 6.7. Another example of *QR*-decomposition is

$$A = \begin{pmatrix} 1 & 1 & 2 \\ 0 & 0 & 1 \\ 1 & 0 & 0 \end{pmatrix} = \begin{pmatrix} 1/\sqrt{2} & 1/\sqrt{2} & 0 \\ 0 & 0 & 1 \\ 1/\sqrt{2} & -1/\sqrt{2} & 0 \end{pmatrix} \begin{pmatrix} \sqrt{2} & 1/\sqrt{2} & \sqrt{2} \\ 0 & 1/\sqrt{2} & \sqrt{2} \\ 0 & 0 & 1 \end{pmatrix}.$$

The QR-decomposition yields a rather efficient and numerically stable method for solving systems of linear equations. Indeed, given a system $Ax = b$, where A is an $n \times n$ invertible matrix, writing $A = QR$, since Q is orthogonal, we get

$$Rx = Q^\top b,$$

and since R is upper triangular, we can solve it by Gaussian elimination, by solving for the last variable x_n first, substituting its value into the system, then solving for x_{n-1}, etc. The QR-decomposition is also very useful in solving least squares problems (we will come back to this later on), and for finding eigenvalues. It can be easily adapted to the case where A is a rectangular $m \times n$ matrix with independent columns (thus, $n \leq m$). In this case, Q is not quite orthogonal. It is an $m \times n$ matrix whose columns are orthogonal, and R is an invertible $n \times n$ upper diagonal matrix with positive diagonal entries. For more on QR, see Strang [23, 24], Golub and Van Loan [9], Demmel [7], Trefethen and Bau [26], or Serre [21].

It should also be said that the Gram–Schmidt orthonormalization procedure that we have presented is not very stable numerically, and instead, one should use the *modified Gram–Schmidt method*. To compute Q'_{k+1}, instead of projecting A_{k+1} onto $Q_1, \ldots, Q_k$ in a single step, it is better to perform k projections. We compute $Q^1_{k+1}, Q^2_{k+1}, \ldots, Q^k_{k+1}$ as follows:

$$Q^1_{k+1} = A_{k+1} - (A_{k+1} \cdot Q_1)\, Q_1,$$
$$Q^{i+1}_{k+1} = Q^i_{k+1} - (Q^i_{k+1} \cdot Q_{i+1})\, Q_{i+1},$$

where $1 \leq i \leq k - 1$. It is easily shown that $Q'_{k+1} = Q^k_{k+1}$. The reader is urged to code this method.

6.6 Some Applications of Euclidean Geometry

Euclidean geometry has applications in computational geometry, in particular Voronoi diagrams and Delaunay triangulations, discussed in Chapter 10. In turn, Voronoi diagrams have applications in motion planning (see O'Rourke [17]).

Euclidean geometry also has applications to matrix analysis. Recall that a real $n \times n$ matrix A is *symmetric* if it is equal to its transpose $A^\top$. One of the most important properties of symmetric matrices is that they have real eigenvalues and that they can be diagonalized by an orthogonal matrix (see Chapter 12). This means that for every symmetric matrix A, there is a diagonal matrix D and an orthogonal matrix P such that

$$A = PDP^\top.$$

Even though it is not always possible to diagonalize an arbitrary matrix, there are various decompositions involving orthogonal matrices that are of great practical interest. For example, for every real matrix A, there is the QR-decomposition, which says that a real matrix A can be expressed as

$$A = QR,$$

where Q is orthogonal and R is an upper triangular matrix. This can be obtained from the Gram–Schmidt orthonormalization procedure, as we saw in Section 6.5, or better, using Householder matrices, as shown in Section 8.3. There is also the *polar decomposition*, which says that a real matrix A can be expressed as

$$A = QS,$$

where Q is orthogonal and S is symmetric positive semidefinite (which means that the eigenvalues of S are nonnegative; see Chapter 12). Such a decomposition is important in continuum mechanics and in robotics, since it separates stretching from rotation. Finally, there is the wonderful *singular value decomposition*, abbreviated as SVD, which says that a real matrix A can be expressed as

$$A = VDU^\top,$$

where U and V are orthogonal and D is a diagonal matrix with nonnegative entries (see Chapter 13). This decomposition leads to the notion of *pseudo-inverse*, which has many applications in engineering (least squares solutions, etc). For an excellent presentation of all these notions, we highly recommend Strang [24, 23], Golub and Van Loan [9], Demmel [7], Serre [21], and Trefethen and Bau [26].

The method of least squares, invented by Gauss and Legendre around 1800, is another great application of Euclidean geometry. Roughly speaking, the method is used to solve inconsistent linear systems $Ax = b$, where the number of equations is greater than the number of variables. Since this is generally impossible, the method of least squares consists in finding a solution x minimizing the Euclidean norm $\|Ax - b\|^2$, that is, the sum of the squares of the "errors." It turns out that there is always a unique solution x^+ of smallest norm minimizing $\|Ax - b\|^2$, and that it is a solution of the square system

$$A^\top Ax = A^\top b,$$

called the system of *normal equations*. The solution x^+ can be found either by using the QR-decomposition in terms of Householder transformations, or by using the notion of pseudo-inverse of a matrix. The pseudo-inverse can be computed using the SVD decomposition. Least squares methods are used extensively in computer vision; see Trucco and Verri [27], or Jain, Katsuri, and Schunck [12]. More details on the method of least squares and pseudo-inverses can be found in Chapter 14.

6.7 Problems

6.1. Prove Lemma 6.2.

6.2. Prove Lemma 6.3.

6.3. Let $(e_1, \ldots, e_n)$ be an orthonormal basis for E. If X and Y are arbitrary $n \times n$ matrices, denoting as usual the jth column of X by X_j, and similarly for Y, show that

$$X^\top Y = (X_i \cdot Y_j)_{1 \le i, j \le n}.$$

Use this to prove that

$$A^\top A = AA^\top = I_n$$

iff the column vectors $(A_1, \ldots, A_n)$ form an orthonormal basis. Show that the conditions $AA^\top = I_n$, $A^\top A = I_n$, and $A^{-1} = A^\top$ are equivalent.

6.4. Given any two linear maps $f \colon E \to F$ and $g \colon F \to E$, where $\dim(E) = n$ and $\dim(F) = m$, prove that

$$(-\lambda)^m \det(g \circ f - \lambda I_n) = (-\lambda)^n \det(f \circ g - \lambda I_m),$$

and thus that $g \circ f$ and $f \circ g$ have the same nonnull eigenvalues.

Hint. If A is an $m \times n$ matrix and B is an $n \times m$ matrix, observe that

$$\begin{vmatrix} AB - XI_m & 0_{m,n} \\ B & -XI_n \end{vmatrix} = \begin{vmatrix} A & XI_m \\ I_n & 0_{n,m} \end{vmatrix} \begin{vmatrix} B & -XI_n \\ -I_m & A \end{vmatrix}$$

and

$$\begin{vmatrix} A & XI_m \\ I_n & 0_{n,m} \end{vmatrix} \begin{vmatrix} B & -XI_n \\ -I_m & A \end{vmatrix} = \begin{vmatrix} BA - XI_n & XB \\ 0_{m,n} & -XI_m \end{vmatrix},$$

where X is a variable.

6.5. (a) Let $\mathscr{C}_1 = (C_1, R_1)$ and $\mathscr{C}_2 = (C_2, R_2)$ be two distinct circles in the plane $\mathbb{E}^2$ (where C_i is the center and R_i is the radius). What is the locus of the centers of all circles tangent to both $\mathscr{C}_1$ and $\mathscr{C}_2$?
Hint. When is it one conic, when is it two conics?

  (b) Repeat question (a) in the case where $\mathscr{C}_2$ is a line.

  (c) Given three pairwise distinct circles $\mathscr{C}_1 = (C_1, R_1)$, $\mathscr{C}_2 = (C_2, R_2)$, and $\mathscr{C}_3 = (C_3, R_3)$ in the plane $\mathbb{E}^2$, prove that there are at most eight circles simultaneously tangent to $\mathscr{C}_1$, $\mathscr{C}_2$, and $\mathscr{C}_3$ (this is known as the *problem of Apollonius*). What happens if the centers C_1, C_2, C_3 of the circles are collinear? In the latter case, show that there are at most two circles exterior and tangent to $\mathscr{C}_1$, $\mathscr{C}_2$, and $\mathscr{C}_3$.
Hint. You may want to use a carefully chosen inversion (see the problems in Section 5.14, especially Problem 5.37).

  (d) Prove that the problem of question (c) reduces to the problem of finding the circles passing through a fixed point and tangent to two given circles. In turn, by inversion, this problem reduces to finding all lines tangent to two circles.

  (e) Given four pairwise distinct spheres $\mathscr{C}_1 = (C_1, R_1)$, $\mathscr{C}_2 = (C_2, R_2)$, $\mathscr{C}_3 = (C_3, R_3)$, and $\mathscr{C}_4 = (C_4, R_4)$, prove that there are at most sixteen spheres simultaneously tangent to $\mathscr{C}_1$, $\mathscr{C}_2$, $\mathscr{C}_3$, and $\mathscr{C}_4$. Prove that this problem reduces to the problem of finding the spheres passing through a fixed point and tangent to three given spheres. In turn, by inversion, this problem reduces to finding all planes tangent to three spheres.

6.6. (a) Given any two circles $\mathscr{C}_1$ and $\mathscr{C}_2$ in $\mathbb{E}^2$ of equations

$$x^2 + y^2 - 2ax - 2by + c = 0 \quad \text{and} \quad x^2 + y^2 - 2a'x - 2'by + c' = 0,$$

we say that $\mathscr{C}_1$ and $\mathscr{C}_2$ are *orthogonal* if they intersect and if the tangents at the intersection points are orthogonal. Prove that $\mathscr{C}_1$ and $\mathscr{C}_2$ are orthogonal iff

$$2(aa' + bb') = c + c'.$$

(b) For any given $c \in \mathbb{R}$ ($c \neq 0$), there is a pencil $\mathscr{F}$ of circles of equations

$$x^2 + y^2 - 2ux - c = 0,$$

where $u \in \mathbb{R}$ is arbitrary. Show that the set of circles orthogonal to all circles in the pencil $\mathscr{F}$ is the pencil $\mathscr{F}^\perp$ of circles of equations

$$x^2 + y^2 - 2vy + c = 0,$$

where $v \in \mathbb{R}$ is arbitrary.

6.7. Let $P = \{p_1, \ldots, p_n\}$ be a finite set of points in $\mathbb{E}^3$. Show that there is a unique point c such that the sum of the squares of the distances from c to each p_i is minimal. Find this point in terms of the p_i.

6.8. (1) Compute the real Fourier coefficients of the function $id(x) = x$ over $[-\pi, \pi]$ and prove that

$$x = 2\left(\frac{\sin x}{1} - \frac{\sin 2x}{2} + \frac{\sin 3x}{3} - \cdots\right).$$

What is the value of the Fourier series at $\pm\pi$? What is the value of the Fourier near $\pm\pi$? Do you find this surprising?

(2) Plot the functions obtained by keeping $1, 2, 4, 5$, and 10 terms. What do you observe around $\pm\pi$?

6.9. The *Dirac delta function* (which is **not** a function!) is the spike function s.t. $\delta(k2\pi) = +\infty$ for all $k \in \mathbb{Z}$, and $\delta(x) = 0$ everywhere else. It has the property that for "well-behaved" functions f (including constant functions and trigonometric functions),

$$\int_{-\pi}^{+\pi} f(t)\delta(t)dt = f(0).$$

(1) Compute the real Fourier coefficients of δ over $[-\pi, \pi]$, and prove that

$$\delta(x) = \frac{1}{2\pi}(1 + 2\cos x + 2\cos 2x + 2\cos 3x + \cdots + 2\cos nx + \cdots).$$

Also compute the complex Fourier coefficients of δ over $[-\pi, \pi]$, and prove that

$$\delta(x) = \frac{1}{2\pi}\left(1 + e^{ix} + e^{-ix} + e^{i2x} + e^{-i2x} + \cdots + e^{inx} + e^{-inx} + \cdots\right).$$

(2) Prove that the partial sum of the first $2n+1$ complex terms is

$$\delta_n(x) = \frac{\sin\left((2n+1)(x/2)\right)}{2\pi\sin\left(x/2\right)}.$$

What is $\delta_n(0)$?

(3) Plot $\delta_n(x)$ for $n = 10, 20$ (over $[-\pi, \pi]$). Prove that the area under the curve δ_n is independent of n. What is it?

6.10. (1) If an upper triangular $n \times n$ matrix R is invertible, prove that its inverse is also upper triangular.

(2) If an upper triangular matrix is orthogonal, prove that it must be a diagonal matrix.

If A is an invertible $n \times n$ matrix and if $A = Q_1 R_1 = Q_2 R_2$, where R_1 and R_2 are upper triangular with positive diagonal entries and Q_1, Q_2 are orthogonal, prove that $Q_1 = Q_2$ and $R_1 = R_2$.

6.11. (1) Review the modified Gram–Schmidt method. Recall that to compute Q'_{k+1}, instead of projecting A_{k+1} onto $Q_1, \ldots, Q_k$ in a single step, it is better to perform k projections. We compute $Q^1_{k+1}, Q^2_{k+1}, \ldots, Q^k_{k+1}$ as follows:

$$Q^1_{k+1} = A_{k+1} - (A_{k+1} \cdot Q_1)\, Q_1,$$
$$Q^{i+1}_{k+1} = Q^i_{k+1} - (Q^i_{k+1} \cdot Q_{i+1})\, Q_{i+1},$$

where $1 \leq i \leq k-1$.

Prove that $Q'_{k+1} = Q^k_{k+1}$.

(2) Write two computer programs to compute the QR-decomposition of an invertible matrix. The first one should use the standard Gram–Schmidt method, and the second one the modified Gram–Schmidt method. Run both on a number of matrices, up to dimension at least 10. Do you observe any difference in their performance in terms of numerical stability?

Run your programs on the Hilbert matrix $H_n = (1/(i+j-1))_{1 \leq i,j \leq n}$. What happens?

Extra Credit. Write a program to solve linear systems of equations $Ax = b$, using your version of the QR-decomposition program, where A is an $n \times n$ matrix.

6.12. Let E be a Euclidean space of finite dimension n, and let $(e_1, \ldots, e_n)$ be an orthonormal basis for E. For any two vectors $u, v \in E$, the linear map $u \otimes v$ is defined such that

$$u \otimes v(x) = (v \cdot x)\, u,$$

for all $x \in E$. If U and V are the column vectors of coordinates of u and v w.r.t. the basis $(e_1, \ldots, e_n)$, prove that $u \otimes v$ is represented by the matrix

$$U^\top V.$$

What sort of linear map is $u \otimes u$ when u is a unit vector?

6.13. Let $\varphi \colon E \times E \to \mathbb{R}$ be a bilinear form on a real vector space E of finite dimension n. Given any basis $(e_1, \ldots, e_n)$ of E, let $A = (\alpha_{ij})$ be the matrix defined such that

$$\alpha_{ij} = \varphi(e_i, e_j),$$

$1 \leq i, j \leq n$. We call A *the matrix of* φ *w.r.t. the basis* $(e_1, \ldots, e_n)$.

(a) For any two vectors x and y, if X and Y denote the column vectors of coordinates of x and y w.r.t. the basis $(e_1, \ldots, e_n)$, prove that

$$\varphi(x, y) = X^\top A Y.$$

(b) Recall that A is a *symmetric* matrix if $A = A^\top$. Prove that φ is symmetric if A is a symmetric matrix.

(c) If $(f_1, \ldots, f_n)$ is another basis of E and P is the change of basis matrix from $(e_1, \ldots, e_n)$ to $(f_1, \ldots, f_n)$, prove that the matrix of φ w.r.t. the basis $(f_1, \ldots, f_n)$ is

$$P^\top A P.$$

The common rank of all matrices representing φ is called the *rank* of φ.

6.14. Let $\varphi \colon E \times E \to \mathbb{R}$ be a symmetric bilinear form on a real vector space E of finite dimension n. Two vectors x and y are said to be *conjugate w.r.t.* φ if $\varphi(x, y) = 0$. The main purpose of this problem is to prove that there is a basis of vectors that are pairwise conjugate w.r.t. φ.

(a) Prove that if $\varphi(x, x) = 0$ for all $x \in E$, then φ is identically null on E.

Otherwise, we can assume that there is some vector $x \in E$ such that $\varphi(x, x) \neq 0$. Use induction to prove that there is a basis of vectors that are pairwise conjugate w.r.t. φ.

For the induction step, proceed as follows. Let $(e_1, e_2, \ldots, e_n)$ be a basis of E, with $\varphi(e_1, e_1) \neq 0$. Prove that there are scalars $\lambda_2, \ldots, \lambda_n$ such that each of the vectors

$$v_i = e_i + \lambda_i e_1$$

is conjugate to e_1 w.r.t. φ, where $2 \leq i \leq n$, and that $(e_1, v_2, \ldots, v_n)$ is a basis.

(b) Let $(e_1, \ldots, e_n)$ be a basis of vectors that are pairwise conjugate w.r.t. φ, and assume that they are ordered such that

$$\varphi(e_i, e_i) = \begin{cases} \theta_i \neq 0 & \text{if } 1 \leq i \leq r, \\ 0 & \text{if } r+1 \leq i \leq n, \end{cases}$$

where r is the rank of φ. Show that the matrix of φ w.r.t. $(e_1, \ldots, e_n)$ is a diagonal matrix, and that

$$\varphi(x, y) = \sum_{i=1}^{r} \theta_i x_i y_i,$$

where $x = \sum_{i=1}^{n} x_i e_i$ and $y = \sum_{i=1}^{n} y_i e_i$.

Prove that for every symmetric matrix A, there is an invertible matrix P such that

$$P^\top AP = D,$$

where D is a diagonal matrix.

(c) Prove that there is an integer p, $0 \leq p \leq r$ (where r is the rank of φ), such that $\varphi(u_i, u_i) > 0$ for exactly p vectors of every basis $(u_1, \ldots, u_n)$ of vectors that are pairwise conjugate w.r.t. φ (*Sylvester's inertia theorem*).

Proceed as follows. Assume that in the basis $(u_1, \ldots, u_n)$, for any $x \in E$, we have

$$\varphi(x,x) = \alpha_1 x_1^2 + \cdots + \alpha_p x_p^2 - \alpha_{p+1} x_{p+1}^2 - \cdots - \alpha_r x_r^2,$$

where $x = \sum_{i=1}^n x_i u_i$, and that in the basis $(v_1, \ldots, v_n)$, for any $x \in E$, we have

$$\varphi(x,x) = \beta_1 y_1^2 + \cdots + \beta_q y_q^2 - \beta_{q+1} y_{q+1}^2 - \cdots - \beta_r y_r^2,$$

where $x = \sum_{i=1}^n y_i v_i$, with $\alpha_i > 0$, $\beta_i > 0$, $1 \leq i \leq r$.

Assume that $p > q$ and derive a contradiction. First, consider x in the subspace F spanned by

$$(u_1, \ldots, u_p, u_{r+1}, \ldots, u_n),$$

and observe that $\varphi(x,x) \geq 0$ if $x \neq 0$. Next, consider x in the subspace G spanned by

$$(v_{q+1}, \ldots, v_r),$$

and observe that $\varphi(x,x) < 0$ if $x \neq 0$. Prove that $F \cap G$ is nontrivial (i.e., contains some nonnull vector), and derive a contradiction. This implies that $p \leq q$. Finish the proof.

The pair $(p, r - p)$ is called the *signature* of φ.

(d) A symmetric bilinear form φ is *definite* if for every $x \in E$, if $\varphi(x,x) = 0$, then $x = 0$.

Prove that a symmetric bilinear form is definite iff its signature is either $(n,0)$ or $(0,n)$. In other words, a symmetric definite bilinear form has rank n and is either positive or negative.

(e) The *kernel* of a symmetric bilinear form φ is the subspace consisting of the vectors that are conjugate to all vectors in E. We say that a symmetric bilinear form φ is *nondegenerate* if its kernel is trivial (i.e., equal to $\{0\}$).

Prove that a symmetric bilinear form φ is nondegenerate iff its rank is n, the dimension of E. Is a definite symmetric bilinear form φ nondegenerate? What about the converse?

Prove that if φ is nondegenerate, then there is a basis of vectors that are pairwise conjugate w.r.t. φ and such that φ is represented by the matrix

$$\begin{pmatrix} I_p & 0 \\ 0 & -I_q \end{pmatrix}$$

where (p, q) is the signature of φ.

(f) Given a nondegenerate symmetric bilinear form φ on E, prove that for every linear map $f\colon E \to E$, there is a unique linear map $f^*\colon E \to E$ such that

$$\varphi(f(u), v) = \varphi(u, f^*(v)),$$

for all $u, v \in E$. The map f^* is called the *adjoint of f (w.r.t. to φ)*. Given any basis $(u_1, \ldots, u_n)$, if Ω is the matrix representing φ and A is the matrix representing f, prove that f^* is represented by $\Omega^{-1} A^\top \Omega$.

Prove that Lemma 6.4 also holds, i.e., the map $\flat\colon E \to E^*$ is a canonical isomorphism.

A linear map $f\colon E \to E$ is an *isometry w.r.t. φ* if

$$\varphi(f(x), f(y)) = \varphi(x, y)$$

for all $x, y \in E$. Prove that a linear map f is an isometry w.r.t. φ iff

$$f^* \circ f = f \circ f^* = \mathrm{id}.$$

Prove that the set of isometries w.r.t. φ is a group. This group is denoted by $\mathbf{O}(\varphi)$, and its subgroup consisting of isometries having determinant $+1$ by $\mathbf{SO}(\varphi)$. Given any basis of E, if Ω is the matrix representing φ and A is the matrix representing f, prove that $f \in \mathbf{O}(\varphi)$ iff

$$A^\top \Omega A = \Omega.$$

Given another nondegenerate symmetric bilinear form ψ on E, we say that φ and ψ are *equivalent* if there is a bijective linear map $h\colon E \to E$ such that

$$\psi(x, y) = \varphi(h(x), h(y)),$$

for all $x, y \in E$. Prove that the groups of isometries $\mathbf{O}(\varphi)$ and $\mathbf{O}(\psi)$ are isomomorphic (use the map $f \mapsto h \circ f \circ h^{-1}$ from $\mathbf{O}(\psi)$ to $\mathbf{O}(\varphi)$).

If φ is a nondegenerate symmetric bilinear form of signature (p, q), prove that the group $\mathbf{O}(\varphi)$ is isomorphic to the group of $n \times n$ matrices A such that

$$A^\top \begin{pmatrix} I_p & 0 \\ 0 & -I_q \end{pmatrix} A = \begin{pmatrix} I_p & 0 \\ 0 & -I_q \end{pmatrix}.$$

Remark: In view of question (f), the groups $\mathbf{O}(\varphi)$ and $\mathbf{SO}(\varphi)$ are also denoted by $\mathbf{O}(p, q)$ and $\mathbf{SO}(p, q)$ when φ has signature (p, q). They are Lie groups. In particular, the group $\mathbf{SO}(3, 1)$, known as the *Lorentz group*, plays an important role in the theory of special relativity.

6.15. (a) Let C be a circle of radius R and center O, and let P be any point in the Euclidean plane $\mathbb{E}^2$. Consider the lines Δ through P that intersect the circle C, generally in two points A and B. Prove that for all such lines,

$$\overrightarrow{PA} \cdot \overrightarrow{PB} = \|\overrightarrow{PO}\|^2 - R^2.$$

Hint. If P is not on C, let B' be the antipodal of B (i.e., $\overrightarrow{OB'} = -\overrightarrow{OB}$). Then $\overrightarrow{AB} \cdot \overrightarrow{AB'} = 0$ and

$$\overrightarrow{PA} \cdot \overrightarrow{PB} = \overrightarrow{PB'} \cdot \overrightarrow{PB} = (\overrightarrow{PO} - \overrightarrow{OB}) \cdot (\overrightarrow{PO} + \overrightarrow{OB}) = \|\overrightarrow{PO}\|^2 - R^2.$$

The quantity $\|\overrightarrow{PO}\|^2 - R^2$ is called the *power of P w.r.t.* C, and it is denoted by $\mathscr{P}(P,C)$.

Show that if Δ is tangent to C, then $A = B$ and

$$\|\overrightarrow{PA}\|^2 = \|\overrightarrow{PO}\|^2 - R^2.$$

Show that P is inside C iff $\mathscr{P}(P,C) < 0$, on C iff $\mathscr{P}(P,C) = 0$, outside C if $\mathscr{P}(P,C) > 0$.

If the equation of C is

$$x^2 + y^2 - 2ax - 2by + c = 0,$$

prove that the power of $P = (x,y)$ w.r.t. C is given by

$$\mathscr{P}(P,C) = x^2 + y^2 - 2ax - 2by + c.$$

(b) Given two nonconcentric circles C and C', show that the set of points having equal power w.r.t. C and C' is a line orthogonal to the line through the centers of C and C'. If the equations of C and C' are

$$x^2 + y^2 - 2ax - 2by + c = 0 \quad \text{and} \quad x^2 + y^2 - 2a'x - 2b'y + c' = 0,$$

show that the equation of this line is

$$2(a - a')x + 2(b - b')y + c' - c = 0.$$

This line is called the *radical axis* of C and C'.

(c) Given three distinct nonconcentric circles C, C', and C'', prove that either the three pairwise radical axes of these circles are parallel or that they intersect in a single point ω that has equal power w.r.t. C, C', and C''. In the first case, the centers of C, C', and C'' are collinear. In the second case, if the power of ω is positive, prove that ω is the center of a circle Γ orthogonal to C, C', and C'', and if the power of ω is negative, ω is inside C, C', and C''.

(d) Given any $k \in \mathbb{R}$ with $k \neq 0$ and any point a, recall that an *inversion of pole a and power k* is a map $h \colon (\mathbb{E}^n - \{a\}) \to \mathbb{E}^n$ defined such that for every $x \in \mathbb{E}^n - \{a\}$,

$$h(x) = a + k\frac{\overrightarrow{ax}}{\|\overrightarrow{ax}\|^2}.$$

For example, when $n = 2$, chosing any orthonormal frame with origin a, h is defined by the map

$$(x,y) \mapsto \left(\frac{kx}{x^2 + y^2}, \frac{ky}{x^2 + y^2} \right).$$

When the centers of C, C' and C'' are not collinear and the power of ω is positive, prove that by a suitable inversion, C, C' and C'' are mapped to three circles whose centers are collinear.

Prove that if three distinct nonconcentric circles C, C', and C'' have collinear centers, then there are at most eight circles simultaneously tangent to C, C', and C'', and at most two for those exterior to C, C', and C''.

(e) Prove that an inversion in $\mathbb{E}^3$ maps a sphere to a sphere or to a plane. Prove that inversions preserve tangency and orthogonality of planes and spheres.

References

1. Emil Artin. *Geometric Algebra*. Wiley Interscience, first edition, 1957.
2. Marcel Berger. *Géométrie 1*. Nathan, 1990. English edition: Geometry 1, Universitext, Springer-Verlag.
3. Marcel Berger. *Géométrie 2*. Nathan, 1990. English edition: Geometry 2, Universitext, Springer-Verlag.
4. G. Cagnac, E. Ramis, and J. Commeau. *Mathématiques Spéciales, Vol. 3, Géométrie*. Masson, 1965.
5. P.G. Ciarlet. *Introduction to Numerical Matrix Analysis and Optimization*. Cambridge University Press, first edition, 1989. French edition: Masson, 1994.
6. H.S.M. Coxeter. *Introduction to Geometry*. Wiley, second edition, 1989.
7. James W. Demmel. *Applied Numerical Linear Algebra*. SIAM Publications, first edition, 1997.
8. Jean Fresnel. *Méthodes Modernes en Géométrie*. Hermann, first edition, 1998.
9. Gene H. Golub and Charles F. Van Loan. *Matrix Computations*. The Johns Hopkins University Press, third edition, 1996.
10. Jacques Hadamard. *Leçons de Géométrie Elémentaire. I Géométrie Plane*. Armand Colin, thirteenth edition, 1947.
11. Jacques Hadamard. *Leçons de Géométrie Elémentaire. II Géométrie dans l'Espace*. Armand Colin, eighth edition, 1949.
12. Ramesh Jain, Rangachar Katsuri, and Brian G. Schunck. *Machine Vision*. McGraw-Hill, first edition, 1995.
13. D. Kincaid and W. Cheney. *Numerical Analysis*. Brooks/Cole Publishing, second edition, 1996.
14. Serge Lang. *Algebra*. Addison-Wesley, third edition, 1993.
15. Serge Lang. *Real and Functional Analysis*. GTM 142. Springer-Verlag, third edition, 1996.
16. Serge Lang. *Undergraduate Analysis*. UTM. Springer-Verlag, second edition, 1997.
17. Joseph O'Rourke. *Computational Geometry in C*. Cambridge University Press, second edition, 1998.
18. Dan Pedoe. *Geometry, A Comprehensive Course*. Dover, first edition, 1988.
19. Eugène Rouché and Charles de Comberousse. *Traité de Géométrie*. Gauthier-Villars, seventh edition, 1900.
20. Walter Rudin. *Real and Complex Analysis*. McGraw-Hill, third edition, 1987.
21. Denis Serre. *Matrices. Theory and Applications*. GTM No. 216. Springer-Verlag, second edition, 2010.
22. Ernst Snapper and Troyer Robert J. *Metric Affine Geometry*. Dover, first edition, 1989.
23. Gilbert Strang. *Introduction to Applied Mathematics*. Wellesley–Cambridge Press, first edition, 1986.
24. Gilbert Strang. *Linear Algebra and Its Applications*. Saunders HBJ, third edition, 1988.

25. Claude Tisseron. *Géométries Affines, Projectives, et Euclidiennes*. Hermann, first edition, 1994.
26. L.N. Trefethen and D. Bau III. *Numerical Linear Algebra*. SIAM Publications, first edition, 1997.
27. Emanuele Trucco and Alessandro Verri. *Introductory Techniques for 3D Computer Vision*. Prentice-Hall, first edition, 1998.

Chapter 7
Separating and Supporting Hyperplanes

7.1 Separation Theorems and Farkas's Lemma

Now that we have a solid background in Euclidean geometry, we can go deeper into our study of convex sets begun in Chapter 3. This chapter is devoted to a thorough study of separating and supporting hyperplanes. We prove two geometric versions of the Hahn–Banach theorem, from which we derive separation results for various kinds of pairs of convex sets (open, closed, compact). We prove various versions of Farkas's lemma, a basic result in the theory of linear programming. We also discuss supporting hyperplanes and prove an important proposition due to Minkowski.

It seems intuitively rather obvious that if A and B are two nonempty disjoint convex sets in $\mathbb{A}^2$, then there is a line H separating them, in the sense that A and B belong to the two (disjoint) open half-planes determined by H. However, this is not always true! For example, this fails if both A and B are closed and unbounded (find an example). Nevertheless, the result is true if both A and B are open, or if the notion of separation is weakened a little bit. The key result, from which most separation results follow, is a geometric version of the *Hahn–Banach theorem*. In the sequel, we restrict our attention to real affine spaces of finite dimension. Then, if X is an affine space of dimension d, there is an affine bijection f between X and $\mathbb{A}^d$.

Now, $\mathbb{A}^d$ is a topological space, under the usual topology on $\mathbb{R}^d$ (in fact, $\mathbb{A}^d$ is a metric space). Recall that if $a = (a_1, \ldots, a_d)$ and $b = (b_1, \ldots, b_d)$ are any two points in $\mathbb{A}^d$, their **Euclidean distance**, $d(a, b)$, is given by

$$d(a,b) = \sqrt{(b_1 - a_1)^2 + \cdots + (b_d - a_d)^2},$$

which is also the *norm* $\|\overrightarrow{ab}\|$ of the vector $\overrightarrow{ab}$, and that for any $\varepsilon > 0$, the *open ball $B(a, \varepsilon)$ of center a and radius ε* is given by

$$B(a,\varepsilon) = \{b \in \mathbb{A}^d \mid d(a,b) < \varepsilon\}.$$

A subset $U \subseteq \mathbb{A}^d$ is *open* (in the *norm topology*) if either U is empty or for every point $a \in U$, there is some (small) open ball $B(a, \varepsilon)$ contained in U. A subset $C \subseteq \mathbb{A}^d$ is *closed* iff $\mathbb{A}^d - C$ is open. For example, the *closed balls* $\overline{B(a, \varepsilon)}$, where

$$\overline{B(a, \varepsilon)} = \{b \in \mathbb{A}^d \mid d(a, b) \leq \varepsilon\},$$

are closed. A subset $W \subseteq \mathbb{A}^d$ is *bounded* iff there is some ball (open or closed) B such that $W \subseteq B$. A subset $W \subseteq \mathbb{A}^d$ is *compact* iff every family $\{U_i\}_{i \in I}$ that is an open cover of W (which means that $W = \bigcup_{i \in I}(W \cap U_i)$, with each U_i an open set) possesses a finite subcover (which means that there is a finite subset $F \subseteq I$ such that $W = \bigcup_{i \in F}(W \cap U_i)$). In $\mathbb{A}^d$, it can be shown that a subset W is compact iff W is closed and bounded. Given a function $f \colon \mathbb{A}^m \to \mathbb{A}^n$, we say that f is *continuous* if $f^{-1}(V)$ is open in $\mathbb{A}^m$ whenever V is open in $\mathbb{A}^n$. If $f \colon \mathbb{A}^m \to \mathbb{A}^n$ is a continuous function, although it is generally *false* that $f(U)$ is open if $U \subseteq \mathbb{A}^m$ is open, it is easily checked that $f(K)$ is compact if $K \subseteq \mathbb{A}^m$ is compact.

An affine space X of dimension d becomes a topological space if we give it the topology for which the open subsets are of the form $f^{-1}(U)$, where U is any open subset of $\mathbb{A}^d$ and $f \colon X \to \mathbb{A}^d$ is an affine bijection.

Given any subset A of a topological space X, the smallest closed set containing A is denoted by $\overline{A}$, and is called the *closure* or *adherence of A*. A subset A of X is *dense in X* if $\overline{A} = X$. The largest open set contained in A is denoted by $\overset{\circ}{A}$, and is called the *interior of A*. The set $\mathrm{Fr}\,A = \overline{A} \cap \overline{X - A}$ is called the *boundary* (or *frontier*) of A. We also denote the boundary of A by ∂A.

In order to prove the Hahn–Banach theorem, we will need two lemmas. Given any two distinct points $x, y \in X$, we let

$$]x, y[= \{(1 - \lambda)x + \lambda y \in X \mid 0 < \lambda < 1\}.$$

Our first lemma (Lemma 7.1) is intuitively quite obvious, so the reader might be puzzled by the length of its proof. However, after proposing several wrong proofs, we realized that its proof is more subtle than it might appear. The proof below is due to Valentine [7]. See whether you can find a shorter (and correct) proof!

Lemma 7.1. *Let S be a nonempty convex set and let $x \in \overset{\circ}{S}$ and $y \in \overline{S}$. Then we have $]x, y[\subseteq \overset{\circ}{S}$.*

Proof. Let $z \in \,]x, y[$, that is, $z = (1 - \lambda)x + \lambda y$, with $0 < \lambda < 1$. Since $x \in \overset{\circ}{S}$, we can find some open subset U contained in S such that $x \in U$. It is easy to check that the central magnification of center z, $H_{z, \frac{\lambda - 1}{\lambda}}$, maps x to y. Then $V = H_{z, \frac{\lambda - 1}{\lambda}}(U)$ is an open subset containing y, and since $y \in \overline{S}$, we have $V \cap S \neq \emptyset$. Let $v \in V \cap S$ be a point of S in this intersection. Now, there is a unique point $u \in U \subseteq S$ such that $H_{z, \frac{\lambda - 1}{\lambda}}(u) = v$, and since S is convex, we deduce that $z = (1 - \lambda)u + \lambda v \in S$. Since U is open, the set

$$W = (1 - \lambda)U + \lambda v = \{(1 - \lambda)w + \lambda v \mid w \in U\} \subseteq S$$

is also open and $z \in W$, which shows that $z \in \overset{\circ}{S}$. $\square$

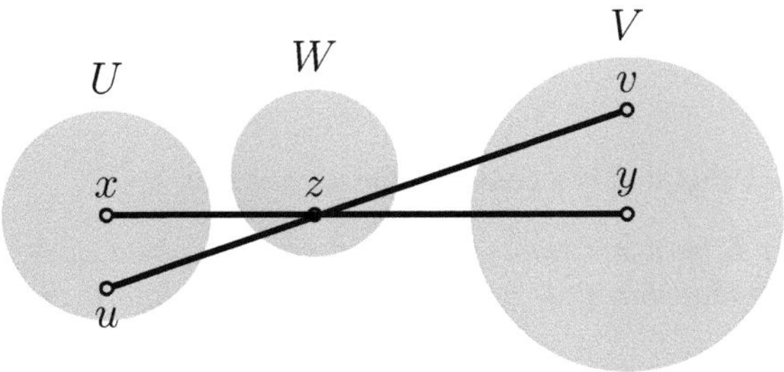

Fig. 7.1 Illustration for the proof of Lemma 7.1.

Corollary 7.1. *If S is convex, then $\overset{\circ}{S}$ is also convex, and we have $\overset{\circ}{\overline{S}} = \overset{\circ}{S}$. Furthermore, if $\overset{\circ}{S} \neq \emptyset$, then $\overline{S} = \overline{\overset{\circ}{S}}$.*

Beware that if S is a closed set, then the convex hull, $\mathrm{conv}(S)$, of S is not necessarily closed! (Find a counterexample.) However, if S is compact, then $\mathrm{conv}(S)$ is also compact and thus closed (see Proposition 3.1).

There is a simple criterion to test whether a convex set has an empty interior, based on the notion of dimension of a convex set (recall that the dimension of a nonempty convex subset is the dimension of its affine hull).

Proposition 7.1. *A nonempty convex set S has a nonempty interior iff* $\dim S = \dim X$.

Proof. Let $d = \dim X$. First, assume that $\overset{\circ}{S} \neq \emptyset$. Then, S contains some open ball of center a_0, and in it, we can find a frame $(a_0, a_1, \ldots, a_d)$ for X. Thus, $\dim S = \dim X$. Conversely, let $(a_0, a_1, \ldots, a_d)$ be a frame of X, with $a_i \in S$, for $i = 0, \ldots, d$. Then we have

$$\frac{a_0 + \cdots + a_d}{d+1} \in \overset{\circ}{S},$$

and $\overset{\circ}{S}$ is nonempty. $\square$

Proposition 7.1 is false in infinite dimension.

We leave the following property as an exercise:

Proposition 7.2. *If S is convex, then $\overline{S}$ is also convex.*

One can also easily prove that convexity is preserved under direct image and inverse image by an affine map.

The next lemma, which seems intuitively obvious, is the core of the proof of the Hahn–Banach theorem. This is the case in which the affine space has dimension two. First, we need to define a convex cone with vertex x.

Definition 7.1. A convex set C is a *convex cone with vertex x* if C is invariant under all central magnifications $H_{x,\lambda}$ of center x and ratio λ, with $\lambda > 0$ (i.e., $H_{x,\lambda}(C) = C$).

Given a convex set S and a point $x \notin S$, we can define

$$\text{cone}_x(S) = \bigcup_{\lambda > 0} H_{x,\lambda}(S).$$

It is easy to check that this is a convex cone with vertex x.

Lemma 7.2. *Let B be a nonempty open and convex subset of $\mathbb{A}^2$, and let O be a point of $\mathbb{A}^2$ such that that $O \notin B$. Then there is some line L through O such that $L \cap B = \emptyset$.*

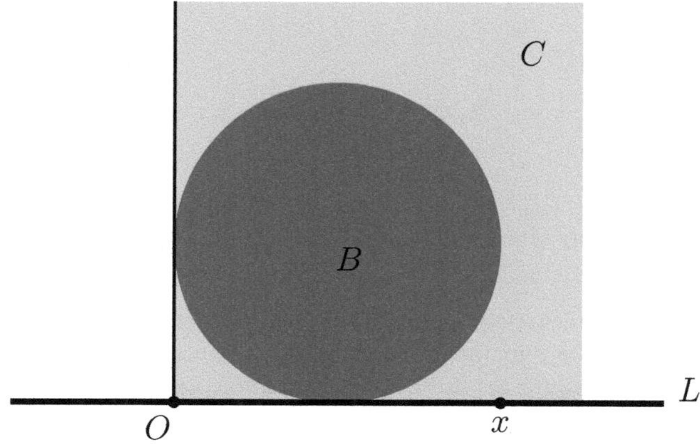

Fig. 7.2 Hahn–Banach theorem in the plane (Lemma 7.2).

Proof. Define the convex cone $C = \text{cone}_O(B)$. Since B is open, it is easy to check that each $H_{O,\lambda}(B)$ is open, and since C is the union of the $H_{O,\lambda}(B)$ (for $\lambda > 0$), which are open, C itself is open. Also, $O \notin C$. We claim that at least one point x of the boundary ∂C of C is distinct from O. Otherwise, $\partial C = \{O\}$, and we claim that $C = \mathbb{A}^2 - \{O\}$, which is not convex, a contradiction. Indeed, since C is convex, it is connected, $\mathbb{A}^2 - \{O\}$ itself is connected, and $C \subseteq \mathbb{A}^2 - \{O\}$. If $C \neq \mathbb{A}^2 - \{O\}$, pick some point $a \neq O$ in $\mathbb{A}^2 - C$ and some point $c \in C$. Now, a basic property of connectivity asserts that every continuous path from a (in the exterior of C) to c (in the interior of C) must intersect the boundary of C, namely $\{O\}$. However, there are plenty of paths from a to c that avoid O, a contradiction. Therefore, $C = \mathbb{A}^2 - \{O\}$.

Since C is open and $x \in \partial C$, we have $x \notin C$. Furthermore, we claim that $y = 2O - x$ (the symmetric of x with respect to O) does not belong to C either. Otherwise, we would have $y \in \overset{\circ}{C} = C$ and $x \in \overline{C}$, and by Lemma 7.1, we would get $O \in C$, a contradiction. Therefore, the line through O and x misses C entirely (since C is a cone), and thus $B \subseteq C$. $\quad\square$

Finally, we come to the Hahn–Banach theorem.

Theorem 7.1. *(Hahn–Banach theorem, geometric form) Let X be a (finite-dimensional) affine space, A a nonempty open and convex subset of X, and L an affine subspace of X such that $A \cap L = \emptyset$. Then there is some hyperplane H containing L that is disjoint from A.*

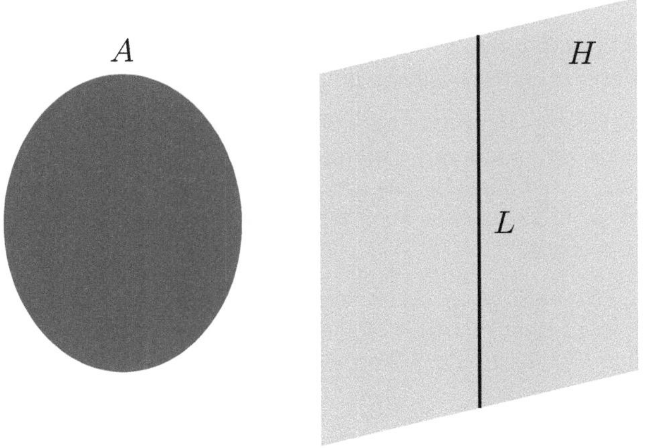

Fig. 7.3 Hahn-Banach theorem, geometric form (Theorem 7.1).

Proof. The case $\dim X = 1$ is trivial. Thus, we may assume that $\dim X \geq 2$. We reduce the proof to the case $\dim X = 2$. Let V be an affine subspace of X of maximal dimension containing L and such that $V \cap A = \emptyset$. Pick an origin $O \in L$ in X, and consider the vector space X_O. We would like to prove that V is a hyperplane, i.e., $\dim V = \dim X - 1$. We proceed by contradiction. Thus, assume that $\dim V \leq \dim X - 2$. In this case, the quotient space X/V has dimension at least 2. We also know that X/V is isomorphic to the orthogonal complement $V^{\perp}$ of V, so we may identify X/V and $V^{\perp}$. The (orthogonal) projection map $\pi\colon X \to V^{\perp}$ is linear and continuous, and we can show that π maps the open subset A to an open subset $\pi(A)$, which is also convex (one way to prove that $\pi(A)$ is open is to observe that for any point $a \in A$, a small open ball of center a contained in A is projected by π to an open ball contained in $\pi(A)$, and since π is surjective, $\pi(A)$ is open). Furthermore, $0 \notin \pi(A)$. Since $V^{\perp}$ has dimension at least 2, there is some plane P (a subspace of dimension 2) intersecting $\pi(A)$, and thus we obtain a nonempty open and convex subset $B = \pi(A) \cap P$ in the plane $P \cong \mathbb{A}^2$. So we can apply Lemma 7.2 to B and the point $O = 0$ in $P \cong \mathbb{A}^2$ to find a line l (in P) through O with $l \cap B = \emptyset$. But then $l \cap \pi(A) = \emptyset$ and $W = \pi^{-1}(l)$ is an affine subspace such that $W \cap A = \emptyset$ and W properly contains V, contradicting the maximality of V. $\square$

Remark: The geometric form of the Hahn–Banach theorem also holds when the dimension of X is infinite, but a slightly more sophisticated proof is required. Actually, all that is needed is to prove that a maximal affine subspace containing L and disjoint from A exists. This can be done using Zorn's lemma. For other proofs, see Bourbaki [3], Chapter 2, Valentine [7], Chapter 2, Barvinok [1], Chapter 2, or Lax [4], Chapter 3.

 Theorem 7.1 is false if we omit the assumption that A is open. For a counter-example, let $A \subseteq \mathbb{A}^2$ be the union of the half-space $y < 0$ with the closed segment $[0,1]$ on the x-axis and let L be the point $(2,0)$ on the boundary of A. It is also false if A is closed! (Find a counterexample).

Theorem 7.1 has many important corollaries. For example, we will eventually prove that for any two nonempty disjoint convex sets A and B, there is a hyperplane separating A and B, but this will take some work (recall the definition of a separating hyperplane given in Definition 3.3). We begin with the following version of the Hahn–Banach theorem:

Theorem 7.2. *(Hahn–Banach, second version) Let X be a (finite-dimensional) affine space, A a nonempty convex subset of X with nonempty interior, and L an affine subspace of X such that $A \cap L = \emptyset$. Then there is some hyperplane H containing L and separating L and A.*

Proof. Since A is convex, by Corollary 7.1, $\overset{\circ}{A}$ is also convex. By hypothesis, $\overset{\circ}{A}$ is nonempty. So we can apply Theorem 7.1 to the nonempty open and convex $\overset{\circ}{A}$ and to the affine subspace L. We get a hyperplane H containing L such that $\overset{\circ}{A} \cap H = \emptyset$. However, $A \subseteq \overline{A} = \overset{\circ}{\overline{A}}$ and $\overset{\circ}{A}$ is contained in the closed half-space (H_+ or H_-) containing $\overset{\circ}{A}$, so H separates A and L. $\square$

Corollary 7.2. *Given an affine space X, let A and B be two nonempty disjoint convex subsets and assume that A has nonempty interior ($\overset{\circ}{A} \neq \emptyset$). Then there is a hyperplane separating A and B.*

Proof. Pick some origin O and consider the vector space X_O. Define $C = A - B$ (a special case of the Minkowski sum) as follows:

$$A - B = \{a - b \mid a \in A, b \in B\} = \bigcup_{b \in B}(A - b).$$

It is easily verified that $C = A - B$ is convex and has nonempty interior (as a union of subsets having a nonempty interior). Furthermore $O \notin C$, since $A \cap B = \emptyset$.[1] (Note

---

[1] Readers who prefer a purely affine argument may define $C = A - B$ as the *affine* subset

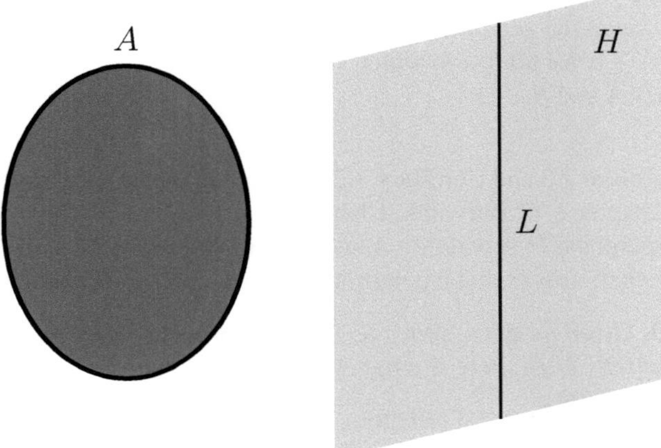

Fig. 7.4 Hahn–Banach theorem, second version (Theorem 7.2).

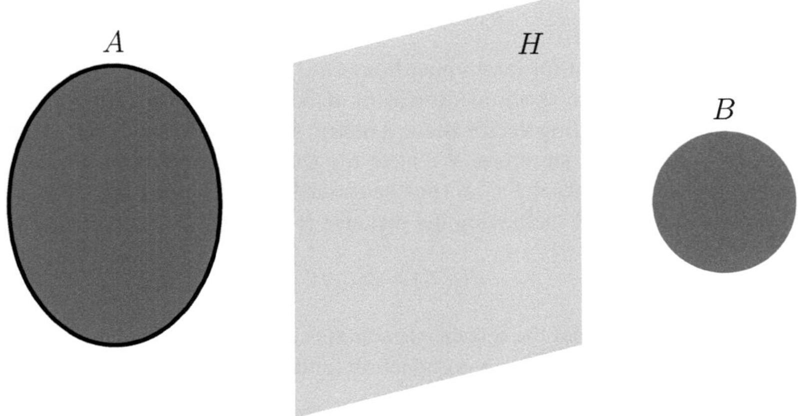

Fig. 7.5 Separation theorem, version 1 (Corollary 7.2).

that the definition depends on the choice of O, but this has no effect on the proof.)

Since $\overset{\circ}{C}$ is nonempty, we can apply Theorem 7.2 to C and to the affine subspace $\{O\}$, and we get a hyperplane H separating C and $\{O\}$. Let f be any linear form defining the hyperplane H. We may assume that $f(a - b) \leq 0$, for all $a \in A$ and

$$A - B = \{O + a - b \mid a \in A, b \in B\}.$$

Again, $O \notin C$ and C is convex. By adjusting O we can pick the affine form f defining a separating hyperplane H of C and $\{O\}$ such that $f(O + a - b) \leq f(O)$, for all $a \in A$ and all $b \in B$, i.e., $f(a) \leq f(b)$.

all $b \in B$, i.e., $f(a) \leq f(b)$. Consequently, if we let $\alpha = \sup\{f(a) \mid a \in A\}$ (which makes sense, since the set $\{f(a) \mid a \in A\}$ is bounded), we have $f(a) \leq \alpha$ for all $a \in A$ and $f(b) \geq \alpha$ for all $b \in B$, which shows that the affine hyperplane defined by $f - \alpha$ separates A and B. $\square$

Remark: Theorem 7.2 and Corollary 7.2 also hold in the infinite-dimensional case; see Lax [4], Chapter 3, or Barvinok, Chapter 3.

Since a hyperplane H separating A and B as in Corollary 7.2 is the boundary of each of the two half-spaces that it determines, we also obtain the following corollary:

Corollary 7.3. *Given an affine space X, let A and B be two nonempty disjoint open and convex subsets. Then there is a hyperplane strictly separating A and B.*

Beware that Corollary 7.3 *fails* for *closed* convex sets. However, Corollary 7.3 holds if we also assume that A (or B) is compact.

We need to review the notion of distance from a point to a subset. Let X be a metric space with distance function d. Given any point $a \in X$ and any nonempty subset B of X, we let

$$d(a,B) = \inf_{b \in B} d(a,b)$$

(where inf is the notation for least upper bound).

Now, if X is an affine space of dimension d, it can be given a metric structure by giving the corresponding vector space a metric structure, for instance, the metric induced by a Euclidean structure. We have the following important property: For any nonempty closed subset $S \subseteq X$ (not necessarily convex) and any point $a \in X$, there is some point $s \in S$ "achieving the distance from a to S," i.e., such that

$$d(a,S) = d(a,s).$$

The proof uses the fact that the distance function is continuous and that a continuous function attains its minimum on a compact set, and is left as an exercise.

Corollary 7.4. *Given an affine space X let A and B be two nonempty disjoint closed and convex subsets, with A compact. Then there is a hyperplane strictly separating A and B.*

Proof. Here is a sketch of the proof. First, we pick an origin O and we give $X_O \cong \mathbb{A}^n$ a Euclidean structure. Let d denote the associated distance. Given any subsets A of X, let

$$A + B(O, \varepsilon) = \{x \in X \mid d(x,A) < \varepsilon\},$$

where $B(a, \varepsilon)$ denotes the open ball $B(a, \varepsilon) = \{x \in X \mid d(a,x) < \varepsilon\}$ of center a and radius $\varepsilon > 0$. Note that

$$A + B(O, \varepsilon) = \bigcup_{a \in A} B(a, \varepsilon),$$

which shows that $A + B(O, \varepsilon)$ is open; furthermore, it is easy to see that if A is convex, then $A + B(O, \varepsilon)$ is also convex. Now, the function $a \mapsto d(a,B)$ (where

$a \in A$) is continuous, and since A is compact, it achieves its minimum, $d(A,B) = \min_{a \in A} d(a,B)$, at some point a of A. Say $d(A,B) = \delta$. Since B is closed, there is some $b \in B$ such that $d(A,B) = d(a,B) = d(a,b)$, and since $A \cap B = \emptyset$, we must have $\delta > 0$. Thus, if we pick $\varepsilon < \delta/2$, we see that

$$(A + B(O,\varepsilon)) \cap (B + B(O,\varepsilon)) = \emptyset.$$

Now, $A + B(O,\varepsilon)$ and $B + B(O,\varepsilon)$ are open, convex, and disjoint, and we conclude by applying Corollary 7.3. $\square$

A "cute" application of Corollary 7.4 is one of the many versions of "Farkas's lemma" (1893–1894, 1902), a basic result in the theory of linear programming. For any vector $x = (x_1, \ldots, x_n) \in \mathbb{R}^n$ and any real $\alpha \in \mathbb{R}$, write $x \geq \alpha$ iff $x_i \geq \alpha$, for $i = 1, \ldots, n$.

Lemma 7.3. *(Farkas's lemma, version I) Given any $d \times n$ real matrix A and any vector $z \in \mathbb{R}^d$, exactly one of the following alternatives occurs:*

(a) The linear system $Ax = z$ has a solution $x = (x_1, \ldots, x_n)$ such that $x \geq 0$ and
$$x_1 + \cdots + x_n = 1.$$
(b) There is some $c \in \mathbb{R}^d$ and some $\alpha \in \mathbb{R}$ such that $c^\top z < \alpha$ and $c^\top A \geq \alpha$.

Proof. Let $A_1, \ldots, A_n \in \mathbb{R}^d$ be the n points corresponding to the columns of A. Then, either $z \in \mathrm{conv}(\{A_1, \ldots, A_n\})$ or $z \notin \mathrm{conv}(\{A_1, \ldots, A_n\})$. In the first case, we have a convex combination

$$z = x_1 A_1 + \cdots + x_n A_n,$$

where $x_i \geq 0$ and $x_1 + \cdots + x_n = 1$, so $x = (x_1, \ldots, x_n)$ is a solution satisfying (a).

In the second case, by Corollary 7.4, there is a hyperplane H strictly separating $\{z\}$ and $\mathrm{conv}(\{A_1, \ldots, A_n\})$, which is obviously closed. In fact, observe that $z \notin \mathrm{conv}(\{A_1, \ldots, A_n\})$ iff there is a hyperplane H such that $z \in \overset{\circ}{H}_-$ and $A_i \in H_+$, or $z \in \overset{\circ}{H}_+$ and $A_i \in H_-$, for $i = 1, \ldots, n$. Since the affine hyperplane H is the zero locus of an equation of the form

$$c_1 y_1 + \cdots + c_d y_d = \alpha,$$

either $c^\top z < \alpha$ and $c^\top A_i \geq \alpha$ for $i = 1, \ldots, n$, that is, $c^\top A \geq \alpha$, or $c^\top z > \alpha$ and $c^\top A \leq \alpha$. In the second case, $(-c)^\top z < -\alpha$ and $(-c)^\top A \geq -\alpha$, so (b) is satisfied by either c and α or by $-c$ and $-\alpha$. $\square$

Remark: If we relax the requirements on solutions of $Ax = z$ and require only $x \geq 0$ ($x_1 + \cdots + x_n = 1$ is no longer required), then in condition (b), we can take $\alpha = 0$. This is another version of Farkas's Lemma. In this case, instead of considering the convex hull of $\{A_1, \ldots, A_n\}$ we are considering the convex cone

$$\mathrm{cone}(A_1, \ldots, A_n) = \{\lambda A_1 + \cdots + \lambda_n A_n \mid \lambda_i \geq 0,\ 1 \leq i \leq n\},$$

that is, we are dropping the condition $\lambda_1 + \cdots + \lambda_n = 1$. For this version of Farkas's lemma we need the following separation lemma:

Proposition 7.3. *Let $C \subseteq \mathbb{E}^d$ be any closed convex cone with vertex O. Then for every point a not in C, there is a hyperplane H passing through O separating a and C with $a \notin H$.*

Proof. Since C is closed and convex and $\{a\}$ is compact and convex, by Corollary 7.4 there is a hyperplane H' strictly separating a and C. Let H be the hyperplane through O parallel to H'. Since C and a lie in the two disjoint open half-spaces determined by H', the point a cannot belong to H. Suppose that some point $b \in C$ lies in the open half-space determined by H and a. Then the line L through O and b intersects H' in some point c, and since C is a cone, the half-line determined by O and b is contained in C. So $c \in C$ would belong to H', a contradiction. Therefore, C is contained in the closed half-space determined by H that does not contain a, as claimed. $\square$

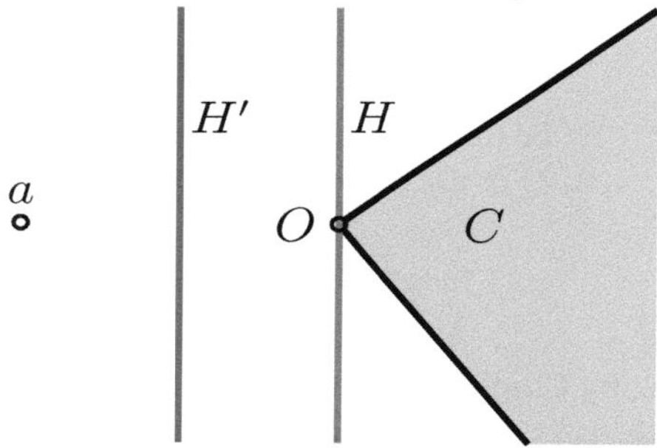

Fig. 7.6 Illustration for the proof of Proposition 7.3.

Lemma 7.4. *(Farkas's lemma, version II) Given any $d \times n$ real matrix A and any vector $z \in \mathbb{R}^d$, exactly one of the following alternatives occurs:*

(a) The linear system $Ax = z$ has a solution x such that $x \geq 0$.
(b) There is some $c \in \mathbb{R}^d$ such that $c^\top z < 0$ and $c^\top A \geq 0$.

Proof. The proof is analogous to the proof of Lemma 7.3 except that it uses Proposition 7.3 instead of Corollary 7.4 and either $z \in \mathrm{cone}(A_1, \ldots, A_n)$ or $z \notin \mathrm{cone}(A_1, \ldots, A_n)$. $\square$

One can show that Farkas II implies Farkas I. Here is another version of Farkas's lemma having to do with a system of inequalities $Ax \leq z$. Although, this version may seem weaker that Farkas II, it is actually equivalent to it!

Lemma 7.5. *(Farkas's lemma, version III) Given any $d \times n$ real matrix A and any vector $z \in \mathbb{R}^d$, exactly one of the following alternatives occurs:*

(a) The system of inequalities $Ax \leq z$ has a solution x.
(b) There is some $c \in \mathbb{R}^d$ such that $c \geq 0$, $c^\top z < 0$, and $c^\top A = 0$.

Proof. We use two tricks from linear programming:

1. We convert the system of inequalities $Ax \leq z$ into a system of equations by introducing a vector of "slack variables" $\gamma = (\gamma_1, \ldots, \gamma_d)$, where the system of equations is

$$(A, I) \begin{pmatrix} x \\ \gamma \end{pmatrix} = z,$$

   with $\gamma \geq 0$.
2. We replace each "unconstrained variable" x_i by $x_i = X_i - Y_i$, with $X_i, Y_i \geq 0$.

Then the original system $Ax \leq z$ has a solution x (unconstrained) iff the system of equations

$$(A, -A, I) \begin{pmatrix} X \\ Y \\ \gamma \end{pmatrix} = z$$

has a solution with $X, Y, \gamma \geq 0$. By Farkas II, this system has no solution iff there exists some $c \in \mathbb{R}^d$ with $c^\top z < 0$ and

$$c^\top (A, -A, I) \geq 0,$$

that is, $c^\top A \geq 0$, $-c^\top A \geq 0$, and $c \geq 0$. However, these four conditions reduce to $c^\top z < 0$, $c^\top A = 0$, and $c \geq 0$. $\square$

These versions of Farkas's lemma are statements of the form $(P \vee Q) \wedge \neg(P \wedge Q)$, which is easily seen to be equivalent to $\neg P \equiv Q$, namely, the logical equivalence of $\neg P$ and Q. Therefore, Farkas-type lemmas can be interpreted as criteria for the unsolvablity of various kinds of systems of linear equations or systems of linear inequalities, in the form of a separation property.

For example, Farkas II (Lemma 7.4) says that a system of linear equations $Ax = z$ does not have any solution $x \geq 0$ iff there is some $c \in \mathbb{R}^d$ such that $c^\top z < 0$ and $c^\top A \geq 0$. This means that there is a hyperplane H of equation $c^\top y = 0$ such that the column vectors A_j forming the matrix A all lie in the positive closed half-space H_+ but z lies in the interior of the other half-space, H_-, determined by H. Therefore, z can't be in the cone spanned by the A_j's.

Farkas III says that a system of linear inequalities $Ax \leq z$ does not have any solution (at all) iff there is some $c \in \mathbb{R}^d$ such that $c \geq 0$, $c^\top z < 0$, and $c^\top A = 0$. This time, there is also a hyperplane of equation $c^\top y = 0$, with $c \geq 0$, such that the column vectors A_j forming the matrix A all lie in H but z lies in the interior of the half-space H_- determined by H. In the "easy" direction, if there are such a vector c and some x satisfying $Ax \leq b$, since $c \geq 0$, we get $c^\top Ax \leq x^\top z$, but $c^\top Ax = 0$ and $x^\top z < 0$, a contradiction.

What is the crirerion for the unsolvability of a system of inequalities $Ax \leq z$ with $x \geq 0$? This problem is equivalent to the unsolvability of the set of inequalities

$$\begin{pmatrix} A \\ -I \end{pmatrix} x \leq \begin{pmatrix} z \\ 0 \end{pmatrix},$$

and by Farkas III, this system has no solution iff there is some vector (c_1, c_2) with $(c_1, c_2) \geq 0$,

$$(c_1^\top, c_2^\top) \begin{pmatrix} A \\ -I \end{pmatrix} = 0, \quad \text{and} \quad (c_1^\top, c_2^\top) \begin{pmatrix} z \\ 0 \end{pmatrix} < 0.$$

The above conditions are equivalent to $c_1 \geq 0$, $c_2 \geq 0$, $c_1^\top A - c_2^\top = 0$, and $c_1^\top z < 0$, which reduce to $c_1 \geq 0$, $c_1^\top A \geq 0$, and $c_1^\top z < 0$.

We can put all these versions together to prove the following version of Farkas's lemma:

Lemma 7.6. *(Farkas's lemma, version IIIb) For any $d \times n$ real matrix A and any vector $z \in \mathbb{R}^d$, the following statements are equivalent:*

(1) The system $Ax = z$ has no solution $x \geq 0$ iff there is some $c \in \mathbb{R}^d$ such that $c^\top A \geq 0$ and $c^\top z < 0$.

(2) The system $Ax \leq z$ has no solution iff there is some $c \in \mathbb{R}^d$ such that $c \geq 0$, $c^\top A = 0$, and $c^\top z < 0$.

(3) The system $Ax \leq z$ has no solution $x \geq 0$ iff there is some $c \in \mathbb{R}^d$ such that $c \geq 0$, $c^\top A \geq 0$, and $c^\top z < 0$.

Proof. We already proved that (1) implies (2) and that (2) implies (3). The proof that (3) implies (1) is left as an easy exercise. $\square$

The reader might wonder whether there is a criterion for the unsolvability of a system $Ax = z$ without any condition on x. However, since the unsolvability of the system $Ax = b$ is equivalent to the unsolvability of the system

$$\begin{pmatrix} A \\ -A \end{pmatrix} x \leq \begin{pmatrix} z \\ -z \end{pmatrix},$$

using (2), the above system is unsolvable iff there is some $(c_1, c_2) \geq (0, 0)$ such that

$$(c_1^\top, c_2^\top) \begin{pmatrix} A \\ -A \end{pmatrix} = 0 \quad \text{and} \quad (c_1^\top, c_2^\top) \begin{pmatrix} z \\ -z \end{pmatrix} < 0,$$

and these are equivalent to $c_1^\top A - c_2^\top A = 0$ and $c_1^\top z - c_2^\top z < 0$, namely, $c^\top A = 0$ and $c^\top z < 0$, where $c = c_1 - c_2 \in \mathbb{R}^d$. However, this simply says that the columns $A_1, \ldots, A_n$ of A are linearly dependent and that z does not belong to the subspace spanned by $A_1, \ldots, A_n$, a criterion that we already knew from linear algebra.

As in Matousek and Gartner [6], we can summarize these various criteria in the following table:

	The system $Ax \leq z$	The system $Ax = z$
has no solution $x \geq 0$ iff	$\exists c \in \mathbb{R}^d$, such that $c \geq 0$, $c^\top A \geq 0$ and $c^\top z < 0$	$\exists c \in \mathbb{R}^d$, such that $c^\top A \geq 0$ and $c^\top z < 0$
has no solution $x \in \mathbb{R}^n$ iff	$\exists c \in \mathbb{R}^d$, such that, $c \geq 0$, $c^\top A = 0$ and $c^\top z < 0$	$\exists c \in \mathbb{R}^d$, such that $c^\top A = 0$ and $c^\top z < 0$

Remark: The strong duality theorem in linear programming can be proved using Lemma 7.6(c).

Finally, we have the separation theorem announced earlier for arbitrary nonempty convex subsets.

Theorem 7.3. *(Separation of disjoint convex sets) Given an affine space X, let A and B be two nonempty disjoint convex subsets. Then there is a hyperplane separating A and B.*

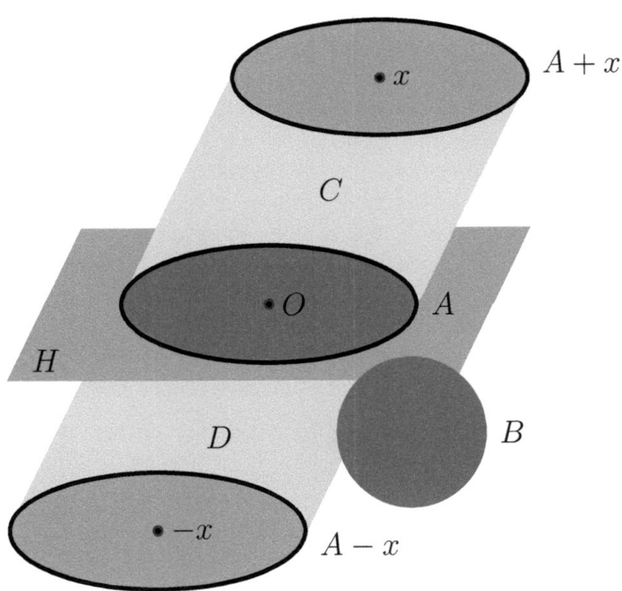

Fig. 7.7 Separation theorem, final version (Theorem 7.3).

Proof. The proof is by descending induction on $n = \dim A$. If $\dim A = \dim X$, we know from Proposition 7.1 that A has nonempty interior, and we conclude using Corollary 7.2. Next, asssume that the induction hypothesis holds if $\dim A \geq n$ and assume $\dim A = n - 1$. Pick an origin $O \in A$ and let H be a hyperplane containing

A. Pick $x \in X$ outside H and define $C = \mathrm{conv}(A \cup \{A + x\})$, where $A + x = \{a + x \mid a \in A\}$ and $D = \mathrm{conv}(A \cup \{A - x\})$, where $A - x = \{a - x \mid a \in A\}$. Note that $C \cup D$ is convex. If $B \cap C \neq \emptyset$ and $B \cap D \neq \emptyset$, then the convexity of B and $C \cup D$ implies that $A \cap B \neq \emptyset$, a contradiction. Without loss of generality, assume that $B \cap C = \emptyset$. Since x is outside H, we have $\dim C = n$, and by the induction hypothesis, there is a hyperplane H_1 separating C and B. Since $A \subseteq C$, we see that H_1 also separates A and B. $\square$

The reader should compare this proof (from Valentine [7], Chapter II) with Berger's proof using compactness of the projective space $\mathbb{P}^d$ [2] (Corollary 11.4.7).

Remarks:

(1) Rather than using the Hahn–Banach theorem to deduce separation results, one may proceed differently and use the following intuitively obvious lemma, as in Valentine [7] (Theorem 2.4):

Lemma 7.7. *If A and B are two nonempty convex sets such that $A \cup B = X$ and $A \cap B = \emptyset$, then $V = \overline{A} \cap \overline{B}$ is a hyperplane.*

One can then deduce Corollary 7.2 and Theorem 7.3. Yet another approach is followed in Barvinok [1].

(2) How can some of the above results be generalized to infinite-dimensional affine spaces, especially Theorem 7.1 and Corollary 7.2? One approach is to simultaneously relax the notion of interior and tighten a little the notion of closure, in a more "linear and less topological" fashion, as in Valentine [7].

Given any subset $A \subseteq X$ (where X may be infinite-dimensional, but is a Hausdorff topological vector space), say that a point $x \in X$ is *linearly accessible from A* if there is some $a \in A$ with $a \neq x$ and $\,]a, x[\,\subseteq A$. We let $\mathrm{lina}\,A$ be the set of all points linearly accessible from A and $\mathrm{lin}\,A = A \cup \mathrm{lina}\,A$.

A point $a \in A$ is a *core point of A* if for every $y \in X$, with $y \neq a$, there is some $z \in \,]a, y[$ such that $[a, z] \subseteq A$. The set of all core points is denoted by $\mathrm{core}\,A$.

It is not difficult to prove that $\mathrm{lin}\,A \subseteq \overline{A}$ and $\overset{\circ}{A} \subseteq \mathrm{core}\,A$. If A has nonempty interior, then $\mathrm{lin}\,A = \overline{A}$ and $\overset{\circ}{A} = \mathrm{core}\,A$. Also, if A is convex, then $\mathrm{core}\,A$ and $\mathrm{lin}\,A$ are convex. Then Lemma 7.7 still holds (where X is not necessarily finite-dimensional) if we redefine V as $V = \mathrm{lin}\,A \cap \mathrm{lin}\,B$ and allow the possibility that V could be X itself. Corollary 7.2 also holds in the general case if we assume that $\mathrm{core}\,A$ is nonempty. For details, see Valentine [7], Chapters I and II.

(3) Yet another approach is to define the notion of an algebraically open convex set, as in Barvinok [1]. A convex set A is *algebraically open* if the intersection of A with every line L is an open interval, possibly empty or infinite at either end (or all of L). An open convex set is algebraically open. Then the Hahn–Banach theorem holds, provided that A is an algebraically open convex set, and similarly, Corollary 7.2 also holds, provided A is algebraically open. For details, see Barvinok [1], Chapters 2 and 3. We do not know how the notion "algebraically open" relates to the concept of core.

(4) Theorems 7.1, 7.2 and Corollary 7.2 are proved in Lax [4] using the notion of *gauge function* in the more general case that A has some core point (but beware that Lax uses the terminology *interior point* instead of core point!).

An important special case of separation is the case that A is convex and $B = \{a\}$, for some point a in A.

7.2 Supporting Hyperplanes and Minkowski's Proposition

Recall the definition of a supporting hyperplane given in Definition 3.4. We have the following important proposition, first proved by Minkowski (1896):

Proposition 7.4. *(Minkowski) Let A be a nonempty, closed, and convex subset. Then for every point $a \in \partial A$, there is a supporting hyperplane to A through a.*

Proof. Let $d = \dim A$. If $d < \dim X$ (i.e., A has empty interior), then A is contained in some affine subspace V of dimension $d < \dim X$, and any hyperplane containing V is a supporting hyperplane for every $a \in A$. Now, assume $d = \dim X$, so that $\overset{\circ}{A} \neq \emptyset$. If $a \in \partial A$, then $\{a\} \cap \overset{\circ}{A} = \emptyset$. By Theorem 7.1, there is a hyperplane H separating $\overset{\circ}{A}$ and $L = \{a\}$. However, by Corollary 7.1, since $\overset{\circ}{A} \neq \emptyset$ and A is closed, we have

$$A = \overline{A} = \overline{\overset{\circ}{A}}.$$

Now, the half-space containing $\overset{\circ}{A}$ is closed, and thus it contains $\overline{\overset{\circ}{A}} = A$. Therefore, H separates A and $\{a\}$. $\square$

Remark: The assumption that A is closed is convenient but unnecessary. Indeed, the proof of Proposition 7.4 shows that the proposition holds for every boundary point $a \in \partial A$ (assuming $\partial A \neq \emptyset$).

Beware that Proposition 7.4 is false when the dimension of X is infinite and when $\overset{\circ}{A} = \emptyset$.

The proposition below gives a sufficient condition for a closed subset to be convex.

Proposition 7.5. *Let A be a closed subset with nonempty interior. If there is a supporting hyperplane for every point $a \in \partial A$, then A is convex.*

Proof. We leave it as an exercise (see Berger [2], Proposition 11.5.4). $\square$

The condition that A have nonempty interior is crucial!

The proposition below characterizes closed convex sets in terms of (closed) half-spaces. It is another intuitive fact whose rigorous proof is nontrivial.

Proposition 7.6. *Let A be a nonempty closed and convex subset. Then A is the intersection of all the closed half-spaces containing it.*

Proof. Let A' be the intersection of all the closed half-spaces containing A. It is immediately checked that A' is closed and convex and that $A \subseteq A'$. Assume that $A' \neq A$, and pick $a \in A' - A$. Then we can apply Corollary 7.4 to $\{a\}$ and A and we find a hyperplane H strictly separating A and $\{a\}$; this shows that A belongs to one of the two half-spaces determined by H, yet a does not belong to the same half-space, contradicting the definition of A'. $\square$

7.3 Problems

7.1. Prove Proposition 7.2.

7.2. Find two closed convex sets such that Corollary 7.3 fails.

7.3. In $\mathbb{E}^3$, consider the closed convex set (cone) A defined by the inequalities

$$x \geq 0, \quad y \geq 0, \quad z \geq 0, \quad z^2 \leq xy,$$

and let D be the line given by $x = 0$, $z = 1$. Prove that $D \cap A = \emptyset$, both A and D are convex and closed, yet every plane containing D meets A. Therefore, A and D give another counterexample to the Hahn–Banach theorem in which A is closed (one cannot relax the hypothesis that A is open).

7.4. Prove Proposition 7.5.

7.5. Let $(v_1, \ldots, v_n)$ be a sequence of n vectors in $\mathbb{R}^d$ and let V be the $d \times n$ matrix whose jth column is v_j. Prove the equivalence of the following two statements:

(a) There is no nontrivial positive linear dependence among the v_j, which means that there is no nonzero vector $y = (y_1, \ldots, y_n) \in \mathbb{R}^n$ with $y_j \geq 0$ for $j = 1, \ldots, n$, so that
$$y_1 v_1 + \cdots + y_n v_n = 0,$$
or equivalently, $Vy = 0$.
(b) There is some vector $c \in \mathbb{R}^d$ such that $c^\top V > 0$, which means that $c^\top v_j > 0$, for $j = 1, \ldots, n$.

References

1. Alexander Barvinok. *A Course in Convexity.* GSM, Vol. 54. AMS, first edition, 2002.

2. Marcel Berger. *Géométrie 2*. Nathan, 1990. English edition: Geometry 2, Universitext, Springer-Verlag.
3. Nicolas Bourbaki. *Espaces Vectoriels Topologiques*. Eléments de Mathématiques. Hermann, 1981.
4. Peter D. Lax. *Functional Analysis*. Wiley, first edition, 2002.
5. Jiri Matousek. *Lectures on Discrete Geometry*. GTM No. 212. Springer-Verlag, first edition, 2002.
6. Jiri Matousek and Bernd Gartner. *Understanding and Using Linear Programming*. Universitext. Springer-Verlag, first edition, 2007.
7. Frederick A. Valentine. *Convex Sets*. McGraw-Hill, first edition, 1964.

Chapter 8
The Cartan–Dieudonné Theorem

8.1 Orthogonal Reflections

In this chapter the structure of the orthogonal group is studied in more depth. In particular, we prove that every isometry in $\mathbf{O}(n)$ is the composition of at most n reflections about hyperplanes (for $n \geq 2$, see Theorem 8.1). This important result is a special case of the "Cartan–Dieudonné theorem" (Cartan [4], Dieudonné [6]). We also prove that every rotation in $\mathbf{SO}(n)$ is the composition of at most n flips (for $n \geq 3$).

Hyperplane reflections are represented by matrices called Householder matrices. These matrices play an important role in numerical methods, for instance for solving systems of linear equations, solving least squares problems, for computing eigenvalues, and for transforming a symmetric matrix into a tridiagonal matrix. We prove a simple geometric lemma that immediately yields the QR-decomposition of arbitrary matrices in terms of Householder matrices.

Affine isometries are defined, and their fixed points are investigated. First, we characterize the set of fixed points of an affine map. Using this characterization, we prove that every affine isometry f can be written uniquely as

$$f = t \circ g, \quad \text{with} \quad t \circ g = g \circ t,$$

where g is an isometry having a fixed point, and t is a translation by a vector τ such that $\overrightarrow{f}(\tau) = \tau$, and with some additional nice properties (see Lemma 8.3). This is a generalization of a classical result of Chasles about (proper) rigid motions in $\mathbb{R}^3$ (screw motions). We also show that the Cartan–Dieudonné theorem can be generalized to affine isometries: Every rigid motion in $\mathbf{Is}(n)$ is the composition of at most n affine reflections if it has a fixed point, or else of at most $n + 2$ affine reflections. We prove that every rigid motion in $\mathbf{SE}(n)$ is the composition of at most n flips (for $n \geq 3$). Finally, the orientation of a Euclidean space is defined, and we discuss volume forms and cross products.

Orthogonal symmetries are a very important example of isometries. First let us review the definition of projections. Given a vector space E, let F and G be subspaces of E that form a direct sum $E = F \oplus G$. Since every $u \in E$ can be written uniquely as $u = v + w$, where $v \in F$ and $w \in G$, we can define the two *projections* $p_F \colon E \to F$ and $p_G \colon E \to G$ such that $p_F(u) = v$ and $p_G(u) = w$. It is immediately verified that p_G and p_F are linear maps, and that $p_F^2 = p_F$, $p_G^2 = p_G$, $p_F \circ p_G = p_G \circ p_F = 0$, and $p_F + p_G = \mathrm{id}$.

Definition 8.1. Given a vector space E, for any two subspaces F and G that form a direct sum $E = F \oplus G$, the *symmetry (or reflection) with respect to F and parallel to G* is the linear map $s \colon E \to E$ defined such that

$$s(u) = 2p_F(u) - u,$$

for every $u \in E$.

Because $p_F + p_G = \mathrm{id}$, note that we also have

$$s(u) = p_F(u) - p_G(u)$$

and

$$s(u) = u - 2p_G(u),$$

$s^2 = \mathrm{id}$, s is the identity on F, and $s = -\mathrm{id}$ on G. We now assume that E is a Euclidean space of finite dimension.

Definition 8.2. Let E be a Euclidean space of finite dimension n. For any two subspaces F and G, if F and G form a direct sum $E = F \oplus G$ and F and G are orthogonal, i.e., $F = G^{\perp}$, the *orthogonal symmetry (or reflection) with respect to F and parallel to G* is the linear map $s \colon E \to E$ defined such that

$$s(u) = 2p_F(u) - u,$$

for every $u \in E$. When F is a hyperplane, we call s a *hyperplane symmetry with respect to F (or reflection about F)*, and when G is a plane (and thus $\dim(F) = n - 2$), we call s a *flip about F*.

A reflection about a hyperplane F is shown in Figure 8.1.

For any two vectors $u, v \in E$, it is easily verified using the bilinearity of the inner product that

$$\|u + v\|^2 - \|u - v\|^2 = 4(u \cdot v).$$

Then, since

$$u = p_F(u) + p_G(u)$$

and

$$s(u) = p_F(u) - p_G(u),$$

since F and G are orthogonal, it follows that

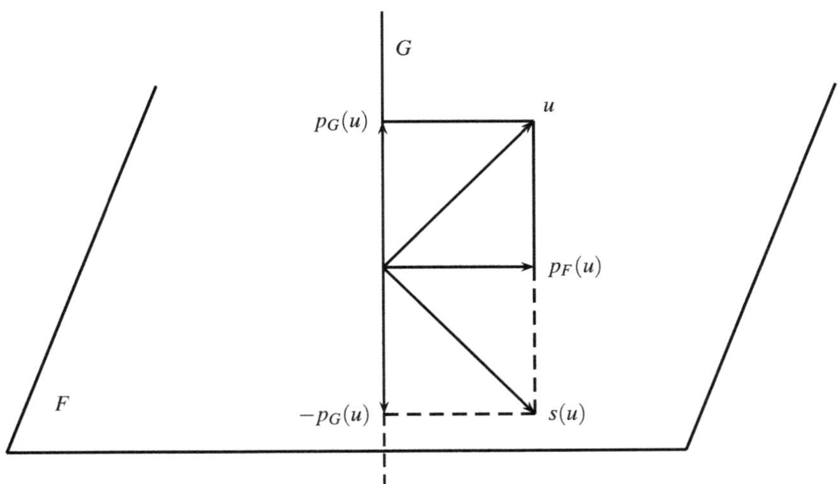

Fig. 8.1 A reflection about a hyperplane F.

$$p_F(u) \cdot p_G(v) = 0,$$

and thus,

$$\|s(u)\| = \|u\|,$$

so that s is an isometry.

Using Lemma 6.7, it is possible to find an orthonormal basis $(e_1,\ldots,e_n)$ of E consisting of an orthonormal basis of F and an orthonormal basis of G. Assume that F has dimension p, so that G has dimension $n - p$. With respect to the orthonormal basis $(e_1,\ldots,e_n)$, the symmetry s has a matrix of the form

$$\begin{pmatrix} I_p & 0 \\ 0 & -I_{n-p} \end{pmatrix}.$$

Thus, $\det(s) = (-1)^{n-p}$, and s is a rotation iff $n - p$ is even. In particular, when F is a hyperplane H, we have $p = n - 1$ and $n - p = 1$, so that s is an improper orthogonal transformation. When $F = \{0\}$, we have $s = -\mathrm{id}$, which is called the *symmetry with respect to the origin*. The symmetry with respect to the origin is a rotation iff n is even, and an improper orthogonal transformation iff n is odd. When n is odd, we observe that every improper orthogonal transformation is the composition of a rotation with the symmetry with respect to the origin. When G is a plane, $p = n - 2$, and $\det(s) = (-1)^2 = 1$, so that a flip about F is a rotation. In particular, when $n = 3$, F is a line, and a flip about the line F is indeed a rotation of measure π.

Remark: Given any two orthogonal subspaces F, G forming a direct sum $E = F \oplus G$, let f be the symmetry with respect to F and parallel to G, and let g be the symmetry with respect to G and parallel to F. We leave as an exercise to show that

$$f \circ g = g \circ f = -\mathrm{id}.$$

When $F = H$ is a hyperplane, we can give an explicit formula for $s(u)$ in terms of any nonnull vector w orthogonal to H. Indeed, from

$$u = p_H(u) + p_G(u),$$

since $p_G(u) \in G$ and G is spanned by w, which is orthogonal to H, we have

$$p_G(u) = \lambda w$$

for some $\lambda \in \mathbb{R}$, and we get

$$u \cdot w = \lambda \|w\|^2,$$

and thus

$$p_G(u) = \frac{(u \cdot w)}{\|w\|^2} \, w.$$

Since

$$s(u) = u - 2p_G(u),$$

we get

$$s(u) = u - 2 \frac{(u \cdot w)}{\|w\|^2} \, w.$$

Such reflections are represented by matrices called *Householder matrices*, and they play an important role in numerical matrix analysis (see Kincaid and Cheney [8] or Ciarlet [5]). Householder matrices are symmetric and orthogonal. It is easily checked that over an orthonormal basis $(e_1, \ldots, e_n)$, a hyperplane reflection about a hyperplane H orthogonal to a nonnull vector w is represented by the matrix

$$H = I_n - 2 \frac{WW^\top}{\|W\|^2} = I_n - 2 \frac{WW^\top}{W^\top W},$$

where W is the column vector of the coordinates of w over the basis $(e_1, \ldots, e_n)$, and I_n is the identity $n \times n$ matrix. Since

$$p_G(u) = \frac{(u \cdot w)}{\|w\|^2} \, w,$$

the matrix representing p_G is

$$\frac{WW^\top}{W^\top W},$$

and since $p_H + p_G = \mathrm{id}$, the matrix representing p_H is

$$I_n - \frac{WW^\top}{W^\top W}.$$

These formulae will be used in Section 9.1 to derive a formula for a rotation of $\mathbb{R}^3$, given the direction w of its axis of rotation and given the angle θ of rotation.

The following fact is the key to the proof that every isometry can be decomposed as a product of reflections.

Lemma 8.1. *Let E be any nontrivial Euclidean space. For any two vectors $u, v \in E$, if $\|u\| = \|v\|$, then there is a hyperplane H such that the reflection s about H maps u to v, and if $u \neq v$, then this reflection is unique.*

Proof. If $u = v$, then any hyperplane containing u does the job. Otherwise, we must have $H = \{v - u\}^{\perp}$, and by the above formula,

$$s(u) = u - 2\frac{(u \cdot (v - u))}{\|(v - u)\|^2}(v - u) = u + \frac{2\|u\|^2 - 2u \cdot v}{\|(v - u)\|^2}(v - u),$$

and since

$$\|(v - u)\|^2 = \|u\|^2 + \|v\|^2 - 2u \cdot v$$

and $\|u\| = \|v\|$, we have

$$\|(v - u)\|^2 = 2\|u\|^2 - 2u \cdot v,$$

and thus, $s(u) = v$. $\square$

If E is a complex vector space and the inner product is Hermitian, Lemma 8.1 is false. The problem is that the vector $v - u$ does not work unless the inner product $u \cdot v$ is real! We will see in the next chapter that the lemma can be salvaged enough to yield the QR-decomposition in terms of Householder transformations.

Using the above property, we can prove a fundamental property of isometries: They are generated by reflections about hyperplanes.

8.2 The Cartan–Dieudonné Theorem for Linear Isometries

The fact that the group $\mathbf{O}(n)$ of linear isometries is generated by the reflections is a special case of a theorem known as the Cartan–Dieudonné theorem. Elie Cartan proved a version of this theorem early in the twentieth century. A proof can be found in his book on spinors [4], which appeared in 1937 (Chapter I, Section 10, pages 10–12). Cartan's version applies to nondegenerate quadratic forms over $\mathbb{R}$ or $\mathbb{C}$. The theorem was generalized to quadratic forms over arbitrary fields by Dieudonné [6]. One should also consult Emil Artin's book [1], which contains an in-depth study of the orthogonal group and another proof of the Cartan–Dieudonné theorem.

First, let us review the notions of eigenvalues and eigenvectors. Recall that given any linear map $f : E \to E$, a vector $u \in E$ is called an *eigenvector, or proper vector, or characteristic vector, of f* if there is some $\lambda \in K$ such that

$$f(u) = \lambda u.$$

In this case, we say that $u \in E$ is an *eigenvector associated with* λ. A scalar $\lambda \in K$ is called an *eigenvalue, or proper value, or characteristic value, of* f if there is some nonnull vector $u \neq 0$ in E such that

$$f(u) = \lambda u,$$

or equivalently if $\mathrm{Ker}\,(f - \lambda\,\mathrm{id}) \neq \{0\}$. Given any scalar $\lambda \in K$, the set of all eigenvectors associated with λ is the subspace $\mathrm{Ker}\,(f - \lambda\,\mathrm{id})$, also denoted by $E_\lambda(f)$ or $E(\lambda, f)$, called the *eigenspace associated with* λ, *or proper subspace associated with* λ.

Theorem 8.1. *Let E be a Euclidean space of dimension $n \geq 1$. Every isometry $f \in \mathbf{O}(E)$ that is not the identity is the composition of at most n reflections. When $n \geq 2$, the identity is the composition of any reflection with itself.*

Proof. We proceed by induction on n. When $n = 1$, every isometry $f \in \mathbf{O}(E)$ is either the identity or $-\mathrm{id}$, but $-\mathrm{id}$ is a reflection about $H = \{0\}$. When $n \geq 2$, we have $\mathrm{id} = s \circ s$ for every reflection s. Let us now consider the case where $n \geq 2$ and f is not the identity. There are two subcases.

Case 1. f admits 1 as an eigenvalue, i.e., there is some nonnull vector w such that $f(w) = w$. In this case, let H be the hyperplane orthogonal to w, so that $E = H \oplus \mathbb{R}w$. We claim that $f(H) \subseteq H$. Indeed, if

$$v \cdot w = 0$$

for any $v \in H$, since f is an isometry, we get

$$f(v) \cdot f(w) = v \cdot w = 0,$$

and since $f(w) = w$, we get

$$f(v) \cdot w = f(v) \cdot f(w) = 0,$$

and thus $f(v) \in H$. Furthermore, since f is not the identity, f is not the identity of H. Since H has dimension $n - 1$, by the induction hypothesis applied to H, there are at most $k \leq n - 1$ reflections $s_1, \ldots, s_k$ about some hyperplanes $H_1, \ldots, H_k$ in H, such that the restriction of f to H is the composition $s_k \circ \cdots \circ s_1$. Each s_i can be extended to a reflection in E as follows: If $H = H_i \oplus L_i$ (where $L_i = H_i^\perp$, the orthogonal complement of H_i in H), $L = \mathbb{R}w$, and $F_i = H_i \oplus L$, since H and L are orthogonal, F_i is indeed a hyperplane, $E = F_i \oplus L_i = H_i \oplus L \oplus L_i$, and for every $u = h + \lambda w \in H \oplus L = E$, since

$$s_i(h) = p_{H_i}(h) - p_{L_i}(h),$$

we can define s_i on E such that

$$s_i(h + \lambda w) = p_{H_i}(h) + \lambda w - p_{L_i}(h),$$

and since $h \in H$, $w \in L$, $F_i = H_i \oplus L$, and $H = H_i \oplus L_i$, we have

$$s_i(h + \lambda w) = p_{F_i}(h + \lambda w) - p_{L_i}(h + \lambda w),$$

which defines a reflection about $F_i = H_i \oplus L$. Now, since f is the identity on $L = \mathbb{R}w$, it is immediately verified that $f = s_k \circ \cdots \circ s_1$, with $k \leq n - 1$.

Case 2. f does not admit 1 as an eigenvalue, i.e., $f(u) \neq u$ for all $u \neq 0$. Pick any $w \neq 0$ in E, and let H be the hyperplane orthogonal to $f(w) - w$. Since f is an isometry, we have $\|f(w)\| = \|w\|$, and by Lemma 8.1, we know that $s(w) = f(w)$, where s is the reflection about H, and we claim that $s \circ f$ leaves w invariant. Indeed, since $s^2 = \mathrm{id}$, we have

$$s(f(w)) = s(s(w)) = w.$$

Since $s^2 = \mathrm{id}$, we cannot have $s \circ f = \mathrm{id}$, since this would imply that $f = s$, where s is the identity on H, contradicting the fact that f is not the identity on any vector. Thus, we are back to Case 1. Thus, there are $k \leq n - 1$ hyperplane reflections such that $s \circ f = s_k \circ \cdots \circ s_1$, from which we get

$$f = s \circ s_k \circ \cdots \circ s_1,$$

with at most $k + 1 \leq n$ reflections. $\square$

Remarks:

(1) A slightly different proof can be given. Either f is the identity, or there is some nonnull vector u such that $f(u) \neq u$. In the second case, proceed as in the second part of the proof, to get back to the case where f admits 1 as an eigenvalue.

(2) Theorem 8.1 still holds if the inner product on E is replaced by a nondegenerate symmetric bilinear form φ, but the proof is a lot harder.

(3) The proof of Theorem 8.1 shows more than stated. If 1 is an eigenvalue of f, for any eigenvector w associated with 1 (i.e., $f(w) = w$, $w \neq 0$), then f is the composition of $k \leq n - 1$ reflections about hyperplanes F_i such that $F_i = H_i \oplus L$, where L is the line $\mathbb{R}w$ and the H_i are subspaces of dimension $n - 2$ all orthogonal to L (the H_i are hyperplanes in H). This situation is illustrated in Figure 8.2.

   If 1 is not an eigenvalue of f, then f is the composition of $k \leq n$ reflections about hyperplanes $H, F_1, \ldots, F_{k-1}$, such that $F_i = H_i \oplus L$, where L is a line intersecting H, and the H_i are subspaces of dimension $n - 2$ all orthogonal to L (the H_i are hyperplanes in $L^{\perp}$). This situation is illustrated in Figure 8.3.

(4) It is natural to ask what is the minimal number of hyperplane reflections needed to obtain an isometry f. This has to do with the dimension of the eigenspace $\mathrm{Ker}\,(f - \mathrm{id})$ associated with the eigenvalue 1. We will prove later that every isometry is the composition of k hyperplane reflections, where

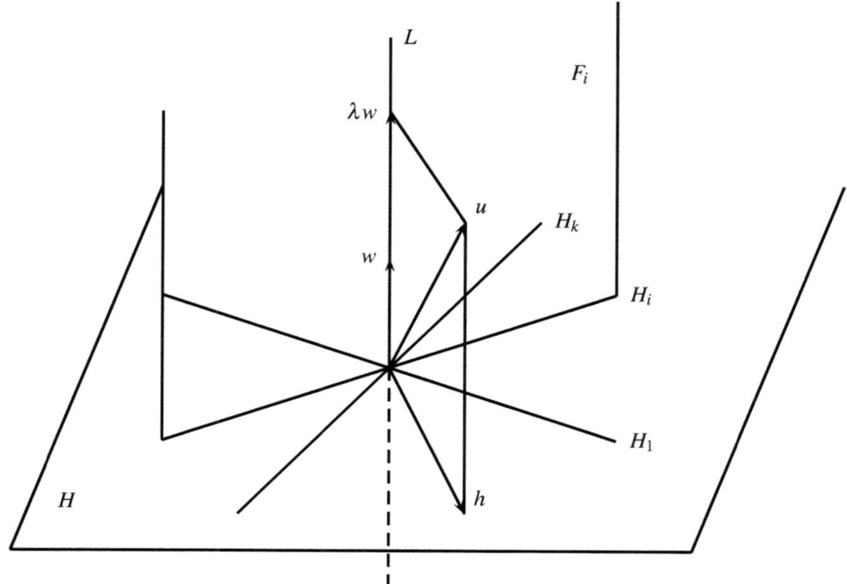

Fig. 8.2 An isometry f as a composition of reflections, when 1 is an eigenvalue of f.

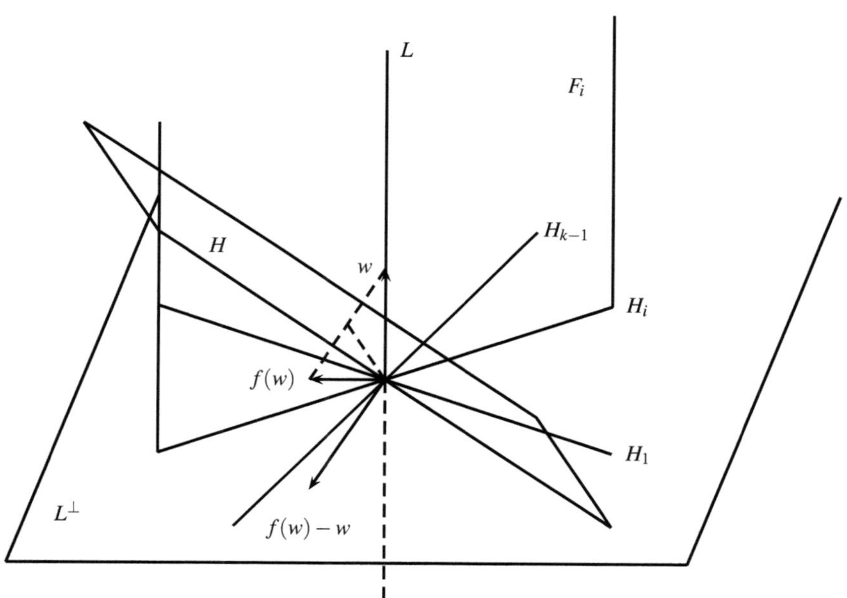

Fig. 8.3 An isometry f as a composition of reflections when 1 is not an eigenvalue of f.

$$k = n - \dim(\operatorname{Ker}(f - \operatorname{id})),$$

and that this number is minimal (where $n = \dim(E)$).

When $n = 2$, a reflection is a reflection about a line, and Theorem 8.1 shows that every isometry in $\mathbf{O}(2)$ is either a reflection about a line or a rotation, and that every rotation is the product of two reflections about some lines. In general, since $\det(s) = -1$ for a reflection s, when $n \geq 3$ is odd, every rotation is the product of an even number less than or equal to $n - 1$ of reflections, and when n is even, every improper orthogonal transformation is the product of an odd number less than or equal to $n - 1$ of reflections.

In particular, for $n = 3$, every rotation is the product of two reflections about planes. When n is odd, we can say more about improper isometries. Indeed, when n is odd, every improper isometry admits the eigenvalue -1. This is because if E is a Euclidean space of finite dimension and $f \colon E \to E$ is an isometry, because $\|f(u)\| = \|u\|$ for every $u \in E$, if λ is any eigenvalue of f and u is an eigenvector associated with λ, then

$$\|f(u)\| = \|\lambda u\| = |\lambda|\|u\| = \|u\|,$$

which implies $|\lambda| = 1$, since $u \neq 0$. Thus, the real eigenvalues of an isometry are either $+1$ or -1. However, it is well known that polynomials of odd degree always have some real root. As a consequence, the characteristic polynomial $\det(f - \lambda\operatorname{id})$ of f has some real root, which is either $+1$ or -1. Since f is an improper isometry, $\det(f) = -1$, and since $\det(f)$ is the product of the eigenvalues, the real roots cannot all be $+1$, and thus -1 is an eigenvalue of f. Going back to the proof of Theorem 8.1, since -1 is an eigenvalue of f, there is some nonnull eigenvector w such that $f(w) = -w$. Using the second part of the proof, we see that the hyperplane H orthogonal to $f(w) - w = -2w$ is in fact orthogonal to w, and thus f is the product of $k \leq n$ reflections about hyperplanes $H, F_1, \ldots, F_{k-1}$ such that $F_i = H_i \oplus L$, where L is a line orthogonal to H, and the H_i are hyperplanes in $H = L^\perp$ orthogonal to L. However, k must be odd, and so $k - 1$ is even, and thus the composition of the reflections about $F_1, \ldots, F_{k-1}$ is a rotation. Thus, when n is odd, an improper isometry is the composition of a reflection about a hyperplane H with a rotation consisting of reflections about hyperplanes $F_1, \ldots, F_{k-1}$ containing a line, L, orthogonal to H. In particular, when $n = 3$, every improper orthogonal transformation is the product of a rotation with a reflection about a plane orthogonal to the axis of rotation.

Using Theorem 8.1, we can also give a rather simple proof of the classical fact that in a Euclidean space of odd dimension, every rotation leaves some nonnull vector invariant, and thus a line invariant.

If λ is an eigenvalue of f, then the following lemma shows that the orthogonal complement $E_\lambda(f)^\perp$ of the eigenspace associated with λ is closed under f.

Lemma 8.2. *Let E be a Euclidean space of finite dimension n, and let $f \colon E \to E$ be an isometry. For any subspace F of E, if $f(F) = F$, then $f(F^\perp) \subseteq F^\perp$ and $E = F \oplus F^\perp$.*

Proof. We just have to prove that if $w \in E$ is orthogonal to every $u \in F$, then $f(w)$ is also orthogonal to every $u \in F$. However, since $f(F) = F$, for every $v \in F$, there is some $u \in F$ such that $f(u) = v$, and we have

$$f(w) \cdot v = f(w) \cdot f(u) = w \cdot u,$$

since f is an isometry. Since we assumed that $w \in E$ is orthogonal to every $u \in F$, we have

$$w \cdot u = 0,$$

and thus

$$f(w) \cdot v = 0,$$

and this for every $v \in F$. Thus, $f(F^{\perp}) \subseteq F^{\perp}$. The fact that $E = F \oplus F^{\perp}$ follows from Lemma 6.8. $\square$

Lemma 8.2 is the starting point of the proof that every orthogonal matrix can be diagonalized over the field of complex numbers. Indeed, if λ is any eigenvalue of f, then $f(E_{\lambda}(f)) = E_{\lambda}(f)$, where $E_{\lambda}(f)$ is the eigenspace associated with λ, and thus the orthogonal $E_{\lambda}(f)^{\perp}$ is closed under f, and $E = E_{\lambda}(f) \oplus E_{\lambda}(f)^{\perp}$. The problem over $\mathbb{R}$ is that there may not be any real eigenvalues. However, when n is odd, the following lemma shows that every rotation admits 1 as an eigenvalue (and similarly, when n is even, every improper orthogonal transformation admits 1 as an eigenvalue).

Lemma 8.3. *Let E be a Euclidean space.*

(1) If E has odd dimension $n = 2m+1$, then every rotation f admits 1 as an eigenvalue and the eigenspace F of all eigenvectors left invariant under f has an odd dimension $2p+1$. Furthermore, there is an orthonormal basis of E, in which f is represented by a matrix of the form

$$\begin{pmatrix} R_{2(m-p)} & 0 \\ 0 & I_{2p+1} \end{pmatrix},$$

where $R_{2(m-p)}$ is a rotation matrix that does not have 1 as an eigenvalue.

(2) If E has even dimension $n = 2m$, then every improper orthogonal transformation f admits 1 as an eigenvalue and the eigenspace F of all eigenvectors left invariant under f has an odd dimension $2p+1$. Furthermore, there is an orthonormal basis of E, in which f is represented by a matrix of the form

$$\begin{pmatrix} S_{2(m-p)-1} & 0 \\ 0 & I_{2p+1} \end{pmatrix},$$

where $S_{2(m-p)-1}$ is an improper orthogonal matrix that does not have 1 as an eigenvalue.

Proof. We prove only (1), the proof of (2) being similar. Since f is a rotation and $n = 2m+1$ is odd, by Theorem 8.1, f is the composition of an even number less than or equal to $2m$ of reflections. From Lemma 2.14, recall the Grassmann relation

$$\dim(M) + \dim(N) = \dim(M + N) + \dim(M \cap N),$$

where M and N are subspaces of E. Now, if M and N are hyperplanes, their dimension is $n - 1$, and thus $\dim(M \cap N) \geq n - 2$. Thus, if we intersect $k \leq n$ hyperplanes, we see that the dimension of their intersection is at least $n - k$. Since each of the reflections is the identity on the hyperplane defining it, and since there are at most $2m = n - 1$ reflections, their composition is the identity on a subspace of dimension at least 1. This proves that 1 is an eigenvalue of f. Let F be the eigenspace associated with 1, and assume that its dimension is q. Let $G = F^{\perp}$ be the orthogonal of F. By Lemma 8.2, G is stable under f, and $E = F \oplus G$. Using Lemma 6.7, we can find an orthonormal basis of E consisting of an orthonormal basis for G and orthonormal basis for F. In this basis, the matrix of f is of the form

$$\begin{pmatrix} R_{2m+1-q} & 0 \\ 0 & I_q \end{pmatrix}.$$

Thus, $\det(f) = \det(R)$, and R must be a rotation, since f is a rotation and $\det(f) = 1$. Now, if f left some vector $u \neq 0$ in G invariant, this vector would be an eigenvector for 1, and we would have $u \in F$, the eigenspace associated with 1, which contradicts $E = F \oplus G$. Thus, by the first part of the proof, the dimension of G must be even, since otherwise, the restriction of f to G would admit 1 as an eigenvalue. Consequently, q must be odd, and R does not admit 1 as an eigenvalue. Letting $q = 2p + 1$, the lemma is established. $\square$

An example showing that Lemma 8.3 fails for n even is the following rotation matrix (when $n = 2$):

$$R = \begin{pmatrix} \cos\theta & -\sin\theta \\ \sin\theta & \cos\theta \end{pmatrix}.$$

The above matrix does not have real eigenvalues for $\theta \neq k\pi$.

It is easily shown that for $n = 2$, with respect to any chosen orthonormal basis (e_1, e_2), every rotation is represented by a matrix of form

$$R = \begin{pmatrix} \cos\theta & -\sin\theta \\ \sin\theta & \cos\theta \end{pmatrix}$$

where $\theta \in [0, 2\pi[$, and that every improper orthogonal transformation is represented by a matrix of the form

$$S = \begin{pmatrix} \cos\theta & \sin\theta \\ \sin\theta & -\cos\theta \end{pmatrix}.$$

In the first case, we call $\theta \in [0, 2\pi[$ the *measure* of the angle of rotation of R w.r.t. the orthonormal basis (e_1, e_2). In the second case, we have a reflection about a line, and it is easy to determine what this line is. It is also easy to see that S is the composition of a reflection about the x-axis with a rotation (of matrix R).

We refrained from calling θ "the angle of rotation," because there are some subtleties involved in defining rigorously the notion of angle of two vectors (or two lines). For example, note that with respect to the "opposite basis" (e_2, e_1), the measure θ must be changed to $2\pi - \theta$ (or $-\theta$ if we consider the quotient set $\mathbb{R}/2\pi$ of the real numbers modulo 2π). We will come back to this point after having defined the notion of orientation (see Section 8.8).

It is easily shown that the group $\mathbf{SO}(2)$ of rotations in the plane is abelian. First, recall that every plane rotation is the product of two reflections (about lines), and that every isometry in $\mathbf{O}(2)$ is either a reflection or a rotation. To alleviate the notation, we will omit the composition operator $\circ$, and write rs instead of $r \circ s$. Now, if r is a rotation and s is a reflection, rs being in $\mathbf{O}(2)$ must be a reflection (since $\det(rs) = \det(r)\det(s) = -1$), and thus $(rs)^2 = \mathrm{id}$, since a reflection is an involution, which implies that

$$srs = r^{-1}.$$

Then, given two rotations r_1 and r_2, writing r_1 as $r_1 = s_2 s_1$ for two reflections s_1, s_2, we have

$$r_1 r_2 r_1^{-1} = s_2 s_1 r_2 (s_2 s_1)^{-1} = s_2 s_1 r_2 s_1^{-1} s_2^{-1} = s_2 s_1 r_2 s_1 s_2 = s_2 r_2^{-1} s_2 = r_2,$$

since $srs = r^{-1}$ for all reflections s and rotations r, and thus $r_1 r_2 = r_2 r_1$.

We can also perform the following calculation, using some elementary trigonometry:

$$\begin{pmatrix} \cos\varphi & \sin\varphi \\ \sin\varphi & -\cos\varphi \end{pmatrix} \begin{pmatrix} \cos\psi & \sin\psi \\ \sin\psi & -\cos\psi \end{pmatrix} = \begin{pmatrix} \cos(\varphi+\psi) & \sin(\varphi+\psi) \\ \sin(\varphi+\psi) & -\cos(\varphi+\psi) \end{pmatrix}.$$

The above also shows that the inverse of a rotation matrix

$$R = \begin{pmatrix} \cos\theta & -\sin\theta \\ \sin\theta & \cos\theta \end{pmatrix}$$

is obtained by changing θ to $-\theta$ (or $2\pi - \theta$). Incidentally, note that in writing a rotation r as the product of two reflections $r = s_2 s_1$, the first reflection s_1 can be chosen arbitrarily, since $s_1^2 = \mathrm{id}$, $r = (rs_1)s_1$, and rs_1 is a reflection.

For $n = 3$, the only two choices for p are $p = 1$, which corresponds to the identity, or $p = 0$, in which case f is a rotation leaving a line invariant. This line D is called the *axis of rotation*. The rotation R behaves like a two-dimensional rotation around the axis of rotation. Thus, the rotation R is the composition of two reflections about planes containing the axis of rotation D and forming an angle $\theta/2$. This is illustrated in Figure 8.4.

The measure of the angle of rotation θ can be determined through its cosine via the formula

$$\cos\theta = u \cdot R(u),$$

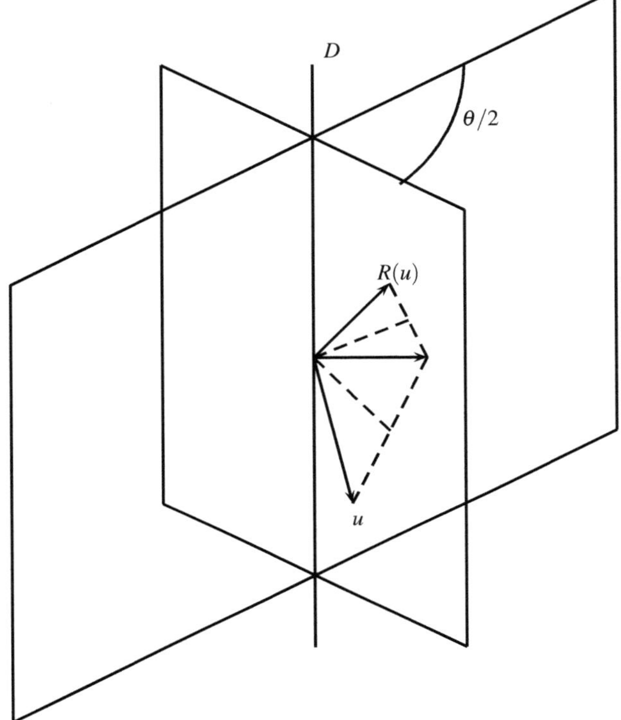

Fig. 8.4 3D rotation as the composition of two reflections.

where u is any unit vector orthogonal to the direction of the axis of rotation. However, this does not determine $\theta \in [0, 2\pi[$ uniquely, since both θ and $2\pi - \theta$ are possible candidates. What is missing is an orientation of the plane (through the origin) orthogonal to the axis of rotation. We will come back to this point in Section 8.8.

In the orthonormal basis of the lemma, a rotation is represented by a matrix of the form

$$R = \begin{pmatrix} \cos\theta & -\sin\theta & 0 \\ \sin\theta & \cos\theta & 0 \\ 0 & 0 & 1 \end{pmatrix}.$$

Remark: For an arbitrary rotation matrix A, since $a_{11} + a_{22} + a_{33}$ (the *trace* of A) is the sum of the eigenvalues of A, and since these eigenvalues are $\cos\theta + i\sin\theta$, $\cos\theta - i\sin\theta$, and 1, for some $\theta \in [0, 2\pi[$, we can compute $\cos\theta$ from

$$1 + 2\cos\theta = a_{11} + a_{22} + a_{33}.$$

It is also possible to determine the axis of rotation (see the problems).

An improper transformation is either a reflection about a plane or the product of three reflections, or equivalently the product of a reflection about a plane with a rotation, and we noted in the discussion following Theorem 8.1 that the axis of rotation is orthogonal to the plane of the reflection. Thus, an improper transformation is represented by a matrix of the form

$$S = \begin{pmatrix} \cos\theta & -\sin\theta & 0 \\ \sin\theta & \cos\theta & 0 \\ 0 & 0 & -1 \end{pmatrix}.$$

When $n \geq 3$, the group of rotations $\mathbf{SO}(n)$ is not only generated by hyperplane reflections, but also by flips (about subspaces of dimension $n-2$). We will also see, in Section 8.4, that every proper affine rigid motion can be expressed as the composition of at most n flips, which is perhaps even more surprising! The proof of these results uses the following key lemma.

Lemma 8.4. *Given any Euclidean space E of dimension $n \geq 3$, for any two reflections h_1 and h_2 about some hyperplanes H_1 and H_2, there exist two flips f_1 and f_2 such that $h_2 \circ h_1 = f_2 \circ f_1$.*

Proof. If $h_1 = h_2$, it is obvious that

$$h_1 \circ h_2 = h_1 \circ h_1 = \mathrm{id} = f_1 \circ f_1$$

for any flip f_1. If $h_1 \neq h_2$, then $H_1 \cap H_2 = F$, where $\dim(F) = n-2$ (by the Grassmann relation). We can pick an orthonormal basis $(e_1, \ldots, e_n)$ of E such that $(e_1, \ldots, e_{n-2})$ is an orthonormal basis of F. We can also extend $(e_1, \ldots, e_{n-2})$ to an orthonormal basis $(e_1, \ldots, e_{n-2}, u_1, v_1)$ of E, where $(e_1, \ldots, e_{n-2}, u_1)$ is an orthonormal basis of H_1, in which case

$$e_{n-1} = \cos\theta_1\, u_1 + \sin\theta_1\, v_1,$$
$$e_n = \sin\theta_1\, u_1 - \cos\theta_1\, v_1,$$

for some $\theta_1 \in [0, 2\pi]$. Since h_1 is the identity on H_1 and v_1 is orthogonal to H_1, it follows that $h_1(u_1) = u_1$, $h_1(v_1) = -v_1$, and we get

$$h_1(e_{n-1}) = \cos\theta_1\, u_1 - \sin\theta_1\, v_1,$$
$$h_1(e_n) = \sin\theta_1\, u_1 + \cos\theta_1\, v_1.$$

After some simple calculations, we get

$$h_1(e_{n-1}) = \cos 2\theta_1\, e_{n-1} + \sin 2\theta_1\, e_n,$$
$$h_1(e_n) = \sin 2\theta_1\, e_{n-1} - \cos 2\theta_1\, e_n.$$

As a consequence, the matrix A_1 of h_1 over the basis $(e_1, \ldots, e_n)$ is of the form

$$A_1 = \begin{pmatrix} I_{n-2} & 0 & 0 \\ 0 & \cos 2\theta_1 & \sin 2\theta_1 \\ 0 & \sin 2\theta_1 & -\cos 2\theta_1 \end{pmatrix}.$$

Similarly, the matrix A_2 of h_2 over the basis $(e_1, \ldots, e_n)$ is of the form

$$A_2 = \begin{pmatrix} I_{n-2} & 0 & 0 \\ 0 & \cos 2\theta_2 & \sin 2\theta_2 \\ 0 & \sin 2\theta_2 & -\cos 2\theta_2 \end{pmatrix}.$$

Observe that both A_1 and A_2 have the eigenvalues -1 and $+1$ with multiplicity $n-1$. The trick is to observe that if we change the last entry in I_{n-2} from $+1$ to -1 (which is possible since $n \geq 3$), we have the following product $A_2 A_1$:

$$\begin{pmatrix} I_{n-3} & 0 & 0 & 0 \\ 0 & -1 & 0 & 0 \\ 0 & 0 & \cos 2\theta_2 & \sin 2\theta_2 \\ 0 & 0 & \sin 2\theta_2 & -\cos 2\theta_2 \end{pmatrix} \begin{pmatrix} I_{n-3} & 0 & 0 & 0 \\ 0 & -1 & 0 & 0 \\ 0 & 0 & \cos 2\theta_1 & \sin 2\theta_1 \\ 0 & 0 & \sin 2\theta_1 & -\cos 2\theta_1 \end{pmatrix}.$$

Now, the two matrices above are clearly orthogonal, and they have the eigenvalues $-1, -1$, and $+1$ with multiplicity $n-2$, which implies that the corresponding isometries leave invariant a subspace of dimension $n-2$ and act as $-\mathrm{id}$ on its orthogonal complement (which has dimension 2). This means that the above two matrices represent two flips f_1 and f_2 such that $h_2 \circ h_1 = f_2 \circ f_1$. $\square$

Using Lemma 8.4 and the Cartan–Dieudonné theorem, we obtain the following characterization of rotations when $n \geq 3$.

Theorem 8.2. *Let E be a Euclidean space of dimension $n \geq 3$. Every rotation $f \in$ $\mathbf{SO}(E)$ is the composition of an even number of flips $f = f_{2k} \circ \cdots \circ f_1$, where $2k \leq n$. Furthermore, if $u \neq 0$ is invariant under f (i.e., $u \in \mathrm{Ker}\,(f - \mathrm{id})$), we can pick the last flip f_{2k} such that $u \in F_{2k}^{\perp}$, where F_{2k} is the subspace of dimension $n-2$ determining f_{2k}.*

Proof. By Theorem 8.1, the rotation f can be expressed as an even number of hyperplane reflections $f = s_{2k} \circ s_{2k-1} \circ \cdots \circ s_2 \circ s_1$, with $2k \leq n$. By Lemma 8.4, every composition of two reflections $s_{2i} \circ s_{2i-1}$ can be replaced by the composition of two flips $f_{2i} \circ f_{2i-1}$ ($1 \leq i \leq k$), which yields $f = f_{2k} \circ \cdots \circ f_1$, where $2k \leq n$.

Assume that $f(u) = u$, with $u \neq 0$. We have already made the remark that in the case where 1 is an eigenvalue of f, the proof of Theorem 8.1 shows that the reflections s_i can be chosen so that $s_i(u) = u$. In particular, if each reflection s_i is a reflection about the hyperplane H_i, we have $u \in H_{2k-1} \cap H_{2k}$. Letting $F = H_{2k-1} \cap H_{2k}$, pick an orthonormal basis $(e_1, \ldots, e_{n-3}, e_{n-2})$ of F, where

$$e_{n-2} = \frac{u}{\|u\|}.$$

The proof of Lemma 8.4 yields two flips f_{2k-1} and f_{2k} such that

$$f_{2k}(e_{n-2}) = -e_{n-2} \quad \text{and} \quad s_{2k} \circ s_{2k-1} = f_{2k} \circ f_{2k-1},$$

since the $(n-2)$th diagonal entry in both matrices is -1, which means that $e_{n-2} \in F_{2k}^{\perp}$, where F_{2k} is the subspace of dimension $n-2$ determining f_{2k}. Since $u = \|u\| e_{n-2}$, we also have $u \in F_{2k}^{\perp}$. $\square$

Remarks:

(1) It is easy to prove that if f is a rotation in $\mathbf{SO}(3)$ and if D is its axis and θ is its angle of rotation, then f is the composition of two flips about lines D_1 and D_2 orthogonal to D and making an angle $\theta/2$.

(2) It is natural to ask what is the minimal number of flips needed to obtain a rotation f (when $n \geq 3$). As for arbitrary isometries, we will prove later that every rotation is the composition of k flips, where

$$k = n - \dim(\operatorname{Ker}(f - \operatorname{id})),$$

and that this number is minimal (where $n = \dim(E)$).

We now show that hyperplane reflections can be used to obtain another proof of the QR-decomposition.

8.3 QR-Decomposition Using Householder Matrices

First, we state the result geometrically. When translated in terms of Householder matrices, we obtain the fact advertised earlier that every matrix (not necessarily invertible) has a QR-decomposition.

Lemma 8.5. *Let E be a nontrivial Euclidean space of dimension n. For any orthonormal basis $(e_1,\ldots, e_n)$ and for any n-tuple of vectors $(v_1, \ldots, v_n)$, there is a sequence of n isometries $h_1,\ldots,h_n$ such that h_i is a hyperplane reflection or the identity, and if $(r_1,\ldots,r_n)$ are the vectors given by*

$$r_j = h_n \circ \cdots \circ h_2 \circ h_1(v_j),$$

then every r_j is a linear combination of the vectors $(e_1,\ldots,e_j)$, $1 \leq j \leq n$. Equivalently, the matrix R whose columns are the components of the r_j over the basis $(e_1,\ldots,e_n)$ is an upper triangular matrix. Furthermore, the h_i can be chosen so that the diagonal entries of R are nonnegative.

Proof. We proceed by induction on n. For $n = 1$, we have $v_1 = \lambda e_1$ for some $\lambda \in \mathbb{R}$. If $\lambda \geq 0$, we let $h_1 = \operatorname{id}$, else if $\lambda < 0$, we let $h_1 = -\operatorname{id}$, the reflection about the origin.

For $n \geq 2$, we first have to find h_1. Let

$$r_{1,1} = \|v_1\|.$$

If $v_1 = r_{1,1}e_1$, we let $h_1 = \mathrm{id}$. Otherwise, there is a unique hyperplane reflection h_1 such that

$$h_1(v_1) = r_{1,1}e_1,$$

defined such that

$$h_1(u) = u - 2\,\frac{(u \cdot w_1)}{\|w_1\|^2}\,w_1$$

for all $u \in E$, where

$$w_1 = r_{1,1}e_1 - v_1.$$

The map h_1 is the reflection about the hyperplane H_1 orthogonal to the vector $w_1 = r_{1,1}e_1 - v_1$. Letting

$$r_1 = h_1(v_1) = r_{1,1}e_1,$$

it is obvious that r_1 belongs to the subspace spanned by e_1, and $r_{1,1} = \|v_1\|$ is non-negative.

Next, assume that we have found k linear maps $h_1, \dots, h_k$, hyperplane reflections or the identity, where $1 \le k \le n-1$, such that if $(r_1, \dots, r_k)$ are the vectors given by

$$r_j = h_k \circ \cdots \circ h_2 \circ h_1(v_j),$$

then every r_j is a linear combination of the vectors $(e_1, \dots, e_j)$, $1 \le j \le k$. The vectors $(e_1, \dots, e_k)$ form a basis for the subspace denoted by U'_k, the vectors $(e_{k+1}, \dots, e_n)$ form a basis for the subspace denoted by U''_k, the subspaces U'_k and U''_k are orthogonal, and $E = U'_k \oplus U''_k$. Let

$$u_{k+1} = h_k \circ \cdots \circ h_2 \circ h_1(v_{k+1}).$$

We can write

$$u_{k+1} = u'_{k+1} + u''_{k+1},$$

where $u'_{k+1} \in U'_k$ and $u''_{k+1} \in U''_k$. Let

$$r_{k+1,k+1} = \|u''_{k+1}\|.$$

If $u''_{k+1} = r_{k+1,k+1}e_{k+1}$, we let $h_{k+1} = \mathrm{id}$. Otherwise, there is a unique hyperplane reflection h_{k+1} such that

$$h_{k+1}(u''_{k+1}) = r_{k+1,k+1}e_{k+1},$$

defined such that

$$h_{k+1}(u) = u - 2\,\frac{(u \cdot w_{k+1})}{\|w_{k+1}\|^2}\,w_{k+1}$$

for all $u \in E$, where

$$w_{k+1} = r_{k+1,k+1}e_{k+1} - u''_{k+1}.$$

The map h_{k+1} is the reflection about the hyperplane H_{k+1} orthogonal to the vector $w_{k+1} = r_{k+1,k+1}e_{k+1} - u''_{k+1}$. However, since $u''_{k+1}, e_{k+1} \in U''_k$ and U'_k is orthogonal to U_k'', the subspace U_k' is contained in H_{k+1}, and thus, the vectors $(r_1, \ldots, r_k)$ and u_{k+1}', which belong to U_k', are invariant under h_{k+1}. This proves that

$$h_{k+1}(u_{k+1}) = h_{k+1}(u_{k+1}') + h_{k+1}(u_{k+1}'') = u_{k+1}' + r_{k+1,k+1}\, e_{k+1}$$

is a linear combination of $(e_1, \ldots, e_{k+1})$. Letting

$$r_{k+1} = h_{k+1}(u_{k+1}) = u_{k+1}' + r_{k+1,k+1}\, e_{k+1},$$

since $u_{k+1} = h_k \circ \cdots \circ h_2 \circ h_1(v_{k+1})$, the vector

$$r_{k+1} = h_{k+1} \circ \cdots \circ h_2 \circ h_1(v_{k+1})$$

is a linear combination of $(e_1, \ldots, e_{k+1})$. The coefficient of r_{k+1} over e_{k+1} is $r_{k+1,k+1} = \|u_{k+1}''\|$, which is nonnegative. This concludes the induction step, and thus the proof. $\quad\square$

Remarks:

(1) Since every h_i is a hyperplane reflection or the identity,

$$\rho = h_n \circ \cdots \circ h_2 \circ h_1$$

   is an isometry.
(2) If we allow negative diagonal entries in R, the last isometry h_n may be omitted.
(3) Instead of picking $r_{k,k} = \|u_k''\|$, which means that

$$w_k = r_{k,k}\, e_k - u_k'',$$

   where $1 \leq k \leq n$, it might be preferable to pick $r_{k,k} = -\|u_k''\|$ if this makes $\|w_k\|^2$ larger, in which case

$$w_k = r_{k,k}\, e_k + u_k''.$$

Indeed, since the definition of h_k involves division by $\|w_k\|^2$, it is desirable to avoid division by very small numbers.
(4) The method also applies to any m-tuple of vectors $(v_1, \ldots, v_m)$, where m is not necessarily equal to n (the dimension of E). In this case, R is an upper triangular $n \times m$ matrix we leave the minor adjustments to the method as an exercise to the reader (if $m > n$, the last $m - n$ vectors are unchanged).

Lemma 8.5 directly yields the QR-decomposition in terms of Householder transformations (see Strang [11, 12], Golub and Van Loan [7], Trefethen and Bau [14], Kincaid and Cheney [8], or Ciarlet [5]).

Lemma 8.6. *For every real $n \times n$ matrix A, there is a sequence $H_1, \ldots, H_n$ of matrices, where each H_i is either a Householder matrix or the identity, and an upper triangular matrix R such that*

$$R = H_n \cdots H_2 H_1 A.$$

As a corollary, there is a pair of matrices Q,R, where Q is orthogonal and R is upper triangular, such that $A = QR$ (a QR-decomposition of A). Furthermore, R can be chosen so that its diagonal entries are nonnegative.

Proof. The jth column of A can be viewed as a vector v_j over the canonical basis $(e_1,\ldots,e_n)$ of $\mathbb{E}^n$ (where $(e_j)_i = 1$ if $i = j$, and 0 otherwise, $1 \le i, j \le n$). Applying Lemma 8.5 to $(v_1,\ldots,v_n)$, there is a sequence of n isometries $h_1,\ldots,h_n$ such that h_i is a hyperplane reflection or the identity, and if $(r_1,\ldots,r_n)$ are the vectors given by

$$r_j = h_n \circ \cdots \circ h_2 \circ h_1(v_j),$$

then every r_j is a linear combination of the vectors $(e_1,\ldots,e_j)$, $1 \le j \le n$. Letting R be the matrix whose columns are the vectors r_j, and H_i the matrix associated with h_i, it is clear that

$$R = H_n \cdots H_2 H_1 A,$$

where R is upper triangular and every H_i is either a Householder matrix or the identity. However, $h_i \circ h_i = \mathrm{id}$ for all i, $1 \le i \le n$, and so

$$v_j = h_1 \circ h_2 \circ \cdots \circ h_n(r_j)$$

for all j, $1 \le j \le n$. But $\rho = h_1 \circ h_2 \circ \cdots \circ h_n$ is an isometry, and by Lemma 6.10, ρ is represented by an orthogonal matrix Q. It is clear that $A = QR$, where R is upper triangular. As we noted in Lemma 8.5, the diagonal entries of R can be chosen to be nonnegative. $\square$

Remarks:

(1) Letting

$$A_{k+1} = H_k \cdots H_2 H_1 A,$$

with $A_1 = A$, $1 \le k \le n$, the proof of Lemma 8.5 can be interpreted in terms of the computation of the sequence of matrices $A_1,\ldots,A_{n+1} = R$. The matrix A_{k+1} has the shape

$$A_{k+1} = \begin{pmatrix} \times & \times & \times & u_1^{k+1} & \times & \times & \times & \times \\ 0 & \times & \vdots & \vdots & \vdots & \vdots & \vdots & \vdots \\ 0 & 0 & \times & u_k^{k+1} & \times & \times & \times & \times \\ 0 & 0 & 0 & u_{k+1}^{k+1} & \times & \times & \times & \times \\ 0 & 0 & 0 & u_{k+2}^{k+1} & \times & \times & \times & \times \\ \vdots & \vdots & \vdots & \vdots & \vdots & \vdots & \vdots & \vdots \\ 0 & 0 & 0 & u_{n-1}^{k+1} & \times & \times & \times & \times \\ 0 & 0 & 0 & u_n^{k+1} & \times & \times & \times & \times \end{pmatrix},$$

where the $(k+1)$th column of the matrix is the vector

$$u_{k+1} = h_k \circ \cdots \circ h_2 \circ h_1(v_{k+1}),$$

and thus

$$u'_{k+1} = \left(u_1^{k+1}, \ldots, u_k^{k+1} \right)$$

and

$$u''_{k+1} = \left(u_{k+1}^{k+1}, u_{k+2}^{k+1}, \ldots, u_n^{k+1} \right).$$

If the last $n - k - 1$ entries in column $k + 1$ are all zero, there is nothing to do, and we let $H_{k+1} = I$. Otherwise, we kill these $n - k - 1$ entries by multiplying A_{k+1} on the left by the Householder matrix H_{k+1} sending

$$\left(0, \ldots, 0, u_{k+1}^{k+1}, \ldots, u_n^{k+1} \right) \quad \text{to} \quad (0, \ldots, 0, r_{k+1,k+1}, 0, \ldots, 0),$$

where $r_{k+1,k+1} = \| (u_{k+1}^{k+1}, \ldots, u_n^{k+1}) \|$.

(2) If A is invertible and the diagonal entries of R are positive, it can be shown that Q and R are unique.

(3) If we allow negative diagonal entries in R, the matrix H_n may be omitted ($H_n = I$).

(4) The method allows the computation of the determinant of A. We have

$$\det(A) = (-1)^m r_{1,1} \cdots r_{n,n},$$

where m is the number of Householder matrices (not the identity) among the H_i.

(5) The "condition number" of the matrix A is preserved (see Strang [12], Golub and Van Loan [7], Trefethen and Bau [14], Kincaid and Cheney [8], or Ciarlet [5]). This is very good for numerical stability.

(6) The method also applies to a rectangular $m \times n$ matrix. In this case, R is also an $m \times n$ matrix (and it is upper triangular).

We now turn to affine isometries.

8.4 Affine Isometries (Rigid Motions)

In the remaining sections we study affine isometries. First, we characterize the set of fixed points of an affine map. Using this characterization, we prove that every affine isometry f can be written uniquely as

$$f = t \circ g, \quad \text{with} \quad t \circ g = g \circ t,$$

where g is an isometry having a fixed point, and t is a translation by a vector τ such that $\overrightarrow{f}(\tau) = \tau$, and with some additional nice properties (see Theorem 8.3). This is a generalization of a classical result of Chasles about (proper) rigid motions in $\mathbb{R}^3$ (screw motions). We prove a generalization of the Cartan–Dieudonné theorem for the affine isometries: Every isometry in $\mathbf{Is}(n)$ can be written as the composition of at most n reflections if it has a fixed point, or else as the composition of at most

$n+2$ reflections. We also prove that every rigid motion in $\mathbf{SE}(n)$ is the composition of at most n flips (for $n \geq 3$). This is somewhat surprising, in view of the previous theorem.

Definition 8.3. Given any two nontrivial Euclidean affine spaces E and F of the same finite dimension n, a function $f\colon E \to F$ is *an affine isometry (or rigid map)* if it is an affine map and

$$\|\overrightarrow{f(a)f(b)}\| = \|\overrightarrow{ab}\|,$$

for all $a, b \in E$. When $E = F$, an affine isometry $f\colon E \to E$ is also called a *rigid motion*.

Thus, an affine isometry is an affine map that preserves the distance. This is a rather strong requirement. In fact, we will show that for any function $f\colon E \to F$, the assumption that

$$\|\overrightarrow{f(a)f(b)}\| = \|\overrightarrow{ab}\|,$$

for all $a, b \in E$, forces f to be an affine map.

Remark: Sometimes, an affine isometry is defined as a *bijective* affine isometry. When E and F are of finite dimension, the definitions are equivalent.

The following simple lemma is left as an exercise.

Lemma 8.7. *Given any two nontrivial Euclidean affine spaces E and F of the same finite dimension n, an affine map $f\colon E \to F$ is an affine isometry iff its associated linear map $\overrightarrow{f}\colon \overrightarrow{E} \to \overrightarrow{F}$ is an isometry. An affine isometry is a bijection.*

Let us now consider affine isometries $f\colon E \to E$. If $\overrightarrow{f}$ is a rotation, we call f a *proper (or direct) affine isometry*, and if $\overrightarrow{f}$ is an improper linear isometry, we call f an *improper (or skew) affine isometry*. It is easily shown that the set of affine isometries $f\colon E \to E$ forms a group, and those for which $\overrightarrow{f}$ is a rotation is a subgroup. The group of affine isometries, or rigid motions, is a subgroup of the affine group $\mathbf{GA}(E)$, denoted by $\mathbf{Is}(E)$ (or $\mathbf{Is}(n)$ when $E = \mathbb{E}^n$). In Snapper and Troyer [10] the group of rigid motions is denoted by $\mathbf{Mo}(E)$. Since we denote the group of affine bijections as $\mathbf{GA}(E)$, perhaps we should denote the group of affine isometries by $\mathbf{IA}(E)$ (or $\mathbf{EA}(E)$!). The subgroup of $\mathbf{Is}(E)$ consisting of the direct rigid motions is also a subgroup of $\mathbf{SA}(E)$, and it is denoted by $\mathbf{SE}(E)$ (or $\mathbf{SE}(n)$, when $E = \mathbb{E}^n$). The translations are the affine isometries f for which $\overrightarrow{f} = \mathrm{id}$, the identity map on $\overrightarrow{E}$. The following lemma is the counterpart of Lemma 6.9 for isometries between Euclidean vector spaces.

Lemma 8.8. *Given any two nontrivial Euclidean affine spaces E and F of the same finite dimension n, for every function $f\colon E \to F$, the following properties are equivalent:*

(1) f is an affine map and $\|\overrightarrow{f(a)f(b)}\| = \|\overrightarrow{ab}\|$, for all $a, b \in E$.
(2) $\|\overrightarrow{f(a)f(b)}\| = \|\overrightarrow{ab}\|$, for all $a, b \in E$.

Proof. Obviously, (1) implies (2). In order to prove that (2) implies (1), we proceed as follows. First, we pick some arbitrary point $\Omega \in E$. We define the map $g \colon \overrightarrow{E} \to \overrightarrow{F}$ such that

$$g(u) = \overrightarrow{f(\Omega)f(\Omega + u)}$$

for all $u \in E$. Since

$$f(\Omega) + g(u) = f(\Omega) + \overrightarrow{f(\Omega)f(\Omega + u)} = f(\Omega + u)$$

for all $u \in \overrightarrow{E}$, f will be affine if we can show that g is linear, and f will be an affine isometry if we can show that g is a linear isometry.

Observe that

$$g(v) - g(u) = \overrightarrow{f(\Omega)f(\Omega + v)} - \overrightarrow{f(\Omega)f(\Omega + u)}$$
$$= \overrightarrow{f(\Omega + u)f(\Omega + v)}.$$

Then, the hypothesis

$$\|\overrightarrow{f(a)f(b)}\| = \|\overrightarrow{ab}\|$$

for all $a, b \in E$, implies that

$$\|g(v) - g(u)\| = \|\overrightarrow{f(\Omega + u)f(\Omega + v)}\| = \|\overrightarrow{(\Omega + u)(\Omega + v)}\| = \|v - u\|.$$

Thus, g preserves the distance. Also, by definition, we have

$$g(0) = 0.$$

Thus, we can apply Lemma 6.9, which shows that g is indeed a linear isometry, and thus f is an affine isometry. $\quad\square$

In order to understand the structure of affine isometries, it is important to investigate the fixed points of an affine map.

8.5 Fixed Points of Affine Maps

Recall that $E\left(1, \overrightarrow{f}\right)$ denotes the eigenspace of the linear map $\overrightarrow{f}$ associated with the scalar 1, that is, the subspace consisting of all vectors $u \in \overrightarrow{E}$ such that $\overrightarrow{f}(u) = u$. Clearly, $\mathrm{Ker}\left(\overrightarrow{f} - \mathrm{id}\right) = E\left(1, \overrightarrow{f}\right)$. Given some origin $\Omega \in E$, since

$$f(a) = f(\Omega + \overrightarrow{\Omega a}) = f(\Omega) + \overrightarrow{f}\left(\overrightarrow{\Omega a}\right),$$

we have $\overrightarrow{f(\Omega)f(a)} = \overrightarrow{f}\,(\overrightarrow{\Omega a})$, and thus

$$\overrightarrow{\Omega f(a)} = \overrightarrow{\Omega f(\Omega)} + \overrightarrow{f}\,(\overrightarrow{\Omega a}).$$

From the above, we get

$$\overrightarrow{\Omega f(a)} - \overrightarrow{\Omega a} = \overrightarrow{\Omega f(\Omega)} + \overrightarrow{f}\,(\overrightarrow{\Omega a}) - \overrightarrow{\Omega a}.$$

Using this, we show the following lemma, which holds for arbitrary affine spaces of finite dimension and for arbitrary affine maps.

Lemma 8.9. *Let E be any affine space of finite dimension. For every affine map $f\colon E \to E$, let $\mathrm{Fix}(f) = \{a \in E \mid f(a) = a\}$ be the set of fixed points of f. The following properties hold:*

(1) If f has some fixed point a, so that $\mathrm{Fix}(f) \neq \emptyset$, then $\mathrm{Fix}(f)$ is an affine subspace of E such that

$$\mathrm{Fix}(f) = a + E\left(1, \overrightarrow{f}\right) = a + \mathrm{Ker}\left(\overrightarrow{f} - \mathrm{id}\right),$$

where $E\left(1, \overrightarrow{f}\right)$ is the eigenspace of the linear map $\overrightarrow{f}$ for the eigenvalue 1.

(2) The affine map f has a unique fixed point iff $E\left(1, \overrightarrow{f}\right) = \mathrm{Ker}\left(\overrightarrow{f} - \mathrm{id}\right) = \{0\}$.

Proof. (1) Since the identity

$$\overrightarrow{\Omega f(b)} - \overrightarrow{\Omega b} = \overrightarrow{\Omega f(\Omega)} + \overrightarrow{f}\,(\overrightarrow{\Omega b}) - \overrightarrow{\Omega b}$$

holds for all $\Omega, b \in E$, if $f(a) = a$, then $\overrightarrow{a f(a)} = 0$, and thus, letting $\Omega = a$, for any $b \in E$,

$$f(b) = b$$

iff

$$\overrightarrow{a f(b)} - \overrightarrow{ab} = 0$$

iff

$$\overrightarrow{f}\,(\overrightarrow{ab}) - \overrightarrow{ab} = 0$$

iff

$$\overrightarrow{ab} \in E\left(1, \overrightarrow{f}\right) = \mathrm{Ker}\left(\overrightarrow{f} - \mathrm{id}\right),$$

which proves that

$$\mathrm{Fix}(f) = a + E\left(1, \overrightarrow{f}\right) = a + \mathrm{Ker}\left(\overrightarrow{f} - \mathrm{id}\right).$$

(2) Again, fix some origin Ω. Some a satisfies $f(a) = a$ iff

$$\overrightarrow{\Omega f(a)} - \overrightarrow{\Omega a} = 0$$

iff

$$\overrightarrow{\Omega f(\Omega)} + \overrightarrow{f}\left(\overrightarrow{\Omega a}\right) - \overrightarrow{\Omega a} = 0,$$

which can be rewritten as

$$\left(\overrightarrow{f} - \mathrm{id}\right)\left(\overrightarrow{\Omega a}\right) = -\overrightarrow{\Omega f(\Omega)}.$$

We have $E\left(1, \overrightarrow{f}\right) = \mathrm{Ker}\left(\overrightarrow{f} - \mathrm{id}\right) = \{0\}$ iff $\overrightarrow{f} - \mathrm{id}$ is injective, and since $\overrightarrow{E}$ has finite dimension, $\overrightarrow{f} - \mathrm{id}$ is also surjective, and thus, there is indeed some $a \in E$ such that

$$\left(\overrightarrow{f} - \mathrm{id}\right)\left(\overrightarrow{\Omega a}\right) = -\overrightarrow{\Omega f(\Omega)},$$

and it is unique, since $\overrightarrow{f} - \mathrm{id}$ is injective. Conversely, if f has a unique fixed point, say a, from

$$\left(\overrightarrow{f} - \mathrm{id}\right)\left(\overrightarrow{\Omega a}\right) = -\overrightarrow{\Omega f(\Omega)},$$

we have $\left(\overrightarrow{f} - \mathrm{id}\right)\left(\overrightarrow{\Omega a}\right) = 0$ iff $f(\Omega) = \Omega$, and since a is the unique fixed point of f, we must have $a = \Omega$, which shows that $\overrightarrow{f} - \mathrm{id}$ is injective. $\square$

Remark: The fact that E has finite dimension is used only to prove (2), and (1) holds in general.

If an isometry f leaves some point fixed, we can take such a point Ω as the origin, and then $f(\Omega) = \Omega$ and we can view f as a rotation or an improper orthogonal transformation, depending on the nature of $\overrightarrow{f}$. Note that it is quite possible that $\mathrm{Fix}(f) = \emptyset$. For example, nontrivial translations have no fixed points. A more interesting example is provided by the composition of a plane reflection about a line composed with a a nontrivial translation parallel to this line.

Otherwise, we will see in Theorem 8.3 that every affine isometry is the (commutative) composition of a translation with an isometry that always has a fixed point.

8.6 Affine Isometries and Fixed Points

Let E be an affine space. Given any two affine subspaces F, G, if F and G are orthogonal complements in E, which means that $\overrightarrow{F}$ and $\overrightarrow{G}$ are orthogonal subspaces of $\overrightarrow{E}$ such that $\overrightarrow{E} = \overrightarrow{F} \oplus \overrightarrow{G}$, for any point $\Omega \in F$, we define $q \colon E \to \overrightarrow{G}$ such that

$$q(a) = p_{\overrightarrow{G}}\left(\overrightarrow{\Omega a}\right).$$

Note that $q(a)$ is independent of the choice of $\Omega \in F$, since we have

$$\overrightarrow{\Omega a} = p_{\overrightarrow{F}}\left(\overrightarrow{\Omega a}\right) + p_{\overrightarrow{G}}\left(\overrightarrow{\Omega a}\right),$$

and for any $\Omega_1 \in F$, we have

$$\overrightarrow{\Omega_1 a} = \overrightarrow{\Omega_1 \Omega} + p_{\overrightarrow{F}}(\overrightarrow{\Omega a}) + p_{\overrightarrow{G}}(\overrightarrow{\Omega a}),$$

and since $\overrightarrow{\Omega_1 \Omega} \in \overrightarrow{F}$, this shows that

$$p_{\overrightarrow{G}}(\overrightarrow{\Omega_1 a}) = p_{\overrightarrow{G}}(\overrightarrow{\Omega a}).$$

Then the map $g \colon E \to E$ such that $g(a) = a - 2q(a)$, or equivalently

$$\overrightarrow{ag(a)} = -2q(a) = -2p_{\overrightarrow{G}}(\overrightarrow{\Omega a}),$$

does not depend on the choice of $\Omega \in F$. If we identify E to $\overrightarrow{E}$ by choosing any origin Ω in F, we note that g is identified with the symmetry with respect to $\overrightarrow{F}$ and parallel to $\overrightarrow{G}$. Thus, the map g is an affine isometry, and it is called the *orthogonal symmetry about F*. Since

$$g(a) = \Omega + \overrightarrow{\Omega a} - 2p_{\overrightarrow{G}}(\overrightarrow{\Omega a})$$

for all $\Omega \in F$ and for all $a \in E$, we note that the linear map $\overrightarrow{g}$ associated with g is the (linear) symmetry about the subspace $\overrightarrow{F}$ (the direction of F), and parallel to $\overrightarrow{G}$ (the direction of G).

Remark: The map $p \colon E \to F$ such that $p(a) = a - q(a)$, or equivalently

$$\overrightarrow{ap(a)} = -q(a) = -p_{\overrightarrow{G}}(\overrightarrow{\Omega a}),$$

is also independent of $\Omega \in F$, and it is called the *orthogonal projection onto F*.

The following amusing lemma shows the extra power afforded by affine orthogonal symmetries: Translations are subsumed! Given two parallel affine subspaces F_1 and F_2 in E, letting $\overrightarrow{F}$ be the common direction of F_1 and F_2 and $\overrightarrow{G} = \overrightarrow{F}^{\perp}$ be its orthogonal complement, for any $a \in F_1$, the affine subspace $a + \overrightarrow{G}$ intersects F_2 in a single point b (see Lemma 2.15). We define the *distance between F_1 and F_2* as $\|\overrightarrow{ab}\|$. It is easily seen that the distance between F_1 and F_2 is independent of the choice of a in F_1, and that it is the minimum of $\|\overrightarrow{xy}\|$ for all $x \in F_1$ and all $y \in F_2$.

Lemma 8.10. *Given any affine space E, if $f \colon E \to E$ and $g \colon E \to E$ are orthogonal symmetries about parallel affine subspaces F_1 and F_2, then $g \circ f$ is a translation defined by the vector $2\overrightarrow{ab}$, where $\overrightarrow{ab}$ is any vector perpendicular to the common direction $\overrightarrow{F}$ of F_1 and F_2 such that $\|\overrightarrow{ab}\|$ is the distance between F_1 and F_2, with $a \in F_1$ and $b \in F_2$. Conversely, every translation by a vector τ is obtained as the composition of two orthogonal symmetries about parallel affine subspaces F_1 and*

F_2 whose common direction is orthogonal to $\tau = \overrightarrow{ab}$, for some $a \in F_1$ and some $b \in F_2$ such that the distance between F_1 and F_2 is $\|\overrightarrow{ab}\|/2$.

Proof. We observed earlier that the linear maps $\overrightarrow{f}$ and $\overrightarrow{g}$ associated with f and g are the linear reflections about the directions of F_1 and F_2. However, F_1 and F_2 have the same direction, and so $\overrightarrow{f} = \overrightarrow{g}$. Since $\overrightarrow{g \circ f} = \overrightarrow{g} \circ \overrightarrow{f}$ and since $\overrightarrow{f} \circ \overrightarrow{g} = \overrightarrow{f} \circ \overrightarrow{f} = \mathrm{id}$, because every reflection is an involution, we have $\overrightarrow{g \circ f} = \mathrm{id}$, proving that $g \circ f$ is a translation. If we pick $a \in F_1$, then $g \circ f(a) = g(a)$, the reflection of $a \in F_1$ about F_2, and it is easily checked that $g \circ f$ is the translation by the vector $\tau = \overrightarrow{ag(a)}$ whose norm is twice the distance between F_1 and F_2. The second part of the lemma is left as an easy exercise. $\quad\square$

We conclude our quick study of affine isometries by proving a result that plays a major role in characterizing the affine isometries. This result may be viewed as a generalization of Chasles's theorem about the direct rigid motions in $\mathbb{E}^3$.

Theorem 8.3. *Let E be a Euclidean affine space of finite dimension n. For every affine isometry $f : E \to E$, there is a unique isometry $g : E \to E$ and a unique translation $t = t_\tau$, with $\overrightarrow{f}(\tau) = \tau$ (i.e., $\tau \in \mathrm{Ker}\left(\overrightarrow{f} - \mathrm{id}\right)$), such that the set $\mathrm{Fix}(g) = \{a \in E \mid g(a) = a\}$ of fixed points of g is a nonempty affine subspace of E of direction*

$$\overrightarrow{G} = \mathrm{Ker}\left(\overrightarrow{f} - \mathrm{id}\right) = E\left(1, \overrightarrow{f}\right),$$

and such that

$$f = t \circ g \quad \text{and} \quad t \circ g = g \circ t.$$

Furthermore, we have the following additional properties:

(a) $f = g$ and $\tau = 0$ iff f has some fixed point, i.e., iff $\mathrm{Fix}(f) \neq \emptyset$.

(b) If f has no fixed points, i.e., $\mathrm{Fix}(f) = \emptyset$, then $\dim\left(\mathrm{Ker}\left(\overrightarrow{f} - \mathrm{id}\right)\right) \geq 1$.

Proof. The proof rests on the following two key facts:

(1) If we can find some $x \in E$ such that $\overrightarrow{xf(x)} = \tau$ belongs to $\mathrm{Ker}\left(\overrightarrow{f} - \mathrm{id}\right)$, we get the existence of g and τ.

(2) $\overrightarrow{E} = \mathrm{Ker}\left(\overrightarrow{f} - \mathrm{id}\right) \oplus \mathrm{Im}\left(\overrightarrow{f} - \mathrm{id}\right)$, and the spaces $\mathrm{Ker}\left(\overrightarrow{f} - \mathrm{id}\right)$ and $\mathrm{Im}\left(\overrightarrow{f} - \mathrm{id}\right)$ are orthogonal. This implies the uniqueness of g and τ.

First, we prove that for every isometry $h : \overrightarrow{E} \to \overrightarrow{E}$, $\mathrm{Ker}\left(h - \mathrm{id}\right)$ and $\mathrm{Im}\left(h - \mathrm{id}\right)$ are orthogonal and that

$$\overrightarrow{E} = \mathrm{Ker}\left(h - \mathrm{id}\right) \oplus \mathrm{Im}\left(h - \mathrm{id}\right).$$

Recall that

$$\dim\left(\overrightarrow{E}\right) = \dim(\mathrm{Ker}\,\varphi) + \dim(\mathrm{Im}\,\varphi),$$

for any linear map $\varphi \colon \overrightarrow{E} \to \overrightarrow{E}$ (for instance, see Lang [9], or Strang [12]). To show that we have a direct sum, we prove orthogonality. Let $u \in \mathrm{Ker}\,(h - \mathrm{id})$, so that $h(u) = u$, let $v \in \overrightarrow{E}$, and compute

$$u \cdot (h(v) - v) = u \cdot h(v) - u \cdot v = h(u) \cdot h(v) - u \cdot v = 0,$$

since $h(u) = u$ and h is an isometry.

Next, assume that there is some $x \in E$ such that $\overrightarrow{xf(x)} = \tau$ belongs to the space $\mathrm{Ker}\,\big(\overrightarrow{f} - \mathrm{id}\big)$. If we define $g \colon E \to E$ such that

$$g = t_{(-\tau)} \circ f,$$

we have

$$g(x) = f(x) - \tau = x,$$

since $\overrightarrow{xf(x)} = \tau$ is equivalent to $x = f(x) - \tau$. As a composition of isometries, g is an isometry, x is a fixed point of g, and since $\tau \in \mathrm{Ker}\,\big(\overrightarrow{f} - \mathrm{id}\big)$, we have

$$\overrightarrow{f}\,(\tau) = \tau,$$

and since

$$g(b) = f(b) - \tau$$

for all $b \in E$, we have $\overrightarrow{g} = \overrightarrow{f}$. Since g has some fixed point x, by Lemma 8.9, $\mathrm{Fix}(g)$ is an affine subspace of E with direction $\mathrm{Ker}\,\big(\overrightarrow{g} - \mathrm{id}\big) = \mathrm{Ker}\,\big(\overrightarrow{f} - \mathrm{id}\big)$. We also have $f(b) = g(b) + \tau$ for all $b \in E$, and thus

$$(g \circ t_\tau)(b) = g(b + \tau) = g(b) + \overrightarrow{g}\,(\tau) = g(b) + \overrightarrow{f}\,(\tau) = g(b) + \tau = f(b),$$

and

$$(t_\tau \circ g)(b) = g(b) + \tau = f(b),$$

which proves that $t \circ g = g \circ t$.

To prove the existence of x as above, pick any arbitrary point $a \in E$. Since

$$\overrightarrow{E} = \mathrm{Ker}\,\big(\overrightarrow{f} - \mathrm{id}\big) \oplus \mathrm{Im}\,\big(\overrightarrow{f} - \mathrm{id}\big),$$

there is a unique vector $\tau \in \mathrm{Ker}\,\big(\overrightarrow{f} - \mathrm{id}\big)$ and some $v \in \overrightarrow{E}$ such that

$$\overrightarrow{af(a)} = \tau + \overrightarrow{f}\,(v) - v.$$

For any $x \in E$, since we also have

$$\overrightarrow{xf(x)} = \overrightarrow{xa} + \overrightarrow{af(a)} + \overrightarrow{f(a)f(x)} = \overrightarrow{xa} + \overrightarrow{af(a)} + \overrightarrow{f}\,(\overrightarrow{ax}),$$

we get

$$\overrightarrow{xf(x)} = \overrightarrow{xa} + \tau + \overrightarrow{f}(v) - v + \overrightarrow{f}(\overrightarrow{ax}),$$

which can be rewritten as

$$\overrightarrow{xf(x)} = \tau + \left(\overrightarrow{f} - \mathrm{id}\right)(v + \overrightarrow{ax}).$$

If we let $\overrightarrow{ax} = -v$, that is, $x = a - v$, we get

$$\overrightarrow{xf(x)} = \tau,$$

with $\tau \in \mathrm{Ker}\left(\overrightarrow{f} - \mathrm{id}\right)$.

Finally, we show that τ is unique. Assume two decompositions (g_1, τ_1) and (g_2, τ_2). Since $\overrightarrow{f} = \overrightarrow{g_1}$, we have $\mathrm{Ker}(\overrightarrow{g_1} - \mathrm{id}) = \mathrm{Ker}\left(\overrightarrow{f} - \mathrm{id}\right)$. Since g_1 has some fixed point b, we get

$$f(b) = g_1(b) + \tau_1 = b + \tau_1,$$

that is, $\overrightarrow{bf(b)} = \tau_1$, and $\overrightarrow{bf(b)} \in \mathrm{Ker}\left(\overrightarrow{f} - \mathrm{id}\right)$, since $\tau_1 \in \mathrm{Ker}\left(\overrightarrow{f} - \mathrm{id}\right)$. Similarly, for some fixed point c of g_2, we get $\overrightarrow{cf(c)} = \tau_2$ and $\overrightarrow{cf(c)} \in \mathrm{Ker}\left(\overrightarrow{f} - \mathrm{id}\right)$. Then we have

$$\tau_2 - \tau_1 = \overrightarrow{cf(c)} - \overrightarrow{bf(b)} = \overrightarrow{cb} - \overrightarrow{f(c)f(b)} = \overrightarrow{cb} - \overrightarrow{f}(\overrightarrow{cb}),$$

which shows that

$$\tau_2 - \tau_1 \in \mathrm{Ker}\left(\overrightarrow{f} - \mathrm{id}\right) \cap \mathrm{Im}\left(\overrightarrow{f} - \mathrm{id}\right),$$

and thus that $\tau_2 = \tau_1$, since we have shown that

$$\overrightarrow{E} = \mathrm{Ker}\left(\overrightarrow{f} - \mathrm{id}\right) \oplus \mathrm{Im}\left(\overrightarrow{f} - \mathrm{id}\right).$$

The fact that (a) holds is a consequence of the uniqueness of g and τ, since f and 0 clearly satisfy the required conditions. That (b) holds follows from Lemma 8.9 (2), since the affine map f has a unique fixed point iff $E\left(1, \overrightarrow{f}\right) = \mathrm{Ker}\left(\overrightarrow{f} - \mathrm{id}\right) = \{0\}$. $\square$

The determination of x is illustrated in Figure 8.5.

Remarks:

(1) Note that $\mathrm{Ker}\left(\overrightarrow{f} - \mathrm{id}\right) = \{0\}$ iff $\tau = 0$, iff $\mathrm{Fix}(g)$ consists of a single element, which is the unique fixed point of f. However, even if f is not a translation, f may not have any fixed points. For example, this happens when E is the affine Euclidean plane and f is the composition of a reflection about a line composed with a nontrivial translation parallel to this line.

(2) The fact that E has finite dimension is used only to prove (b).

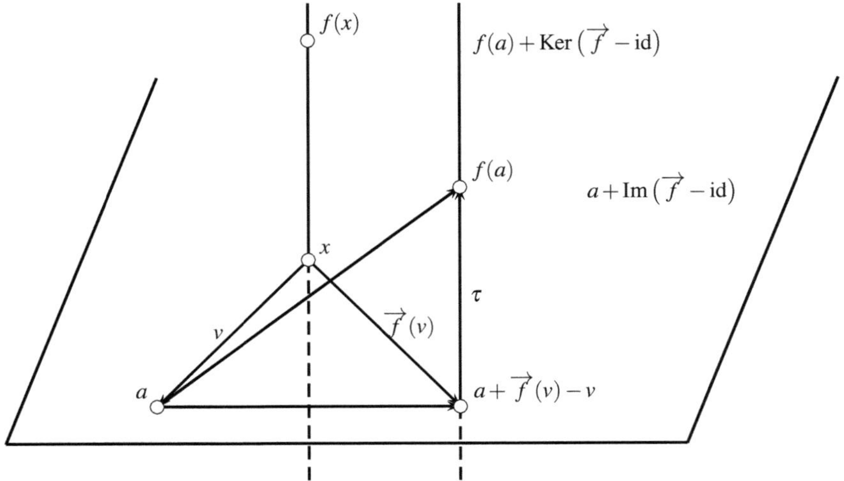

Fig. 8.5 Rigid motion as $f = t \circ g$, where g has some fixed point x.

(3) It is easily checked that $\mathrm{Fix}(g)$ consists of the set of points x such that $\|\overrightarrow{xf(x)}\|$ is minimal.

In the affine Euclidean plane it is easy to see that the affine isometries (besides the identity) are classified as follows. An isometry f that has a fixed point is a rotation if it is a direct isometry; otherwise, it is a reflection about a line. If f has no fixed point, then it is either a nontrivial translation or the composition of a reflection about a line with a nontrivial translation parallel to this line.

In an affine space of dimension 3 it is easy to see that the affine isometries (besides the identity) are classified as follows. There are three kinds of isometries that have a fixed point. A proper isometry with a fixed point is a rotation around a line D (its set of fixed points), as illustrated in Figure 8.6.

An improper isometry with a fixed point is either a reflection about a plane H (the set of fixed points) or the composition of a rotation followed by a reflection about a plane H orthogonal to the axis of rotation D, as illustrated in Figures 8.7 and 8.8. In the second case, there is a single fixed point $O = D \cap H$.

There are three types of isometries with no fixed point. The first kind is a nontrivial translation. The second kind is the composition of a rotation followed by a nontrivial translation parallel to the axis of rotation D. Such a rigid motion is proper, and is called a *screw motion*. A screw motion is illustrated in Figure 8.9.

The third kind is the composition of a reflection about a plane followed by a nontrivial translation by a vector parallel to the direction of the plane of the reflection, as illustrated in Figure 8.10.

This last transformation is an improper isometry.

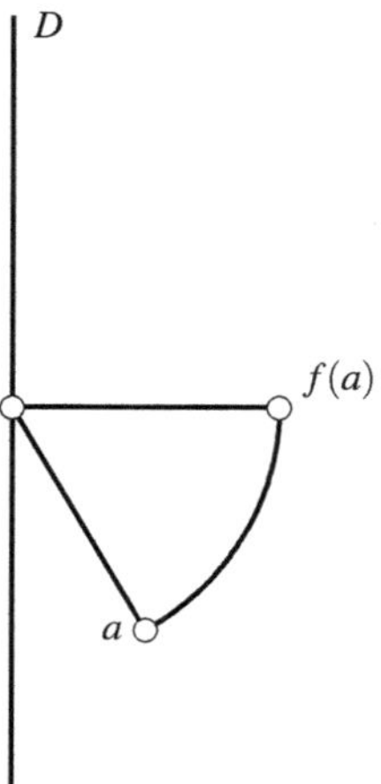

Fig. 8.6 3D proper rigid motion with line D of fixed points (rotation).

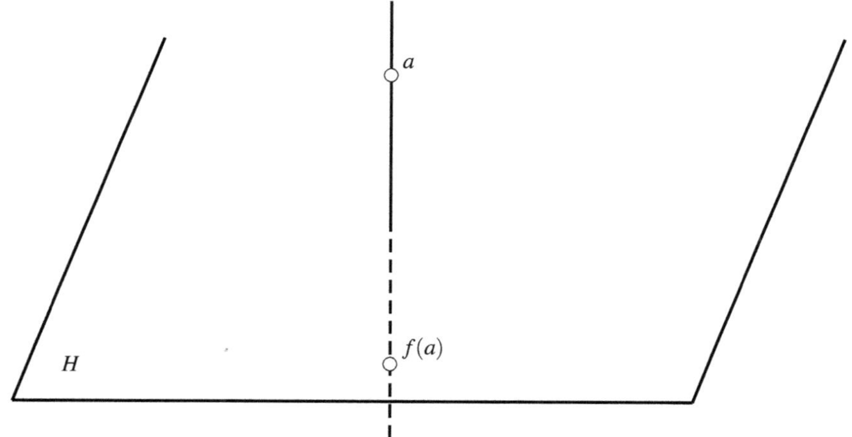

Fig. 8.7 3D improper rigid motion with a plane H of fixed points (reflection).

8.7 The Cartan–Dieudonné Theorem for Affine Isometries

The Cartan–Dieudonné theorem also holds for affine isometries, with a small twist due to translations. The reader is referred to Berger [2], Snapper and Troyer [10], or Tisseron [13] for a detailed treatment of the Cartan–Dieudonné theorem and its variants.

Theorem 8.4. *Let E be an affine Euclidean space of dimension $n \geq 1$. Every isometry $f \in \mathbf{Is}(E)$ that has a fixed point and is not the identity is the composition of at most n reflections. Every isometry $f \in \mathbf{Is}(E)$ that has no fixed point is the composi-*

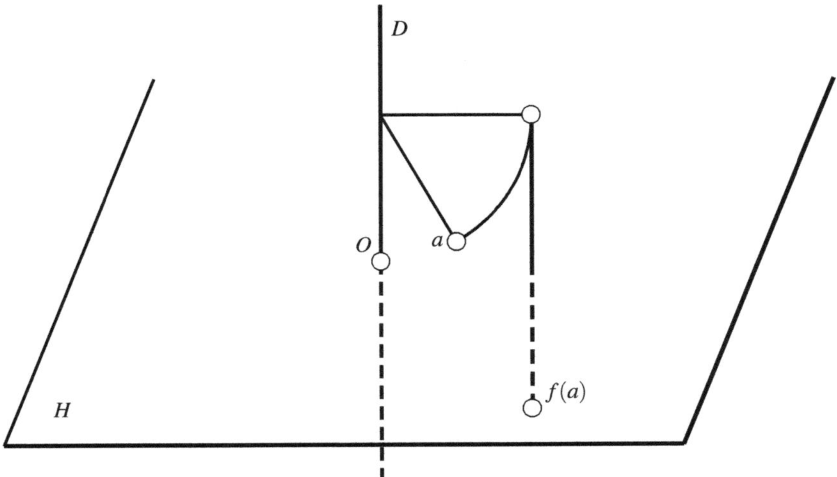

Fig. 8.8 3*D* improper rigid motion with a unique fixed point.

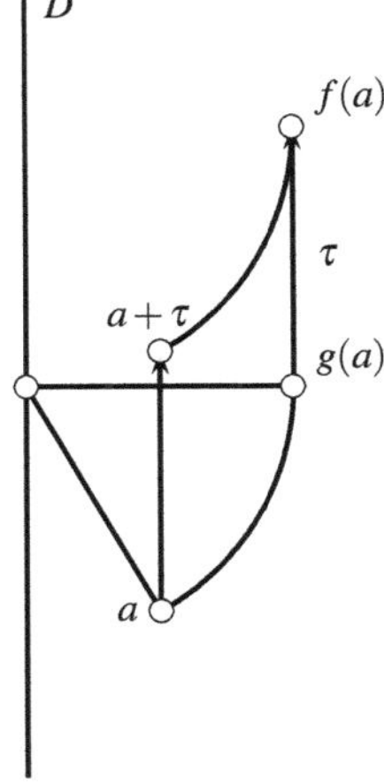

Fig. 8.9 3D proper rigid motion with no fixed point (screw motion).

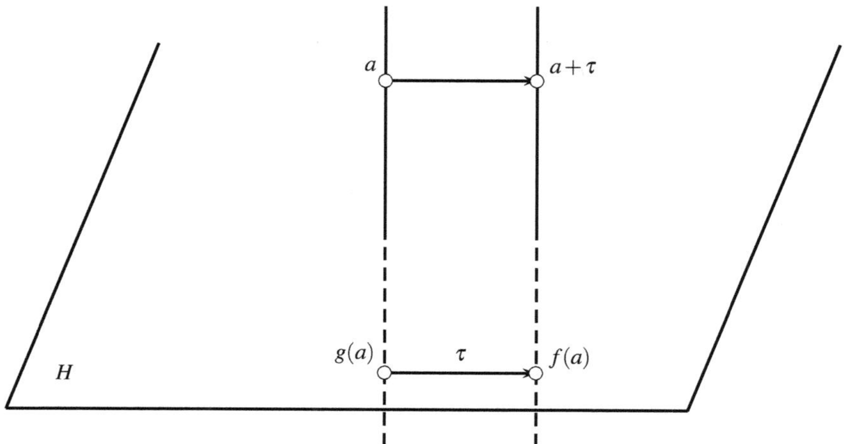

Fig. 8.10 3D improper rigid motion with no fixed points.

tion of at most $n+2$ reflections. When $n \geq 2$, the identity is the composition of any reflection with itself.

Proof. First, we use Theorem 8.3. If f has a fixed point Ω, we choose Ω as an origin and work in the vector space E_Ω. Since f behaves as a linear isometry, the result follows from Theorem 8.1. More specifically, we can write $\overrightarrow{f} = \overrightarrow{s_k} \circ \cdots \circ \overrightarrow{s_1}$ for $k \leq n$ hyperplane reflections $\overrightarrow{s_i}$. We define the affine reflections s_i such that

$$s_i(a) = \Omega + \overrightarrow{s_i}(\overrightarrow{\Omega a})$$

for all $a \in E$, and we note that $f = s_k \circ \cdots \circ s_1$, since

$$f(a) = \Omega + \overrightarrow{s_k} \circ \cdots \circ \overrightarrow{s_1}(\overrightarrow{\Omega a})$$

for all $a \in E$. If f has no fixed point, then $f = t \circ g$ for some isometry g that has a fixed point Ω and some translation $t = t_\tau$, with $\overrightarrow{f}(\tau) = \tau$. By the argument just given, we can write $g = s_k \circ \cdots \circ s_1$ for some affine reflections (at most n). However, by Lemma 8.10, the translation $t = t_\tau$ can be achieved by two reflections about parallel hyperplanes, and thus $f = s_{k+2} \circ \cdots \circ s_1$, for some affine reflections (at most $n+2$). □

When $n \geq 3$, we can also characterize the affine isometries in $\mathbf{SE}(n)$ in terms of flips. Remarkably, not only we can do without translations, but we can even bound the number of flips by n.

Theorem 8.5. *Let E be a Euclidean affine space of dimension $n \geq 3$. Every rigid motion $f \in \mathbf{SE}(E)$ is the composition of an even number of flips $f = f_{2k} \circ \cdots \circ f_1$, where $2k \leq n$.*

Proof. As in the proof of Theorem 8.4, we distinguish between the two cases where f has some fixed point or not. If f has a fixed point Ω, we apply Theorem 8.2. More specifically, we can write $\overrightarrow{f} = \overrightarrow{f_{2k}} \circ \cdots \circ \overrightarrow{f_1}$ for some flips $\overrightarrow{f_i}$. We define the affine flips f_i such that

$$f_i(a) = \Omega + \overrightarrow{f_i}(\overrightarrow{\Omega a})$$

for all $a \in E$, and we note that $f = f_{2k} \circ \cdots \circ f_1$, since

$$f(a) = \Omega + \overrightarrow{f_{2k}} \circ \cdots \circ \overrightarrow{f_1}(\overrightarrow{\Omega a})$$

for all $a \in E$.

If f does not have a fixed point, as in the proof of Theorem 8.4, we get

$$f = t_\tau \circ f_{2k} \circ \cdots \circ f_1,$$

for some affine flips f_i. We need to get rid of the translation. However, $\overrightarrow{f}(\tau) = \tau$, and by the second part of Theorem 8.2, we can assume that $\tau \in \overrightarrow{F_{2k}}^{\perp}$, where $\overrightarrow{F_{2k}}$ is the direction of the affine subspace defining the affine flip f_{2k}. Finally, appealing to Lemma 8.10, since $\tau \in \overrightarrow{F_{2k}}^{\perp}$, the translation t_τ can be expressed as the composition $f'_{2k} \circ f'_{2k-1}$ of two flips f'_{2k-1} and f'_{2k} about the two parallel subspaces $\Omega + \overrightarrow{F_{2k}}$ and $\Omega + \tau/2 + \overrightarrow{F_{2k}}$, whose distance is $\|\tau\|/2$. However, since f'_{2k-1} and f_{2k} are both the identity on $\Omega + \overrightarrow{F_{2k}}$, we must have $f'_{2k-1} = f_{2k}$, and thus

$$\begin{aligned}
f &= t_\tau \circ f_{2k} \circ f_{2k-1} \circ \cdots \circ f_1 \\
&= f'_{2k} \circ f'_{2k-1} \circ f_{2k} \circ f_{2k-1} \circ \cdots \circ f_1 \\
&= f'_{2k} \circ f_{2k-1} \circ \cdots \circ f_1,
\end{aligned}$$

since $f'_{2k-1} = f_{2k}$ and $f'_{2k-1} \circ f_{2k} = f_{2k} \circ f_{2k} = \mathrm{id}$, since f_{2k} is a symmetry. $\quad\square$

Remark: It is easy to prove that if f is a screw motion in $\mathbf{SE}(3)$, D its axis, θ is its angle of rotation, and τ the translation along the direction of D, then f is the composition of two flips about lines D_1 and D_2 orthogonal to D, at a distance $\|\tau\|/2$ and making an angle $\theta/2$.

There is one more topic that we would like to cover, since it is often useful in practice: The concept of *cross product of vectors*, also called vector product. But first, we need to discuss the question of orientation of bases.

8.8 Orientations of a Euclidean Space, Angles

In this section we return to vector spaces. In order to deal with the notion of orientation correctly, it is important to assume that every family $(u_1, \ldots, u_n)$ of vectors is ordered (by the natural ordering on $\{1, 2, \ldots, n\}$). Thus, we will assume that all families $(u_1, \ldots, u_n)$ of vectors, in particular bases and orthonormal bases, are ordered.

Let E be a vector space of finite dimension n over $\mathbb{R}$, and let $(u_1, \ldots, u_n)$ and $(v_1, \ldots, v_n)$ be any two bases for E. Recall that the change of basis matrix from $(u_1, \ldots, u_n)$ to $(v_1, \ldots, v_n)$ is the matrix P whose columns are the coordinates of the vectors v_j over the basis $(u_1, \ldots, u_n)$. It is immediately verified that the set of alternating n-linear forms on E is a vector space, which we denote by $\Lambda(E)$ (see Lang [9]).

We now show that $\Lambda(E)$ has dimension 1. For any alternating n-linear form $\varphi \colon E \times \cdots \times E \to K$ and any two sequences of vectors $(u_1, \ldots, u_n)$ and $(v_1, \ldots, v_n)$, if

$$(v_1, \ldots, v_n) = (u_1, \ldots, u_n)P,$$

then

$$\varphi(v_1, \ldots, v_n) = \det(P)\varphi(u_1, \ldots, u_n).$$

In particular, if we consider nonnull alternating n-linear forms $\varphi \colon E \times \cdots \times E \to K$, we must have $\varphi(u_1, \ldots, u_n) \neq 0$ for every basis $(u_1, \ldots, u_n)$. Since for any two alternating n-linear forms φ and ψ we have

$$\varphi(v_1, \ldots, v_n) = \det(P)\varphi(u_1, \ldots, u_n)$$

and

$$\psi(v_1, \ldots, v_n) = \det(P)\psi(u_1, \ldots, u_n),$$

we get

$$\varphi(u_1, \ldots, u_n)\psi(v_1, \ldots, v_n) - \psi(u_1, \ldots, u_n)\varphi(v_1, \ldots, v_n) = 0.$$

Fixing $(u_1, \ldots, u_n)$ and letting $(v_1, \ldots, v_n)$ vary, this shows that φ and ψ are linearly dependent, and since $\Lambda(E)$ is nontrivial, it has dimension 1.

We now define an equivalence relation on $\Lambda(E) - \{0\}$ (where we let 0 denote the null alternating n-linear form):

φ and ψ are equivalent if $\psi = \lambda \varphi$ for some $\lambda > 0$.

It is immediately verified that the above relation is an equivalence relation. Furthermore, it has exactly two equivalence classes O_1 and O_2.

The first way of defining an *orientation of* E is to pick one of these two equivalence classes, say O $(O \in \{O_1, O_2\})$. Given such a choice of a class O, we say that a basis $(w_1, \ldots, w_n)$ has *positive orientation* iff $\varphi(w_1, \ldots, w_n) > 0$ for any alternating n-linear form $\varphi \in O$. Note that this makes sense, since for any other $\psi \in O$, $\varphi = \lambda \psi$ for some $\lambda > 0$.

According to the previous definition, two bases $(u_1, \ldots, u_n)$ and $(v_1, \ldots, v_n)$ have the same orientation iff $\varphi(u_1, \ldots, u_n)$ and $\varphi(v_1, \ldots, v_n)$ have the same sign for all $\varphi \in \Lambda(E) - \{0\}$. From

$$\varphi(v_1, \ldots, v_n) = \det(P)\varphi(u_1, \ldots, u_n)$$

we must have $\det(P) > 0$. Conversely, if $\det(P) > 0$, the same argument shows that $(u_1, \ldots, u_n)$ and $(v_1, \ldots, v_n)$ have the same orientation. This leads us to an equivalent and slightly less contorted definition of the notion of orientation. We define a relation between bases of E as follows: Two bases $(u_1, \ldots, u_n)$ and $(v_1, \ldots, v_n)$ are related if $\det(P) > 0$, where P is the change of basis matrix from $(u_1, \ldots, u_n)$ to $(v_1, \ldots, v_n)$.

Since $\det(PQ) = \det(P)\det(Q)$, and since change of basis matrices are invertible, the relation just defined is indeed an equivalence relation, and it has two equivalence classes. Furthermore, from the discussion above, any nonnull alternating n-linear form φ will have the same sign on any two equivalent bases.

The above discussion motivates the following definition.

Definition 8.4. Given any vector space E of finite dimension over $\mathbb{R}$, we define an *orientation of E* as the choice of one of the two equivalence classes of the equivalence relation on the set of bases defined such that $(u_1, \ldots, u_n)$ and $(v_1, \ldots, v_n)$ have the same orientation iff $\det(P) > 0$, where P is the change of basis matrix from $(u_1, \ldots, u_n)$ to $(v_1, \ldots, v_n)$. A basis in the chosen class is said to have *positive orientation, or to be positive*. An *orientation of a Euclidean affine space E* is an orientation of its underlying vector space $\overrightarrow{E}$.

In practice, to give an orientation, one simply picks a fixed basis considered as having positive orientation. The orientation of every other basis is determined by the sign of the determinant of the change of basis matrix.

Having the notation of orientation at hand, we wish to go back briefly to the concept of (oriented) angle. Let E be a Euclidean space of dimension $n = 2$, and assume a given orientation. In any given positive orthonormal basis for E, every rotation r is represented by a matrix

$$R = \begin{pmatrix} \cos\theta & -\sin\theta \\ \sin\theta & \cos\theta \end{pmatrix}.$$

Actually, we claim that the matrix R representing the rotation r is the same in *all* orthonormal positive bases. This is because the change of basis matrix from one positive orthonormal basis to another positive orthonormal basis is a rotation represented by some matrix of the form

$$P = \begin{pmatrix} \cos\psi & -\sin\psi \\ \sin\psi & \cos\psi \end{pmatrix}$$

and that we have

$$P^{-1} = \begin{pmatrix} \cos(-\psi) & -\sin(-\psi) \\ \sin(-\psi) & \cos(-\psi) \end{pmatrix},$$

and after calculations, we find that PRP^{-1} is the rotation matrix associated with $\psi + \theta - \psi = \theta$. We can choose $\theta \in [0, 2\pi[$, and we call θ the *measure of the angle of rotation of r (and R)*. If the orientation is changed, the measure changes to $2\pi - \theta$.

We now let E be a Euclidean space of dimension $n = 2$, but we do not assume any orientation. It is easy to see that given any two unit vectors $u_1, u_2 \in E$ (unit means that $\|u_1\| = \|u_2\| = 1$), there is a unique rotation r such that

$$r(u_1) = u_2.$$

It is also possible to define an equivalence relation of pairs of unit vectors such that

$$\langle u_1, u_2 \rangle \equiv \langle u_3, u_4 \rangle$$

iff there is some rotation r such that $r(u_1) = u_3$ and $r(u_2) = u_4$.

Then the equivalence class of $\langle u_1, u_2 \rangle$ can be taken as the definition of the (oriented) *angle of $\langle u_1, u_2 \rangle$*, which is denoted by $\widehat{u_1 u_2}$.

Furthermore, it can be shown that there is a rotation mapping the pair $\langle u_1, u_2 \rangle$ to the pair $\langle u_3, u_4 \rangle$ iff there is a rotation mapping the pair $\langle u_1, u_3 \rangle$ to the pair $\langle u_2, u_4 \rangle$ (all vectors being unit vectors), as illustrated in Figure 8.11.

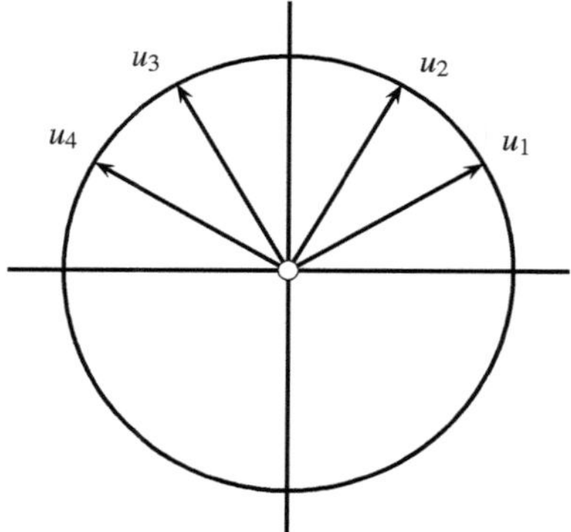

Fig. 8.11 Defining angles.

As a consequence of all this, since for any pair $\langle u_1, u_2 \rangle$ of unit vectors there is a unique rotation r mapping u_1 to u_2, the angle $\widehat{u_1 u_2}$ of $\langle u_1, u_2 \rangle$ corresponds bijectively to the rotation r, and there is a bijection between the set of angles of pairs of unit vectors and the set of rotations in the plane. As a matter of fact, the set of angles forms an abelian group isomorphic to the (abelian) group of rotations in the plane.

Thus, even though we can consider angles as oriented, note that the notion of orientation is not necessary to define angles. However, to define the *measure of an angle*, the notion of orientation is needed.

If we now assume that an orientation of E (still a Euclidean plane) is given, the unique rotation r associated with an angle $\widehat{u_1 u_2}$ corresponds to a unique matrix

$$R = \begin{pmatrix} \cos\theta & -\sin\theta \\ \sin\theta & \cos\theta \end{pmatrix}.$$

The number θ is defined up to $2k\pi$ (with $k \in \mathbb{Z}$) and is called a *measure of the angle* $\widehat{u_1 u_2}$. There is a unique $\theta \in [0, 2\pi[$ that is a measure of the angle $\widehat{u_1 u_2}$. It is also immediately seen that

$$\cos\theta = u_1 \cdot u_2.$$

In fact, since $\cos\theta = \cos(2\pi - \theta) = \cos(-\theta)$, the quantity $\cos\theta$ does not depend on the orientation.

Now, still considering a Euclidean plane, given any pair $\langle u_1, u_2 \rangle$ of nonnull vectors, we define their angle as the angle of the unit vectors $u_1/\|u_1\|$ and $u_2/\|u_2\|$, and if E is oriented, we define the *measure* θ of this angle as the measure of the angle of these unit vectors. Note that

$$\cos\theta = \frac{u_1 \cdot u_2}{\|u_1\|\|u_2\|},$$

and this independently of the orientation.

Finally, if E is a Euclidean space of dimension $n \geq 2$, we define the angle of a pair $\langle u_1, u_2 \rangle$ of nonnull vectors as the angle of this pair in the Euclidean plane spanned by $\langle u_1, u_2 \rangle$ if they are linearly independent, or any Euclidean plane containing u_1 if they are collinear.

If E is a Euclidean affine space of dimension $n \geq 2$, for any two pairs $\langle a_1, b_1 \rangle$ and $\langle a_2, b_2 \rangle$ of points in E, where $a_1 \neq b_1$ and $a_2 \neq b_2$, we define the angle of the pair $\langle \langle a_1, b_1 \rangle, \langle a_2, b_2 \rangle \rangle$ as the angle of the pair $\langle \overrightarrow{a_1 b_1}, \overrightarrow{a_2 b_2} \rangle$.

As for the issue of measure of an angle when $n \geq 3$, all we can do is to define the measure of the angle $\widehat{u_1 u_2}$ as either θ or $2\pi - \theta$, where $\theta \in [0, 2\pi[$. For a detailed treatment, see Berger [2] or Cagnac, Ramis, and Commeau [3]. In particular, when $n = 3$, one should note that it is not enough to give a line D through the origin (the axis of rotation) and an angle θ to specify a rotation! The problem is that depending on the orientation of the plane H (through the origin) orthogonal to D, we get two different rotations: one of angle θ, the other of angle $2\pi - \theta$. Thus, to specify a rotation, we also need to give an orientation of the plane orthogonal to the axis of rotation. This can be done by specifying an orientation of the axis of rotation by some unit vector ω, and chosing the basis (e_1, e_2, ω) (where (e_1, e_2) is a basis of H) such that it has positive orientation w.r.t. the chosen orientation of E.

We now return to alternating multilinear forms on a Euclidean space.

When E is a Euclidean space, we have an interesting situation regarding the value of determinants over orthornormal bases described by the following lemma. Given any basis $B = (u_1, \ldots, u_n)$ for E, for any sequence $(w_1, \ldots, w_n)$ of n vectors, we denote by $\det_B(w_1, \ldots, w_n)$ the determinant of the matrix whose columns are the coordinates of the w_j over the basis $B = (u_1, \ldots, u_n)$.

Lemma 8.11. *Let E be a Euclidean space of dimension n, and assume that an orientation of E has been chosen. For any sequence $(w_1, \ldots, w_n)$ of n vectors and any two orthonormal bases $B_1 = (u_1, \ldots, u_n)$ and $B_2 = (v_1, \ldots, v_n)$ of positive orientation, we have*

$$\det_{B_1}(w_1, \ldots, w_n) = \det_{B_2}(w_1, \ldots, w_n).$$

Proof. Let P be the change of basis matrix from $B_1 = (u_1, \ldots, u_n)$ to $B_2 = (v_1, \ldots, v_n)$. Since $B_1 = (u_1, \ldots, u_n)$ and $B_2 = (v_1, \ldots, v_n)$ are orthonormal, P is orthogonal, and we must have $\det(P) = +1$, since the bases have positive orientation. Let U_1 be the matrix whose columns are the coordinates of the w_j over the basis $B_1 = (u_1, \ldots, u_n)$, and let U_2 be the matrix whose columns are the coordinates of the w_j over the basis $B_2 = (v_1, \ldots, v_n)$. Then, by definition of P, we have

$$(w_1, \ldots, w_n) = (u_1, \ldots, u_n)U_2 P,$$

that is,

$$U_1 = U_2 P.$$

Then, we have

$$\det_{B_1}(w_1, \ldots, w_n) = \det(U_1) = \det(U_2 P) = \det(U_2)\det(P)$$
$$= \det_{B_2}(w_1, \ldots, w_n)\det(P) = \det_{B_2}(w_1, \ldots, w_n),$$

since $\det(P) = +1$. $\square$

By Lemma 8.11, the determinant $\det_B(w_1, \ldots, w_n)$ is independent of the base B, provided that B is orthonormal and of positive orientation. Thus, Lemma 8.11 suggests the following definition.

8.9 Volume Forms, Cross Products

In this section we generalize the familiar notion of cross product of vectors in $\mathbb{R}^3$ to Euclidean spaces of any finite dimension. First, we define the mixed product, or volume form.

Definition 8.5. Given any Euclidean space E of finite dimension n over $\mathbb{R}$ and any orientation of E, for any sequence $(w_1, \ldots, w_n)$ of n vectors in E, the common value $\lambda_E(w_1, \ldots, w_n)$ of the determinant $\det_B(w_1, \ldots, w_n)$ over all positive orthonormal bases B of E is called the *mixed product (or volume form) of* $(w_1, \ldots, w_n)$.

The mixed product $\lambda_E(w_1, \ldots, w_n)$ will also be denoted by $(w_1, \ldots, w_n)$, even though the notation is overloaded. The following properties hold.

- The mixed product $\lambda_E(w_1, \ldots, w_n)$ changes sign when the orientation changes.
- The mixed product $\lambda_E(w_1, \ldots, w_n)$ is a scalar, and Definition 8.5 really defines an alternating multilinear form from e^n to $\mathbb{R}$.

- $\lambda_E(w_1, \ldots, w_n) = 0$ iff $(w_1, \ldots, w_n)$ is linearly dependent.
- A basis $(u_1, \ldots, u_n)$ is positive or negative iff $\lambda_E(u_1, \ldots, u_n)$ is positive or negative.
- $\lambda_E(w_1, \ldots, w_n)$ is invariant under every isometry f such that $\det(f) = 1$.

The terminology "volume form" is justified because $\lambda_E(w_1, \ldots, w_n)$ is indeed the volume of some geometric object. Indeed, viewing E as an affine space, the *parallelotope defined by* $(w_1, \ldots, w_n)$ is the set of points

$$\{\lambda_1 w_1 + \cdots + \lambda_n w_n \mid 0 \leq \lambda_i \leq 1, 1 \leq i \leq n\}.$$

Then, it can be shown (see Berger [2], Section 9.12) that the volume of the parallelotope defined by $(w_1, \ldots, w_n)$ is indeed $\lambda_E(w_1, \ldots, w_n)$. If $(E, \overrightarrow{E})$ is a Euclidean affine space of dimension n, given any $n+1$ affinely independent points $(a_0, \ldots, a_n)$, the set

$$\{a_0 + \lambda_1 \overrightarrow{a_0 a_1} + \cdots + \lambda_n \overrightarrow{a_0 a_n} \mid \text{where } 0 \leq \lambda_i \leq 1, 1 \leq i \leq n\}$$

is called the *parallelotope spanned by* $(a_0, \ldots, a_n)$. Then the volume of the parallelotope spanned by $(a_0, \ldots, a_n)$ is $\lambda_{\overrightarrow{E}}(\overrightarrow{a_0 a_1}, \ldots, \overrightarrow{a_0 a_n})$. It can also be shown that the volume $\mathrm{vol}(a_0, \ldots, a_n)$ of the n-simplex $(a_0, \ldots, a_n)$ is

$$\mathrm{vol}(a_0, \ldots, a_n) = \frac{1}{n!} \lambda_{\overrightarrow{E}}(\overrightarrow{a_0 a_1}, \ldots, \overrightarrow{a_0 a_n}).$$

Now, given a sequence $(w_1, \ldots, w_{n-1})$ of $n-1$ vectors in E, the map

$$x \mapsto \lambda_E(w_1, \ldots, w_{n-1}, x)$$

is a linear form. Thus, by Lemma 6.4, there is a unique vector $u \in E$ such that

$$\lambda_E(w_1, \ldots, w_{n-1}, x) = u \cdot x$$

for all $x \in E$. The vector u has some interesting properties that motivate the next definition.

Definition 8.6. Given any Euclidean space E of finite dimension n over $\mathbb{R}$, for any orientation of E and any sequence $(w_1, \ldots, w_{n-1})$ of $n-1$ vectors in E, the unique vector $w_1 \times \cdots \times w_{n-1}$ such that

$$\lambda_E(w_1, \ldots, w_{n-1}, x) = w_1 \times \cdots \times w_{n-1} \cdot x$$

for all $x \in E$ is the *cross product, or vector product, of* $(w_1, \ldots, w_{n-1})$.

The following properties hold.

- The cross product $w_1 \times \cdots \times w_{n-1}$ changes sign when the orientation changes.
- The cross product $w_1 \times \cdots \times w_{n-1}$ is a vector, and Definition 8.6 really defines an alternating multilinear map from e^{n-1} to E.

- $w_1 \times \cdots \times w_{n-1} = 0$ iff $(w_1, \ldots, w_{n-1})$ is linearly dependent. This is because

$$w_1 \times \cdots \times w_{n-1} = 0$$

  iff

$$\lambda_E(w_1, \ldots, w_{n-1}, x) = 0$$

  for all $x \in E$, and thus if $(w_1, \ldots, w_{n-1})$ were linearly independent, we could find a vector $x \in E$ to complete $(w_1, \ldots, w_{n-1})$ into a basis of E, and we would have

$$\lambda_E(w_1, \ldots, w_{n-1}, x) \neq 0.$$

- The cross product $w_1 \times \cdots \times w_{n-1}$ is orthogonal to each of the w_j.
- If $(w_1, \ldots, w_{n-1})$ is linearly independent, then the sequence

$$(w_1, \ldots, w_{n-1}, w_1 \times \cdots \times w_{n-1})$$

  is a positive basis of E.

We now show how to compute the coordinates of $u_1 \times \cdots \times u_{n-1}$ over an orthonormal basis.

Given an orthonormal basis $(e_1, \ldots, e_n)$, for any sequence $(u_1, \ldots, u_{n-1})$ of $n-1$ vectors in E, if

$$u_j = \sum_{i=1}^{n} u_{i,j} e_i,$$

where $1 \leq j \leq n-1$, for any $x = x_1 e_1 + \cdots + x_n e_n$, consider the determinant

$$\lambda_E(u_1, \ldots, u_{n-1}, x) = \begin{vmatrix} u_{11} & \cdots & u_{1\,n-1} & x_1 \\ u_{21} & \cdots & u_{2\,n-1} & x_2 \\ \vdots & \vdots & \ddots & \vdots \\ u_{n1} & \cdots & u_{n\,n-1} & x_n \end{vmatrix}.$$

Calling the underlying matrix above A, we can expand $\det(A)$ according to the last column, using the Laplace formula (see Strang [12]), where A_{ij} is the $(n-1) \times (n-1)$ matrix obtained from A by deleting row i and column j, and we get

$$\begin{vmatrix} u_{11} & \cdots & u_{1\,n-1} & x_1 \\ u_{21} & \cdots & u_{2\,n-1} & x_2 \\ \vdots & \vdots & \ddots & \vdots \\ u_{n1} & \cdots & u_{n\,n-1} & x_n \end{vmatrix} = (-1)^{n+1} x_1 \det(A_{1n}) + \cdots + x_n \det(A_{nn}).$$

Each $(-1)^{i+n} \det(A_{in})$ is called the *cofactor of x_i*. We note that $\det(A)$ is in fact the inner product

$$\det(A) = ((-1)^{n+1} \det(A_{1n}) e_1 + \cdots + (-1)^{n+n} \det(A_{nn}) e_n) \cdot x.$$

Since the cross product $u_1 \times \cdots \times u_{n-1}$ is the unique vector u such that

$$u \cdot x = \lambda_E(u_1, \ldots, u_{n-1}, x),$$

for all $x \in E$, the coordinates of the cross product $u_1 \times \cdots \times u_{n-1}$ must be

$$((-1)^{n+1} \det(A_{1n}), \ldots, (-1)^{n+n} \det(A_{nn})),$$

the sequence of cofactors of the x_i in the determinant $\det(A)$.

For example, when $n = 3$, the coordinates of the cross product $u \times v$ are given by the cofactors of x_1, x_2, x_3, in the determinant

$$\begin{vmatrix} u_1 & v_1 & x_1 \\ u_2 & v_2 & x_2 \\ u_3 & v_3 & x_3 \end{vmatrix},$$

or, more explicitly, by

$$(-1)^{3+1} \begin{vmatrix} u_2 & v_2 \\ u_3 & v_3 \end{vmatrix}, \qquad (-1)^{3+2} \begin{vmatrix} u_1 & v_1 \\ u_3 & v_3 \end{vmatrix}, \qquad (-1)^{3+3} \begin{vmatrix} u_1 & v_1 \\ u_2 & v_2 \end{vmatrix},$$

that is,

$$(u_2 v_3 - u_3 v_2, \; u_3 v_1 - u_1 v_3, \; u_1 v_2 - u_2 v_1).$$

It is also useful to observe that if we let U be the matrix

$$U = \begin{pmatrix} 0 & -u_3 & u_2 \\ u_3 & 0 & -u_1 \\ -u_2 & u_1 & 0 \end{pmatrix},$$

then the coordinates of the cross product $u \times v$ are given by

$$\begin{pmatrix} 0 & -u_3 & u_2 \\ u_3 & 0 & -u_1 \\ -u_2 & u_1 & 0 \end{pmatrix} \begin{pmatrix} v_1 \\ v_2 \\ v_3 \end{pmatrix} = \begin{pmatrix} u_2 v_3 - u_3 v_2 \\ u_3 v_1 - u_1 v_3 \\ u_1 v_2 - u_2 v_1 \end{pmatrix}.$$

We finish our discussion of cross products by mentioning without proof a few more of their properties, in the case $n = 3$. Firstly, the following so-called *Lagrange identity* holds:

$$(u \cdot v)^2 + \|u \times v\|^2 = \|u\|^2 \|v\|^2.$$

If u and v are linearly independent, and if θ (or $2\pi - \theta$) is a measure of the angle $\widehat{uv}$, then

$$|\sin \theta| = \frac{\|u \times v\|}{\|u\| \|v\|}.$$

It can also be shown that $u \times v$ is the only vector w such that the following properties hold:

(1) $w \cdot u = 0$, and $w \cdot v = 0$.
(2) $\lambda_E(u, v, w) \geq 0$.
(3) $(u \cdot v)^2 + \|w\|^2 = \|u\|^2 \|v\|^2$.

Recall that the mixed product $\lambda_E(w_1, w_1, w_3)$ is also denoted by (w_1, w_2, w_3), and that

$$w_1 \cdot (w_2 \times w_3) = (w_1, w_2, w_3).$$

8.10 Problems

8.1. Prove Lemma 8.7.

8.2. This problem is a warm-up for the next problem. Consider the set of matrices of the form

$$\begin{pmatrix} 0 & -a \\ a & 0 \end{pmatrix},$$

where $a \in \mathbb{R}$.

(a) Show that these matrices are invertible when $a \neq 0$ (give the inverse explicitly). Given any two such matrices A, B, show that $AB = BA$. Describe geometrically the action of such a matrix on points in the affine plane $\mathbb{A}^2$, with its usual Euclidean inner product. Verify that this set of matrices is a vector space isomorphic to $(\mathbb{R}, +)$. This vector space is denoted by $\mathfrak{so}(2)$.

(b) Given an $n \times n$ matrix A, we define the *exponential* e^A as

$$e^A = I_n + \sum_{k \geq 1} \frac{A^k}{k!},$$

where I_n denotes the $n \times n$ identity matrix. It can be shown rigorously that this power series is indeed convergent for every A (over $\mathbb{R}$ or $\mathbb{C}$), so that e^A makes sense (and you do not have to prove it!).

Given any matrix

$$A = \begin{pmatrix} 0 & -\theta \\ \theta & 0 \end{pmatrix},$$

prove that

$$e^A = \cos\theta \begin{pmatrix} 1 & 0 \\ 0 & 1 \end{pmatrix} + \sin\theta \begin{pmatrix} 0 & -1 \\ 1 & 0 \end{pmatrix} = \begin{pmatrix} \cos\theta & -\sin\theta \\ \sin\theta & \cos\theta \end{pmatrix}.$$

Hint. Check that

$$\begin{pmatrix} 0 & -\theta \\ \theta & 0 \end{pmatrix} = \theta \begin{pmatrix} 0 & -1 \\ 1 & 0 \end{pmatrix} \quad \text{and} \quad \begin{pmatrix} 0 & -\theta \\ \theta & 0 \end{pmatrix}^2 = -\theta^2 \begin{pmatrix} 1 & 0 \\ 0 & 1 \end{pmatrix},$$

and use the power series for $\cos\theta$ and $\sin\theta$. Conclude that the exponential map provides a surjective map $\exp: \mathfrak{so}(2) \to \mathbf{SO}(2)$ from $\mathfrak{so}(2)$ onto the group $\mathbf{SO}(2)$ of plane rotations. Is this map injective? How do you need to restrict θ to get an injective map?

Remark: By the way, $\mathfrak{so}(2)$ is the *Lie algebra* of the (Lie) group $\mathbf{SO}(2)$.

(c) Consider the set $\mathbf{U}(1)$ of complex numbers of the form $\cos\theta + i\sin\theta$. Check that this is a group under multiplication. Assuming that we use the standard affine frame for the affine plane $\mathbb{A}^2$, every point (x,y) corresponds to the complex number $z = x + iy$, and this correspondence is a bijection. Then, every $\alpha = \cos\theta + i\sin\theta \in \mathbf{U}(1)$ induces the map $R_\alpha \colon \mathbb{A}^2 \to \mathbb{A}^2$ defined such that

$$R_\alpha(z) = \alpha z.$$

Prove that R_α is the rotation of matrix

$$\begin{pmatrix} \cos\theta & -\sin\theta \\ \sin\theta & \cos\theta \end{pmatrix}.$$

Prove that the map $R \colon \mathbf{U}(1) \to \mathbf{SO}(2)$ defined such that $R(\alpha) = R_\alpha$ is an isomorphism. Deduce that topologically, $\mathbf{SO}(2)$ is a circle. Using the exponential map from $\mathbb{R}$ to $\mathbf{U}(1)$ defined such that $\theta \mapsto e^{i\theta} = \cos\theta + i\sin\theta$, prove that there is a surjective homomorphism from $(R,+)$ to $\mathbf{SO}(2)$. What is the connection with the exponential map from $\mathfrak{so}(2)$ to $\mathbf{SO}(2)$?

8.3. (a) Recall that the coordinates of the cross product $u \times v$ of two vectors $u = (u_1, u_2, u_3)$ and $v = (v_1, v_2, v_3)$ in $\mathbb{R}^3$ are

$$(u_2 v_3 - u_3 v_2, \; u_3 v_1 - u_1 v_3, \; u_1 v_2 - u_2 v_1).$$

Letting U be the matrix

$$U = \begin{pmatrix} 0 & -u_3 & u_2 \\ u_3 & 0 & -u_1 \\ -u_2 & u_1 & 0 \end{pmatrix},$$

check that the coordinates of the cross product $u \times v$ are given by

$$\begin{pmatrix} 0 & -u_3 & u_2 \\ u_3 & 0 & -u_1 \\ -u_2 & u_1 & 0 \end{pmatrix} \begin{pmatrix} v_1 \\ v_2 \\ v_3 \end{pmatrix} = \begin{pmatrix} u_2 v_3 - u_3 v_2 \\ u_3 v_1 - u_1 v_3 \\ u_1 v_2 - u_2 v_1 \end{pmatrix}.$$

(b) Show that the set of matrices of the form

$$U = \begin{pmatrix} 0 & -u_3 & u_2 \\ u_3 & 0 & -u_1 \\ -u_2 & u_1 & 0 \end{pmatrix}$$

is a vector space isomorphic to $(\mathbb{R}^3 +)$. This vector space is denoted by $\mathfrak{so}(3)$. Show that such matrices are never invertible. Find the kernel of the linear map associated with a matrix U. Describe geometrically the action of the linear map defined by a matrix U. Show that when restricted to the plane orthogonal to $u = (u_1, u_2, u_3)$ through the origin, it is a rotation by $\pi/2$.

(c) Consider the map $\psi \colon (\mathbb{R}^3, \times) \to \mathfrak{so}(3)$ defined by the formula

$$\psi(u_1, u_2, u_3) = \begin{pmatrix} 0 & -u_3 & u_2 \\ u_3 & 0 & -u_1 \\ -u_2 & u_1 & 0 \end{pmatrix}.$$

For any two matrices $A, B \in \mathfrak{so}(3)$, defining $[A, B]$ as

$$[A, B] = AB - BA,$$

verify that

$$\psi(u \times v) = [\psi(u), \psi(v)].$$

Show that $[-, -]$ is not associative. Show that $[A, A] = 0$, and that the so-called *Jacobi identity* holds:

$$[A, [B, C]] + [C, [A, B]] + [B, [C, A]] = 0.$$

Show that $[AB]$ is bilinear (linear in both A and B).

Remark: $[A, B]$ is called a *Lie bracket*, and under this operation, the vector space $\mathfrak{so}(3)$ is called a *Lie algebra*. In fact, it is the Lie algebra of the (Lie) group $\mathbf{SO}(3)$.

(d) For any matrix

$$A = \begin{pmatrix} 0 & -c & b \\ c & 0 & -a \\ -b & a & 0 \end{pmatrix},$$

letting $\theta = \sqrt{a^2 + b^2 + c^2}$ and

$$B = \begin{pmatrix} a^2 & ab & ac \\ ab & b^2 & bc \\ ac & bc & c^2 \end{pmatrix},$$

prove that

$$A^2 = -\theta^2 I + B,$$
$$AB = BA = 0.$$

From the above, deduce that

$$A^3 = -\theta^2 A,$$

and for any $k \geq 0$,

$$A^{4k+1} = \theta^{4k} A,$$
$$A^{4k+2} = \theta^{4k} A^2,$$
$$A^{4k+3} = -\theta^{4k+2} A,$$
$$A^{4k+4} = -\theta^{4k+2} A^2.$$

Then prove that the exponential map $\exp\colon \mathfrak{so}(3) \to \mathbf{SO}(3)$ is given by

$$\exp A = e^A = \cos\theta\, I_3 + \frac{\sin\theta}{\theta} A + \frac{(1-\cos\theta)}{\theta^2} B,$$

or, equivalently, by

$$e^A = I_3 + \frac{\sin\theta}{\theta} A + \frac{(1-\cos\theta)}{\theta^2} A^2,$$

if $\theta \neq k2\pi$ ($k \in \mathbb{Z}$), with $\exp(0_3) = I_3$.

Remark: This formula is known as Rodrigues's formula (1840).

(e) Prove that $\exp A$ is a rotation of axis (a,b,c) and of angle $\theta = \sqrt{a^2 + b^2 + c^2}$. *Hint.* Check that e^A is an orthogonal matrix of determinant $+1$, etc., or look up any textbook on kinematics or classical dynamics!

(f) Prove that the exponential map $\exp\colon \mathfrak{so}(3) \to \mathbf{SO}(3)$ is surjective. Prove that if R is a rotation matrix different from I_3, letting $\omega = (a,b,c)$ be a unit vector defining the axis of rotation, if $\mathrm{tr}(R) = -1$, then

$$(\exp(R))^{-1} = \left\{ \pm\pi \begin{pmatrix} 0 & -c & b \\ c & 0 & -a \\ -b & a & 0 \end{pmatrix} \right\},$$

and if $\mathrm{tr}(R) \neq -1$, then

$$(\exp(R))^{-1} = \left\{ \frac{\theta}{2\sin\theta}(R - R^T) \;\middle|\; 1 + 2\cos\theta = \mathrm{tr}(R) \right\}.$$

(Recall that $\mathrm{tr}(R) = r_{11} + r_{22} + r_{33}$, the *trace* of the matrix R). Note that both θ and $2\pi - \theta$ yield the same matrix $\exp(R)$.

8.4. Prove that for any plane isometry f such that $\overrightarrow{f}$ is a reflection, f is the composition of a reflection about a line with a translation (possibly null) parallel to this line.

8.5. (1) Given a unit vector $(-\sin\theta, \cos\theta)$, prove that the Householder matrix determined by the vector $(-\sin\theta, \cos\theta)$ is

$$\begin{pmatrix} \cos 2\theta & \sin 2\theta \\ \sin 2\theta & -\cos 2\theta \end{pmatrix}.$$

Give a geometric interpretation (i.e., why the choice $(-\sin\theta, \cos\theta)$?).

(2) Given any matrix

$$A = \begin{pmatrix} a & b \\ c & d \end{pmatrix},$$

prove that there is a Householder matrix H such that AH is lower triangular, i.e.,

$$AH = \begin{pmatrix} a' & 0 \\ c' & d' \end{pmatrix}$$

for some $a', c', d' \in \mathbb{R}$.

8.6. Given a Euclidean space E of dimension n, if h is a reflection about some hyperplane orthogonal to a nonnull vector u and f is any isometry, prove that $f \circ h \circ f^{-1}$ is the reflection about the hyperplane orthogonal to $f(u)$.

8.7. Let E be a Euclidean space of dimension $n = 2$. Prove that given any two unit vectors $u_1, u_2 \in E$ (unit means that $\|u_1\| = \|u_2\| = 1$), there is a unique rotation r such that

$$r(u_1) = u_2.$$

Prove that there is a rotation mapping the pair $\langle u_1, u_2 \rangle$ to the pair $\langle u_3, u_4 \rangle$ iff there is a rotation mapping the pair $\langle u_1, u_3 \rangle$ to the pair $\langle u_2, u_4 \rangle$ (all vectors being unit vectors).

8.8. (1) Recall that

$$\det(v_1, \ldots, v_n) = \begin{vmatrix} v_{11} & v_{12} & \cdots & v_{1n} \\ v_{21} & v_{22} & \cdots & v_{2n} \\ \vdots & \vdots & \ddots & \vdots \\ v_{n1} & v_{n2} & \cdots & v_{nn} \end{vmatrix},$$

where v_i has coordinates $(v_{i1}, \ldots, v_{in})$ with respect to a basis $(e_1, \ldots, e_n)$. Prove that the volume of the parallelotope spanned by $(a_0, \ldots, a_n)$ is given by

$$\lambda_E(a_0, \ldots, a_n) = (-1)^n \begin{vmatrix} a_{01} & a_{02} & \cdots & a_{0n} & 1 \\ a_{11} & a_{12} & \cdots & a_{1n} & 1 \\ \vdots & \vdots & \ddots & \vdots & \vdots \\ a_{n1} & a_{n2} & \cdots & a_{nn} & 1 \end{vmatrix},$$

and letting $\lambda_E(a_0, \ldots, a_n) = \lambda_{\overrightarrow{E}}(\overrightarrow{a_0a_1}, \ldots, \overrightarrow{a_0a_n})$, that

$$\lambda_E(a_0, \ldots, a_n) = \begin{vmatrix} a_{11} - a_{01} & a_{12} - a_{02} & \cdots & a_{1n} - a_{0n} \\ a_{21} - a_{01} & a_{22} - a_{02} & \cdots & a_{2n} - a_{0n} \\ \vdots & \vdots & \ddots & \vdots \\ a_{n1} - a_{01} & a_{n2} - a_{02} & \cdots & a_{nn} - a_{0n} \end{vmatrix},$$

where a_i has coordinates $(a_{i1}, \ldots, a_{in})$ with respect to the affine frame $(O, (e_1, \ldots, e_n))$.

(2) Prove that the volume $\mathrm{vol}(a_0,\ldots,a_n)$ of the n-simplex $(a_0,\ldots,a_n)$ is

$$\mathrm{vol}(a_0,\ldots,a_n) = \frac{1}{n!}\lambda_{\overrightarrow{E}}(\overrightarrow{a_0a_1},\ldots,\overrightarrow{a_0a_n}).$$

8.9. Prove that the so-called *Lagrange identity* holds:

$$(u\cdot v)^2 + \|u\times v\|^2 = \|u\|^2\|v\|^2.$$

8.10. Given p vectors $(u_1,\ldots,u_p)$ in a Euclidean space E of dimension $n \geq p$, the *Gram determinant (or Gramian)* of the vectors $(u_1,\ldots,u_p)$ is the determinant

$$\mathrm{Gram}(u_1,\ldots,u_p) = \begin{vmatrix} \|u_1\|^2 & \langle u_1,u_2\rangle & \cdots & \langle u_1,u_p\rangle \\ \langle u_2,u_1\rangle & \|u_2\|^2 & \cdots & \langle u_2,u_p\rangle \\ \vdots & \vdots & \ddots & \vdots \\ \langle u_p,u_1\rangle & \langle u_p,u_2\rangle & \cdots & \|u_p\|^2 \end{vmatrix}.$$

(1) Prove that

$$\mathrm{Gram}(u_1,\ldots,u_n) = \lambda_E(u_1,\ldots,u_n)^2.$$

Hint. By a previous problem, if $(e_1,\ldots,e_n)$ is an orthonormal basis of E and A is the matrix of the vectors $(u_1,\ldots,u_n)$ over this basis,

$$\det(A)^2 = \det(A^\top A) = \det(A_i\cdot A_j),$$

where A_i denotes the ith column of the matrix A, and $(A_i\cdot A_j)$ denotes the $n\times n$ matrix with entries $A_i\cdot A_j$.

(2) Prove that

$$\|u_1\times\cdots\times u_{n-1}\|^2 = \mathrm{Gram}(u_1,\ldots,u_{n-1}).$$

Hint. Letting $w = u_1\times\cdots\times u_{n-1}$, observe that

$$\lambda_E(u_1,\ldots,u_{n-1},w) = \langle w,w\rangle = \|w\|^2,$$

and show that

$$\|w\|^4 = \lambda_E(u_1,\ldots,u_{n-1},w)^2 = \mathrm{Gram}(u_1,\ldots,u_{n-1},w)$$
$$= \mathrm{Gram}(u_1,\ldots,u_{n-1})\|w\|^2.$$

8.11. Given a Euclidean space E, let U be a nonempty affine subspace of E, and let a be any point in E. We define the *distance* $d(a,U)$ of a to U as

$$d(a,U) = \inf\{\|\overrightarrow{ab}\| \mid b \in U\}.$$

(a) Prove that the affine subspace $U_a^\perp$ defined such that

$$U_a^\perp = a + \overrightarrow{U}^\perp$$

intersects U in a single point b such that $d(a,U) = \|\overrightarrow{ab}\|$.

Hint. Recall the discussion after Lemma 2.15.

(b) Let $(a_0,\ldots,a_p)$ be a frame for U (not necessarily orthonormal). Prove that

$$d(a,U)^2 = \frac{\mathrm{Gram}(\overrightarrow{a_0a},\overrightarrow{a_0a_1},\ldots,\overrightarrow{a_0a_p})}{\mathrm{Gram}(\overrightarrow{a_0a_1},\ldots,\overrightarrow{a_0a_p})}.$$

Hint. Gram is unchanged when a linear combination of other vectors is added to one of the vectors, and thus

$$\mathrm{Gram}(\overrightarrow{a_0a},\overrightarrow{a_0a_1},\ldots,\overrightarrow{a_0a_p}) = \mathrm{Gram}(\overrightarrow{ba},\overrightarrow{a_0a_1},\ldots,\overrightarrow{a_0a_p}),$$

where b is the unique point defined in question (a).

(c) If D and D' are two lines in E that are not coplanar, $a,b \in D$ are distinct points on D, and $a',b' \in D'$ are distinct points on D', prove that if $d(D,D')$ is the shortest distance between D and D' (why does it exist?), then

$$d(D,D')^2 = \frac{\mathrm{Gram}(\overrightarrow{aa'},\overrightarrow{ab},\overrightarrow{a'b'})}{\mathrm{Gram}(\overrightarrow{ab},\overrightarrow{a'b'})}.$$

8.12. Given a hyperplane H in $\mathbb{E}^n$ of equation

$$u_1x_1 + \cdots + u_nx_n - v = 0,$$

for any point $a = (a_1,\ldots,a_n)$, prove that the distance $d(a,H)$ of a to H (see problem 8.11) is given by

$$d(a,H) = \frac{|u_1a_1 + \cdots + u_na_n - v|}{\sqrt{u_1^2 + \cdots + u_n^2}}.$$

8.13. Given a Euclidean space E, let U and V be two nonempty affine subspaces such that $U \cap V = \emptyset$. We define the *distance $d(U,V)$ of U and V* as

$$d(U,V) = \inf\{\|\overrightarrow{ab}\| \mid a \in U, b \in V\}.$$

(a) Prove that $\dim(\overrightarrow{U} + \overrightarrow{V}) \leq \dim(\overrightarrow{E}) - 1$, and that $\overrightarrow{U}^\perp \cap \overrightarrow{V}^\perp = (\overrightarrow{U} + \overrightarrow{V})^\perp \neq \{0\}$.

Hint. Recall the discussion after Lemma 2.15 in Chapter 2.

(b) Let $\overrightarrow{W} = \overrightarrow{U}^\perp \cap \overrightarrow{V}^\perp = (\overrightarrow{U} + \overrightarrow{V})^\perp$. Prove that $U' = U + \overrightarrow{W}$ is an affine subspace with direction $\overrightarrow{U} \oplus \overrightarrow{W}$, $V' = V + \overrightarrow{W}$ is an affine subspace with direction $\overrightarrow{V} \oplus \overrightarrow{W}$, and that $W' = U' \cap V'$ is a nonempty affine subspace with direction $(\overrightarrow{U} \cap$

$\overrightarrow{V}) \oplus \overrightarrow{W}$ such that $U \cap W' \neq \emptyset$ and $V \cap W' \neq \emptyset$. Prove that $U \cap W'$ and $V \cap W'$ are parallel affine subspaces such that

$$\overrightarrow{U \cap W'} = \overrightarrow{V \cap W'} = \overrightarrow{U} \cap \overrightarrow{V}.$$

Prove that if $a, c \in U$, $b, d \in V$, and $\overrightarrow{ab}, \overrightarrow{cd} \in (\overrightarrow{U} + \overrightarrow{V})^{\perp}$, then $\overrightarrow{ab} = \overrightarrow{cd}$ and $\overrightarrow{ac} = \overrightarrow{bd}$.

Prove that if $c \in W'$, then $c + (\overrightarrow{U} + \overrightarrow{V})^{\perp}$ intersects $U \cap W'$ and $V \cap W'$ in unique points $a \in U \cap W'$ and $b \in V \cap W'$ such that $\overrightarrow{ab} \in (\overrightarrow{U} + \overrightarrow{V})^{\perp}$.

Prove that for all $a \in U \cap W'$ and all $b \in V \cap W'$,

$$d(U, V) = \|\overrightarrow{ab}\| \quad \text{iff} \quad \overrightarrow{ab} \in (\overrightarrow{U} + \overrightarrow{V})^{\perp}.$$

Prove that $a \in U$ and $b \in V$ as above are unique iff $\overrightarrow{U} \cap \overrightarrow{V} = \{0\}$.

(c) If $m = \dim(\overrightarrow{U} + \overrightarrow{V})$, $(e_1, \ldots, e_m)$ is any basis of $\overrightarrow{U} + \overrightarrow{V}$, and $a_0 \in U$ and $b_0 \in V$ are any two points, prove that

$$d(U, V)^2 = \frac{\mathrm{Gram}(\overrightarrow{a_0 b_0}, e_1, \ldots, e_m)}{\mathrm{Gram}(e_1, \ldots, e_m)}.$$

8.14. Let E be a real vector space of dimension n, and let $\varphi \colon E \times E \to \mathbb{R}$ be a symmetric bilinear form. Recall that φ is *nondegenerate* if for every $u \in E$,

$$\text{if} \quad \varphi(u, v) = 0 \quad \text{for all } v \in E, \quad \text{then} \quad u = 0.$$

A linear map $f \colon E \to E$ is an *isometry w.r.t.* φ if

$$\varphi(f(x), f(y)) = \varphi(x, y)$$

for all $x, y \in E$. The purpose of this problem is to prove that the Cartan–Dieudonné theorem still holds when φ is nondegenerate. The difficulty is that there may be *isotropic vectors*, i.e., nonnull vectors u such that $\varphi(u, u) = 0$. A vector u is called *nonisotropic* if $\varphi(u, u) \neq 0$. Of course, a nonisotropic vector is nonnull.

(a) Assume that φ is nonnull and that f is an isometry w.r.t. φ. Prove that $f(u) - u$ and $f(u) + u$ are conjugate w.r.t. φ, i.e.,

$$\varphi(f(u) - u, f(u) + u) = 0.$$

Prove that there is some nonisotropic vector $u \in E$ such that either $f(u) - u$ or $f(u) + u$ is nonisotropic.

(b) Let φ be nondegenerate. Prove the following version of the Cartan–Dieudonné theorem:

Every isometry $f \in \mathbf{O}(\varphi)$ that is not the identity is the composition of at most $2n - 1$ reflections w.r.t. hyperplanes. When $n \geq 2$, the identity is the composition of any reflection with itself.

Proceed by induction. In the induction step, consider the following three cases:

(1) f admits 1 as an eigenvalue.
(2) f admits -1 as an eigenvalue.
(3) $f(u) \neq u$ and $f(u) \neq -u$ for every nonnull vector $u \in E$.

Argue that there is some nonisotropic vector u such that either $f(u) - u$ or $f(u) + u$ is nonisotropic, and use a suitable reflection s about the hyperplane orthogonal to $f(u) - u$ or $f(u) + u$, such that $s \circ f$ admits 1 or -1 as an eigenvalue.

(c) What goes wrong with the argument in (b) if φ is nonnull but possibly degenerate? Is $\mathbf{O}(\varphi)$ still a group?

Remark: A stronger version of the Cartan–Dieudonné theorem holds: in fact, at most n reflections are needed, but the proof is much harder (for instance, see Dieudonné [6]).

References

1. Emil Artin. *Geometric Algebra*. Wiley Interscience, first edition, 1957.
2. Marcel Berger. *Géométrie 1*. Nathan, 1990. English edition: Geometry 1, Universitext, Springer-Verlag.
3. G. Cagnac, E. Ramis, and J. Commeau. *Mathématiques Spéciales, Vol. 3, Géométrie*. Masson, 1965.
4. Élie Cartan. *Theory of Spinors*. Dover, first edition, 1966.
5. P.G. Ciarlet. *Introduction to Numerical Matrix Analysis and Optimization*. Cambridge University Press, first edition, 1989. French edition: Masson, 1994.
6. Jean Dieudonné. *Sur les Groupes Classiques*. Hermann, third edition, 1967.
7. Gene H. Golub and Charles F. Van Loan. *Matrix Computations*. The Johns Hopkins University Press, third edition, 1996.
8. D. Kincaid and W. Cheney. *Numerical Analysis*. Brooks/Cole Publishing, second edition, 1996.
9. Serge Lang. *Algebra*. Addison-Wesley, third edition, 1993.
10. Ernst Snapper and Troyer Robert J. *Metric Affine Geometry*. Dover, first edition, 1989.
11. Gilbert Strang. *Introduction to Applied Mathematics*. Wellesley–Cambridge Press, first edition, 1986.
12. Gilbert Strang. *Linear Algebra and Its Applications*. Saunders HBJ, third edition, 1988.
13. Claude Tisseron. *Géométries Affines, Projectives, et Euclidiennes*. Hermann, first edition, 1994.
14. L.N. Trefethen and D. Bau III. *Numerical Linear Algebra*. SIAM Publications, first edition, 1997.

Chapter 9
The Quaternions and the Spaces S^3, $\mathbf{SU}(2)$, $\mathbf{SO}(3)$, and $\mathbb{RP}^3$

9.1 The Algebra $\mathbb{H}$ of Quaternions

In this chapter, we discuss the representation of rotations of $\mathbb{R}^3$ in terms of quaternions. Such a representation is not only concise and elegant, it also yields a very efficient way of handling composition of rotations. It also tends to be numerically more stable than the representation in terms of orthogonal matrices.

The group of rotations $\mathbf{SO}(2)$ is isomorphic to the group $\mathbf{U}(1)$ of complex numbers $\mathrm{e}^{\mathrm{i}\theta} = \cos\theta + \mathrm{i}\sin\theta$ of unit length. This follows immediately from the fact that the map

$$\mathrm{e}^{\mathrm{i}\theta} \mapsto \begin{pmatrix} \cos\theta & -\sin\theta \\ \sin\theta & \cos\theta \end{pmatrix}$$

is a group isomorphism. Geometrically, observe that $\mathbf{U}(1)$ is the unit circle S^1. We can identify the plane $\mathbb{R}^2$ with the complex plane $\mathbb{C}$, letting $z = x + \mathrm{i}y \in \mathbb{C}$ represent $(x, y) \in \mathbb{R}^2$. Then every plane rotation ρ_θ by an angle θ is represented by multiplication by the complex number $\mathrm{e}^{\mathrm{i}\theta} \in \mathbf{U}(1)$, in the sense that for all $z, z' \in \mathbb{C}$,

$$z' = \rho_\theta(z) \quad \text{iff} \quad z' = \mathrm{e}^{\mathrm{i}\theta} z.$$

In some sense, the quaternions generalize the complex numbers in such a way that rotations of $\mathbb{R}^3$ are represented by multiplication by quaternions of unit length. This is basically true with some twists. For instance, quaternion multiplication is not commutative, and a rotation in $\mathbf{SO}(3)$ requires conjugation with a quaternion for its representation. Instead of the unit circle S^1, we need to consider the sphere S^3 in $\mathbb{R}^4$, and $\mathbf{U}(1)$ is replaced by $\mathbf{SU}(2)$.

Recall that the 3-sphere S^3 is the set of points $(x, y, z, t) \in \mathbb{R}^4$ such that

$$x^2 + y^2 + z^2 + t^2 = 1,$$

and that the real projective space $\mathbb{RP}^3$ is the quotient of S^3 modulo the equivalence relation that identifies antipodal points (where (x, y, z, t) and $(-x, -y, -z, -t)$ are antipodal points). The group $\mathbf{SO}(3)$ of rotations of $\mathbb{R}^3$ is intimately related to the 3-sphere S^3 and to the real projective space $\mathbb{RP}^3$. The key to this relationship is the fact that rotations can be represented by quaternions, discovered by Hamilton in 1843. Historically, the quaternions were the first instance of a skew field. As we shall see, quaternions represent rotations in $\mathbb{R}^3$ very concisely.

It will be convenient to define the quaternions as certain 2×2 complex matrices. We write a complex number z as $z = a + ib$, where $a, b \in \mathbb{R}$, and the *conjugate* $\bar{z}$ of z is $\bar{z} = a - ib$. Let $\mathbf{1}, \mathbf{i}, \mathbf{j}$, and $\mathbf{k}$ be the following matrices:

$$\mathbf{1} = \begin{pmatrix} 1 & 0 \\ 0 & 1 \end{pmatrix}, \qquad\qquad \mathbf{i} = \begin{pmatrix} i & 0 \\ 0 & -i \end{pmatrix},$$

$$\mathbf{j} = \begin{pmatrix} 0 & 1 \\ -1 & 0 \end{pmatrix}, \qquad\qquad \mathbf{k} = \begin{pmatrix} 0 & i \\ i & 0 \end{pmatrix}.$$

Definition 9.1. Let $\mathbb{H}$ be the set of all matrices of the form

$$a\mathbf{1} + b\mathbf{i} + c\mathbf{j} + d\mathbf{k},$$

where $(a, b, c, d) \in \mathbb{R}^4$. Thus, every matrix in $\mathbb{H}$ is of the form

$$A = \begin{pmatrix} x & y \\ -\bar{y} & \bar{x} \end{pmatrix},$$

where $x = a + ib$ and $y = c + id$. The matrices in $\mathbb{H}$ are called *quaternions*. The null quaternion is denoted by 0 (or $\mathbf{0}$, if confusion may arise). Quaternions of the form $b\mathbf{i} + c\mathbf{j} + d\mathbf{k}$ are called *pure quaternions*. The set of pure quaternions is denoted by $\mathbb{H}_p$.

Note that the rows (and columns) of matrices in $\mathbb{H}$ are vectors in $\mathbb{C}^2$ that are orthogonal with respect to the Hermitian inner product of $\mathbb{C}^2$ given by

$$(x_1, y_1) \cdot (x_2, y_2) = x_1 \bar{x_2} + y_1 \bar{y_2}.$$

Furthermore, their norm is

$$\sqrt{x\bar{x} + y\bar{y}} = \sqrt{a^2 + b^2 + c^2 + d^2},$$

and the determinant of A is $a^2 + b^2 + c^2 + d^2$.

It is easily seen that the following famous identities (discovered by Hamilton) hold:

$$\mathbf{i}^2 = \mathbf{j}^2 = \mathbf{k}^2 = \mathbf{ijk} = -\mathbf{1},$$
$$\mathbf{ij} = -\mathbf{ji} = \mathbf{k},$$
$$\mathbf{jk} = -\mathbf{kj} = \mathbf{i},$$
$$\mathbf{ki} = -\mathbf{ik} = \mathbf{j}.$$

Using these identities, it can be verified that $\mathbb{H}$ is a ring (with multiplicative identity **1**) and a real vector space of dimension 4 with basis $(\mathbf{1}, \mathbf{i}, \mathbf{j}, \mathbf{k})$. In fact, the quaternions form an associative algebra. For details, see Berger [3], Veblen and Young [22], Dieudonné [5], Bertin [4].

The quaternions $\mathbb{H}$ are often defined as the real algebra generated by the four elements **1**, **i**, **j**, **k**, and satisfying the identities just stated above. The problem with such a definition is that it is not obvious that the algebraic structure $\mathbb{H}$ actually exists. A rigorous justification requires the notions of freely generated algebra and of quotient of an algebra by an ideal. Our definition in terms of matrices makes the existence of $\mathbb{H}$ trivial (but requires showing that the identities hold, which is an easy matter).

Given any two quaternions $X = a\mathbf{1} + b\mathbf{i} + c\mathbf{j} + d\mathbf{k}$ and $Y = a'\mathbf{1} + b'\mathbf{i} + c'\mathbf{j} + d'\mathbf{k}$, it can be verified that

$$XY = (aa' - bb' - cc' - dd')\mathbf{1} + (ab' + ba' + cd' - dc')\mathbf{i}$$
$$+ (ac' + ca' + db' - bd')\mathbf{j} + (ad' + da' + bc' - cb')\mathbf{k}.$$

It is worth noting that these formulae were discovered independently by Olinde Rodrigues in 1840, a few years before Hamilton (Veblen and Young [22]). However, Rodrigues was working with a different formalism, homogeneous transformations, and he did not discover the quaternions. The map from $\mathbb{R}$ to $\mathbb{H}$ defined such that $a \mapsto a\mathbf{1}$ is an injection that allows us to view $\mathbb{R}$ as a subring $\mathbb{R}\mathbf{1}$ (in fact, a field) of $\mathbb{H}$. Similarly, the map from $\mathbb{R}^3$ to $\mathbb{H}$ defined such that $(b, c, d) \mapsto b\mathbf{i} + c\mathbf{j} + d\mathbf{k}$ is an injection that allows us to view $\mathbb{R}^3$ as a subspace of $\mathbb{H}$, in fact, the hyperplane $\mathbb{H}_p$.

Given a quaternion $X = a\mathbf{1} + b\mathbf{i} + c\mathbf{j} + d\mathbf{k}$, we define its *conjugate* $\overline{X}$ as

$$\overline{X} = a\mathbf{1} - b\mathbf{i} - c\mathbf{j} - d\mathbf{k}.$$

It is easily verified that

$$X\overline{X} = (a^2 + b^2 + c^2 + d^2)\mathbf{1}.$$

The quantity $a^2 + b^2 + c^2 + d^2$, also denoted by $N(X)$, is called the *reduced norm* of X.

Clearly, X is nonnull iff $N(X) \neq 0$, in which case $\overline{X}/N(X)$ is the multiplicative inverse of X. Thus, $\mathbb{H}$ is a skew field. Since $X + \overline{X} = 2a\mathbf{1}$, we also call $2a$ the *reduced trace* of X, and we denote it by $\mathrm{Tr}(X)$. A quaternion X is a pure quaternion iff $\overline{X} = -X$ iff $\mathrm{Tr}(X) = 0$.

The following identities can be shown (see Berger [3], Dieudonné [5], Bertin [4]):

$$\overline{XY} = \overline{Y}\,\overline{X},$$

$$\mathrm{Tr}(XY) = \mathrm{Tr}(YX),$$

$$N(XY) = N(X)N(Y),$$

$$\mathrm{Tr}(ZXZ^{-1}) = \mathrm{Tr}(X),$$

whenever $Z \neq 0$.

If $X = b\mathbf{i} + c\mathbf{j} + d\mathbf{k}$ and $Y = b'\mathbf{i} + c'\mathbf{j} + d'\mathbf{k}$ are pure quaternions, identifying X and Y with the corresponding vectors in $\mathbb{R}^3$, the inner product $X \cdot Y$ and the cross product $X \times Y$ make sense, and letting $[0, X \times Y]$ denote the quaternion whose first component is 0 and whose last three components are those of $X \times Y$, we have the remarkable identity

$$XY = -(X \cdot Y)\mathbf{1} + [0, X \times Y].$$

More generally, given a quaternion $X = a\mathbf{1} + b\mathbf{i} + c\mathbf{j} + d\mathbf{k}$, we can write it as

$$X = [a, (b, c, d)],$$

where a is called the *scalar part* of X and (b, c, d) the *pure part* of X. Then, if $X = [a, U]$ and $Y = [a', U']$, it is easily seen that the quaternion product XY can be expressed as

$$XY = [aa' - U \cdot U',\ aU' + a'U + U \times U'].$$

The above formula for quaternion multiplication allows us to show the following fact. Let $Z \in \mathbb{H}$, and assume that $ZX = XZ$ for all $X \in \mathbb{H}$. We claim that the pure part of Z is null, i.e., $Z = a\mathbf{1}$ for some $a \in \mathbb{R}$. Indeed, writing $Z = [a, U]$, if $U \neq 0$, there is at least one nonnull pure quaternion $X = [0, V]$ such that $U \times V \neq 0$ (for example, take any nonnull vector V in the orthogonal complement of U). Then

$$ZX = [-U \cdot V,\ aV + U \times V], \quad XZ = [-V \cdot U,\ aV + V \times U],$$

and since $V \times U = -(U \times V)$ and $U \times V \neq 0$, we have $XZ \neq ZX$, a contradiction. Conversely, it is trivial that if $Z = [a, 0]$, then $XZ = ZX$ for all $X \in \mathbb{H}$. Thus, the set of quaternions that commute with all quaternions is $\mathbb{R}\mathbf{1}$.

Remark: It is easy to check that for arbitrary quaternions $X = [a, U]$ and $Y = [a', U']$,

$$XY - YX = [0, 2(U \times U')],$$

and that for pure quaternions $X, Y \in \mathbb{H}_p$,

$$2(X \cdot Y)\mathbf{1} = -(XY + YX).$$

Since quaternion multiplication is bilinear, for a given X, the map $Y \mapsto XY$ is linear, and similarly for a given Y, the map $X \mapsto XY$ is linear. It is immediate that if the matrix of the first map is L_X and the matrix of the second map is R_Y, then

$$XY = L_X Y = \begin{pmatrix} a & -b & -c & -d \\ b & a & -d & c \\ c & d & a & -b \\ d & -c & b & a \end{pmatrix} \begin{pmatrix} a' \\ b' \\ c' \\ d' \end{pmatrix}$$

and

$$XY = R_Y X = \begin{pmatrix} a' & -b' & -c' & -d' \\ b' & a' & d' & -c' \\ c' & -d' & a' & b' \\ d' & c' & -b' & a' \end{pmatrix} \begin{pmatrix} a \\ b \\ c \\ d \end{pmatrix}.$$

Observe that the columns (and the rows) of the above matrices are orthogonal. Thus, when X and Y are unit quaternions, both L_X and R_Y are orthogonal matrices. Furthermore, it is obvious that $L_{\overline{X}} = L_X^\top$, the transpose of L_X, and similarly, $R_{\overline{Y}} = R_Y^\top$. Since $X\overline{X} = N(X)$, the matrix $L_X L_X^\top$ is the diagonal matrix $N(X)I$ (where I is the identity 4×4 matrix), and similarly the matrix $R_Y R_Y^\top$ is the diagonal matrix $N(Y)I$. Since L_X and $L_X^\top$ have the same determinant, we deduce that $\det(L_X)^2 = N(X)^4$, and thus $\det(L_X) = \pm N(X)^2$. However, it is obvious that one of the terms in $\det(L_X)$ is a^4, and thus

$$\det(L_X) = (a^2 + b^2 + c^2 + d^2)^2.$$

This shows that when X is a unit quaternion, L_X is a rotation matrix, and similarly when Y is a unit quaternion, R_Y is a rotation matrix (see Veblen and Young [22]).

Define the map $\varphi \colon \mathbb{H} \times \mathbb{H} \to \mathbb{R}$ as follows:

$$\varphi(X,Y) = \frac{1}{2}\mathrm{Tr}(X\overline{Y}) = aa' + bb' + cc' + dd'.$$

It is easily verified that φ is bilinear, symmetric, and definite positive. Thus, the quaternions form a Euclidean space under the inner product defined by φ (see Berger [3], Dieudonné [5], Bertin [4]).

It is immediate that under this inner product, the norm of a quaternion X is just $\sqrt{N(X)}$. As a Euclidean space, $\mathbb{H}$ is isomorphic to $\mathbb{E}^4$. It is also immediate that the subspace $\mathbb{H}_p$ of pure quaternions is orthogonal to the space of "real quaternions" $\mathbb{R}\mathbf{1}$. The subspace $\mathbb{H}_p$ of pure quaternions inherits a Euclidean structure, and this subspace is isomorphic to the Euclidean space $\mathbb{E}^3$. Since $\mathbb{H}$ and $\mathbb{E}^4$ are isomorphic Euclidean spaces, their groups of rotations $\mathbf{SO}(\mathbb{H})$ and $\mathbf{SO}(4)$ are isomorphic, and we will identify them. Similarly, we will identify $\mathbf{SO}(\mathbb{H}_p)$ and $\mathbf{SO}(3)$.

9.2 Quaternions and Rotations in **SO**(3)

We have just observed that for any nonnull quaternion X, both maps $Y \mapsto XY$ and $Y \mapsto YX$ (where $Y \in \mathbb{H}$) are linear maps, and that when $N(X) = 1$, these linear maps are in $\mathbf{SO}(4)$. This suggests looking at maps $\rho_{Y,Z} \colon \mathbb{H} \to \mathbb{H}$ of the form $X \mapsto YXZ$, where $Y, Z \in \mathbb{H}$ are any two fixed nonnull quaternions such that $N(Y)N(Z) = 1$. In view of the identity $N(UV) = N(U)N(V)$ for all $U, V \in \mathbb{H}$, we see that $\rho_{Y,Z}$ is an isometry. In fact, since $\rho_{Y,Z} = \rho_{Y,1} \circ \rho_{1,Z}$, and since $\rho_{Y,1}$ is the map $X \mapsto YX$ and $\rho_{1,Z}$ is the map $X \mapsto XZ$, which are both rotations, $\rho_{Y,Z}$ itself is a rotation, i.e., $\rho_{Y,Z} \in \mathbf{SO}(4)$. We will prove that every rotation in $\mathbf{SO}(4)$ arises in this fashion.

When $Z = Y^{-1}$, the map $\rho_{Y,Y^{-1}}$ is denoted more simply by ρ_Y. In this case, it is easy to check that ρ_Y is the identity on $\mathbf{1}\mathbb{R}$, and maps $\mathbb{H}_p$ into itself. Indeed (renaming Y as Z), observe that

$$\rho_Z(X + Y) = \rho_Z(X) + \rho_Z(Y).$$

It is also easy to check that

$$\rho_Z(\overline{X}) = \overline{\rho_Z(X)}.$$

Then we have

$$\rho_Z(X + \overline{X}) = \rho_Z(X) + \rho_Z(\overline{X}) = \rho_Z(X) + \overline{\rho_Z(X)},$$

and since if $X = [a, U]$, then $X + \overline{X} = 2a\mathbf{1}$, where a is the real part of X, if X is pure, i.e., $X + \overline{X} = 0$, then $\rho_Z(X) + \overline{\rho_Z(X)} = 0$, i.e., $\rho_Z(X)$ is also pure. Thus, $\rho_Z \in \mathbf{SO}(3)$, i.e., ρ_Z is a rotation of $\mathbb{E}^3$. We will prove that every rotation in $\mathbf{SO}(3)$ arises in this fashion.

Remark: If a bijective map $\rho \colon \mathbb{H} \to \mathbb{H}$ satisfies the three conditions

$$\rho(X + Y) = \rho(X) + \rho(Y),$$
$$\rho(\lambda X) = \lambda \rho(X),$$
$$\rho(XY) = \rho(X)\rho(Y),$$

for all quaternions $X, Y \in \mathbb{H}$ and all $\lambda \in \mathbb{R}$, i.e., ρ is a linear automorphism of $\mathbb{H}$, it can be shown that $\rho(\overline{X}) = \overline{\rho(X)}$ and $N(\rho(X)) = N(X)$. In fact, ρ must be of the form ρ_Z for some nonnull $Z \in \mathbb{H}$.

The quaternions of norm 1, also called *unit quaternions*, are in bijection with points of the real 3-sphere S^3. It is easy to verify that the unit quaternions form a subgroup of the multiplicative group $\mathbb{H}^*$ of nonnull quaternions. In terms of complex matrices, the unit quaternions correspond to the group of unitary complex 2×2 matrices of determinant 1 (i.e., $x\overline{x} + y\overline{y} = 1$),

$$A = \begin{pmatrix} x & y \\ -\overline{y} & \overline{x} \end{pmatrix},$$

with respect to the Hermitian inner product in $\mathbb{C}^2$. This group is denoted by $\mathbf{SU}(2)$. The obvious bijection between $\mathbf{SU}(2)$ and S^3 is in fact a homeomorphism, and it can be used to transfer the group structure on $\mathbf{SU}(2)$ to S^3, which becomes a topological group isomorphic to the topological group $\mathbf{SU}(2)$ of unit quaternions. Incidentally, it is easy to see that the group $\mathbf{U}(2)$ of all unitary complex 2×2 matrices consists of all matrices of the form

$$A = \begin{pmatrix} \lambda x & y \\ -\lambda \bar{y} & \bar{x} \end{pmatrix},$$

with $x\bar{x} + y\bar{y} = 1$, and where λ is a complex number of modulus 1 ($\lambda \bar{\lambda} = 1$). It should also be noted that the fact that the sphere S^3 has a group structure is quite exceptional. As a matter of fact, the only spheres for which a continuous group structure is definable are S^1 and S^3. The algebraic structure of the groups $\mathbf{SU}(2)$ and $\mathbf{SO}(3)$, and their relationship to S^3, is explained very clearly in Chapter 8 of Artin [1], which we highly recommend as a general reference on algebra.

One of the most important properties of the quaternions is that they can be used to represent rotations of $\mathbb{R}^3$, as stated in the following lemma. Our proof is inspired by Berger [3], Dieudonné [5], and Bertin [4].

Lemma 9.1. *For every quaternion $Z \neq 0$, the map*

$$\rho_Z \colon X \mapsto ZXZ^{-1}$$

(where $X \in \mathbb{H}$) is a rotation in $\mathbf{SO}(\mathbb{H}) = \mathbf{SO}(4)$ whose restriction to the space $\mathbb{H}_p$ of pure quaternions is a rotation in $\mathbf{SO}(\mathbb{H}_p) = \mathbf{SO}(3)$. Conversely, every rotation in $\mathbf{SO}(3)$ is of the form

$$\rho_Z \colon X \mapsto ZXZ^{-1},$$

for some quaternion $Z \neq 0$ and for all $X \in \mathbb{H}_p$. Furthermore, if two nonnull quaternions Z and Z' represent the same rotation, then $Z' = \lambda Z$ for some $\lambda \neq 0$ in $\mathbb{R}$.

Proof. We have already observed that $\rho_Z \in \mathbf{SO}(3)$. We have to prove that every rotation is of the form ρ_Z. First, it is easily seen that

$$\rho_{YX} = \rho_Y \circ \rho_X.$$

By Theorem 8.1, every rotation that is not the identity is the composition of an even number of reflections (in the three-dimensional case, two reflections), and thus it is enough to show that for every reflection σ of $\mathbb{H}_p$ about a plane H, there is some pure quaternion $Z \neq 0$ such that $\sigma(X) = -ZXZ^{-1}$ for all $X \in \mathbb{H}_p$. If Z is a pure quaternion orthogonal to the plane H, we know that

$$\sigma(X) = X - 2\frac{(X \cdot Z)}{(Z \cdot Z)} Z$$

for all $X \in \mathbb{H}_p$. However, for pure quaternions $Y, Z \in \mathbb{H}_p$, we have

$$2(Y \cdot Z)\mathbf{1} = -(YZ + ZY).$$

Then $(Z \cdot Z)\mathbf{1} = -Z^2$, and we have

$$\sigma(X) = X - 2\frac{(X \cdot Z)}{(Z \cdot Z)} Z = X + 2(X \cdot Z)Z^{-1}$$
$$= X - (XZ + ZX)Z^{-1} = -ZXZ^{-1},$$

which shows that $\sigma(X) = -ZXZ^{-1}$ for all $X \in \mathbb{H}_p$, as desired.

If $\rho(Z_1) = \rho(Z_2)$, then

$$Z_1 X Z_1^{-1} = Z_2 X Z_2^{-1}$$

for all $X \in \mathbb{H}$, which is equivalent to

$$Z_2^{-1} Z_1 X = X Z_2^{-1} Z_1$$

for all $X \in \mathbb{H}$. However, we showed earlier that $Z_2^{-1} Z_1 = a\mathbf{1}$ for some $a \in \mathbb{R}$, and since Z_1 and Z_2 are nonnull, we get $Z_2 = (1/a)Z_1$, where $a \neq 0$. $\square$

As a corollary of

$$\rho_{YX} = \rho_Y \circ \rho_X,$$

it is easy to show that the map $\rho : \mathbf{SU}(2) \to \mathbf{SO}(3)$ defined such that $\rho(Z) = \rho_Z$ is a surjective and continuous homomorphism whose kernel is $\{\mathbf{1}, -\mathbf{1}\}$. Since $\mathbf{SU}(2)$ and S^3 are homeomorphic as topological spaces, this shows that $\mathbf{SO}(3)$ is homeomorphic to the quotient of the sphere S^3 modulo the antipodal map. But the real projective space $\mathbb{RP}^3$ is defined precisely this way in terms of the antipodal map $\pi : S^3 \to \mathbb{RP}^3$, and thus $\mathbf{SO}(3)$ and $\mathbb{RP}^3$ are homeomorphic. This homeomorphism can then be used to transfer the group structure on $\mathbf{SO}(3)$ to $\mathbb{RP}^3$, which becomes a topological group. Moreover, it can be shown that $\mathbf{SO}(3)$ and $\mathbb{RP}^3$ are diffeomorphic manifolds (see Marsden and Ratiu [15]). Thus, $\mathbf{SO}(3)$ and $\mathbb{RP}^3$ are at the same time groups, topological spaces, and manifolds, and in fact they are Lie groups (see Marsden and Ratiu [15] or Bryant [6]).

The axis and the angle of a rotation can also be extracted from a quaternion representing that rotation. The proof of the following lemma is adapted from Berger [3] and Dieudonné [5].

Lemma 9.2. *For every quaternion $Z = a\mathbf{1} + t$ where t is a nonnull pure quaternion, the axis of the rotation ρ_Z associated with Z is determined by the vector in $\mathbb{R}^3$ corresponding to t, and the angle of rotation θ is equal to π when $a = 0$, or when $a \neq 0$, given a suitable orientation of the plane orthogonal to the axis of rotation, the angle is given by*

$$\tan \frac{\theta}{2} = \frac{\sqrt{N(t)}}{|a|},$$

with $0 < \theta \leq \pi$.

Proof. A simple calculation shows that the line of direction t is invariant under the rotation ρ_Z, and thus it is the axis of rotation. Note that for any two nonnull vectors $X, Y \in \mathbb{R}^3$ such that $N(X) = N(Y)$, there is some rotation ρ such that $\rho(X) = Y$. If $X = Y$, we use the identity, and if $X \neq Y$, we use the rotation of axis determined by

$X \times Y$ rotating X to Y in the plane containing X and Y. Thus, given any two nonnull pure quaternions X, Y such that $N(X) = N(Y)$, there is some nonnull quaternion W such that $Y = WXW^{-1}$. Furthermore, given any two nonnull quaternions Z, W, we claim that the angle of the rotation ρ_Z is the same as the angle of the rotation $\rho_{WZW^{-1}}$. This can be shown as follows. First, letting $Z = a\mathbf{1} + t$ where t is a pure nonnull quaternion, we show that the axis of the rotation $\rho_{WZW^{-1}}$ is $WtW^{-1} = \rho_W(t)$. Indeed, it is easily checked that WtW^{-1} is pure, and

$$WZW^{-1} = W(a\mathbf{1} + t)W^{-1} = Wa\mathbf{1}W^{-1} + WtW^{-1} = a\mathbf{1} + WtW^{-1}.$$

Second, given any pure nonnull quaternion X orthogonal to t, the angle of the rotation Z is the angle between X and $\rho_Z(X)$. Since rotations preserve orientation (since they preserve the cross product), the angle θ between two vectors X and Y is preserved under rotation. Since rotations preserve the inner product, if $X \cdot t = 0$, we have $\rho_W(X) \cdot \rho_W(t) = 0$, and the angle of the rotation $\rho_{WZW^{-1}} = \rho_W \circ \rho_Z \circ (\rho_W)^{-1}$ is the angle between the two vectors $\rho_W(X)$ and $\rho_{WZW^{-1}}(\rho_W(X))$. Since

$$\rho_{WZW^{-1}}(\rho_W(X)) = (\rho_W \circ \rho_Z \circ (\rho_W)^{-1} \circ \rho_W)(X)$$
$$= (\rho_W \circ \rho_Z)(X) = \rho_W(\rho_Z(X)),$$

the angle of the rotation $\rho_{WZW^{-1}}$ is the angle between the two vectors $\rho_W(X)$ and $\rho_W(\rho_Z(X))$. Since rotations preserves angles, this is also the angle between the two vectors X and $\rho_Z(X)$, which is the angle of the rotation ρ_Z, as claimed. Thus, given any quaternion $Z = a\mathbf{1} + t$, where t is a nonnull pure quaternion, since there is some nonnull quaternion W such that $WtW^{-1} = \sqrt{N(t)}\,\mathbf{i}$ and $WZW^{-1} = a\mathbf{1} + \sqrt{N(t)}\,\mathbf{i}$, it is enough to figure out the angle of rotation for a quaternion Z of the form $a\mathbf{1} + b\mathbf{i}$ (a rotation of axis $\mathbf{i}$). It suffices to find the angle between $\mathbf{j}$ and $\rho_Z(\mathbf{j})$, and since

$$\rho_Z(\mathbf{j}) = (a\mathbf{1} + b\mathbf{i})\mathbf{j}(a\mathbf{1} + b\mathbf{i})^{-1},$$

we get

$$\rho_Z(\mathbf{j}) = \frac{1}{a^2 + b^2}(a\mathbf{1} + b\mathbf{i})\mathbf{j}(a\mathbf{1} - b\mathbf{i}) = \frac{a^2 - b^2}{a^2 + b^2}\mathbf{j} + \frac{2ab}{a^2 + b^2}\mathbf{k}.$$

Then if $a \neq 0$, we must have

$$\tan\theta = \frac{2ab}{a^2 - b^2} = \frac{2(b/a)}{1 - (b/a)^2},$$

and since

$$\tan\theta = \frac{2\tan(\theta/2)}{1 - \tan^2(\theta/2)},$$

under a suitable orientation of the plane orthogonal to the axis of rotation, we get

$$\tan\frac{\theta}{2} = \frac{b}{|a|} = \frac{\sqrt{N(t)}}{|a|}.$$

If $a = 0$, we get

$$\rho_Z(\mathbf{j}) = -\mathbf{j},$$

and $\theta = \pi$. $\square$

Note that if Z is a unit quaternion, then since

$$\cos\theta = \frac{1 - \tan^2(\theta/2)}{1 + \tan^2(\theta/2)}$$

and $a^2 + N(t) = N(Z) = 1$, we get $\cos\theta = a^2 - N(t) = 2a^2 - 1$, and since $\cos\theta = 2\cos^2(\theta/2) - 1$, under a suitable orientation we have

$$\cos\frac{\theta}{2} = |a|.$$

Now, since $a^2 + N(t) = N(Z) = 1$, we can write the unit quaternion Z as

$$Z = \left[\cos\frac{\theta}{2}, \sin\frac{\theta}{2}V\right],$$

where V is the unit vector $\frac{t}{\sqrt{N(t)}}$ (with $-\pi \leq \theta \leq \pi$). Also note that $VV = -\mathbf{1}$, and thus, formally, every unit quaternion looks like a complex number $\cos\varphi + i\sin\varphi$, except that i is replaced by a unit vector, and multiplication is quaternion multiplication.

In order to explain the homomorphism $\rho: \mathbf{SU}(2) \to \mathbf{SO}(3)$ more concretely, we now derive the formula for the rotation matrix of a rotation ρ whose axis D is determined by the nonnull vector w and whose angle of rotation is θ. For simplicity, we may assume that w is a unit vector. Letting $W = (b, c, d)$ be the column vector representing w and H be the plane orthogonal to w, recall from the discussion just before Lemma 8.1 that the matrices representing the projections p_D and p_H are

$$WW^\top \quad \text{and} \quad I - WW^\top.$$

Given any vector $u \in \mathbb{R}^3$, the vector $\rho(u)$ can be expressed in terms of the vectors $p_D(u)$, $p_H(u)$, and $w \times p_H(u)$ as

$$\rho(u) = p_D(u) + \cos\theta\, p_H(u) + \sin\theta\, w \times p_H(u).$$

However, it is obvious that

$$w \times p_H(u) = w \times u,$$

so that

$$\rho(u) = p_D(u) + \cos\theta\, p_H(u) + \sin\theta\, w \times u,$$
$$\rho(u) = (u \cdot w)w + \cos\theta\,(u - (u \cdot w)w) + \sin\theta\, w \times u,$$

and we know from Section 8.9 that the cross product $w \times u$ can be expressed in terms of the multiplication on the left by the matrix

$$A = \begin{pmatrix} 0 & -d & c \\ d & 0 & -b \\ -c & b & 0 \end{pmatrix}.$$

Then, letting

$$B = WW^\top = \begin{pmatrix} b^2 & bc & bd \\ bc & c^2 & cd \\ bd & cd & d^2 \end{pmatrix},$$

the matrix R representing the rotation ρ is

$$\begin{aligned} R &= WW^\top + \cos\theta(I - WW^\top) + \sin\theta A, \\ &= \cos\theta\, I + \sin\theta A + (1 - \cos\theta)WW^\top, \\ &= \cos\theta\, I + \sin\theta A + (1 - \cos\theta)B. \end{aligned}$$

It is immediately verified that

$$A^2 = B - I,$$

and thus R is also given by

$$R = I + \sin\theta A + (1 - \cos\theta)A^2.$$

Then the nonnull unit quaternion

$$Z = \left[\cos\frac{\theta}{2}, \ \sin\frac{\theta}{2}\, V \right],$$

where $V = (b, c, d)$ is a unit vector, corresponds to the rotation ρ_Z of matrix

$$R = I + \sin\theta A + (1 - \cos\theta)A^2.$$

Remark: A related formula known as Rodrigues's formula (1840) gives an expression for a rotation matrix in terms of the exponential of a matrix (the exponential map). Indeed, given $(b, c, d) \in \mathbb{R}^3$, letting $\theta = \sqrt{b^2 + c^2 + d^2}$, we have

$$e^A = \cos\theta\, I + \frac{\sin\theta}{\theta}A + \frac{(1 - \cos\theta)}{\theta^2}B,$$

with A and B as above, but (b, c, d) not necessarily a unit vector. We will study exponential maps later on.

Using the matrices L_X and R_Y introduced earlier, since $XY = L_X Y = R_Y X$, from $Y = ZXZ^{-1} = ZX\overline{Z}/N(Z)$, we get

$$Y = \frac{1}{N(Z)} L_Z R_{\overline{Z}} X.$$

Thus, if we want to see the effect of the rotation specified by the quaternion Z in terms of matrices, we simply have to compute the matrix

$$R(Z) = \frac{1}{N(Z)} L_Z R_{\overline{Z}} = v \begin{pmatrix} a & -b & -c & -d \\ b & a & -d & c \\ c & d & a & -b \\ d & -c & b & a \end{pmatrix} \begin{pmatrix} a & b & c & d \\ -b & a & -d & c \\ -c & d & a & -b \\ -d & -c & b & a \end{pmatrix},$$

where

$$N(Z) = a^2 + b^2 + c^2 + d^2 \quad \text{and} \quad v = \frac{1}{N(Z)},$$

which yields

$$v \begin{pmatrix} N(Z) & 0 & 0 & 0 \\ 0 & a^2 + b^2 - c^2 - d^2 & 2bc - 2ad & 2ac + 2bd \\ 0 & 2bc + 2ad & a^2 - b^2 + c^2 - d^2 & -2ab + 2cd \\ 0 & -2ac + 2bd & 2ab + 2cd & a^2 - b^2 - c^2 + d^2 \end{pmatrix}.$$

But since every pure quaternion X is a vector whose first component is 0, we see that the rotation matrix $R(Z)$ associated with the quaternion Z is

$$\frac{1}{N(Z)} \begin{pmatrix} a^2 + b^2 - c^2 - d^2 & 2bc - 2ad & 2ac + 2bd \\ 2bc + 2ad & a^2 - b^2 + c^2 - d^2 & -2ab + 2cd \\ -2ac + 2bd & 2ab + 2cd & a^2 - b^2 - c^2 + d^2 \end{pmatrix}.$$

This expression for a rotation matrix is due to Euler (see Veblen and Young [22]). It is quite remarkable that this matrix contains only quadratic polynomials in a, b, c, d. This makes it possible to compute easily a quaternion from a rotation matrix.

From a computational point of view, it is worth noting that computing the composition of two rotations ρ_Y and ρ_Z specified by two quaternions Y, Z using quaternion multiplication (i.e., $\rho_Y \circ \rho_Z = \rho_{YZ}$) is cheaper than using rotation matrices and matrix multiplication. On the other hand, computing the image of a point X under a rotation ρ_Z is more expensive in terms of quaternions (it requires computing ZXZ^{-1}) than it is in terms of rotation matrices (where only AX needs to be computed, where A is a rotation matrix). Thus, if many points need to be rotated and the rotation is specified by a quaternion, it is advantageous to precompute the Euler matrix.

9.3 Quaternions and Rotations in SO(4)

For every nonnull quaternion Z, the map $X \mapsto ZXZ^{-1}$ (where X is a pure quaternion) defines a rotation of $\mathbb{H}_p$, and conversely, every rotation of $\mathbb{H}_p$ is of the above form. What happens if we consider a map of the form

$$X \mapsto YXZ,$$

where $X \in \mathbb{H}$ and $N(Y)N(Z) = 1$? Remarkably, it turns out that we get all the rotations of $\mathbb{H}$. The proof of the following lemma is inspired by Berger [3], Dieudonné [5], and Tisseron [21].

Lemma 9.3. *For every pair (Y, Z) of quaternions such that $N(Y)N(Z) = 1$, the map*

$$\rho_{Y,Z} \colon X \mapsto YXZ$$

(where $X \in \mathbb{H}$) is a rotation in $\mathbf{SO}(\mathbb{H}) = \mathbf{SO}(4)$. Conversely, every rotation in $\mathbf{SO}(4)$ is of the form

$$\rho_{Y,Z} \colon X \mapsto YXZ,$$

for some quaternions Y, Z such that $N(Y)N(Z) = 1$. Furthermore, if two nonnull pairs of quaternions (Y, Z) and (Y', Z') represent the same rotation, then $Y' = \lambda Y$ and $Z' = \lambda^{-1}Z$, for some $\lambda \neq 0$ in $\mathbb{R}$.

Proof. We have already shown that $\rho_{Y,Z} \in \mathbf{SO}(4)$. It remains to prove that every rotation in $\mathbf{SO}(4)$ is of this form.

It is easily seen that

$$\rho_{(Y'Y, ZZ')} = \rho_{Y',Z'} \circ \rho_{Y,Z}.$$

Let $\rho \in \mathbf{SO}(4)$ be a rotation, and let $Z_0 = \rho(1)$ and $g = \rho_{Z_0^{-1}, 1}$. Since ρ is an isometry, $Z_0 = \rho(1)$ is a unit quaternion, and thus $g \in \mathbf{SO}(4)$. Observe that

$$g(\rho(1)) = 1,$$

which implies that $F = \mathbb{R}1$ is invariant under $g \circ \rho$. Since $F^{\perp} = \mathbb{H}_p$, by Lemma 8.2, $g \circ \rho(\mathbb{H}_p) \subseteq \mathbb{H}_p$, which shows that the restriction of $g \circ \rho$ to $\mathbb{H}_p$ is a rotation. By Lemma 9.1, there is some nonnull quaternion Z such that $g \circ \rho = \rho_Z$ on $\mathbb{H}_p$, but since both $g \circ \rho$ and ρ_Z are the identity on $\mathbb{R}1$, we must have $g \circ \rho = \rho_Z$ on $\mathbb{H}$. Finally, a trivial calculation shows that

$$\rho = g^{-1} \circ \rho_Z = \rho_{Z_0, 1}\rho_Z = \rho_{Z_0, 1}\rho_{Z, Z^{-1}} = \rho_{Z_0 Z, Z^{-1}}.$$

If $\rho_{Y,Z} = \rho_{Y',Z'}$, then

$$YXZ = Y'XZ'$$

for all $X \in \mathbb{H}$, that is,

$$Y^{-1}Y'XZ'Z^{-1} = X$$

for all $X \in \mathbb{H}$. Letting $X = (Y^{-1}Y')^{-1}$, we get $Z'Z^{-1} = (Y^{-1}Y')^{-1}$. From

$$Y^{-1}Y'X(Y^{-1}Y')^{-1} = X$$

for all $Z \in \mathbb{H}$, by a previous remark, we must have $Y^{-1}Y' = \lambda 1$ for some $\lambda \neq 0$ in $\mathbb{R}$, so that $Y' = \lambda Y$, and since $Z'Z^{-1} = (Y^{-1}Y')^{-1}$, we get $Z'Z^{-1} = \lambda^{-1}1$, i.e. $Z' = \lambda^{-1}Z$. $\qquad \square$

Since

$$\rho_{(Y'Y,ZZ')} = \rho_{Y',Z'} \circ \rho_{Y,Z},$$

it is easy to show that the map $\eta \colon S^3 \times S^3 \to \mathbf{SO}(4)$ defined by $\eta(Y,Z) = \rho_{Y,\overline{Z}}$ is a surjective homomorphism whose kernel is $\{(\mathbf{1},\mathbf{1}),(-\mathbf{1},-\mathbf{1})\}$.

Remark: Note that it is necessary to define $\eta \colon S^3 \times S^3 \to \mathbf{SO}(4)$ such that

$$\eta(Y,Z)(X) = YX\overline{Z},$$

where the conjugate $\overline{Z}$ of Z is used rather than Z, to compensate for the switch between Z and Z' in

$$\rho_{(Y'Y,ZZ')} = \rho_{Y',Z'} \circ \rho_{Y,Z}.$$

Otherwise, η would not be a homomorphism from the product group $S^3 \times S^3$ to $\mathbf{SO}(4)$.

We conclude this section on the quaternions with a mention of the exponential map, since it has applications to quaternion interpolation, which, in turn, has applications to motion interpolation.

Observe that the quaternions $\mathbf{i}, \mathbf{j}, \mathbf{k}$ can also be written as

$$\mathbf{i} = \begin{pmatrix} i & 0 \\ 0 & -i \end{pmatrix} = i \begin{pmatrix} 1 & 0 \\ 0 & -1 \end{pmatrix},$$

$$\mathbf{j} = \begin{pmatrix} 0 & 1 \\ -1 & 0 \end{pmatrix} = i \begin{pmatrix} 0 & -i \\ i & 0 \end{pmatrix},$$

$$\mathbf{k} = \begin{pmatrix} 0 & i \\ i & 0 \end{pmatrix} = i \begin{pmatrix} 0 & 1 \\ 1 & 0 \end{pmatrix},$$

so that if we define the matrices $\sigma_1, \sigma_2, \sigma_3$ such that

$$\sigma_1 = \begin{pmatrix} 0 & 1 \\ 1 & 0 \end{pmatrix}, \quad \sigma_2 = \begin{pmatrix} 0 & -i \\ i & 0 \end{pmatrix}, \quad \sigma_3 = \begin{pmatrix} 1 & 0 \\ 0 & -1 \end{pmatrix},$$

we can write

$$Z = a\mathbf{1} + b\mathbf{i} + c\mathbf{j} + d\mathbf{k} = a\mathbf{1} + i(d\sigma_1 + c\sigma_2 + b\sigma_3).$$

The matrices $\sigma_1, \sigma_2, \sigma_3$ are called the *Pauli spin matrices*. Note that their traces are null and that they are Hermitian (recall that a complex matrix is Hermitian if it is equal to the transpose of its conjugate, i.e., $A^* = A$). The somewhat unfortunate order reversal of b, c, d has to do with the traditional convention for listing the Pauli matrices. If we let $e_0 = a$, $e_1 = d$, $e_2 = c$, and $e_3 = b$, then Z can be written as

$$Z = e_0\mathbf{1} + i(e_1\sigma_1 + e_2\sigma_2 + e_3\sigma_3),$$

and e_0, e_1, e_2, e_3 are called the *Euler parameters* of the rotation specified by Z. If $N(Z) = 1$, then we can also write

$$Z = \cos\frac{\theta}{2}\,\mathbf{1} + \mathrm{i}\sin\frac{\theta}{2}\,(\beta\sigma_3 + \gamma\sigma_2 + \delta\sigma_1),$$

where

$$(\beta,\gamma,\delta) = \frac{1}{\sin\frac{\theta}{2}}\,(b,c,d).$$

Letting $A = \beta\sigma_3 + \gamma\sigma_2 + \delta\sigma_1$, it can be shown that

$$\mathrm{e}^{\mathrm{i}\theta A} = \cos\theta\,\mathbf{1} + \mathrm{i}\sin\theta\,A,$$

where the exponential is the usual exponential of matrices, i.e., for a square $n \times n$ matrix M,

$$\exp(M) = I_n + \sum_{k \geq 1}\frac{M^k}{k!}.$$

Note that since A is Hermitian of null trace, iA is skew Hermitian of null trace.

The above formula turns out to define the exponential map from the Lie algebra of $\mathbf{SU}(2)$ to $\mathbf{SU}(2)$. The Lie algebra of $\mathbf{SU}(2)$ is a real vector space having $\mathrm{i}\sigma_1$, $\mathrm{i}\sigma_2$, and $\mathrm{i}\sigma_3$ as a basis. Now, the vector space $\mathbb{R}^3$ is a Lie algebra if we define the Lie bracket on $\mathbb{R}^3$ as the usual cross product $u \times v$ of vectors. Then the Lie algebra of $\mathbf{SU}(2)$ is isomorphic to $(\mathbb{R}^3, \times)$, and the exponential map can be viewed as a map $\exp\colon (\mathbb{R}^3, \times) \to \mathbf{SU}(2)$ given by the formula

$$\exp(\theta v) = \left[\cos\frac{\theta}{2}, \sin\frac{\theta}{2}\,v\right],$$

for every vector θv, where v is a unit vector in $\mathbb{R}^3$ and $\theta \in \mathbb{R}$.

The exponential map can be used for quaternion interpolation. Given two unit quaternions X, Y, suppose we want to find a quaternion Z "interpolating" between X and Y. Of course, we have to clarify what this means. Since $\mathbf{SU}(2)$ is topologically the same as the sphere S^3, we define an *interpolant* of X and Y as a quaternion Z on the great circle (on the sphere S^3) determined by the intersection of S^3 with the (2-)plane defined by the two points X and Y (viewed as points on S^3) and the origin $(0,0,0,0)$.

Then the points (quaternions) on this great circle can be defined by first rotating X and Y so that X goes to $\mathbf{1}$ and Y goes to $X^{-1}Y$, by multiplying (on the left) by X^{-1}. Letting

$$X^{-1}Y = [\cos\Omega, \sin\Omega\,w],$$

where $-\pi < \Omega \leq \pi$, the points on the great circle from $\mathbf{1}$ to $X^{-1}Y$ are given by the quaternions

$$(X^{-1}Y)^\lambda = [\cos\lambda\Omega, \sin\lambda\Omega\,w],$$

where $\lambda \in \mathbb{R}$. This is because $X^{-1}Y = \exp(2\Omega w)$, and since an interpolant between $(0,0,0)$ and $2\Omega w$ is $2\lambda\Omega w$ in the Lie algebra of $\mathbf{SU}(2)$, the corresponding quaternion is indeed

$$\exp(2\lambda\Omega) = [\cos\lambda\Omega, \sin\lambda\Omega\,w].$$

We cannot justify all this here, but it is indeed correct.

If $\Omega \neq \pi$, then the shortest arc between X and Y is unique, and it corresponds to those λ such that $0 \leq \lambda \leq 1$ (it is a geodesic arc). However, if $\Omega = \pi$, then X and Y are antipodal, and there are infinitely many half circles from X to Y. In this case, w can be chosen arbitrarily.

Finally, having the arc of great circle between $\mathbf{1}$ and $X^{-1}Y$ (assuming $\Omega \neq \pi$), we get the arc of interpolants $Z(\lambda)$ between X and Y by performing the inverse rotation from $\mathbf{1}$ to X and from $X^{-1}Y$ to Y, i.e., by multiplying (on the left) by X, and we get

$$Z(\lambda) = X(X^{-1}Y)^\lambda.$$

Note how the geometric reasoning immediately shows that

$$Z(\lambda) = X(X^{-1}Y)^\lambda = (YX^{-1})^\lambda X.$$

It is remarkable that a closed-form formula for $Z(\lambda)$ can be given, as shown by Shoemake [19, 20]. If $X = [\cos\theta, \sin\theta\, u]$ and $Y = [\cos\varphi, \sin\varphi\, v]$ (where u and v are unit vectors in $\mathbb{R}^3$), letting

$$\cos\Omega = \cos\theta \cos\varphi + \sin\theta \sin\varphi\,(u \cdot v)$$

be the inner product of X and Y viewed as vectors in $\mathbb{R}^4$, it is a bit laborious to show that

$$Z(\lambda) = \frac{\sin(1-\lambda)\Omega}{\sin\Omega} X + \frac{\sin\lambda\Omega}{\sin\Omega} Y.$$

The above formula is quite remarkable, since if $X = \cos\theta + i\sin\theta$ and $Y = \cos\varphi + i\sin\varphi$ are two points on the unit circle S^1 (given as complex numbers of unit length), letting $\Omega = \varphi - \theta$, the interpolating point $\cos((1-\lambda)\theta + \lambda\varphi) + i\sin((1-\lambda)\theta + \lambda\varphi)$ on S^1 is given by the same formula

$$\cos((1-\lambda)\theta + \lambda\varphi) + i\sin((1-\lambda)\theta + \lambda\varphi) = \frac{\sin(1-\lambda)\Omega}{\sin\Omega} X + \frac{\sin\lambda\Omega}{\sin\Omega} Y.$$

9.4 Applications of Euclidean Geometry to Motion Interpolation

Euclidean geometry has a number applications including computer vision, computer graphics, kinematics, and robotics. The motion of a rigid body in space can be described using rigid motions. Given a fixed Euclidean frame $(O, (e_1, e_2, e_3))$, we can assume that some moving frame $(C, (u_1, u_2, u_3))$ is attached (say glued) to a rigid body B (for example, at the center of gravity of B) so that the position and orientation of B in space are completely (and uniquely) determined by some rigid motion (R, U), where U specifies the position of C w.r.t. O, and R is a rotation matrix specifying the orientation of B w.r.t. the fixed frame $(O, (e_1, e_2, e_3))$. For simplicity, we can separate the motion of the center of gravity C of B from the rotation of B around its center of gravity. Then a motion of B in space corresponds to two curves: The trajectory of the center of gravity and a curve in $\mathbf{SO}(3)$ representing the various orientations of B. Given a sequence of "snapshots" of B, say $B_0, B_1, \ldots, B_m$, we may want to find an interpolating motion passing through the given snapshots. Furthermore, in most cases, it desirable that the curve be invariant with respect to a change of coordinates and to rescaling. Often, one looks for an energy minimizing motion. The problem is not as simple as it looks, because the space of rotations $\mathbf{SO}(3)$ is topologically rather complex, and in particular, it is curved.

The problem of motion interpolation has been studied quite extensively both in the robotics and computer graphics communities. Since rotations in $\mathbf{SO}(3)$ can be represented by quaternions (see Chapter 9), the problem of quaternion interpolation has been investigated, an approach apparently initiated by Shoemake [19, 20], who extended the de Casteljau algorithm to the 3-sphere. Related work was done by Barr, Currin, Gabriel, and Hughes [2]. Kim, M.-J., Kim, M.-S. and Shin [12, 13] corrected bugs in Shoemake and introduced various kinds of splines on S^3, using the exponential map. Motion interpolation and rational motions have been investigated by Jüttler [8, 9], Jüttler and Wagner [10, 11], Horsch and Jüttler [7], and Röschel [18]. Park and Ravani [16, 17] also investigated Bézier curves on Riemannian manifolds and Lie groups, $\mathbf{SO}(3)$ in particular. More generally, the problem of interpolating curves on surfaces or higher-dimensional manifolds in an efficient way remains an open problem. A very interesting book on the quaternions and their applications to a number of engineering problems, including aerospace systems, is the book by Kuipers [14], which we highly recommend.

9.5 Problems

9.1. Prove the following identities about quaternion multiplication (discovered by Hamilton):

$$\mathbf{i}^2 = \mathbf{j}^2 = \mathbf{k}^2 = \mathbf{ijk} = -1,$$
$$\mathbf{ij} = -\mathbf{ji} = \mathbf{k},$$
$$\mathbf{jk} = -\mathbf{kj} = \mathbf{i},$$
$$\mathbf{ki} = -\mathbf{ik} = \mathbf{j}.$$

9.2. Given any two quaternions $X = a\mathbf{1} + b\mathbf{i} + c\mathbf{j} + d\mathbf{k}$ and $Y = a'\mathbf{1} + b'\mathbf{i} + c'\mathbf{j} + d'\mathbf{k}$, prove that

$$XY = (aa' - bb' - cc' - dd')\mathbf{1} + (ab' + ba' + cd' - dc')\mathbf{i}$$
$$+ (ac' + ca' + db' - bd')\mathbf{j} + (ad' + da' + bc' - cb')\mathbf{k}.$$

Also prove that if $X = [a, U]$ and $Y = [a', U']$, the quaternion product XY can be expressed as
$$XY = [aa' - U \cdot U', aU' + a'U + U \times U'].$$

9.3. Show that there is a very simple method for producing an orthonormal frame in $\mathbb{R}^4$ whose first vector is any given nonnull vector (a,b,c,d).

9.4. Prove that

$$\rho_Z(XY) = \rho_Z(X)\rho_Z(Y),$$
$$\rho_Z(X+Y) = \rho_Z(X)+\rho_Z(Y),$$

for any nonnull quaternion Z and any two quaternions X,Y (i.e., ρ_Z is an automorphism of $\mathbb{H}$), and that

$$XY - YX = [0, 2(U \times U')]$$

for arbitrary quaternions $X = [a,U]$ and $Y = [a',U']$.

9.5. Give an algorithm to find a quaternion Z corresponding to a rotation matrix R using the Euler form of a rotation matrix $R(Z)$:

$$\frac{1}{N(Z)} \begin{pmatrix} a^2+b^2-c^2-d^2 & 2bc-2ad & 2ac+2bd \\ 2bc+2ad & a^2-b^2+c^2-d^2 & -2ab+2cd \\ -2ac+2bd & 2ab+2cd & a^2-b^2-c^2+d^2 \end{pmatrix}.$$

What about the choice of the sign of Z?

9.6. Let i, j, and k, be the unit vectors of coordinates $(1,0,0)$, $(0,1,0)$, and $(0,0,1)$ in $\mathbb{R}^3$.

(i) Describe geometrically the rotations defined by the following quaternions:

$$p = (0,i), \quad q = (0,j).$$

Prove that the interpolant $Z(\lambda) = p(p^{-1}q)^\lambda$ is given by

$$Z(\lambda) = (0, \, \cos(\lambda\pi/2)i + \sin(\lambda\pi/2)j).$$

Describe geometrically what this rotation is.

(ii) Repeat question (i) with the rotations defined by the quaternions

$$p = \left(\frac{1}{2}, \frac{\sqrt{3}}{2}i\right), \quad q = (0,j).$$

Prove that the interpolant $Z(\lambda)$ is given by

$$Z(\lambda) = \left(\frac{1}{2}\cos(\lambda\pi/2), \, \frac{\sqrt{3}}{2}\cos(\lambda\pi/2)i + \sin(\lambda\pi/2)j\right).$$

Describe geometrically what this rotation is.

(iii) Repeat question (i) with the rotations defined by the quaternions

$$p = \left(\frac{1}{\sqrt{2}}, \frac{1}{\sqrt{2}}i\right), \quad q = \left(0, \frac{1}{\sqrt{2}}(i+j)\right).$$

Prove that the interpolant $Z(\lambda)$ is given by

$$Z(\lambda) = \left(\frac{1}{\sqrt{2}} \cos(\lambda \pi/3) - \frac{1}{\sqrt{6}} \sin(\lambda \pi/3), \right.$$

$$\left. (1/\sqrt{2}\cos(\lambda \pi/3) + 1/\sqrt{6}\sin(\lambda \pi/3))i + \frac{2}{\sqrt{6}} \sin(\lambda \pi/3)j \right).$$

(iv) Prove that

$$w \times (u \times v) = (w \cdot v)u - (u \cdot w)v.$$

Conclude that

$$u \times (u \times v) = (u \cdot v)u - (u \cdot u)v.$$

(v) Let

$$p = (\cos \theta, \sin \theta u), \quad q = (\cos \varphi, \sin \varphi v),$$

where u and v are unit vectors in $\mathbb{R}^3$. If

$$\cos \Omega = \cos \theta \cos \varphi + \sin \theta \sin \varphi (u \cdot v)$$

is the inner product of X and Y viewed as vectors in $\mathbb{R}^4$, assuming that $\Omega \neq k\pi$, prove that

$$Z(\lambda) = \frac{\sin(1-\lambda)\Omega}{\sin \Omega} p + \frac{\sin \lambda \Omega}{\sin \Omega} q.$$

References

1. Michael Artin. *Algebra*. Prentice-Hall, first edition, 1991.
2. A.H. Barr, B. Currin, S. Gabriel, and J.F. Hughes. Smooth Interpolation of Orientations with Angular Velocity Constraints using Quaternions. In *Computer Graphics Proceedings, Annual Conference Series*, pages 313–320. ACM, 1992.
3. Marcel Berger. *Géométrie 1*. Nathan, 1990. English edition: Geometry 1, Universitext, Springer-Verlag.
4. J.E. Bertin. *Algèbre Linéaire et Géométrie Classique*. Masson, first edition, 1981.
5. Jean Dieudonné. *Algèbre Linéaire et Géométrie Elémentaire*. Hermann, second edition, 1965.
6. R.L. Bryant. An introduction to Lie groups and symplectic geometry. In D.S. Freed and K.K. Uhlenbeck, editors, *Geometry and Quantum Field Theory*, pages 5–181. AMS, Providence, RI, 1995.
7. Thomas Horsch and Bert Jüttler. Cartesian spline interpolation for industrial robots. *Computer-Aided Design*, 30(3):217–224, 1998.
8. Bert Jüttler. Visualization of moving objects using dual quaternion curves. *Computers & Graphics*, 18(3):315–326, 1994.
9. Bert Jüttler. An osculating motion with second order contact for spacial Euclidean motions. *Mech. Mach. Theory*, 32(7):843–853, 1997.
10. Bert Jüttler and M.G. Wagner. Computer-aided design with spacial rational *B*-spline motions. *Journal of Mechanical Design*, 118:193–201, 1996.
11. Bert Jüttler and M.G. Wagner. Rational motion-based surface generation. *Computer-Aided Design*, 31:203–213, 1999.

12. M.J. Kim, M.S. Kim, and S.Y. Shin. A general construction scheme for unit quaternion curves with simple high-order derivatives. In *Computer Graphics Proceedings, Annual Conference Series*, pages 369–376. ACM, 1995.

13. M.J. Kim, M.S. Kim, and S.Y. Shin. A compact differential formula for the first derivative of a unit quaternion curve. *Journal of Visualization and Computer Animation*, 7:43–57, 1996.

14. Jack Kuipers. *Quaternion and Rotation Sequences*. Princeton University Press, first edition, 1999.

15. Jerrold E. Marsden and T.S. Ratiu. *Introduction to Mechanics and Symmetry*. TAM, Vol. 17. Springer-Verlag, first edition, 1994.

16. F.C. Park and B. Ravani. Bézier curves on Riemannian manifolds and Lie groups with kinematic applications. *ASME J. Mech. Des.*, 117:36–40, 1995.

17. F.C. Park and B. Ravani. Smooth invariant interpolation of rotations. *ACM Transactions on Graphics*, 16:277–295, 1997.

18. Otto Röschel. Rational motion design: A survey. *Computer-Aided Design*, 30(3):169–178, 1998.

19. Ken Shoemake. Animating rotation with quaternion curves. In *ACM SIGGRAPH'85*, volume 19, pages 245–254. ACM, 1985.

20. Ken Shoemake. Quaternion calculus for animation. In *Math for SIGGRAPH*, pages 1–19. ACM, 1991. Course Note No. 2.

21. Claude Tisseron. *Géométries Affines, Projectives, et Euclidiennes*. Hermann, first edition, 1994.

22. O. Veblen and J. W. Young. *Projective Geometry, Vol. 2*. Ginn, first edition, 1946.

Chapter 10
Dirichlet–Voronoi Diagrams and Delaunay Triangulations

10.1 Dirichlet–Voronoi Diagrams

In this chapter we present the concepts of a Voronoi diagram and of a Delaunay triangulation. These are important tools in computational geometry, and Delaunay triangulations are important in problems where it is necessary to fit 3D data using surface splines. It is usually useful to compute a good mesh for the projection of this set of data points onto the xy-plane, and a Delaunay triangulation is a good candidate. Our presentation will be rather sketchy. We are primarily interested in defining these concepts and stating their most important properties without proofs. For a comprehensive exposition of Voronoi diagrams, Delaunay triangulations, and more topics in computational geometry, our readers may consult O'Rourke [10], Preparata and Shamos [11], Boissonnat and Yvinec [2], de Berg, Van Kreveld, Overmars, and Schwarzkopf [1], or Risler [12]. The survey by Graham and Yao [7] contains a very gentle and lucid introduction to computational geometry. Some practical applications of Voronoi diagrams and Delaunay triangulations are briefly discussed in Section 10.5.

Let $\mathscr{E}$ be a Euclidean space of finite dimension, that is, an affine space $\mathscr{E}$ whose underlying vector space $\overrightarrow{\mathscr{E}}$ is equipped with an inner product (and has finite dimension). For concreteness, one may safely assume that $\mathscr{E} = \mathbb{E}^m$, although what follows applies to any Euclidean space of finite dimension. Given a set $P = \{p_1, \ldots, p_n\}$ of n points in $\mathscr{E}$, it is often useful to find a partition of the space $\mathscr{E}$ into regions each containing a single point of P and having some nice properties. It is also often useful to find triangulations of the convex hull of P having some nice properties. We shall see that this can be done and that the two problems are closely related. In order to solve the first problem, we need to introduce bisector lines and bisector planes.

For simplicity, let us first assume that $\mathscr{E}$ is a plane i.e., has dimension 2. Given any two distinct points $a, b \in \mathscr{E}$, the line orthogonal to the line segment (a, b) and passing through the midpoint of this segment is the locus of all points having equal distance to a and b. It is called the *bisector line of a and b*. The bisector line of two points is illustrated in Figure 10.1.

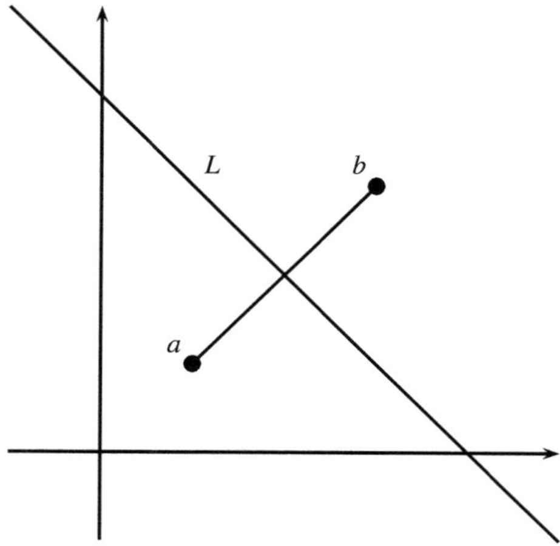

Fig. 10.1 The bisector line L of a and b.

If $h = \frac{1}{2}a + \frac{1}{2}b$ is the midpoint of the line segment (a,b), letting m be an arbitrary point on the bisector line, the equation of this line can be found by writing that $\overrightarrow{hm}$ is orthogonal to $\overrightarrow{ab}$. In any orthogonal frame, letting $m = (x,y)$, $a = (a_1, a_2)$, $b = (b_1, b_2)$, the equation of this line is

$$(b_1 - a_1)(x - (a_1 + b_1)/2) + (b_2 - a_2)(y - (a_2 + b_2)/2) = 0,$$

which can also be written as

$$(b_1 - a_1)x + (b_2 - a_2)y = (b_1^2 + b_2^2)/2 - (a_1^2 + a_2^2)/2.$$

The closed half-plane $H(a,b)$ containing a and with boundary the bisector line is the locus of all points such that

$$(b_1 - a_1)x + (b_2 - a_2)y \le (b_1^2 + b_2^2)/2 - (a_1^2 + a_2^2)/2,$$

and the closed half-plane $H(b,a)$ containing b and with boundary the bisector line is the locus of all points such that

$$(b_1 - a_1)x + (b_2 - a_2)y \ge (b_1^2 + b_2^2)/2 - (a_1^2 + a_2^2)/2.$$

The closed half-plane $H(a,b)$ is the set of all points whose distance to a is less that or equal to the distance to b, and vice versa for $H(b,a)$. Thus, points in the closed half-plane $H(a,b)$ are closer to a than they are to b.

We now consider a problem called the *post office problem* by Graham and Yao [7]. Given any set $P = \{p_1, \ldots, p_n\}$ of n points in the plane (considered as *post offices* or *sites*), for any arbitrary point x, find out which post office is closest to x.

Since x can be arbitrary, it seems desirable to precompute the sets $V(p_i)$ consisting of all points that are closer to p_i than to any other point $p_j \neq p_i$. Indeed, if the sets $V(p_i)$ are known, the answer is any post office p_i such that $x \in V(p_i)$. Thus, it remains to compute the sets $V(p_i)$. For this, if x is closer to p_i than to any other point $p_j \neq p_i$, then x is on the same side as p_i with respect to the bisector line of p_i and p_j for every $j \neq i$, and thus

$$V(p_i) = \bigcap_{j \neq i} H(p_i, p_j).$$

If $\mathscr{E}$ has dimension 3, the locus of all points having equal distance to a and b is a plane. It is called the *bisector plane of a and b*. The equation of this plane is also found by writing that $\overrightarrow{hm}$ is orthogonal to $\overrightarrow{ab}$. The equation of this plane is

$$(b_1 - a_1)(x - (a_1 + b_1)/2) + (b_2 - a_2)(y - (a_2 + b_2)/2)$$
$$+ (b_3 - a_3)(z - (a_3 + b_3)/2) = 0,$$

which can also be written as

$$(b_1 - a_1)x + (b_2 - a_2)y + (b_3 - a_3)z = (b_1^2 + b_2^2 + b_3^2)/2 - (a_1^2 + a_2^2 + a_3^2)/2.$$

The closed half-space $H(a, b)$ containing a and with boundary the bisector plane is the locus of all points such that

$$(b_1 - a_1)x + (b_2 - a_2)y + (b_3 - a_3)z \leq (b_1^2 + b_2^2 + b_3^2)/2 - (a_1^2 + a_2^2 + a_3^2)/2,$$

and the closed half-space $H(b, a)$ containing b and with boundary the bisector plane is the locus of all points such that

$$(b_1 - a_1)x + (b_2 - a_2)y + (b_3 - a_3)z \geq (b_1^2 + b_2^2 + b_3^2)/2 - (a_1^2 + a_2^2 + a_3^2)/2.$$

The closed half-space $H(a, b)$ is the set of all points whose distance to a is less that or equal to the distance to b, and vice versa for $H(b, a)$. Again, points in the closed half-space $H(a, b)$ are closer to a than they are to b.

Given any set $P = \{p_1, \ldots, p_n\}$ of n points in $\mathscr{E}$ (of dimension $m = 2, 3$), it is often useful to find for every point p_i the region consisting of all points that are closer to p_i than to any other point $p_j \neq p_i$, that is, the set

$$V(p_i) = \{x \in \mathscr{E} \mid d(x, p_i) \leq d(x, p_j), \text{ for all } j \neq i\},$$

where $d(x, y) = (\overrightarrow{xy} \cdot \overrightarrow{xy})^{1/2}$, the Euclidean distance associated with the inner product $\cdot$ on $\mathscr{E}$. From the definition of the bisector line (or plane), it is immediate that

$$V(p_i) = \bigcap_{j \neq i} H(p_i, p_j).$$

Families of sets of the form $V(p_i)$ were investigated by Dirichlet [4] (1850) and Voronoi [13] (1908). Voronoi diagrams also arise in crystallography (Gilbert [6]). Other applications, including facility location and path planning, are discussed in O'Rourke [10]. For simplicity, we also denote the set $V(p_i)$ by V_i, and we introduce the following definition.

Definition 10.1. Let $\mathscr{E}$ be a Euclidean space of dimension $m = 2, 3$. Given any set $P = \{p_1, \ldots, p_n\}$ of n points in $\mathscr{E}$, the *Dirichlet–Voronoi diagram* $\mathscr{V}(P)$ of $P = \{p_1, \ldots, p_n\}$ is the family of subsets of $\mathscr{E}$ consisting of the sets $V_i = \bigcap_{j \neq i} H(p_i, p_j)$ and of all of their intersections.

Dirichlet–Voronoi diagrams are also called *Voronoi diagrams*, *Voronoi tessellations*, or *Thiessen polygons*. Following common usage, we will use the terminology *Voronoi diagram*. As intersections of convex sets (closed half-planes or closed half-spaces), the *Voronoi regions* $V(p_i)$ are convex sets. In dimension two, the boundaries of these regions are convex polygons, and in dimension three, the boundaries are convex polyhedra.

Whether a region $V(p_i)$ is bounded or not depends on the location of p_i. If p_i belongs to the boundary of the convex hull of the set P, then $V(p_i)$ is unbounded, and otherwise bounded. In dimension two, the convex hull is a convex polygon, and in dimension three, the convex hull is a convex polyhedron. As we will see later, there is an intimate relationship between convex hulls and Voronoi diagrams.

Generally, if $\mathscr{E}$ is a Euclidean space of dimension m, given any two distinct points $a, b \in \mathscr{E}$, the locus of all points having equal distance to a and b is a hyperplane. It is called the *bisector hyperplane of a and b*. The equation of this hyperplane is still found by writing that $\overrightarrow{hm}$ is orthogonal to $\overrightarrow{ab}$. The equation of this hyperplane is

$$(b_1 - a_1)(x_1 - (a_1 + b_1)/2) + \cdots + (b_m - a_m)(x_m - (a_m + b_m)/2) = 0,$$

which can also be written as

$$(b_1 - a_1)x_1 + \cdots + (b_m - a_m)x_m = (b_1^2 + \cdots + b_m^2)/2 - (a_1^2 + \cdots + a_m^2)/2.$$

The closed half-space $H(a, b)$ containing a and with boundary the bisector hyperplane is the locus of all points such that

$$(b_1 - a_1)x_1 + \cdots + (b_m - a_m)x_m \leq (b_1^2 + \cdots + b_m^2)/2 - (a_1^2 + \cdots + a_m^2)/2,$$

and the closed half-space $H(b, a)$ containing b and with boundary the bisector hyperplane is the locus of all points such that

$$(b_1 - a_1)x_1 + \cdots + (b_m - a_m)x_m \geq (b_1^2 + \cdots + b_m^2)/2 - (a_1^2 + \cdots + a_m^2)/2.$$

The closed half-space $H(a, b)$ is the set of all points whose distance to a is less than or equal to the distance to b, and vice versa for $H(b, a)$.

Figure 10.2 shows the Voronoi diagram of a set of twelve points. In the general case where $\mathscr{E}$ has dimension m, the definition of the Voronoi diagram $\mathscr{V}(P)$ of P is

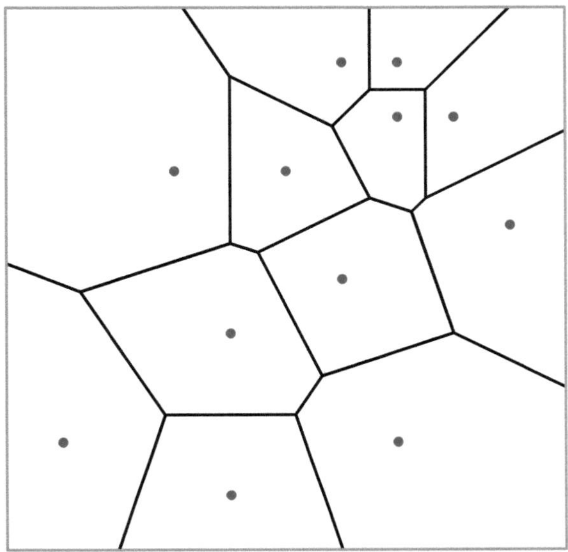

Fig. 10.2 A Voronoi diagram.

the same as Definition 10.1, except that $H(p_i, p_j)$ is the closed half-space containing p_i and having the bisector hyperplane of a and b as boundary. Also, observe that the convex hull of P is a convex polytope.

We will now state a lemma listing the main properties of Voronoi diagrams. It turns out that certain degenerate situations can be avoided if we assume that if P is a set of points in an affine space of dimension m, then no $m + 2$ points from P belong to the same $(m - 1)$-sphere. We will say that the points of P are in *general position*. Thus when $m = 2$, no 4 points in P are cocyclic, and when $m = 3$, no 5 points in P are on the same sphere.

Lemma 10.1. *Given a set $P = \{p_1, \ldots, p_n\}$ of n points in some Euclidean space $\mathscr{E}$ of dimension m (say $\mathbb{E}^m$), if the points in P are in general position and not in a common hyperplane then the Voronoi diagram of P satisfies the following conditions:*

(1) Each region V_i is convex and contains p_i in its interior.
(2) Each vertex of V_i belongs to $m + 1$ regions V_j and to $m + 1$ edges.
(3) The region V_i is unbounded iff p_i belongs to the boundary of the convex hull of P.
(4) If p is a vertex that belongs to the regions $V_1, \ldots, V_{m+1}$, then p is the center of the $(m - 1)$-sphere $S(p)$ determined by $p_1, \ldots, p_{m+1}$. Furthermore, no point in P is inside the sphere $S(p)$ (i.e., in the open ball associated with the sphere $S(p)$).

(5) If p_j is a nearest neighbor of p_i, then one of the faces of V_i is contained in the bisector hyperplane of (p_i, p_j).

(6)

$$\bigcup_{i=1}^{n} V_i = \mathscr{E}, \quad \text{and} \quad \overset{\circ}{V}_i \cap \overset{\circ}{V}_j = \emptyset, \quad \text{for all } i, j, \text{ with } i \neq j,$$

where $\overset{\circ}{V}_i$ denotes the interior of V_i.

Proof. We prove only some of the statements, leaving the others as an exercise (or see Risler [12]).

(1) Since $V_i = \bigcap_{j \neq i} H(p_i, p_j)$ and each half-space $H(p_i, p_j)$ is convex, as an intersection of convex sets, V_i is convex. Also, since p_i belongs to the interior of each $H(p_i, p_j)$, the point p_i belongs to the interior of V_i.

(2) Let $F_{i,j}$ denote $V_i \cap V_j$. Any vertex p of the Vononoi diagram of P must belong to r faces $F_{i,j}$. Now, given a vector space E and any two subspaces M and N of E, recall that we have the *Grassmann relation* (see Lemma 2.14)

$$\dim(M) + \dim(N) = \dim(M + N) + \dim(M \cap N).$$

Then since p belongs to the intersection of the hyperplanes that form the boundaries of the V_i, and since a hyperplane has dimension $m - 1$, by the Grassmann relation, we must have $r \geq m$. For simplicity of notation, let us denote these faces by $F_{1,2}, F_{2,3}, \ldots, F_{r,r+1}$. Since $F_{i,j} = V_i \cap V_j$, we have

$$F_{i,j} = \{p \mid d(p, p_i) = d(p, p_j) \leq d(p, p_k), \text{ for all } k \neq i, j\},$$

and since $p \in F_{1,2} \cap F_{2,3} \cap \cdots \cap F_{r,r+1}$, we have

$$d(p, p_1) = \cdots = d(p, p_{r+1}) < d(p, p_k) \text{ for all } k \notin \{1, \ldots, r+1\}.$$

This means that p is the center of a sphere passing through $p_1, \ldots, p_{r+1}$ and containing no other point in P. By the assumption that points in P are in general position, we must have $r \leq m$, and thus $r = m$. Thus, p belongs to $V_1 \cap \cdots \cap V_{m+1}$, but to no other V_j with $j \notin \{1, \ldots, m+1\}$. Furthermore, every edge of the Voronoi diagram containing p is the intersection of m of the regions $V_1, \ldots, V_{m+1}$, and so there are $m + 1$ of them. $\square$

For simplicity, let us again consider the case where $\mathscr{E}$ is a plane. It should be noted that certain Voronoi regions, although closed, may extend very far. Figure 10.3 shows such an example.

It is also possible for certain unbounded regions to have parallel edges.

There are a number of methods for computing Voronoi diagrams. A fairly simple (although not very efficient) method is to compute each Voronoi region $V(p_i)$ by intersecting the half-planes $H(p_i, p_j)$. One way to do this is to construct successive convex polygons that converge to the boundary of the region. At every step we intersect the current convex polygon with the bisector line of p_i and p_j. There are at most two intersection points. We also need a starting polygon, and for this we can

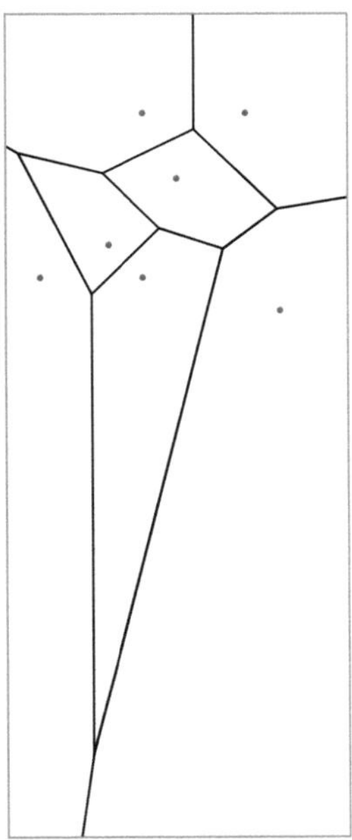

Fig. 10.3 Another Voronoi diagram.

pick a square containing all the points. A naive implementation will run in $O(n^3)$. However, the intersection of half-planes can be done in $O(n\log n)$, using the fact that the vertices of a convex polygon can be sorted. Thus, the above method runs in $O(n^2\log n)$. Actually, there are faster methods (see Preparata and Shamos [11] or O'Rourke [10]), and it is possible to design algorithms running in $O(n\log n)$. The most direct method to obtain fast algorithms is to use the "lifting method" discussed in Section 10.4, whereby the original set of points is lifted onto a paraboloid, and to use fast algorithms for finding a convex hull.

A very interesting (undirected) graph can be obtained from the Voronoi diagram as follows: The vertices of this graph are the points p_i (each corresponding to a unique region of $\mathscr{V}(P)$), and there is an edge between p_i and p_j iff the regions V_i and V_j share an edge. The resulting graph is called a *Delaunay triangulation* of the convex hull of P, after Delaunay, who invented this concept in 1934. Such triangulations have remarkable properties.

Figure 10.4 shows the Delaunay triangulation associated with the earlier Voronoi diagram of a set of twelve points.

One has to be careful to make sure that all the Voronoi vertices have been computed before computing a Delaunay triangulation, since otherwise, some edges could be missed. In Figure 10.5 illustrating such a situation, if the lowest Voronoi vertex had not been computed (not shown on the diagram!), the lowest edge of the

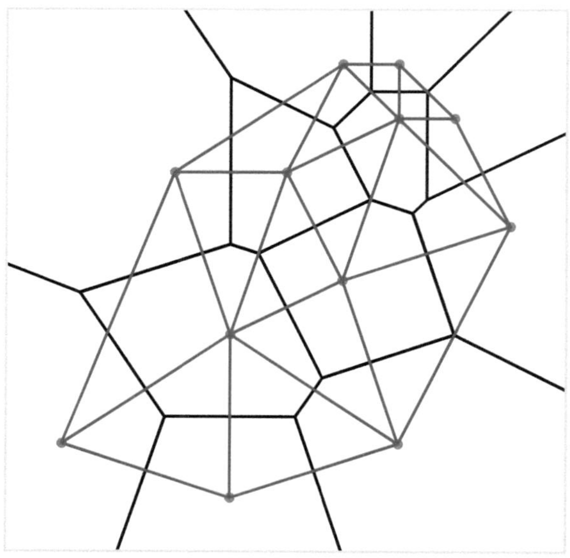

Fig. 10.4 Delaunay triangulation associated with a Voronoi diagram.

Delaunay triangulation would be missing. The concept of a triangulation can be generalized to dimension 3, or even to any dimension m. But first, we need to define a triangulation precisely, and for this, we need to review what is a simplicial complex.

10.2 Simplicial Complexes and Triangulations

A simplex is just the convex hull of a finite number of affinely independent points, but we also need to define faces, the boundary, and the interior of a simplex.

Definition 10.2. Let $\mathscr{E}$ be any normed affine space, say $\mathscr{E} = \mathbb{E}^m$ with its usual Euclidean norm. Given any $n+1$ affinely independent points $a_0, \dots, a_n$ in $\mathscr{E}$, the *n-simplex (or simplex)* σ *defined by* $a_0, \dots, a_n$ is the convex hull of the points $a_0, \dots, a_n$, that is, the set of all convex combinations $\lambda_0 a_0 + \cdots + \lambda_n a_n$, where $\lambda_0 + \cdots + \lambda_n = 1$ and $\lambda_i \geq 0$ for all i, $0 \leq i \leq n$. We call n the *dimension* of the *n-simplex* σ, and the points $a_0, \dots, a_n$ are the *vertices* of σ. Given any subset $\{a_{i_0}, \dots, a_{i_k}\}$ of $\{a_0, \dots, a_n\}$ (where $0 \leq k \leq n$), the *k*-simplex generated by $a_{i_0}, \dots, a_{i_k}$ is called a *face* of σ. A face s of σ is a *proper face* if $s \neq \sigma$ (we agree that the empty set is a face of any simplex). For any vertex a_i, the face generated by

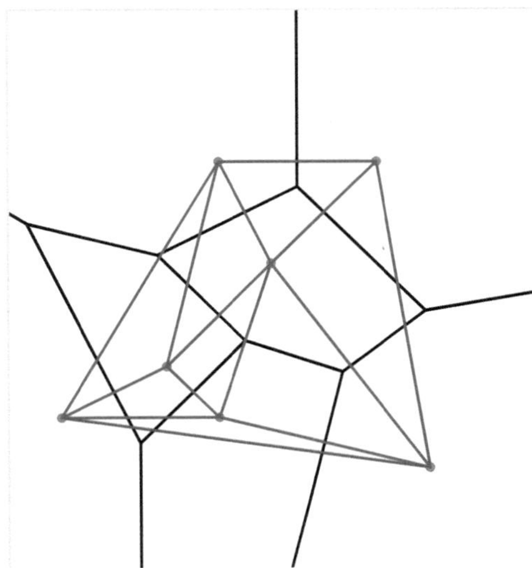

Fig. 10.5 Another Delaunay triangulation associated with a Voronoi diagram.

$a_0, \ldots, a_{i-1}, a_{i+1}, \ldots, a_n$ (i.e., omitting a_i) is called the *face opposite* a_i. Every face that is an $(n-1)$-simplex is called a *boundary face*. The union of the boundary faces is the *boundary of* σ, denoted by $\partial\sigma$, and the complement of $\partial\sigma$ in σ is the *interior* $\mathrm{Int}\,\sigma = \sigma - \partial\sigma$ of σ. The interior $\mathrm{Int}\,\sigma$ of σ is sometimes called an *open simplex*.

It should be noted that for a 0-simplex consisting of a single point $\{a_0\}$, $\partial\{a_0\} = \emptyset$, and $\mathrm{Int}\{a_0\} = \{a_0\}$. Of course, a 0-simplex is a single point, a 1-simplex is the line segment (a_0, a_1), a 2-simplex is a triangle (a_0, a_1, a_2) (with its interior), and a 3-simplex is a tetrahedron (a_0, a_1, a_2, a_3) (with its interior); see Figure 10.6.

We now state a number of properties of simplices, whose proofs are left as an exercise. Clearly, a point x belongs to the boundary $\partial\sigma$ of σ iff at least one of its barycentric coordinates $(\lambda_0, \ldots, \lambda_n)$ is zero, and a point x belongs to the interior $\mathrm{Int}\,\sigma$ of σ iff all of its barycentric coordinates $(\lambda_0, \ldots, \lambda_n)$ are positive, i.e., $\lambda_i > 0$ for all i, $0 \leq i \leq n$. Then, for every $x \in \sigma$, there is a unique face s such that $x \in \mathrm{Int}\,s$, the face generated by those points a_i for which $\lambda_i > 0$, where $(\lambda_0, \ldots, \lambda_n)$ are the barycentric coordinates of x.

A simplex σ is convex, arcwise connected, compact, and closed. The interior $\mathrm{Int}\,\sigma$ of a complex is convex, arcwise connected, open, and σ is the closure of $\mathrm{Int}\,\sigma$.

We now need to put simplices together to form more complex shapes, following Munkres [9].

Definition 10.3. A *simplicial complex in* $\mathbb{E}^m$ (for short, a *complex in* $\mathbb{E}^m$) is a set K consisting of a (finite or infinite) set of simplices in $\mathbb{E}^m$ satisfying the following conditions:

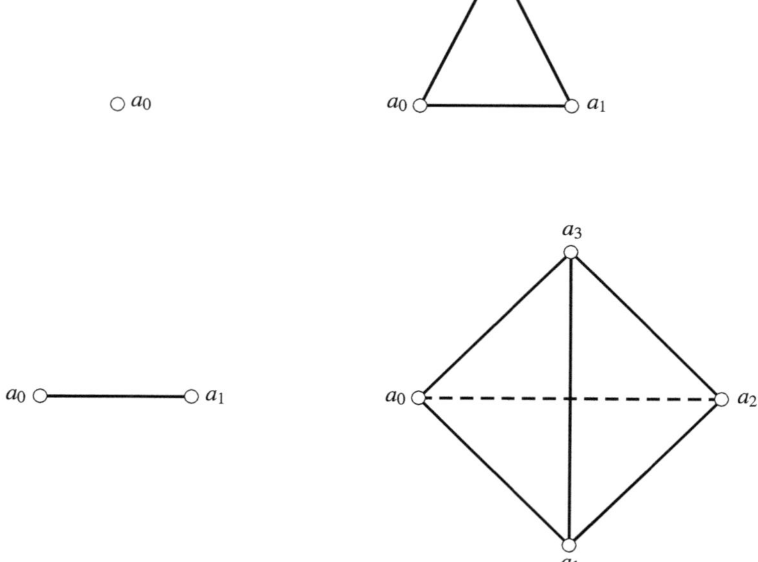

Fig. 10.6 Examples of simplices.

(1) Every face of a simplex in K also belongs to K.

(2) For any two simplices σ_1 and σ_2 in K, if $\sigma_1 \cap \sigma_2 \neq \emptyset$, then $\sigma_1 \cap \sigma_2$ is a common face of both σ_1 and σ_2.

If $\sigma \in K$ is a simplex of $n+1$ elements, then its dimension is n, and it is called an *n-simplex*. A 0-simplex $\{x\}$ is called a *vertex*. The *dimension* of the simplicial complex K is the maximum of the dimensions of simplices in K.

Condition (2) guarantees that the various simplices forming a complex are glued nicely. It can be shown that the following condition is equivalent to condition (2):

(2′) For any two distinct simplices σ_1, σ_2, $\mathrm{Int}\,\sigma_1 \cap \mathrm{Int}\,\sigma_2 = \emptyset$.

The union K_g of all the simplices in K is a subset of $\mathbb{E}^m$. We can define a topology on K_g by defining a subset F of K_g to be closed iff $F \cap \sigma$ is closed in σ for every simplex $\sigma \in K$. It is immediately verified that the axioms of a topological space are indeed satisfied. The resulting topological space K_g is called the *geometric realization of K*. A *polytope* is the geometric realization of some simplicial complex. A polytope of dimension 1 is usually called a *polygon*, and a polytope of dimension 2 is usually called a *polyhedron*. It can be checked that each region V_i of a Voronoi diagram is a (convex) polytope.

In the sequel, we will consider only finite simplicial complexes, that is, complexes K consisting of a finite number of simplices. In this case, the topology of K_g defined above is identical to the topology induced from $\mathbb{E}^m$. In this case, for any simplex σ in K, $\operatorname{Int}\sigma$ coincides with the interior $\overset{\circ}{\sigma}$ of σ in the topological sense, and $\partial\sigma$ coincides with the boundary of σ in the topological sense. We can now define triangulations.

First, assume that $\mathscr{E} = \mathbb{E}^2$. Given a subset S of $\mathbb{E}^2$, a *triangulation of S* is a finite complex K of dimension 2 such that S is the union of the 2-simplices in K. Equivalently, S is the union of the (closed) triangles in K. Thus, a triangulation of S specifies a way of cutting up S into a collection of (closed) triangles that intersect nicely. Next, if $\mathscr{E} = \mathbb{E}^3$, given a subset S of $\mathbb{E}^3$, a triangulation of S is a finite complex K of dimension 3 such that S is the union of the 3-simplices in K. Equivalently, S is the union of the (closed) tetrahedra in K. Thus, a triangulation of S specifies a way of cutting up S into a collection of (closed) tetrahedra that intersect nicely. In general, we have the following definition.

Definition 10.4. Given a subset S of $\mathbb{E}^m$ (where $m \geq 2$), a *d-triangulation of S* (where $d \leq m$) is a finite complex K such that

$$S = \bigcup_{\substack{\sigma \in K \\ \dim(\sigma)=d}} \sigma,$$

i.e., such that S is the union of all d-simplices in K.

Given a finite set P of n points in the plane, and given a triangulation of the convex hull of P having P as its set of vertices, observe that the boundary of P is a convex polygon. Similarly, given a finite set P of points in 3-space, and given a triangulation of the convex hull of P having P as its set of vertices, observe that the boundary of P is a convex polyhedron. It is interesting to know how many triangulations exist for a set of n points (in the plane or in 3-space), and it is also interesting to know the number of edges and faces in terms of the number of vertices in P. These questions can be settled using the Euler–Poincaré characteristic. We say that a polygon in the plane is a *simple polygon* iff it is a connected closed polygon such that no two edges intersect (except at a common vertex).

Lemma 10.2.

(1) For any triangulation of a region of the plane whose boundary is a simple polygon, letting v be the number of vertices, e the number of edges, and f the number of triangles, we have the "Euler formula"

$$v - e + f = 1.$$

(2) For any polytope S homeomorphic to a closed ball in $\mathbb{E}^3$ and any triangulation of S, letting v be the number of vertices, e the number of edges, f the number of triangles, and t the number of tetrahedra, we have the "Euler formula"

$$v - e + f - t = 1.$$

(3) Furthermore, for any triangulation of the polyhedron $B(S)$ that is the boundary of S, letting v' be the number of vertices, e' the number of edges, and f' the number of triangles, we have the "Euler formula"

$$v' - e' + f' = 2.$$

Proof. We only sketch the proof. More details can be found in O'Rourke [10], Risler [12], or books on algebraic topology, such as Massey [8] or Munkres [9]. The proof of (1) is by induction on the number f of triangles. The proof of (2) is by induction on the number t of tetrahedra. The proof of (3) consists in first flattening the polyhedron into a planar graph in the plane. This can be done by removing some face and then by deformation. The boundary of this planar graph is a simple polygon, and the region outside this boundary corresponds to the removed face. Then by (1) we get the formula, remembering that there is one more face (this is why we get 2 instead of 1). $\square$

It is now easy to see that in case (1), the number of edges and faces is a linear function of the number of vertices and boundary edges, and that in case (3), the number of edges and faces is a linear function of the number of vertices. Indeed, in the case of a planar triangulation, each face has 3 edges, and if there are e_b edges in the boundary and e_i edges not in the boundary, each nonboundary edge is shared by two faces, and thus $3f = e_b + 2e_i$. Since $v - e_b - e_i + f = 1$, we get

$$v - e_b - e_i + e_b/3 + 2e_i/3 = 1,$$
$$2e_b/3 + e_i/3 = v - 1,$$

and thus $e_i = 3v - 3 - 2e_b$. Since $f = e_b/3 + 2e_i/3$, we have $f = 2v - 2 - e_b$.

Similarly, since $v' - e' + f' = 2$ and $3f' = 2e'$, we easily get $e = 3v - 6$ and $f = 2v - 4$. Thus, given a set P of n points, the number of triangles (and edges) for any triangulation of the convex hull of P using the n points in P for its vertices is fixed.

Case (2) is trickier, but it can be shown that

$$v - 3 \leq t \leq (v-1)(v-2)/2.$$

Thus, there can be different numbers of tetrahedra for different triangulations of the convex hull of P.

Remark: The numbers of the form $v - e + f$ and $v - e + f - t$ are called *Euler–Poincaré characteristics*. They are topological invariants, in the sense that they are the same for all triangulations of a given polytope. This is a fundamental fact of algebraic topology.

We shall now investigate triangulations induced by Voronoi diagrams.

10.3 Delaunay Triangulations

Given a set $P = \{p_1, \ldots, p_n\}$ of n points in the plane and the Voronoi diagram $\mathscr{V}(P)$ for P, we explained in Section 10.1 how to define an (undirected) graph: The vertices of this graph are the points p_i (each corresponding to a unique region of $\mathscr{V}(P)$), and there is an edge between p_i and p_j iff the regions V_i and V_j share an edge. The resulting graph turns out to be a triangulation of the convex hull of P having P as its set of vertices. Such a complex can be defined in general. For any set $P = \{p_1, \ldots, p_n\}$ of n points in $\mathbb{E}^m$, we say that a triangulation of the convex hull of P is *associated with P* if its set of vertices is the set P.

Definition 10.5. Let $P = \{p_1, \ldots, p_n\}$ be a set of n points in $\mathbb{E}^m$, and let $\mathscr{V}(P)$ be the Voronoi diagram of P. We define a complex $\mathscr{D}(P)$ as follows. The complex $\mathscr{D}(P)$ contains the k-simplex $\{p_1, \ldots, p_{k+1}\}$ iff $V_1 \cap \cdots \cap V_{k+1} \neq \emptyset$, where $0 \leq k \leq m$. The complex $\mathscr{D}(P)$ is called the *Delaunay triangulation of the convex hull of P*.

Thus, $\{p_i, p_j\}$ is an edge iff $V_i \cap V_j \neq \emptyset$, $\{p_i, p_j, p_h\}$ is a triangle iff $V_i \cap V_j \cap V_h \neq \emptyset$, $\{p_i, p_j, p_h, p_k\}$ is a tetrahedron iff $V_i \cap V_j \cap V_h \cap V_k \neq \emptyset$, etc.

For simplicity, we often write $\mathscr{D}$ instead of $\mathscr{D}(P)$. A Delaunay triangulation for a set of twelve points is shown in Figure 10.7.

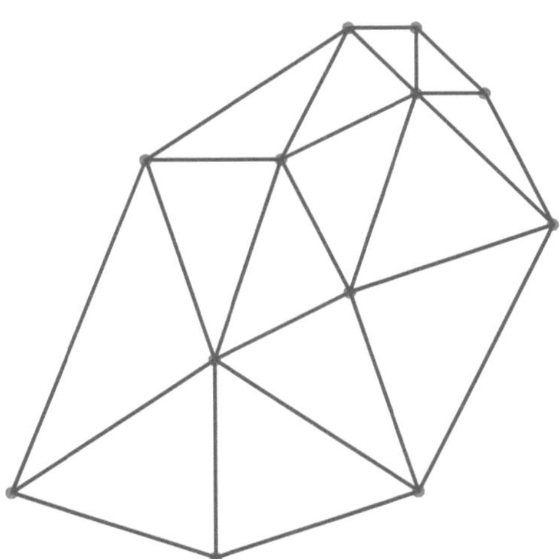

Fig. 10.7 A Delaunay triangulation.

Actually, it is not obvious that $\mathscr{D}(P)$ is a triangulation of the convex hull of P, but this can be shown, as well as the properties listed in the following lemma.

Lemma 10.3. *Let $P = \{p_1, \ldots, p_n\}$ be a set of n points in $\mathbb{E}^m$, and assume that they are in general position. Then the Delaunay triangulation of the convex hull of P is indeed a triangulation associated with P, and it satisfies the following properties:*

(1) The boundary of $\mathscr{D}(P)$ is the convex hull of P.
(2) A triangulation T associated with P is the Delaunay triangulation $\mathscr{D}(P)$ iff every $(m-1)$-sphere $S(\sigma)$ circumscribed about an m-simplex σ of T contains no other point from P (i.e., the open ball associated with $S(\sigma)$ contains no point from P).

The proof can be found in Risler [12] and O'Rourke [10]. In the case of a planar set P, it can also be shown that the Delaunay triangulation has the property that it maximizes the minimum angle of the triangles involved in any triangulation of P. However, this does not characterize the Delaunay triangulation. Given a connected graph in the plane, it can also be shown that any minimal spanning tree is contained in the Delaunay triangulation of the convex hull of the set of vertices of the graph (O'Rourke [10]).

We will now explore briefly the connection between Delaunay triangulations and convex hulls.

10.4 Delaunay Triangulations and Convex Hulls

In this section we show that there is an intimate relationship between convex hulls and Delaunay triangulations. We will see that given a set P of points in the Euclidean space $\mathbb{E}^m$ of dimension m, we can "lift" these points onto a paraboloid living in the space $\mathbb{E}^{m+1}$ of dimension $m+1$, and that the Delaunay triangulation of P is the projection of the downward-facing faces of the convex hull of the set of lifted points. This remarkable connection was first discovered by Brown [3], and refined by Edelsbrunner and Seidel [5]. For simplicity, we consider the case of a set P of points in the plane $\mathbb{E}^2$, and we assume that they are in general position.

Consider the paraboloid of revolution of equation $z = x^2 + y^2$. A point $p = (x, y)$ in the plane is lifted to the point $l(p) = (X, Y, Z)$ in $\mathbb{E}^3$, where $X = x$, $Y = y$, and $Z = x^2 + y^2$.

The first crucial observation is that a circle in the plane is lifted into a plane curve (an ellipse). Indeed, if such a circle C is defined by the equation

$$x^2 + y^2 + ax + by + c = 0,$$

since $X = x$, $Y = y$, and $Z = x^2 + y^2$, by eliminating $x^2 + y^2$ we get

$$Z = -ax - by - c,$$

and thus X, Y, Z satisfy the linear equation

$$aX + bY + Z + c = 0,$$

which is the equation of a plane. Thus, the intersection of the cylinder of revolution consisting of the lines parallel to the z-axis and passing through a point of the circle C with the paraboloid $z = x^2 + y^2$ is a planar curve (an ellipse).

We can compute the convex hull of the set of lifted points. Let us focus on the downward-facing faces of this convex hull. Let $(l(p_1), l(p_2), l(p_3))$ be such a face. The points p_1, p_2, p_3 belong to the set P. We claim that no other point from P is inside the circle C. Indeed, a point p inside the circle C would lift to a point $l(p)$ on the paraboloid. Since no four points are cocyclic, one of the four points p_1, p_2, p_3, p is further from O than the others; say this point is p_3. Then, the face $(l(p_1), l(p_2), l(p))$ would be below the face $(l(p_1), l(p_2), l(p_3))$, contradicting the fact that $(l(p_1), l(p_2), l(p_3))$ is one of the downward-facing faces of the convex hull of P. But then, by property (2) of Lemma 10.3, the triangle (p_1, p_2, p_3) would belong to the Delaunay triangulation of P.

Therefore, we have shown that *the projection of the part of the convex hull of the lifted set $l(P)$ consisting of the downward-facing faces is the Delaunay triangulation of P.* Figure 10.8 shows the lifting of the Delaunay triangulation shown earlier.

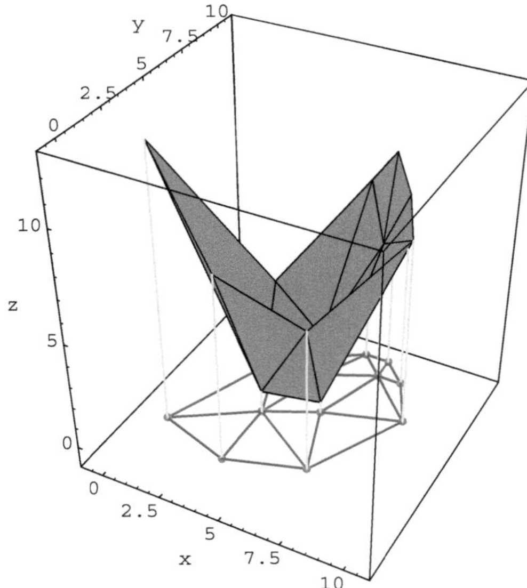

Fig. 10.8 A Delaunay triangulation and its lifting to a paraboloid.

Another example of the lifting of a Delaunay triangulation is shown in Figure 10.9. The fact that a Delaunay triangulation can be obtained by projecting a lower convex hull can be used to find efficient algorithms for computing a Delaunay triangulation. It also holds for higher dimensions.

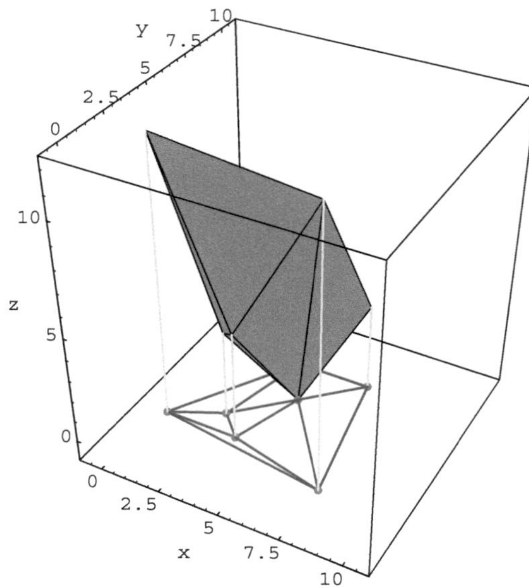

Fig. 10.9 Another Delaunay triangulation and its lifting to a paraboloid.

The Voronoi diagram itself can also be obtained from the lifted set $l(P)$. However, this time, we need to consider tangent planes to the paraboloid at the lifted points. It is fairly obvious that the tangent plane at the lifted point $(a, b, a^2 + b^2)$ is

$$z = 2ax + 2by - (a^2 + b^2).$$

Given two distinct lifted points $(a_1, b_1, a_1^2 + b_1^2)$ and $(a_2, b_2, a_2^2 + b_2^2)$, the intersection of the tangent planes at these points is a line belonging to the plane of equation

$$(b_1 - a_1)x + (b_2 - a_2)y = (b_1^2 + b_2^2)/2 - (a_1^2 + a_2^2)/2.$$

This is precisely the equation of the bisector line of the two points (a_1, b_1) and (a_2, b_2). Therefore, *if we look at the paraboloid from $z = +\infty$ (with the paraboloid transparent), the projection of the tangent planes at the lifted points is the Voronoi diagram*!

It should be noted that the "duality" between the Delaunay triangulation, which is the projection of the convex hull of the lifted set $l(P)$ viewed from $z = -\infty$, and the Voronoi diagram, which is the projection of the tangent planes at the lifted set $l(P)$ viewed from $z = +\infty$, is reminiscent of the polar duality with respect to a quadric.

The reader interested in algorithms for finding Voronoi diagrams and Delaunay triangulations is referred to O'Rourke [10], Preparata and Shamos [11], Boissonnat and Yvinec [2], de Berg, Van Kreveld, Overmars, and Schwarzkopf [1], and Risler

[12]. We conclude our brief presentation of Voronoi diagrams and Delaunay triangulations with a short section on applications.

10.5 Applications of Voronoi Diagrams and Delaunay Triangulations

The examples below are taken from O'Rourke [10]. Other examples can be found in Preparata and Shamos [11], Boissonnat and Yvinec [2], and de Berg, Van Kreveld, Overmars, and Schwarzkopf [1].

The first example is the *nearest neighbors* problem. There are actually two subproblems: *Nearest neighbor queries* and *all nearest neighbors*.

The nearest neighbor queries problem is as follows. Given a set P of points and a query point q, find the nearest neighbor(s) of q in P. This problem can be solved by computing the Voronoi diagram of P and determining in which Voronoi region q falls. This last problem, called *point location*, has been heavily studied (see O'Rourke [10]). The all neighbors problem is as follows: Given a set P of points, find the nearest neighbor(s) to all points in P. This problem can be solved by building a graph, the *nearest neighbor graph*, for short *nng*. The nodes of this undirected graph are the points in P, and there is an arc from p to q iff p is a nearest neighbor of q or vice versa. Then it can be shown that this graph is contained in the Delaunay triangulation of P.

The second example is the *largest empty circle*. Some practical applications of this problem are to locate a new store (to avoid competition), or to locate a nuclear plant as far as possible from a set of towns. More precisely, the problem is as follows. Given a set P of points, find a largest empty circle whose center is in the (closed) convex hull of P, empty in that it contains no points from P inside it, and largest in the sense that there is no other circle with strictly larger radius. The Voronoi diagram of P can be used to solve this problem. It can be shown that if the center p of a largest empty circle is strictly inside the convex hull of P, then p coincides with a Voronoi vertex. However, not every Voronoi vertex is a good candidate. It can also be shown that if the center p of a largest empty circle lies on the boundary of the convex hull of P, then p lies on a Voronoi edge.

The third example is the *minimum spanning tree*. Given a graph G, a minimum spanning tree of G is a subgraph of G that is a tree, contains every vertex of the graph G, and minimizes the sum of the lengths of the tree edges. It can be shown that a minimum spanning tree is a subgraph of the Delaunay triangulation of the vertices of the graph. This can be used to improve algorithms for finding minimum spanning trees, for example Kruskal's algorithm (see O'Rourke [10]).

We conclude by mentioning that Voronoi diagrams have applications to *motion planning*. For example, consider the problem of moving a disk on a plane while avoiding a set of polygonal obstacles. If we "extend" the obstacles by the diameter of the disk, the problem reduces to finding a collision–free path between two points in the extended obstacle space. One needs to generalize the notion of a Voronoi diagram. Indeed, we need to define the distance to an object, and medial curves (consisting of points equidistant to two objects) may no longer be straight lines. A collision–free path with maximal clearance from the obstacles can be found by moving along the edges of the generalized Voronoi diagram. This is an active area of research in robotics. For more on this topic, see O'Rourke [10].

10.6 Problems

10.1. Investigate the different shapes of the Voronoi diagram for a set of 3 points, and then for a set of 4 points.

10.2. Prove (3)–(6) of Lemma 10.1.

10.3. Show that the intersection of n half-planes can be done in $O(n \log n)$, using the fact that the vertices of a convex polygon can be sorted.

10.4. Write a computer program computing the Voronoi diagram of a set of points in the plane. Can you do it in time $O(n^2 \log n)$?

10.5. Let σ be a simplex. (i) Prove that a point x belongs to the boundary $\partial \sigma$ of σ iff at least one of its barycentric coordinates $(\lambda_0, \ldots, \lambda_n)$ is zero, and a point x belongs to the interior $\mathrm{Int}\,\sigma$ of σ iff all of its barycentric coordinates $(\lambda_0, \ldots, \lambda_n)$ are positive, i.e., $\lambda_i > 0$ for all $i, 0 \leq i \leq n$. Prove that for every $x \in \sigma$, there is a unique face s such that $x \in \mathrm{Int}\,s$, the face generated by those points a_i for which $\lambda_i > 0$, where $(\lambda_0, \ldots, \lambda_n)$ are the barycentric coordinates of x.

(ii) Prove that a simplex σ is convex, arcwise connected, compact, and closed. The interior $\mathrm{Int}\,\sigma$ of a complex is convex, arcwise connected, open, and σ is the closure of $\mathrm{Int}\,\sigma$.

10.6. Prove that condition (2) of Definition 10.3 is equivalent to condition:
(2′) For any two distinct simplices σ_1, σ_2, $\mathrm{Int}\,\sigma_1 \cap \mathrm{Int}\,\sigma_2 = \emptyset$.

10.7. Complete the proof of (1) in Lemma 10.2 (use induction).

10.8. Prove that a sphere does not have any triangulation in which every vertex belongs to six triangles. Conclude that a sphere cannot be triangulated by regular hexagons. Look at a golf ball!

10.9. Given a connected graph in the plane, show that any minimal spanning tree is contained in the Delaunay triangulation of the convex hull of the set of vertices of the graph.

10.10. Write a computer program computing the Delaunay triangulation of a finite set of points in the plane using the method of lifting to a paraboloid.

10.11. Prove Lemma 10.3.

10.12. Let $\{p_1, \ldots, p_n\}$ be a finite set of points contained in a given square S. Consider the following path-planning problem. Given an initial position s and a final position t both on the boundary on the given square S, find a C^2-continuous path from s to t staying inside S with the property that at any given time, a point moving on the path is as far as possible from the nearest point p_i. You may think of the points p_i as radar stations and the moving particle as a flying airplane. The airplane is trying to maximize the minimum distance from the radars.

Solve the above problem as best as you can using Voronoi diagrams and B-splines.

References

1. M. Berg, M. Van Kreveld, M. Overmars, and O. Schwarzkopf. *Computational Geometry. Algorithms and Applications.* Springer-Verlag, first edition, 1997.
2. J.-D. Boissonnat and M. Yvinec. *Géométrie Algorithmique.* Ediscience International, first edition, 1995.
3. K.Q. Brown. Voronoi diagrams from convex hulls. *Inform. Process. Lett.*, 9:223–228, 1979.
4. G.L. Dirichlet. Über die Reduktion der positiven quadratischen Formen mit drei unbestimmten ganzen Zahlen. *Journal für die reine und angewandte Mathematik*, 40:209–227, 1850.
5. H. Edelsbrunner and R. Seidel. Voronoi diagrams and arrangements. *Discrete Computational Geometry*, 1:25–44, 1986.
6. E.N. Gilbert. Random subdivisions of space into crystals. *Annals of Math. Stat.*, 33:958–972, 1962.
7. R. Graham and F. Yao. A whirlwind tour of computational geometry. *American Mathematical Monthly*, 97(8):687–701, 1990.
8. William S. Massey. *A Basic Course in Algebraic Topology.* GTM No. 127. Springer-Verlag, first edition, 1991.
9. James R. Munkres. *Elements of Algebraic Topology.* Addison-Wesley, first edition, 1984.
10. Joseph O'Rourke. *Computational Geometry in C.* Cambridge University Press, second edition, 1998.
11. F.P. Preparata and M.I. Shamos. *Computational Geometry: An Introduction.* Springer-Verlag, first edition, 1988.
12. J.-J. Risler. *Mathematical Methods for CAD.* Masson, first edition, 1992.
13. M.G. Voronoi. Nouvelles applications des paramètres continus à la théorie des formes quadratiques. *J. Reine u. Agnew. Math.*, 134:198–287, 1908.

Chapter 11
Basics of Hermitian Geometry

11.1 Sesquilinear and Hermitian Forms, Pre-Hilbert Spaces and Hermitian Spaces

In this chapter we generalize the basic results of Euclidean geometry presented in Chapter 6 to vector spaces over the complex numbers. Such a generalization is inevitable, and not simply a luxury. For example, linear maps may not have real eigenvalues, but they always have complex eigenvalues. Furthermore, some very important classes of linear maps can be diagonalized if they are extended to the complexification of a real vector space. This is the case for orthogonal matrices, and, more generally, normal matrices. Also, complex vector spaces are often the natural framework in physics or engineering, and they are more convenient for dealing with Fourier series. However, some complications arise due to complex conjugation. Recall that for any complex number $z \in \mathbb{C}$, if $z = x + iy$ where $x, y \in \mathbb{R}$, we let $\Re z = x$, the real part of z, and $\Im z = y$, the imaginary part of z. We also denote the conjugate of $z = x + iy$ by $\bar{z} = x - iy$, and the absolute value (or length, or modulus) of z by $|z|$. Recall that $|z|^2 = z\bar{z} = x^2 + y^2$. There are many natural situations where a map $\varphi \colon E \times E \to \mathbb{C}$ is linear in its first argument and only semilinear in its second argument, which means that $\varphi(u, \mu v) = \bar{\mu}\varphi(u, v)$, as opposed to $\varphi(u, \mu v) = \mu\varphi(u, v)$. For example, the natural inner product to deal with functions $f \colon \mathbb{R} \to \mathbb{C}$, especially Fourier series, is

$$\langle f, g \rangle = \int_{-\pi}^{\pi} f(x)\overline{g(x)}dx,$$

which is semilinear (but not linear) in g. Thus, when generalizing a result from the real case of a Euclidean space to the complex case, we always have to check very carefully that our proofs do not rely on linearity in the second argument. Otherwise, we need to revise our proofs, and sometimes the result is simply wrong!

Before defining the natural generalization of an inner product, it is convenient to define semilinear maps.

Definition 11.1. Given two vector spaces E and F over the complex field $\mathbb{C}$, a function $f \colon E \to F$ is *semilinear* if

$$f(u+v) = f(u) + f(v),$$
$$f(\lambda u) = \overline{\lambda} f(u),$$

for all $u, v \in E$ and all $\lambda \in \mathbb{C}$. The set of all semilinear maps $f \colon E \to \mathbb{C}$ is denoted by $\overline{E}^*$.

It is trivially verified that $\overline{E}^*$ is a vector space over $\mathbb{C}$. It is not quite the dual space E^* of E.

Remark: Instead of defining semilinear maps, we could have defined the vector space $\overline{E}$ as the vector space with the same carrier set E whose addition is the same as that of E, but whose multiplication by a complex number is given by

$$(\lambda, u) \mapsto \overline{\lambda} u.$$

Then it is easy to check that a function $f \colon E \to \mathbb{C}$ is semilinear iff $f \colon \overline{E} \to \mathbb{C}$ is linear. If E has finite dimension n, it is easy to see that $\overline{E}^*$ has the same dimension n (if $(e_1, \ldots, e_n)$ is a basis for E, check that the semilinear maps $(\overline{e_1}, \ldots, \overline{e_n})$ defined such that

$$\overline{e_i}\left(\sum_{j=1}^{n} \lambda_j e_j \right) = \overline{\lambda_i},$$

form a basis of $\overline{E}^*$.)

We can now define sesquilinear forms and Hermitian forms.

Definition 11.2. Given a complex vector space E, a function $\varphi \colon E \times E \to \mathbb{C}$ is a *sesquilinear form* if it is linear in its first argument and semilinear in its second argument, which means that

$$\varphi(u_1 + u_2, v) = \varphi(u_1, v) + \varphi(u_2, v),$$
$$\varphi(u, v_1 + v_2) = \varphi(u, v_1) + \varphi(u, v_2),$$
$$\varphi(\lambda u, v) = \lambda \varphi(u, v),$$
$$\varphi(u, \mu v) = \overline{\mu} \varphi(u, v),$$

for all $u, v, u_1, u_2, v_1, v_2 \in E$, and all $\lambda, \mu \in \mathbb{C}$. A function $\varphi \colon E \times E \to \mathbb{C}$ is a *Hermitian form* if it is sesquilinear and if

$$\varphi(v, u) = \overline{\varphi(u, v)}$$

for all all $u, v \in E$.

Obviously, $\varphi(0, v) = \varphi(u, 0) = 0$. Also note that if $\varphi \colon E \times E \to \mathbb{C}$ is sesquilinear, we have

$$\varphi(\lambda u + \mu v, \lambda u + \mu v) = |\lambda|^2 \varphi(u, u) + \lambda \overline{\mu} \varphi(u, v) + \overline{\lambda} \mu \varphi(v, u) + |\mu|^2 \varphi(v, v),$$

and if $\varphi\colon E\times E\to\mathbb{C}$ is Hermitian, we have

$$\varphi(\lambda u+\mu v,\lambda u+\mu v)=|\lambda|^2\varphi(u,u)+2\Re(\lambda\overline{\mu}\varphi(u,v))+|\mu|^2\varphi(v,v).$$

Note that restricted to real coefficients, a sesquilinear form is bilinear (we sometimes say $\mathbb{R}$-bilinear). The function $\Phi\colon E\to\mathbb{C}$ defined such that $\Phi(u)=\varphi(u,u)$ for all $u\in E$ is called the *quadratic form* associated with φ.

The standard example of a Hermitian form on $\mathbb{C}^n$ is the map φ defined such that

$$\varphi((x_1,\ldots,x_n),(y_1,\ldots,y_n))=x_1\overline{y_1}+x_2\overline{y_2}+\cdots+x_n\overline{y_n}.$$

This map is also positive definite, but before dealing with these issues, we show the following useful lemma.

Lemma 11.1. *Given a complex vector space E, the following properties hold:*

(1) A sesquilinear form $\varphi\colon E\times E\to\mathbb{C}$ is a Hermitian form iff $\varphi(u,u)\in\mathbb{R}$ for all $u\in E$.

(2) If $\varphi\colon E\times E\to\mathbb{C}$ is a sesquilinear form, then

$$\begin{aligned}
4\varphi(u,v)&=\varphi(u+v,u+v)-\varphi(u-v,u-v)\\
&\quad+i\varphi(u+iv,u+iv)-i\varphi(u-iv,u-iv),
\end{aligned}$$

and

$$2\varphi(u,v)=(1+i)(\varphi(u,u)+\varphi(v,v))-\varphi(u-v,u-v)-i\varphi(u-iv,u-iv).$$

These are called polarization identities.

Proof. (1) If φ is a Hermitian form, then

$$\varphi(v,u)=\overline{\varphi(u,v)}$$

implies that

$$\varphi(u,u)=\overline{\varphi(u,u)},$$

and thus $\varphi(u,u)\in\mathbb{R}$. If φ is sesquilinear and $\varphi(u,u)\in\mathbb{R}$ for all $u\in E$, then

$$\varphi(u+v,u+v)=\varphi(u,u)+\varphi(u,v)+\varphi(v,u)+\varphi(v,v),$$

which proves that

$$\varphi(u,v)+\varphi(v,u)=\alpha,$$

where α is real, and changing u to iu, we have

$$i(\varphi(u,v)-\varphi(v,u))=\beta,$$

where β is real, and thus

$$\varphi(u,v)=\frac{\alpha-i\beta}{2}\quad\text{and}\quad\varphi(v,u)=\frac{\alpha+i\beta}{2},$$

proving that φ is Hermitian.

(2) These identities are verified by expanding the right-hand side, and we leave them as an exercise. $\quad\square$

Lemma 11.1 shows that a sesquilinear form is completely determined by the quadratic form $\Phi(u) = \varphi(u,u)$, even if φ is not Hermitian. This is false for a real bilinear form, unless it is symmetric. For example, the bilinear form $\varphi\colon \mathbb{R}\times\mathbb{R}\to\mathbb{R}$ defined such that

$$\varphi((x_1,y_1),(x_2,y_2)) = x_1 y_2 - x_2 y_1$$

is not identically zero, and yet it is null on the diagonal. However, a real symmetric bilinear form is indeed determined by its values on the diagonal, as we saw in Chapter 11.

As in the Euclidean case, Hermitian forms for which $\varphi(u,u) \geq 0$ play an important role.

Definition 11.3. Given a complex vector space E, a Hermitian form $\varphi\colon E\times E\to\mathbb{C}$ is *positive* if $\varphi(u,u) \geq 0$ for all $u \in E$, and *positive definite* if $\varphi(u,u) > 0$ for all $u \neq 0$. A pair $\langle E,\varphi\rangle$ where E is a complex vector space and φ is a Hermitian form on E is called a *pre-Hilbert space* if φ is positive, and a *Hermitian (or unitary) space* if φ is positive definite.

We warn our readers that some authors, such as Lang [3], define a pre-Hilbert space as what we define as a Hermitian space. We prefer following the terminology used in Schwartz [5] and Bourbaki [1]. The quantity $\varphi(u,v)$ is usually called the *Hermitian product* of u and v. We will occasionally call it the inner product of u and v.

Given a pre-Hilbert space $\langle E,\varphi\rangle$, as in the case of a Euclidean space, we also denote $\varphi(u,v)$ by

$$u\cdot v \quad\text{or}\quad \langle u,v\rangle \quad\text{or}\quad (u|v),$$

and $\sqrt{\Phi(u)}$ by $\|u\|$.

Example 11.1. The complex vector space $\mathbb{C}^n$ under the Hermitian form

$$\varphi((x_1,\ldots,x_n),(y_1,\ldots,y_n)) = x_1\overline{y_1} + x_2\overline{y_2} + \cdots + x_n\overline{y_n}$$

is a Hermitian space.

Example 11.2. Let l^2 denote the set of all countably infinite sequences $x = (x_i)_{i\in\mathbb{N}}$ of complex numbers such that $\sum_{i=0}^{\infty} |x_i|^2$ is defined (i.e., the sequence $\sum_{i=0}^{n} |x_i|^2$ converges as $n\to\infty$). It can be shown that the map $\varphi\colon l^2\times l^2\to\mathbb{C}$ defined such that

$$\varphi((x_i)_{i\in\mathbb{N}},(y_i)_{i\in\mathbb{N}}) = \sum_{i=0}^{\infty} x_i\overline{y_i}$$

is well defined, and l^2 is a Hermitian space under φ. Actually, l^2 is even a Hilbert space (Chapter 26 on the web site, see `http://www.cis.upenn.edu/~jean/gbooks/geom2.html`).

Example 11.3. Let $\mathscr{C}_{\text{piece}}[a,b]$ be the set of piecewise bounded continuous functions $f: [a,b] \to \mathbb{C}$ under the Hermitian form

$$\langle f,g \rangle = \int_a^b f(x)\overline{g(x)}dx.$$

It is easy to check that this Hermitian form is positive, but it is not definite. Thus, under this Hermitian form, $\mathscr{C}_{\text{piece}}[a,b]$ is only a pre-Hilbert space.

Example 11.4. Let $\mathscr{C}[-\pi,\pi]$ be the set of complex-valued continuous functions $f: [-\pi,\pi] \to \mathbb{C}$ under the Hermitian form

$$\langle f,g \rangle = \int_a^b f(x)\overline{g(x)}dx.$$

It is easy to check that this Hermitian form is positive definite. Thus, $\mathscr{C}[-\pi,\pi]$ is a Hermitian space.

The Cauchy–Schwarz inequality and the Minkowski inequalities extend to pre-Hilbert spaces and to Hermitian spaces.

Lemma 11.2. *Let $\langle E, \varphi \rangle$ be a pre-Hilbert space with associated quadratic form Φ. For all $u,v \in E$, we have the* Cauchy–Schwarz *inequality*

$$|\varphi(u,v)| \le \sqrt{\Phi(u)}\sqrt{\Phi(v)}.$$

Furthermore, if $\langle E, \varphi \rangle$ is a Hermitian space, the equality holds iff u and v are linearly dependent.
We also have the Minkowski *inequality*

$$\sqrt{\Phi(u+v)} \le \sqrt{\Phi(u)} + \sqrt{\Phi(v)}.$$

Furthermore, if $\langle E, \varphi \rangle$ is a Hermitian space, the equality holds iff u and v are linearly dependent, where in addition, if $u \ne 0$ and $v \ne 0$, then $u = \lambda v$ for some real λ such that $\lambda > 0$.

Proof. For all $u,v \in E$ and all $\mu \in \mathbb{C}$, we have observed that

$$\varphi(u+\mu v, u+\mu v) = \varphi(u,u) + 2\Re(\overline{\mu}\varphi(u,v)) + |\mu|^2\varphi(v,v).$$

Let $\varphi(u,v) = \rho e^{i\theta}$, where $|\varphi(u,v)| = \rho$ ($\rho \ge 0$). Let $F: \mathbb{R} \to \mathbb{R}$ be the function defined such that
$$F(t) = \Phi(u+te^{i\theta}v),$$

for all $t \in \mathbb{R}$. The above shows that

$$F(t) = \varphi(u,u) + 2t|\varphi(u,v)| + t^2\varphi(v,v) = \Phi(u) + 2t|\varphi(u,v)| + t^2\Phi(v).$$

Since φ is assumed to be positive, we have $F(t) \ge 0$ for all $t \in \mathbb{R}$. If $\Phi(v) = 0$, we must have $\varphi(u,v) = 0$, since otherwise, $F(t)$ could be made negative by choosing

t negative and small enough. If $\Phi(v) > 0$, in order for $F(t)$ to be nonnegative, the equation

$$\Phi(u) + 2t|\varphi(u,v)| + t^2 \Phi(v) = 0$$

must not have distinct real roots, which is equivalent to

$$|\varphi(u,v)|^2 \leq \Phi(u)\Phi(v).$$

Taking the square root on both sides yields the Cauchy–Schwarz inequality.

For the second part of the claim, if φ is positive definite, we argue as follows. If u and v are linearly dependent, it is immediately verified that we get an equality. Conversely, if

$$|\varphi(u,v)|^2 = \Phi(u)\Phi(v),$$

then the equation

$$\Phi(u) + 2t|\varphi(u,v)| + t^2 \Phi(v) = 0$$

has a double root t_0, and thus

$$\Phi(u + t_0 e^{i\theta} v) = 0.$$

Since φ is positive definite, we must have

$$u + t_0 e^{i\theta} v = 0,$$

which shows that u and v are linearly dependent.

If we square the Minkowski inequality, we get

$$\Phi(u+v) \leq \Phi(u) + \Phi(v) + 2\sqrt{\Phi(u)}\sqrt{\Phi(v)}.$$

However, we observed earlier that

$$\Phi(u+v) = \Phi(u) + \Phi(v) + 2\Re(\varphi(u,v)).$$

Thus, it is enough to prove that

$$\Re(\varphi(u,v)) \leq \sqrt{\Phi(u)}\sqrt{\Phi(v)},$$

but this follows from the Cauchy–Schwarz inequality

$$|\varphi(u,v)| \leq \sqrt{\Phi(u)}\sqrt{\Phi(v)}$$

and the fact that $\Re z \leq |z|$.

If φ is positive definite and u and v are linearly dependent, it is immediately verified that we get an equality. Conversely, if equality holds in the Minkowski inequality, we must have

$$\Re(\varphi(u,v)) = \sqrt{\Phi(u)}\sqrt{\Phi(v)},$$

which implies that

$$|\varphi(u,v)| = \sqrt{\Phi(u)}\sqrt{\Phi(v)},$$

since otherwise, by the Cauchy–Schwarz inequality, we would have

$$\Re(\varphi(u,v)) \leq |\varphi(u,v)| < \sqrt{\Phi(u)}\sqrt{\Phi(v)}.$$

Thus, equality holds in the Cauchy–Schwarz inequality, and

$$\Re(\varphi(u,v)) = |\varphi(u,v)|.$$

But then, we proved in the Cauchy–Schwarz case that u and v are linearly dependent. Since we also just proved that $\varphi(u,v)$ is real and nonnegative, the coefficient of proportionality between u and v is indeed nonnegative. $\square$

As in the Euclidean case, if $\langle E, \varphi \rangle$ is a Hermitian space, the Minkowski inequality

$$\sqrt{\Phi(u+v)} \leq \sqrt{\Phi(u)} + \sqrt{\Phi(v)}$$

shows that the map $u \mapsto \sqrt{\Phi(u)}$ is a *norm* on E. The norm induced by φ is called the *Hermitian norm induced by* φ. We usually denote $\sqrt{\Phi(u)}$ by $\|u\|$, and the Cauchy–Schwarz inequality is written as

$$|u \cdot v| \leq \|u\|\|v\|.$$

Since a Hermitian space is a normed vector space, it is a topological space under the topology induced by the norm (a basis for this topology is given by the open balls $B_0(u,\rho)$ of center u and radius $\rho > 0$, where

$$B_0(u,\rho) = \{v \in E \mid \|v - u\| < \rho\}.$$

If E has finite dimension, every linear map is continuous; see Lang [3, 4], Dixmier [2], or Schwartz [5, 6]. The Cauchy–Schwarz inequality

$$|u \cdot v| \leq \|u\|\|v\|$$

shows that $\varphi \colon E \times E \to \mathbb{C}$ is continuous, and thus, that $\|\ \|$ is continuous.

If $\langle E, \varphi \rangle$ is only pre-Hilbertian, $\|u\|$ is called a *seminorm*. In this case, the condition

$$\|u\| = 0 \quad \text{implies} \quad u = 0$$

is not necessarily true. However, the Cauchy–Schwarz inequality shows that if $\|u\| = 0$, then $u \cdot v = 0$ for all $v \in E$.

We will now basically mirror the presentation of Euclidean geometry given in Chapter 6 rather quickly, leaving out most proofs, except when they need to be seriously amended. This will be the case for the Cartan–Dieudonné theorem.

11.2 Orthogonality, Duality, Adjoint of a Linear Map

In this section we assume that we are dealing with Hermitian spaces. We denote the Hermitian inner product by $u \cdot v$ or $\langle u, v \rangle$. The concepts of orthogonality, orthogonal family of vectors, orthonormal family of vectors, and orthogonal complement of a set of vectors are unchanged from the Euclidean case (Definition 6.2).

For example, the set $\mathscr{C}[-\pi, \pi]$ of continuous functions $f \colon [-\pi, \pi] \to \mathbb{C}$ is a Hermitian space under the product

$$\langle f, g \rangle = \int_{-\pi}^{\pi} f(x)\overline{g(x)}dx,$$

and the family $(e^{ikx})_{k \in \mathbb{Z}}$ is orthogonal.

Lemma 6.2 and 6.3 hold without any changes. It is easy to show that

$$\left\| \sum_{i=1}^{n} u_i \right\|^2 = \sum_{i=1}^{n} \|u_i\|^2 + \sum_{1 \le i < j \le n} 2\Re(u_i \cdot u_j).$$

Analogously to the case of Euclidean spaces of finite dimension, the Hermitian product induces a canonical bijection (i.e., independent of the choice of bases) between the vector space E and the space E^*. This is one of the places where conjugation shows up, but in this case, troubles are minor.

Given a Hermitian space E, for any vector $u \in E$, let $\varphi_u^l \colon E \to \mathbb{C}$ be the map defined such that

$$\varphi_u^l(v) = u \cdot v,$$

for all $v \in E$. Similarly, for any vector $v \in E$, let $\varphi_v^r \colon E \to \mathbb{C}$ be the map defined such that

$$\varphi_v^r(u) = u \cdot v,$$

for all $u \in E$.

Since the Hermitian product is linear in its first argument u, the map φ_v^r is a linear form in E^*, and since it is semilinear in its second argument v, the map φ_u^l is a semilinear form in $\overline{E}^*$. Thus, we have two maps $\flat^l \colon E \to \overline{E}^*$ and $\flat^r \colon E \to E^*$, defined such that

$$\flat^l(u) = \varphi_u^l, \quad \text{and} \quad \flat^r(v) = \varphi_v^r.$$

Lemma 11.3. *let E be a Hermitian space E.*

(1) The map $\flat^l \colon E \to \overline{E}^*$ defined such that

$$\flat^l(u) = \varphi_u^l$$

is linear and injective.

(2) The map $\flat^r \colon E \to E^*$ defined such that

$$\flat^r(v) = \varphi_v^r$$

is semilinear and injective.

When E is also of finite dimension, the maps $\flat^l : E \to \overline{E}^*$ and $\flat^r : E \to E^*$ are canonical isomorphisms.

Proof. (1) That $\flat^l : E \to \overline{E}^*$ is a linear map follows immediately from the fact that the Hermitian product is linear in its first argument. If $\varphi_u^l = \varphi_v^l$, then $\varphi_u^l(w) = \varphi_v^l(w)$ for all $w \in E$, which by definition of φ_u^l means that

$$u \cdot w = v \cdot w$$

for all $w \in E$, which by linearity on the left is equivalent to

$$(v - u) \cdot w = 0$$

for all $w \in E$, which implies that $u = v$, since the Hermitian product is positive definite. Thus, $\flat^l : E \to \overline{E}^*$ is injective. Finally, when E is of finite dimension n, $\overline{E}^*$ is also of dimension n, and then $\flat^l : E \to \overline{E}^*$ is bijective.

The proof of (2) is essentially the same as the proof of (1), except that the Hermitian product is semilinear in its second argument. $\square$

The inverse of the isomorphism $\flat^l : E \to \overline{E}^*$ is denoted by $\sharp^l : \overline{E}^* \to E$, and the inverse of the isomorphism $\flat^r : E \to E^*$ is denoted by $\sharp^r : E^* \to E$.

As a corollary of the isomorphism $\flat^r : E \to E^*$, if E is a Hermitian space of finite dimension, then every linear form $f \in E^*$ corresponds to a unique $v \in E$, such that

$$f(u) = u \cdot v,$$

for every $u \in E$. In particular, if f is not the null form, the kernel of f, which is a hyperplane H, is precisely the set of vectors that are orthogonal to v.

Remark: The "musical map" $\flat^r : E \to E^*$ is not surjective when E has infinite dimension. This result can be salvaged by restricting our attention to continuous linear maps, and by assuming that the vector space E is a *Hilbert space*.

The existence of the isomorphism $\flat^l : E \to \overline{E}^*$ is crucial to the existence of adjoint maps. Indeed, Lemma 11.3 allows us to define the adjoint of a linear map on a Hermitian space. Let E be a Hermitian space of finite dimension n, and let $f : E \to E$ be a linear map. For every $u \in E$, the map

$$v \mapsto u \cdot f(v)$$

is clearly a semilinear form in $\overline{E}^*$, and by Lemma 11.3, there is a unique vector in E denoted by $f^*(u)$ such that

$$f^*(u) \cdot v = u \cdot f(v),$$

for every $v \in E$. The following lemma shows that the map f^* is linear.

Lemma 11.4. *Given a Hermitian space E of finite dimension, for every linear map $f\colon E \to E$ there is a unique linear map $f^*\colon E \to E$ such that*

$$f^*(u) \cdot v = u \cdot f(v),$$

for all $u, v \in E$. The map f^* is called the adjoint of f (w.r.t. to the Hermitian product).

Proof. Careful inspection of the proof of lemma 6.5 reveals that it applies unchanged. The only potential problem is in proving that $f^*(\lambda u) = \lambda f^*(u)$, but everything takes place in the first argument of the Hermitian product, and there, we have linearity. $\square$

The fact that

$$v \cdot u = \overline{u \cdot v}$$

implies that the adjoint f^* of f is also characterized by

$$f(u) \cdot v = u \cdot f^*(v),$$

for all $u, v \in E$. It is also obvious that $f^{**} = f$.

Given two Hermitian spaces E and F, where the Hermitian product on E is denoted by $\langle -, - \rangle_1$ and the Hermitian product on F is denoted by $\langle -, - \rangle_2$, given any linear map $f\colon E \to F$, it is immediately verified that the proof of Lemma 11.4 can be adapted to show that there is a unique linear map $f^*\colon F \to E$ such that

$$\langle f(u), v \rangle_2 = \langle u, f^*(v) \rangle_1$$

for all $u \in E$ and all $v \in F$. The linear map f^* is also called the adjoint of f.

As in the Euclidean case, Lemma 11.3 can be used to show that any Hermitian space of finite dimension has an orthonormal basis. The proof is unchanged.

Lemma 11.5. *Given any nontrivial Hermitian space E of finite dimension $n \geq 1$, there is an orthonormal basis $(u_1, \ldots, u_n)$ for E.*

The *Gram–Schmidt orthonormalization procedure* also applies to Hermitian spaces of finite dimension, without any changes from the Euclidean case!

Lemma 11.6. *Given a nontrivial Hermitian space E of finite dimension $n \geq 1$, from any basis $(e_1, \ldots, e_n)$ for E we can construct an orthonormal basis $(u_1, \ldots, u_n)$ for E with the property that for every k, $1 \leq k \leq n$, the families $(e_1, \ldots, e_k)$ and $(u_1, \ldots, u_k)$ generate the same subspace.*

Remark: The remarks made after Lemma 6.7 also apply here, except that in the QR-decomposition, Q is a unitary matrix.

As a consequence of Lemma 6.6 (or Lemma 11.6), given any Hermitian space of finite dimension n, if $(e_1, \ldots, e_n)$ is an orthonormal basis for E, then for any two vectors $u = u_1 e_1 + \cdots + u_n e_n$ and $v = v_1 e_1 + \cdots + v_n e_n$, the Hermitian product $u \cdot v$ is expressed as

$$u \cdot v = (u_1 e_1 + \cdots + u_n e_n) \cdot (v_1 e_1 + \cdots + v_n e_n) = \sum_{i=1}^{n} u_i \overline{v_i},$$

and the norm $\|u\|$ as

$$\|u\| = \|u_1 e_1 + \cdots + u_n e_n\| = \sqrt{\sum_{i=1}^{n} |u_i|^2}.$$

Lemma 6.8 also holds unchanged.

Lemma 11.7. *Given any nontrivial Hermitian space E of finite dimension $n \geq 1$, for any subspace F of dimension k, the orthogonal complement $F^{\perp}$ of F has dimension $n - k$, and $E = F \oplus F^{\perp}$. Furthermore, we have $F^{\perp\perp} = F$.*

Affine Hermitian spaces are defined just as affine Euclidean spaces, except that we modify Definition 6.3 to require that the complex vector space $\overrightarrow{E}$ be a Hermitian space. We denote by $\mathbb{E}_{\mathbb{C}}^m$ the Hermitian affine space obtained from the affine space $\mathbb{A}_{\mathbb{C}}^m$ by defining on the vector space $\mathbb{C}^m$ the standard Hermitian product

$$(x_1, \ldots, x_m) \cdot (y_1, \ldots, y_m) = x_1 \overline{y_1} + \cdots + x_m \overline{y_m}.$$

The corresponding Hermitian norm is

$$\|(x_1, \ldots, x_m)\| = \sqrt{|x_1|^2 + \cdots + |x_m|^2}.$$

Lemma 8.2 also holds for Hermitian spaces, and the proof is the same.

Lemma 11.8. *Let E be a Hermitian space of finite dimension n, and let $f \colon E \to E$ be an isometry. For any subspace F of E, if $f(F) = F$, then $f(F^{\perp}) \subseteq F^{\perp}$ and $E = F \oplus F^{\perp}$.*

11.3 Linear Isometries (Also Called Unitary Transformations)

In this section we consider linear maps between Hermitian spaces that preserve the Hermitian norm. All definitions given for Euclidean spaces in Section 6.3 extend to Hermitian spaces, except that orthogonal transformations are called unitary transformation, but Lemma 6.9 extends only with a modified condition (2). Indeed, the old proof that (2) implies (3) does not work, and the implication is in fact false! It can be repaired by strengthening condition (2). For the sake of completeness, we state the Hermitian version of Definition 6.4.

Definition 11.4. Given any two nontrivial Hermitian spaces E and F of the same finite dimension n, a function $f \colon E \to F$ is *a unitary transformation, or a linear isometry*, if it is linear and

$$\|f(u)\| = \|u\|,$$

for all $u \in E$.

Lemma 6.9 can be salvaged by strengthening condition (2).

Lemma 11.9. *Given any two nontrivial Hermitian spaces E and F of the same finite dimension n, for every function $f \colon E \to F$, the following properties are equivalent:*

(1) f is a linear map and $\|f(u)\| = \|u\|$, for all $u \in E$;
(2) $\|f(v) - f(u)\| = \|v - u\|$ and $f(iu) = if(u)$, for all $u, v \in E$.
(3) $f(u) \cdot f(v) = u \cdot v$, for all $u, v \in E$.

Furthermore, such a map is bijective.

Proof. The proof that (2) implies (3) given in Lemma 6.9 needs to be revised as follows. We use the polarization identity

$$2\varphi(u,v) = (1 + i)(\|u\|^2 + \|v\|^2) - \|u - v\|^2 - i\|u - iv\|^2.$$

Since $f(iv) = if(v)$, we get $f(0) = 0$ by setting $v = 0$, so the function f preserves distance and norm, and we get

$$\begin{aligned}
2\varphi(f(u), f(v)) &= (1 + i)(\|f(u)\|^2 + \|f(v)\|^2) - \|f(u) - f(v)\|^2 \\
&\quad - i\|f(u) - if(v)\|^2 \\
&= (1 + i)(\|f(u)\|^2 + \|f(v)\|^2) - \|f(u) - f(v)\|^2 \\
&\quad - i\|f(u) - f(iv)\|^2 \\
&= (1 + i)(\|u\|^2 + \|v\|^2) - \|u - v\|^2 - i\|u - iv\|^2 \\
&= 2\varphi(u, v),
\end{aligned}$$

which shows that f preserves the Hermitian inner product, as desired. The rest of the proof is unchanged. $\square$

Remarks:

(i) In the Euclidean case, we proved that the assumption

$$\|f(v) - f(u)\| = \|v - u\| \quad \text{for all } u, v \in E \text{ and } f(0) = 0 \tag{$2'$}$$

implies (3). For this we used the polarization identity

$$2u \cdot v = \|u\|^2 + \|v\|^2 - \|u - v\|^2.$$

In the Hermitian case the polarization identity involves the complex number i. In fact, the implication (2′) implies (3) is false in the Hermitian case! Conjugation $z \mapsto \bar{z}$ satisfies (2′) since

$$|\bar{z}_2 - \bar{z}_1| = |\overline{z_2 - z_1}| = |z_2 - z_1|,$$

and yet, it is not linear!

(ii) If we modify (2) by changing the second condition by now requiring that there be some $\tau \in E$ such that

$$f(\tau + iu) = f(\tau) + i(f(\tau + u) - f(\tau))$$

for all $u \in E$, then the function $g \colon E \to E$ defined such that

$$g(u) = f(\tau + u) - f(\tau)$$

satisfies the old conditions of (2), and the implications (2) $\to$ (3) and (3) $\to$ (1) prove that g is linear, and thus that f is affine. In view of the first remark, some condition involving i is needed on f, in addition to the fact that f is distance-preserving.

11.4 The Unitary Group, Unitary Matrices

In this section, as a mirror image of our treatment of the isometries of a Euclidean space, we explore some of the fundamental properties of the unitary group and of unitary matrices. As an immediate corollary of the Gram–Schmidt orthonormalization procedure, we obtain the QR-decomposition for invertible matrices. In the Hermitian framework, the matrix of the adjoint of a linear map is not given by the transpose of the original matrix, but by its conjugate.

Definition 11.5. Given a complex $m \times n$ matrix A, the *transpose* $A^\top$ *of* A is the $n \times m$ matrix $A^\top = \left(a^\top_{i,j} \right)$ defined such that

$$a^\top_{i,j} = a_{j,i},$$

and the *conjugate* $\overline{A}$ *of* A is the $m \times n$ matrix $\overline{A} = (b_{i,j})$ defined such that

$$b_{i,j} = \bar{a}_{i,j}$$

for all i, j, $1 \le i \le m$, $1 \le j \le n$. The *adjoint* A^* *of* A is the matrix defined such that

$$A^* = \overline{(A^\top)} = \left(\overline{A} \right)^\top.$$

Lemma 11.10. *Let E be any Hermitian space of finite dimension n, and let $f\colon E \to E$ be any linear map. The following properties hold:*

(1) The linear map $f\colon E \to E$ is an isometry iff

$$f \circ f^* = f^* \circ f = \mathrm{id}.$$

(2) For every orthonormal basis $(e_1,\ldots,e_n)$ of E, if the matrix of f is A, then the matrix of f^* is the adjoint A^* of A, and f is an isometry iff A satisfies the identities

$$AA^* = A^*A = I_n,$$

where I_n denotes the identity matrix of order n, iff the columns of A form an orthonormal basis of E, iff the rows of A form an orthonormal basis of E.

Proof. (1) The proof is identical to that of Lemma 6.10 (1).

(2) If $(e_1,\ldots,e_n)$ is an orthonormal basis for E, let $A = (a_{i,j})$ be the matrix of f, and let $B = (b_{i,j})$ be the matrix of f^*. Since f^* is characterized by

$$f^*(u) \cdot v = u \cdot f(v)$$

for all $u, v \in E$, using the fact that if $w = w_1 e_1 + \cdots + w_n e_n$, we have $w_k = w \cdot e_k$, for all k, $1 \leq k \leq n$; letting $u = e_i$ and $v = e_j$, we get

$$b_{j,i} = f^*(e_i) \cdot e_j = e_i \cdot f(e_j) = \overline{f(e_j) \cdot e_i} = \overline{a_{i,j}},$$

for all i, j, $1 \leq i, j \leq n$. Thus, $B = A^*$. Now, if X and Y are arbitrary matrices over the basis $(e_1,\ldots,e_n)$, denoting as usual the jth column of X by X_j, and similarly for Y, a simple calculation shows that

$$Y^*X = (X_j \cdot Y_i)_{1 \leq i, j \leq n}.$$

Then it is immediately verified that if $X = Y = A$, then $A^*A = AA^* = I_n$ iff the column vectors $(A_1,\ldots,A_n)$ form an orthonormal basis. Thus, from (1), we see that (2) is clear. $\square$

Lemma 6.10 shows that the inverse of an isometry f is its adjoint f^*. Lemma 6.10 also motivates the following definition.

Definition 11.6. A complex $n \times n$ matrix is a *unitary matrix* if

$$AA^* = A^*A = I_n.$$

Remarks:

(1) The conditions $AA^* = I_n$, $A^*A = I_n$, and $A^{-1} = A^*$ are equivalent. Given any two orthonormal bases $(u_1,\ldots,u_n)$ and $(v_1,\ldots,v_n)$, if P is the change of basis

matrix from $(u_1, \ldots, u_n)$ to $(v_1, \ldots, v_n)$, it is easy to show that the matrix P is unitary. The proof of Lemma 11.9 (3) also shows that if f is an isometry, then the image of an orthonormal basis $(u_1, \ldots, u_n)$ is an orthonormal basis.

(2) If f is unitary and A is its matrix with respect to any orthonormal basis, the characteristic polynomial $D(A - \lambda I)$ of A is a polynomial with complex coefficients, and thus it has n (complex) roots (counting multiplicities). If u is an eigenvector of f for λ, then from $f(u) = \lambda u$ and the fact that f is an isometry we get

$$\|u\| = \|f(u)\| = \|\lambda u\| = |\lambda| \|u\|,$$

which shows that $|\lambda| = 1$. Since the determinant $D(A)$ of f is the product of the eigenvalues of f, we have $|D(A)| = 1$. It is clear that the isometries of a Hermitian space of dimension n form a group, and that the isometries of determinant $+1$ form a subgroup.

This leads to the following definition.

Definition 11.7. Given a Hermitian space E of dimension n, the set of isometries $f \colon E \to E$ forms a subgroup of $\mathbf{GL}(E, \mathbb{C})$ denoted by $\mathbf{U}(E)$, or $\mathbf{U}(n)$ when $E = \mathbb{C}^n$, called the *unitary group (of E)*. For every isometry f we have $|D(f)| = 1$, where $D(f)$ denotes the determinant of f. The isometries such that $D(f) = 1$ are called *rotations, or proper isometries, or proper unitary transformations*, and they form a subgroup of the special linear group $\mathbf{SL}(E, \mathbb{C})$ (and of $\mathbf{U}(E)$), denoted by $\mathbf{SU}(E)$, or $\mathbf{SU}(n)$ when $E = \mathbb{C}^n$, called the *special unitary group (of E)*. The isometries such that $D(f) \neq 1$ are called *improper isometries, or improper unitary transformations, or flip transformations*.

A very important example of unitary matrices is provided by Fourier matrices (up to a factor of $\sqrt{n}$), matrices that arise in the various versions of the discrete Fourier transform. For more on this topic, see the problems, and Strang [7, 8].

Now that we have the definition of a unitary matrix, we can explain how the Gram–Schmidt orthonormalization procedure immediately yields the QR-decomposition for matrices.

Lemma 11.11. *Given any $n \times n$ complex matrix A, if A is invertible, then there is a unitary matrix Q and an upper triangular matrix R with positive diagonal entries such that $A = QR$.*

The proof is absolutely the same as in the real case!

Due to space limitations, we will not study the isometries of a Hermitian space in this chapter. However, the reader will find such a study in the supplements on the web site (Chapter 25, see `http://www.cis.upenn.edu/~jean/gbooks/geom2.html`).

11.5 Problems

11.1. Given a complex vector space E of finite dimension n, prove that $\overline{E}^*$ also has dimension n.

Hint. If $(e_1, \ldots, e_n)$ is a basis for E, check that the semilinear maps $\overline{e_i}$ defined such that

$$\overline{e_i}\left(\sum_{j=1}^{n} \lambda_j e_j\right) = \overline{\lambda_i}$$

form a basis of $\overline{E}^*$.

11.2. Prove the polarization identities in Lemma 11.1 (2).

11.3. Given a Hermitian space E, for any orthonormal basis $(e_1, \ldots, e_n)$, if X and Y are arbitrary matrices over the basis $(e_1, \ldots, e_n)$, denoting as usual the jth column of X by X_j, and similarly for Y, prove that

$$Y^*X = (X_j \cdot Y_i)_{1 \leq i,j \leq n}.$$

Then prove that

$$A^*A = AA^* = I_n$$

iff the column vectors $(A_1, \ldots, A_n)$ form an orthonormal basis.

11.4. Given a Hermitian space E, prove that if f is an isometry, then f maps any orthonormal basis of E to an orthonormal basis.

11.5. Given p vectors $(u_1, \ldots, u_p)$ in a Hermitian space E of dimension $n \geq p$, the *Gram determinant (or Gramian)* of the vectors $(u_1, \ldots, u_p)$ is the determinant

$$\mathrm{Gram}(u_1, \ldots, u_p) = \begin{vmatrix} \|u_1\|^2 & \langle u_1, u_2 \rangle & \ldots & \langle u_1, u_p \rangle \\ \langle u_2, u_1 \rangle & \|u_2\|^2 & \ldots & \langle u_2, u_p \rangle \\ \vdots & \vdots & \ddots & \vdots \\ \langle u_p, u_1 \rangle & \langle u_p, u_2 \rangle & \ldots & \|u_p\|^2 \end{vmatrix}.$$

(1) Prove that

$$\mathrm{Gram}(u_1, \ldots, u_n) = \lambda_E(u_1, \ldots, u_n)^2.$$

Hint. By Problem 11.3, if $(e_1, \ldots, e_n)$ is an orthonormal basis of E and A is the matrix of the vectors $(u_1, \ldots, u_n)$ over this basis, then

$$\det(A)^2 = \det(A^*A) = \det(A_i \cdot A_j),$$

where A_i denotes the ith column of the matrix A, and $(A_i \cdot A_j)$ denotes the $n \times n$ matrix with entries $A_i \cdot A_j$.

11.6. Let F_n be the symmetric $n \times n$ matrix (with complex coefficients)

$$F_n = \left(e^{i2\pi kl/n}\right)_{\substack{0 \le k \le n-1 \\ 0 \le l \le n-1}}$$

assuming that we index the entries in F_n over $[0, 1, \ldots, n-1] \times [0, 1, \ldots, n-1]$, the standard kth row now being indexed by $k-1$ and the standard lth column now being indexed by $l-1$. The matrix F_n is called a *Fourier matrix*.

(1) Letting $\overline{F_n} = \left(e^{-i2\pi kl/n}\right)_{\substack{0 \le k \le n-1 \\ 0 \le l \le n-1}}$ be the conjugate of F_n, prove that

$$F_n\overline{F_n} = \overline{F_n}F_n = nI_n.$$

The above shows that $F_n/\sqrt{n}$ is unitary.

(2) Define the *discrete Fourier transform* $\widehat{f}$ of a sequence $f = (f_0, \ldots, f_{n-1}) \in \mathbb{C}^n$ as

$$\widehat{f} = \overline{F_n}f.$$

Define the *inverse discrete Fourier transform* (taking c back to f) as

$$\overline{\overline{c}} = F_n c,$$

where $c = (c_0, \ldots, c_{n-1}) \in \mathbb{C}^n$. Define the *circular shift matrix S_n (of order n)* as the matrix

$$S_n = \begin{pmatrix} 0 & 0 & 0 & 0 & \cdots & 0 & 1 \\ 1 & 0 & 0 & 0 & \cdots & 0 & 0 \\ 0 & 1 & 0 & 0 & \cdots & 0 & 0 \\ 0 & 0 & 1 & 0 & \cdots & 0 & 0 \\ \vdots & \vdots & \vdots & \vdots & \vdots & \vdots & \vdots \\ 0 & 0 & 0 & 0 & \cdots & 1 & 0 \end{pmatrix}$$

consisting of cyclic permutations of its first column. For any sequence $f = (f_0, \ldots, f_{n-1}) \in \mathbb{C}^n$, we define the *circulant matrix $H(f)$* as

$$H(f) = \sum_{j=0}^{n-1} f_j S_n^j,$$

where $S_n^0 = I_n$, as usual.

Prove that

$$H(f)F_n = F_n\widehat{f}.$$

The above shows that the columns of the Fourier matrix F_n are the eigenvectors of the circulant matrix $H(f)$, and that the eigenvalue associated with the lth eigenvector is $(\widehat{f})_l$, the lth component of the Fourier transform $\widehat{f}$ of f (counting from 0).

Hint. Prove that

$$S_n F_n = F_n \operatorname{diag}(v^1)$$

where $\operatorname{diag}(v^1)$ is the diagonal matrix with the following entries on the diagonal:

$$v^1 = \left(1, e^{-i2\pi/n}, \ldots, e^{-ik2\pi/n}, \ldots, e^{-i(n-1)2\pi/n}\right).$$

(3) If the sequence $f = (f_0, \ldots, f_{n-1})$ is even, which means that $f_{-j} = f_j$ for all $j \in \mathbb{Z}$ (viewed as a periodic sequence), or equivalently that $f_{n-j} = f_j$ for all j, $0 \leq j \leq n-1$, prove that the Fourier transform $\widehat{f}$ is expressed as

$$\widehat{f}(k) = \sum_{j=0}^{n-1} f_j \cos\left(2\pi jk/n\right),$$

and that the inverse Fourier transform (taking c back to f) is expressed as

$$\overline{\widehat{c}}(k) = \sum_{j=0}^{n-1} c_j \cos\left(2\pi jk/n\right),$$

for every k, $0 \leq k \leq n-1$.

(4) Define the *convolution* $f \star g$ of two sequences $f = (f_0, \ldots, f_{n-1})$ and $g = (g_0, \ldots, g_{n-1})$ as

$$f \star g = H(f)\,g,$$

viewing f and g as column vectors.

Prove the *(circular) convolution rule*

$$\widehat{f \star g} = \widehat{f}\,\widehat{g},$$

where the multiplication on the right-hand side is just the inner product of the vectors $\widehat{f}$ and $\widehat{g}$.

11.7. Let $\varphi \colon E \times E \to \mathbb{C}$ be a sesquilinear form on a complex vector space E of finite dimension n. Given any basis $(e_1, \ldots, e_n)$ of E, let $A = (\alpha_{ij})$ be the matrix defined such that

$$\alpha_{ij} = \varphi(e_i, e_j),$$

$1 \leq i, j \leq n$. We call A *the matrix of φ w.r.t. the basis* $(e_1, \ldots, e_n)$.

(a) For any two vectors x and y, if X and Y denote the column vectors of coordinates of x and y w.r.t. the basis $(e_1, \ldots, e_n)$, prove that

$$\varphi(x,y) = X^{\top} A \overline{Y}.$$

(b) Recall that A is a *Hermitian* matrix if $A = A^* = \overline{A^{\top}}$. Prove that φ is Hermitian iff A is a Hermitian matrix. When is it true that

$$\varphi(x,y) = Y^* A X?$$

(c) If $(f_1, \ldots, f_n)$ is another basis of E and P is the change of basis matrix from $(e_1, \ldots, e_n)$ to $(f_1, \ldots, f_n)$, prove that the matrix of φ w.r.t. the basis $(f_1, \ldots, f_n)$ is

$$P^{\top} A \overline{P}.$$

The common rank of all matrices representing φ is called the *rank* of φ.

11.8. Let $\varphi\colon E \times E \to \mathbb{C}$ be a Hermitian form on a complex vector space E of finite dimension n. Two vectors x and y are said to be *conjugate w.r.t.* φ if $\varphi(x,y) = 0$. The main purpose of this problem is to prove that there is a basis of vectors that are pairwise conjugate w.r.t. φ.

(a) Prove that if $\varphi(x,x) = 0$ for all $x \in E$, then φ is identically null on E. For this, compute $\varphi(ix+y, ix+y)$ and $i\varphi(x+y, x+y)$, and conclude that $\varphi(x,y) = 0$.

Otherwise, we can assume that there is some vector $x \in E$ such that $\varphi(x,x) \neq 0$. Use induction to prove that there is a basis of vectors that are pairwise conjugate w.r.t. φ.

For the induction step, proceed as follows. Let $(e_1, e_2, \ldots, e_n)$ be a basis of E, with $\varphi(e_1, e_1) \neq 0$. Prove that there are scalars $\lambda_2, \ldots, \lambda_n$ such that each of the vectors

$$v_i = e_i + \lambda_i e_1,$$

is conjugate to e_1 w.r.t. φ, where $2 \leq i \leq n$, and that $(e_1, v_2, \ldots, v_n)$ is a basis.

(b) Let $(e_1, \ldots, e_n)$ be a basis of vectors that are pairwise conjugate w.r.t. φ, and assume that they are ordered such that

$$\varphi(e_i, e_i) = \begin{cases} \theta_i \neq 0 & \text{if } 1 \leq i \leq r, \\ 0 & \text{if } r+1 \leq i \leq n, \end{cases}$$

where r is the rank of φ. Show that the matrix of φ w.r.t. $(e_1, \ldots, e_n)$ is a diagonal matrix, and that

$$\varphi(x,y) = \sum_{i=1}^{r} \theta_i x_i \overline{y_i},$$

where $x = \sum_{i=1}^{n} x_i e_i$ and $y = \sum_{i=1}^{n} y_i e_i$.

Prove that for every Hermitian matrix A there is an invertible matrix P such that

$$P^\top A \overline{P} = D,$$

where D is a diagonal matrix.

(c) Prove that there is an integer p, $0 \leq p \leq r$ (where r is the rank of φ), such that $\varphi(u_i, u_i) > 0$ for exactly p vectors of every basis $(u_1, \ldots, u_n)$ of vectors that are pairwise conjugate w.r.t. φ (*Sylvester's inertia theorem*).

Proceed as follows. Assume that in the basis $(u_1, \ldots, u_n)$, for any $x \in E$, we have

$$\varphi(x,x) = \alpha_1 |x_1|^2 + \cdots + \alpha_p |x_p|^2 - \alpha_{p+1} |x_{p+1}|^2 - \cdots - \alpha_r |x_r|^2,$$

where $x = \sum_{i=1}^{n} x_i u_i$, and that in the basis $(v_1, \ldots, v_n)$, for any $x \in E$, we have

$$\varphi(x,x) = \beta_1 |y_1|^2 + \cdots + \beta_q |y_q|^2 - \beta_{q+1} |y_{q+1}|^2 - \cdots - \beta_r |y_r|^2,$$

where $x = \sum_{i=1}^{n} y_i v_i$, with $\alpha_i > 0$, $\beta_i > 0$, $1 \leq i \leq r$.

Assume that $p > q$ and derive a contradiction. First, consider x in the subspace F spanned by

$$(u_1, \ldots, u_p, u_{r+1}, \ldots, u_n),$$

and observe that $\varphi(x,x) \geq 0$ if $x \neq 0$. Next, consider x in the subspace G spanned by

$$(v_{q+1}, \ldots, v_r),$$

and observe that $\varphi(x,x) < 0$ if $x \neq 0$. Prove that $F \cap G$ is nontrivial (i.e., contains some nonnull vector), and derive a contradiction. This implies that $p \leq q$. Finish the proof.

The pair $(p, r - p)$ is called the *signature* of φ.

(d) A Hermitian form φ is *definite* if for every $x \in E$, if $\varphi(x,x) = 0$, then $x = 0$.

Prove that a Hermitian form is definite iff its signature is either $(n, 0)$ or $(0, n)$. In other words, a Hermitian definite form has rank n and is either positive or negative.

(e) The *kernel* of a Hermitian form φ is the subspace consisting of the vectors that are conjugate to all vectors in E. We say that a Hermitian form φ is *nondegenerate* if its kernel is trivial (i.e., reduced to $\{0\}$).

Prove that a Hermitian form φ is nondegenerate iff its rank is n, the dimension of E. Is a definite Hermitian form φ nondegenerate? What about the converse?

Prove that if φ is nondegenerate, then there is a basis of vectors that are pairwise conjugate w.r.t. φ and such that φ is represented by the matrix

$$\begin{pmatrix} I_p & 0 \\ 0 & -I_q \end{pmatrix},$$

where (p, q) is the signature of φ.

(f) Given a nondegenerate Hermitian form φ on E, prove that for every linear map $f \colon E \to E$, there is a unique linear map $f^* \colon E \to E$ such that

$$\varphi(f(u), v) = \varphi(u, f^*(v)),$$

for all $u, v \in E$. The map f^* is called the *adjoint of f (w.r.t. to φ)*. Given any basis $(u_1, \ldots, u_n)$, if Ω is the matrix representing φ and A is the matrix representing f, prove that f^* is represented by $(\Omega^\top)^{-1} A^* \Omega^\top$.

Prove that Lemma 11.3 also holds, i.e., the maps $\flat^l \colon E \to \overline{E}^*$ and $\flat^r \colon E \to E^*$ are canonical isomorphisms.

A linear map $f \colon E \to E$ is an *isometry w.r.t.* φ if

$$\varphi(f(x), f(y)) = \varphi(x, y)$$

for all $x, y \in E$. Prove that a linear map f is an isometry w.r.t. φ iff

$$f^* \circ f = f \circ f^* = \text{id}.$$

Prove that the set of isometries w.r.t. φ is a group. This group is denoted by $\mathbf{U}(\varphi)$, and its subgroup consisting of isometries having determinant $+1$ by $\mathbf{SU}(\varphi)$. Given any basis of E, if Ω is the matrix representing φ and A is the matrix representing f, prove that $f \in \mathbf{U}(\varphi)$ iff

$$A^* \Omega^\top A = \Omega^\top.$$

Given another nondegenerate Hermitian form ψ on E, we say that φ and ψ are *equivalent* if there is a bijective linear map $h \colon E \to E$ such that

$$\psi(x, y) = \varphi(h(x), h(y)),$$

for all $x, y \in E$. Prove that the groups of isometries $\mathbf{U}(\varphi)$ and $\mathbf{U}(\psi)$ are isomomorphic (use the map $f \mapsto h \circ f \circ h^{-1}$ from $\mathbf{U}(\psi)$ to $\mathbf{U}(\varphi)$).

If φ is a nondegenerate Hermitian form of signature (p, q), prove that the group $\mathbf{U}(\varphi)$ is isomorphic to the group of $n \times n$ matrices A such that

$$A^\top \begin{pmatrix} I_p & 0 \\ 0 & -I_q \end{pmatrix} \overline{A} = \begin{pmatrix} I_p & 0 \\ 0 & -I_q \end{pmatrix}.$$

Remark: In view of question (f), the groups $\mathbf{U}(\varphi)$ and $\mathbf{SU}(\varphi)$ are also denoted by $\mathbf{U}(p, q)$ and $\mathbf{SU}(p, q)$ when φ has signature (p, q). They are Lie groups.

11.9. (a) If A is a real symmetric $n \times n$ matrix and B is a real skew symmetric $n \times n$ matrix, then $A + iB$ is Hermitian. Conversely, every Hermitian matrix can be written as $A + iB$, where A is real symmetric and B is real skew symmetric.

(b) Every complex $n \times n$ matrix can be written as $A + iB$, for some Hermitian matrices A, B.

11.10. (a) Given a complex $n \times n$ matrix A, prove that

$$\sum_{i,j=1}^{n} |a_{i,j}|^2 = \operatorname{tr}(A^* A) = \operatorname{tr}(A A^*).$$

(b) Prove that $\|A\| = \sqrt{\operatorname{tr}(A^* A)}$ defines a norm on matrices. Prove that

$$\|AB\| \le \|A\| \, \|B\|.$$

(c) When A is Hermitian, prove that

$$\|A\|^2 = \sum_{i=1}^{n} \lambda_i^2,$$

where the λ_i are the (real) eigenvalues of A.

11.11. Given a Hermitian matrix A, prove that $I_n + iA$ and $I_n - iA$ are invertible. Prove that $(I_n + iA)(I_n - iA)^{-1}$ is a unitary matrix.

11.12. Let E be a Hermitian space of dimension n. For any basis $(e_1, \ldots, e_n)$ of E, orthonormal or not, let G be the Gram matrix associated with $(e_1, \ldots, e_n)$, i.e., the matrix

$$G = (e_i \cdot e_j).$$

Given any linear map $f\colon E \to E$, if A is the matrix of f w.r.t. $(e_1,\ldots,e_n)$, prove that f is self-adjoint ($f^* = f$) iff

$$G^\top A = A^* G^\top .$$

References

1. Nicolas Bourbaki. *Espaces Vectoriels Topologiques*. Eléments de Mathématiques. Hermann, 1981.
2. Jacques Dixmier. *General Topology*. UTM. Springer-Verlag, first edition, 1984.
3. Serge Lang. *Real and Functional Analysis*. GTM 142. Springer-Verlag, third edition, 1996.
4. Serge Lang. *Undergraduate Analysis*. UTM. Springer-Verlag, second edition, 1997.
5. Laurent Schwartz. *Analyse I. Théorie des Ensembles et Topologie*. Collection Enseignement des Sciences. Hermann, 1991.
6. Laurent Schwartz. *Analyse II. Calcul Différentiel et Equations Différentielles*. Collection Enseignement des Sciences. Hermann, 1992.
7. Gilbert Strang. *Introduction to Applied Mathematics*. Wellesley–Cambridge Press, first edition, 1986.
8. Gilbert Strang and Nguyen Truong. *Wavelets and Filter Banks*. Wellesley–Cambridge Press, second edition, 1997.

Chapter 12
Spectral Theorems in Euclidean and Hermitian Spaces

12.1 Introduction: What's with Lie Groups and Lie Algebras?

The goal of this chapter is to show that there are nice normal forms for symmetric matrices, skew-symmetric matrices, orthogonal matrices, and normal matrices. The spectral theorem for symmetric matrices states that symmetric matrices have real eigenvalues and that they can be diagonalized over an orthonormal basis. The spectral theorem for Hermitian matrices states that Hermitian matrices also have real eigenvalues and that they can be diagonalized over a complex orthonormal basis. Normal matrices can be block diagonalized over an orthonormal basis with blocks having size at most two, and there are refinements of this normal form for skew-symmetric and orthogonal matrices.

One of the main purposes of this book is to give a concrete introduction to Lie groups and Lie algebras. Our ulterior motive is to present some beautiful mathematical concepts that can also be used as tools for solving practical problems arising in computer science, more specifically in robotics, motion planning, computer vision, and computer graphics.

Most texts on Lie groups and Lie algebras begin with prerequisites in differential geometry that are often formidable to average computer scientists (or average scientists, whatever that means!). We also struggled for a long time, trying to figure out what Lie groups and Lie algebras are all about, but this can be done! A good way to sneak into the wonderful world of Lie groups and Lie algebras is to play with explicit matrix groups such as the group of rotations in $\mathbb{R}^2$ (or $\mathbb{R}^3$) and with the exponential map. After actually computing the exponential $A = e^B$ of a 2×2 skew-symmetric matrix B and observing that it is a rotation matrix, and similarly for a 3×3 skew-symmetric matrix B, one begins to suspect that there is something deep going on. Similarly, after the discovery that every real invertible $n \times n$ matrix A can be written as $A = RP$, where R is an orthogonal matrix and P is a positive definite symmetric matrix, and that P can be written as $P = e^S$ for some symmetric matrix S, one begins to appreciate the exponential map.

Our goal is to give an elementary and concrete introduction to Lie groups and Lie algebras by studying a number of the so-called *classical groups*, such as the general linear group $\mathbf{GL}(n,\mathbb{R})$, the special linear group $\mathbf{SL}(n,\mathbb{R})$, the orthogonal group $\mathbf{O}(n)$, the special orthogonal group $\mathbf{SO}(n)$, and the group of affine rigid motions $\mathbf{SE}(n)$, and their Lie algebras $\mathfrak{gl}(n,\mathbb{R})$ (all matrices), $\mathfrak{sl}(n,\mathbb{R})$ (matrices with null trace), $\mathfrak{o}(n)$, and $\mathfrak{so}(n)$ (skew-symmetric matrices). We also consider the corresponding groups of complex matrices and their Lie algebras. Whenever possible, we show that the exponential map is surjective. For this, all we need is some results of linear algebra about various normal forms for symmetric matrices and skew-symmetric matrices. Thus, we begin by proving that there are nice normal forms (block diagonal matrices whith blocks having size at most two) for normal matrices and other special cases (symmetric matrices, skew-symmetric matrices, orthogonal matrices). We also prove the spectral theorem for complex normal matrices.

12.2 Normal Linear Maps

We begin by studying normal maps, to understand the structure of their eigenvalues and eigenvectors. This section and the next two were inspired by Lang [4], Artin [1], Mac Lane and Birkhoff [5], Berger [2], and Bertin [3].

Definition 12.1. Given a Euclidean space E, a linear map $f : E \to E$ is *normal* if

$$f \circ f^* = f^* \circ f.$$

A linear map $f : E \to E$ is *self-adjoint* if $f = f^*$, *skew-self-adjoint* if $f = -f^*$, and *orthogonal* if $f \circ f^* = f^* \circ f = \mathrm{id}$.

Obviously, a self-adjoint, skew-self-adjoint, or orthogonal linear map is a normal linear map. Our first goal is to show that for every normal linear map $f : E \to E$, there is an orthonormal basis (w.r.t. $\langle -, - \rangle$) such that the matrix of f over this basis has an especially nice form: It is a block diagonal matrix in which the blocks are either one-dimensional matrices (i.e., single entries) or two-dimensional matrices of the form

$$\begin{pmatrix} \lambda & \mu \\ -\mu & \lambda \end{pmatrix}.$$

This normal form can be further refined if f is self-adjoint, skew-self-adjoint, or orthogonal. As a first step, we show that f and f^* have the same kernel when f is normal.

Lemma 12.1. *Given a Euclidean space E, if $f : E \to E$ is a normal linear map, then* $\mathrm{Ker}\, f = \mathrm{Ker}\, f^*$.

Proof. First, let us prove that

$$\langle f(u), f(v) \rangle = \langle f^*(u), f^*(v) \rangle$$

for all $u, v \in E$. Since f^* is the adjoint of f and $f \circ f^* = f^* \circ f$, we have

$$\begin{aligned}
\langle f(u), f(u) \rangle &= \langle u, (f^* \circ f)(u) \rangle, \\
&= \langle u, (f \circ f^*)(u) \rangle, \\
&= \langle f^*(u), f^*(u) \rangle.
\end{aligned}$$

Since $\langle -, - \rangle$ is positive definite,

$$\begin{aligned}
\langle f(u), f(u) \rangle = 0 \quad &\text{iff} \quad f(u) = 0, \\
\langle f^*(u), f^*(u) \rangle = 0 \quad &\text{iff} \quad f^*(u) = 0,
\end{aligned}$$

and since

$$\langle f(u), f(u) \rangle = \langle f^*(u), f^*(u) \rangle,$$

we have

$$f(u) = 0 \quad \text{iff} \quad f^*(u) = 0.$$

Consequently, $\operatorname{Ker} f = \operatorname{Ker} f^*$. $\square$

The next step is to show that for every linear map $f \colon E \to E$ there is some subspace W of dimension 1 or 2 such that $f(W) \subseteq W$. When $\dim(W) = 1$, the subspace W is actually an eigenspace for some real eigenvalue of f. Furthermore, when f is normal, there is a subspace W of dimension 1 or 2 such that $f(W) \subseteq W$ and $f^*(W) \subseteq W$. The difficulty is that the eigenvalues of f are not necessarily real. One way to get around this problem is to complexify both the vector space E and the inner product $\langle -, - \rangle$.

In Section 5.11 it was explained how a real vector space E is embedded into a complex vector space $E_{\mathbb{C}}$, and how a linear map $f \colon E \to E$ is extended to a linear map $f_{\mathbb{C}} \colon E_{\mathbb{C}} \to E_{\mathbb{C}}$. For the sake of convenience, we repeat the definition of $E_{\mathbb{C}}$.

Definition 12.2. Given a real vector space E, let $E_{\mathbb{C}}$ be the structure $E \times E$ under the addition operation

$$(u_1, u_2) + (v_1, v_2) = (u_1 + v_1, u_2 + v_2),$$

and let multiplication by a complex scalar $z = x + iy$ be defined such that

$$(x + iy) \cdot (u, v) = (xu - yv, yu + xv).$$

It is convenient to write $u + iv$ for (u, v).

A linear map $f \colon E \to E$ is extended to the linear map $f_{\mathbb{C}} \colon E_{\mathbb{C}} \to E_{\mathbb{C}}$ defined such that

$$f_{\mathbb{C}}(u + iv) = f(u) + if(v).$$

Next, we need to extend the inner product on E to an inner product on $E_{\mathbb{C}}$.

The inner product $\langle -, - \rangle$ on a Euclidean space E is extended to the Hermitian positive definite form $\langle -, - \rangle_{\mathbb{C}}$ on $E_{\mathbb{C}}$ as follows:

$$\langle u_1 + iv_1, u_2 + iv_2 \rangle_{\mathbb{C}} = \langle u_1, u_2 \rangle + \langle v_1, v_2 \rangle + i(\langle u_2, v_1 \rangle - \langle u_1, v_2 \rangle).$$

It is easily verified that $\langle -, - \rangle_{\mathbb{C}}$ is indeed a Hermitian form that is positive definite, and it is clear that $\langle -, - \rangle_{\mathbb{C}}$ agrees with $\langle -, - \rangle$ on real vectors. Then, given any linear map $f \colon E \to E$, it is easily verified that the map $f_{\mathbb{C}}^*$ defined such that

$$f_{\mathbb{C}}^*(u + iv) = f^*(u) + if^*(v)$$

for all $u, v \in E$ is the adjoint of $f_{\mathbb{C}}$ w.r.t. $\langle -, - \rangle_{\mathbb{C}}$.

Assuming again that E is a Hermitian space, observe that Lemma 12.1 also holds. We have the following crucial lemma relating the eigenvalues of f and f^*.

Lemma 12.2. *Given a Hermitian space E, for any normal linear map $f \colon E \to E$, a vector u is an eigenvector of f for the eigenvalue λ (in $\mathbb{C}$) iff u is an eigenvector of f^* for the eigenvalue $\overline{\lambda}$.*

Proof. First, it is immediately verified that the adjoint of $f - \lambda \, \mathrm{id}$ is $f^* - \overline{\lambda} \, \mathrm{id}$. Furthermore, $f - \lambda \, \mathrm{id}$ is normal. Indeed,

$$
\begin{aligned}
(f - \lambda \, \mathrm{id}) \circ (f - \lambda \, \mathrm{id})^* &= (f - \lambda \, \mathrm{id}) \circ (f^* - \overline{\lambda} \, \mathrm{id}), \\
&= f \circ f^* - \overline{\lambda} f - \lambda f^* + \lambda \overline{\lambda} \, \mathrm{id}, \\
&= f^* \circ f - \lambda f^* - \overline{\lambda} f + \overline{\lambda} \lambda \, \mathrm{id}, \\
&= (f^* - \overline{\lambda} \, \mathrm{id}) \circ (f - \lambda \, \mathrm{id}), \\
&= (f - \lambda \, \mathrm{id})^* \circ (f - \lambda \, \mathrm{id}).
\end{aligned}
$$

Applying Lemma 12.1 to $f - \lambda \, \mathrm{id}$, for every nonnull vector u, we see that

$$(f - \lambda \, \mathrm{id})(u) = 0 \quad \text{iff} \quad (f^* - \overline{\lambda} \, \mathrm{id})(u) = 0,$$

which is exactly the statement of the lemma. $\quad \square$

The next lemma shows a very important property of normal linear maps: Eigenvectors corresponding to distinct eigenvalues are orthogonal.

Lemma 12.3. *Given a Hermitian space E, for any normal linear map $f \colon E \to E$, if u and v are eigenvectors of f associated with the eigenvalues λ and μ (in $\mathbb{C}$) where $\lambda \neq \mu$, then $\langle u, v \rangle = 0$.*

Proof. Let us compute $\langle f(u), v \rangle$ in two different ways. Since v is an eigenvector of f for μ, by Lemma 12.2, v is also an eigenvector of f^* for $\overline{\mu}$, and we have

$$\langle f(u), v \rangle = \langle \lambda u, v \rangle = \lambda \langle u, v \rangle$$

and

$$\langle f(u), v \rangle = \langle u, f^*(v) \rangle = \langle u, \overline{\mu} v \rangle = \mu \langle u, v \rangle,$$

where the last identity holds because of the semilinearity in the second argument, and thus

$$\lambda \langle u, v \rangle = \mu \langle u, v \rangle,$$

that is,

$$(\lambda - \mu)\langle u, v \rangle = 0,$$

which implies that $\langle u, v \rangle = 0$, since $\lambda \neq \mu$. $\square$

We can also show easily that the eigenvalues of a self-adjoint linear map are real.

Lemma 12.4. *Given a Hermitian space E, the eigenvalues of any self-adjoint linear map $f \colon E \to E$ are real.*

Proof. Let z (in $\mathbb{C}$) be an eigenvalue of f and let u be an eigenvector for z. We compute $\langle f(u), u \rangle$ in two different ways. We have

$$\langle f(u), u \rangle = \langle zu, u \rangle = z\langle u, u \rangle,$$

and since $f = f^*$, we also have

$$\langle f(u), u \rangle = \langle u, f^*(u) \rangle = \langle u, f(u) \rangle = \langle u, zu \rangle = \overline{z}\langle u, u \rangle.$$

Thus,

$$z\langle u, u \rangle = \overline{z}\langle u, u \rangle,$$

which implies that $z = \overline{z}$, since $u \neq 0$, and z is indeed real. $\square$

Given any subspace W of a Hermitian space E, recall that the *orthogonal complement* $W^\perp$ of W is the subspace defined such that

$$W^\perp = \{u \in E \mid \langle u, w \rangle = 0, \text{ for all } w \in W \}.$$

Recall from Lemma 11.7 that that $E = W \oplus W^\perp$ (this can be easily shown, for example, by constructing an orthonormal basis of E using the Gram–Schmidt orthonormalization procedure). The same result also holds for Euclidean spaces (see Lemma 6.8). The following lemma provides the key to the induction that will allow us to show that a normal linear map can be diagonalized. It actually holds for any linear map. We found the inspiration for this lemma in Berger [2].

Lemma 12.5. *Given a Hermitian space E, for any linear map $f \colon E \to E$, if W is any subspace of E such that $f(W) \subseteq W$ and $f^*(W) \subseteq W$, then $f(W^\perp) \subseteq W^\perp$ and $f^*(W^\perp) \subseteq W^\perp$.*

Proof. If $u \in W^\perp$, then

$$\langle u, w \rangle = 0$$

for all $w \in W$. However,

$$\langle f(u), w \rangle = \langle u, f^*(w) \rangle,$$

and since $f^*(W) \subseteq W$, we have $f^*(w) \in W$, and since $u \in W^\perp$, we get

$$\langle u, f^*(w) \rangle = 0,$$

which shows that

$$\langle f(u), w \rangle = 0$$

for all $w \in W$, that is, $f(u) \in W^\perp$. Thus, $f(W^\perp) \subseteq W^\perp$. The proof that $f^*(W^\perp) \subseteq W^\perp$ is analogous. $\square$

The above lemma also holds for Euclidean spaces. Although we are ready to prove that for every normal linear map f (over a Hermitian space) there is an orthonormal basis of eigenvectors, we now return to real Euclidean spaces.

If $f \colon E \to E$ is a linear map and $w = u + iv$ is an eigenvector of $f_\mathbb{C} \colon E_\mathbb{C} \to E_\mathbb{C}$ for the eigenvalue $z = \lambda + i\mu$, where $u, v \in E$ and $\lambda, \mu \in \mathbb{R}$, since

$$f_\mathbb{C}(u + iv) = f(u) + if(v)$$

and

$$f_\mathbb{C}(u + iv) = (\lambda + i\mu)(u + iv) = \lambda u - \mu v + i(\mu u + \lambda v),$$

we have

$$f(u) = \lambda u - \mu v \quad \text{and} \quad f(v) = \mu u + \lambda v,$$

from which we immediately obtain

$$f_\mathbb{C}(u - iv) = (\lambda - i\mu)(u - iv),$$

which shows that $\overline{w} = u - iv$ is an eigenvector of $f_\mathbb{C}$ for $\overline{z} = \lambda - i\mu$. Using this fact, we can prove the following lemma.

Lemma 12.6. *Given a Euclidean space E, for any normal linear map $f \colon E \to E$, if $w = u + iv$ is an eigenvector of $f_\mathbb{C}$ associated with the eigenvalue $z = \lambda + i\mu$ (where $u, v \in E$ and $\lambda, \mu \in \mathbb{R}$), if $\mu \neq 0$ (i.e., z is not real) then $\langle u, v \rangle = 0$ and $\langle u, u \rangle = \langle v, v \rangle$, which implies that u and v are linearly independent, and if W is the subspace spanned by u and v, then $f(W) = W$ and $f^*(W) = W$. Furthermore, with respect to the (orthogonal) basis (u, v), the restriction of f to W has the matrix*

$$\begin{pmatrix} \lambda & \mu \\ -\mu & \lambda \end{pmatrix}.$$

If $\mu = 0$, then λ is a real eigenvalue of f, and either u or v is an eigenvector of f for λ. If W is the subspace spanned by u if $u \neq 0$, or spanned by $v \neq 0$ if $u = 0$, then $f(W) \subseteq W$ and $f^*(W) \subseteq W$.

Proof. Since $w = u + iv$ is an eigenvector of $f_\mathbb{C}$, by definition it is nonnull, and either $u \neq 0$ or $v \neq 0$. From the fact stated just before Lemma 12.6, $u - iv$ is an eigenvector of $f_\mathbb{C}$ for $\lambda - i\mu$. It is easy to check that $f_\mathbb{C}$ is normal. However, if $\mu \neq 0$, then $\lambda + i\mu \neq \lambda - i\mu$, and from Lemma 12.3, the vectors $u + iv$ and $u - iv$ are orthogonal w.r.t. $\langle -, - \rangle_\mathbb{C}$, that is,

$$\langle u + iv, u - iv \rangle_\mathbb{C} = \langle u, u \rangle - \langle v, v \rangle + 2i\langle u, v \rangle = 0.$$

Thus, we get $\langle u,v\rangle = 0$ and $\langle u,u\rangle = \langle v,v\rangle$, and since $u \neq 0$ or $v \neq 0$, u and v are linearly independent. Since

$$f(u) = \lambda u - \mu v \quad \text{and} \quad f(v) = \mu u + \lambda v$$

and since by Lemma 12.2 $u + iv$ is an eigenvector of f^* for $\lambda - i\mu$, we have

$$f^*(u) = \lambda u + \mu v \quad \text{and} \quad f^*(v) = -\mu u + \lambda v,$$

and thus $f(W) = W$ and $f^*(W) = W$, where W is the subspace spanned by u and v.

When $\mu = 0$, we have

$$f(u) = \lambda u \quad \text{and} \quad f(v) = \lambda v,$$

and since $u \neq 0$ or $v \neq 0$, either u or v is an eigenvector of f for λ. If W is the subspace spanned by u if $u \neq 0$, or spanned by v if $u = 0$, it is obvious that $f(W) \subseteq W$ and $f^*(W) \subseteq W$. Note that $\lambda = 0$ is possible, and this is why $\subseteq$ cannot be replaced by $=$. $\square$

The beginning of the proof of Lemma 12.6 actually shows that for every linear map $f \colon E \to E$ there is some subspace W such that $f(W) \subseteq W$, where W has dimension 1 or 2. In general, it doesn't seem possible to prove that $W^\perp$ is invariant under f. However, this happens when f is normal.

We can finally prove our first main theorem.

Theorem 12.1. *Given a Euclidean space E of dimension n, for every normal linear map $f \colon E \to E$ there is an orthonormal basis $(e_1,\ldots,e_n)$ such that the matrix of f w.r.t. this basis is a block diagonal matrix of the form*

$$\begin{pmatrix} A_1 & & \cdots & \\ & A_2 & \cdots & \\ \vdots & \vdots & \ddots & \vdots \\ & & \cdots & A_p \end{pmatrix}$$

such that each block A_i is either a one-dimensional matrix (i.e., a real scalar) or a two-dimensional matrix of the form

$$A_i = \begin{pmatrix} \lambda_i & -\mu_i \\ \mu_i & \lambda_i \end{pmatrix},$$

where $\lambda_i, \mu_i \in \mathbb{R}$, with $\mu_i > 0$.

Proof. We proceed by induction on the dimension n of E as follows. If $n = 1$, the result is trivial. Assume now that $n \geq 2$. First, since $\mathbb{C}$ is algebraically closed (i.e., every polynomial has a root in $\mathbb{C}$), the linear map $f_\mathbb{C} \colon E_\mathbb{C} \to E_\mathbb{C}$ has some eigenvalue $z = \lambda + i\mu$ (where $\lambda, \mu \in \mathbb{R}$). Let $w = u + iv$ be some eigenvector of $f_\mathbb{C}$ for $\lambda + i\mu$ (where $u, v \in E$). We can now apply Lemma 12.6.

If $\mu = 0$, then either u or v is an eigenvector of f for $\lambda \in \mathbb{R}$. Let W be the subspace of dimension 1 spanned by $e_1 = u/\|u\|$ if $u \neq 0$, or by $e_1 = v/\|v\|$ otherwise. It is obvious that $f(W) \subseteq W$ and $f^*(W) \subseteq W$. The orthogonal $W^\perp$ of W has dimension $n-1$, and by Lemma 12.5, we have $f(W^\perp) \subseteq W^\perp$. But the restriction of f to $W^\perp$ is also normal, and we conclude by applying the induction hypothesis to $W^\perp$.

If $\mu \neq 0$, then $\langle u, v \rangle = 0$ and $\langle u, u \rangle = \langle v, v \rangle$, and if W is the subspace spanned by $u/\|u\|$ and $v/\|v\|$, then $f(W) = W$ and $f^*(W) = W$. We also know that the restriction of f to W has the matrix

$$\begin{pmatrix} \lambda & \mu \\ -\mu & \lambda \end{pmatrix}$$

with respect to the basis $(u/\|u\|, v/\|v\|)$. If $\mu < 0$, we let $\lambda_1 = \lambda$, $\mu_1 = -\mu$, $e_1 = u/\|u\|$, and $e_2 = v/\|v\|$. If $\mu > 0$, we let $\lambda_1 = \lambda$, $\mu_1 = \mu$, $e_1 = v/\|v\|$, and $e_2 = u/\|u\|$. In all cases, it is easily verified that the matrix of the restriction of f to W w.r.t. the orthonormal basis (e_1, e_2) is

$$A_1 = \begin{pmatrix} \lambda_1 & -\mu_1 \\ \mu_1 & \lambda_1 \end{pmatrix},$$

where $\lambda_1, \mu_1 \in \mathbb{R}$, with $\mu_1 > 0$. However, $W^\perp$ has dimension $n-2$, and by Lemma 12.5, $f(W^\perp) \subseteq W^\perp$. Since the restriction of f to $W^\perp$ is also normal, we conclude by applying the induction hypothesis to $W^\perp$. $\square$

After this relatively hard work, we can easily obtain some nice normal forms for the matrices of self-adjoint, skew-self-adjoint, and orthogonal linear maps. However, for the sake of completeness (and since we have all the tools to so do), we go back to the case of a Hermitian space and show that normal linear maps can be diagonalized with respect to an orthonormal basis.

Theorem 12.2. *Given a Hermitian space E of dimension n, for every normal linear map $f: E \to E$ there is an orthonormal basis $(e_1, \ldots, e_n)$ of eigenvectors of f such that the matrix of f w.r.t. this basis is a diagonal matrix*

$$\begin{pmatrix} \lambda_1 & & & \cdots & \\ & \lambda_2 & & \cdots & \\ \vdots & \vdots & \ddots & \vdots \\ & & & \cdots & \lambda_n \end{pmatrix},$$

where $\lambda_i \in \mathbb{C}$.

Proof. We proceed by induction on the dimension n of E as follows. If $n = 1$, the result is trivial. Assume now that $n \geq 2$. Since $\mathbb{C}$ is algebraically closed (i.e., every polynomial has a root in $\mathbb{C}$), the linear map $f: E \to E$ has some eigenvalue $\lambda \in \mathbb{C}$, and let w be some eigenvector for λ. Let W be the subspace of dimension 1 spanned by w. Clearly, $f(W) \subseteq W$. By Lemma 12.2, w is an eigenvector of f^* for $\overline{\lambda}$, and thus $f^*(W) \subseteq W$. By Lemma 12.5, we also have $f(W^\perp) \subseteq W^\perp$. The restriction of f

to $W^\perp$ is still normal, and we conclude by applying the induction hypothesis to $W^\perp$ (whose dimension is $n-1$). $\quad\square$

Thus, in particular, self-adjoint, skew-self-adjoint, and orthogonal linear maps can be diagonalized with respect to an orthonormal basis of eigenvectors. In this latter case, though, an orthogonal map is called a *unitary* map. Also, Lemma 12.4 shows that the eigenvalues of a self-adjoint linear map are real. It is easily shown that skew-self-adjoint maps have eigenvalues that are pure imaginary or null, and that unitary maps have eigenvalues of absolute value 1.

Remark: There is a converse to Theorem 12.2, namely, if there is an orthonormal basis $(e_1,\ldots,e_n)$ of eigenvectors of f, then f is normal. We leave the easy proof as an exercise.

12.3 Self-Adjoint, Skew-Self-Adjoint, and Orthogonal Linear Maps

We begin with self-adjoint maps.

Theorem 12.3. *Given a Euclidean space E of dimension n, for every self-adjoint linear map $f\colon E \to E$, there is an orthonormal basis $(e_1,\ldots,e_n)$ of eigenvectors of f such that the matrix of f w.r.t. this basis is a diagonal matrix*

$$\begin{pmatrix} \lambda_1 & & \cdots & \\ & \lambda_2 & \cdots & \\ \vdots & \vdots & \ddots & \vdots \\ & & \cdots & \lambda_n \end{pmatrix},$$

where $\lambda_i \in \mathbb{R}$.

Proof. The case $n=1$ is trivial. If $n \geq 2$, we need to show that $f\colon E \to E$ has some real eigenvalue. There are several ways to do so. One method is to observe that the linear map $f_{\mathbb{C}}\colon E_{\mathbb{C}} \to E_{\mathbb{C}}$ is also self-adjoint, and by Lemma 12.4 the eigenvalues of $f_{\mathbb{C}}$ are all real. This implies that f itself has some real eigenvalue, and in fact, all eigenvalues of f are real. We now give a more direct method not involving the complexification of $\langle -, - \rangle$ and Lemma 12.4.

Since $\mathbb{C}$ is algebraically closed, $f_{\mathbb{C}}$ has some eigenvalue $\lambda + i\mu$, and let $u + iv$ be some eigenvector of $f_{\mathbb{C}}$ for $\lambda + i\mu$, where $\lambda, \mu \in \mathbb{R}$ and $u, v \in E$. We saw in the proof of Lemma 12.6 that

$$f(u) = \lambda u - \mu v \quad \text{and} \quad f(v) = \mu u + \lambda v.$$

Since $f = f^*$,

$$\langle f(u), v \rangle = \langle u, f(v) \rangle$$

for all $u, v \in E$. Applying this to

$$f(u) = \lambda u - \mu v \quad \text{and} \quad f(v) = \mu u + \lambda v,$$

we get

$$\langle f(u), u \rangle = \langle \lambda u - \mu v, v \rangle = \lambda \langle u, v \rangle - \mu \langle v, v \rangle$$

and

$$\langle u, f(v) \rangle = \langle u, \mu u + \lambda v \rangle = \mu \langle u, u \rangle + \lambda \langle u, v \rangle,$$

and thus we get

$$\lambda \langle u, v \rangle - \mu \langle v, v \rangle = \mu \langle u, u \rangle + \lambda \langle u, v \rangle,$$

that is,

$$\mu (\langle u, u \rangle + \langle v, v \rangle) = 0,$$

which implies $\mu = 0$, since either $u \neq 0$ or $v \neq 0$. Therefore, λ is a real eigenvalue of f.

Now, going back to the proof of Theorem 12.1, only the case where $\mu = 0$ applies, and the induction shows that all the blocks are one-dimensional. $\quad \square$

Theorem 12.3 implies that if $\lambda_1, \ldots, \lambda_p$ are the distinct real eigenvalues of f, and E_i is the eigenspace associated with λ_i, then

$$E = E_1 \oplus \cdots \oplus E_p,$$

where E_i and E_j are orthogonal for all $i \neq j$.

Remark: Another way to prove that a self-adjoint map has a real eigenvalue is to use a little bit of calculus. We learned such a proof from Herman Gluck. The idea is to consider the real-valued function $\Phi \colon E \to \mathbb{R}$ defined such that

$$\Phi(u) = \langle f(u), u \rangle$$

for every $u \in E$. This function is C^∞, and if we represent f by a matrix A over some orthonormal basis, it is easy to compute the gradient vector

$$\nabla \Phi(X) = \left(\frac{\partial \Phi}{\partial x_1}(X), \ldots, \frac{\partial \Phi}{\partial x_n}(X) \right)$$

of Φ at X. Indeed, we find that

$$\nabla \Phi(X) = (A + A^\top) X,$$

where X is a column vector of size n. But since f is self-adjoint, $A = A^\top$, and thus

$$\nabla \Phi(X) = 2AX.$$

The next step is to find the maximum of the function Φ on the sphere

$$S^{n-1} = \{(x_1, \ldots, x_n) \in \mathbb{R}^n \mid x_1^2 + \cdots + x_n^2 = 1\}.$$

Since S^{n-1} is compact and Φ is continuous, and in fact C^∞, Φ takes a maximum at some X on S^{n-1}. But then it is well known that at an extremum X of Φ we must have

$$d\Phi_X(Y) = \langle \nabla\Phi(X), Y \rangle = 0$$

for all tangent vectors Y to S^{n-1} at X, and so $\nabla\Phi(X)$ is orthogonal to the tangent plane at X, which means that

$$\nabla\Phi(X) = \lambda X$$

for some $\lambda \in \mathbb{R}$. Since $\nabla\Phi(X) = 2AX$, we get

$$2AX = \lambda X,$$

and thus $\lambda/2$ is a real eigenvalue of A (i.e., of f).

Next, we consider skew-self-adjoint maps.

Theorem 12.4. *Given a Euclidean space E of dimension n, for every skew-self-adjoint linear map $f \colon E \to E$ there is an orthonormal basis $(e_1, \ldots, e_n)$ such that the matrix of f w.r.t. this basis is a block diagonal matrix of the form*

$$\begin{pmatrix} A_1 & & \cdots & \\ & A_2 & \cdots & \\ \vdots & \vdots & \ddots & \vdots \\ & & \cdots & A_p \end{pmatrix}$$

such that each block A_i is either 0 or a two-dimensional matrix of the form

$$A_i = \begin{pmatrix} 0 & -\mu_i \\ \mu_i & 0 \end{pmatrix},$$

where $\mu_i \in \mathbb{R}$, with $\mu_i > 0$. In particular, the eigenvalues of $f_\mathbb{C}$ are pure imaginary of the form $\pm i\mu_i$ or 0.

Proof. The case where $n = 1$ is trivial. As in the proof of Theorem 12.1, $f_\mathbb{C}$ has some eigenvalue $z = \lambda + i\mu$, where $\lambda, \mu \in \mathbb{R}$. We claim that $\lambda = 0$. First, we show that

$$\langle f(w), w \rangle = 0$$

for all $w \in E$. Indeed, since $f = -f^*$, we get

$$\langle f(w), w \rangle = \langle w, f^*(w) \rangle = \langle w, -f(w) \rangle = -\langle w, f(w) \rangle = -\langle f(w), w \rangle,$$

since $\langle -, - \rangle$ is symmetric. This implies that

$$\langle f(w), w \rangle = 0.$$

Applying this to u and v and using the fact that

$$f(u) = \lambda u - \mu v \quad \text{and} \quad f(v) = \mu u + \lambda v,$$

we get

$$0 = \langle f(u), u \rangle = \langle \lambda u - \mu v, u \rangle = \lambda \langle u, u \rangle - \mu \langle u, v \rangle$$

and

$$0 = \langle f(v), v \rangle = \langle \mu u + \lambda v, v \rangle = \mu \langle u, v \rangle + \lambda \langle v, v \rangle,$$

from which, by addition, we get

$$\lambda (\langle v, v \rangle + \langle v, v \rangle) = 0.$$

Since $u \neq 0$ or $v \neq 0$, we have $\lambda = 0$.

Then, going back to the proof of Theorem 12.1, unless $\mu = 0$, the case where u and v are orthogonal and span a subspace of dimension 2 applies, and the induction shows that all the blocks are two-dimensional or reduced to 0. $\square$

Remark: One will note that if f is skew-self-adjoint, then $i f_\mathbb{C}$ is self-adjoint w.r.t. $\langle -, - \rangle_\mathbb{C}$. By Lemma 12.4, the map $i f_\mathbb{C}$ has real eigenvalues, which implies that the eigenvalues of $f_\mathbb{C}$ are pure imaginary or 0.

Finally, we consider orthogonal linear maps.

Theorem 12.5. *Given a Euclidean space E of dimension n, for every orthogonal linear map $f \colon E \to E$ there is an orthonormal basis $(e_1, \dots, e_n)$ such that the matrix of f w.r.t. this basis is a block diagonal matrix of the form*

$$\begin{pmatrix} A_1 & & \cdots & \\ & A_2 & \cdots & \\ \vdots & \vdots & \ddots & \vdots \\ & & \cdots & A_p \end{pmatrix}$$

such that each block A_i is either 1, -1, or a two-dimensional matrix of the form

$$A_i = \begin{pmatrix} \cos \theta_i & -\sin \theta_i \\ \sin \theta_i & \cos \theta_i \end{pmatrix}$$

where $0 < \theta_i < \pi$. In particular, the eigenvalues of $f_\mathbb{C}$ are of the form $\cos \theta_i \pm i \sin \theta_i$, 1, or -1.

Proof. The case where $n = 1$ is trivial. As in the proof of Theorem 12.1, $f_\mathbb{C}$ has some eigenvalue $z = \lambda + i\mu$, where $\lambda, \mu \in \mathbb{R}$. Since $f \circ f^* = f^* \circ f = \text{id}$, the map f is invertible. In fact, the eigenvalues of f have absolute value 1. Indeed, if z (in $\mathbb{C}$) is an eigenvalue of f, and u is an eigenvector for z, we have

$$\langle f(u), f(u) \rangle = \langle zu, zu \rangle = z\bar{z} \langle u, u \rangle$$

and

$$\langle f(u), f(u) \rangle = \langle u, (f^* \circ f)(u) \rangle = \langle u, u \rangle,$$

from which we get

$$z\bar{z} \langle u, u \rangle = \langle u, u \rangle.$$

Since $u \neq 0$, we have $z\bar{z} = 1$, i.e., $|z| = 1$. As a consequence, the eigenvalues of $f_{\mathbb{C}}$ are of the form $\cos \theta \pm i \sin \theta$, 1, or -1. The theorem then follows immediately from Theorem 12.1, where the condition $\mu > 0$ implies that $\sin \theta_i > 0$, and thus, $0 < \theta_i < \pi$. $\square$

It is obvious that we can reorder the orthonormal basis of eigenvectors given by Theorem 12.5, so that the matrix of f w.r.t. this basis is a block diagonal matrix of the form

$$\begin{pmatrix} A_1 \; \cdots & & & \\ \vdots & \ddots & \vdots & & \vdots \\ & \cdots & A_r & & \\ & & & -I_q & \\ \cdots & & & & I_p \end{pmatrix}$$

where each block A_i is a two-dimensional rotation matrix $A_i \neq \pm I_2$ of the form

$$A_i = \begin{pmatrix} \cos \theta_i & -\sin \theta_i \\ \sin \theta_i & \cos \theta_i \end{pmatrix}$$

with $0 < \theta_i < \pi$.

The linear map f has an eigenspace $E(1, f) = \mathrm{Ker}\,(f - \mathrm{id})$ of dimension p for the eigenvalue 1, and an eigenspace $E(-1, f) = \mathrm{Ker}\,(f + \mathrm{id})$ of dimension q for the eigenvalue -1. If $\det(f) = +1$ (f is a rotation), the dimension q of $E(-1, f)$ must be even, and the entries in $-I_q$ can be paired to form two-dimensional blocks, if we wish. In this case, every rotation in $\mathbf{SO}(n)$ has a matrix of the form

$$\begin{pmatrix} A_1 \; \cdots & & & \\ \vdots & \ddots & \vdots & \\ & \cdots & A_m & \\ \cdots & & & I_{n-2m} \end{pmatrix}$$

where the first m blocks A_i are of the form

$$A_i = \begin{pmatrix} \cos \theta_i & -\sin \theta_i \\ \sin \theta_i & \cos \theta_i \end{pmatrix}$$

with $0 < \theta_i \leq \pi$.

Theorem 12.5 can be used to prove a sharper version of the Cartan–Dieudonné theorem, as claimed in remark (3) after Theorem 8.1.

Theorem 12.6. *Let E be a Euclidean space of dimension $n \geq 2$. For every isometry $f \in \mathbf{O}(E)$, if $p = \dim(E(1,f)) = \dim(\mathrm{Ker}\,(f - \mathrm{id}))$, then f is the composition of $n - p$ reflections, and $n - p$ is minimal.*

Proof. From Theorem 12.5 there are r subspaces $F_1, \ldots, F_r$, each of dimension 2, such that

$$E = E(1,f) \oplus E(-1,f) \oplus F_1 \oplus \cdots \oplus F_r,$$

and all the summands are pairwise orthogonal. Furthermore, the restriction r_i of f to each F_i is a rotation $r_i \neq \pm\mathrm{id}$. Each 2D rotation r_i can be written a the composition $r_i = s'_i \circ s_i$ of two reflections s_i and s'_i about lines in F_i (forming an angle $\theta_i/2$). We can extend s_i and s'_i to hyperplane reflections in E by making them the identity on $F_i^\perp$. Then,

$$s'_r \circ s_r \circ \cdots \circ s'_1 \circ s_1$$

agrees with f on $F_1 \oplus \cdots \oplus F_r$ and is the identity on $E(1,f) \oplus E(-1,f)$. If $E(-1,f)$ has an orthonormal basis of eigenvectors $(v_1, \ldots, v_q)$, letting s''_j be the reflection about the hyperplane $(v_j)^\perp$, it is clear that

$$s''_q \circ \cdots \circ s''_1$$

agrees with f on $E(-1,f)$ and is the identity on $E(1,f) \oplus F_1 \oplus \cdots \oplus F_r$. But then,

$$f = s''_q \circ \cdots \circ s''_1 \circ s'_r \circ s_r \circ \cdots \circ s'_1 \circ s_1,$$

the composition of $2r + q = n - p$ reflections.

If

$$f = s_t \circ \cdots \circ s_1,$$

for t reflections s_i, it is clear that

$$F = \bigcap_{i=1}^{t} E(1,s_i) \subseteq E(1,f),$$

where $E(1,s_i)$ is the hyperplane defining the reflection s_i. By the Grassmann relation, if we intersect $t \leq n$ hyperplanes, the dimension of their intersection is at least $n - t$. Thus, $n - t \leq p$, that is, $t \geq n - p$, and $n - p$ is the smallest number of reflections composing f. $\quad\square$

The theorems of this section and of the previous section can be immediately applied to matrices.

12.4 Normal, Symmetric, Skew-Symmetric, Orthogonal, Hermitian, Skew-Hermitian, and Unitary Matrices

First, we consider real matrices. Recall the following definitions.

Definition 12.3. Given a real $m \times n$ matrix A, the *transpose* $A^\top$ of A is the $n \times m$ matrix $A^\top = (a_{i,j}^\top)$ defined such that

$$a_{i,j}^\top = a_{j,i}$$

for all i, j, $1 \leq i \leq m$, $1 \leq j \leq n$. A real $n \times n$ matrix A is

- *normal* if
$$AA^\top = A^\top A,$$

- *symmetric* if
$$A^\top = A,$$

- *skew-symmetric* if
$$A^\top = -A,$$

- *orthogonal* if
$$AA^\top = A^\top A = I_n.$$

Recall from Lemma 6.10 that when E is a Euclidean space and $(e_1, \ldots, e_n)$ is an orthonormal basis for E, if A is the matrix of a linear map $f \colon E \to E$ w.r.t. the basis $(e_1, \ldots, e_n)$, then $A^\top$ is the matrix of the adjoint f^* of f. Consequently, a normal linear map has a normal matrix, a self-adjoint linear map has a symmetric matrix, a skew-self-adjoint linear map has a skew-symmetric matrix, and an orthogonal linear map has an orthogonal matrix. Similarly, if E and F are Euclidean spaces, $(u_1, \ldots, u_n)$ is an orthonormal basis for E, and $(v_1, \ldots, v_m)$ is an orthonormal basis for F, if a linear map $f \colon E \to F$ has the matrix A w.r.t. the bases $(u_1, \ldots, u_n)$ and $(v_1, \ldots, v_m)$, then its adjoint f^* has the matrix $A^\top$ w.r.t. the bases $(v_1, \ldots, v_m)$ and $(u_1, \ldots, u_n)$.

Furthermore, if $(u_1, \ldots, u_n)$ is another orthonormal basis for E and P is the change of basis matrix whose columns are the components of the u_i w.r.t. the basis $(e_1, \ldots, e_n)$, then P is orthogonal, and for any linear map $f \colon E \to E$, if A is the matrix of f w.r.t $(e_1, \ldots, e_n)$ and B is the matrix of f w.r.t. $(u_1, \ldots, u_n)$, then

$$B = P^\top A P.$$

As a consequence, Theorems 12.1 and 12.3–12.5 can be restated as follows.

Theorem 12.7. *For every normal matrix A there is an orthogonal matrix P and a block diagonal matrix D such that $A = PDP^\top$, where D is of the form*

$$D = \begin{pmatrix} D_1 & & \cdots & \\ & D_2 & \cdots & \\ \vdots & \vdots & \ddots & \vdots \\ & & \cdots & D_p \end{pmatrix}$$

such that each block D_i is either a one-dimensional matrix (i.e., a real scalar) or a two-dimensional matrix of the form

$$D_i = \begin{pmatrix} \lambda_i & -\mu_i \\ \mu_i & \lambda_i \end{pmatrix},$$

where $\lambda_i, \mu_i \in \mathbb{R}$, with $\mu_i > 0$.

Theorem 12.8. *For every symmetric matrix A there is an orthogonal matrix P and a diagonal matrix D such that $A = PDP^\top$, where D is of the form*

$$D = \begin{pmatrix} \lambda_1 & & & \cdots \\ & \lambda_2 & & \cdots \\ \vdots & \vdots & \ddots & \vdots \\ & & \cdots & \lambda_n \end{pmatrix},$$

where $\lambda_i \in \mathbb{R}$.

Theorem 12.9. *For every skew-symmetric matrix A there is an orthogonal matrix P and a block diagonal matrix D such that $A = PDP^\top$, where D is of the form*

$$D = \begin{pmatrix} D_1 & & & \cdots \\ & D_2 & & \cdots \\ \vdots & \vdots & \ddots & \vdots \\ & & \cdots & D_p \end{pmatrix}$$

such that each block D_i is either 0 or a two-dimensional matrix of the form

$$D_i = \begin{pmatrix} 0 & -\mu_i \\ \mu_i & 0 \end{pmatrix},$$

where $\mu_i \in \mathbb{R}$, with $\mu_i > 0$. In particular, the eigenvalues of A are pure imaginary of the form $\pm i\mu_i$, or 0.

Theorem 12.10. *For every orthogonal matrix A there is an orthogonal matrix P and a block diagonal matrix D such that $A = PDP^\top$, where D is of the form*

$$D = \begin{pmatrix} D_1 & & & \cdots \\ & D_2 & & \cdots \\ \vdots & \vdots & \ddots & \vdots \\ & & \cdots & D_p \end{pmatrix}$$

such that each block D_i is either 1, -1, or a two-dimensional matrix of the form

$$D_i = \begin{pmatrix} \cos\theta_i & -\sin\theta_i \\ \sin\theta_i & \cos\theta_i \end{pmatrix}$$

where $0 < \theta_i < \pi$. *In particular, the eigenvalues of* A *are of the form* $\cos\theta_i \pm i\sin\theta_i$,
1, or -1.

We now consider complex matrices.

Definition 12.4. Given a complex $m \times n$ matrix A, the *transpose* $A^\top$ *of* A is the $n \times m$ matrix $A^\top = \left(a_{i,j}^\top\right)$ defined such that

$$a_{i,j}^\top = a_{j,i}$$

for all i, j, $1 \leq i \leq m$, $1 \leq j \leq n$. The *conjugate* $\overline{A}$ *of* A is the $m \times n$ matrix $\overline{A} = (b_{i,j})$ defined such that

$$b_{i,j} = \overline{a}_{i,j}$$

for all i, j, $1 \leq i \leq m$, $1 \leq j \leq n$. Given an $m \times n$ complex matrix A, the *adjoint* A^* *of* A is the matrix defined such that

$$A^* = \overline{(A^\top)} = (\overline{A})^\top.$$

A complex $n \times n$ matrix A is

- *normal* if
$$AA^* = A^*A,$$

- *Hermitian* if
$$A^* = A,$$

- *skew-Hermitian* if
$$A^* = -A,$$

- *unitary* if
$$AA^* = A^*A = I_n.$$

Recall from Lemma 11.10 that when E is a Hermitian space and $(e_1, \ldots, e_n)$ is an orthonormal basis for E, if A is the matrix of a linear map $f: E \to E$ w.r.t. the basis $(e_1, \ldots, e_n)$, then A^* is the matrix of the adjoint f^* of f. Consequently, a normal linear map has a normal matrix, a self-adjoint linear map has a Hermitian matrix, a skew-self-adjoint linear map has a skew-Hermitian matrix, and a unitary linear map has a unitary matrix. Similarly, if E and F are Hermitian spaces, $(u_1, \ldots, u_n)$ is an orthonormal basis for E, and $(v_1, \ldots, v_m)$ is an orthonormal basis for F, if a linear map $f: E \to F$ has the matrix A w.r.t. the bases $(u_1, \ldots, u_n)$ and $(v_1, \ldots, v_m)$, then its adjoint f^* has the matrix A^* w.r.t. the bases $(v_1, \ldots, v_m)$ and $(u_1, \ldots, u_n)$.

Furthermore, if $(u_1, \ldots, u_n)$ is another orthonormal basis for E and P is the change of basis matrix whose columns are the components of the u_i w.r.t. the basis

$(e_1, \ldots, e_n)$, then P is unitary, and for any linear map $f \colon E \to E$, if A is the matrix of f w.r.t $(e_1, \ldots, e_n)$ and B is the matrix of f w.r.t. $(u_1, \ldots, u_n)$, then

$$B = P^*AP.$$

Theorem 12.2 can be restated in terms of matrices as follows. We can also say a little more about eigenvalues (easy exercise left to the reader).

Theorem 12.11. *For every complex normal matrix A there is a unitary matrix U and a diagonal matrix D such that $A = UDU^*$. Furthermore, if A is Hermitian, then D is a real matrix; if A is skew-Hermitian, then the entries in D are pure imaginary or null; and if A is unitary, then the entries in D have absolute value 1.*

We now have all the tools to present the important *singular value decomposition* (SVD) and the *polar form* of a matrix.

12.5 Problems

12.1. Given a Hermitian space of finite dimension n, for any linear map $f \colon E \to E$, prove that if there is an orthonormal basis $(e_1, \ldots, e_n)$ of eigenvectors of f, then f is normal.

12.2. The purpose of this problem is to prove that given any self-adjoint linear map $f \colon E \to E$ (i.e., such that $f^* = f$), where E is a Euclidean space of dimension $n \geq 3$, given an orthonormal basis $(e_1, \ldots, e_n)$, there are $n - 2$ isometries h_i, hyperplane reflections or the identity, such that the matrix of

$$h_{n-2} \circ \cdots \circ h_1 \circ f \circ h_1 \circ \cdots \circ h_{n-2}$$

is a symmetric tridiagonal matrix.

(1) Prove that for any isometry $f \colon E \to E$ we have $f = f^* = f^{-1}$ iff $f \circ f = \mathrm{id}$.

Prove that if f and h are self-adjoint linear maps ($f^* = f$ and $h^* = h$), then $h \circ f \circ h$ is a self-adjoint linear map.

(2) Proceed by induction, taking inspiration from the proof of the triangular decomposition given in Chapter 8. Let V_k be the subspace spanned by $(e_{k+1}, \ldots, e_n)$. For the base case, proceed as follows.

Let

$$f(e_1) = a_1^0 e_1 + \cdots + a_n^0 e_n,$$

and let

$$r_{1,2} = \|a_2^0 e_2 + \cdots + a_n^0 e_n\|.$$

Find an isometry h_1 (reflection or id) such that

$$h_1(f(e_1) - a_1^0 e_1) = r_{1,2} e_2.$$

Observe that

$$w_1 = r_{1,2}\, e_2 + a_1^0 e_1 - f(e_1) \in V_1,$$

and prove that $h_1(e_1) = e_1$, so that

$$h_1 \circ f \circ h_1(e_1) = a_1^0 e_1 + r_{1,2}\, e_2.$$

Let $f_1 = h_1 \circ f \circ h_1$.

Assuming by induction that

$$f_k = h_k \circ \cdots \circ h_1 \circ f \circ h_1 \circ \cdots \circ h_k$$

has a tridiagonal matrix up to the kth row and column, $1 \le k \le n-3$, let

$$f_k(e_{k+1}) = a_k^k e_k + a_{k+1}^k e_{k+1} + \cdots + a_n^k e_n,$$

and let

$$r_{k+1,k+2} = \|a_{k+2}^k e_{k+2} + \cdots + a_n^k e_n\|.$$

Find an isometry h_{k+1} (reflection or id) such that

$$h_{k+1}(f_k(e_{k+1}) - a_k^k e_k - a_{k+1}^k e_{k+1}) = r_{k+1,k+2}\, e_{k+2}.$$

Observe that

$$w_{k+1} = r_{k+1,k+2}\, e_{k+2} + a_k^k e_k + a_{k+1}^k e_{k+1} - f_k(e_{k+1}) \in V_{k+1},$$

and prove that $h_{k+1}(e_k) = e_k$ and $h_{k+1}(e_{k+1}) = e_{k+1}$, so that

$$h_{k+1} \circ f_k \circ h_{k+1}(e_{k+1}) = a_k^k e_k + a_{k+1}^k e_{k+1} + r_{k+1,k+2}\, e_{k+2}.$$

Let $f_{k+1} = h_{k+1} \circ f_k \circ h_{k+1}$, and finish the proof.

Do f and f_{n-2} have the same eigenvalues? If so, explain why.

(3) Prove that given any symmetric $n \times n$ matrix A, there are $n - 2$ matrices $H_1, \ldots, H_{n-2}$, Householder matrices or the identity, such that

$$B = H_{n-2} \cdots H_1 A H_1 \cdots H_{n-2}$$

is a symmetric tridiagonal matrix.

12.3. Write a computer program implementing the method of Problem 12.2(3).

12.4. Let A be a symmetric tridiagonal $n \times n$ matrix

$$A = \begin{pmatrix} b_1 & c_1 & & & & \\ c_1 & b_2 & c_2 & & & \\ & c_2 & b_3 & c_3 & & \\ & & \ddots & \ddots & \ddots & \\ & & & c_{n-2} & b_{n-1} & c_{n-1} \\ & & & & c_{n-1} & b_n \end{pmatrix},$$

where it is assumed that $c_i \neq 0$ for all i, $1 \leq i \leq n-1$, and let A_k be the $k \times k$ submatrix consisting of the first k rows and columns of A, $1 \leq k \leq n$. We define the polynomials $P_k(x)$ as follows: $(0 \leq k \leq n)$.

$$P_0(x) = 1,$$
$$P_1(x) = b_1 - x,$$
$$P_k(x) = (b_k - x)P_{k-1}(x) - c_{k-1}^2 P_{k-2}(x),$$

where $2 \leq k \leq n$.

(1) Prove the following properties:

(i) $P_k(x)$ is the characteristic polynomial of A_k, where $1 \leq k \leq n$.

(ii) $\lim_{x \to -\infty} P_k(x) = +\infty$, where $1 \leq k \leq n$.

(iii) If $P_k(x) = 0$, then $P_{k-1}(x)P_{k+1}(x) < 0$, where $1 \leq k \leq n-1$.

(iv) $P_k(x)$ has k distinct real roots that separate the $k+1$ roots of P_{k+1}, where $1 \leq k \leq n-1$.

(2) Given any real number $\mu > 0$, for every k, $1 \leq k \leq n$, define the function $\mathrm{sg}_k(\mu)$ as follows:

$$\mathrm{sg}_k(\mu) = \begin{cases} \text{sign of } P_k(\mu) & \text{if } P_k(\mu) \neq 0, \\ \text{sign of } P_{k-1}(\mu) & \text{if } P_k(\mu) = 0. \end{cases}$$

We encode the sign of a positive number as $+$, and the sign of a negative number as $-$. Then let $E(k,\mu)$ be the ordered list

$$E(k,\mu) = \langle +, \mathrm{sg}_1(\mu), \mathrm{sg}_2(\mu), \ldots, \mathrm{sg}_k(\mu) \rangle,$$

and let $N(k,\mu)$ be the number changes of sign between consecutive signs in $E(k,\mu)$.

Prove that $\mathrm{sg}_k(\mu)$ is well defined, and that $N(k,\mu)$ is the number of roots λ of $P_k(x)$ such that $\lambda < \mu$.

Remark: The above can be used to compute the eigenvalues of a (tridiagonal) symmetric matrix (the method of Givens–Householder).

12.5. Let $A = (a_{ij})$ be a real or complex $n \times n$ matrix.

(1) If λ is an eigenvalue of A, prove that there is some eigenvector $u = (u_1, \ldots, u_n)$ of A for λ such that

$$\max_{1 \leq i \leq n} |u_i| = 1.$$

(2) If $u = (u_1, \ldots, u_n)$ is an eigenvector of A for λ as in (1), assuming that i, $1 \leq i \leq n$, is an index such that $|u_i| = 1$, prove that

$$(\lambda - a_{ii})u_i = \sum_{\substack{j=1 \\ j \neq i}}^{n} a_{ij}u_j,$$

and thus that

$$|\lambda - a_{ii}| \leq \sum_{\substack{j=1 \\ j \neq i}}^{n} |a_{ij}|.$$

Conclude that the eigenvalues of A are inside the union of the closed disks D_i defined such that

$$D_i = \left\{ z \in \mathbb{C} \mid |z - a_{ii}| \leq \sum_{\substack{j=1 \\ j \neq i}}^{n} |a_{ij}| \right\}.$$

Remark: This result is known as *Gershgorin's theorem*.

12.6. (a) Given a rotation matrix

$$R = \begin{pmatrix} \cos\theta & -\sin\theta \\ \sin\theta & \cos\theta \end{pmatrix},$$

where $0 < \theta < \pi$, prove that there is a skew-symmetric matrix B such that

$$R = (I - B)(I + B)^{-1}.$$

(b) If B is a skew-symmetric $n \times n$ matrix, prove that $\lambda I_n - B$ and $\lambda I_n + B$ are invertible for all $\lambda \neq 0$, and that they commute.

(c) Prove that

$$R = (\lambda I_n - B)(\lambda I_n + B)^{-1}$$

is a rotation matrix that does not admit -1 as an eigenvalue.

(d) Given any rotation matrix R that does not admit -1 as an eigenvalue, prove that there is a skew-symmetric matrix B such that

$$R = (I_n - B)(I_n + B)^{-1} = (I_n + B)^{-1}(I_n - B).$$

This is known as the *Cayley representation* of rotations (Cayley, 1846).

(e) Given any rotation matrix R, prove that there is a skew-symmetric matrix B such that

$$R = \left((I_n - B)(I_n + B)^{-1} \right)^2.$$

12.7. Given a Euclidean space E, let $\varphi \colon E \times E \to \mathbb{R}$ be a symmetric bilinear form on E. Prove that there is an orthonormal basis of E w.r.t. which φ is represented by a diagonal matrix. Given any basis $(e_1, \ldots, e_n)$ of E, recall that for any two vectors x and y, if X and Y denote the column vectors of coordinates of x and y w.r.t.

$(e_1, \ldots, e_n)$, then

$$\varphi(x,y) = X^\top AY,$$

for some symmetric matrix A; see Chapter 6, Problem 6.13.

Hint. Let A be the symmetric matrix representing φ over $(e_1, \ldots, e_n)$. Use the fact that there is an orthogonal matrix P and a (real) diagonal matrix D such that

$$A = PDP^\top.$$

12.8. Given a Hermitian space E, let $\varphi \colon E \times E \to \mathbb{C}$ be a Hermitian form on E. Prove that there is an orthonormal basis of E w.r.t. which φ is represented by a diagonal matrix. Given any basis $(e_1, \ldots, e_n)$ of E, recall that for any two vectors x and y, if X and Y denote the column vectors of coordinates of x and y w.r.t. $(e_1, \ldots, e_n)$, then

$$\varphi(x,y) = X^\top A\overline{Y},$$

for some Hermitian matrix A; see Chapter 11, Problem 11.7.

Hint. Let A be the Hermitian matrix representing φ over $(e_1, \ldots, e_n)$. Use the fact that there is a unitary matrix P and a (real) diagonal matrix D such that

$$A^\top = PDP^*.$$

12.9. Let E be a Euclidean space of dimension n. For any linear map $f \colon E \to E$, we define the *Rayleigh–Ritz ratio* of f as the function $R_f \colon (E - \{0\}) \to \mathbb{R}$ defined such that

$$R_f(x) = \frac{f(x) \cdot x}{x \cdot x},$$

for all $x \neq 0$.

(a) Prove that

$$R_f(x) = R_f(\lambda x)$$

for all $\lambda \in \mathbb{R}$, $\lambda \neq 0$. As a consequence, show that it can be assumed that R_f is defined on the unit sphere

$$S^{n-1} = \{x \in E \mid \|x\| = 1\}.$$

(b) Assume that f is self-adjoint, and let $\lambda_1 \leq \lambda_2 \leq \cdots \leq \lambda_n$ be the (real) eigenvalues of f listed in nondecreasing order. Prove that there is an orthonormal basis $(e_1, \ldots, e_n)$ such that, letting $V_k = S^{n-1} \cap E_k$ be the intersection of S^{n-1} with the subspace E_k spanned by $\{e_1, \ldots, e_k\}$, the following properties hold for all k, $1 \leq k \leq n$:

(1) $\lambda_k = R_f(e_k)$;
(2) $\lambda_k = \max_{x \in V_k} R_f(x)$.

(c) Letting $\mathcal{V}_k$ denote the set of all sets of the form $W \cap S^{n-1}$, where W is any subspace of dimension $k \geq 1$, prove that

(3) $\lambda_k = \min_{W \in \mathcal{V}_k} \max_{x \in W} R_f(x)$.

Hint. You will need to prove that if W is any subspace of dimension k, then

$$\dim(W \cap E_k^\perp) \geq 1.$$

The formula given in (3) is usually called the *Courant–Fischer* formula.
(d) Prove that

$$R_f(S^{n-1}) = [\lambda_1, \lambda_n].$$

References

1. Michael Artin. *Algebra*. Prentice-Hall, first edition, 1991.
2. Marcel Berger. *Géométrie 1*. Nathan, 1990. English edition: Geometry 1, Universitext, Springer-Verlag.
3. J.E. Bertin. *Algèbre Linéaire et Géométrie Classique*. Masson, first edition, 1981.
4. Serge Lang. *Algebra*. Addison-Wesley, third edition, 1993.
5. Saunders Mac Lane and Garrett Birkhoff. *Algebra*. Macmillan, first edition, 1967.

Chapter 13
Singular Value Decomposition (SVD) and Polar Form

13.1 Polar Form

In this section we assume that we are dealing with a real Euclidean space E. Let $f\colon E \to E$ be any linear map. In general, it may not be possible to diagonalize f. We show that every linear map can be diagonalized if we are willing to use *two* orthonormal bases. This is the celebrated *singular value decomposition (SVD)*. A close cousin of the SVD is the *polar form* of a linear map, which shows how a linear map can be decomposed into its purely rotational component (perhaps with a flip) and its purely stretching part.

The key observation is that $f^* \circ f$ is self-adjoint, since

$$\langle (f^* \circ f)(u), v \rangle = \langle f(u), f(v) \rangle = \langle u, (f^* \circ f)(v) \rangle.$$

Similarly, $f \circ f^*$ is self-adjoint.

The fact that $f^* \circ f$ and $f \circ f^*$ are self-adjoint is very important, because it implies that $f^* \circ f$ and $f \circ f^*$ can be diagonalized and that they have real eigenvalues. In fact, these eigenvalues are all nonnegative. Indeed, if u is an eigenvector of $f^* \circ f$ for the eigenvalue λ, then

$$\langle (f^* \circ f)(u), u \rangle = \langle f(u), f(u) \rangle$$

and

$$\langle (f^* \circ f)(u), u \rangle = \lambda \langle u, u \rangle,$$

and thus

$$\lambda \langle u, u \rangle = \langle f(u), f(u) \rangle,$$

which implies that $\lambda \geq 0$, since $\langle -, - \rangle$ is positive definite. A similar proof applies to $f \circ f^*$. Thus, the eigenvalues of $f^* \circ f$ are of the form $\mu_1^2, \ldots, \mu_r^2$ or 0, where $\mu_i > 0$, and similarly for $f \circ f^*$. The situation is even better, since we will show shortly that $f^* \circ f$ and $f \circ f^*$ have the same eigenvalues.

Remark: Given any two linear maps $f\colon E \to F$ and $g\colon F \to E$, where $\dim(E) = n$ and $\dim(F) = m$, it can be shown that

$$(-\lambda)^m \det(g \circ f - \lambda\, I_n) = (-\lambda)^n \det(f \circ g - \lambda\, I_m),$$

and thus $g \circ f$ and $f \circ g$ always have the same nonnull eigenvalues!

Definition 13.1. The square roots $\mu_i > 0$ of the positive eigenvalues of $f^* \circ f$ (and $f \circ f^*$) are called the *singular values of* f.

Definition 13.2. A self-adjoint linear map $f\colon E \to E$ whose eigenvalues are nonnegative is called *positive semidefinite* (or *positive*), and if f is also invertible, f is said to be *positive definite*. In the latter case, every eigenvalue of f is strictly positive.

We just showed that $f^* \circ f$ and $f \circ f^*$ are positive semidefinite self-adjoint linear maps. This fact has the remarkable consequence that every linear map has two important decompositions:

1. The polar form.
2. The singular value decomposition (SVD).

The wonderful thing about the singular value decomposition is that there exist two orthonormal bases $(u_1, \ldots, u_n)$ and $(v_1, \ldots, v_n)$ such that with respect to these bases, f is a diagonal matrix consisting of the singular values of f, or 0. Thus, in some sense, f can always be diagonalized with respect to *two* orthonormal bases. The SVD is also a useful tool for solving overdetermined linear systems in the least squares sense and for data analysis, as we show later on.

First, we show some useful relationships between the kernels and the images of f, f^*, $f^* \circ f$, and $f \circ f^*$. Recall that if $f\colon E \to F$ is a linear map, the *image* $\operatorname{Im} f$ *of* f is the subspace $f(E)$ of F, and the *rank of* f is the dimension $\dim(\operatorname{Im} f)$ of its image. Also recall that

$$\dim\left(\operatorname{Ker} f\right) + \dim\left(\operatorname{Im} f\right) = \dim\left(E\right),$$

and that for every subspace W of E,

$$\dim\left(W\right) + \dim\left(W^\perp\right) = \dim\left(E\right).$$

Lemma 13.1. *Given any two Euclidean spaces E and F, where E has dimension n and F has dimension m, for any linear map $f\colon E \to F$, we have*

$$\operatorname{Ker} f = \operatorname{Ker}\left(f^* \circ f\right),$$
$$\operatorname{Ker} f^* = \operatorname{Ker}\left(f \circ f^*\right),$$
$$\operatorname{Ker} f = \left(\operatorname{Im} f^*\right)^\perp,$$
$$\operatorname{Ker} f^* = \left(\operatorname{Im} f\right)^\perp,$$
$$\dim(\operatorname{Im} f) = \dim(\operatorname{Im} f^*),$$
$$\dim(\operatorname{Ker} f) = \dim(\operatorname{Ker} f^*),$$

and f, f^*, $f^* \circ f$, and $f \circ f^*$ have the same rank.

Proof. To simplify the notation, we will denote the inner products on E and F by the same symbol $\langle -, - \rangle$ (to avoid subscripts). If $f(u) = 0$, then $(f^* \circ f)(u) = f^*(f(u)) = f^*(0) = 0$, and so $\operatorname{Ker} f \subseteq \operatorname{Ker}(f^* \circ f)$. By definition of f^*, we have

$$\langle f(u), f(u) \rangle = \langle (f^* \circ f)(u), u \rangle$$

for all $u \in E$. If $(f^* \circ f)(u) = 0$, since $\langle -, - \rangle$ is positive definite, we must have $f(u) = 0$, and so $\operatorname{Ker}(f^* \circ f) \subseteq \operatorname{Ker} f$. Therefore,

$$\operatorname{Ker} f = \operatorname{Ker}(f^* \circ f).$$

The proof that $\operatorname{Ker} f^* = \operatorname{Ker}(f \circ f^*)$ is similar.

By definition of f^*, we have

$$\langle f(u), v \rangle = \langle u, f^*(v) \rangle$$

for all $u \in E$ and all $v \in F$. This immediately implies that

$$\operatorname{Ker} f = (\operatorname{Im} f^*)^{\perp} \quad \text{and} \quad \operatorname{Ker} f^* = (\operatorname{Im} f)^{\perp}.$$

Since

$$\dim(\operatorname{Im} f) = n - \dim(\operatorname{Ker} f)$$

and

$$\dim((\operatorname{Im} f^*)^{\perp}) = n - \dim(\operatorname{Im} f^*),$$

from

$$\operatorname{Ker} f = (\operatorname{Im} f^*)^{\perp}$$

we also have

$$\dim(\operatorname{Ker} f) = \dim((\operatorname{Im} f^*)^{\perp}),$$

from which we obtain

$$\dim(\operatorname{Im} f) = \dim(\operatorname{Im} f^*).$$

The above immediately implies that $\dim(\operatorname{Ker} f) = \dim(\operatorname{Ker} f^*)$. From all this we easily deduce that

$$\dim(\operatorname{Im} f) = \dim(\operatorname{Im}(f^* \circ f)) = \dim(\operatorname{Im}(f \circ f^*)),$$

i.e., f, f^*, $f^* \circ f$, and $f \circ f^*$ have the same rank. $\square$

The next lemma shows a very useful property of positive semidefinite self-adjoint linear maps.

Lemma 13.2. *Given a Euclidean space E of dimension n, for any positive semidefinite self-adjoint linear map $f\colon E \to E$ there is a unique positive semidefinite self-adjoint linear map $h\colon E \to E$ such that $f = h^2 = h \circ h$. Furthermore, $\operatorname{Ker} f = \operatorname{Ker} h$, and if $\mu_1, \ldots, \mu_p$ are the distinct eigenvalues of h and E_i is the eigenspace associated* *with μ_i, then $\mu_1^2, \ldots, \mu_p^2$ are the distinct eigenvalues of f, and E_i is the eigenspace associated with μ_i^2.*

Proof. Since f is self-adjoint, by Theorem 12.3 there is an orthonormal basis $(u_1, \ldots, u_n)$ consisting of eigenvectors of f, and if $\lambda_1, \ldots, \lambda_n$ are the eigenvalues of f, we know that $\lambda_i \in \mathbb{R}$. Since f is assumed to be positive semidefinite, we have $\lambda_i \geq 0$, and we can write $\lambda_i = \mu_i^2$, where $\mu_i \geq 0$. If we define $h \colon E \to E$ by its action on the basis $(u_1, \ldots, u_n)$, so that

$$h(u_i) = \mu_i u_i,$$

it is obvious that $f = h^2$ and that h is positive semidefinite self-adjoint (since its matrix over the orthonormal basis $(u_1, \ldots, u_n)$ is diagonal, thus symmetric). It remains to prove that h is uniquely determined by f. Let $g \colon E \to E$ be any positive semidefinite self-adjoint linear map such that $f = g^2$. Then there is an orthonormal basis $(v_1, \ldots, v_n)$ of eigenvectors of g, and let $\mu_1, \ldots, \mu_n$ be the eigenvalues of g, where $\mu_i \geq 0$. Note that

$$f(v_i) = g^2(v_i) = g(g(v_i)) = \mu_i^2 v_i,$$

so that v_i is an eigenvector of f for the eigenvalue μ_i^2. If $\mu_1, \ldots, \mu_p$ are the distinct eigenvalues of g and $E_1, \ldots, E_p$ are the corresponding eigenspaces, the above argument shows that each E_i is a subspace of the eigenspace U_i of f associated with μ_i^2. However, we observed (just after Theorem 12.3) that

$$E = E_1 \oplus \cdots \oplus E_p,$$

where E_i and E_j are orthogonal if $i \neq j$, and thus we must have $E_i = U_i$. Since $\mu_i, \mu_j \geq 0$ and $\mu_i \neq \mu_j$ implies that $\mu_i^2 \neq \mu_j^2$, the values $\mu_1^2, \ldots, \mu_p^2$ are the distinct eigenvalues of f, and the corresponding eigenspaces are also $E_1, \ldots, E_p$. This shows that $g = h$, and h is unique. Also, as a consequence, $\operatorname{Ker} f = \operatorname{Ker} h$, and if $\mu_1, \ldots, \mu_p$ are the distinct eigenvalues of h, then $\mu_1^2, \ldots, \mu_p^2$ are the distinct eigenvalues of f, and the corresponding eigenspaces are identical. $\square$

There are now two ways to proceed. We can prove directly the singular value decomposition, as Strang does [8, 7], or prove the so-called *polar decomposition* theorem. The proofs are of roughly the same difficulty. We have chosen the second approach, since it is less common in textbook presentations, and since it also yields a little more, namely uniqueness when f is invertible. It is somewhat disconcerting that the next two theorems are given only as an exercise in Bourbaki [1] (Algèbre, Chapter 9, Problem 14, page 127). Yet, the SVD decomposition is of great practical importance. This is probably typical of the attitude of "pure mathematicians." However, the proof hinted at in Bourbaki is quite elegant.

The early history of the singular value decomposition is described in a fascinating paper by Stewart [6]. The SVD is due to Beltrami and Camille Jordan independently (1873, 1874). Gauss is the grandfather of all this, for his work on least squares (1809, 1823) (but Legendre also published a paper on least squares!). Then come Sylvester, Schmidt, and Hermann Weyl. Sylvester's work was apparently "opaque." He gave a computational method to find an SVD. Schimdt's work really has to do with integral equations and symmetric and asymmetric kernels (1907). Weyl's work has to do with perturbation theory (1912). Autonne came up with the polar decomposition (1902, 1915). Eckart and Young extended SVD to rectangular matrices (1936, 1939).

The next three theorems deal with a linear map $f \colon E \to E$ over a Euclidean space E. We will show later on how to generalize these results to linear maps $f \colon E \to F$ between two Euclidean spaces E and F.

Theorem 13.1. *Given a Euclidean space E of dimension n, for any linear map $f \colon E \to E$ there are two positive semidefinite self-adjoint linear maps $h_1 \colon E \to E$ and $h_2 \colon E \to E$ and an orthogonal linear map $g \colon E \to E$ such that*

$$f = g \circ h_1 = h_2 \circ g.$$

Furthermore, if f has rank r, the maps h_1 and h_2 have the same positive eigenvalues $\mu_1, \ldots, \mu_r$, which are the singular values of f, i.e., the positive square roots of the nonnull eigenvalues of both $f^* \circ f$ and $f \circ f^*$. Finally, h_1, h_2 are unique, g is unique if f is invertible, and $h_1 = h_2$ if f is normal.

Proof. By Lemma 13.2 there are two (unique) positive semidefinite self-adjoint linear maps $h_1 \colon E \to E$ and $h_2 \colon E \to E$ such that $f^* \circ f = h_1^2$ and $f \circ f^* = h_2^2$. Note that

$$\langle f(u), f(v) \rangle = \langle h_1(u), h_1(v) \rangle$$

for all $u, v \in E$, since

$$\langle f(u), f(v) \rangle = \langle u, (f^* \circ f)(v) \rangle = \langle u, (h_1 \circ h_1)(v) \rangle = \langle h_1(u), h_1(v) \rangle,$$

because $f^* \circ f = h_1^2$ and $h_1 = h_1^*$ (h_1 is self-adjoint). From Lemma 13.1, $\mathrm{Ker}\, f = \mathrm{Ker}\,(f^* \circ f)$, and from Lemma 13.2, $\mathrm{Ker}\,(f^* \circ f) = \mathrm{Ker}\, h_1$. Thus,

$$\mathrm{Ker}\, f = \mathrm{Ker}\, h_1.$$

If r is the rank of f, then since h_1 is self-adjoint, by Theorem 12.3 there is an orthonormal basis $(u_1, \ldots, u_n)$ of eigenvectors of h_1, and by reordering these vectors if necessary, we can assume that $(u_1, \ldots, u_r)$ are associated with the strictly positive eigenvalues $\mu_1, \ldots, \mu_r$ of h_1 (the singular values of f), and that $\mu_{r+1} = \cdots = \mu_n = 0$. Observe that $(u_{r+1}, \ldots, u_n)$ is an orthonormal basis of $\mathrm{Ker}\, f = \mathrm{Ker}\, h_1$, and that $(u_1, \ldots, u_r)$ is an orthonormal basis of $(\mathrm{Ker}\, f)^{\perp} = \mathrm{Im}\, f^*$. Note that

$$\langle f(u_i), f(u_j) \rangle = \langle h_1(u_i), h_1(u_j) \rangle = \mu_i \mu_j \langle u_i, u_j \rangle = \mu_i^2 \delta_{ij}$$

when $1 \le i, j \le n$ (recall that $\delta_{ij} = 1$ if $i = j$, and $\delta_{ij} = 0$ if $i \ne j$). Letting

$$v_i = \frac{f(u_i)}{\mu_i}$$

when $1 \le i \le r$, observe that

$$\langle v_i, v_j \rangle = \delta_{ij}$$

when $1 \leq i, j \leq r$. Using the Gram–Schmidt orthonormalization procedure, we can extend $(v_1, \ldots, v_r)$ to an orthonormal basis $(v_1, \ldots, v_n)$ of E (even when $r = 0$). Also note that $(v_1, \ldots, v_r)$ is an orthonormal basis of $\mathrm{Im}\, f$, and $(v_{r+1}, \ldots, v_n)$ is an orthonormal basis of $\mathrm{Im}\, f^{\perp} = \mathrm{Ker}\, f^*$.

We define the linear map $g \colon E \to E$ by its action on the basis $(u_1, \ldots, u_n)$ as follows:

$$g(u_i) = v_i$$

for all i, $1 \leq i \leq n$. We have

$$(g \circ h_1)(u_i) = g(h_1(u_i)) = g(\mu_i u_i) = \mu_i g(u_i) = \mu_i v_i = \mu_i \frac{f(u_i)}{\mu_i} = f(u_i)$$

when $1 \leq i \leq r$, and

$$(g \circ h_1)(u_i) = g(h_1(u_i)) = g(0) = 0$$

when $r + 1 \leq i \leq n$ (since $(u_{r+1}, \ldots, u_n)$ is a basis for $\mathrm{Ker}\, f = \mathrm{Ker}\, h_1$), which shows that $f = g \circ h_1$. The fact that g is orthogonal follows easily from the fact that it maps the orthonormal basis $(u_1, \ldots, u_n)$ to the orthonormal basis $(v_1, \ldots, v_n)$.

We can show that $f = h_2 \circ g$ as follows. Notice that

$$\begin{aligned}
h_2^2(v_i) &= (f \circ f^*)\left(\frac{f(u_i)}{\mu_i}\right), \\
&= (f \circ (f^* \circ f))\left(\frac{u_i}{\mu_i}\right), \\
&= \frac{1}{\mu_i}(f \circ h_1^2)(u_i), \\
&= \frac{1}{\mu_i} f(h_1^2(u_i)), \\
&= \frac{1}{\mu_i} f(\mu_i^2 u_i), \\
&= \mu_i f(u_i), \\
&= \mu_i^2 v_i
\end{aligned}$$

when $1 \leq i \leq r$, and

$$h_2^2(v_i) = (f \circ f^*)(v_i) = f(f^*(v_i)) = 0$$

when $r + 1 \leq i \leq n$, since $(v_{r+1}, \ldots, v_n)$ is a basis for $\mathrm{Ker}\, f^* = (\mathrm{Im}\, f)^{\perp}$. Since h_2 is positive semidefinite self-adjoint, so is h_2^2, and by Lemma 13.2, we must have

$$h_2(v_i) = \mu_i v_i$$

when $1 \le i \le r$, and

$$h_2(v_i) = 0$$

when $r+1 \le i \le n$. This shows that $(v_1, \ldots, v_n)$ are eigenvectors of h_2 for $\mu_1, \ldots, \mu_n$ (since $\mu_{r+1} = \cdots = \mu_n = 0$), and thus h_1 and h_2 have the same eigenvalues $\mu_1, \ldots, \mu_n$.

As a consequence,

$$(h_2 \circ g)(u_i) = h_2(g(u_i)) = h_2(v_i) = \mu_i v_i = f(u_i)$$

when $1 \le i \le n$. Since $h_1, h_2, f^* \circ f$, and $f \circ f^*$ are positive semidefinite self-adjoint, $f^* \circ f = h_1^2$, $f \circ f^* = h_2^2$, and $\mu_1, \ldots, \mu_r$ are the eigenvalues of both h_1 and h_2, it follows that $\mu_1, \ldots, \mu_r$ are the singular values of f, i.e., the positive square roots of the nonnull eigenvalues of both $f^* \circ f$ and $f \circ f^*$.

Finally, since

$$f^* \circ f = h_1^2 \quad \text{and} \quad f \circ f^* = h_2^2,$$

by Lemma 13.2, h_1 and h_2 are unique and if f is invertible, then h_1 and h_2 are invertible and thus g is also unique, since $g = f \circ h_1^{-1}$. If h is normal, then $f^* \circ f = f \circ f^*$ and $h_1 = h_2$. $\square$

In matrix form, Theorem 13.1 can be stated as follows. For every real $n \times n$ matrix A, there is some orthogonal matrix R and some positive semidefinite symmetric matrix S such that

$$A = RS.$$

Furthermore, R, S are unique if A is invertible.

Definition 13.3. A pair (R, S) such that $A = RS$ wih R orthogonal and A symmetric positive semi-defnite is called a *polar decomposition of* A.

For example, the matrix

$$A = \frac{1}{2} \begin{pmatrix} 1 & 1 & 1 & 1 \\ 1 & 1 & -1 & -1 \\ 1 & -1 & 1 & -1 \\ 1 & -1 & -1 & 1 \end{pmatrix}$$

is both orthogonal and symmetric, and $A = RS$ with $R = A$ and $S = I$, which implies that some of the eigenvalues of A are negative.

Remark: If E is a Hermitian space, Theorem 13.1 also holds, but the orthogonal linear map g becomes a unitary map. In terms of matrices, the polar decomposition states that for every complex $n \times n$ matrix A, there is some unitary matrix U and some positive semidefinite Hermitian matrix H such that

$$A = UH.$$

13.2 Singular Value Decomposition (SVD)

The proof of Theorem 13.1 shows that there are two orthonormal bases $(u_1, \ldots, u_n)$ and $(v_1, \ldots, v_n)$, where $(u_1, \ldots, u_n)$ are eigenvectors of h_1 and $(v_1, \ldots, v_n)$ are eigenvectors of h_2. Furthermore, $(u_1, \ldots, u_r)$ is an orthonormal basis of $\mathrm{Im}\, f^*$, $(u_{r+1}, \ldots, u_n)$ is an orthonormal basis of $\mathrm{Ker}\, f$, $(v_1, \ldots, v_r)$ is an orthonormal basis of $\mathrm{Im}\, f$, and $(v_{r+1}, \ldots, v_n)$ is an orthonormal basis of $\mathrm{Ker}\, f^*$. Using this, we immediately obtain the singular value decomposition theorem. Note that the singular value decomposition for linear maps of determinant $+1$ is called the *Cartan decomposition* (after Elie Cartan)!

Theorem 13.2. *Given a Euclidean space E of dimension n, for every linear map $f : E \to E$ there are two orthonormal bases $(u_1, \ldots, u_n)$ and $(v_1, \ldots, v_n)$ such that if r is the rank of f, the matrix of f w.r.t. these two bases is a diagonal matrix of the form*

$$\begin{pmatrix} \mu_1 & & & \cdots & \\ & \mu_2 & & \cdots & \\ \vdots & \vdots & \ddots & \vdots & \\ & & & \cdots & \mu_n \end{pmatrix},$$

where $\mu_1, \ldots, \mu_r$ are the singular values of f, i.e., the (positive) square roots of the nonnull eigenvalues of $f^* \circ f$ and $f \circ f^*$, and $\mu_{r+1} = \cdots = \mu_n = 0$. Furthermore, $(u_1, \ldots, u_n)$ are eigenvectors of $f^* \circ f$, $(v_1, \ldots, v_n)$ are eigenvectors of $f \circ f^*$, and $f(u_i) = \mu_i v_i$ when $1 \le i \le n$.

Proof. Going back to the proof of Theorem 13.2, there are two orthonormal bases $(u_1, \ldots, u_n)$ and $(v_1, \ldots, v_n)$, where $(u_1, \ldots, u_n)$ are eigenvectors of h_1, $(v_1, \ldots, v_n)$ are eigenvectors of h_2, $f(u_i) = \mu_i v_i$ when $1 \le i \le r$, and $f(u_i) = 0$ when $r+1 \le i \le n$. But now, with respect to the orthonormal bases $(u_1, \ldots, u_n)$ and $(v_1, \ldots, v_n)$, the matrix of f is indeed

$$\begin{pmatrix} \mu_1 & & & \cdots & \\ & \mu_2 & & \cdots & \\ \vdots & \vdots & \ddots & \vdots & \\ & & & \cdots & \mu_n \end{pmatrix},$$

where $\mu_1, \ldots, \mu_r$ are the singular values of f and $\mu_{r+1} = \cdots = \mu_n = 0$. $\square$

Note that $\mu_i > 0$ for all i ($1 \le i \le n$) iff f is invertible. Given an orientation of the Euclidean space E specified by some orthonormal basis $(e_1, \ldots, e_n)$ taken as direct, if $\det(f) \ge 0$, we can always make sure that the two orthonormal bases $(u_1, \ldots, u_n)$ and $(v_1, \ldots, v_n)$ are oriented positively. Indeed, if $\det(f) = 0$, we just have to flip u_n to $-u_n$ if necessary, and v_n to $-v_n$ if necessary. If $\det(f) > 0$, since $\mu_i > 0$ for all i, $1 \le i \le n$, the orthogonal matrices U and V whose columns are the u_i's and the v_i's have determinants of the same sign. Since $f(u_n) = \mu_n v_n$ and $\mu_n > 0$, we just have to flip u_n to $-u_n$ if necessary, since v_n will also be flipped. Theorem 13.2 can be restated in terms of (real) matrices as follows.

Theorem 13.3. *For every real $n \times n$ matrix A there are two orthogonal matrices U and V and a diagonal matrix D such that $A = VDU^\top$, where D is of the form*

$$D = \begin{pmatrix} \mu_1 & & \cdots & \\ & \mu_2 & \cdots & \\ \vdots & \vdots & \ddots & \vdots \\ & & \cdots & \mu_n \end{pmatrix},$$

where $\mu_1, \ldots, \mu_r$ are the singular values of f, i.e., the (positive) square roots of the nonnull eigenvalues of $A^\top A$ and $AA^\top$, and $\mu_{r+1} = \cdots = \mu_n = 0$. The columns of U are eigenvectors of $A^\top A$, and the columns of V are eigenvectors of $AA^\top$. Furthermore, if $\det(A) \geq 0$, it is possible to choose U and V such that $\det(U) = \det(V) = +1$, i.e., U and V are rotation matrices.

Definition 13.4. A triple (U, D, V) such that $A = VDU^\top$ where U and V are orthogonal and D is a diagonal matrix whose entries are nonnegative (it is positive semidefinite) is called a *singular value decomposition (SVD) of A*.

Remarks:

(1) In Strang [8] the matrices U, V, D are denoted by $U = Q_2$, $V = Q_1$, and $D = \Sigma$, and an SVD is written as $A = Q_1 \Sigma Q_2^\top$. This has the advantage that Q_1 comes before Q_2 in $A = Q_1 \Sigma Q_2^\top$. This has the disadvantage that A maps the columns of Q_2 (eigenvectors of $A^\top A$) to multiples of the columns of Q_1 (eigenvectors of $AA^\top$).

(2) Algorithms for actually computing the SVD of a matrix are presented in Golub and Van Loan [4], Demmel [3], and Trefethen and Bau [9], where the SVD and its applications are also discussed quite extensively.

(3) The SVD also applies to complex matrices. In this case, for every complex $n \times n$ matrix A, there are two unitary matrices U and V and a diagonal matrix D such that

$$A = VDU^*,$$

where D is a diagonal matrix consisting of real entries $\mu_1, \ldots, \mu_n$, where $\mu_1, \ldots, \mu_r$ are the singular values of f, i.e., the positive square roots of the nonnull eigenvalues of A^*A and AA^*, and $\mu_{r+1} = \ldots = \mu_n = 0$.

It is easy to go from the polar form to the SVD, and conversely. Indeed, given a polar decomposition $A = R_1 S$, where R_1 is orthogonal and S is positive semidefinite symmetric, there is an orthogonal matrix R_2 and a positive semidefinite diagonal matrix D such that $S = R_2 D R_2^\top$, and thus

$$A = R_1 R_2 D R_2^\top = VDU^\top,$$

where $V = R_1 R_2$ and $U = R_2$ are orthogonal.

Going the other way, given an SVD decomposition $A = VDU^\top$, let $R = VU^\top$ and $S = UDU^\top$. It is clear that R is orthogonal and that S is positive semidefinite symmetric, and

$$RS = VU^\top UDU^\top = VDU^\top = A.$$

Note that it is possible to require that $\det(R) = +1$ when $\det(A) \geq 0$.

Theorem 13.3 can be easily extended to rectangular $m \times n$ matrices (see Strang [8] or Golub and Van Loan [4], Demmel [3], Trefethen and Bau [9]). As a matter of fact, both Theorem 13.1 and Theorem 13.2 can be generalized to linear maps $f \colon E \to F$ between two Euclidean spaces E and F. In order to do so, we need to define the analogue of the notion of an orthogonal linear map for linear maps $f \colon E \to F$. By definition, the adjoint $f^* \colon F \to E$ of a linear map $f \colon E \to F$ is the unique linear map such that

$$\langle f(u), v \rangle_2 = \langle u, f^*(v) \rangle_1$$

for all $u \in E$ and all $v \in F$. Then we have

$$\langle f(u), f(v) \rangle_2 = \langle u, (f^* \circ f)(v) \rangle_1$$

for all $u, v \in E$. Letting $n = \dim(E)$, $m = \dim(F)$, and $p = \min(m,n)$, if f has rank p and if for every p orthonormal vectors $(u_1, \ldots, u_p)$ in $(\operatorname{Ker} f)^\perp$ the vectors $(f(u_1), \ldots, f(u_p))$ are also orthonormal in F, then

$$f^* \circ f = \operatorname{id}$$

on $(\operatorname{Ker} f)^\perp$. The converse is immediately proved. Thus, we will say that a linear map $f \colon E \to F$ is *weakly orthogonal* if is has rank $p = \min(m,n)$ and if

$$f^* \circ f = \operatorname{id}$$

on $(\operatorname{Ker} f)^\perp$. Of course, $f^* \circ f = 0$ on $\operatorname{Ker} f$. In terms of matrices, we will say that a real $m \times n$ matrix A is weakly orthogonal if its first $p = \min(m,n)$ columns are orthonormal, the remaining ones (if any) being null columns. This is equivalent to saying that

$$A^\top A = I_n$$

if $m \geq n$, and that

$$A^\top A = \begin{pmatrix} I_m & 0_{m,n-m} \\ 0_{n-m,m} & 0_{n-m,n-m} \end{pmatrix}$$

if $n > m$. In this latter case ($n > m$), it is immediately shown that

$$AA^\top = I_m,$$

and $A^\top$ is also weakly orthogonal. The main difference with orthogonal matrices is that $AA^\top$ is usually not a nice matrix of the above form when $m \geq n$ (unless $m = n$). Weakly unitary linear maps are defined analogously.

Theorem 13.4. *Given any two Euclidean spaces E and F, where E has dimension n and F has dimension m, for every linear map $f \colon E \to F$ there are two positive* *semidefinite self-adjoint linear maps $h_1 \colon E \to E$ and $h_2 \colon F \to F$ and a weakly orthogonal linear map $g \colon E \to F$ such that*

$$f = g \circ h_1 = h_2 \circ g.$$

Furthermore, if f has rank r, the maps h_1 and h_2 have the same positive eigenvalues $\mu_1, \ldots, \mu_r$, which are the singular values of f, i.e., the positive square roots of the nonnull eigenvalues of both $f^* \circ f$ and $f \circ f^*$. Finally, h_1, h_2 are unique, g is unique if $\mathrm{rank}(f) = \min(m,n)$ and $h_1 = h_2$ if f is normal.

Proof. By Lemma 13.2 there are two (unique) positive semidefinite self-adjoint linear maps $h_1 \colon E \to E$ and $h_2 \colon F \to F$ such that $f^* \circ f = h_1^2$ and $f \circ f^* = h_2^2$. As in the proof of Theorem 13.1,

$$\mathrm{Ker}\, f = \mathrm{Ker}\, h_1,$$

and letting r be the rank of f, there is an orthonormal basis $(u_1, \ldots, u_n)$ of eigenvectors of h_1 such that $(u_1, \ldots, u_r)$ are associated with the strictly positive eigenvalues $\mu_1, \ldots, \mu_r$ of h_1 (the singular values of f). The vectors $(u_{r+1}, \ldots, u_n)$ form an orthonormal basis of $\mathrm{Ker}\, f = \mathrm{Ker}\, h_1$, and the vectors $(u_1, \ldots, u_r)$ form an orthonormal basis of $(\mathrm{Ker}\, f)^{\perp} = \mathrm{Im}\, f^*$. Furthermore, letting

$$v_i = \frac{f(u_i)}{\mu_i}$$

when $1 \le i \le r$, using the Gram–Schmidt orthonormalization procedure, we can extend $(v_1, \ldots, v_r)$ to an orthonormal basis $(v_1, \ldots, v_m)$ of F (even when $r = 0$). Also note that $(v_1, \ldots, v_r)$ is an orthonormal basis of $\mathrm{Im}\, f$, and $(v_{r+1}, \ldots, v_m)$ is an orthonormal basis of $\mathrm{Im}\, f^{\perp} = \mathrm{Ker}\, f^*$.

Letting $p = \min(m,n)$, we define the linear map $g \colon E \to F$ by its action on the basis $(u_1, \ldots, u_n)$ as follows:

$$g(u_i) = v_i$$

for all i, $1 \le i \le p$, and

$$g(u_i) = 0$$

for all i, $p+1 \le i \le n$. Note that $r \le p$. Just as in the proof of Theorem 13.1, we have

$$(g \circ h_1)(u_i) = f(u_i)$$

when $1 \le i \le r$, and

$$(g \circ h_1)(u_i) = g(h_1(u_i)) = g(0) = 0$$

when $r+1 \le i \le n$ (since $(u_{r+1}, \ldots, u_n)$ is a basis for $\mathrm{Ker}\, f = \mathrm{Ker}\, h_1$), which shows that $f = g \circ h_1$. The fact that g is weakly orthogonal follows easily from the fact that it maps the orthonormal vectors $(u_1, \ldots, u_p)$ to the orthonormal vectors $(v_1, \ldots, v_p)$.

We can show that $f = h_2 \circ g$ as follows. Just as in the proof of Theorem 13.1,

$$h_2^2(v_i) = \mu_i^2 v_i$$

when $1 \leq i \leq r$, and

$$h_2^2(v_i) = (f \circ f^*)(v_i) = f(f^*(v_i)) = 0$$

when $r + 1 \leq i \leq m$, since $(v_{r+1}, \ldots, v_m)$ is a basis for $\operatorname{Ker} f^* = (\operatorname{Im} f)^{\perp}$. Since h_2 is positive semidefinite self-adjoint, so is h_2^2, and by Lemma 13.2, we must have

$$h_2(v_i) = \mu_i v_i$$

when $1 \leq i \leq r$, and

$$h_2(v_i) = 0$$

when $r + 1 \leq i \leq m$. This shows that $(v_1, \ldots, v_m)$ are eigenvectors of h_2 for $\mu_1, \ldots, \mu_m$ (letting $\mu_{r+1} = \cdots = \mu_m = 0$), and thus h_1 and h_2 have the same nonnull eigenvalues $\mu_1, \ldots, \mu_r$. As a consequence,

$$(h_2 \circ g)(u_i) = h_2(g(u_i)) = h_2(v_i) = \mu_i v_i = f(u_i)$$

when $1 \leq i \leq m$. Since $h_1, h_2, f^* \circ f$, and $f \circ f^*$ are positive semidefinite self-adjoint, $f^* \circ f = h_1^2$, $f \circ f^* = h_2^2$, and $\mu_1, \ldots, \mu_r$ are the eigenvalues of both h_1 and h_2, it follows that $\mu_1, \ldots, \mu_r$ are the singular values of f, i.e., the positive square roots of the nonnull eigenvalues of both $f^* \circ f$ and $f \circ f^*$.

Finally, if $m \geq n$ and $\operatorname{rank}(f) = n$, then $\operatorname{Ker} h_1 = \operatorname{Ker} f = (0)$ and h_1 is invertible and if $n \geq m$ and $\operatorname{rank}(f) = m$, then $\operatorname{Ker} h_2 = \operatorname{Ker} f^* = (0)$ and h_2 is invertible. By Lemma 13.2 h_1 and h_2 are unique and since

$$f = g \circ h_1 \quad \text{and} \quad f = h_2 \circ g,$$

if h_1 is invertible then $g = f \circ h_1^{-1}$ and if h_2 is invertible then $g = h_2^{-1} \circ f$, and thus g is also unique. If h is normal, then $f^* \circ f = f \circ f^*$ and $h_1 = h_2$. $\square$

In matrix form, Theorem 13.4 can be stated as follows. For every real $m \times n$ matrix A, there is some weakly orthogonal $m \times n$ matrix R and some positive semidefinite symmetric $n \times n$ matrix S such that

$$A = RS.$$

The proof also shows that if $n > m$, the last $n - m$ columns of R are zero vectors. A pair (R, S) such that $A = RS$ is called a *polar decomposition of* A.

Remark: If E is a Hermitian space, Theorem 13.4 also holds, but the weakly orthogonal linear map g becomes a weakly unitary map. In terms of matrices, the polar decomposition states that for every complex $m \times n$ matrix A, there is some weakly unitary $m \times n$ matrix U and some positive semidefinite Hermitian $n \times n$ matrix H such that

$$A = UH.$$

The proof of Theorem 13.4 shows that there are two orthonormal bases $(u_1, \ldots, u_n)$ and $(v_1, \ldots, v_m)$ for E and F, respectively, where $(u_1, \ldots, u_n)$ are eigenvectors of

h_1 and $(v_1, \ldots, v_m)$ are eigenvectors of h_2. Furthermore, $(u_1, \ldots, u_r)$ is an orthonormal basis of $\operatorname{Im} f^*$, $(u_{r+1}, \ldots, u_n)$ is an orthonormal basis of $\operatorname{Ker} f$, $(v_1, \ldots, v_r)$ is an orthonormal basis of $\operatorname{Im} f$, and $(v_{r+1}, \ldots, v_m)$ is an orthonormal basis of $\operatorname{Ker} f^*$. Using this, we immediately obtain the singular value decomposition theorem for linear maps $f \colon E \to F$, where E and F can have different dimensions.

Theorem 13.5. *Given any two Euclidean spaces E and F, where E has dimension n and F has dimension m, for every linear map $f \colon E \to F$ there are two orthonormal bases $(u_1, \ldots, u_n)$ and $(v_1, \ldots, v_m)$ such that if r is the rank of f, the matrix of f w.r.t. these two bases is a $m \times n$ matrix D of the form*

$$
D = \begin{pmatrix}
\mu_1 & & & \cdots & \\
& \mu_2 & & \cdots & \\
\vdots & \vdots & \ddots & \vdots \\
& & & \cdots & \mu_n \\
0 & \vdots & \cdots & 0 \\
\vdots & \vdots & \ddots & \vdots \\
0 & \vdots & \cdots & 0
\end{pmatrix}
\quad \text{or} \quad
D = \begin{pmatrix}
\mu_1 & & \cdots & & 0 & \cdots & 0 \\
& \mu_2 & \cdots & & 0 & \cdots & 0 \\
\vdots & \vdots & \ddots & \vdots & 0 & \vdots & 0 \\
& & \cdots & \mu_m & 0 & \cdots & 0
\end{pmatrix},
$$

where $\mu_1, \ldots, \mu_r$ are the singular values of f, i.e., the (positive) square roots of the nonnull eigenvalues of $f^* \circ f$ and $f \circ f^*$, and $\mu_{r+1} = \cdots = \mu_p = 0$, where $p = \min(m, n)$. Furthermore, $(u_1, \ldots, u_n)$ are eigenvectors of $f^* \circ f$, $(v_1, \ldots, v_m)$ are eigenvectors of $f \circ f^*$, and $f(u_i) = \mu_i v_i$ when $1 \le i \le p = \min(m, n)$.

Even though the matrix D is an $m \times n$ rectangular matrix, since its only nonzero entries are on the descending diagonal, we still say that D is a diagonal matrix. Theorem 13.5 can be restated in terms of (real) matrices as follows.

Theorem 13.6. *For every real $m \times n$ matrix A, there are two orthogonal matrices U $(n \times n)$ and V $(m \times m)$ and a diagonal $m \times n$ matrix D such that $A = VDU^\top$, where D is of the form*

$$
D = \begin{pmatrix}
\mu_1 & & & \cdots & \\
& \mu_2 & & \cdots & \\
\vdots & \vdots & \ddots & \vdots \\
& & & \cdots & \mu_n \\
0 & \vdots & \cdots & 0 \\
\vdots & \vdots & \ddots & \vdots \\
0 & \vdots & \cdots & 0
\end{pmatrix}
\quad \text{or} \quad
D = \begin{pmatrix}
\mu_1 & & \cdots & & 0 & \cdots & 0 \\
& \mu_2 & \cdots & & 0 & \cdots & 0 \\
\vdots & \vdots & \ddots & \vdots & 0 & \vdots & 0 \\
& & \cdots & \mu_m & 0 & \cdots & 0
\end{pmatrix},
$$

where $\mu_1, \ldots, \mu_r$ are the singular values of f, i.e. the (positive) square roots of the nonnull eigenvalues of $A^\top A$ and $AA^\top$, and $\mu_{r+1} = \ldots = \mu_p = 0$, where $p = \min(m, n)$. The columns of U are eigenvectors of $A^\top A$, and the columns of V are eigenvectors of $AA^\top$.

A triple (U,D,V) such that $A = VDU^\top$ is called a *singular value decomposition (SVD) of A*.

The SVD of matrices can be used to define the pseudo-inverse of a rectangular matrix; see Strang [8], Demmel [3], Trefethen and Bau [9], or Golub and Van Loan [4] for a thorough presentation.

Remark: The matrix form of Theorem 13.4 also yields a variant of the singular value decomposition. First, assume that $m \geq n$. Given an $m \times n$ matrix A, there is a weakly orthogonal $m \times n$ matrix R_1 and a positive semidefinite symmetric $n \times n$ matrix S such that

$$A = R_1 S.$$

Since S is positive semidefinite symmetric, there is an orthogonal $n \times n$ matrix R_2 and a diagonal $n \times n$ matrix D with nonnegative entries such that

$$S = R_2 D R_2^\top.$$

Thus, we can write

$$A = R_1 R_2 D R_2^\top.$$

We claim that $R_1 R_2$ is weakly orthogonal. Indeed,

$$(R_1 R_2)^\top (R_1 R_2) = R_2^\top (R_1^\top R_1) R_2,$$

and if $m \geq n$, we have

$$R_1^\top R_1 = I_n,$$

so that

$$(R_1 R_2)^\top (R_1 R_2) = I_n.$$

Thus, $R_1 R_2$ is indeed weakly orthogonal. Let us now consider the case $n > m$. From the version of SVD in which

$$A = VDU^\top$$

where U is $n \times n$ orthogonal, V is $m \times m$ orthogonal, and D is $m \times n$ diagonal with nonnegative diagonal entries, letting V' be the $m \times n$ matrix obtained from V by adding $n - m$ zero columns and D' be the $n \times n$ matrix obtained from D by adding $n - m$ zero rows, it is immediately verified that

$$V'D' = VD,$$

and thus when $n > m$, we also have

$$A = V'D'U^\top,$$

where U is $n \times n$ orthogonal, V' is $m \times n$ weakly orthogonal, and D' is $n \times n$ diagonal with nonnegative diagonal entries. As a consequence, in both cases we have shown that there exists a weakly orthogonal $m \times n$ matrix V, an orthogonal $n \times n$ matrix U, and a diagonal $n \times n$ matrix D with nonnegative entries such that

$$A = VDU^\top.$$

There is yet another alternative when $n > m$. Given an $m \times n$ matrix A, there is a positive semidefinite symmetric $m \times m$ matrix S and a weakly orthogonal $m \times n$ matrix R_1, such that

$$A = SR_1.$$

Since S is positive semidefinite symmetric, there is an orthogonal $m \times m$ matrix R_2 and a diagonal $m \times m$ matrix D with nonnegative entries such that

$$S = R_2 D R_2^\top.$$

Thus, we can write

$$A = R_2 D R_2^\top R_1.$$

We claim that $R_2^\top R_1$ is weakly orthogonal. Indeed,

$$(R_2^\top R_1)^\top R_2^\top R_1 = R_1^\top (R_2 R_2^\top) R_1 = R_1^\top R_1,$$

since R_2 is orthogonal, and if $n > m$, we have

$$R_1^\top R_1 = \begin{pmatrix} I_m & 0_{m,n-m} \\ 0_{n-m,m} & 0_{n-m,n-m} \end{pmatrix},$$

so that

$$(R_2^\top R_1)^\top R_2^\top R_1 = \begin{pmatrix} I_m & 0_{m,n-m} \\ 0_{n-m,m} & 0_{n-m,n-m} \end{pmatrix},$$

and $R_2^\top R_1$ is weakly orthogonal. Since $n > m$, $(R_2^\top R_1)^\top = R_1^\top R_2$ is also weakly orthogonal. As a consequence, we have shown that when $m \geq n$, there exists a weakly orthogonal $m \times n$ matrix V, an orthogonal $n \times n$ matrix U, and a diagonal $n \times n$ matrix D with nonnegative entries such that

$$A = VDU^\top,$$

and when $n > m$, there exists an orthogonal $m \times m$ matrix V, a weakly orthogonal $m \times n$ matrix $U^\top$ (with U also weakly orthogonal), and a diagonal $m \times m$ matrix D with nonnegative entries, such that

$$A = VDU^\top.$$

In both cases,

$$V^\top A U = D.$$

One of the spectral theorems states that a symmetric matrix can be diagonalized by an orthogonal matrix. There are several numerical methods to compute the eigenvalues of a symmetric matrix A. One method consists in *tridiagonalizing* A, which means that there exists some orthogonal matrix P and some symmetric tridiagonal matrix T such that $A = PTP^\top$. In fact, this can be done using Householder transformations. It is then possible to compute the eigenvalues of T using a bisection method based on Sturm sequences. One can also use Jacobi's method. For details, see Golub and Van Loan [4], Chapter 8, Demmel [3], Trefethen and Bau [9], Lecture 26, or Ciarlet [2]. Computing the SVD of a matrix A is more involved. Most methods begin by finding orthogonal matrices U and V and a *bidiagonal* matrix B such that $A = VBU^\top$. This can also be done using Householder transformations. Observe that $B^\top B$ is symmetric tridiagonal. Thus, in principle, the previous method to diagonalize a symmetric tridiagonal matrix can be applied. However, it is unwise to compute $B^\top B$ explicitly, and more subtle methods are used for this last step. Again, see Golub and Van Loan [4], Chapter 8, Demmel [3], and Trefethen and Bau [9], Lecture 31.

The polar form has applications in continuum mechanics. Indeed, in any deformation it is important to separate stretching from rotation. This is exactly what QS achieves. The orthogonal part Q corresponds to rotation (perhaps with an additional reflection), and the symmetric matrix S to stretching (or compression). The real eigenvalues $\sigma_1, \ldots, \sigma_r$ of S are the stretch factors (or compression factors) (see Marsden and Hughes [5]). The fact that S can be diagonalized by an orthogonal matrix corresponds to a natural choice of axes, the principal axes.

The SVD has applications to data compression, for instance in image processing. The idea is to retain only singular values whose magnitudes are significant enough. The SVD can also be used to determine the rank of a matrix when other methods such as Gaussian elimination produce very small pivots. One of the main applications of the SVD is the computation of the pseudo-inverse. Pseudo-inverses are the key to the solution of various optimization problems, in particular the method of least squares. This topic is discussed in the next chapter (Chapter 14). Applications of the material of this chapter can be found in Strang [8, 7]; Ciarlet [2]; Golub and Van Loan [4], which contains many other references; Demmel [3]; and Trefethen and Bau [9].

13.3 Problems

13.1. (1) Given a matrix

$$A = \begin{pmatrix} a & b \\ c & d \end{pmatrix}$$

prove that there are Householder matrices G, H such that

$$GAH = \begin{pmatrix} \cos\theta & \sin\theta \\ \sin\theta & -\cos\theta \end{pmatrix} \begin{pmatrix} a & b \\ c & d \end{pmatrix} \begin{pmatrix} \cos\varphi & \sin\varphi \\ \sin\varphi & -\cos\varphi \end{pmatrix} = D,$$

where D is a diagonal matrix, iff the following equations hold:

$$(b+c)\cos(\theta + \varphi) = (a-d)\sin(\theta + \varphi),$$
$$(c-b)\cos(\theta - \varphi) = (a+d)\sin(\theta - \varphi).$$

(2) Discuss the solvability of the system. Consider the following cases:

1. $a-d = a+d = 0$.
2a. $a-d = b+c = 0$, $a+d \neq 0$.
2b. $a-d = 0$, $b+c \neq 0$, $a+d \neq 0$.
3a. $a+d = c-b = 0$, $a-d \neq 0$.
3b. $a+d = 0$, $c-b \neq 0$, $a-d \neq 0$.
4. $a+d \neq 0$, $a-d \neq 0$. Show that the solution in this case is

$$\theta = \frac{1}{2}\left[\arctan\left(\frac{b+c}{a-d}\right) + \arctan\left(\frac{c-b}{a+d}\right)\right],$$
$$\varphi = \frac{1}{2}\left[\arctan\left(\frac{b+c}{a-d}\right) - \arctan\left(\frac{c-b}{a+d}\right)\right].$$

If $b = 0$, show that the discussion is simpler: Basically, consider $c = 0$ or $c \neq 0$.

(3) Expressing everything in terms of $u = \cot\theta$ and $v = \cot\varphi$, show that the equations of question (1) become

$$(b+c)(uv-1) = (u+v)(a-d),$$
$$(c-b)(uv+1) = (-u+v)(a+d).$$

Remark: I was unable to find an *elegant* solution for this system.

13.2. The purpose of this problem is to prove that given any linear map $f \colon E \to E$, where E is a Euclidean space of dimension $n \geq 2$ and an orthonormal basis $(e_1, \ldots, e_n)$, there are isometries g_i, h_i, hyperplane reflections or the identity, such that the matrix of

$$g_n \circ \cdots \circ g_1 \circ f \circ h_1 \circ \cdots \circ h_n$$

is a lower bidiagonal matrix, which means that the nonzero entries (if any) are on the main descending diagonal and on the diagonal below it.

(1) Prove that for any isometry $f \colon E \to E$ we have $f = f^* = f^{-1}$ iff $f \circ f = \mathrm{id}$.

(2) Proceed by induction, taking inspiration from the proof of the triangular decomposition given in Chapter 6. Let U_k' be the subspace spanned by $(e_1, \ldots, e_k)$ and U_k'' be the subspace spanned by $(e_{k+1}, \ldots, e_n)$, $1 \leq k \leq n-1$. For the base case, proceed as follows.

Let $v_1 = f^*(e_1)$ and $r_{1,1} = \|v_1\|$. Find an isometry h_1 (reflection or id) such that

$$h_1(f^*(e_1)) = r_{1,1}e_1.$$

Observe that $h_1(f^*(e_1)) \in U_1'$, so that

$$\langle h_1(f^*(e_1)), e_j \rangle = 0$$

for all j, $2 \le j \le n$, and conclude that

$$\langle e_1, f \circ h_1(e_j) \rangle = 0$$

for all j, $2 \le j \le n$.

Next, let

$$u_1 = f \circ h_1(e_1) = u_1' + u_1'',$$

where $u_1' \in U_1'$ and $u_1'' \in U_1''$, and let $r_{2,1} = \|u_1''\|$. Find an isometry g_1 (reflection or id) such that

$$g_1(u_1'') = r_{2,1} e_2.$$

Show that $g_1(e_1) = e_1$,

$$g_1 \circ f \circ h_1(e_1) = u_1' + r_{2,1} e_2,$$

and that

$$\langle e_1, g_1 \circ f \circ h_1(e_j) \rangle = 0$$

for all j, $2 \le j \le n$. At the end of this stage, show that $g_1 \circ f \circ h_1$ has a matrix such that all entries on its first row except perhaps the first are null, and that all entries on the first column, except perhaps the first two, are null.

Assume by induction that some isometries $g_1, \ldots, g_k$ and $h_1, \ldots, h_k$ have been found, either reflections or the identity, and such that

$$f_k = g_k \circ \cdots \circ g_1 \circ f \circ h_1 \circ \cdots \circ h_k$$

has a matrix that is lower bidiagonal up to and including row and column k, where $1 \le k \le n-2$.

Let

$$v_{k+1} = f_k^*(e_{k+1}) = v_{k+1}' + v_{k+1}'',$$

where $v_{k+1}' \in U_k'$ and $v_{k+1}'' \in U_k''$, and let $r_{k+1,k+1} = \|v_{k+1}''\|$. Find an isometry h_{k+1} (reflection or id) such that

$$h_{k+1}(v_{k+1}'') = r_{k+1,k+1} e_{k+1}.$$

Show that if h_{k+1} is a reflection, then $U_k' \subseteq H_{k+1}$, where H_{k+1} is the hyperplane defining the reflection h_{k+1}. Deduce that $h_{k+1}(v_{k+1}') = v_{k+1}'$, and that

$$h_{k+1}(f_k^*(e_{k+1})) = v_{k+1}' + r_{k+1,k+1} e_{k+1}.$$

Observe that $h_{k+1}(f_k^*(e_{k+1})) \in U_{k+1}'$, so that

$$\langle h_{k+1}(f_k^*(e_{k+1})), e_j \rangle = 0$$

for all j, $k+2 \le j \le n$, and thus

$$\langle e_{k+1}, f_k \circ h_{k+1}(e_j) \rangle = 0$$

for all j, $k+2 \le j \le n$.

Next, let

$$u_{k+1} = f_k \circ h_{k+1}(e_{k+1}) = u'_{k+1} + u''_{k+1},$$

where $u'_{k+1} \in U'_{k+1}$ and $u''_{k+1} \in U''_{k+1}$, and let $r_{k+2,k+1} = \|u''_{k+1}\|$. Find an isometry g_{k+1} (reflection or id) such that

$$g_{k+1}(u''_{k+1}) = r_{k+2,k+1} e_{k+2}.$$

Show that if g_{k+1} is a reflection, then $U'_{k+1} \subseteq G_{k+1}$, where G_{k+1} is the hyperplane defining the reflection g_{k+1}. Deduce that $g_{k+1}(e_i) = e_i$ for all i, $1 \le i \le k+1$, and that

$$g_{k+1} \circ f_k \circ h_{k+1}(e_{k+1}) = u'_{k+1} + r_{k+2,k+1} e_{k+2}.$$

Since by induction hypothesis

$$\langle e_i, f_k \circ h_{k+1}(e_j) \rangle = 0$$

for all i, j, $1 \le i \le k+1$, $k+2 \le j \le n$, and since $g_{k+1}(e_i) = e_i$ for all i, $1 \le i \le k+1$, conclude that

$$\langle e_i, g_{k+1} \circ f_k \circ h_{k+1}(e_j) \rangle = 0$$

for all i, j, $1 \le i \le k+1$, $k+2 \le j \le n$. Finish the proof.

13.3. Write a computer program implementing the method of Problem 13.2 to convert an $n \times n$ matrix to bidiagonal form.

References

1. Nicolas Bourbaki. *Algèbre, Chapitre 9*. Eléments de Mathématiques. Hermann, 1968.
2. P.G. Ciarlet. *Introduction to Numerical Matrix Analysis and Optimization*. Cambridge University Press, first edition, 1989. French edition: Masson, 1994.
3. James W. Demmel. *Applied Numerical Linear Algebra*. SIAM Publications, first edition, 1997.
4. Gene H. Golub and Charles F. Van Loan. *Matrix Computations*. The Johns Hopkins University Press, third edition, 1996.
5. Jerrold E. Marsden and Thomas J.R. Hughes. *Mathematical Foundations of Elasticity*. Dover, first edition, 1994.
6. G.W. Stewart. On the early history of the singular value decomposition. *SIAM Review*, 35(4):551–566, 1993.
7. Gilbert Strang. *Introduction to Applied Mathematics*. Wellesley–Cambridge Press, first edition, 1986.
8. Gilbert Strang. *Linear Algebra and Its Applications*. Saunders HBJ, third edition, 1988.
9. L.N. Trefethen and D. Bau III. *Numerical Linear Algebra*. SIAM Publications, first edition, 1997.

Chapter 14
Applications of SVD and Pseudo-inverses

De tous les principes qu'on peut proposer pour cet objet, je pense qu'il n'en est pas de plus général, de plus exact, ni d'une application plus facile, que celui dont nous avons fait usage dans les recherches pécédentes, et qui consiste à rendre *minimum* la somme des carrés des erreurs. Par ce moyen il s'établit entre les erreurs une sorte d'équilibre qui, empêchant les extrêmes de prévaloir, est très propre à faire connaitre l'état du système le plus proche de la vérité.

—**Legendre, 1805,** *Nouvelles Méthodes pour la détermination des Orbites des Comètes*

14.1 Least Squares Problems and the Pseudo-inverse

This chapter presents several applications of SVD. The first one is the pseudo-inverse, which plays a crucial role in solving linear systems by the method of least squares. The second application is data compression. The third application is principal component analysis (PCA), whose purpose is to identify patterns in data and understand the variance–covariance structure of the data. The fourth application is the best affine approximation of a set of data, a problem closely related to PCA.

The method of least squares is a way of "solving" an overdetermined system of linear equations

$$Ax = b,$$

i.e., a system in which A is a rectangular $m \times n$ matrix with more equations than unknowns (when $m > n$). Historically, the method of least squares was used by Gauss and Legendre to solve problems in astronomy and geodesy. The method was first published by Legendre in 1805 in a paper on methods for determining the orbits of comets. However, Gauss had already used the method of least squares as early as 1801 to determine the orbit of the asteroid Ceres, and he published a paper about it in 1810 after the discovery of the asteroid Pallas. Incidentally, it is in that same paper that Gaussian elimination using pivots is introduced.

The reason why more equations than unknowns arise in such problems is that repeated measurements are taken to minimize errors. This produces an overdetermined and often inconsistent system of linear equations. For example, Gauss solved a system of eleven equations in six unknowns to determine the orbit of the asteroid Pallas. As a concrete illustration, suppose that we observe the motion of a small object, assimilated to a point, in the plane. From our observations, we suspect that this point moves along a straight line, say of equation $y = dx + c$. Suppose that we observed the moving point at three different locations (x_1, y_1), (x_2, y_2), and (x_3, y_3). Then we should have

$$c + dx_1 = y_1,$$
$$c + dx_2 = y_2,$$
$$c + dx_3 = y_3.$$

If there were no errors in our measurements, these equations would be compatible, and c and d would be determined by only two of the equations. However, in the presence of errors, the system may be inconsistent. Yet we would like to find c and d!

The idea of the method of least squares is to determine (c, d) such that it minimizes the sum of the squares of the errors, namely,

$$(c + dx_1 - y_1)^2 + (c + dx_2 - y_2)^2 + (c + dx_3 - y_3)^2.$$

In general, for an overdetermined $m \times n$ system $Ax = b$, what Gauss and Legendre discovered is that there are solutions x minimizing

$$\|Ax - b\|^2$$

(where $\|u\|^2 = u_1^2 + \cdots + u_n^2$, the square of the Euclidean norm of the vector $u = (u_1, \ldots, u_n)$), and that these solutions are given by the square $n \times n$ system

$$A^\top Ax = A^\top b,$$

called the *normal equations*. Furthermore, when the columns of A are linearly independent, it turns out that $A^\top A$ is invertible, and so x is unique and given by

$$x = (A^\top A)^{-1} A^\top b.$$

Note that $A^\top A$ is a symmetric matrix, one of the nice features of the normal equations of a least squares problem. For instance, the normal equations for the above problem are

$$\begin{pmatrix} 3 & x_1 + x_2 + x_3 \\ x_1 + x_2 + x_3 & x_1^2 + x_2^2 + x_3^2 \end{pmatrix} \begin{pmatrix} c \\ d \end{pmatrix} = \begin{pmatrix} y_1 + y_2 + y_3 \\ x_1 y_1 + x_2 y_2 + x_3 y_3 \end{pmatrix}.$$

In fact, given any real $m \times n$ matrix A, there is always a unique x^+ of minimum norm that minimizes $\|Ax - b\|^2$, even when the columns of A are linearly dependent. How do we prove this, and how do we find x^+?

Theorem 14.1. *Every linear system $Ax = b$, where A is an $m \times n$ matrix, has a unique least squares solution x^+ of smallest norm.*

Proof. Geometry offers a nice proof of the existence and uniqueness of x^+. Indeed, we can interpret b as a point in the Euclidean (affine) space $\mathbb{R}^m$, and the image subspace of A (also called the column space of A) as a subspace U of $\mathbb{R}^m$ (passing through the origin). Then, we claim that x minimizes $\|Ax - b\|^2$ iff Ax is the orthogonal projection p of b onto the subspace U, which is equivalent to $\overrightarrow{pb} = b - Ax$ being orthogonal to U.

First of all, if $U^\perp$ is the vector space orthogonal to U, the affine space $b + U^\perp$ intersects U in a unique point p (this follows from Lemma 2.15 (2)). Next, for any point $y \in U$, the vectors $\overrightarrow{py}$ and $\overrightarrow{bp}$ are orthogonal, which implies that

$$\|\overrightarrow{by}\|^2 = \|\overrightarrow{bp}\|^2 + \|\overrightarrow{py}\|^2.$$

Thus, p is indeed the unique point in U that minimizes the distance from b to any point in U.

To show that there is a unique x^+ of minimum norm minimizing the (square) error $\|Ax - b\|^2$, we use the fact that

$$\mathbb{R}^n = \operatorname{Ker}A \oplus (\operatorname{Ker}A)^\perp.$$

Indeed, every $x \in \mathbb{R}^n$ can be written uniquely as $x = u + v$, where $u \in \operatorname{Ker}A$ and $v \in (\operatorname{Ker}A)^\perp$, and since u and v are orthogonal,

$$\|x\|^2 = \|u\|^2 + \|v\|^2.$$

Furthermore, since $u \in \operatorname{Ker}A$, we have $Au = 0$, and thus $Ax = p$ iff $Av = p$, which shows that the solutions of $Ax = p$ for which x has minimum norm must belong to $(\operatorname{Ker}A)^\perp$. However, the restriction of A to $(\operatorname{Ker}A)^\perp$ is injective. This is because if $Av_1 = Av_2$, where $v_1, v_2 \in (\operatorname{Ker}A)^\perp$, then $A(v_2 - v_2) = 0$, which implies $v_2 - v_1 \in \operatorname{Ker}A$, and since $v_1, v_2 \in (\operatorname{Ker}A)^\perp$, we also have $v_2 - v_1 \in (\operatorname{Ker}A)^\perp$, and consequently, $v_2 - v_1 = 0$. This shows that there is a unique x of minimum norm minimizing $\|Ax - b\|^2$, and that it must belong to $(\operatorname{Ker}A)^\perp$. $\square$

The proof also shows that x minimizes $\|Ax - b\|^2$ iff $\overrightarrow{pb} = b - Ax$ is orthogonal to U, which can be expressed by saying that $b - Ax$ is orthogonal to every column of A. However, this is equivalent to

$$A^\top (b - Ax) = 0, \quad \text{i.e.,} \quad A^\top A x = A^\top b.$$

Finally, it turns out that the minimum norm least squares solution x^+ can be found in terms of the pseudo-inverse A^+ of A, which is itself obtained from any SVD of A.

Definition 14.1. Given any $m \times n$ matrix A, if $A = VDU^\top$ is an SVD of A with

$$D = \mathrm{diag}(\lambda_1, \ldots, \lambda_r, 0, \ldots, 0),$$

where D is an $m \times n$ matrix and $\lambda_i > 0$, if we let

$$D^+ = \mathrm{diag}(1/\lambda_1, \ldots, 1/\lambda_r, 0, \ldots, 0),$$

an $n \times m$ matrix, the *pseudo-inverse* of A is defined by

$$A^+ = UD^+V^\top.$$

Actually, it seems that A^+ depends on the specific choice of U and V in an SVD (U, D, V) for A, but the next theorem shows that this is not so.

Theorem 14.2. *The least squares solution of smallest norm of the linear system* $Ax = b$, *where A is an $m \times n$ matrix, is given by*

$$x^+ = A^+b = UD^+V^\top b.$$

Proof. First, assume that A is a (rectangular) diagonal matrix D, as above. Then, since x minimizes $\|Dx - b\|^2$ iff Dx is the projection of b onto the image subspace F of D, it is fairly obvious that $x^+ = D^+b$. Otherwise, we can write

$$A = VDU^\top,$$

where U and V are orthogonal. However, since V is an isometry,

$$\|Ax - b\| = \|VDU^\top x - b\| = \|DU^\top x - V^\top b\|.$$

Letting $y = U^\top x$, we have $\|x\| = \|y\|$, since U is an isometry, and since U is surjective, $\|Ax - b\|$ is minimized iff $\|Dy - V^\top b\|$ is minimized, and we have shown that the least solution is

$$y^+ = D^+V^\top b.$$

Since $y = U^\top x$, with $\|x\| = \|y\|$, we get

$$x^+ = UD^+V^\top b = A^+b.$$

Thus, the pseudo-inverse provides the optimal solution to the least squares problem. $\square$

By Lemma 14.2 and Theorem 14.1, A^+b is uniquely defined by every b, and thus A^+ depends only on A.

Let $A = U\Sigma V^\top$ be an SVD for A. It is easy to check that

$$AA^+A = A,$$
$$A^+AA^+ = A^+,$$

and both AA^+ and A^+A are symmetric matrices. In fact,

$$AA^+ = U\Sigma V^\top V\Sigma^+ U^\top = U\Sigma\Sigma^+ U^\top = U \begin{pmatrix} I_r & 0 \\ 0 & 0_{n-r} \end{pmatrix} U^\top$$

and

$$A^+A = V\Sigma^+ U^\top U\Sigma V^\top = V\Sigma^+\Sigma V^\top = V \begin{pmatrix} I_r & 0 \\ 0 & 0_{n-r} \end{pmatrix} V^\top.$$

We immediately get

$$(AA^+)^2 = AA^+,$$
$$(A^+A)^2 = A^+A,$$

so both AA^+ and A^+A are orthogonal projections (since they are both symmetric). *We claim that AA^+ is the orthogonal projection onto the range of A and A^+A is the orthogonal projection onto* $\mathrm{Ker}(A)^\perp = \mathrm{Im}(A^\top)$, *the range of $A^\top$.*

Obviously, we have $\mathrm{range}(AA^+) \subseteq \mathrm{range}(A)$, and for any $y = Ax \in \mathrm{range}(A)$, since $AA^+A = A$, we have

$$AA^+y = AA^+Ax = Ax = y,$$

so the image of AA^+ is indeed the range of A. It is also clear that $\mathrm{Ker}(A) \subseteq \mathrm{Ker}(A^+A)$, and since $AA^+A = A$, we also have $\mathrm{Ker}(A^+A) \subseteq \mathrm{Ker}(A)$, and so

$$\mathrm{Ker}(A^+A) = \mathrm{Ker}(A).$$

Since A^+A is Hermitian, $\mathrm{range}(A^+A) = \mathrm{Ker}(A^+A)^\perp = \mathrm{Ker}(A)^\perp$, as claimed.

It will also be useful to see that $\mathrm{range}(A) = \mathrm{range}(AA^+)$ consists of all vectors $y \in \mathbb{R}^n$ such that

$$U^\top y = \begin{pmatrix} z \\ 0 \end{pmatrix},$$

with $z \in \mathbb{R}^r$.

Indeed, if $y = Ax$, then

$$U^\top y = U^\top Ax = U^\top U\Sigma V^\top x = \Sigma V^\top x = \begin{pmatrix} \Sigma_r & 0 \\ 0 & 0_{n-r} \end{pmatrix} V^\top x = \begin{pmatrix} z \\ 0 \end{pmatrix},$$

where Σ_r is the $r \times r$ diagonal matrix $\mathrm{diag}(\sigma_1, \ldots, \sigma_r)$. Conversely, if $U^\top y = \begin{pmatrix} z \\ 0 \end{pmatrix}$, then $y = U \begin{pmatrix} z \\ 0 \end{pmatrix}$, and

$$AA^+ y = U \begin{pmatrix} I_r & 0 \\ 0 & 0_{n-r} \end{pmatrix} U^\top y$$

$$= U \begin{pmatrix} I_r & 0 \\ 0 & 0_{n-r} \end{pmatrix} U^\top U \begin{pmatrix} z \\ 0 \end{pmatrix}$$

$$= U \begin{pmatrix} I_r & 0 \\ 0 & 0_{n-r} \end{pmatrix} \begin{pmatrix} z \\ 0 \end{pmatrix}$$

$$= U \begin{pmatrix} z \\ 0 \end{pmatrix} = y,$$

which shows that y belongs to the range of A.

Similarly, we claim that $\mathrm{range}(A^+A) = \mathrm{Ker}(A)^\perp$ consists of all vectors $y \in \mathbb{R}^n$ such that

$$V^\top y = \begin{pmatrix} z \\ 0 \end{pmatrix},$$

with $z \in \mathbb{R}^r$.

If $y = A^+Au$, then

$$y = A^+Au = V \begin{pmatrix} I_r & 0 \\ 0 & 0_{n-r} \end{pmatrix} V^\top u = V \begin{pmatrix} z \\ 0 \end{pmatrix},$$

for some $z \in \mathbb{R}^r$. Conversely, if $V^\top y = \begin{pmatrix} z \\ 0 \end{pmatrix}$, then $y = V \begin{pmatrix} z \\ 0 \end{pmatrix}$, and so

$$A^+AV \begin{pmatrix} z \\ 0 \end{pmatrix} = V \begin{pmatrix} I_r & 0 \\ 0 & 0_{n-r} \end{pmatrix} V^\top V \begin{pmatrix} z \\ 0 \end{pmatrix}$$

$$= V \begin{pmatrix} I_r & 0 \\ 0 & 0_{n-r} \end{pmatrix} \begin{pmatrix} z \\ 0 \end{pmatrix}$$

$$= V \begin{pmatrix} z \\ 0 \end{pmatrix} = y,$$

which shows that $y \in \mathrm{range}(A^+A)$.

If A is a symmetric matrix, then in general, there is no SVD $U\Sigma V^\top$ of A with $U = V$. However, if A is positive semidefinite, then the eigenvalues of A are nonnegative, and so the nonzero eigenvalues of A are equal to the singular values of A and SVDs of A are of the form

$$A = U\Sigma U^\top.$$

Analogous results hold for complex matrices, but in this case, U and V are unitary matrices and AA^+ and A^+A are Hermitian orthogonal projections.

If A is a normal matrix, which means that $AA^\top = A^\top A$, then there is an intimate relationship between SVD's of A and block diagonalizations of A. As a consequence, the pseudo-inverse of a normal matrix A can be obtained directly from a block diagonalization of A.

If A is a (real) normal matrix, then we know from Theorem 12.7 that A can be block diagonalized with respect to an orthogonal matrix U as

$$A = U\Lambda U^\top,$$

where Λ is the (real) block diagonal matrix

$$\Lambda = \operatorname{diag}(B_1,\ldots,B_n),$$

consisting either of 2×2 blocks of the form

$$B_j = \begin{pmatrix} \lambda_j & -\mu_j \\ \mu_j & \lambda_j \end{pmatrix}$$

with $\mu_j \neq 0$, or of one-dimensional blocks $B_k = (\lambda_k)$. Then we have the following proposition:

Proposition 14.1. *For any (real) normal matrix A and any block diagonalization $A = U\Lambda U^\top$ of A as above, the pseudo-inverse of A is given by*

$$A^+ = U\Lambda^+ U^\top,$$

where Λ^+ is the pseudo-inverse of Λ. Furthermore, if

$$\Lambda = \begin{pmatrix} \Lambda_r & 0 \\ 0 & 0 \end{pmatrix},$$

where Λ_r has rank r, then

$$\Lambda^+ = \begin{pmatrix} \Lambda_r^{-1} & 0 \\ 0 & 0 \end{pmatrix}.$$

Proof. Assume that $B_1,\ldots,B_p$ are 2×2 blocks and that $\lambda_{2p+1},\ldots,\lambda_n$ are the scalar entries. We know that the numbers $\lambda_j \pm i\mu_j$, and the λ_{2p+k} are the eigenvalues of A. Let $\rho_{2j-1} = \rho_{2j} = \sqrt{\lambda_j^2 + \mu_j^2}$ for $j = 1,\ldots,p$, $\rho_{2p+j} = \lambda_j$ for $j = 1,\ldots,n-2p$, and assume that the blocks are ordered so that $\rho_1 \geq \rho_2 \geq \cdots \geq \rho_n$. Then it is easy to see that

$$UU^\top = U^\top U = U\Lambda U^\top U\Lambda^\top U^\top = U\Lambda\Lambda^\top U^\top,$$

with

$$\Lambda\Lambda^\top = \operatorname{diag}(\rho_1^2,\ldots,\rho_n^2),$$

so the singular values $\sigma_1 \geq \sigma_2 \geq \cdots \geq \sigma_n$ of A, which are the nonnegative square roots of the eigenvalues of $AA^\top$, are such that

$$\sigma_j = \rho_j, \quad 1 \leq j \leq n.$$

We can define the diagonal matrices

$$\Sigma = \operatorname{diag}(\sigma_1,\ldots,\sigma_r,0,\ldots,0),$$

where $r = \mathrm{rank}(A)$, $\sigma_1 \geq \cdots \geq \sigma_r > 0$ and

$$\Theta = \mathrm{diag}(\sigma_1^{-1} B_1, \ldots, \sigma_{2p}^{-1} B_p, 1, \ldots, 1),$$

so that Θ is an orthogonal matrix and

$$\Lambda = \Theta \Sigma = (B_1, \ldots, B_p, \lambda_{2p+1}, \ldots, \lambda_r, 0, \ldots, 0).$$

But then we can write

$$A = U \Lambda U^\top = U \Theta \Sigma U^\top,$$

and we if let $V = U\Theta$, since U is orthogonal and Θ is also orthogonal, V is also orthogonal and $A = V \Sigma U^\top$ *is an SVD for A.* Now we get

$$A^+ = U \Sigma^+ V^\top = U \Sigma^+ \Theta^\top U^\top.$$

However, since Θ is an orthogonal matrix, $\Theta^\top = \Theta^{-1}$, and a simple calculation shows that

$$\Sigma^+ \Theta^\top = \Sigma^+ \Theta^{-1} = \Lambda^+,$$

which yields the formula

$$A^+ = U \Lambda^+ U^\top.$$

Also observe that if we write

$$\Lambda_r = (B_1, \ldots, B_p, \lambda_{2p+1}, \ldots, \lambda_r),$$

then Λ_r is invertible and

$$\Lambda^+ = \begin{pmatrix} \Lambda_r^{-1} & 0 \\ 0 & 0 \end{pmatrix}.$$

Therefore, the pseudo-inverse of a normal matrix can be computed directly from any block diagonalization of A, as claimed. $\square$

The following properties, due to Penrose, characterize the pseudo-inverse of a matrix. We have already proved that the pseudo-inverse satisfies these equations. For a proof of the converse, see Kincaid and Cheney [6].

Lemma 14.1. *Given any $m \times n$ matrix A (real or complex), the pseudo-inverse A^+ of A is the unique $n \times m$ matrix satisfying the following properties:*

$$AA^+A = A,$$
$$A^+AA^+ = A^+,$$
$$(AA^+)^\top = AA^+,$$
$$(A^+A)^\top = A^+A.$$

If A is an $m \times n$ matrix of rank n (and so $m \geq n$), it is immediately shown that the QR-decomposition in terms of Householder transformations applies as follows:

There are n $m \times m$ matrices $H_1, \ldots, H_n$, Householder matrices or the identity, and an upper triangular $m \times n$ matrix R of rank n such that

$$A = H_1 \cdots H_n R.$$

Then, because each H_i is an isometry,

$$\|Ax - b\| = \|Rx - H_n \cdots H_1 b\|,$$

and the least squares problem $Ax = b$ is equivalent to the system

$$Rx = H_n \cdots H_1 b.$$

Now, the system

$$Rx = H_n \cdots H_1 b$$

is of the form

$$\begin{pmatrix} R_1 \\ 0_{m-n} \end{pmatrix} x = \begin{pmatrix} c \\ d \end{pmatrix},$$

where R_1 is an invertible $n \times n$ matrix (since A has rank n), $c \in \mathbb{R}^n$, and $d \in \mathbb{R}^{m-n}$, and the least squares solution of smallest norm is

$$x^+ = R_1^{-1} c.$$

Since R_1 is a triangular matrix, it is very easy to invert R_1.

The method of least squares is one of the most effective tools of the mathematical sciences. There are entire books devoted to it. Readers are advised to consult Strang [7], Golub and Van Loan [4], Demmel [1], and Trefethen and Bau [8], where extensions and applications of least squares (such as weighted least squares and recursive least squares) are described. Golub and Van Loan [4] also contains a very extensive bibliography, including a list of books on least squares.

14.2 Data Compression and SVD

Among the many applications of SVD, a very useful one is *data compression*, notably for images. In order to make precise the notion of closeness of matrices, we review briefly the notion of *matrix norm*. We assume that the reader is familiar with the concept of a norm in a vector space. The concept of a norm is defined in Section 21.2 of the appendix, and the reader may want to review it before reading any further.

A familiar example of a norm on $\mathbb{R}^n$ (resp. $\mathbb{C}^n$) is the l_p *norm*,

$$\|u\|_p = \left(\sum_{i=1}^n |u_i|^p \right)^{1/p},$$

where $p \geq 1$. When $p = 1$, we have

$$\|u\|_1 = \sum_{i=1}^{n} |u_i|;$$

when $p = 2$, we have the *Euclidean norm*,

$$\|u\|_2 = \sqrt{\sum_{i=1}^{n} |u_i|^2};$$

and when $p = \infty$, we have

$$\|u\|_\infty = \max_{1 \leq i \leq n} |u_i|.$$

Now let E and and F be two normed vector spaces (we will use the same notation, $\|\ \|$, for the norms on E and F). If $A \colon E \to F$ is a linear map, we say that A is *bounded* iff there is some constant $c \geq 0$ such that

$$\|Au\| \leq c \|u\|,$$

for all $u \in E$.

It is well known that *a linear map is continuous iff it is bounded*. Also, if E is finite-dimensional, then a linear map is always bounded. The norms on E and F induce a norm on bounded linear maps as follows:

Definition 14.2. Given two normed vector spaces E and F, for any linear map $A \colon E \to F$, we define $\|A\|$ by

$$\|A\| = \sup_{u \neq 0} \frac{\|Au\|}{\|u\|} = \sup_{\|u\|=1} \|Au\|.$$

Proposition 14.2. *Given two normed vector spaces E and F, the quantity $\|A\|$ is a norm on bounded linear maps $A \colon E \to F$. Furthermore, $\|Au\| \leq \|A\| \|u\|$ for all $u \in E$.*

The norm $\|A\|$ on (bounded) linear maps defined as above is called an *operator norm* or *induced norm* or *subordinate norm*. From Proposition 14.2, we deduce that if $A \colon E \to F$ and $B \colon F \to G$ are bounded linear maps, where E, F, G are normed vector spaces, then

$$\|BA\| \leq \|A\| \|B\|.$$

Let us now consider $m \times n$ matrices. A *matrix norm* is simply a norm on $\mathbb{R}^{mn}$ (or $\mathbb{C}^{mn}$). Some authors require a matrix norm to satisfy $\|AB\| \leq \|A\| \|B\|$ whenever AB makes sense. We immediately have the subordinate matrix norms induced by the l_p norms, but there are also useful matrix norms that are not subordinate norms.

For example, we have the *Frobenius norm* (also known as *Schur norm* or *Hilbert norm*) defined so that if $A = (a_{ij})$ is an $m \times n$ matrix, then

$$\|A\|_F = \sqrt{\sum_{ij} |a_{ij}|^2}.$$

We leave the following useful proposition as an exercise:

Proposition 14.3. *Let A be an $m \times n$ matrix (over $\mathbb{R}$ or $\mathbb{C}$) and let $\sigma_1 \geq \sigma_2 \geq \cdots \geq \sigma_p$ be its singular values (where $p = \min(m,n)$). Then the following properties hold:*

1. *$\|Au\| \leq \|A\|\,\|u\|$, where $\|A\|$ is a subordinate norm and $\|Au\|_2 \leq \|A\|_F\,\|u\|_2$, where $\|A\|_F$ is the Frobenius norm.*
2. *$\|AB\| \leq \|A\|\,\|B\|$, for a subordinate norm or the Frobenius norm.*
3. *$\|UAV\| = \|A\|$ if U and V are orthogonal (or unitary) and $\|\ \|$ is the Frobenius norm or the subordinate norm $\|\ \|_2$.*
4. *$\|A\|_\infty = \max_i \sum_j |a_{ij}|$.*
5. *$\|A\|_1 = \max_j \sum_i |a_{ij}|$.*
6. *$\|A\|_2 = \sigma_1 = \sqrt{\lambda_{\max}(A^*A)}$, where $\lambda_{\max}(A^*A)$ is the largest eigenvalue of A^*A.*
7. *$\|A\|_F = \sqrt{\sum_{i=1}^{p} \sigma_i^2}$, where $p = \min(m,n)$.*
8. *$\|A\|_2 \leq \|A\|_F \leq \sqrt{p}\,\|A\|_2$.*

In (4), (5), (6), (8), the matrix norms are the subordinate norms induced by the corresponding norms ($\|\ \|_\infty$, $\|\ \|_1$ and $\|\ \|_2$) on $\mathbb{R}^m$ and $\mathbb{R}^n$.

Having all this, given an $m \times n$ matrix of rank r, we would like to find a best approximation of A by a matrix B of rank $k \leq r$ (actually, $k < r$) so that $\|A - B\|_2$ (or $\|A - B\|_F$) is minimized.

Proposition 14.4. *Let A be an $m \times n$ matrix of rank r and let $VDU^\top = A$ be an SVD for A. Write u_i for the columns of U, v_i for the columns of V, and $\sigma_1 \geq \sigma_2 \geq \cdots \geq \sigma_p$ for the singular values of A ($p = \min(m,n)$). Then a matrix of rank $k < r$ closest to A (in the $\|\ \|_2$ norm) is given by*

$$A_k = \sum_{i=1}^{k} \sigma_i v_i u_i^\top = V \mathrm{diag}(\sigma_1, \ldots, \sigma_k) U^\top$$

and $\|A - A_k\|_2 = \sigma_{k+1}$.

Proof. By construction, A_k has rank k, and we have

$$\|A - A_k\|_2 = \left\| \sum_{i=k+1}^{p} \sigma_i v_i u_i^\top \right\|_2 = \left\| V \mathrm{diag}(0, \ldots, 0, \sigma_{k+1}, \ldots, \sigma_p) U^\top \right\|_2 = \sigma_{k+1}.$$

It remains to show that $\|A - B\|_2 \geq \sigma_{k+1}$ for all rank-k matrices B. Let B be any rank-k matrix, so its kernel has dimension $p - k$. The subspace V_{k+1} spanned by $(v_1, \ldots, v_{k+1})$ has dimension $k + 1$, and because the sum of the dimensions of the kernel of B and of V_{k+1} is $(p - k) + k + 1 = p + 1$, these two subspaces must intersect in a subspace of dimension at least 1. Pick any unit vector h in $\mathrm{Ker}(B) \cap V_{k+1}$. Then since $Bh = 0$, we have

$$\|A - B\|_2^2 \geq \|(A - B)h\|_2^2 = \|Ah\|_2^2 = \left\|VDU^\top h\right\|_2^2 \geq \sigma_{k+1}^2 \left\|U^\top h\right\|_2^2 = \sigma_{k+1}^2,$$

which proves our claim. $\square$

Note that A_k can be stored using $(m+n)k$ entries, as opposed to mn entries. When $k \ll m$, this is a substantial gain.

A nice example of the use of Proposition 14.4 in image compression is given in Demmel [1], Chapter 3, Section 3.2.3, pages 113–115; see the Matlab demo.

An interesting topic that we have not addressed is the actual computation of an SVD. This is a very interesting but tricky subject. Most methods reduce the computation of an SVD to the diagonalization of a well-chosen symmetric matrix (which is not $A^\top A$). Interested readers should read Section 5.4 of Demmel's excellent book [1], which contains an overview of most known methods and an extensive list of references.

14.3 Principal Components Analysis (PCA)

Suppose we have a set of data consisting of n points $X_1, \ldots, X_n$, with each $X_i \in \mathbb{R}^d$ *viewed as a row vector*.

Think of the X_i's as persons, and if $X_i = (x_{i1}, \ldots, x_{id})$, each x_{ij} is the value of some *feature* (or *attribute*) of that person. For example, the X_i's could be mathematicians, $d = 2$, and the first component, x_{i1}, of X_i could be the year that X_i was born, and the second component, x_{i2}, the length of the beard of X_i in centimeters. Here is a small data set:

Name	year	length
Carl Friedrich Gauss	1777	0
Camille Jordan	1838	12
Adrien-Marie Legendre	1752	0
Bernhard Riemann	1826	15
David Hilbert	1862	2
Henri Poincaré	1854	5
Emmy Noether	1882	0
Karl Weierstrass	1815	0
Eugenio Beltrami	1835	2
Hermann Schwarz	1843	20

We usually form the $n \times d$ matrix X whose ith row is X_i, with $1 \leq i \leq n$. Then the jth column is denoted by C_j $(1 \leq j \leq d)$. It is sometimes called a *feature vector*, but this terminology is far from being universally accepted. In fact, many people in computer vision call the data points X_i feature vectors!

The purpose of *principal components analysis*, for short *PCA*, is to identify patterns in data and understand the *variance–covariance* structure of the data. This is useful for the following tasks:

1. Data reduction: Often much of the variability of the data can be accounted for by a smaller number of *principal components*.
2. Interpretation: PCA can show relationships that were not previously suspected.

Given a vector (a *sample* of measurements) $x = (x_1, \ldots, x_n) \in \mathbb{R}^n$, recall that the *mean* (or *average*) $\bar{x}$ of x is given by

$$\bar{x} = \frac{\sum_{i=1}^{n} x_i}{n}.$$

We let $x - \bar{x}$ denote the *centered data point*

$$x - \bar{x} = (x_1 - \bar{x}, \ldots, x_n - \bar{x}).$$

In order to *measure the spread* of the x_i's around the mean, we define the *sample variance* (for short, *variance*) $\mathrm{var}(x)$ (or s^2) of the sample x by

$$\mathrm{var}(x) = \frac{\sum_{i=1}^{n} (x_i - \bar{x})^2}{n - 1}.$$

There is a reason for using $n - 1$ instead of n. The above definition makes $\mathrm{var}(x)$ an unbiased estimator of the variance of the random variable being sampled. However, we don't need to worry about this. Curious readers will find an explanation of these peculiar definitions in Epstein [2] (Chapter 14, Section 14.5), or in any decent statistics book.

Given two vectors $x = (x_1, \ldots, x_n)$ and $y = (y_1, \ldots, y_n)$, the *sample covariance* (for short, *covariance*) of x and y is given by

$$\mathrm{cov}(x, y) = \frac{\sum_{i=1}^{n} (x_i - \bar{x})(y_i - \bar{y})}{n - 1}.$$

The covariance of x and y measures how x and y vary from the mean with respect to each other. Obviously, $\mathrm{cov}(x, y) = \mathrm{cov}(y, x)$ and $\mathrm{cov}(x, x) = \mathrm{var}(x)$.

Note that

$$\mathrm{cov}(x, y) = \frac{(x - \bar{x})^\top (y - \bar{y})}{n - 1}.$$

We say that x and y are *uncorrelated* iff $\mathrm{cov}(x, y) = 0$.

Finally, given an $n \times d$ matrix X of n points X_i, for PCA to be meaningful, it will be necessary to translate the origin to the *centroid* (or *center of gravity*) μ of the X_i's, defined by

$$\mu = \frac{1}{n}(X_1 + \cdots + X_n).$$

Observe that if $\mu = (\mu_1, \ldots, \mu_d)$, then μ_j is the mean of the vector C_j (the jth column of X).

We let $X - \mu$ denote the *matrix* whose ith row is the centered data point $X_i - \mu$ ($1 \leq i \leq n$). Then, the *sample covariance matrix* (for short, *covariance matrix*) of X

is the $d \times d$ symmetric matrix

$$\Sigma = \frac{1}{n-1}(X-\mu)^\top(X-\mu) = (\text{cov}(C_i,C_j)).$$

Remark: The factor $\frac{1}{n-1}$ is irrelevant for our purposes and can be ignored. Here is the matrix $X - \mu$ in the case of our bearded mathematicians: Since

$$\mu_1 = 1828.4, \quad \mu_2 = 5.6,$$

we get

Name	year	length
Carl Friedrich Gauss	−51.4	−5.6
Camille Jordan	9.6	6.4
Adrien-Marie Legendre	−76.4	−5.6
Bernhard Riemann	−2.4	9.4
David Hilbert	33.6	−3.6
Henri Poincaré	25.6	−0.6
Emmy Noether	53.6	−5.6
Karl Weierstrass	13.4	−5.6
Eugenio Beltrami	6.6	−3.6
Hermann Schwarz	14.6	14.4

We can think of the vector C_j as representing the features of X in the direction e_j (the jth canonical basis vector in $\mathbb{R}^d$, namely $e_j = (0,\ldots,1,\ldots 0)$, with a 1 in the jth position).

If $v \in \mathbb{R}^d$ is a unit vector, we wish to consider the projection of the data points $X_1,\ldots,X_n$ onto the line spanned by v. Recall from Euclidean geometry that if $x \in \mathbb{R}^d$ is any vector and $v \in \mathbb{R}^d$ is a unit vector, the projection of x onto the line spanned by v is

$$\langle x,v \rangle v.$$

Thus, with respect to the basis v, the projection of x has coordinate $\langle x,v \rangle$. If x is represented by a row vector and v by a column vector, then

$$\langle x,v \rangle = xv.$$

Therefore, the vector $Y \in \mathbb{R}^n$ consisting of the coordinates of the projections of $X_1,\ldots,X_n$ onto the line spanned by v is given by $Y = Xv$, and this is the linear combination

$$Xv = v_1 C_1 + \cdots + v_d C_d$$

of the columns of X (with $v = (v_1,\ldots,v_d)$).

Observe that because μ_j is the mean of the vector C_j (the jth column of X), we get

$$\overline{Y} = \overline{Xv} = v_1\mu_1 + \cdots + v_d\mu_d,$$

and so the centered point $Y - \overline{Y}$ is given by

$$Y - \overline{Y} = v_1(C_1 - \mu_1) + \cdots + v_d(C_d - \mu_d) = (X - \mu)v.$$

Furthermore, if $Y = Xv$ and $Z = Xw$, then

$$\begin{aligned}
\operatorname{cov}(Y,Z) &= \frac{((X-\mu)v)^\top(X-\mu)w}{n-1} \\
&= v^\top \frac{1}{n-1}(X-\mu)^\top(X-\mu)w \\
&= v^\top \Sigma w,
\end{aligned}$$

where Σ is the covariance matrix of X. Since $Y - \overline{Y}$ has zero mean, we have

$$\operatorname{var}(Y) = \operatorname{var}(Y - \overline{Y}) = v^\top \frac{1}{n-1}(X-\mu)^\top(X-\mu)v.$$

The above suggests that we should move the origin to the centroid μ of the X_i's and consider the matrix $X - \mu$ of the centered data points $X_i - \mu$.

From now on, beware that we denote the columns of $X - \mu$ by $C_1, \ldots, C_d$ and that Y denotes the *centered* point $Y = (X - \mu)v = \sum_{j=1}^{d} v_j C_j$, where v is a unit vector.

Basic idea of PCA: The principal components of X are *uncorrelated* projections Y of the data points $X_1, \ldots, X_n$ onto some directions v (where the v's are unit vectors) such that $\operatorname{var}(Y)$ is maximal.

This suggests the following definition:

Definition 14.3. Given an $n \times d$ matrix X of data points $X_1, \ldots, X_n$, if μ is the centroid of the X_i's, then a *first principal component of X (first PC)* is a centered point $Y_1 = (X - \mu)v_1$, the projection of $X_1, \ldots, X_n$ onto a direction v_1 such that $\operatorname{var}(Y_1)$ is maximized, where v_1 is a unit vector (recall that $Y_1 = (X - \mu)v_1$ is a linear combination of the C_j's, the columns of $X - \mu$).

More generally, if $Y_1, \ldots, Y_k$ are k principal components of X along some unit vectors $v_1, \ldots, v_k$, where $1 \le k < d$, a *$(k+1)$th principal component of X $((k+1)$th PC)* is a centered point $Y_{k+1} = (X - \mu)v_{k+1}$, the projection of $X_1, \ldots, X_n$ onto some direction v_{k+1} such that $\operatorname{var}(Y_{k+1})$ is maximized, subject to $\operatorname{cov}(Y_h, Y_{k+1}) = 0$ for all h with $1 \le h \le k$, and where v_{k+1} is a unit vector (recall that $Y_h = (X - \mu)v_h$ is a linear combination of the C_j's). The v_h are called *principal directions*.

The following lemma is the key to the main result about PCA:

Lemma 14.2. *If A is a symmetric $d \times d$ matrix with eigenvalues $\lambda_1 \ge \lambda_2 \ge \cdots \ge \lambda_d$ and if $(u_1, \ldots, u_d)$ is any orthonormal basis of eigenvectors of A, where u_i is a unit eigenvector associated with λ_i, then*

$$\max_{x \ne 0} \frac{x^\top A x}{x^\top x} = \lambda_1$$

(with the maximum attained for $x = u_1$) and

$$\max_{x \neq 0, x \in \{u_1,\dots,u_k\}^\perp} \frac{x^\top A x}{x^\top x} = \lambda_{k+1}$$

(with the maximum attained for $x = u_{k+1}$), where $1 \leq k \leq d-1$.

Proof. First, observe that

$$\max_{x \neq 0} \frac{x^\top A x}{x^\top x} = \max_x \{x^\top A x \mid x^\top x = 1\},$$

and similarly,

$$\max_{x \neq 0, x \in \{u_1,\dots,u_k\}^\perp} \frac{x^\top A x}{x^\top x} = \max_x \left\{ x^\top A x \mid (x \in \{u_1,\dots,u_k\}^\perp) \wedge (x^\top x = 1) \right\}.$$

Since A is a symmetric matrix, its eigenvalues are real and it can be diagonalized with respect to an orthonormal basis of eigenvectors, so let $(u_1,\dots,u_d)$ be such a basis. If we write

$$x = \sum_{i=1}^d x_i u_i,$$

a simple computation shows that

$$x^\top A x = \sum_{i=1}^d \lambda_i x_i^2.$$

If $x^\top x = 1$, then $\sum_{i=1}^d x_i^2 = 1$, and since we assumed that $\lambda_1 \geq \lambda_2 \geq \cdots \geq \lambda_d$, we get

$$x^\top A x = \sum_{i=1}^d \lambda_i x_i^2 \leq \lambda_1 \left(\sum_{i=1}^d x_i^2 \right) = \lambda_1.$$

Thus,

$$\max_x \left\{ x^\top A x \mid x^\top x = 1 \right\} \leq \lambda_1,$$

and since this maximum is achieved for $e_1 = (1,0,\dots,0)$, we conclude that

$$\max_x \left\{ x^\top A x \mid x^\top x = 1 \right\} = \lambda_1.$$

Next, observe that $x \in \{u_1,\dots,u_k\}^\perp$ and $x^\top x = 1$ iff $x_1 = \cdots = x_k = 0$ and $\sum_{i=1}^d x_i = 1$. Consequently, for such an x, we have

$$x^\top A x = \sum_{i=k+1}^d \lambda_i x_i^2 \leq \lambda_{k+1} \left(\sum_{i=k+1}^d x_i^2 \right) = \lambda_{k+1}.$$

Thus,

$$\max_x \left\{ x^\top A x \mid (x \in \{u_1, \ldots, u_k\}^\perp) \wedge (x^\top x = 1) \right\} \leq \lambda_{k+1},$$

and since this maximum is achieved for $e_{k+1} = (0, \ldots, 0, 1, 0, \ldots, 0)$ with a 1 in position $k+1$, we conclude that

$$\max_x \left\{ x^\top A x \mid (x \in \{u_1, \ldots, u_k\}^\perp) \wedge (x^\top x = 1) \right\} = \lambda_{k+1},$$

as claimed. $\quad\square$

The quantity

$$\frac{x^\top A x}{x^\top x}$$

is known as the *Rayleigh–Ritz ratio* and Lemma 14.2 is often known as part of the *Rayleigh–Ritz theorem*.

Lemma 14.2 also holds if A is a Hermitian matrix and if we replace $x^\top A x$ by $x^* A x$ and $x^\top x$ by $x^* x$. The proof is unchanged, since a Hermitian matrix has real eigenvalues and is diagonalized with respect to an orthonormal basis of eigenvectors (with respect to the Hermitian inner product).

We then have the following fundamental result showing how *the SVD of X yields the PCs*:

Theorem 14.3. *(SVD yields PCA) Let X be an $n \times d$ matrix of data points $X_1, \ldots, X_n$, and let μ be the centroid of the X_i's. If $X - \mu = V D U^\top$ is an SVD decomposition of $X - \mu$ and if the main diagonal of D consists of the singular values $\sigma_1 \geq \sigma_2 \geq \cdots \geq \sigma_d$, then the centered points $Y_1, \ldots, Y_d$, where*

$$Y_k = (X - \mu) u_k = \text{kth column of } V D$$

and u_k is the kth column of U, are d principal components of X. Furthermore,

$$\text{var}(Y_k) = \frac{\sigma_k^2}{n - 1}$$

and $\text{cov}(Y_h, Y_k) = 0$, whenever $h \neq k$ and $1 \leq k, h \leq d$.

Proof. Recall that for any unit vector v, the centered projection of the points $X_1, \ldots, X_n$ onto the line of direction v is $Y = (X - \mu)v$ and that the variance of Y is given by

$$\text{var}(Y) = v^\top \frac{1}{n-1} (X - \mu)^\top (X - \mu) v.$$

Since $X - \mu = V D U^\top$, we get

$$\operatorname{var}(Y) = v^\top \frac{1}{(n-1)}(X-\mu)^\top(X-\mu)v$$

$$= v^\top \frac{1}{(n-1)}UDV^\top VDU^\top v$$

$$= v^\top U \frac{1}{(n-1)}D^2 U^\top v.$$

Similarly, if $Y = (X-\mu)v$ and $Z = (X-\mu)w$, then the covariance of Y and Z is given by

$$\operatorname{cov}(Y,Z) = v^\top U \frac{1}{(n-1)}D^2 U^\top w.$$

Obviously, $U\frac{1}{(n-1)}D^2 U^\top$ is a symmetric matrix whose eigenvalues are $\frac{\sigma_1^2}{n-1} \geq \cdots \geq \frac{\sigma_d^2}{n-1}$, and the columns of U form an orthonormal basis of unit eigenvectors.

We proceed by induction on k. For the base case, $k = 1$, maximizing $\operatorname{var}(Y)$ is equivalent to maximizing

$$v^\top U \frac{1}{(n-1)}D^2 U^\top v,$$

where v is a unit vector. By Lemma 14.2, the maximum of the above quantity is the largest eigenvalue of $U\frac{1}{(n-1)}D^2 U^\top$, namely $\frac{\sigma_1^2}{n-1}$, and it is achieved for u_1, the first columnn of U. Now we get

$$Y_1 = (X-\mu)u_1 = VDU^\top u_1,$$

and since the columns of U form an orthonormal basis, $U^\top u_1 = e_1 = (1,0,\ldots,0)$, and so Y_1 is indeed the first column of VD.

By the induction hypothesis, the centered points $Y_1,\ldots,Y_k$, where $Y_h = (X-\mu)u_h$ and $u_1,\ldots,u_k$ are the first k columns of U, are k principal components of X. Because

$$\operatorname{cov}(Y,Z) = v^\top U \frac{1}{(n-1)}D^2 U^\top w,$$

where $Y = (X-\mu)v$ and $Z = (X-\mu)w$, the condition $\operatorname{cov}(Y_h,Z) = 0$ for $h = 1,\ldots,k$ is equivalent to the fact that w belongs to the orthogonal complement of the subspace spanned by $\{u_1,\ldots,u_k\}$, and maximizing $\operatorname{var}(Z)$ subject to $\operatorname{cov}(Y_h,Z) = 0$ for $h = 1,\ldots,k$ is equivalent to maximizing

$$w^\top U \frac{1}{(n-1)}D^2 U^\top w,$$

where w is a unit vector orthogonal to the subspace spanned by $\{u_1,\ldots,u_k\}$. By Lemma 14.2, the maximum of the above quantity is the $(k+1)$th eigenvalue of $U\frac{1}{(n-1)}D^2 U^\top$, namely $\frac{\sigma_{k+1}^2}{n-1}$, and it is achieved for u_{k+1}, the $(k+1)$th columnn of U. Now we get

$$Y_{k+1} = (X-\mu)u_{k+1} = VDU^\top u_{k+1},$$

and since the columns of U form an orthonormal basis, $U^\top u_{k+1} = e_{k+1}$, and Y_{k+1} is indeed the $(k+1)$th column of VD, which completes the proof of the induction step. $\quad\square$

The d columns $u_1,\ldots,u_d$ of U are usually called the *principal directions* of $X - \mu$ (and X). We note that not only do we have $\mathrm{cov}(Y_h, Y_k) = 0$ whenever $h \neq k$, but the directions $u_1,\ldots,u_d$ along which the data are projected are mutually orthogonal.

We know from our study of SVD that $\sigma_1^2,\ldots,\sigma_d^2$ are the eigenvalues of the symmetric positive semidefinite matrix $(X - \mu)^\top (X - \mu)$ and that $u_1,\ldots,u_d$ are corresponding eigenvectors. Numerically, it is preferable to use SVD on $X - \mu$ rather than to compute explicitly $(X - \mu)^\top (X - \mu)$ and then diagonalize it. Indeed, the explicit computation of $A^\top A$ from a matrix A can be numerically quite unstable, and good SVD algorithms avoid computing $A^\top A$ explicitly.

In general, since an SVD of X is not unique, *the principal directions $u_1,\ldots,u_d$ are not unique*. This can happen when a data set has some *rotational symmetries*, and in such a case, PCA is not a very good method for analyzing the data set.

14.4 Best Affine Approximation

A problem very close to PCA (and based on least squares) is to *best approximate a data set of n points $X_1,\ldots,X_n$, with $X_i \in \mathbb{R}^d$, by a p-dimensional affine subspace A of $\mathbb{R}^d$, with $1 \leq p \leq d-1$* (the terminology rank $d - p$ is also used).

First, consider $p = d - 1$. Then $A = A_1$ is an affine hyperplane (in $\mathbb{R}^d$), and it is given by an equation of the form

$$a_1 x_1 + \cdots + a_d x_d + c = 0.$$

By *best approximation*, we mean that $(a_1,\ldots,a_d,c)$ solves the homogeneous linear system

$$\begin{pmatrix} x_{11} & \cdots & x_{1d} & 1 \\ \vdots & \vdots & \vdots & \vdots \\ x_{n1} & \cdots & x_{nd} & 1 \end{pmatrix} \begin{pmatrix} a_1 \\ \vdots \\ a_d \\ c \end{pmatrix} = \begin{pmatrix} 0 \\ \vdots \\ 0 \\ 0 \end{pmatrix}$$

in the *least squares sense, subject to the condition that $a = (a_1,\ldots,a_d)$ is a unit vector*, that is, $a^\top a = 1$, where $X_i = (x_{i1},\cdots,x_{id})$.

If we form the symmetric matrix

$$\begin{pmatrix} x_{11} & \cdots & x_{1d} & 1 \\ \vdots & \vdots & \vdots & \vdots \\ x_{n1} & \cdots & x_{nd} & 1 \end{pmatrix}^\top \begin{pmatrix} x_{11} & \cdots & x_{1d} & 1 \\ \vdots & \vdots & \vdots & \vdots \\ x_{n1} & \cdots & x_{nd} & 1 \end{pmatrix}$$

involved in the normal equations, we see that the bottom row (and last column) of that matrix is

$$n\mu_1 \quad \cdots \quad n\mu_d \quad n,$$

where $n\mu_j = \sum_{i=1}^n x_{ij}$ is n times the mean of the column C_j of X.

Therefore, if $(a_1,\ldots,a_d,c)$ is a least squares solution, that is, a solution of the normal equations, we must have

$$n\mu_1 a_1 + \cdots + n\mu_d a_d + nc = 0,$$

that is,

$$a_1\mu_1 + \cdots + a_d\mu_d + c = 0,$$

which means that the *hyperplane A_1 must pass through the centroid μ of the data points $X_1,\ldots,X_n$.* Then we can rewrite the original system with respect to the centered data $X_i - \mu$, and we find that the variable c drops out and we get the system

$$(X - \mu)a = 0,$$

where $a = (a_1,\ldots,a_d)$.

Thus, we are looking for a unit vector a solving $(X - \mu)a = 0$ in the least squares sense, that is, some a such that $a^\top a = 1$ minimizing

$$a^\top (X - \mu)^\top (X - \mu)a.$$

Compute some SVD $VDU^\top$ of $X - \mu$, where the main diagonal of D consists of the singular values $\sigma_1 \geq \sigma_2 \geq \cdots \geq \sigma_d$ of $X - \mu$ arranged in descending order. Then

$$a^\top (X - \mu)^\top (X - \mu)a = a^\top UD^2 U^\top a,$$

where $D^2 = \mathrm{diag}(\sigma_1^2,\ldots,\sigma_d^2)$ is a diagonal matrix, so pick a to be *the last column in U* (corresponding to the smallest eigenvalue σ_d^2 of $(X - \mu)^\top (X - \mu)$). This is a solution to our best fit problem.

Therefore, if U_{d-1} is the linear hyperplane defined by a, that is,

$$U_{d-1} = \{u \in \mathbb{R}^d \mid \langle u,a \rangle = 0\},$$

where a is the last column in U for some SVD $VDU^\top$ of $X - \mu$, we have shown that the affine hyperplane $A_1 = \mu + U_{d-1}$ is a best approximation of the data set $X_1,\ldots,X_n$ in the least squares sense.

Is is easy to show that this hyperplane $A_1 = \mu + U_{d-1}$ minimizes the sum of the square distances of each X_i to its orthogonal projection onto A_1. Also, since U_{d-1} is the orthogonal complement of a, the last column of U, we see that U_{d-1} is spanned by the first $d - 1$ columns of U, that is, the first $d - 1$ principal directions of $X - \mu$.

All this can be generalized to a *best $(d-k)$-dimensional affine subspace A_k approximating $X_1,\ldots,X_n$ in the least squares sense* $(1 \leq k \leq d - 1)$. Such an affine subspace A_k is cut out by k independent hyperplanes H_i (with $1 \leq i \leq k$), each given by some equation

$$a_{i1}x_1 + \cdots + a_{id}x_d + c_i = 0.$$

If we write $a_i = (a_{i1}, \cdots, a_{id})$, to say that the H_i are independent means that $a_1, \ldots, a_k$ are linearly independent. In fact, we may assume that $a_1, \ldots, a_k$ form an *orthonormal system*.

Then, finding a best $(d-k)$-dimensional affine subspace A_k amounts to solving the homogeneous linear system

$$
\begin{pmatrix} X & \mathbf{1} & 0 & \cdots & 0 & 0 & 0 \\ \vdots & \vdots & \vdots & \ddots & \vdots & \vdots & \vdots \\ 0 & 0 & 0 & \cdots & 0 & X & \mathbf{1} \end{pmatrix}
\begin{pmatrix} a_1 \\ c_1 \\ \vdots \\ a_k \\ c_k \end{pmatrix}
=
\begin{pmatrix} 0 \\ \vdots \\ 0 \end{pmatrix},
$$

in the least squares sense, subject to the conditions $a_i^\top a_j = \delta_{ij}$, for all i, j with $1 \le i, j \le k$, where the matrix of the system is a block diagonal matrix consisting of k diagonal blocks $(X, \mathbf{1})$, where $\mathbf{1}$ denotes the column vector $(1, \ldots, 1) \in \mathbb{R}^n$.

Again, it is easy to see that each hyperplane H_i *must pass through the centroid μ* of $X_1, \ldots, X_n$, and by switching to the centered data $X_i - \mu$ we get the system

$$
\begin{pmatrix} X - \mu & 0 & \cdots & 0 \\ \vdots & \vdots & \ddots & \vdots \\ 0 & 0 & \cdots & X - \mu \end{pmatrix}
\begin{pmatrix} a_1 \\ \vdots \\ a_k \end{pmatrix}
=
\begin{pmatrix} 0 \\ \vdots \\ 0 \end{pmatrix},
$$

with $a_i^\top a_j = \delta_{ij}$ for all i, j with $1 \le i, j \le k$.

If $VDU^\top = X - \mu$ is an SVD decomposition, it is easy to see that a least squares solution of this system is given by the last k columns of U, assuming that the main diagonal of D consists of the singular values $\sigma_1 \ge \sigma_2 \ge \cdots \ge \sigma_d$ of $X - \mu$ arranged in descending order. But now the $(d-k)$-dimensional subspace U_{d-k} cut out by the hyperplanes defined by $a_1, \ldots, a_k$ is simply the orthogonal complement of $(a_1, \ldots, a_k)$, which is the subspace spanned by the first $d-k$ columns of U.

So the best $(d-k)$-dimensional affine subpsace A_k approximating $X_1, \ldots, X_n$ in the least squares sense is

$$
A_k = \mu + U_{d-k},
$$

where U_{d-k} is the linear subspace spanned by the first $d-k$ principal directions of $X - \mu$, that is, the first $d-k$ columns of U. Consequently, we get the following interesting interpretation of PCA (actually, principal directions):

Theorem 14.4. *Let X be an $n \times d$ matrix of data points $X_1, \ldots, X_n$, and let μ be the centroid of the X_i's. If $X - \mu = VDU^\top$ is an SVD decomposition of $X - \mu$ and if the main diagonal of D consists of the singular values $\sigma_1 \ge \sigma_2 \ge \cdots \ge \sigma_d$, then a best $(d-k)$-dimensional affine approximation A_k of $X_1, \ldots, X_n$ in the least squares sense is given by*

$$
A_k = \mu + U_{d-k},
$$

where U_{d-k} is the linear subspace spanned by the first $d-k$ columns of U, the first $d-k$ principal directions of $X - \mu$ $(1 \le k \le d-1)$.

There are many applications of PCA to data compression, dimension reduction, and pattern analysis. The basic idea is that in many cases, given a data set $X_1, \ldots, X_n$, with $X_i \in \mathbb{R}^d$, only a "small" subset of $m < d$ of the features is needed to describe the data set accurately.

If $u_1, \ldots, u_d$ are the principal directions of $X - \mu$, then the first m projections of the data (the first m principal components, i.e., the first m columns of VD) onto the first m principal directions represent the data without much loss of information. Thus, instead of using the original data points $X_1, \ldots, X_n$, with $X_i \in \mathbb{R}^d$, we can use their projections onto the first m principal directions $Y_1, \ldots, Y_m$, where $Y_i \in \mathbb{R}^m$ and $m < d$, obtaining a compressed version of the original data set.

For example, PCA is used in computer vision for *face recognition*. Sirovitch and Kirby (1987) seem to be the first to have had the idea of using PCA to compress facial images. They introduced the term *eigenpicture* to refer to the principal directions, u_i. However, an explicit face recognition algorithm was given only later, by Turk and Pentland (1991). They renamed eigenpictures as *eigenfaces*.

For details on the topic of eigenfaces, see Forsyth and Ponce [3] (Chapter 22, Section 22.3.2), where you will also find exact references to Turk and Pentland's papers.

Another interesting application of PCA is to the *recognition of handwritten digits*. Such an application is described in Hastie, Tibshirani, and Friedman, [5] (Chapter 14, Section 14.5.1).

14.5 Problems

14.1. We observe m positions $((x_1, y_1), \ldots, (x_m, y_m))$ of a point moving in the plane ($m \geq 2$), and assume that they are roughly on a straight line. Prove that the line $y = c + dx$ that minimizes the error

$$(c + dx_1 - y_1)^2 + \cdots + (c + dx_m - y_m)^2$$

is the line of equation

$$y = \bar{y} + d(x - \bar{x}),$$

where

$$\bar{x} = \frac{x_1 + \cdots + x_m}{m},$$

$$\bar{y} = \frac{y_1 + \cdots + y_m}{m},$$

$$d = \frac{\sum_{i=1}^{m} (x_i - \bar{x}) y_i}{\sum_{i=1}^{m} (x_i - \bar{x})^2}.$$

14.2. Find the least squares solution to the problem

$$\begin{pmatrix} 2 & -1 \\ 2 & 2 \\ -1 & 2 \end{pmatrix} \begin{pmatrix} x_1 \\ x_2 \end{pmatrix} = \begin{pmatrix} 1 \\ 1 \\ 1 \end{pmatrix}.$$

Do the problem again with the right-hand sides

$$\begin{pmatrix} 2 \\ -1 \\ 2 \end{pmatrix} \quad \text{and} \quad \begin{pmatrix} 2 \\ 2 \\ -1 \end{pmatrix}.$$

14.3. Given m real numbers $(y_1, \ldots, y_m)$, prove that the constant function c that minimizes the error

$$e = (y_1 - c)^2 + \cdots + (y_m - c)^2$$

is the *mean* $\bar{y}$ of the data,

$$\bar{y} = \frac{y_1 + \cdots + y_m}{m}.$$

Note that the corresponding error is the *variance* of the data.

14.4. Given the four points $(-1, 2)$, $(0, 0)$, $(1, -3)$, $(2, -5)$, find (in the least squares sense)

(i) The best horizontal line $y = c$;
(ii) The best line $y = c + dx$;
(iii) The best parabola $y = c + dx + ex^2$.

14.5. Given the four points $(1, 1, 3)$, $(0, 3, 6)$, $(2, 1, 5)$, $(0, 0, 0)$, find the best plane (in the least squares sense)

$$z = c + dx + ey$$

that fits the four points.

14.6. (a) Prove that if A has independent columns, then its pseudo-inverse is $(A^\top A)^{-1} A^\top$, which is also the left inverse of A.

(b) Prove that if A has independent rows, then its pseudo-inverse is $A^\top (AA^\top)^{-1}$, which is also the right inverse of A.

14.7. Prove Proposition 14.2.

14.8. Prove Proposition 14.3.

14.9. Let A be any invertible (real) $n \times n$ matrix.

(a) Prove that for every SVD $A = VDU^\top$ of A, the product $VU^\top$ is the same (i.e., if $V_1 DU_1^\top = V_2 DU_2^\top$, then $V_1 U_1^\top = V_2 U_2^\top$). What does $VU^\top$ have to do with the polar form of A?

(b) Given any invertible (real) $n \times n$ matrix A, prove that there is a unique orthogonal matrix $Q \in \mathbf{O}(n)$ such that $\|A - Q\|_F$ is minimal (under the Frobenius norm). In fact, prove that $Q = VU^\top$, where $A = VDU^\top$ is an SVD of A. Moreover, if $\det(A) > 0$, show that $Q \in \mathbf{SO}(n)$.

What can you say if A is singular (i.e., noninvertible)?

References

1. James W. Demmel. *Applied Numerical Linear Algebra.* SIAM Publications, first edition, 1997.
2. Charles L. Epstein. *Introduction to the Mathematics of Medical Imaging.* SIAM, second edition, 2007.
3. David A. Forsyth and Jean Ponce. *Computer Vision: A Modern Approach.* Prentice Hall, first edition, 2002.
4. Gene H. Golub and Charles F. Van Loan. *Matrix Computations.* The Johns Hopkins University Press, third edition, 1996.
5. Trevor Hastie, Robert Tibshirani, and Jerome Friedman. *The Elements of Statistical Learning: Data Mining, Inference, and Prediction.* Springer, second edition, 2009.
6. D. Kincaid and W. Cheney. *Numerical Analysis.* Brooks/Cole Publishing, second edition, 1996.
7. Gilbert Strang. *Linear Algebra and Its Applications.* Saunders HBJ, third edition, 1988.
8. L.N. Trefethen and D. Bau III. *Numerical Linear Algebra.* SIAM Publications, first edition, 1997.

Chapter 15
Quadratic Optimization Problems

15.1 Quadratic Optimization: The Positive Definite Case

In this chapter, we consider two classes of quadratic optimization problems that appear frequently in engineering and in computer science (especially in computer vision):

1. Minimizing

$$f(x) = \frac{1}{2}x^\top Ax + x^\top b$$

over all $x \in \mathbb{R}^n$, or subject to linear or affine constraints.

2. Minimizing

$$f(x) = \frac{1}{2}x^\top Ax + x^\top b$$

over the unit sphere.

In both cases, A is a symmetric matrix. We also seek necessary and sufficient conditions for f to have a global minimum.

Many problems in physics and engineering can be stated as the minimization of some energy function, with or without constraints. Indeed, it is a fundamental principle of mechanics that nature acts so as to minimize energy. Furthermore, if a physical system is in a stable state of equilibrium, then the energy in that state should be minimal. For example, a small ball placed on top of a sphere is in an unstable equilibrium position. A small motion causes the ball to roll down. On the other hand, a ball placed inside and at the bottom of a sphere is in a stable equilibrium position, because the potential energy is minimal.

The simplest kind of energy function is a quadratic function. Such functions can be conveniently defined in the form

$$P(x) = x^\top Ax - x^\top b,$$

where A is a symmetric $n \times n$ matrix, and x, b, are vectors in $\mathbb{R}^n$, viewed as column vectors. Actually, for reasons that will be clear shortly, it is preferable to put a factor

$\frac{1}{2}$ in front of the quadratic term, so that

$$P(x) = \frac{1}{2}x^\top Ax - x^\top b.$$

The question is, under what conditions (on A) does $P(x)$ have a global minimum, preferably unique?

We give a complete answer to the above question in two stages:

1. In this section, we show that if A is symmetric positive definite, then $P(x)$ has a unique global minimum precisely when

$$Ax = b.$$

2. In Section 15.2, we give necessary and sufficient conditions in the general case, in terms of the pseudo-inverse of A.

We begin with the matrix version of Definition 13.2.

Definition 15.1. A symmetric *positive definite matrix* is a matrix whose eigenvalues are strictly positive, and a symmetric *positive semidefinite matrix* is a matrix whose eigenvalues are nonnegative.

Equivalent criteria are given in the following lemma.

Lemma 15.1. *Given any Euclidean space E of dimension n, the following properties hold:*

(1) Every self-adjoint linear map $f \colon E \to E$ is positive definite iff

$$\langle x, f(x) \rangle > 0$$

for all $x \in E$ with $x \neq 0$.
(2) Every self-adjoint linear map $f \colon E \to E$ is positive semidefinite iff

$$\langle x, f(x) \rangle \geq 0$$

for all $x \in E$.

Proof. (1) First, assume that f is positive definite. Recall that every self-adjoint linear map has an orthonormal basis $(e_1, \ldots, e_n)$ of eigenvectors, and let $\lambda_1, \ldots, \lambda_n$ be the corresponding eigenvalues. With respect to this basis, for every $x = x_1 e_1 + \cdots + x_n e_n \neq 0$, we have

$$\langle x, f(x) \rangle = \left\langle \sum_{i=1}^n x_i e_i, f\left(\sum_{i=1}^n x_i e_i\right) \right\rangle = \left\langle \sum_{i=1}^n x_i e_i, \sum_{i=1}^n \lambda_i x_i e_i \right\rangle = \sum_{i=1}^n \lambda_i x_i^2,$$

which is strictly positive, since $\lambda_i > 0$ for $i = 1, \ldots, n$, and $x_i^2 > 0$ for some i, since $x \neq 0$.

Conversely, assume that

$$\langle x, f(x)\rangle > 0$$

for all $x \neq 0$. Then for $x = e_i$, we get

$$\langle e_i, f(e_i)\rangle = \langle e_i, \lambda_i e_i\rangle = \lambda_i,$$

and thus $\lambda_i > 0$ for all $i = 1, \ldots, n$.

(2) As in (1), we have

$$\langle x, f(x)\rangle = \sum_{i=1}^{n} \lambda_i x_i^2,$$

and since $\lambda_i \geq 0$ for $i = 1, \ldots, n$ because f is positive semidefinite, we have $\langle x, f(x)\rangle \geq 0$, as claimed. The converse is as in (1) except that we get only $\lambda_i \geq 0$ since $\langle e_i, f(e_i)\rangle \geq 0$. $\square$

Some special notation is customary (especially in the field of convex optimization) to express that a symmetric matrix is positive definite or positive semidefinite.

Definition 15.2. Given any $n \times n$ symmetric matrix A we write $A \succeq 0$ if A is positive semidefinite and we write $A \succ 0$ if A is positive definite.

It should be noted that we can define the relation

$$A \succeq B$$

between any two $n \times n$ matrices (symmetric or not) iff $A - B$ is symmetric positive semidefinite. It is easy to check that this relation is actually a partial order on matrices, called the *positive semidefinite cone ordering*; for details, see Boyd and Vandenberghe [1], Section 2.4.

If A is symmetric positive definite, it is easily checked that A^{-1} is also symmetric positive definite. Also, if C is a symmetric positive definite $m \times m$ matrix and A is an $m \times n$ matrix of rank n (and so $m \geq n$), then $A^{\top}CA$ is symmetric positive definite.

We can now prove that

$$P(x) = \frac{1}{2}x^{\top}Ax - x^{\top}b$$

has a global minimum when A is symmetric positive definite.

Lemma 15.2. *Given a quadratic function*

$$P(x) = \frac{1}{2}x^{\top}Ax - x^{\top}b,$$

if A is symmetric positive definite, then $P(x)$ has a unique global minimum for the solution of the linear system $Ax = b$. The minimum value of $P(x)$ is

$$P(A^{-1}b) = -\frac{1}{2}b^{\top}A^{-1}b.$$

Proof. Since A is positive definite, it is invertible, since its eigenvalues are all strictly positive. Let $x = A^{-1}b$, and compute $P(y) - P(x)$ for any $y \in \mathbb{R}^n$. Since $Ax = b$, we get

$$
\begin{aligned}
P(y) - P(x) &= \frac{1}{2}y^\top Ay - y^\top b - \frac{1}{2}x^\top Ax + x^\top b \\
&= \frac{1}{2}y^\top Ay - y^\top Ax + \frac{1}{2}x^\top Ax \\
&= \frac{1}{2}(y - x)^\top A(y - x).
\end{aligned}
$$

Since A is positive definite, the last expression is nonnegative, and thus

$$
P(y) \geq P(x)
$$

for all $y \in \mathbb{R}^n$, which proves that $x = A^{-1}b$ is a global minimum of $P(x)$. A simple computation yields

$$
P(A^{-1}b) = -\frac{1}{2}b^\top A^{-1}b.
$$

$\square$

Remarks:

(1) The quadratic function $P(x)$ is also given by

$$
P(x) = \frac{1}{2}x^\top Ax - b^\top x,
$$

but the definition using $x^\top b$ is more convenient for the proof of Lemma 15.2.

(2) If $P(x)$ contains a constant term $c \in \mathbb{R}$, so that

$$
P(x) = \frac{1}{2}x^\top Ax - x^\top b + c,
$$

the proof of Lemma 15.2 still shows that $P(x)$ has a unique global minimum for $x = A^{-1}b$, but the minimal value is

$$
P(A^{-1}b) = -\frac{1}{2}b^\top A^{-1}b + c.
$$

Thus, when the energy function $P(x)$ of a system is given by a quadratic function

$$
P(x) = \frac{1}{2}x^\top Ax - x^\top b,
$$

where A is symmetric positive definite, finding the global minimum of $P(x)$ is equivalent to solving the linear system $Ax = b$. Sometimes, it is useful to recast a linear problem $Ax = b$ as a variational problem (finding the minimum of some energy function). However, very often, a minimization problem comes with extra constraints that must be satisfied for all admissible solutions. For instance, we may want to minimize the quadratic function

$$Q(y_1, y_2) = \frac{1}{2}\left(y_1^2 + y_2^2\right)$$

subject to the constraint

$$2y_1 - y_2 = 5.$$

The solution for which $Q(y_1, y_2)$ is minimum is no longer $(y_1, y_2) = (0, 0)$, but instead, $(y_1, y_2) = (2, -1)$, as will be shown later.

Geometrically, the graph of the function defined by $z = Q(y_1, y_2)$ in $\mathbb{R}^3$ is a paraboloid of revolution P with axis of revolution Oz. The constraint

$$2y_1 - y_2 = 5$$

corresponds to the vertical plane H parallel to the z-axis and containing the line of equation $2y_1 - y_2 = 5$ in the xy-plane. Thus, the constrained minimum of Q is located on the parabola that is the intersection of the paraboloid P with the plane H.

A nice way to solve constrained minimization problems of the above kind is to use the method of *Lagrange multipliers*. But first, let us define precisely what kind of minimization problems we intend to solve.

Definition 15.3. The *quadratic constrained minimization problem* consists in minimizing a quadratic function

$$Q(y) = \frac{1}{2}y^\top C^{-1}y - b^\top y$$

subject to the linear constraints

$$A^\top y = f,$$

where C^{-1} is an $m \times m$ symmetric positive definite matrix, A is an $m \times n$ matrix of rank n (so that $m \geq n$), and where $b, y \in \mathbb{R}^m$ (viewed as column vectors), and $f \in \mathbb{R}^n$ (viewed as a column vector).

The reason for using C^{-1} instead of C is that the constrained minimization problem has an interpretation as a set of equilibrium equations in which the matrix that arises naturally is C (see Strang [10]). Since C and C^{-1} are both symmetric positive definite, this doesn't make any difference, but it seems preferable to stick to Strang's notation.

The method of Lagrange consists in incorporating the n constraints $A^\top y = f$ into the quadratic function $Q(y)$, by introducing extra variables $\lambda = (\lambda_1, \ldots, \lambda_n)$ called *Lagrange multipliers*, one for each constraint. We form the *Lagrangian*

$$L(y, \lambda) = Q(y) + \lambda^\top (A^\top y - f) = \frac{1}{2}y^\top C^{-1}y - (b - A\lambda)^\top y - \lambda^\top f.$$

We shall prove that our constrained minimization problem has a unique solution given by the system of linear equations

$$C^{-1}y + A\lambda = b,$$
$$A^\top y = f,$$

which can be written in matrix form as

$$\begin{pmatrix} C^{-1} & A \\ A^\top & 0 \end{pmatrix} \begin{pmatrix} y \\ \lambda \end{pmatrix} = \begin{pmatrix} b \\ f \end{pmatrix}.$$

Note that the matrix of this system is symmetric. Eliminating y from the first equation

$$C^{-1}y + A\lambda = b,$$

we get

$$y = C(b - A\lambda),$$

and substituting into the second equation, we get

$$A^\top C(b - A\lambda) = f,$$

that is,

$$A^\top C A\lambda = A^\top C b - f.$$

However, by a previous remark, since C is symmetric positive definite and the columns of A are linearly independent, $A^\top C A$ is symmetric positive definite, and thus invertible. Note that this way of solving the system requires solving for the Lagrange multipliers first.

Letting $e = b - A\lambda$, we also note that the system

$$\begin{pmatrix} C^{-1} & A \\ A^\top & 0 \end{pmatrix} \begin{pmatrix} y \\ \lambda \end{pmatrix} = \begin{pmatrix} b \\ f \end{pmatrix}$$

is equivalent to the system

$$e = b - A\lambda,$$
$$y = Ce,$$
$$A^\top y = f.$$

The latter system is called the *equilibrium equations* by Strang [10]. Indeed, Strang shows that the equilibrium equations of many physical systems can be put in the above form. This includes spring-mass systems, electrical networks, and trusses, which are structures built from elastic bars. In each case, y, e, b, C, λ, f, and $K = A^\top C A$ have a physical interpretation. The matrix $K = A^\top C A$ is usually called the *stiffness matrix*. Again, the reader is referred to Strang [10].

In order to prove that our constrained minimization problem has a unique solution, we proceed to prove that the constrained minimization of $Q(y)$ subject to $A^\top y = f$ is equivalent to the unconstrained maximization of another function $-P(\lambda)$. We get $P(\lambda)$ by minimizing the Lagrangian $L(y, \lambda)$ treated as a function of y alone. Since C^{-1} is symmetric positive definite and

$$L(y, \lambda) = \frac{1}{2} y^\top C^{-1} y - (b - A\lambda)^\top y - \lambda^\top f,$$

by Lemma 15.2 the global minimum (with respect to y) of $L(y, \lambda)$ is obtained for the solution y of

$$C^{-1} y = b - A\lambda,$$

that is, when

$$y = C(b - A\lambda),$$

and the minimum of $L(y, \lambda)$ is

$$\min_y L(y, \lambda) = -\frac{1}{2}(A\lambda - b)^\top C(A\lambda - b) - \lambda^\top f.$$

Letting

$$P(\lambda) = \frac{1}{2}(A\lambda - b)^\top C(A\lambda - b) + \lambda^\top f,$$

we claim that the solution of the constrained minimization of $Q(y)$ subject to $A^\top y = f$ is equivalent to the unconstrained maximization of $-P(\lambda)$. Of course, since we minimized $L(y, \lambda)$ with respect to y, we have

$$L(y, \lambda) \geq -P(\lambda)$$

for all y and all λ. However, when the constraint $A^\top y = f$ holds, $L(y, \lambda) = Q(y)$, and thus for any admissible y, which means that $A^\top y = f$, we have

$$\min_y Q(y) \geq \max_\lambda -P(\lambda).$$

In order to prove that the unique minimum of the constrained problem $Q(y)$ subject to $A^\top y = f$ is the unique maximum of $-P(\lambda)$, we compute $Q(y) + P(\lambda)$.

Lemma 15.3. *The quadratic constrained minimization problem of Definition 15.3 has a unique solution (y, λ) given by the system*

$$\begin{pmatrix} C^{-1} & A \\ A^\top & 0 \end{pmatrix} \begin{pmatrix} y \\ \lambda \end{pmatrix} = \begin{pmatrix} b \\ f \end{pmatrix}.$$

Furthermore, the component λ of the above solution is the unique value for which $-P(\lambda)$ is maximum.

Proof. As we suggested earlier, let us compute $Q(y) + P(\lambda)$, assuming that the constraint $A^\top y = f$ holds. Eliminating f, since $b^\top y = y^\top b$ and $\lambda^\top A^\top y = y^\top A\lambda$, we get

$$Q(y) + P(\lambda) = \frac{1}{2} y^\top C^{-1} y - b^\top y + \frac{1}{2}(A\lambda - b)^\top C(A\lambda - b) + \lambda^\top f$$

$$= \frac{1}{2}(C^{-1}y + A\lambda - b)^\top C(C^{-1}y + A\lambda - b).$$

Since C is positive definite, the last expression is nonnegative. In fact, it is null iff

$$C^{-1}y + A\lambda - b = 0,$$

that is,

$$C^{-1}y + A\lambda = b.$$

But then the unique constrained minimum of $Q(y)$ subject to $A^\top y = f$ is equal to the unique maximum of $-P(\lambda)$ exactly when $A^\top y = f$ and $C^{-1}y + A\lambda = b$, which proves the lemma. $\square$

Remarks:

(1) There is a form of duality going on in this situation. The constrained minimization of $Q(y)$ subject to $A^\top y = f$ is called the *primal problem*, and the unconstrained maximization of $-P(\lambda)$ is called the *dual problem*. Duality is the fact stated slightly loosely as

$$\min_y Q(y) = \max_\lambda -P(\lambda).$$

Recalling that $e = b - A\lambda$, since

$$P(\lambda) = \frac{1}{2}(A\lambda - b)^\top C(A\lambda - b) + \lambda^\top f,$$

we can also write

$$P(\lambda) = \frac{1}{2}e^\top Ce + \lambda^\top f.$$

This expression often represents the total potential energy of a system. Again, the optimal solution is the one that minimizes the potential energy (and thus maximizes $-P(\lambda)$).

(2) It is immediately verified that the equations of Lemma 15.3 are equivalent to the equations stating that the partial derivatives of the Lagrangian $L(y, \lambda)$ are null:

$$\frac{\partial L}{\partial y_i} = 0, \quad i = 1, \ldots, m,$$

$$\frac{\partial L}{\partial \lambda_j} = 0, \quad j = 1, \ldots, n.$$

Thus, the constrained minimum of $Q(y)$ subject to $A^\top y = f$ is an extremum of the Lagrangian $L(y, \lambda)$. As we showed in Lemma 15.3, this extremum corresponds to simultaneously minimizing $L(y,\lambda)$ with respect to y and maximizing $L(y,\lambda)$ with respect to λ. Geometrically, such a point is a *saddle point* for $L(y,\lambda)$.

(3) The Lagrange multipliers sometimes have a natural physical meaning. For example, in the spring-mass system they correspond to node displacements. In some general sense, Lagrange multipliers are correction terms needed to satisfy equilibrium equations and the price paid for the constraints. For more details, see Strang [10].

Going back to the constrained minimization of $Q(y_1,y_2) = \frac{1}{2}(y_1^2 + y_2^2)$ subject to

$$2y_1 - y_2 = 5,$$

the Lagrangian is

$$L(y_1,y_2,\lambda) = \frac{1}{2}(y_1^2 + y_2^2) + \lambda(2y_1 - y_2 - 5),$$

and the equations stating that the Lagrangian has a saddle point are

$$y_1 + 2\lambda = 0,$$
$$y_2 - \lambda = 0,$$
$$2y_1 - y_2 - 5 = 0.$$

We obtain the solution $(y_1,y_2,\lambda) = (2,-1,-1)$.

Much more should be said about the use of Lagrange multipliers in optimization or variational problems. This is a vast topic. Least squares methods and Lagrange multipliers are used to tackle many problems in computer graphics and computer vision; see Trucco and Verri [11], Metaxas [9], Jain, Katsuri, and Schunck [8], Faugeras [4], and Foley, van Dam, Feiner, and Hughes [5]. For a lucid introduction to optimization methods, see Ciarlet [2].

15.2 Quadratic Optimization: The General Case

In this section, we complete the study initiated in Section 15.1 and give necessary and sufficient conditions for the quadratic function $\frac{1}{2}x^\top A x + x^\top b$ to have a global minimum. We begin with the following simple fact:

Proposition 15.1. *If A is an invertible symmetric matrix, then the function*

$$f(x) = \frac{1}{2}x^\top A x + x^\top b$$

has a minimum value iff $A \succeq 0$, in which case this optimal value is obtained for a unique value of x, namely $x^* = -A^{-1}b$, and with

$$f(A^{-1}b) = -\frac{1}{2}b^\top A^{-1}b.$$

Proof. Observe that

$$\frac{1}{2}(x+A^{-1}b)^\top A(x+A^{-1}b) = \frac{1}{2}x^\top Ax + x^\top b + \frac{1}{2}b^\top A^{-1}b.$$

Thus,

$$f(x) = \frac{1}{2}x^\top Ax + x^\top b = \frac{1}{2}(x+A^{-1}b)^\top A(x+A^{-1}b) - \frac{1}{2}b^\top A^{-1}b.$$

If A has some negative eigenvalue, say $-\lambda$ (with $\lambda > 0$), if we pick any eigenvector u of A associated with λ, then for any $\alpha \in \mathbb{R}$ with $\alpha \neq 0$, if we let $x = \alpha u - A^{-1}b$, then since $Au = -\lambda u$, we get

$$\begin{aligned}
f(x) &= \frac{1}{2}(x+A^{-1}b)^\top A(x+A^{-1}b) - \frac{1}{2}b^\top A^{-1}b \\
&= \frac{1}{2}\alpha u^\top A\alpha u - \frac{1}{2}b^\top A^{-1}b \\
&= -\frac{1}{2}\alpha^2 \lambda \|u\|_2^2 - \frac{1}{2}b^\top A^{-1}b,
\end{aligned}$$

and since α can be made as large as we want and $\lambda > 0$, we see that f has no minimum. Consequently, in order for f to have a minimum, we must have $A \succeq 0$. In this case, since $(x+A^{-1}b)^\top A(x+A^{-1}b) \geq 0$, it is clear that the minimum value of f is achieved when $x+A^{-1}b = 0$, that is, $x = -A^{-1}b$. $\square$

Let us now consider the case of an arbitrary symmetric matrix A.

Proposition 15.2. *If A is a symmetric matrix, then the function*

$$f(x) = \frac{1}{2}x^\top Ax + x^\top b$$

has a minimum value iff $A \succeq 0$ and $(I - AA^+)b = 0$, in which case this minimum value is

$$p^* = -\frac{1}{2}b^\top A^+ b.$$

Furthermore, if $A = U^\top \Sigma U$ is an SVD of A, then the optimal value is achieved by all $x \in \mathbb{R}^n$ of the form

$$x = -A^+ b + U^\top \begin{pmatrix} 0 \\ z \end{pmatrix},$$

for any $z \in \mathbb{R}^{n-r}$, where r is the rank of A.

Proof. The case that A is invertible is taken care of by Proposition 15.1, so we may assume that A is singular. If A has rank $r < n$, then we can diagonalize A as

$$A = U^\top \begin{pmatrix} \Sigma_r & 0 \\ 0 & 0 \end{pmatrix} U,$$

where U is an orthogonal matrix and where Σ_r is an $r \times r$ diagonal invertible matrix. Then we have

$$f(x) = \frac{1}{2} x^\top U^\top \begin{pmatrix} \Sigma_r & 0 \\ 0 & 0 \end{pmatrix} U x + x^\top U^\top U b$$

$$= \frac{1}{2}(Ux)^\top \begin{pmatrix} \Sigma_r & 0 \\ 0 & 0 \end{pmatrix} U x + (Ux)^\top U b.$$

If we write

$$Ux = \begin{pmatrix} y \\ z \end{pmatrix} \quad \text{and} \quad Ub = \begin{pmatrix} c \\ d \end{pmatrix},$$

with $y, c \in \mathbb{R}^r$ and $z, d \in \mathbb{R}^{n-r}$, we get

$$f(x) = \frac{1}{2}(Ux)^\top \begin{pmatrix} \Sigma_r & 0 \\ 0 & 0 \end{pmatrix} U x + (Ux)^\top U b$$

$$= \frac{1}{2}(y^\top, z^\top) \begin{pmatrix} \Sigma_r & 0 \\ 0 & 0 \end{pmatrix} \begin{pmatrix} y \\ z \end{pmatrix} + (y^\top, z^\top) \begin{pmatrix} c \\ d \end{pmatrix}$$

$$= \frac{1}{2} y^\top \Sigma_r y + y^\top c + z^\top d.$$

For $y = 0$, we get

$$f(x) = z^\top d,$$

so if $d \neq 0$, the function f has no minimum. Therefore, if f has a minimum, then $d = 0$. However, $d = 0$ means that

$$Ub = \begin{pmatrix} c \\ 0 \end{pmatrix},$$

and we know from Section 14.1 that b is in the range of A (here, U is $U^\top$), which is equivalent to $(I - AA^+)b = 0$. If $d = 0$, then

$$f(x) = \frac{1}{2} y^\top \Sigma_r y + y^\top c,$$

and since Σ_r is invertible, by Proposition 15.1, the function f has a minimum iff $\Sigma_r \succeq 0$, which is equivalent to $A \succeq 0$.

Therefore, we have proved that if f has a minimum, then $(I - AA^+)b = 0$ and $A \succeq 0$. Conversely, if $(I - AA^+)b = 0$ and $A \succeq 0$, what we just did proves that f does have a minimum.

When the above conditions hold, the minimum is achieved if $y = -\Sigma_r^{-1}c$, $z = 0$ and $d = 0$, that is, for x^* given by

$$Ux^* = \begin{pmatrix} -\Sigma_r^{-1}c \\ 0 \end{pmatrix} \quad \text{and} \quad Ub = \begin{pmatrix} c \\ 0 \end{pmatrix},$$

from which we deduce that

$$x^* = -U^\top \begin{pmatrix} \Sigma_r^{-1}c \\ 0 \end{pmatrix} = -U^\top \begin{pmatrix} \Sigma_r^{-1}c & 0 \\ 0 & 0 \end{pmatrix} \begin{pmatrix} c \\ 0 \end{pmatrix} = -U^\top \begin{pmatrix} \Sigma_r^{-1}c & 0 \\ 0 & 0 \end{pmatrix} Ub = -A^+ b$$

and the minimum value of f is

$$f(x^*) = -\frac{1}{2}b^\top A^+ b.$$

For any $x \in \mathbb{R}^n$ of the form

$$x = -A^+ b + U^\top \begin{pmatrix} 0 \\ z \end{pmatrix},$$

for any $z \in \mathbb{R}^{n-r}$, our previous calculations show that $f(x) = -\frac{1}{2}b^\top A^+ b$. $\square$

The case in which we add either linear constraints of the form $C^\top x = 0$ or affine constraints of the form $C^\top x = t$ (where $t \neq 0$) can be reduced to the unconstrained case using a *QR*-decomposition of C or N. Let us show how to do this for linear constraints of the form $C^\top x = 0$.

If we use a *QR* decomposition of C, by permuting the columns, we may assume that

$$C = Q^\top \begin{pmatrix} R & S \\ 0 & 0 \end{pmatrix} \Pi,$$

where R is an $r \times r$ invertible upper triangular matrix and S is an $r \times (m-r)$ matrix (C has rank r). Then, if we let

$$x = Q^\top \begin{pmatrix} y \\ z \end{pmatrix},$$

where $y \in \mathbb{R}^r$ and $z \in \mathbb{R}^{n-r}$, then $C^\top x = 0$ becomes

$$\Pi^\top \begin{pmatrix} R & 0 \\ S & 0 \end{pmatrix} Qx = \Pi^\top \begin{pmatrix} R^\top & 0 \\ S^\top & 0 \end{pmatrix} \begin{pmatrix} y \\ z \end{pmatrix} = 0,$$

which implies $y = 0$, and every solution of $C^\top x = 0$ is of the form

$$x = Q^\top \begin{pmatrix} 0 \\ z \end{pmatrix}.$$

Our original problem becomes

$$\text{minimize} \quad \frac{1}{2}(y^\top, z^\top)QAQ^\top \begin{pmatrix} y \\ z \end{pmatrix} + (y^\top, z^\top)Qb$$

$$\text{subject to} \quad y = 0, \, y \in \mathbb{R}^r, \, z \in \mathbb{R}^{n-r}.$$

Thus, the constraint $C^\top x = 0$ has been eliminated, and if we write

$$QAQ^\top = \begin{pmatrix} G_{11} & G_{12} \\ G_{21} & G_{22} \end{pmatrix}$$

and

$$Qb = \begin{pmatrix} b_1 \\ b_2 \end{pmatrix}, \quad b_1 \in \mathbb{R}^r, \, b_2 \in \mathbb{R}^{n-r},$$

our problem becomes

$$\text{minimize } \frac{1}{2} z^\top G_{22} z + z^\top b_2, \quad z \in \mathbb{R}^{n-r},$$

the problem solved in Proposition 15.2.

Constraints of the form $C^\top x = t$ (where $t \neq 0$) can be handled in a similar fashion. In this case, we may assume that C is an $n \times m$ matrix with full rank (so that $m \leq n$) and $t \in \mathbb{R}^m$. Then we use a QR-decomposition of the form

$$C = P \begin{pmatrix} R \\ 0 \end{pmatrix},$$

where P is an orthogonal matrix and R is an $m \times m$ invertible upper triangular matrix. If we write

$$x = P \begin{pmatrix} y \\ z \end{pmatrix},$$

where $y \in \mathbb{R}^m$ and $z \in \mathbb{R}^{n-m}$, the equation $C^\top x = t$ becomes

$$(R^\top, 0)P^\top x = t,$$

that is,

$$(R^\top, 0) \begin{pmatrix} y \\ z \end{pmatrix} = t,$$

which yields

$$R^\top y = t.$$

Since R is invertible, we get $y = (R^\top)^{-1}t$, and then it is easy to see that our original problem reduces to an unconstrained problem in terms of the matrix $P^\top A P$; the details are left as an exercise.

15.3 Maximizing a Quadratic Function on the Unit Sphere

In this section we discuss various quadratic optimization problems mostly arising from computer vision (image segmentation and contour grouping). These problems can be reduced to the following basic optimization problem: Given an $n \times n$ real symmetric matrix A

$$\text{maximize}\ \ x^\top A x$$
$$\text{subject to}\ \ x^\top x = 1, x \in \mathbb{R}^n.$$

In view of Lemma 14.2, the maximum value of $x^\top A x$ on the unit sphere is equal to the largest eigenvalue λ_1 of the matrix A, and it is achieved for any unit eigenvector u_1 associated with λ_1.

A variant of the above problem often encountered in computer vision consists in minimizing $x^\top A x$ on the ellipsoid given by an equation of the form

$$x^\top B x = 1,$$

where B is a symmetric positive definite matrix. Since B is positive definite, it can be diagonalized as

$$B = QDQ^\top,$$

where Q is an orthogonal matrix and D is a diagonal matrix,

$$D = \text{diag}(d_1, \ldots, d_n),$$

with $d_i > 0$, for $i = 1, \ldots, n$. If we define the matrices $B^{1/2}$ and $B^{-1/2}$ by

$$B^{1/2} = Q \, \text{diag}\left(\sqrt{d_1}, \ldots, \sqrt{d_n}\right) Q^\top$$

and

$$B^{-1/2} = Q \, \text{diag}\left(1/\sqrt{d_1}, \ldots, 1/\sqrt{d_n}\right) Q^\top,$$

it is clear that these matrices are symmetric, that $B^{1/2} B B^{-1/2} = I$, and that $B^{1/2}$ and $B^{-1/2}$ are mutual inverses. Then, if we make the change of variable

$$x = B^{-1/2}y,$$

the equation $x^\top B x = 1$ becomes $y^\top y = 1$, and the optimization problem

$$\text{maximize}\ \ x^\top A x$$
$$\text{subject to}\ \ x^\top B x = 1, x \in \mathbb{R}^n,$$

is equivalent to the problem

$$\text{maximize}\ \ y^\top B^{-1/2} A B^{-1/2} y$$
$$\text{subject to}\ \ y^\top y = 1, y \in \mathbb{R}^n,$$

where $y = B^{1/2}x$ and where $B^{-1/2} A B^{-1/2}$ is symmetric.

We will see in Chapter 17 that the complex version of our basic optimization problem in which A is a Hermitian matrix also arises, namely, given an $n \times n$ complex Hermitian matrix A,

$$\text{maximize} \quad x^*Ax$$
$$\text{subject to} \quad x^*x = 1, x \in \mathbb{C}^n.$$

Again by Lemma 14.2, the maximum value of x^*Ax on the unit sphere is equal to the largest eigenvalue λ_1 of the matrix A and it is achieved for any unit eigenvector u_1 associated with λ_1.

It is worth pointing out (and we will use this fact in Section 17.5) that if A is a *skew-Hermitian* matrix, that is, if $A^* = -A$, then x^*Ax is *pure imaginary or zero*.

Indeed, since $z = x^*Ax$ is a scalar, we have $z^* = \bar{z}$ (the conjugate of z), so we have

$$\overline{x^*Ax} = (x^*Ax)^* = x^*A^*x = -x^*Ax,$$

so $\overline{x^*Ax} + x^*Ax = 2\text{Re}(x^*Ax) = 0$, which means that x^*Ax is pure imaginary or zero.

In particular, if A is a real matrix and if A is *skew-symmetric*, then

$$x^\top Ax = 0.$$

Thus, for any real matrix (symmetric or not),

$$x^\top Ax = x^\top H(A)x,$$

where $H(A) = (A + A^\top)/2$, the symmetric part of A.

There are situations in which it is necessary to add linear constraints to the problem of maximizing a quadratic function on the sphere. This problem was completely solved by Golub [7] (1973). The problem is the following: Given an $n \times n$ real symmetric matrix A and an $n \times p$ matrix C,

$$\text{minimize} \quad x^\top Ax$$
$$\text{subject to} \quad x^\top x = 1, C^\top x = 0, x \in \mathbb{R}^n.$$

Golub shows that the linear constraint $C^\top x = 0$ can be eliminated as follows: If we use a *QR* decomposition of C, by permuting the columns, we may assume that

$$C = Q^\top \begin{pmatrix} R & S \\ 0 & 0 \end{pmatrix} \Pi,$$

where R is an $r \times r$ invertible upper triangular matrix and S is an $r \times (p-r)$ matrix (assuming C has rank r). Then if we let

$$x = Q^\top \begin{pmatrix} y \\ z \end{pmatrix},$$

where $y \in \mathbb{R}^r$ and $z \in \mathbb{R}^{n-r}$, then $C^\top x = 0$ becomes

$$\Pi^\top \begin{pmatrix} R^\top & 0 \\ S^\top & 0 \end{pmatrix} Qx = \Pi^\top \begin{pmatrix} R^\top & 0 \\ S^\top & 0 \end{pmatrix} \begin{pmatrix} y \\ z \end{pmatrix} = 0,$$

which implies $y = 0$, and every solution of $C^\top x = 0$ is of the form

$$x = Q^\top \begin{pmatrix} 0 \\ z \end{pmatrix}.$$

Our original problem becomes

$$\text{minimize} \quad (y^\top, z^\top) QAQ^\top \begin{pmatrix} y \\ z \end{pmatrix}$$
$$\text{subject to} \quad z^\top z = 1, z \in \mathbb{R}^{n-r},$$
$$y = 0, y \in \mathbb{R}^r.$$

Thus, the constraint $C^\top x = 0$ has been eliminated, and if we write

$$QAQ^\top = \begin{pmatrix} G_{11} & G_{12} \\ G_{12}^\top & G_{22} \end{pmatrix},$$

our problem becomes

$$\text{minimize} \quad z^\top G_{22} z$$
$$\text{subject to} \quad z^\top z = 1, z \in \mathbb{R}^{n-r},$$

a standard eigenvalue problem. Observe that if we let

$$J = \begin{pmatrix} 0 & 0 \\ 0 & I_{n-r} \end{pmatrix},$$

then

$$JQAQ^\top J = \begin{pmatrix} 0 & 0 \\ 0 & G_{22} \end{pmatrix},$$

and if we set

$$P = Q^\top JQ,$$

then

$$PAP = Q^\top JQAQ^\top JQ.$$

Now, $Q^\top JQAQ^\top JQ$ and $JQAQ^\top J$ have the same eigenvalues, so PAP and $JQAQ^\top J$ also have the same eigenvalues. It follows that the solutions of our optimization problem are among the eigenvalues of $K = PAP$, and at least r of those are 0. Using the fact that CC^+ is the projection onto the range of C, where C^+ is the pseudo-inverse of C, it can also be shown that

$$P = I - CC^+,$$

the projection onto the kernel of $C^\top$. In particular, when $n \geq p$ and C has full rank (the columns of C are linearly independent), then we know that $C^+ = (C^\top C)^{-1} C^\top$ and

$$P = I - C(C^\top C)^{-1} C^\top.$$

This fact is used by Cour and Shi [3] and implicitly by Yu and Shi [12].

The problem of adding affine constraints of the form $N^\top x = t$, where $t \neq 0$, also comes up in practice. At first glance, this problem may not seem harder than the linear problem in which $t = 0$, but it is. This problem was extensively studied in a paper by Gander, Golub, and von Matt [6] (1989).

Gander, Golub, and von Matt consider the following problem: Given an $(n + m) \times (n + m)$ real symmetric matrix A (with $n > 0$), an $(n + m) \times m$ matrix N with full rank, and a nonzero vector $t \in \mathbb{R}^m$ with $\left\| (N^\top)^\dagger t \right\| < 1$ (where $(N^\top)^\dagger$ denotes the pseudo-inverse of $N^\top$),

$$\begin{aligned}
&\text{minimize} \quad x^\top A x \\
&\text{subject to} \quad x^\top x = 1,\ N^\top x = t,\ x \in \mathbb{R}^{n+m}.
\end{aligned}$$

The condition $\left\| (N^\top)^\dagger t \right\| < 1$ ensures that the problem has a solution and is not trivial. The authors begin by proving that the affine constraint $N^\top x = t$ can be eliminated. One way to do so is to use a QR decomposition of N. If

$$N = P \begin{pmatrix} R \\ 0 \end{pmatrix},$$

where P is an orthogonal matrix and R is an $m \times m$ invertible upper triangular matrix, then if we observe that

$$\begin{aligned}
x^\top A x &= x^\top P P^\top A P P^\top x, \\
N^\top x &= (R^\top, 0) P^\top x = t, \\
x^\top x &= x^\top P P^\top x = 1,
\end{aligned}$$

and if we write

$$P^\top A P = \begin{pmatrix} B & \Gamma^\top \\ \Gamma & C \end{pmatrix}$$

and

$$P^\top x = \begin{pmatrix} y \\ z \end{pmatrix},$$

then we get

$$x^\top Ax = y^\top By + 2z^\top \Gamma y + z^\top Cz,$$
$$R^\top y = t,$$
$$y^\top y + z^\top z = 1.$$

Thus

$$y = (R^\top)^{-1}t,$$

and if we write

$$s^2 = 1 - y^\top y > 0$$

and

$$b = \Gamma y,$$

we get the simplified problem

$$\text{minimize} \quad z^\top Cz + 2z^\top b$$
$$\text{subject to} \quad z^\top z = s^2, \; z \in \mathbb{R}^m.$$

Unfortunately, if $b \neq 0$, Lemma 14.2 is no longer applicable. It is still possible to find the minimum of the function $z^\top Cz + 2z^\top b$ using Lagrange multipliers, but such a solution is too involved to be presented here. Interested readers will find a thorough discussion in Gander, Golub, and von Matt [6].

15.4 Problems

15.1. If A is symmetric positive definite, prove that A^{-1} is also symmetric positive definite. If C is a symmetric positive definite $m \times m$ matrix and A is an $m \times n$ matrix of rank n (and so $m \geq n$), prove that $A^\top CA$ is symmetric positive definite.

15.2. Minimize

$$Q = \frac{1}{2}\left(y_1^2 + \frac{1}{3}y_2^2\right)$$

subject to $y_1 + y_2 = 1$.

15.3. Find the nearest point to the origin on the hyperplane

$$y_1 + \cdots + y_m = 1.$$

15.4. (i) Find the minimum of

$$Q = \frac{1}{2}\left(y_1^2 + 2y_1 y_2\right) - y_2$$

subject to $y_1 + y_2 = 0$.
  (ii) Find the minimum of

$$Q = \frac{1}{2}\left(y_1^2 + y_2^2 + y_3^2\right)$$

subject to $y_1 - y_2 = 1$ and $y_2 - y_3 = 2$.

15.5. Find the rectangle with corners at points $(\pm y_1, \pm y_2)$ on the ellipse $y_1^2 + 4y_2^2 = 1$ such that the perimeter $4y_1 + 4y_2$ is maximized.

15.6. What is the minimum-length least squares solution to

$$\begin{pmatrix} 1 & 0 & 0 \\ 1 & 0 & 0 \\ 1 & 1 & 1 \end{pmatrix} \begin{pmatrix} c \\ d \\ e \end{pmatrix} = \begin{pmatrix} 0 \\ 2 \\ 2 \end{pmatrix}.$$

15.7. Give the details of the proof showing that minimizing a quadratic function

$$f(x) = \frac{1}{2}x^\top A x + x^\top b$$

subject to constraints of the form $C^\top x = t$, where $t \neq 0$ and C is an $n \times m$ matrix with full rank, reduces to a similar unconstrained problem.
Hint. Use a *QR*-decomposition of the form

$$C = P \begin{pmatrix} R \\ 0 \end{pmatrix},$$

where P is an orthogonal matrix and R is an $m \times m$ invertible upper triangular matrix, and write

$$x = P \begin{pmatrix} y \\ z \end{pmatrix},$$

where $y \in \mathbb{R}^m$ and $z \in \mathbb{R}^{n-m}$.

15.8. Let A be any symmetric $n \times n$ matrix, let $b \in \mathbb{R}^n$, and let $c \in \mathbb{R}$.
  (a) Prove that if $A \succeq 0$, then the set

$$S = \{x \in \mathbb{R}^n \mid x^\top A x + b^\top x + c \leq 0\}$$

is convex.
Hint. Intersect S with an arbitrary line determined by a point p and a unit vector u.
  (b) Prove that if S as above is convex, then $A \succeq 0$.
  (c) Let H be an affine hyperplane defined by an equation of the form $g^\top x + h = 0$, where we may assume that g is a unit vector. Prove that

$$H = \{z \in \mathbb{R}^n \mid z = -hg + (I - gg^\top)x, \, x \in \mathbb{R}^n\}$$

and that $(g^\top)^+ = g$ (where $(g^\top)^+$ is the pseudo-inverse of $g^\top$). Prove that $S \cap H$ is convex (where S is defined in (a)) iff

$$(I - gg^\top)A(I - gg^\top) \succeq 0.$$

Prove that if there is some $\lambda \in \mathbb{R}$ such that $A + \lambda gg^\top \succeq 0$, then $S \cap H$ is convex but that the converse is false.

References

1. Stephen Boyd and Lieven Vandenberghe. *Convex Optimization*. Cambridge University Press, first edition, 2004.
2. P.G. Ciarlet. *Introduction to Numerical Matrix Analysis and Optimization*. Cambridge University Press, first edition, 1989. French edition: Masson, 1994.
3. Timothée Cour and Jianbo Shi. Solving Markov random fields with spectral relaxation. In Marita Meila and Xiaotong Shen, editors, *Artifical Intelligence and Statistics*. Society for Artificial Intelligence and Statistics, 2007.
4. Olivier Faugeras. *Three-Dimensional Computer Vision, A Geometric Viewpoint*. MIT Press, first edition, 1996.
5. James Foley, Andries van Dam, Steven Feiner, and John Hughes. *Computer Graphics. Principles and Practice*. Addison-Wesley, second edition, 1993.
6. Walter Gander, Gene H. Golub, and Urs von Matt. A constrained eigenvalue problem. *Linear Algebra and Its Applications*, 114/115:815–839, 1989.
7. Gene H. Golub. Some modified eigenvalue problems. *SIAM Review*, 15(2):318–334, 1973.
8. Ramesh Jain, Rangachar Katsuri, and Brian G. Schunck. *Machine Vision*. McGraw-Hill, first edition, 1995.
9. Dimitris N. Metaxas. *Physics-Based Deformable Models*. Kluwer Academic Publishers, first edition, 1997.
10. Gilbert Strang. *Introduction to Applied Mathematics*. Wellesley–Cambridge Press, first edition, 1986.
11. Emanuele Trucco and Alessandro Verri. *Introductory Techniques for 3D Computer Vision*. Prentice-Hall, first edition, 1998.
12. Stella X. Yu and Jianbo Shi. Grouping with bias. In Thomas G. Dietterich, Sue Becker, and Zoubin Ghahramani, editors, *Neural Information Processing Systems, Vancouver, Canada, 3–8 Dec. 2001*. MIT Press, 2001.

Chapter 16
Schur Complements and Applications

16.1 Schur Complements

Schur complements arise naturally in the process of inverting block matrices of the form

$$M = \begin{pmatrix} A & B \\ C & D \end{pmatrix}$$

and in characterizing when symmetric versions of these matrices are positive definite or positive semidefinite. These characterizations come up in various quadratic optimization problems; see Boyd and Vandenberghe [1], especially Appendix B. In the most general case, pseudo-inverses are also needed.

In this chapter we introduce Schur complements and describe several interesting ways in which they are used. Along the way we provide some details and proofs of some results from Appendix A.5 (especially Section A.5.5) of Boyd and Vandenberghe [1].

Let M be an $n \times n$ matrix written as a 2×2 block matrix

$$M = \begin{pmatrix} A & B \\ C & D \end{pmatrix},$$

where A is a $p \times p$ matrix and D is a $q \times q$ matrix, with $n = p + q$ (so B is a $p \times q$ matrix and C is a $q \times p$ matrix). We can try to solve the linear system

$$\begin{pmatrix} A & B \\ C & D \end{pmatrix} \begin{pmatrix} x \\ y \end{pmatrix} = \begin{pmatrix} c \\ d \end{pmatrix},$$

that is,

$$Ax + By = c,$$
$$Cx + Dy = d,$$

by mimicking Gaussian elimination. If we assume that D is invertible, then we first solve for y, getting

$$y = D^{-1}(d - Cx),$$

and after substituting this expression for y in the first equation, we get

$$Ax + B(D^{-1}(d - Cx)) = c,$$

that is,

$$(A - BD^{-1}C)x = c - BD^{-1}d.$$

If the matrix $A - BD^{-1}C$ is invertible, then we obtain the solution to our system

$$x = (A - BD^{-1}C)^{-1}(c - BD^{-1}d),$$
$$y = D^{-1}(d - C(A - BD^{-1}C)^{-1}(c - BD^{-1}d)).$$

If A is invertible, then by eliminating x first using the first equation, we obtain analogous formulas involving the matrix $D - CA^{-1}B$. The above formulas suggest that the matrices $A - BD^{-1}C$ and $D - CA^{-1}B$ play a special role and suggest the followig definition:

Definition 16.1. Given any block matrix of the form

$$M = \begin{pmatrix} A & B \\ C & D \end{pmatrix},$$

if D is invertible, then the matrix $A - BD^{-1}C$ is called the *Schur complement* of D in M. If A is invertible, then the matrix $D - CA^{-1}B$ is called the *Schur complement* of A in M.

The above equations written as

$$x = (A - BD^{-1}C)^{-1}c - (A - BD^{-1}C)^{-1}BD^{-1}d,$$
$$y = -D^{-1}C(A - BD^{-1}C)^{-1}c$$
$$\qquad + (D^{-1} + D^{-1}C(A - BD^{-1}C)^{-1}BD^{-1})d,$$

yield a formula for the inverse of M in terms of the Schur complement of D in M, namely

$$\begin{pmatrix} A & B \\ C & D \end{pmatrix}^{-1} = \begin{pmatrix} (A - BD^{-1}C)^{-1} & -(A - BD^{-1}C)^{-1}BD^{-1} \\ -D^{-1}C(A - BD^{-1}C)^{-1} & D^{-1} + D^{-1}C(A - BD^{-1}C)^{-1}BD^{-1} \end{pmatrix}.$$

A moment of reflection reveals that

$$\begin{pmatrix} A & B \\ C & D \end{pmatrix}^{-1} = \begin{pmatrix} (A - BD^{-1}C)^{-1} & 0 \\ -D^{-1}C(A - BD^{-1}C)^{-1} & D^{-1} \end{pmatrix} \begin{pmatrix} I & -BD^{-1} \\ 0 & I \end{pmatrix},$$

and then

$$\begin{pmatrix} A & B \\ C & D \end{pmatrix}^{-1} = \begin{pmatrix} I & 0 \\ -D^{-1}C & I \end{pmatrix} \begin{pmatrix} (A - BD^{-1}C)^{-1} & 0 \\ 0 & D^{-1} \end{pmatrix} \begin{pmatrix} I & -BD^{-1} \\ 0 & I \end{pmatrix}.$$

It follows that

$$\begin{pmatrix} A & B \\ C & D \end{pmatrix} = \begin{pmatrix} I & BD^{-1} \\ 0 & I \end{pmatrix} \begin{pmatrix} A - BD^{-1}C & 0 \\ 0 & D \end{pmatrix} \begin{pmatrix} I & 0 \\ D^{-1}C & I \end{pmatrix}.$$

The above expression can be checked directly and has the advantage of requiring only the invertibility of D.

Remark: If A is invertible, then we can use the Schur complement $D - CA^{-1}B$ of A to obtain the following factorization of M:

$$\begin{pmatrix} A & B \\ C & D \end{pmatrix} = \begin{pmatrix} I & 0 \\ CA^{-1} & I \end{pmatrix} \begin{pmatrix} A & 0 \\ 0 & D - CA^{-1}B \end{pmatrix} \begin{pmatrix} I & A^{-1}B \\ 0 & I \end{pmatrix}.$$

If $D - CA^{-1}B$ is invertible, we can invert all three matrices above, and we get another formula for the inverse of M in terms of $(D - CA^{-1}B)$, namely,

$$\begin{pmatrix} A & B \\ C & D \end{pmatrix}^{-1} = \begin{pmatrix} A^{-1} + A^{-1}B(D - CA^{-1}B)^{-1}CA^{-1} & -A^{-1}B(D - CA^{-1}B)^{-1} \\ -(D - CA^{-1}B)^{-1}CA^{-1} & (D - CA^{-1}B)^{-1} \end{pmatrix}.$$

If A, D and both Schur complements $A - BD^{-1}C$ and $D - CA^{-1}B$ are all invertible, by comparing the two expressions for M^{-1}, we get the (nonobvious) formula

$$(A - BD^{-1}C)^{-1} = A^{-1} + A^{-1}B(D - CA^{-1}B)^{-1}CA^{-1}.$$

Using this formula, we obtain another expression for the inverse of M involving the Schur complements of A and D (see Horn and Johnson [2]):

$$\begin{pmatrix} A & B \\ C & D \end{pmatrix}^{-1} = \begin{pmatrix} (A - BD^{-1}C)^{-1} & -A^{-1}B(D - CA^{-1}B)^{-1} \\ -(D - CA^{-1}B)^{-1}CA^{-1} & (D - CA^{-1}B)^{-1} \end{pmatrix}.$$

If we set $D = I$ and change B to $-B$, we get

$$(A + BC)^{-1} = A^{-1} - A^{-1}B(I - CA^{-1}B)^{-1}CA^{-1},$$

a formula known as the *matrix inversion lemma* (see Boyd and Vandenberghe [1], Appendix C.4, especially C.4.3).

16.2 Symmetric Positive Definite Matrices and Schur Complements

If we assume that our block matrix M is symmetric, so that A, D are symmetric and $C = B^\top$, then we see that M is expressed as

$$M = \begin{pmatrix} A & B \\ B^\top & D \end{pmatrix} = \begin{pmatrix} I & BD^{-1} \\ 0 & I \end{pmatrix} \begin{pmatrix} A - BD^{-1}B^\top & 0 \\ 0 & D \end{pmatrix} \begin{pmatrix} I & BD^{-1} \\ 0 & I \end{pmatrix}^\top,$$

which shows that M is similar to a block diagonal matrix (obviously, the Schur complement, $A - BD^{-1}B^\top$, is symmetric). As a consequence, we have the following version of "Schur's trick" to check whether $M \succ 0$ for a symmetric matrix.

Proposition 16.1. *For any symmetric matrix M of the form*

$$M = \begin{pmatrix} A & B \\ B^\top & C \end{pmatrix},$$

if C is invertible, then the following properties hold:

(1) $M \succ 0$ iff $C \succ 0$ and $A - BC^{-1}B^\top \succ 0$.
(2) If $C \succ 0$, then $M \succeq 0$ iff $A - BC^{-1}B^\top \succeq 0$.

Proof. (1) Observe that

$$\begin{pmatrix} I & BD^{-1} \\ 0 & I \end{pmatrix}^{-1} = \begin{pmatrix} I & -BD^{-1} \\ 0 & I \end{pmatrix},$$

and we know that for any symmetric matrix T and any invertible matrix N, the matrix T is positive definite ($T \succ 0$) iff $NTN^\top$ (which is obviously symmetric) is positive definite ($NTN^\top \succ 0$). But a block diagonal matrix is positive definite iff each diagonal block is positive definite, which concludes the proof.

(2) This is because for any symmetric matrix T and any invertible matrix N, we have $T \succeq 0$ iff $NTN^\top \succeq 0$. $\square$

Another version of Proposition 16.1 using the Schur complement of A instead of the Schur complement of C also holds. The proof uses the factorization of M using the Schur complement of A (see Section 16.1).

Proposition 16.2. *For any symmetric matrix M of the form*

$$M = \begin{pmatrix} A & B \\ B^\top & C \end{pmatrix},$$

if A is invertible then the following properties hold:

(1) $M \succ 0$ iff $A \succ 0$ and $C - B^\top A^{-1}B \succ 0$.

(2) If $A \succ 0$, then $M \succeq 0$ iff $C - B^\top A^{-1} B \succeq 0$.

When C is singular (or A is singular), it is still possible to characterize when a symmetric matrix M as above is positive semidefinite, but this requires using a version of the Schur complement involving the pseudo-inverse of C, namely $A - BC^+ B^\top$ (or the Schur complement, $C - B^\top A^+ B$, of A). We use the crierion of Proposition 15.2, which tells us when a quadratic function of the form $\frac{1}{2} x^\top P x + x^\top b$ has a minimum and what this optimum value is (where P is a symmetric matrix).

16.3 Symmetric Positive Semidefinite Matrices and Schur Complements

We now return to our original problem, characterizing when a symmetric matrix

$$M = \begin{pmatrix} A & B \\ B^\top & C \end{pmatrix}$$

is positive semidefinite.

Thus, we want to know when the function

$$f(x,y) = (x^\top, y^\top) \begin{pmatrix} A & B \\ B^\top & C \end{pmatrix} \begin{pmatrix} x \\ y \end{pmatrix} = x^\top A x + 2 x^\top B y + y^\top C y$$

has a minimum with respect to both x and y. If we hold y constant, Proposition 15.2 implies that $f(x,y)$ has a minimum iff $A \succeq 0$ and $(I - AA^+)By = 0$, and then the minimum value is

$$f(x^*, y) = -y^\top B^\top A^+ B y + y^\top C y = y^\top (C - B^\top A^+ B) y.$$

Since we want $f(x,y)$ to be uniformly bounded from below for all x, y, we must have $(I - AA^+)B = 0$. Now, $f(x^*, y)$ has a minimum iff $C - B^\top A^+ B \succeq 0$. Therefore, we have established that $f(x,y)$ has a minimum over all x, y iff

$$A \succeq 0, \quad (I - AA^+)B = 0, \quad C - B^\top A^+ B \succeq 0.$$

Similar reasoning applies if we first minimize with respect to y and then with respect to x, but this time, the Schur complement $A - BC^+ B^\top$ of C is involved. Putting all these facts together, we get our main result:

Theorem 16.1. *Given any symmetric matrix*

$$M = \begin{pmatrix} A & B \\ B^\top & C \end{pmatrix}$$

the following conditions are equivalent:

(1) $M \succeq 0$ (M is positive semidefinite).
(2) $A \succeq 0$, $(I - AA^+)B = 0$, $C - B^\top A^+ B \succeq 0$.
(3) $C \succeq 0$, $(I - CC^+)B^\top = 0$, $A - BC^+ B^\top \succeq 0$.

If $M \succeq 0$ as in Theorem 16.1, then it is easy to check that we have the following factorizations (using the fact that $A^+ AA^+ = A^+$ and $C^+ CC^+ = C^+$):

$$\begin{pmatrix} A & B \\ B^\top & C \end{pmatrix} = \begin{pmatrix} I & BC^+ \\ 0 & I \end{pmatrix} \begin{pmatrix} A - BC^+ B^\top & 0 \\ 0 & C \end{pmatrix} \begin{pmatrix} I & 0 \\ C^+ B^\top & I \end{pmatrix}$$

and

$$\begin{pmatrix} A & B \\ B^\top & C \end{pmatrix} = \begin{pmatrix} I & 0 \\ B^\top A^+ & I \end{pmatrix} \begin{pmatrix} A & 0 \\ 0 & C - B^\top A^+ B \end{pmatrix} \begin{pmatrix} I & A^+ B \\ 0 & I \end{pmatrix}.$$

16.4 Problems

16.1. Supply the details of the argument showing that if D is invertible, then

$$\begin{pmatrix} A & B \\ C & D \end{pmatrix} = \begin{pmatrix} I & BD^{-1} \\ 0 & I \end{pmatrix} \begin{pmatrix} A - BD^{-1}C & 0 \\ 0 & D \end{pmatrix} \begin{pmatrix} I & 0 \\ D^{-1}C & I \end{pmatrix}.$$

16.2. Let X be a symmetric $n \times n$ matrix and let $x \in \mathbb{R}^n$ be a vector. Prove that

$$X \succeq xx^\top$$

iff

$$\begin{pmatrix} X & x \\ x^\top & 1 \end{pmatrix} \succeq 0.$$

16.3. Consider the following quadratic optimization problem with quadratic constraints

$$\begin{aligned} \text{minimize} \quad & x^\top A_1 x + 2b_1^\top x \\ \text{subject to} \quad & x^\top A_2 x + 2b_2^\top x \leq 0, \end{aligned}$$

where A_1, A_2 are symmetric $n \times n$ matrices and $b_1, b_2 \in \mathbb{R}^n$. Using the fact that $\text{tr}(Axx^\top) = x^\top Ax$, prove that the above problem is equivalent to the problem

$$\begin{aligned} \text{minimize} \quad & \text{tr}(A_1 X) + 2b_1^\top x \\ \text{subject to} \quad & \text{tr}(A_2 X) + 2b_2^\top x \leq 0, \ X = xx^\top, \end{aligned}$$

where A_1, A_2 are symmetric $n \times n$ matrices, X is a symmetric matrix with $X \succeq 0$, and $b_1, b_2 \in \mathbb{R}^n$.

The above problem is hard to solve, but it can be relaxed to the following problem:

$$\text{minimize} \quad \text{tr}(A_1 X) + 2b_1^\top x$$
$$\text{subject to} \quad \text{tr}(A_2 X) + 2b_2^\top x \leq 0,\ X \succeq xx^\top,$$

where A_1, A_2, X are symmetric $n \times n$ matrices, and $b_1, b_2 \in \mathbb{R}^n$.

Show that the relaxed problem is equivalent to the problem

$$\text{minimize} \quad \text{tr}(A_1 X) + 2b_1^\top x$$
$$\text{subject to} \quad \text{tr}(A_2 X) + 2b_2^\top x \leq 0,\ \begin{pmatrix} X & x \\ x^\top & 1 \end{pmatrix} \succeq 0,$$

where A_1, A_2, X are symmetric $n \times n$ matrices, and $b_1, b_2 \in \mathbb{R}^n$.

The above is an SDP program, and a number of methods are available to solve it; see Boyd and Vandenberghe [1].

References

1. Stephen Boyd and Lieven Vandenberghe. *Convex Optimization*. Cambridge University Press, first edition, 2004.
2. Roger A. Horn and Charles R. Johnson. *Matrix Analysis*. Cambridge University Press, first edition, 1990.

Chapter 17
Quadratic Optimization and Contour Grouping

17.1 Formulation of the Problem

This chapter presents a new and exciting application of quadratic optimization methods to the problem of contour grouping in computer vision. It turns out that this problem leads to finding the local maxima of a Hermitian matrix depending on a parameter. We are thus led to the problem of finding the derivative of an eigenvalue and the derivative of some eigenvector associated with this eigenvalue, in the case of a normal matrix. The problem also leads naturally to the consideration of the field of values of a matrix, a concept studied as early as 1918 by Toeplitz and Hausdorff. We prove that the field of values is convex, a theorem due to Toeplitz and Hausdorff. This fact is helpful in improving the search for local maxima.

Many problems in computer vision can be cast as quadratic optimization problems. In a seminal paper, Shi and Malik [5] showed how image segmentation can be performed using certain types of graph cuts called *normalized cuts*. Inspired by this work, Jianbo Shi and his students Qihui Zhu and Gang Song investigated the problem of *contour grouping* in 2D images [6]. Recently, this method was significantly improved and a better optimization function was introduced; see Kennedy, Gallier, and Shi [3]. We present a method using the new optimization function but for simplicity, we use the older method described in [6]. The problem is to find 1D (closed) curve-like structures in images. The goal is to find cycles linking small edges called *edgels*.

The method uses a directed graph in which the nodes are edgels and the edges connect pairs of edgels within some distance. Every edge has a *weight* W_{ij} measuring the (directed) collinearity of two edgels using the elastic energy between these edgels.

Given a weighted directed graph $G = (V, E, W)$, we seek a set of edges $S \subseteq V$ (a *cut*) and an *ordering* $\mathscr{O}$ on S that maximizes a certain objective function,

$$C(S, \mathscr{O}, k) = \frac{1 - E_{\mathrm{cut}}(S) - I_{\mathrm{cut}}(S, \mathscr{O}, k)}{T(k)},$$

where

1. $E_{\mathrm{cut}}(S)$ measures how strongly S is separated from its surrounding background (*external cut*);
2. $I_{\mathrm{cut}}(S, \mathcal{O}, k)$ is a measure of the *entanglement* of the edges between the nodes in S (*internal cut*);
3. $T(k)$ is the *tube size* of the cut; it depends on the *thickness factor* k (in fact, $T(k) = k/|S|$).

Maximizing $C(S, \mathcal{O}, k)$ is a hard combinatorial problem, so Shi, Zhu, and Gong had the idea of converting the orginal problem to a simpler problem using a *circular embedding*.

The main idea is that a cycle is an image of the unit circle. Thus, we try to map the nodes of the graph onto the unit circle, but nodes not in a cycle will be mapped to the origin. A point on the unit circle has coordinates

$$(\cos\theta, \sin\theta),$$

which are conveniently encoded as the complex number

$$z = \cos\theta + i\sin\theta = e^{i\theta}.$$

The nodes in a cycle will be mapped to the complex numbers

$$z_j = e^{i\theta_j}, \qquad \theta_j = \frac{2\pi j}{|S|}.$$

The *maximum jumping angle* $\theta_{\max}$ will also play a role; this is the maximum of the angle between two consecutive nodes. Then, Shi and Zhu proved that maximizing $C(S, \mathcal{O}, k)$ is equivalent to maximizing the *circular embedding score*

$$C_e(r, \theta, \theta_{\max}) = \frac{1}{\theta_{\max}} \sum_{\substack{\theta_i < \theta_j \le \theta_i + \theta_{\max} \\ r_i > 0, \, r_j > 0}} P_{ij}/|S|,$$

where:

1. The matrix $P = (P_{ij})$ is obtained from the weight matrix W (of the graph $G = (V, E, W)$) by a suitable *normalization*;
2. $r_j \in \{0, 1\}$;
3. θ_j is an angle specifying the ordering of the nodes in the cycle;
4. $\theta_{\max}$ is the maximum jumping angle.

This optimization problem is still hard to solve. Consequently, Shi and Zhu considered a *continuous relaxation* of the probem by allowing r_j to be any real number in the interval $[0, 1]$ and θ_j to be any angle (within a suitable range). In the circular embedding, a node in then represented by the complex number

$$x_j = r_j e^{i\theta_j}.$$

We also introduce the *average jumping angle*

$$\Delta\theta = \overline{\theta_k - \theta_j}.$$

Then it is not hard to see that the numerator of $C_e(r,\theta,\theta_{\max})$ is well approximated by the expression

$$\sum_{j,k} P_{jk}\cos(\theta_k - \theta_j - \Delta\theta) = \sum_{j,k}\operatorname{Re}(x_j^* x_k \cdot e^{-i\Delta\theta}).$$

Thus, $C_e(r,\theta,\theta_{\max})$ is well approximated by

$$\frac{1}{\theta_{\max}}\frac{\sum_{j,k}\operatorname{Re}(x_j^* x_k \cdot e^{-i\Delta\theta})}{\sum_j |x_j|^2}.$$

This term can be written in terms of the matrix P as

$$C_e(r,\theta,\theta_{\max}) \approx \frac{1}{\theta_{\max}}\frac{\operatorname{Re}(x^* P x \cdot e^{-i\Delta\theta})}{x^* x},$$

where $x \in \mathbb{C}^n$ is the vector $x = (x_1,\ldots,x_n)$. The matrix P is a real matrix, but in general, it is neither symmetric nor normal ($PP^* = P^*P$). If we write $\delta = \Delta\theta$ and if we assume that $0 < \delta_{\min} \le \delta \le \delta_{\max}$, we need to solve the following quadratic optimization problem:

$$\begin{aligned} &\text{maximize} \quad \operatorname{Re}(x^* e^{-i\delta} P x)\\ &\text{subject to} \quad x^* x = 1,\, x \in \mathbb{C}^n;\ \delta_{\min} \le \delta \le \delta_{\max}. \end{aligned}$$

At first glance, this problem does not look like any of the standard quadratic optimization problems on the unit sphere. Nevertheless, we show that it reduces to a standard quadratic optimization problem involving a Hermitian matrix.

Let

$$c = e^{-i\delta} = a + ib,$$

with $a = \cos\delta$ and $b = -\sin\delta$. Following Horn and Johnson [1], for any (real or complex) $n \times n$ matrix P let $H(P)$ be the *Hermitian part* of P and let $S(P)$ be the *skew-Hermitian part* of P, where $H(P)$ and $S(P)$ are given by

$$H(P) = \frac{P + P^*}{2} \quad\text{and}\quad S(P) = \frac{P - P^*}{2}.$$

Obviously,

$$H(P)^* = H(P), \quad S(P)^* = -S(P), \quad\text{and}\quad P = H(P) + S(P).$$

Observe that $\frac{1}{i}S(P) = -iS(P)$ is Hermitian. If P is a real matrix, then $H(P)$ is said to be *symmetric* and $S(P)$ is said to be *skew-symmetric*. In this case,

$$H(P) = \frac{P + P^\top}{2} \quad \text{and} \quad S(P) = \frac{P - P^\top}{2}.$$

For every complex number $z = x + iy$, recall that

$$\mathrm{Re}(z) = y = \frac{1}{2}(z + \bar{z}).$$

Now, $z = x^* cPx$ is a complex number and $\bar{z} = z^*$, viewing z as a 1×1 matrix, so

$$\mathrm{Re}(x^* cPx) = \frac{1}{2}(x^* cPx + (x^* cPx)^*) = \frac{1}{2}(x^* cPx + x^* \bar{c}P^\top x) = x^* \frac{1}{2}(cP + \bar{c}P^\top)x.$$

The matrix $cP + \bar{c}P^\top$ is clearly Hermitian, and in fact,

$$\frac{1}{2}(cP + \bar{c}P^\top) = \frac{1}{2}((a + ib)P + (a - ib)P^\top)$$

$$= a\frac{1}{2}(P + P^\top) + ib\frac{1}{2}(P - P^\top) = aH(P) + ibS(P).$$

Define the Hermitian matrix H as

$$H = aH(P) + ibS(P) = \frac{1}{2}(cP + \bar{c}P^\top).$$

Observe that

$$H = \frac{1}{2}(cP + \bar{c}P^\top) = \frac{1}{2}(cP + (cP)^*) = H(cP),$$

the Hermitian part of cP, and since $c = e^{-i\delta}$, we have

$$H = H(e^{-i\delta}P).$$

In view of the above, our optimimization problem can also be stated as

$$\text{maximize} \quad x^* H(\delta)x$$
$$\text{subject to} \quad x^* x = 1,\, x \in \mathbb{C}^n;\, \delta_{\min} \leq \delta \leq \delta_{\max},$$

with

$$H(\delta) = H(e^{-i\delta}P) = \cos\delta\, H(P) - i\sin\delta\, S(P),$$

a Hermitian matrix.

By Lemma 14.2, the optimal value is the largest eigenvalue λ_1 of $H(\delta)$ over all δ such that $\delta_{\min} \leq \delta \leq \delta_{\max}$, and it is attained for the associated complex eigenvector $x = x_{\mathrm{re}} + ix_{\mathrm{im}}$.

To study the variation of the eigenvalues of $H(\delta)$, we will need to compute the derivative of $H(\delta)$ with respect to δ, denoted by $H'(\delta)$. We have

$$H'(\delta) = -\sin\delta\, H(P) - i\cos\delta\, S(P).$$

17.2 Derivatives of Eigenvalues and Eigenvectors for Normal Matrices

Let $X(\delta)$ be a normal matrix that depends differentiably on δ, let λ be some eigenvalue of X, which we assume to be simple (it has algebraic multiplicity 1), and let u be the corresponding unit eigenvector. We are going to derive formulas for the derivative of λ and the derivative of u. We adapt the derivation given by Peter Lax [4] (Chapter 9, Section 2) to normal matrices. The step missing in Lax is the application of the pseudo-inverse. However, Lax's derivation applies to arbitrary matrices X. A similar derivation is given in a blog by Terence Tao, assuming that the matrix X has only simple eigenvalues. The simplification afforded by normal matrices is that there is no need to deal with the dual space, since $Xu = \lambda u$ iff $X^*u = \overline{\lambda}u$ iff $u^*X = \lambda u^*$. When the eigenvalues are all simple, we can use a basis of eigenvectors $(u_1, \ldots, u_n)$ and its dual basis $(u_1^*, \ldots, u_n^*)$, because $Xu_i = \lambda u_i$ iff $u_i^*X = \lambda u_i^*$ (where $u_i^*(u_j) = \delta_{ij}$).

It is proved in Lax [4] (Chapter 9, Theorem 7 and Theorem 8) that if λ is a simple eigenvalue of $X(\delta)$ for $\delta = \delta_0$ and if u is a unit eigenvector associated with λ, then in a small open interval around δ_0, the matrix $X(\delta)$ has a simple eigenvalue $\lambda(\delta)$ that is differentiable (with $\lambda(\delta_0) = \lambda$) and that there is a choice of an eigenvector $u(t)$ associated with $\lambda(t)$, so that $u(t)$ is also differentiable (with $u(\delta_0) = u$). In the case of an eigenvalue, the proof uses the implicit function theorem applied to the characteristic polynomial $\det(\lambda I - X(\delta)) = f(\lambda, \delta)$. The proof of differentiability for an eigenvector is more involved and uses the nonvanishing of some principal minor of $\det(\lambda I - X(\delta))$.

Since explicit formulas (for normal matrices) for the derivative of a simple eigenvalue and the derivative of the corresponding unit eigenvector are not so easily found in the literature, we will prove the following proposition in full detail:

Proposition 17.1. *Let $X(\delta)$ be a normal matrix that depends differentiably on δ. If λ is any simple eigenvalue of X at δ_0 (it has algebraic multiplicity 1) and if u is the corresponding unit eigenvector, then the derivatives at $\delta = \delta_0$ of $\lambda(\delta)$ and $u(\delta)$ are given by*

$$\lambda' = u^*X'u,$$
$$u' = (\lambda I - X)^+X'u,$$

where $(\lambda I - X)^+$ is the pseudo-inverse of $\lambda I - X$, X' is the derivative of X at $\delta = \delta_0$, and u' is orthogonal to u.

Proof. If X is a normal matrix, then by Lemma 12.2, we known that $Xu = \lambda u$ iff $X^*u = \overline{\lambda}u$, and so if $Xu = \lambda u$, then

$$u^*X = \lambda u^*.$$

Taking the derivative of $Xu = \lambda u$ and using the chain rule, we get

$$X'u + Xu' = \lambda'u + \lambda u'.$$

By taking the inner product with u^*, we get

$$u^*X'u + u^*Xu' = \lambda'u^*u + \lambda u^*u'.$$

However, $u^*X = \lambda u^*$, so

$$u^*Xu' = \lambda u^*u',$$

and since u is a unit vector, $u^*u = 1$, so

$$u^*X'u + \lambda u^*u' = \lambda' + \lambda u^*u',$$

that is,

$$\lambda' = u^*X'u.$$

Let us rewrite the equation

$$X'u + Xu' = \lambda'u + \lambda u'$$

as

$$(\lambda I - X)u' = (X' - \lambda'I)u.$$

We need to show that this equation has a solution, and for this, it is enough to prove that $(X' - \lambda'I)u$ is in the range of $\lambda I - X$. However, the range of $\lambda I - X$ is equal to the orthogonal complement of the kernel of its adjoint $(\lambda I - X)^* = \overline{\lambda}I - X^*$, and since λ is a simple eigenvalue of X, $\overline{\lambda}$ is also a simple eigenvalue of X^* and $\mathrm{Ker}(\lambda I - X) = \mathrm{Ker}(\overline{\lambda}I - X^*) = \mathbb{C}u$, the one-dimensional space spanned by the unit eigenvector u. Thus, $(X' - \lambda'I)u$ is in the range of $\lambda I - X$ iff it is orthogonal to $\mathrm{Ker}(\overline{\lambda}I - X^*) = \mathbb{C}u$ iff

$$u^*(X' - \lambda'I)u = 0$$

iff

$$u^*X'u - \lambda'u^*u = 0,$$

that is, $\lambda' = u^*X'u$, which we have just proved.

Therefore the set of solutions of the linear equation

$$(\lambda I - X)u' = (X' - \lambda'I)u$$

is an affine line whose direction is the one-dimensional subspace spanned by the unit eigenvector u.

By Theorem 14.2, the pseudo-inverse of $\lambda I - X$ yields a solution of minimum norm belonging to the orthogonal complement of the kernel of $\lambda I - X$, that is, a solution orthogonal to the unit vector u, given by

$$u' = (\lambda I - X)^+(X' - \lambda'I)u.$$

Actually, because X is normal, we claim that

$$(\lambda I - X)^+ u = 0,$$

and so

$$u' = (\lambda I - X)^+ X' u.$$

For this, it is enough to prove that if X is a normal matrix and if $Xu = 0$, then $X^+ u = 0$. Indeed, since $\lambda I - X$ is also normal and since $(\lambda I - X)u = 0$, the above fact implies that $(\lambda I - X)^+ u = 0$.

Now, since X is a (real) normal matrix, by Theorem 12.7 it can be block diagonalized with respect to an orthogonal matrix U as

$$X = U\Lambda U^\top,$$

where Λ is the (real) block diagonal matrix

$$\Lambda = \mathrm{diag}(B_1, \ldots, B_n),$$

consisting either of 2×2 blocks of the form

$$B_j = \begin{pmatrix} \lambda_j & -\mu_j \\ \mu_j & \lambda_j \end{pmatrix}$$

with $\mu_j \neq 0$ or of one-dimensional blocks $B_k = (\lambda_k)$. If we write

$$\Lambda = \begin{pmatrix} \Lambda_r & 0 \\ 0 & 0 \end{pmatrix},$$

where Λ is invertible (with rank r) and all the other entries are zero, then by Proposition 14.1, the pseudo-inverse of X is given by

$$X^+ = U\Lambda^+ U^\top = U \begin{pmatrix} \Lambda_r^{-1} & 0 \\ 0 & 0 \end{pmatrix} U^\top.$$

Now, $Xu = 0$ implies that

$$\begin{pmatrix} \Lambda_r & 0 \\ 0 & 0 \end{pmatrix} U^\top u = 0,$$

which means that

$$U^\top u = \begin{pmatrix} 0 \\ y \end{pmatrix}$$

with $\dim(y) = n - r$. Then we have

$$X^+ u = U \begin{pmatrix} \Lambda_r^{-1} & 0 \\ 0 & 0 \end{pmatrix} U^\top u = U \begin{pmatrix} \Lambda_r^{-1} & 0 \\ 0 & 0 \end{pmatrix} \begin{pmatrix} 0 \\ y \end{pmatrix} = 0,$$

as claimed. $\square$

Applying the above to the Hermitian matrix $H(\delta)$, we get

$$\lambda'(\delta) = u^*H(\delta)u = -u^*(\sin\delta H(P) + i\cos\delta S(P))u.$$

The derivative $u'(\delta)$ orthogonal to u is given by

$$u'(\delta) = (\lambda(\delta)I - H(\delta))^+ H'(\delta)u(\delta),$$

where $(\lambda(\delta)I - H(\delta))^+$ is the pseudo-inverse of $\lambda(\delta)I - H(\delta)$.

17.3 Relationship between the Eigenvectors of P and $H(\delta)$

Experimental evidence suggests that there is a close relationship between the eigenvectors of the real matrix P and the eigenvectors of the Hermitian matrix $H(\delta)$. If P is a normal matrix, we can indeed prove such a relationship.

Recall that a matrix P is *normal* if P commutes with its transpose, that is,

$$PP^\top = P^\top P.$$

Proposition 17.2. *For any normal matrix P if $u + iv$ is an eigenvector of P for the eigenvalue $\lambda + i\mu$, then $u + iv$ is also an eigenvector of $H = aH(P) + ibS(P)$ for the real eigenvalue $a\lambda - b\mu$. Furthermore, all the eigenvalues of $H = aH(P) + ibS(P)$ are of the form $a\lambda - b\mu$, where $\lambda + i\mu$ is an eigenvalue of P.*

Proof. If P is a normal matrix, then by Lemma 12.2 we know that a complex vector $u + iv$ is an eigenvector of P for the eigenvalue $\lambda + i\mu$ iff $u + iv$ is an eigenvector of $P^\top$ for the conjugate eigenvalue, $\lambda - i\mu$.

As a consequence,

$$\begin{aligned}
H(u + iv) &= \frac{1}{2}(cP + \overline{c}P^\top)(u + iv) \\
&= \frac{1}{2}(cP(u + iv) + \overline{c}P^\top(u + iv)) \\
&= \frac{1}{2}(c(\lambda + i\mu) + \overline{c}(\lambda - i\mu))(u + iv) \\
&= \mathrm{Re}(c(\lambda + i\mu))(u + iv) \\
&= (a\lambda - b\mu)(u + iv),
\end{aligned}$$

since $c = a + ib$. The last statement holds because a normal matrix is diagonalizable (over an orthonormal basis with respect to the Hermitian inner product). $\square$

With the values of a and b as in Section 17.1,

$$a\lambda - b\mu = \lambda\cos\delta + \mu\sin\delta = \lambda(\delta).$$

If we write

$$\lambda + i\mu = \rho(\cos\varphi + i\sin\varphi),$$

then we get

$$\lambda(\delta) = \lambda \cos\delta + \mu \sin\delta = \cos\varphi \cos\delta + \sin\varphi \sin\delta = \rho \cos(\varphi - \delta).$$

The function $\delta \mapsto \rho \cos(\varphi - \delta)$ has a maximum for $\delta = \varphi$.

This confirms the experimental evidence that the numerator $\rho \cos(\varphi - \delta)$ of the eigenvalue $\lambda(\delta)$ of $H(\delta)$ associated with $\lambda + i\mu$ has a local maximum exactly when

$$\mathrm{Re}(u^*Pu \cdot c)$$

(subject to $u^*u = 1$) has a local maximum, which also happens for $\delta = \varphi$. It appears that these results still hold as long as P is not "too distant" from a normal matrix.

It would be desirable to measure how far a matrix A is from being normal. According to Horn and Johnson [1] (Chapter 3, Problem 18, page 156), this can be done using the *defect from normality.*

Definition 17.1. If $\sigma_1, \ldots, \sigma_n$ are the singular values of A listed in decreasing order and $\lambda_1, \ldots, \lambda_n$ are the eigenvalues of A listed so that $|\lambda_1| \geq \cdots \geq |\lambda_n|$, then the *defect from normality of A with respect to the Frobenius norm* is defined by

$$\delta_F(A) = \sqrt{\sum_{i=1}^{n}(\sigma_i^2 - |\lambda_i|^2)}.$$

Recall that the singular values $\sigma_1, \ldots, \sigma_n$ of A are the nonnegative square roots of the eigenvalues of A^*A (and AA^*), so that the Frobenius norm of A is given by

$$\|A\|_F = \sqrt{\mathrm{tr}(A^*A)} = \sqrt{\sum_{i=1}^{n}\sigma_i^2}.$$

For any upper triangular Schur decomposition of A,

$$A = U(D+T)U^*,$$

where U is a unitary matrix, D is the diagonal matrix $D = \mathrm{diag}(\lambda_1, \ldots, \lambda_n)$ and T is a strictly upper triangular matrix, since the Frobenius norm is unitarily invariant, which means that

$$\|A\|_F = \|D+T\|_F .$$

Since D is a diagonal matrix, a straightforward computation shows that

$$\|A\|_F^2 = \|D\|_F^2 + \|F\|_F^2 .$$

However,

$$\|D\|_F^2 = \sum_{i=1}^{n}|\lambda_i|^2,$$

and since $\|A\|_F^2 = \sum_{i=1}^{n}\sigma_i^2$, we conclude that

$$\|T\|_F^2 = \sum_{i=1}^{n} (\sigma_i^2 - |\lambda_i|^2).$$

Therefore, the quantity $\sum_{i=1}^{n}(\sigma_i^2 - |\lambda_i|^2)$ is always nonnegative, and moreover, $\|T\|_F$ *has the same value for all upper triangular Schur decompositions of A*, namely, the defect from normality of A,

$$\delta_F(A) = \sqrt{\sum_{i=1}^{n}(\sigma_i^2 - |\lambda_i|^2)}.$$

We have also proved that A is normal iff

$$\sum_{i=1}^{n}(\sigma_i^2 - |\lambda_i|^2) = 0,$$

which implies that

$$\sigma_i = |\lambda_i|$$

for $i = 1,\ldots,n$. Indeed, if A is normal, then A can be diagonalized with respect to a unitary matrix U, so that $A = U\Lambda U^*$ and then

$$U^*U = U\Lambda^*\Lambda U^*,$$

which proves that the singular values of A are indeed $|\lambda_1|,\ldots,|\lambda_n|$.

Conversely, if

$$\sigma_i = |\lambda_i|$$

for $i = 1,\ldots,n$, then

$$\sum_{i=1}^{n}(\sigma_i^2 - |\lambda_i|^2) = 0,$$

which, as we proved above, implies that A is normal. Thus, we have just proved the following proposition:

Proposition 17.3. *A matrix A is normal iff*

(1) $\sum_{i=1}^{n}\sigma_i^2 = \sum_{i=1}^{n}|\lambda_i|^2$, *or*
(2) $\sigma_i = |\lambda_i|$, *for* $i = 1,\ldots,n$.

The quantity $\delta_F(A) = \sqrt{\sum_{i=1}^{n}(\sigma_i^2 - |\lambda_i|^2)}$ measures the defect from normality of A. If $\delta_F(A)$ is "small," then A behaves much like a normal matrix.

Also observe that if $\|\ \|$ is *any* unitarily invariant matrix norm, then (following Horn and Johnson [1] (Chapter 3, Problem 31, page 192) we can define the *defect from normality of A with respect to the norm* $\|\ \|$ by

$$\delta(A,\|\ \|) = \inf\{\|T\| \mid A = U(D+T)U^*\},$$

where $U(D+T)U^*$ is any upper triangular Schur decomposition of A. Intuitively, A is "almost normal" iff $\delta(A,\|\ \|)$ is small. In the case of the Frobenius norm, we proved that

$$\delta(A, \|\ \|_F) = \delta_F(A) = \sqrt{\sum_{i=1}^{n}(\sigma_i^2 - |\lambda_i|^2)}.$$

17.4 Study of the Continuous Relaxation of the Problem

In this section, we study the variations of the the objective function

$$f(x, \delta) = x^*(\cos \delta \, H(P) - i \sin \delta \, S(P))x,$$

where $x \in \mathbb{C}^n$ with $\|x\| = 1$, and $0 \leq \delta \leq 2\pi$.

Figures 17.1, 17.2, and 17.3 show plots of the eigenvalues of various matrices as functions of $\delta \in [0, 2\pi)$ and were produced by Ryan Kennedy. Figure 17.4 corresponds to an actual image.

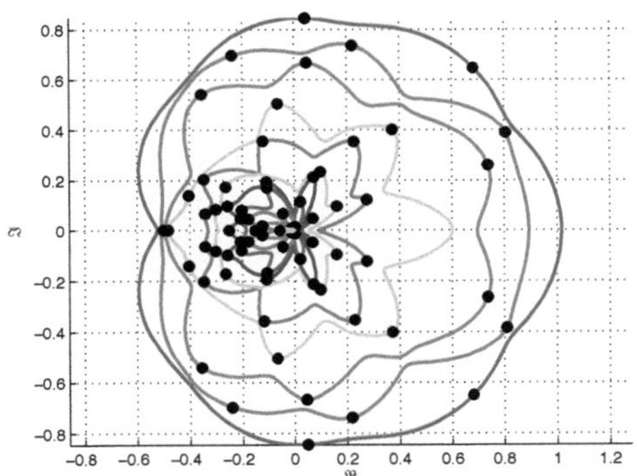

Fig. 17.1 The eigenvalues of a matrix $H(\delta)$ that is not normal.

It turns out that $x^*H(\delta)x \leq |x^*Px|$ for all x and all δ, and this has some important implications regarding the local maxima of these two functions.

Proposition 17.4. *For any (real) matrix P if we write $x^*Px = |x^*Px|(\cos \varphi + i \sin \varphi)$ and $H(\delta) = \cos \delta \, H(P) - i \sin \delta \, S(P)$ (as usual), then*

$$x^*H(\delta)x = |x^*Px|\cos(\delta - \varphi).$$

Proof. First, let us compute x^*Hx and $|x^*Px|$. We can write

$$P = H(P) + S(P) = H(P) + i(-iS(P)) = H_1 + iH_2,$$

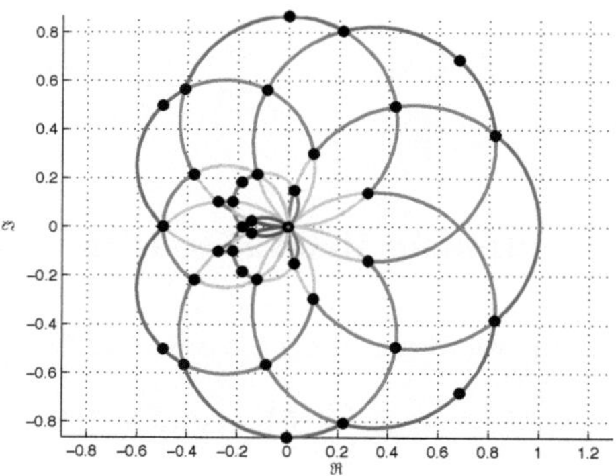

Fig. 17.2 The eigenvalues of a normal matrix $H(\delta)$.

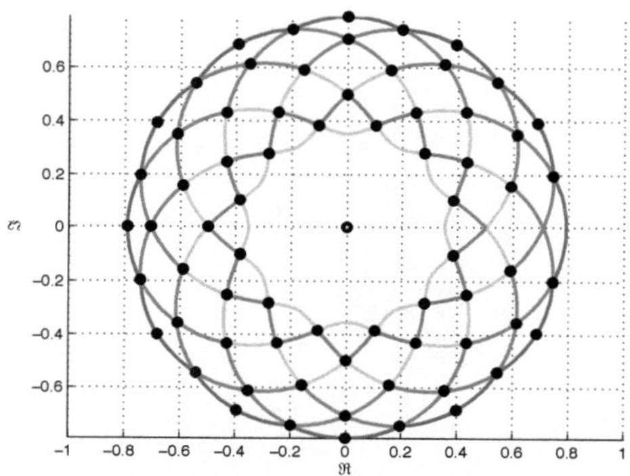

Fig. 17.3 The eigenvalues of a matrix $H(\delta)$ that is near normal.

with $H_1 = H(P)$ and $H_2 = -iS(P)$. Recall that H_1 and H_2 are Hermitian, so $\alpha = x^*H_1x$ and $\beta = x^*H_2x$ are both real, and we have

$$x^*Px = x^*H_1x + ix^*H_2x = \alpha + i\beta.$$

Now,

$$H = \cos\delta\, H(P) - i\sin\delta\, S(P) = \cos\delta\, H_1 + \sin\delta\, H_2,$$

and so

$$x^*Hx = \cos\delta\, x^*H_1x + \sin\delta\, x^*H_2x = \cos\delta\, \alpha + \sin\delta\, \beta.$$

In summary,

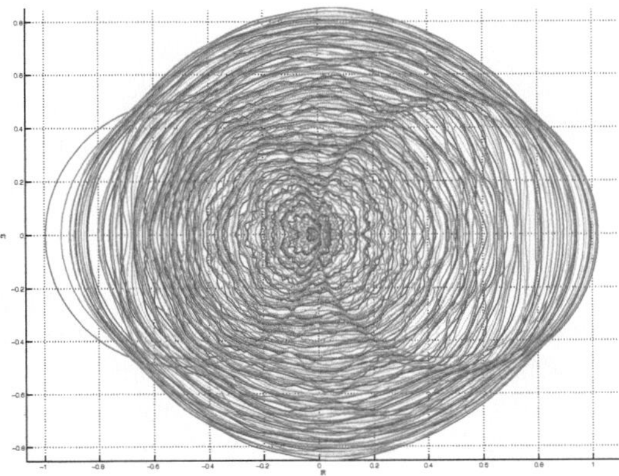

Fig. 17.4 The eigenvalues of the matrix for an actual image.

$$x^*Hx = \cos\delta\,\alpha + \sin\delta\,\beta,$$
$$x^*Px = \alpha + i\beta.$$

Since $x^*Px = |x^*Px|(\cos\varphi + i\sin\varphi)$, we have $\alpha = |x^*Px|\cos\varphi$ and $\beta = |x^*Px|\sin\varphi$ and we get

$$x^*Hx = \cos\delta\,\alpha + \sin\delta\,\beta = |x^*Px|(\cos\delta\cos\varphi + \sin\delta\sin\varphi) = |x^*Px|\cos(\delta - \varphi).$$

$\square$

The equation

$$x^*Hx = |x^*Px|\cos(\delta - \varphi)$$

implies that

$$x^*Hx \le |x^*Px|$$

for all $x \in \mathbb{C}^n$ and all δ ($0 \le \delta \le 2\pi$), with equality iff

$$\delta = \varphi,$$

the argument (phase angle) of x^*Px. In particular, for x fixed, $f(x,\delta) = x^*Hx$ has a local optimum when $\delta = \varphi$, and in this case, $x^*Hx = |x^*Px|$.

The inequality $x^*Hx \le |x^*Px|$ also implies that if $|x^*Px|$ achieves a local maximum for some vector x, then $f(x,\delta) = x^*Hx$ achieves a local maximum equal to $|x^*Px|$ for $\delta = \varphi$ and for the same x (where φ is the argument of x^*Px).

Indeed, we know that $f(x,\varphi) = |x^*Px|$, and if $f(x,\varphi)$ were not a local maximum at (x,φ), then for every open set $U \subseteq \mathbb{C}^n \times [0, 2\pi]$ with $(x,\varphi) \in U$, there would be some pair $(y,\eta) \in U$ such that

$$f(y, \eta) > f(x, \varphi) = |x^*Px|,$$

and since

$$|x^*Px| < f(y, \eta) \leq |y^*Py|,$$

we would have $|y^*Py| > |x^*Px|$. In particular, we can pick the open set $U \subseteq \mathbb{C}^n \times [0, 2\pi]$ to be a product $U = \Omega \times (\delta - \varepsilon, \delta + \varepsilon)$, where Ω is some arbitrary open subset of $\mathbb{C}^n$, and the above reasoning shows that $|y^*Py| > |x^*Px|$ for some $y \in \Omega$, contradicting the fact that x is a local maximum of $|x^*Px|$.

Now, since H is a Hermitian matrix, for δ fixed, we know that if $f(x, \delta) = x^*Hx$ has a local maximum for x, then x must be an eigenvector of H. Therefore, we proved that if $|x^*Px|$ achieves a local maximum for some unit vector x, then x must be an eigenvector of $H(\delta)$ for some δ, namely, the argument of x^*Px.

Generally, if $f(x, \delta) = x^*Hx$ is a local maximum of f at (x, δ), then $|x^*Px|$ is *not* necessarily a local maximum at x.

However, we can show that if $f(x, \delta) = x^*Hx$ is a local maximum of f at (x, δ), then $\delta = \varphi$, the phase angle of $|x^*Px|$, and so $x^*Hx = |x^*Px|$.

This is because

$$x^*Hx = |x^*Px|\cos(\delta - \varphi),$$

and for every open subset $U \subseteq \mathbb{C}^n \times [0, 2\pi]$ with $(x, \delta) \in U$, we can find some η small enough that $(x, \delta + \eta) \in U$ and $|\delta + \eta - \varphi| < |\delta - \varphi|$, and thus

$$x^*H(\delta + \eta)x > x^*H(\delta)x,$$

contradicting the fact that (x, δ) is a local maximum.

Unfortunately, this does not seem to help much in finding for which δ the function $f(x, \delta)$ has local maxima.

17.5 The Field of Values

The determination of the local extrema of $|x^*Px|$ (with $x^*x = 1$) is closely related to the structure of the set of complex numbers

$$F(P) = \{x^*Px \in \mathbb{C} \mid x \in \mathbb{C}^n, x^*x = 1\},$$

known as the *field of values* of P or the *numerical range* of P; see Horn and Johnson [2] (Chapter 1).

The notation $W(P)$ is also commonly used, corresponding to the German terminology "Wertvorrat" or "Wertevorrat." This set was studied as early as 1918 by Toeplitz and Hausdorff. Toeplitz proved that the boundary of $F(P)$ is convex, and Hausdorff proved the remarkable fact that $F(P)$ itself is convex. The quantity

$$r(P) = \max\{|z| \mid z \in F(P)\}$$

is called the *numerical radius* of P. It is obviously of interest to us, since it corresponds to the maximum of $|x^*Px|$ over all unit vectors x.

Figures 17.5, 17.6, and 17.7 give examples of numerical ranges and were produced by Ryan Kennedy.

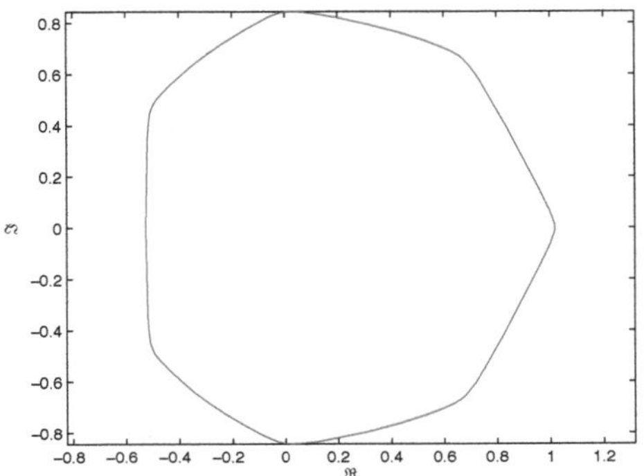

Fig. 17.5 Numerical range of a matrix that is not normal.

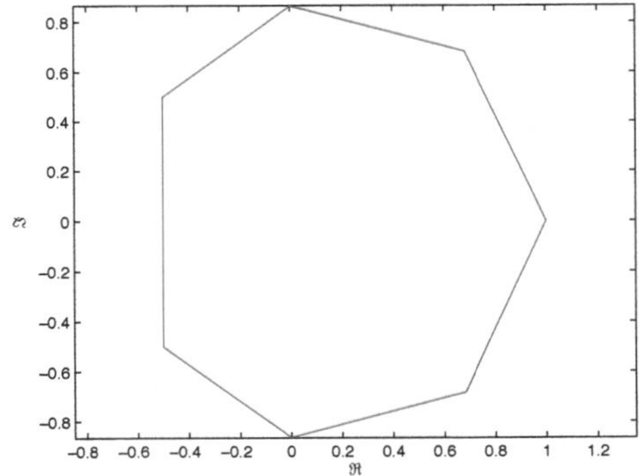

Fig. 17.6 Numerical range of a normal matrix.

Here is a summary of properties of the field of values relevant to our problem (assuming P is an $n \times n$ matrix):

(1) The set $F(P)$ is a compact subset of the complex plane $\mathbb{C}$.
(2) $F(P)$ is convex.
(3) Every eigenvalue of P belongs to $F(P)$.

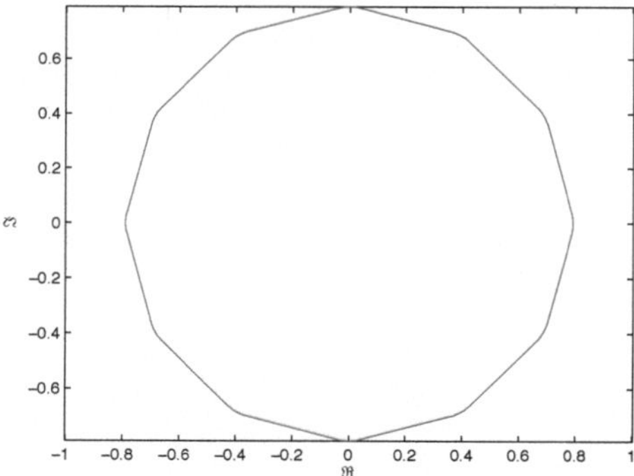

Fig. 17.7 Numerical range of a matrix that is near normal.

 (4) If P is normal, then $F(P)$ is the convex hull of its eigenvalues.
 (5) If P is Hermitian, then $F(P)$ is a real closed interval $[\alpha_n, \alpha_1]$, where α_n is the smallest eigenvalue of P and α_1 is the largest eigenvalue of P.
 (6) $F(U^*PU) = F(P)$, for every unitary matrix U.
 (7) $F(H(P)) = \mathrm{Re}(F(P))$ and $F(S(P)) = \mathrm{Im}(F(P))$.
 (8) $F(P + \alpha I) = F(P) + \alpha$.
 (9) $F(\alpha P) = \alpha F(P)$.
(10) If α is a sharp point of the boundary $\partial F(P)$ of $F(P)$, then α is an eigenvalue of P.
(11) The boundary $\partial F(P)$ of $F(P)$ has at most n sharp points.
(12) If the boundary $\partial F(P)$ of $F(P)$ is C^1 (does not have any sharp point), then every eigenvalue of P is in the interior of $F(P)$.
(13) The boundary $\partial F(P)$ of P is a piecewise algebraic curve.
(14) If $\lambda \in \partial F(P)$ for some eigenvalue λ of P, then

> a. Every eigenvector associated with λ is orthogonal to every eigenvector associated with every eigenvalue $\mu \neq \lambda$ of P.
> b. The dimension of the eigenspace associated with λ is equal to the algebraic multiplicity of λ.

(15) If P is a real matrix, then $F(P)$ is symmetric with respect to the x-axis.

Let us prove (2), since it is the main property of the field of values. Rather than following the proof given in Horn and Johnson [2] (Chapter 1, Section 1.3), which reduces the general case to the two-dimensional case, we give a proof much closer to Hausdorff's original proof based on a path connectivity argument (a similar proof is outlined in Horn and Johnson [2]; Section 1.3, Problem 7).

Lemma 17.1. *If A is any Hermitian matrix, then for any $\lambda \in \mathbb{C}$, the set*

$$L_A(\lambda) = \{x \in \mathbb{C}^n \mid x^*Ax = \lambda, x^*x = 1\}$$

is path connected (which means that there is a continuous curve contained in $L_A(\lambda)$ joining any two points in $L_A(\lambda)$). Furthermore, for any $\alpha \in \mathbb{C}$, the set $L_{\alpha A}(\lambda)$ is also path connected.

Proof. Because A is Hermitian, x^*Ax is real, its eigenvalues are real, and it can be diagonalized with respect to a unitary matrix. Thus $\lambda \in \mathbb{R}$ and, by properties (6) and (8) above, we may assume that $\lambda = 0$ and that A is a real diagonal matrix, $A = \mathrm{diag}(a_1, \ldots, a_n)$. In this case,

$$F(A) = \left\{ \sum_{j=1}^n a_j |x_j|^2 \mid (x_1, \ldots, x_n) \in \mathbb{C}^n, \ \sum_{j=1}^n |x_j|^2 = 1 \right\}.$$

Let $x, y \in \mathbb{C}^n$ be two unit vectors $x, y \in L_A(0)$, that is, such that

$$\sum_{j=1}^n a_j |x_j|^2 = \sum_{j=1}^n a_j |y_j|^2.$$

If we write $x_j = r_j e^{i\theta_j}$, with $r_j \in \mathbb{R}$, $r_j \geq 0$, and $\theta_j \in [0, 2\pi)$, it is clear that the points $(r_1 e^{i\theta_1}, \ldots, r_n e^{i\theta_n}) \in L_A(0)$ and $(r_1, \ldots, r_n) \in L_A(0)$ are connected by the continuous curve

$$\gamma_1(t) = (r_1 e^{i\theta_1(1-t)}, \ldots, r_n e^{i\theta_n(1-t)}),$$

where $\gamma_1(t) \in L_A(0)$ for all $t \in [0, 1]$. Therefore, it is enough to prove that any two points $x, y \in L_A(0)$, with $x_j, y_j \in \mathbb{R}$ and $x_j, y_j \geq 0$, are path connected. This is indeed the case, since the continuous curve

$$\gamma_2(t) = \left(\sqrt{(1-t)x_1^2 + ty_1^2}, \ldots, \sqrt{(1-t)x_n^2 + ty_n^2} \right)$$

stays in $L_A(0)$ and connects x and y. The second part of the lemma follows from the fact that $F(\alpha P) = \alpha F(P)$ (property (9)). $\square$

We can now prove property (2). For any complex matrix A, the matrix $H(A)$ is Hermitian and $S(A)$ is skew-Hermitian. However, if S is skew-Hermitian, then $S = -i\,iS$ where iS is Hermitian.

Theorem 17.1. *(Toeplitz and Hausdorff) For every complex matrix A, the field of values $F(A)$ is convex.*

Proof. We need to prove that for any two distinct complex numbers $\alpha, \beta \in \mathbb{C}$, if $\alpha, \beta \in F(A)$, then $(1-t)\alpha + t\beta \in F(A)$ for all $t \in [0, 1]$. By Properties (8) and (9), we may assume that $\alpha = 0$ and $\beta = 1$. Let $x, y \in \mathbb{C}^n$ be unit vectors such that $x^*Ax = 0$ and $y^*Ay = 1$. Since the skew-Hermitian part $S(A)$ of A is a scalar multiple of the Hermitian matrix $iS(A)$, by Lemma 17.1, the set

$$L_{S(A)}(0) = \{ x \in \mathbb{C}^n \mid x^*S(A)x = 0, \ x^*x = 1 \}$$

is path connected. Because $x^*Ax = 0$ and $y^*Ay = 1$ are real and because $S(A)$ is skew-Hermitian, as remarked in Section 15.3, $x^*S(A)x$ and $y^*S(A)y$ are pure imaginary or zero, so we must have $x^*S(A)x = y^*S(A)y = 0$. Therefore, $x, y \in L_{S(A)}(0)$, and there is some continuous curve $\gamma(t)$ in $L_{S(A)}(0)$ such that $\gamma(0) = x$ and $\gamma(1) = y$. Consequently, since $H(A)$ is Hermitian, $\gamma(t)^*H(A)\gamma(t) \in \mathbb{R}$, and the function

$$\gamma(t)^*A\gamma(t) = \gamma(t)^*H(A)\gamma(t) + \gamma(t)^*S(A)\gamma(t) = \gamma(t)^*H(A)\gamma(t)$$

is a real and continuous function from $\gamma(0)^*A\gamma(0) = x^*Ax = 0$ to $\gamma(1)^*A\gamma(1) = y^*Ay = 1$, which proves that $[0, 1] \subseteq F(A)$. Therefore, $F(A)$ is indeed convex. $\square$

Property (12) shows that in general, the eigenvalues of P do not yield the local maxima of $|x^*Px|$. Property (14) shows that if some eigenvalue λ of P belongs to $\partial F(P)$, then λ behaves like an eigenvalue of a normal matrix. Property (9) implies that

$$F(\mathrm{e}^{-\mathrm{i}\delta}P) = \mathrm{e}^{-\mathrm{i}\delta}F(P),$$

and so

$$F(P) = \mathrm{e}^{\mathrm{i}\delta}F(\mathrm{e}^{-\mathrm{i}\delta}P).$$

Geometrically, this means that $F(P)$ is obtained from $F(\mathrm{e}^{-\mathrm{i}\delta}P)$ by rotating it by δ. This with (5) and (7) yields a nice way of finding supporting lines for the convex set $F(P)$. To show this, we use a proposition from Horn and Johnson [2], whose proof is quite simple:

Proposition 17.5. *For any $n \times n$ matrix P and any unit vector $x \in \mathbb{C}^n$, the following properties are equivalent:*

(1) $\mathrm{Re}(x^*Px) = \max\{\mathrm{Re}(z) \mid z \in F(P)\}$.
(2) $x^*H(P)x = \max\{r \mid r \in F(H(P))\}$.
(3) The vector x is an eigenvector of $H(P)$ corresponding to the largest eigenvalue λ_1 of $H(P)$.

In fact, Proposition 17.5 immediately implies that

$$\max\{\mathrm{Re}(z) \mid z \in F(P)\} = \max\{r \mid r \in F(H(P))\} = \lambda_1.$$

As a consequence, for every angle $\delta \in [0, 2\pi)$, if we let λ_δ be the largest eigenvalue of the matrix $H(\mathrm{e}^{-\mathrm{i}\delta}P)$ and if x_δ is a corresponding unit eigenvector, then $z_\delta = x_\delta^*Px_\delta$ is on the boundary $\partial F(P)$ of $F(P)$, and the line L_δ given by

$$L_\delta = \left\{\mathrm{e}^{\mathrm{i}\delta}(\lambda_\delta + t\mathrm{i}) \mid t \in \mathbb{R}\right\} = \left\{(x, y) \in \mathbb{R}^2 \mid \cos\delta\, x + \sin\delta\, y - \lambda_\delta = 0\right\}$$

is a supporting line of $F(P)$ at z_δ. This is because by Proposition 17.5, the vertical line through the real point λ_δ is the supporting line to $F(P)$ at λ_δ, and the line L_δ is obtained by rotating by δ.

Observe that the triple $(\cos\delta, \sin\delta, \lambda_\delta)$ satisfies the equation

$$\det(wI - uH_1 - vH_2) = 0,$$

in the variables u, v, w, since λ_δ is the largest eigenvalue of the matrix $H(\mathrm{e}^{-\mathrm{i}\delta}P) = \cos\delta\, H_1 + \sin\delta\, H_2$ (recall that $H_1 = H(P)$ and $H_2 = -\mathrm{i}S(P)$). We can extend the domain of the variables u, v, w to be $\mathbb{C}$, in which case, the equation

$$\det(uH_1 + vH_2 + wI) = 0$$

defines a set of projective lines in the complex projective plane $\mathbb{CP}^2$, each one given by the equation

$$ux + vy + wz = 0,$$

in homogeneous coordinates $(x : y : z)$, and this set of lines is the set of tangent lines of a complex projective algebraic curve $C(P)$.

The above equation is the so-called *equation in line coordinates* of the curve $C(P)$. Since all supporting lines of $F(P)$ have line coordinates of the form $(\cos\delta, \sin\delta, -\lambda_\delta)$, they are among such tangent lines, and it is easy to see that the convex hull of the set of real points of the curve $C(P)$ is $F(P)$. The curve $C(P)$ first introduced and studied by Rudolph Kippenhahn in 1951 (note: Francis Murnaghan also briefly discussed this curve in 1932), is called the *boundary generating curve of P*. It is an algebraic curve of *class n*, which means that n tangent lines to the curve pass that through any "general" point. The degree of this curve is $n(n-1)$ minus the number of multiple tangents counted with their multiplicity.

17.6 Problems

17.1. Prove properties (2)–(9) of the field of values.

17.2. Prove Proposition 17.5.

References

1. Roger A. Horn and Charles R. Johnson. *Matrix Analysis*. Cambridge University Press, first edition, 1990.
2. Roger A. Horn and Charles R. Johnson. *Topics in Matrix Analysis*. Cambridge University Press, first edition, 1994.
3. Ryan Kennedy, Jean Gallier, and Jianbo Shi. Contour cuts: identifying salient contours in images by solving a hermitian eigenvalue problem. In *CVPR 2011, Colorado Springs, June 21-23, 2011*, pages 2065–2072. IEEE, 2011.
4. Peter D. Lax. *Linear Algebra and Its Applications*. Wiley, second edition, 2007.
5. Jianbo Shi and Jitendra Malik. Normalized cuts and image segmentation. *IEEE Transations on Pattern Analysis and Machine Intelligence*, 22(8):888–905, 2000.
6. Qihui Zhu, Gang Song, and Jianbo Shi. Untangling cycles for countour grouping. In *Eleventh IEEE International Conference on Computer Vision, Rio de Janeiro, Brazil, October 14–20, 2007*, pages 1–8. IEEE, 2007.

Chapter 18
Basics of Manifolds and Classical Lie Groups: The Exponential Map, Lie Groups, and Lie Algebras

Le rôle prépondérant de la théorie des groupes en mathématiques a été longtemps insoupçonné; il y a quatre-vingts ans, le nom même de groupe était ignoré. C'est Galois qui, le premier, en a eu une notion claire, mais c'est seulement depuis les travaux de Klein et surtout de Lie que l'on a commencé à voir qu'il n'y a presque aucune théorie mathématique où cette notion ne tienne une place importante.

—**Henri Poincaré**

18.1 The Exponential Map

This chapter is an introduction to manifolds, Lie groups, and Lie algebras.

The inventors of Lie groups and Lie algebras (starting with Lie!) regarded Lie groups as groups of symmetries of various topological or geometric objects. Lie algebras were viewed as the "infinitesimal transformations" associated with the symmetries in the Lie group. For example, the group $\mathbf{SO}(n)$ of rotations is the group of orientation-preserving isometries of the Euclidean space $\mathbb{E}^n$. The Lie algebra $\mathfrak{so}(n, \mathbb{R})$ consisting of real skew-symmetric $n \times n$ matrices is the corresponding set of infinitesimal rotations. The geometric link between a Lie group and its Lie algebra is the fact that the Lie algebra can be viewed as the tangent space to the Lie group at the identity. There is a map from the tangent space to the Lie group, called the *exponential map*. The Lie algebra can be considered as a linearization of the Lie group (near the identity element), and the exponential map provides the "delinearization," i.e., it takes us back to the Lie group. These concepts have a concrete realization in the case of groups of matrices, and for this reason we begin by studying the behavior of the exponential maps on matrices.

We begin by defining the exponential map on matrices and proving some of its properties. The exponential map allows us to "linearize" certain algebraic properties of matrices. It also plays a crucial role in the theory of linear differential equations with constant coefficients. But most of all, as we mentioned earlier, it is a stepping-stone to Lie groups and Lie algebras. On the way to Lie algebras, we derive the

459

classical "Rodrigues-like" formulae for rotations and for rigid motions in $\mathbb{R}^2$ and $\mathbb{R}^3$. We give an elementary proof that the exponential map is surjective for both $\mathbf{SO}(n)$ and $\mathbf{SE}(n)$, not using any topology, just our normal forms for matrices.

The last section gives a quick introduction to Lie groups and Lie algebras. We define manifolds as embedded submanifolds of $\mathbb{R}^N$, and we define linear Lie groups, using the famous result of Cartan (apparently actually due to Von Neumann) that a closed subgroup of $\mathbf{GL}(n, \mathbb{R})$ is a manifold, and thus a Lie group. This way, Lie algebras can be "computed" using tangent vectors to curves of the form $t \mapsto A(t)$, where $A(t)$ is a matrix. This section is inspired from Artin [6], Chevalley [12], Marsden and Ratiu [33], Curtis [14], Howe [23], and Sattinger and Weaver [42].

Given an $n \times n$ (real or complex) matrix $A = (a_{ij})$, we would like to define the exponential e^A of A as the sum of the series

$$e^A = I_n + \sum_{p \geq 1} \frac{A^p}{p!} = \sum_{p \geq 0} \frac{A^p}{p!},$$

letting $A^0 = I_n$. The problem is, Why is it well-defined? The following lemma shows that the above series is indeed absolutely convergent.

Lemma 18.1. *Let $A = (a_{ij})$ be a (real or complex) $n \times n$ matrix, and let*

$$\mu = \max\{|a_{ij}| \mid 1 \leq i, j \leq n\}.$$

If $A^p = (a_{ij}^{(p)})$, then

$$\left|a_{ij}^{(p)}\right| \leq (n\mu)^p$$

for all i, j, $1 \leq i, j \leq n$. As a consequence, the n^2 series

$$\sum_{p \geq 0} \frac{a_{ij}^{(p)}}{p!}$$

converge absolutely, and the matrix

$$e^A = \sum_{p \geq 0} \frac{A^p}{p!}$$

is a well-defined matrix.

Proof. The proof is by induction on p. For $p = 0$, we have $A^0 = I_n$, $(n\mu)^0 = 1$, and the lemma is obvious. Assume that

$$|a_{ij}^{(p)}| \leq (n\mu)^p$$

for all i, j, $1 \leq i, j \leq n$. Then we have

$$\left|a_{ij}^{(p+1)}\right| = \left|\sum_{k=1}^{n} a_{ik}^{(p)} a_{kj}\right| \leq \sum_{k=1}^{n} |a_{ik}^{(p)}| |a_{kj}| \leq \mu \sum_{k=1}^{n} |a_{ik}^{(p)}| \leq n\mu(n\mu)^p$$

and so, $\left|a_{ij}^{(p+1)}\right| \le (n\mu)^{p+1}$ for all i, j, $1 \le i, j \le n$. For every pair (i, j) such that $1 \le i, j \le n$, since

$$\left|a_{ij}^{(p)}\right| \le (n\mu)^p,$$

the series

$$\sum_{p\ge 0} \frac{\left|a_{ij}^{(p)}\right|}{p!}$$

is bounded by the convergent series

$$e^{n\mu} = \sum_{p\ge 0} \frac{(n\mu)^p}{p!},$$

and thus it is absolutely convergent. This shows that

$$e^A = \sum_{k\ge 0} \frac{A^k}{k!}$$

is well defined. $\square$

It is instructive to compute explicitly the exponential of some simple matrices. As an example, let us compute the exponential of the real skew-symmetric matrix

$$A = \begin{pmatrix} 0 & -\theta \\ \theta & 0 \end{pmatrix}.$$

We need to find an inductive formula expressing the powers A^n. Let us observe that

$$\begin{pmatrix} 0 & -\theta \\ \theta & 0 \end{pmatrix} = \theta \begin{pmatrix} 0 & -1 \\ 1 & 0 \end{pmatrix} \quad \text{and} \quad \begin{pmatrix} 0 & -\theta \\ \theta & 0 \end{pmatrix}^2 = -\theta^2 \begin{pmatrix} 1 & 0 \\ 0 & 1 \end{pmatrix}.$$

Then, letting

$$J = \begin{pmatrix} 0 & -1 \\ 1 & 0 \end{pmatrix},$$

we have

$$A^{4n} = \theta^{4n} I_2,$$
$$A^{4n+1} = \theta^{4n+1} J,$$
$$A^{4n+2} = -\theta^{4n+2} I_2,$$
$$A^{4n+3} = -\theta^{4n+3} J,$$

and so

$$e^A = I_2 + \frac{\theta}{1!} J - \frac{\theta^2}{2!} I_2 - \frac{\theta^3}{3!} J + \frac{\theta^4}{4!} I_2 + \frac{\theta^5}{5!} J - \frac{\theta^6}{6!} I_2 - \frac{\theta^7}{7!} J + \cdots.$$

Rearranging the order of the terms, we have

$$e^A = \left(1 - \frac{\theta^2}{2!} + \frac{\theta^4}{4!} - \frac{\theta^6}{6!} + \cdots\right) I_2 + \left(\frac{\theta}{1!} - \frac{\theta^3}{3!} + \frac{\theta^5}{5!} - \frac{\theta^7}{7!} + \cdots\right) J.$$

We recognize the power series for $\cos\theta$ and $\sin\theta$, and thus

$$e^A = \cos\theta I_2 + \sin\theta J,$$

that is

$$e^A = \begin{pmatrix} \cos\theta & -\sin\theta \\ \sin\theta & \cos\theta \end{pmatrix}.$$

Thus, e^A is a rotation matrix! This is a general fact. If A is a skew-symmetric matrix, then e^A is an orthogonal matrix of determinant $+1$, i.e., a rotation matrix. Furthermore, every rotation matrix is of this form; i.e., the exponential map from the set of skew-symmetric matrices to the set of rotation matrices is surjective. In order to prove these facts, we need to establish some properties of the exponential map. But before that, let us work out another example showing that the exponential map is not always surjective. Let us compute the exponential of a real 2×2 matrix with null trace of the form

$$A = \begin{pmatrix} a & b \\ c & -a \end{pmatrix}.$$

We need to find an inductive formula expressing the powers A^n. Observe that

$$A^2 = (a^2 + bc)I_2 = -\det(A)I_2.$$

If $a^2 + bc = 0$, we have

$$e^A = I_2 + A.$$

If $a^2 + bc < 0$, let $\omega > 0$ be such that $\omega^2 = -(a^2 + bc)$. Then, $A^2 = -\omega^2 I_2$. We get

$$e^A = I_2 + \frac{A}{1!} - \frac{\omega^2}{2!}I_2 - \frac{\omega^2}{3!}A + \frac{\omega^4}{4!}I_2 + \frac{\omega^4}{5!}A - \frac{\omega^6}{6!}I_2 - \frac{\omega^6}{7!}A + \cdots.$$

Rearranging the order of the terms, we have

$$e^A = \left(1 - \frac{\omega^2}{2!} + \frac{\omega^4}{4!} - \frac{\omega^6}{6!} + \cdots\right) I_2 + \frac{1}{\omega}\left(\omega - \frac{\omega^3}{3!} + \frac{\omega^5}{5!} - \frac{\omega^7}{7!} + \cdots\right) A.$$

We recognize the power series for $\cos\omega$ and $\sin\omega$, and thus

$$e^A = \cos\omega I_2 + \frac{\sin\omega}{\omega}A.$$

If $a^2 + bc > 0$, let $\omega > 0$ be such that $\omega^2 = (a^2 + bc)$. Then $A^2 = \omega^2 I_2$. We get

$$e^A = I_2 + \frac{A}{1!} + \frac{\omega^2}{2!}I_2 + \frac{\omega^2}{3!}A + \frac{\omega^4}{4!}I_2 + \frac{\omega^4}{5!}A + \frac{\omega^6}{6!}I_2 + \frac{\omega^6}{7!}A + \cdots.$$

Rearranging the order of the terms, we have

$$e^A = \left(1 + \frac{\omega^2}{2!} + \frac{\omega^4}{4!} + \frac{\omega^6}{6!} + \cdots\right) I_2 + \frac{1}{\omega}\left(\omega + \frac{\omega^3}{3!} + \frac{\omega^5}{5!} + \frac{\omega^7}{7!} + \cdots\right) A.$$

If we recall that $\cosh\omega = \left(e^\omega + e^{-\omega}\right)/2$ and $\sinh\omega = \left(e^\omega - e^{-\omega}\right)/2$, we recognize the power series for $\cosh\omega$ and $\sinh\omega$, and thus

$$e^A = \cosh\omega\, I_2 + \frac{\sinh\omega}{\omega} A.$$

It immediately verified that in all cases,

$$\det\left(e^A\right) = 1.$$

This shows that the exponential map is a function from the set of 2×2 matrices with null trace to the set of 2×2 matrices with determinant 1. This function is not surjective. Indeed, $tr(e^A) = 2\cos\omega$ when $a^2 + bc < 0$, $tr(e^A) = 2\cosh\omega$ when $a^2 + bc > 0$, and $tr(e^A) = 2$ when $a^2 + bc = 0$. As a consequence, for any matrix A with null trace,

$$tr\left(e^A\right) \geq -2,$$

and any matrix B with determinant 1 and whose trace is less than -2 is not the exponential e^A of any matrix A with null trace. For example,

$$B = \begin{pmatrix} a & 0 \\ 0 & a^{-1} \end{pmatrix},$$

where $a < 0$ and $a \neq -1$, is not the exponential of any matrix A with null trace.

A fundamental property of the exponential map is that if $\lambda_1, \ldots, \lambda_n$ are the eigenvalues of A, then the eigenvalues of e^A are $e^{\lambda_1}, \ldots, e^{\lambda_n}$. For this we need two lemmas.

Lemma 18.2. *Let A and U be (real or complex) matrices, and assume that U is invertible. Then*

$$e^{UAU^{-1}} = Ue^A U^{-1}.$$

Proof. A trivial induction shows that

$$UA^p U^{-1} = (UAU^{-1})^p,$$

and thus

$$e^{UAU^{-1}} = \sum_{p\geq 0} \frac{(UAU^{-1})^p}{p!} = \sum_{p\geq 0} \frac{UA^p U^{-1}}{p!}$$

$$= U\left(\sum_{p\geq 0} \frac{A^p}{p!}\right) U^{-1} = Ue^A U^{-1}.$$

$\square$

Say that a square matrix A is an *upper triangular matrix* if it has the following shape,

$$
\begin{pmatrix}
a_{11} & a_{12} & a_{13} & \dots & a_{1\,n-1} & a_{1n} \\
0 & a_{22} & a_{23} & \dots & a_{2\,n-1} & a_{2n} \\
0 & 0 & a_{33} & \dots & a_{3\,n-1} & a_{3n} \\
\vdots & \vdots & \vdots & \ddots & \vdots & \vdots \\
0 & 0 & 0 & \dots & a_{n-1\,n-1} & a_{n-1\,n} \\
0 & 0 & 0 & \dots & 0 & a_{nn}
\end{pmatrix},
$$

i.e., $a_{ij} = 0$ whenever $j < i$, $1 \leq i, j \leq n$.

Lemma 18.3. *Given any complex $n \times n$ matrix A, there is an invertible matrix P and an upper triangular matrix T such that*

$$
A = PTP^{-1}.
$$

Proof. We prove by induction on n that if $f \colon \mathbb{C}^n \to \mathbb{C}^n$ is a linear map, then there is a basis $(u_1, \dots, u_n)$ with respect to which f is represented by an upper triangular matrix. For $n = 1$ the result is obvious. If $n > 1$, since $\mathbb{C}$ is algebraically closed, f has some eigenvalue $\lambda_1 \in \mathbb{C}$, and let u_1 be an eigenvector for λ_1. We can find $n - 1$ vectors $(v_2, \dots, v_n)$ such that $(u_1, v_2, \dots, v_n)$ is a basis of $\mathbb{C}^n$, and let W be the subspace of dimension $n - 1$ spanned by $(v_2, \dots, v_n)$. In the basis $(u_1, v_2 \dots, v_n)$, the matrix of f is of the form

$$
\begin{pmatrix}
a_{11} & a_{12} & \dots & a_{1n} \\
0 & a_{22} & \dots & a_{2n} \\
\vdots & \vdots & \ddots & \vdots \\
0 & a_{n2} & \dots & a_{nn}
\end{pmatrix},
$$

since its first column contains the coordinates of $\lambda_1 u_1$ over the basis $(u_1, v_2, \dots, v_n)$. Letting $p \colon \mathbb{C}^n \to W$ be the projection defined such that $p(u_1) = 0$ and $p(v_i) = v_i$ when $2 \leq i \leq n$, the linear map $g \colon W \to W$ defined as the restriction of $p \circ f$ to W is represented by the $(n-1) \times (n-1)$ matrix $(a_{ij})_{2 \leq i, j \leq n}$ over the basis $(v_2, \dots, v_n)$. By the induction hypothesis, there is a basis $(u_2, \dots, u_n)$ of W such that g is represented by an upper triangular matrix $(b_{ij})_{1 \leq i, j \leq n-1}$.

However,

$$
\mathbb{C}^n = \mathbb{C}u_1 \oplus W,
$$

and thus $(u_1, \dots, u_n)$ is a basis for $\mathbb{C}^n$. Since p is the projection from $\mathbb{C}^n = \mathbb{C}u_1 \oplus W$ onto W and $g \colon W \to W$ is the restriction of $p \circ f$ to W, we have

$$
f(u_1) = \lambda_1 u_1
$$

and

$$
f(u_{i+1}) = a_{1i} u_1 + \sum_{j=1}^{n-1} b_{ij} u_{j+1}
$$

for some $a_{1i} \in \mathbb{C}$, when $1 \le i \le n-1$. But then the matrix of f with respect to $(u_1, \ldots, u_n)$ is upper triangular. Thus, there is a change of basis matrix P such that $A = PTP^{-1}$ where T is upper triangular. $\quad\square$

Remark: If E is a Hermitian space, the proof of Lemma 18.3 can be easily adapted to prove that there is an *orthonormal* basis $(u_1, \ldots, u_n)$ with respect to which the matrix of f is upper triangular. In terms of matrices, this means that there is a unitary matrix U and an upper triangular matrix T such that $A = UTU^*$. This is usually known as *Schur's lemma*. Using this result, we can immediately rederive the fact that if A is a Hermitian matrix, then there is a unitary matrix U and a real diagonal matrix D such that $A = UDU^*$.

If $A = PTP^{-1}$ where T is upper triangular, note that the diagonal entries on T are the eigenvalues $\lambda_1, \ldots, \lambda_n$ of A. Indeed, A and T have the same characteristic polynomial. This is because if A and B are any two matrices such that $A = PBP^{-1}$, then

$$
\begin{aligned}
\det(A - \lambda I) &= \det(PBP^{-1} - \lambda \, PIP^{-1}), \\
&= \det(P(B - \lambda I)P^{-1}), \\
&= \det(P)\det(B - \lambda I)\det(P^{-1}), \\
&= \det(P)\det(B - \lambda I)\det(P)^{-1}, \\
&= \det(B - \lambda I).
\end{aligned}
$$

Furthermore, it is well known that the determinant of a matrix of the form

$$
\begin{pmatrix}
\lambda_1 - \lambda & a_{12} & a_{13} & \cdots & a_{1n-1} & a_{1n} \\
0 & \lambda_2 - \lambda & a_{23} & \cdots & a_{2n-1} & a_{2n} \\
0 & 0 & \lambda_3 - \lambda & \cdots & a_{3n-1} & a_{3n} \\
\vdots & \vdots & \vdots & \ddots & \vdots & \vdots \\
0 & 0 & 0 & \cdots & \lambda_{n-1} - \lambda & a_{n-1n} \\
0 & 0 & 0 & \cdots & 0 & \lambda_n - \lambda
\end{pmatrix}
$$

is $(\lambda_1 - \lambda) \cdots (\lambda_n - \lambda)$, and thus the eigenvalues of $A = PTP^{-1}$ are the diagonal entries of T. We use this property to prove the following lemma.

Lemma 18.4. *Given any complex $n \times n$ matrix A, if $\lambda_1, \ldots, \lambda_n$ are the eigenvalues of A, then $e^{\lambda_1}, \ldots, e^{\lambda_n}$ are the eigenvalues of e^A. Furthermore, if u is an eigenvector of A for λ_i, then u is an eigenvector of e^A for e^{λ_i}.*

Proof. By Lemma 18.3 there is an invertible matrix P and an upper triangular matrix T such that

$$
A = PTP^{-1}.
$$

By Lemma 18.2,

$$e^{PTP^{-1}} = Pe^T P^{-1}.$$

However, we showed that A and T have the same eigenvalues, which are the diagonal entries $\lambda_1, \ldots, \lambda_n$ of T, and $e^A = e^{PTP^{-1}} = Pe^T P^{-1}$ and e^T have the same eigenvalues, which are the diagonal entries of e^T. Clearly, the diagonal entries of e^T are $e^{\lambda_1}, \ldots, e^{\lambda_n}$. Now, if u is an eigenvector of A for the eigenvalue λ, a simple induction shows that u is an eigenvector of A^n for the eigenvalue λ^n, from which is follows that u is an eigenvector of e^A for e^λ. $\square$

As a consequence, we can show that

$$\det(e^A) = e^{\mathrm{tr}(A)},$$

where $\mathrm{tr}(A)$ is the *trace of A*, i.e., the sum $a_{11} + \cdots + a_{nn}$ of its diagonal entries, which is also equal to the sum of the eigenvalues of A. This is because the determinant of a matrix is equal to the product of its eigenvalues, and if $\lambda_1, \ldots, \lambda_n$ are the eigenvalues of A, then by Lemma 18.4, $e^{\lambda_1}, \ldots, e^{\lambda_n}$ are the eigenvalues of e^A, and thus

$$\det\left(e^A\right) = e^{\lambda_1} \cdots e^{\lambda_n} = e^{\lambda_1 + \cdots + \lambda_n} = e^{\mathrm{tr}(A)}.$$

This shows that e^A is always an invertible matrix, since e^z is never null for every $z \in \mathbb{C}$. In fact, the inverse of e^A is e^{-A}, but we need to prove another lemma. This is because it is generally not true that

$$e^{A+B} = e^A e^B,$$

unless A and B commute, i.e., $AB = BA$. We need to prove this last fact.

Lemma 18.5. *Given any two complex $n \times n$ matrices A, B, if $AB = BA$, then*

$$e^{A+B} = e^A e^B.$$

Proof. Since $AB = BA$, we can expand $(A+B)^p$ using the binomial formula:

$$(A+B)^p = \sum_{k=0}^{p} \binom{p}{k} A^k B^{p-k},$$

and thus

$$\frac{1}{p!}(A+B)^p = \sum_{k=0}^{p} \frac{A^k B^{p-k}}{k!(p-k)!}.$$

Note that for any integer $N \geq 0$, we can write

$$\sum_{p=0}^{2N} \frac{1}{p!}(A+B)^p = \sum_{p=0}^{2N} \sum_{k=0}^{p} \frac{A^k B^{p-k}}{k!(p-k)!}$$

$$= \left(\sum_{p=0}^{N} \frac{A^p}{p!} \right) \left(\sum_{p=0}^{N} \frac{B^p}{p!} \right) + \sum_{\substack{\max(k,l) > N \\ k+l \leq 2N}} \frac{A^k}{k!} \frac{B^l}{l!},$$

where there are $N(N+1)$ pairs (k,l) in the second term. Letting

$$\|A\| = \max\{|a_{ij}| \mid 1 \le i, j \le n\}, \quad \|B\| = \max\{|b_{ij}| \mid 1 \le i, j \le n\},$$

and $\mu = \max(\|A\|, \|B\|)$, note that for every entry c_{ij} in $\left(A^k/k!\right)\left(B^l/l!\right)$ we have

$$|c_{ij}| \le n \frac{(n\mu)^k}{k!} \frac{(n\mu)^l}{l!} \le \frac{(n^2\mu)^{2N}}{N!}.$$

As a consequence, the absolute value of every entry in

$$\sum_{\substack{\max(k,l) > N \\ k+l \le 2N}} \frac{A^k}{k!} \frac{B^l}{l!}$$

is bounded by

$$N(N+1) \frac{(n^2\mu)^{2N}}{N!},$$

which goes to 0 as $N \mapsto \infty$. From this, it immediately follows that

$$e^{A+B} = e^A e^B.$$

$\square$

Now, using Lemma 18.5, since A and $-A$ commute, we have

$$e^A e^{-A} = e^{A+-A} = e^{0_n} = I_n,$$

which shows that the inverse of e^A is e^{-A}.

We will now use the properties of the exponential that we have just established to show how various matrices can be represented as exponentials of other matrices.

18.2 The Lie Groups $\mathbf{GL}(n, \mathbb{R})$, $\mathbf{SL}(n, \mathbb{R})$, $\mathbf{O}(n)$, $\mathbf{SO}(n)$, the Lie Algebras $\mathfrak{gl}(n, \mathbb{R})$, $\mathfrak{sl}(n, \mathbb{R})$, $\mathfrak{o}(n)$, $\mathfrak{so}(n)$, and the Exponential Map

First, we recall some basic facts and definitions. The set of real invertible $n \times n$ matrices forms a group under multiplication, denoted by $\mathbf{GL}(n, \mathbb{R})$. The subset of $\mathbf{GL}(n, \mathbb{R})$ consisting of those matrices having determinant $+1$ is a subgroup of $\mathbf{GL}(n, \mathbb{R})$, denoted by $\mathbf{SL}(n, \mathbb{R})$. It is also easy to check that the set of real $n \times n$ orthogonal matrices forms a group under multiplication, denoted by $\mathbf{O}(n)$. The subset of $\mathbf{O}(n)$ consisting of those matrices having determinant $+1$ is a subgroup of $\mathbf{O}(n)$, denoted by $\mathbf{SO}(n)$. We will also call matrices in $\mathbf{SO}(n)$ *rotation matrices*. Staying with easy things, we can check that the set of real $n \times n$ matrices with null trace forms a vector space under addition, and similarly for the set of skew-symmetric matrices.

Definition 18.1. The group $\mathbf{GL}(n,\mathbb{R})$ is called the *general linear group*, and its subgroup $\mathbf{SL}(n,\mathbb{R})$ is called the *special linear group*. The group $\mathbf{O}(n)$ of orthogonal matrices is called the *orthogonal group*, and its subgroup $\mathbf{SO}(n)$ is called the *special orthogonal group* (or *group of rotations*). The vector space of real $n \times n$ matrices with null trace is denoted by $\mathfrak{sl}(n,\mathbb{R})$, and the vector space of real $n \times n$ skew-symmetric matrices is denoted by $\mathfrak{so}(n)$.

Remark: The notation $\mathfrak{sl}(n,\mathbb{R})$ and $\mathfrak{so}(n)$ is rather strange and deserves some explanation. The groups $\mathbf{GL}(n,\mathbb{R})$, $\mathbf{SL}(n,\mathbb{R})$, $\mathbf{O}(n)$, and $\mathbf{SO}(n)$ are more than just groups. They are also topological groups, which means that they are topological spaces (viewed as subspaces of $\mathbb{R}^{n^2}$) and that the multiplication and the inverse operations are continuous (in fact, smooth). Furthermore, they are smooth real manifolds.[1] Such objects are called *Lie groups*. The real vector spaces $\mathfrak{sl}(n)$ and $\mathfrak{so}(n)$ are what is called *Lie algebras*. However, we have not defined the algebra structure on $\mathfrak{sl}(n,\mathbb{R})$ and $\mathfrak{so}(n)$ yet. The algebra structure is given by what is called the *Lie bracket*, which is defined as

$$[A, B] = AB - BA.$$

Lie algebras are associated with Lie groups. What is going on is that the Lie algebra of a Lie group is its tangent space at the identity, i.e., the space of all tangent vectors at the identity (in this case, I_n). In some sense, the Lie algebra achieves a "linearization" of the Lie group. The exponential map is a map from the Lie algebra to the Lie group, for example,

$$\exp\colon \mathfrak{so}(n) \to \mathbf{SO}(n)$$

and

$$\exp\colon \mathfrak{sl}(n,\mathbb{R}) \to \mathbf{SL}(n,\mathbb{R}).$$

The exponential map often allows a parametrization of the Lie group elements by simpler objects, the Lie algebra elements.

One might ask, What happened to the Lie algebras $\mathfrak{gl}(n,\mathbb{R})$ and $\mathfrak{o}(n)$ associated with the Lie groups $\mathbf{GL}(n,\mathbb{R})$ and $\mathbf{O}(n)$? We will see later that $\mathfrak{gl}(n,\mathbb{R})$ is the set of *all* real $n \times n$ matrices, and that $\mathfrak{o}(n) = \mathfrak{so}(n)$.

The properties of the exponential map play an important role in studying a Lie group. For example, it is clear that the map

$$\exp\colon \mathfrak{gl}(n,\mathbb{R}) \to \mathbf{GL}(n,\mathbb{R})$$

---

[1] We refrain from defining manifolds right now, not to interupt the flow of intuitive ideas.

is well-defined, but since every matrix of the form e^A has a positive determinant, exp is not surjective. Similarly, since

$$\det(e^A) = e^{\mathrm{tr}(A)},$$

the map

$$\exp\colon \mathfrak{sl}(n,\mathbb{R}) \to \mathbf{SL}(n,\mathbb{R})$$

is well-defined. However, we showed in Section 18.1 that it is not surjective either. As we will see in the next theorem, the map

$$\exp\colon \mathfrak{so}(n) \to \mathbf{SO}(n)$$

is well-defined and surjective. The map

$$\exp\colon \mathfrak{o}(n) \to \mathbf{O}(n)$$

is well-defined, but it is not surjective, since there are matrices in $\mathbf{O}(n)$ with determinant -1.

Remark: The situation for matrices over the field $\mathbb{C}$ of complex numbers is quite different, as we will see later.

We now show the fundamental relationship between $\mathbf{SO}(n)$ and $\mathfrak{so}(n)$.

Theorem 18.1. *The exponential map*

$$\exp\colon \mathfrak{so}(n) \to \mathbf{SO}(n)$$

is well-defined and surjective.

Proof. First, we need to prove that if A is a skew-symmetric matrix, then e^A is a rotation matrix. For this, first check that

$$\left(e^A\right)^\top = e^{A^\top}.$$

Then, since $A^\top = -A$, we get

$$\left(e^A\right)^\top = e^{A^\top} = e^{-A},$$

and so

$$\left(e^A\right)^\top e^A = e^{-A}e^A = e^{-A+A} = e^{0_n} = I_n,$$

and similarly,

$$e^A \left(e^A\right)^\top = I_n,$$

showing that e^A is orthogonal. Also,

$$\det\left(e^A\right) = e^{\mathrm{tr}(A)},$$

and since A is real skew-symmetric, its diagonal entries are 0, i.e., $\operatorname{tr}(A) = 0$, and so $\det(e^A) = +1$.

For the surjectivity, we will use Theorem 12.9 and Theorem 12.10. Theorem 12.9 says that for every skew-symmetric matrix A there is an orthogonal matrix P such that $A = PDP^\top$, where D is a block diagonal matrix of the form

$$
D = \begin{pmatrix} D_1 & & \cdots & \\ & D_2 & \cdots & \\ \vdots & \vdots & \ddots & \vdots \\ & & \cdots & D_p \end{pmatrix}
$$

such that each block D_i is either 0 or a two-dimensional matrix of the form

$$
D_i = \begin{pmatrix} 0 & -\theta_i \\ \theta_i & 0 \end{pmatrix}
$$

where $\theta_i \in \mathbb{R}$, with $\theta_i > 0$. Theorem 12.10 says that for every orthogonal matrix R there is an orthogonal matrix P such that $R = PEP^\top$, where E is a block diagonal matrix of the form

$$
E = \begin{pmatrix} E_1 & & \cdots & \\ & E_2 & \cdots & \\ \vdots & \vdots & \ddots & \vdots \\ & & \cdots & E_p \end{pmatrix}
$$

such that each block E_i is either 1, -1, or a two-dimensional matrix of the form

$$
E_i = \begin{pmatrix} \cos\theta_i & -\sin\theta_i \\ \sin\theta_i & \cos\theta_i \end{pmatrix} .
$$

If R is a rotation matrix, there is an even number of -1's and they can be grouped into blocks of size 2 associated with $\theta = \pi$. Let D be the block matrix associated with E in the obvious way, where an entry 0 in D is associated with a 1 in E and where

$$
D_i = \begin{pmatrix} 0 & -\theta_i \\ \theta_i & 0 \end{pmatrix}
$$

is associated with the rotation matrix

$$
E_i = \begin{pmatrix} \cos\theta_i & -\sin\theta_i \\ \sin\theta_i & \cos\theta_i \end{pmatrix} .
$$

Since by Lemma 18.2

$$
e^A = e^{PDP^{-1}} = Pe^D P^{-1},
$$

and since D is a block diagonal matrix, we can compute e^D by computing the exponentials of its blocks. If $D_i = 0$, we get $E_i = e^0 = +1$, and if

$$D_i = \begin{pmatrix} 0 & -\theta_i \\ \theta_i & 0 \end{pmatrix},$$

we showed earlier that

$$e^{D_i} = \begin{pmatrix} \cos\theta_i & -\sin\theta_i \\ \sin\theta_i & \cos\theta_i \end{pmatrix},$$

exactly the block E_i. Thus, $E = e^D$, and as a consequence,

$$e^A = e^{PDP^{-1}} = Pe^D P^{-1} = PEP^{-1} = PEP^\top = R.$$

This shows the surjectivity of the exponential. $\square$

When $n = 3$ (and A is skew-symmetric), it is possible to work out an explicit formula for e^A. For any 3×3 real skew-symmetric matrix

$$A = \begin{pmatrix} 0 & -c & b \\ c & 0 & -a \\ -b & a & 0 \end{pmatrix},$$

letting $\theta = \sqrt{a^2 + b^2 + c^2}$ and

$$B = \begin{pmatrix} a^2 & ab & ac \\ ab & b^2 & bc \\ ac & bc & c^2 \end{pmatrix},$$

we have the following result known as *Rodrigues's formula* (1840).

Lemma 18.6. *The exponential map* $\exp\colon \mathfrak{so}(3) \to \mathbf{SO}(3)$ *is given by*

$$e^A = \cos\theta\, I_3 + \frac{\sin\theta}{\theta}A + \frac{(1-\cos\theta)}{\theta^2}B,$$

or, equivalently, by

$$e^A = I_3 + \frac{\sin\theta}{\theta}A + \frac{(1-\cos\theta)}{\theta^2}A^2$$

if $\theta \neq 0$, *with* $e^{0_3} = I_3$.

Proof. Here is a proof sketch. First, prove that

$$A^2 = -\theta^2 I + B,$$
$$AB = BA = 0.$$

From the above, deduce that

$$A^3 = -\theta^2 A,$$

and for any $k \geq 0$,

$$A^{4k+1} = \theta^{4k}A,$$
$$A^{4k+2} = \theta^{4k}A^2,$$
$$A^{4k+3} = -\theta^{4k+2}A,$$
$$A^{4k+4} = -\theta^{4k+2}A^2.$$

Then prove the desired result by writing the power series for e^A and regrouping terms so that the power series for cos and sin show up. $\square$

The above formulae are the well-known formulae expressing a rotation of axis specified by the vector (a,b,c) and angle θ. Since the exponential is surjective, it is possible to write down an explicit formula for its inverse (but it is a multivalued function!). This has applications in kinematics, robotics, and motion interpolation.

18.3 Symmetric Matrices, Symmetric Positive Definite Matrices, and the Exponential Map

Recall that a real symmetric matrix is called *positive* (or *positive semidefinite*) if its eigenvalues are all positive or null, and *positive definite* if its eigenvalues are all strictly positive. We denote the vector space of real symmetric $n \times n$ matrices by $\mathbf{S}(n)$, the set of symmetric positive matrices by $\mathbf{SP}(n)$, and the set of symmetric positive definite matrices by $\mathbf{SPD}(n)$.

The next lemma shows that every symmetric positive definite matrix A is of the form e^B for some unique symmetric matrix B. The set of symmetric matrices is a vector space, but it is not a Lie algebra because the Lie bracket $[A,B]$ is not symmetric unless A and B commute, and the set of symmetric (positive) definite matrices is not a multiplicative group, so this result is of a different flavor as Theorem 18.1.

Lemma 18.7. *For every symmetric matrix B, the matrix e^B is symmetric positive definite. For every symmetric positive definite matrix A, there is a unique symmetric matrix B such that $A = e^B$.*

Proof. We showed earlier that

$$\left(e^B\right)^{\top} = e^{B^{\top}}.$$

If B is a symmetric matrix, then since $B^{\top} = B$, we get

$$\left(e^B\right)^{\top} = e^{B^{\top}} = e^B,$$

and e^B is also symmetric. Since the eigenvalues $\lambda_1, \ldots, \lambda_n$ of the symmetric matrix B are real and the eigenvalues of e^B are $e^{\lambda_1}, \ldots, e^{\lambda_n}$, and since $e^\lambda > 0$ if $\lambda \in \mathbb{R}$, e^B is positive definite.

If A is symmetric positive definite, by Theorem 12.8 there is an orthogonal matrix P such that $A = PDP^{\top}$, where D is a diagonal matrix

$$
D = \begin{pmatrix} \lambda_1 & & \cdots & \\ & \lambda_2 & \cdots & \\ \vdots & \vdots & \ddots & \vdots \\ & & \cdots & \lambda_n \end{pmatrix},
$$

where $\lambda_i > 0$, since A is positive definite. Letting

$$
L = \begin{pmatrix} \log \lambda_1 & & \cdots & \\ & \log \lambda_2 & \cdots & \\ \vdots & \vdots & \ddots & \vdots \\ & & \cdots & \log \lambda_n \end{pmatrix},
$$

it is obvious that $e^L = D$, with $\log \lambda_i \in \mathbb{R}$, since $\lambda_i > 0$.

Let

$$
B = PLP^\top.
$$

By Lemma 18.2, we have

$$
e^B = e^{PLP^\top} = e^{PLP^{-1}} = Pe^L P^{-1} = Pe^L P^\top = PDP^\top = A.
$$

Finally, we prove that if B_1 and B_2 are symmetric and $A = e^{B_1} = e^{B_2}$, then $B_1 = B_2$. Since B_1 is symmetric, there is an orthonormal basis $(u_1, \ldots, u_n)$ of eigenvectors of B_1. Let $\mu_1, \ldots, \mu_n$ be the corresponding eigenvalues. Similarly, there is an orthonormal basis $(v_1, \ldots, v_n)$ of eigenvectors of B_2. We are going to prove that B_1 and B_2 agree on the basis $(v_1, \ldots, v_n)$, thus proving that $B_1 = B_2$.

Let μ be some eigenvalue of B_2, and let $v = v_i$ be some eigenvector of B_2 associated with μ. We can write

$$
v = \alpha_1 u_1 + \cdots + \alpha_n u_n.
$$

Since v is an eigenvector of B_2 for μ and $A = e^{B_2}$, by Lemma 18.4

$$
A(v) = e^\mu v = e^\mu \alpha_1 u_1 + \cdots + e^\mu \alpha_n u_n.
$$

On the other hand,

$$
A(v) = A(\alpha_1 u_1 + \cdots + \alpha_n u_n) = \alpha_1 A(u_1) + \cdots + \alpha_n A(u_n),
$$

and since $A = e^{B_1}$ and $B_1(u_i) = \mu_i u_i$, by Lemma 18.4 we get

$$
A(v) = e^{\mu_1} \alpha_1 u_1 + \cdots + e^{\mu_n} \alpha_n u_n.
$$

Therefore, $\alpha_i = 0$ if $\mu_i \neq \mu$. Letting

$$
I = \{i \mid \mu_i = \mu, \, i \in \{1, \ldots, n\}\},
$$

we have

$$v = \sum_{i \in I} \alpha_i u_i.$$

Now,

$$B_1(v) = B_1\left(\sum_{i \in I} \alpha_i u_i\right) = \sum_{i \in I} \alpha_i B_1(u_i) = \sum_{i \in I} \alpha_i \mu_i u_i$$

$$= \sum_{i \in I} \alpha_i \mu u_i = \mu\left(\sum_{i \in I} \alpha_i u_i\right) = \mu v,$$

since $\mu_i = \mu$ when $i \in I$. Since v is an eigenvector of B_2 for μ,

$$B_2(v) = \mu v,$$

which shows that

$$B_1(v) = B_2(v).$$

Since the above holds for every eigenvector v_i, we have $B_1 = B_2$. $\square$

Lemma 18.7 can be reformulated as stating that the map $\exp\colon \mathbf{S}(n) \to \mathbf{SPD}(n)$ is a bijection. It can be shown that it is a homeomorphism.

It should be noted that Lemma 18.7 is a key ingredient in the *log-Euclidean framework* due to Arsigny, Fillard, Pennec and Ayache, which has important applications to medical imaging, especially diffusion tensor imaging (DTI) [2, 3, 4, 5].

In the case of invertible matrices, the polar form theorem can be reformulated as stating that there is a bijection between the topological space $\mathbf{GL}(n, \mathbb{R})$ of real $n \times n$ invertible matrices (also a group) and $\mathbf{O}(n) \times \mathbf{SPD}(n)$.

As a corollary of the polar form theorem (Theorem 13.1) and Lemma 18.7, we have the following result: For every invertible matrix A there is a unique orthogonal matrix R and a unique symmetric matrix S such that

$$A = R e^S.$$

Thus, we have a bijection between $\mathbf{GL}(n, \mathbb{R})$ and $\mathbf{O}(n) \times \mathbf{S}(n)$. But $\mathbf{S}(n)$ itself is isomorphic to $\mathbb{R}^{n(n+1)/2}$. Thus, there is a bijection between $\mathbf{GL}(n, \mathbb{R})$ and $\mathbf{O}(n) \times \mathbb{R}^{n(n+1)/2}$. It can also be shown that this bijection is a homeomorphism. This is an interesting fact. Indeed, this homeomorphism essentially reduces the study of the topology of $\mathbf{GL}(n, \mathbb{R})$ to the study of the topology of $\mathbf{O}(n)$. This is nice, since it can be shown that $\mathbf{O}(n)$ is compact.

In $A = R e^S$, if $\det(A) > 0$, then R must be a rotation matrix (i.e., $\det(R) = +1$), since $\det\left(e^S\right) > 0$. In particular, if $A \in \mathbf{SL}(n, \mathbb{R})$, since $\det(A) = \det(R) = +1$, the symmetric matrix S must have a null trace, i.e., $S \in \mathbf{S}(n) \cap \mathfrak{sl}(n, \mathbb{R})$. Thus, we have a bijection between $\mathbf{SL}(n, \mathbb{R})$ and $\mathbf{SO}(n) \times (\mathbf{S}(n) \cap \mathfrak{sl}(n, \mathbb{R}))$.

We can also use the results of Section 12.4 to show that the exponential map is a surjective map from the skew-Hermitian matrices to the unitary matrices.

18.4 The Lie Groups $\mathbf{GL}(n,\mathbb{C})$, $\mathbf{SL}(n,\mathbb{C})$, $\mathbf{U}(n)$, $\mathbf{SU}(n)$, the Lie Algebras $\mathfrak{gl}(n,\mathbb{C})$, $\mathfrak{sl}(n,\mathbb{C})$, $\mathfrak{u}(n)$, $\mathfrak{su}(n)$, and the Exponential Map

The set of complex invertible $n \times n$ matrices forms a group under multiplication, denoted by $\mathbf{GL}(n,\mathbb{C})$. The subset of $\mathbf{GL}(n,\mathbb{C})$ consisting of those matrices having determinant $+1$ is a subgroup of $\mathbf{GL}(n,\mathbb{C})$, denoted by $\mathbf{SL}(n,\mathbb{C})$. It is also easy to check that the set of complex $n \times n$ unitary matrices forms a group under multiplication, denoted by $\mathbf{U}(n)$. The subset of $\mathbf{U}(n)$ consisting of those matrices having determinant $+1$ is a subgroup of $\mathbf{U}(n)$, denoted by $\mathbf{SU}(n)$. We can also check that the set of complex $n \times n$ matrices with null trace forms a real vector space under addition, and similarly for the set of skew-Hermitian matrices and the set of skew-Hermitian matrices with null trace.

Definition 18.2. The group $\mathbf{GL}(n,\mathbb{C})$ is called the *general linear group*, and its subgroup $\mathbf{SL}(n,\mathbb{C})$ is called the *special linear group*. The group $\mathbf{U}(n)$ of unitary matrices is called the *unitary group*, and its subgroup $\mathbf{SU}(n)$ is called the *special unitary group*. The real vector space of complex $n \times n$ matrices with null trace is denoted by $\mathfrak{sl}(n,\mathbb{C})$, the real vector space of skew-Hermitian matrices is denoted by $\mathfrak{u}(n)$, and the real vector space $\mathfrak{u}(n) \cap \mathfrak{sl}(n,\mathbb{C})$ is denoted by $\mathfrak{su}(n)$.

Remarks:

(1) As in the real case, the groups $\mathbf{GL}(n,\mathbb{C})$, $\mathbf{SL}(n,\mathbb{C})$, $\mathbf{U}(n)$, and $\mathbf{SU}(n)$ are also topological groups (viewed as subspaces of $\mathbb{R}^{2n^2}$), and in fact, smooth real manifolds. Such objects are called *(real) Lie groups*. The real vector spaces $\mathfrak{sl}(n,\mathbb{C})$, $\mathfrak{u}(n)$, and $\mathfrak{su}(n)$ are *Lie algebras* associated with $\mathbf{SL}(n,\mathbb{C})$, $\mathbf{U}(n)$, and $\mathbf{SU}(n)$. The algebra structure is given by the *Lie bracket*, which is defined as

$$[A, B] = AB - BA.$$

(2) It is also possible to define complex Lie groups, which means that they are topological groups and smooth *complex* manifolds. It turns out that $\mathbf{GL}(n,\mathbb{C})$ and $\mathbf{SL}(n,\mathbb{C})$ are complex manifolds, but not $\mathbf{U}(n)$ and $\mathbf{SU}(n)$.

One should be very careful to observe that even though the Lie algebras $\mathfrak{sl}(n,\mathbb{C})$, $\mathfrak{u}(n)$, and $\mathfrak{su}(n)$ consist of matrices with complex coefficients, we view them as *real* vector spaces. The Lie algebra $\mathfrak{sl}(n,\mathbb{C})$ is also a complex vector space, but $\mathfrak{u}(n)$ and $\mathfrak{su}(n)$ are not! Indeed, if A is a skew-Hermitian matrix, iA is *not* skew-Hermitian, but Hermitian!

Again the Lie algebra achieves a "linearization" of the Lie group. In the complex case, the Lie algebras $\mathfrak{gl}(n,\mathbb{C})$ is the set of *all* complex $n \times n$ matrices, but $\mathfrak{u}(n) \neq \mathfrak{su}(n)$, because a skew-Hermitian matrix does not necessarily have a null trace.

The properties of the exponential map also play an important role in studying complex Lie groups. For example, it is clear that the map

$$\exp\colon \mathfrak{gl}(n,\mathbb{C}) \to \mathbf{GL}(n,\mathbb{C})$$

is well-defined, but this time, it is surjective! One way to prove this is to use the Jordan normal form. Similarly, since

$$\det\left(e^A\right) = e^{\operatorname{tr}(A)},$$

the map

$$\exp\colon \mathfrak{sl}(n,\mathbb{C}) \to \mathbf{SL}(n,\mathbb{C})$$

is well-defined, but it is not surjective! As we will see in the next theorem, the maps

$$\exp\colon \mathfrak{u}(n) \to \mathbf{U}(n)$$

and

$$\exp\colon \mathfrak{su}(n) \to \mathbf{SU}(n)$$

are well-defined and surjective.

Theorem 18.2. *The exponential maps*

$$\exp\colon \mathfrak{u}(n) \to \mathbf{U}(n) \quad \textit{and} \quad \exp\colon \mathfrak{su}(n) \to \mathbf{SU}(n)$$

are well-defined and surjective.

Proof. First, we need to prove that if A is a skew-Hermitian matrix, then e^A is a unitary matrix. For this, first check that

$$\left(e^A\right)^* = e^{A^*}.$$

Then, since $A^* = -A$, we get

$$\left(e^A\right)^* = e^{A^*} = e^{-A},$$

and so

$$\left(e^A\right)^* e^A = e^{-A} e^A = e^{-A+A} = e^{0_n} = I_n,$$

and similarly, $e^A \left(e^A\right)^* = I_n$, showing that e^A is unitary. Since

$$\det\left(e^A\right) = e^{\operatorname{tr}(A)},$$

if A is skew-Hermitian and has null trace, then $\det(e^A) = +1$.

For the surjectivity we will use Theorem 12.11. First, assume that A is a unitary matrix. By Theorem 12.11, there is a unitary matrix U and a diagonal matrix D such that $A = UDU^*$. Furthermore, since A is unitary, the entries $\lambda_1,\ldots,\lambda_n$ in D (the eigenvalues of A) have absolute value $+1$. Thus, the entries in D are of the form $\cos\theta + i\sin\theta = e^{i\theta}$. Thus, we can assume that D is a diagonal matrix of the form

$$
D = \begin{pmatrix} e^{i\theta_1} & & \cdots & \\ & e^{i\theta_2} & \cdots & \\ \vdots & \vdots & \ddots & \vdots \\ & & \cdots & e^{i\theta_p} \end{pmatrix}.
$$

If we let E be the diagonal matrix

$$
E = \begin{pmatrix} i\theta_1 & & \cdots & \\ & i\theta_2 & \cdots & \\ \vdots & \vdots & \ddots & \vdots \\ & & \cdots & i\theta_p \end{pmatrix}
$$

it is obvious that E is skew-Hermitian and that

$$
e^E = D.
$$

Then, letting $B = UEU^*$, we have

$$
e^B = A,
$$

and it is immediately verified that B is skew-Hermitian, since E is.

If A is a unitary matrix with determinant $+1$, since the eigenvalues of A are $e^{i\theta_1}, \ldots, e^{i\theta_p}$ and the determinant of A is the product

$$
e^{i\theta_1} \cdots e^{i\theta_p} = e^{i(\theta_1 + \cdots + \theta_p)}
$$

of these eigenvalues, we must have

$$
\theta_1 + \cdots + \theta_p = 0,
$$

and so, E is skew-Hermitian and has zero trace. As above, letting

$$
B = UEU^*,
$$

we have

$$
e^B = A,
$$

where B is skew-Hermitian and has null trace. $\quad\square$

We now extend the result of Section 18.3 to Hermitian matrices.

18.5 Hermitian Matrices, Hermitian Positive Definite Matrices, and the Exponential Map

Recall that a Hermitian matrix is called *positive* (or *positive semidefinite*) if its eigenvalues are all positive or null, and *positive definite* if its eigenvalues are all strictly positive. We denote the real vector space of Hermitian $n \times n$ matrices by $\mathbf{H}(n)$, the set of Hermitian positive matrices by $\mathbf{HP}(n)$, and the set of Hermitian positive definite matrices by $\mathbf{HPD}(n)$.

The next lemma shows that every Hermitian positive definite matrix A is of the form e^B for some unique Hermitian matrix B. As in the real case, the set of Hermitian matrices is a real vector space, but it is not a Lie algebra because the Lie bracket $[A,B]$ is not Hermitian unless A and B commute, and the set of Hermitian (positive) definite matrices is not a multiplicative group.

Lemma 18.8. *For every Hermitian matrix B, the matrix e^B is Hermitian positive definite. For every Hermitian positive definite matrix A, there is a unique Hermitian matrix B such that $A = e^B$.*

Proof. It is basically the same as the proof of Theorem 18.8, except that a Hermitian matrix can be written as $A = UDU^*$, where D is a real diagonal matrix and U is unitary instead of orthogonal. $\square$

Lemma 18.8 can be reformulated as stating that the map $\exp\colon \mathbf{H}(n) \to \mathbf{HPD}(n)$ is a bijection. In fact, it can be shown that it is a homeomorphism. In the case of complex invertible matrices, the polar form theorem can be reformulated as stating that there is a bijection between the topological space $\mathbf{GL}(n,\mathbb{C})$ of complex $n \times n$ invertible matrices (also a group) and $\mathbf{U}(n) \times \mathbf{HPD}(n)$. As a corollary of the polar form theorem and Lemma 18.8, we have the following result: For every complex invertible matrix A, there is a unique unitary matrix U and a unique Hermitian matrix S such that

$$A = U e^S.$$

Thus, we have a bijection between $\mathbf{GL}(n,\mathbb{C})$ and $\mathbf{U}(n) \times \mathbf{H}(n)$. But $\mathbf{H}(n)$ itself is isomorphic to $\mathbb{R}^{n^2}$, and so there is a bijection between $\mathbf{GL}(n,\mathbb{C})$ and $\mathbf{U}(n) \times \mathbb{R}^{n^2}$. It can also be shown that this bijection is a homeomorphism. This is an interesting fact. Indeed, this homeomorphism essentially reduces the study of the topology of $\mathbf{GL}(n,\mathbb{C})$ to the study of the topology of $\mathbf{U}(n)$. This is nice, since it can be shown that $\mathbf{U}(n)$ is compact (as a real manifold).

In the polar decomposition $A = U e^S$, we have $|\det(U)| = 1$, since U is unitary, and $\mathrm{tr}(S)$ is real, since S is Hermitian (since it is the sum of the eigenvalues of S, which are real), so that $\det(e^S) > 0$. Thus, if $\det(A) = 1$, we must have $\det(e^S) = 1$, which implies that $S \in \mathbf{H}(n) \cap \mathfrak{sl}(n,\mathbb{C})$. Thus, we have a bijection between $\mathbf{SL}(n,\mathbb{C})$ and $\mathbf{SU}(n) \times (\mathbf{H}(n) \cap \mathfrak{sl}(n,\mathbb{C}))$.

In the next section we study the group $\mathbf{SE}(n)$ of affine maps induced by orthogonal transformations, also called rigid motions, and its Lie algebra. We will show that the exponential map is surjective. The groups $\mathbf{SE}(2)$ and $\mathbf{SE}(3)$ play play a fundamental role in robotics, dynamics, and motion planning.

18.6 The Lie Group $\mathbf{SE}(n)$ and the Lie Algebra $\mathfrak{se}(n)$

First, we review the usual way of representing affine maps of $\mathbb{R}^n$ in terms of $(n+1) \times (n+1)$ matrices.

Definition 18.3. The set of affine maps ρ of $\mathbb{R}^n$, defined such that

$$\rho(X) = RX + U,$$

where R is a rotation matrix ($R \in \mathbf{SO}(n)$) and U is some vector in $\mathbb{R}^n$, is a group under composition called the group of *direct affine isometries, or rigid motions*, denoted by $\mathbf{SE}(n)$.

Every rigid motion can be represented by the $(n+1) \times (n+1)$ matrix

$$\begin{pmatrix} R & U \\ 0 & 1 \end{pmatrix}$$

in the sense that

$$\begin{pmatrix} \rho(X) \\ 1 \end{pmatrix} = \begin{pmatrix} R & U \\ 0 & 1 \end{pmatrix} \begin{pmatrix} X \\ 1 \end{pmatrix}$$

iff

$$\rho(X) = RX + U.$$

Definition 18.4. The vector space of real $(n+1) \times (n+1)$ matrices of the form

$$A = \begin{pmatrix} \Omega & U \\ 0 & 0 \end{pmatrix},$$

where Ω is a skew-symmetric matrix and U is a vector in $\mathbb{R}^n$, is denoted by $\mathfrak{se}(n)$.

Remark: The group $\mathbf{SE}(n)$ is a Lie group, and its Lie algebra turns out to be $\mathfrak{se}(n)$.

We will show that the exponential map $\exp \colon \mathfrak{se}(n) \to \mathbf{SE}(n)$ is surjective. First, we prove the following key lemma.

Lemma 18.9. *Given any $(n+1) \times (n+1)$ matrix of the form*

$$A = \begin{pmatrix} \Omega & U \\ 0 & 0 \end{pmatrix}$$

where Ω is any matrix and $U \in \mathbb{R}^n$,

$$A^k = \begin{pmatrix} \Omega^k & \Omega^{k-1}U \\ 0 & 0 \end{pmatrix},$$

where $\Omega^0 = I_n$. As a consequence,

$$e^A = \begin{pmatrix} e^\Omega & VU \\ 0 & 1 \end{pmatrix},$$

where

$$V = I_n + \sum_{k \geq 1} \frac{\Omega^k}{(k+1)!}.$$

Proof. A trivial induction on k shows that

$$A^k = \begin{pmatrix} \Omega^k & \Omega^{k-1}U \\ 0 & 0 \end{pmatrix}.$$

Then we have

$$\begin{aligned}
e^A &= \sum_{k \geq 0} \frac{A^k}{k!}, \\
&= I_{n+1} + \sum_{k \geq 1} \frac{1}{k!} \begin{pmatrix} \Omega^k & \Omega^{k-1}U \\ 0 & 0 \end{pmatrix}, \\
&= \begin{pmatrix} I_n + \sum_{k \geq 0} \frac{\Omega^k}{k!} & \sum_{k \geq 1} \frac{\Omega^{k-1}}{k!}U \\ 0 & 1 \end{pmatrix}, \\
&= \begin{pmatrix} e^\Omega & VU \\ 0 & 1 \end{pmatrix}.
\end{aligned}$$

$\square$

We can now prove our main theorem. We will need to prove that V is invertible when Ω is a skew-symmetric matrix. It would be tempting to write V as

$$V = \Omega^{-1}(e^\Omega - I).$$

Unfortunately, for odd n, a skew-symmetric matrix of order n is not invertible! Thus, we have to find another way of proving that V is invertible. However, observe that we have the following useful fact:

$$V = I_n + \sum_{k \geq 1} \frac{\Omega^k}{(k+1)!} = \int_0^1 e^{\Omega t} dt.$$

This is what we will use in Theorem 18.3 to prove surjectivity.

Theorem 18.3. *The exponential map*

$$\exp\colon \mathfrak{se}(n) \to \mathbf{SE}(n)$$

is well-defined and surjective.

Proof. Since Ω is skew-symmetric, e^{Ω} is a rotation matrix, and by Theorem 18.1, the exponential map

$$\exp\colon \mathfrak{so}(n) \to \mathbf{SO}(n)$$

is surjective. Thus, it remains to prove that for every rotation matrix R, there is some skew-symmetric matrix Ω such that $R = e^{\Omega}$ and

$$V = I_n + \sum_{k \geq 1} \frac{\Omega^k}{(k+1)!}$$

is invertible. By Theorem 12.9, for every skew-symmetric matrix Ω there is an orthogonal matrix P such that $\Omega = PDP^{\top}$, where D is a block diagonal matrix of the form

$$D = \begin{pmatrix} D_1 & & \cdots & \\ & D_2 & \cdots & \\ \vdots & \vdots & \ddots & \vdots \\ & & \cdots & D_p \end{pmatrix}$$

such that each block D_i is either 0 or a two-dimensional matrix of the form

$$D_i = \begin{pmatrix} 0 & -\theta_i \\ \theta_i & 0 \end{pmatrix}$$

where $\theta_i \in \mathbb{R}$, with $\theta_i > 0$. Actually, we can assume that $\theta_i \neq k2\pi$ for all $k \in \mathbb{Z}$, since when $\theta_i = k2\pi$ we have $e^{D_i} = I_2$, and D_i can be replaced by two one-dimensional blocks each consisting of a single zero. To compute V, since $\Omega = PDP^{\top} = PDP^{-1}$, observe that

$$V = I_n + \sum_{k \geq 1} \frac{\Omega^k}{(k+1)!}$$

$$= I_n + \sum_{k \geq 1} \frac{PD^k P^{-1}}{(k+1)!}$$

$$= P\left(I_n + \sum_{k \geq 1} \frac{D^k}{(k+1)!}\right) P^{-1}$$

$$= PWP^{-1},$$

where

$$W = I_n + \sum_{k \geq 1} \frac{D^k}{(k+1)!}.$$

We can compute

$$W = I_n + \sum_{k \geq 1} \frac{D^k}{(k+1)!} = \int_0^1 e^{Dt}\,dt,$$

by computing

$$W = \begin{pmatrix} W_1 & & & \cdots \\ & W_2 & & \cdots \\ \vdots & \vdots & \ddots & \vdots \\ & & & \cdots & W_p \end{pmatrix}$$

by blocks. Since

$$e^{D_i} = \begin{pmatrix} \cos\theta_i & -\sin\theta_i \\ \sin\theta_i & \cos\theta_i \end{pmatrix}$$

when D_i is a 2×2 skew-symmetric matrix and $W_i = \int_0^1 e^{D_i t}\,dt$, we get

$$W_i = \begin{pmatrix} \int_0^1 \cos(\theta_i t)dt & \int_0^1 -\sin(\theta_i t)dt \\ \int_0^1 \sin(\theta_i t)dt & \int_0^1 \cos(\theta_i t)dt \end{pmatrix} = \frac{1}{\theta_i}\begin{pmatrix} \sin(\theta_i t)\,|_0^1 & \cos(\theta_i t)\,|_0^1 \\ -\cos(\theta_i t)\,|_0^1 & \sin(\theta_i t)\,|_0^1 \end{pmatrix},$$

that is,

$$W_i = \frac{1}{\theta_i}\begin{pmatrix} \sin\theta_i & -(1-\cos\theta_i) \\ 1-\cos\theta_i & \sin\theta_i \end{pmatrix},$$

and $W_i = 1$ when $D_i = 0$. Now, in the first case, the determinant is

$$\frac{1}{\theta_i^2}\left((\sin\theta_i)^2 + (1-\cos\theta_i)^2\right) = \frac{2}{\theta_i^2}(1-\cos\theta_i),$$

which is nonzero, since $\theta_i \neq k2\pi$ for all $k \in \mathbb{Z}$. Thus, each W_i is invertible, and so is W, and thus, $V = PWP^{-1}$ is invertible. $\square$

In the case $n = 3$, given a skew-symmetric matrix

$$\Omega = \begin{pmatrix} 0 & -c & b \\ c & 0 & -a \\ -b & a & 0 \end{pmatrix},$$

letting $\theta = \sqrt{a^2 + b^2 + c^2}$, it it easy to prove that if $\theta = 0$, then

$$e^A = \begin{pmatrix} I_3 & U \\ 0 & 1 \end{pmatrix},$$

and that if $\theta \neq 0$ (using the fact that $\Omega^3 = -\theta^2\Omega$), then

$$e^\Omega = I_3 + \frac{\sin\theta}{\theta}\Omega + \frac{(1-\cos\theta)}{\theta^2}\Omega^2$$

and

$$V = I_3 + \frac{(1-\cos\theta)}{\theta^2}\Omega + \frac{(\theta-\sin\theta)}{\theta^3}\Omega^2.$$

Our next goal is to define *embedded submanifolds* and (linear) Lie groups. Before doing this, we believe that some readers might appreciate a review of the notion of the *derivative* of a function between two normed vector spaces.

18.7 The Derivative of a Function Between Normed Vector Spaces, a Review

In this brief section, we review some basic notions of differential calculus, in particular, the *derivative* of a function $f\colon E \to F$, where E and F are normed vector spaces. In most cases, $E = \mathbb{R}^n$ and $F = \mathbb{R}^m$. However, if we need to deal with infinite-dimensional manifolds, then it is necessary to allow E and F to be infinite-dimensional. This section can be omitted by readers already familiar with this standard material. We omit all proofs and refer the reader to standard analysis textbooks such as Lang [29, 28], Munkres [38], Abraham and Marsden [1], Choquet-Bruhat [13], Schwartz [43], or Cartan [10] for a complete exposition.

Let E and F be two *normed vector spaces*, let $A \subseteq E$ be some open subset of A, and let $a \in A$ be some element of A. Even though a is a vector, we may also call it a point.

The idea behind the derivative of the function f at a is that it is a *linear approximation* (actually, an *affine approximation*) of f in a small open set around a. The difficulty is to make sense of the quotient

$$\frac{f(a+h) - f(a)}{h},$$

where h is a vector. We circumvent this difficulty in two stages.

A first possibility is to consider the *directional derivative* with respect to a vector $u \neq 0$ in E.

We can consider the vector $f(a+tu) - f(a)$, where $t \in \mathbb{R}$ (or $t \in \mathbb{C}$). Now

$$\frac{f(a+tu) - f(a)}{t}$$

makes sense.

The idea is that in E, the points of the form $a + tu$, for t in some small closed interval $[r, s] \subseteq A$ containing a, form a line segment and that the image of this line segment defines a small curve segment on $f(A)$. This curve (segment) is defined by the map $t \mapsto f(a+tu)$, from $[r, s]$ to F, and the directional derivative $\mathrm{D}_u f(a)$ defines the direction of the tangent line at a to this curve.

Definition 18.5. Let E and F be two normed spaces, let A be a nonempty open subset of E, and let $f\colon A \to F$ be any function. For any $a \in A$, for any $u \neq 0$ in E, the *directional derivative of f at a with respect to the vector u*, denoted by $\mathrm{D}_u f(a)$, is the limit (if it exists)

$$\lim_{t \to 0,\, t \in U} \frac{f(a+tu) - f(a)}{t},$$

where $U = \{t \in \mathbb{R} \mid a+tu \in A,\, t \neq 0\}$ (or $U = \{t \in \mathbb{C} \mid a+tu \in A,\, t \neq 0\}$).

Since the map $t \mapsto a + tu$ is continuous, and since $A - \{a\}$ is open, the inverse image U of $A - \{a\}$ under the above map is open, and the definition of the limit in Definition 18.5 makes sense.

Remark: Since the notion of limit is purely topological, the existence and value of a directional derivative is independent of the choice of norms in E and F, as long as they are equivalent norms.

The directional derivative is sometimes called the *Gâteaux derivative*.

In the special case that $E = \mathbb{R}$, $F = \mathbb{R}$ and we let $u = 1$ (i.e., the real number 1, viewed as a vector), it is immediately verified that $D_1 f(a) = f'(a)$. When $E = \mathbb{R}$ (or $E = \mathbb{C}$) and F is any normed vector space, the derivative $D_1 f(a)$, also denoted by $f'(a)$, provides a suitable generalization of the notion of derivative.

However, when E has dimension ≥ 2, directional derivatives present a serious problem, which is that their definition is not sufficiently uniform. Indeed, there is no reason to believe that the directional derivatives with respect to all nonzero vectors u share something in common. As a consequence, a function can have all directional derivatives at a, and yet not be continuous at a. Two functions may have all directional derivatives in some open sets, and yet their composition may not. Thus, we introduce a more uniform notion.

Definition 18.6. Let E and F be two normed spaces, let A be a nonempty open subset of E, and let $f \colon A \to F$ be any function. For any $a \in A$, we say that f is *differentiable at* $a \in A$ if there are a linear continuous map $L \colon E \to F$ and a function $\varepsilon(h)$ such that

$$f(a+h) = f(a) + L(h) + \varepsilon(h)\|h\|$$

for every $a + h \in A$, where

$$\lim_{h \to 0, h \in U} \varepsilon(h) = 0,$$

with $U = \{h \in E \mid a + h \in A, h \neq 0\}$. The linear map L is denoted by $\mathrm{D}f(a)$, or $\mathrm{D}f_a$, or $df(a)$, or df_a, or $f'(a)$, and it is called the *Fréchet derivative*, or *total derivative*, or *derivative*, or *total differential*, or *differential*, of f at a.

Since the map $h \mapsto a + h$ from E to E is continuous, and since A is open in E, the inverse image U of $A - \{a\}$ under the above map is open in E, and it makes sense to say that

$$\lim_{h \to 0, h \in U} \varepsilon(h) = 0.$$

Note that for every $h \in U$, since $h \neq 0$, $\varepsilon(h)$ is uniquely determined, since

$$\varepsilon(h) = \frac{f(a+h) - f(a) - L(h)}{\|h\|},$$

and the value $\varepsilon(0)$ plays absolutely no role in this definition. It does no harm to assume that $\varepsilon(0) = 0$, and we will assume this from now on.

Remark: Since the notion of limit is purely topological, the existence and value of a derivative is independent of the choice of norms in E and F, as long as they are equivalent norms.

The following proposition shows that our new definition is consistent with the definition of the directional derivative and that *the continuous linear map L is unique*, if it exists.

Proposition 18.1. *Let E and F be two normed spaces, let A be a nonempty open subset of E, and let $f: A \to F$ be any function. For any $a \in A$, if $Df(a)$ is defined, then f is continuous at a and f has a directional derivative $D_u f(a)$ for every $u \neq 0$ in E. Furthermore,*

$$D_u f(a) = Df(a)(u),$$

and thus $Df(a)$ is uniquely defined.

Proof. If $L = Df(a)$ exists, then for any nonzero vector $u \in E$, because A is open, for any $t \in \mathbb{R} - \{0\}$ (or $t \in \mathbb{C} - \{0\}$) small enough, $a + tu \in A$, so

$$f(a+tu) = f(a) + L(tu) + \varepsilon(tu)\|tu\| = f(a) + tL(u) + |t|\varepsilon(tu)\|u\|,$$

which implies that

$$L(u) = \frac{f(a+tu) - f(a)}{t} - \frac{|t|}{t}\varepsilon(tu)\|u\|,$$

and since $\lim_{t \mapsto 0} \varepsilon(tu) = 0$, we deduce that

$$L(u) = Df(a)(u) = D_u f(a).$$

Because

$$f(a+h) = f(a) + L(h) + \varepsilon(h)\|h\|$$

for all h such that $\|h\|$ is small enough, L is continuous, and $\lim_{h \mapsto 0} \varepsilon(h)\|h\| = 0$, we have $\lim_{h \mapsto 0} f(a+h) = f(a)$, that is, f is continuous at a. $\quad\square$

Observe that the uniqueness of $Df(a)$ follows from Proposition 18.1. Also, when E is of finite dimension, it is easily shown that every linear map is continuous, and this assumption is then redundant.

If $Df(a)$ exists for every $a \in A$, we get a map

$$Df: A \to \mathscr{L}(E;F),$$

called the *derivative of f on A*, and also denoted by df. Here $\mathscr{L}(E;F)$ denotes the vector space of continuous linear maps from E to F.

When E is of finite dimension n, for any basis $(u_1, \ldots, u_n)$ of E, we can define the directional derivatives with respect to the vectors in the basis $(u_1, \ldots, u_n)$ (actually, we can also do it for an infinite basis). In this way, we obtain the definition of partial derivatives, as follows:

Definition 18.7. For any two normed spaces E and F, if E is of finite dimension n, then for every basis $(u_1, \ldots, u_n)$ for E, for every $a \in E$, for every function $f: E \to F$, the directional derivatives $D_{u_j} f(a)$ (if they exist) are called the *partial derivatives* *of f with respect to the basis* $(u_1, \ldots, u_n)$. The partial derivative $D_{u_j} f(a)$ is also denoted by $\partial_j f(a)$, or $\frac{\partial f}{\partial x_j}(a)$.

The notation $\frac{\partial f}{\partial x_j}(a)$ for a partial derivative, although customary and going back to Leibniz, is a "logical obscenity." Indeed, the variable x_j really has nothing to do with the formal definition. This is just another of these situations in which tradition is just too hard to overthrow!

We now consider a number of standard results about derivatives.

Proposition 18.2. *Given two normed spaces E and F, if $f\colon E \to F$ is a constant function, then $\mathrm{D}f(a) = 0$, for every $a \in E$. If $f\colon E \to F$ is a continuous affine map, then $\mathrm{D}f(a) = \overrightarrow{f}$, for every $a \in E$, where $\overrightarrow{f}$ denotes the linear map associated with f.*

Proposition 18.3. *Given a normed space E and a normed vector space F, for any two functions $f, g\colon E \to F$, for every $a \in E$, if $\mathrm{D}f(a)$ and $\mathrm{D}g(a)$ exist, then $\mathrm{D}(f + g)(a)$ and $\mathrm{D}(\lambda f)(a)$ exist, and*

$$\mathrm{D}(f+g)(a) = \mathrm{D}f(a) + \mathrm{D}g(a),$$
$$\mathrm{D}(\lambda f)(a) = \lambda \mathrm{D}f(a).$$

Proposition 18.4. *Given three normed vector spaces E_1, E_2, and F, for any continuous bilinear map $f\colon E_1 \times E_2 \to F$, for every $(a,b) \in E_1 \times E_2$, $\mathrm{D}f(a,b)$ exists, and for every $u \in E_1$ and $v \in E_2$,*

$$\mathrm{D}f(a,b)(u,v) = f(u,b) + f(a,v).$$

We now state the very useful *chain rule*.

Theorem 18.4. *Given three normed spaces E, F, and G, let A be an open set in E, and let B an open set in F. For any functions $f\colon A \to F$ and $g\colon B \to G$ such that $f(A) \subseteq B$, for any $a \in A$, if $\mathrm{D}f(a)$ exists and $\mathrm{D}g(f(a))$ exists, then $\mathrm{D}(g \circ f)(a)$ exists, and*

$$\mathrm{D}(g \circ f)(a) = \mathrm{D}g(f(a)) \circ \mathrm{D}f(a).$$

Theorem 18.4 has many interesting consequences. We mention two corollaries.

Proposition 18.5. *Given two normed spaces E and F, let A be some open subset in E, let B be some open subset in F, let $f\colon A \to B$ be a bijection from A to B, and assume that $\mathrm{D}f$ exists on A and that $\mathrm{D}f^{-1}$ exists on B. Then for every $a \in A$,*

$$\mathrm{D}f^{-1}(f(a)) = (\mathrm{D}f(a))^{-1}.$$

Proposition 18.5 has the remarkable consequence that the two vector spaces E and F have the same dimension. In other words, a local property, the existence of a bijection f between an open set A of E and an open set B of F such that f is differentiable on A and f^{-1} is differentiable on B, implies a global property, that the two vector spaces E and F have the same dimension.

If both E and F are of finite dimension, for any basis $(u_1, \ldots, u_n)$ of E and any basis $(v_1, \ldots, v_m)$ of F, every function $f\colon E \to F$ is determined by m functions $f_i\colon E \to \mathbb{R}$ (or $f_i\colon E \to \mathbb{C}$), where

$$f(x) = f_1(x)v_1 + \cdots + f_m(x)v_m,$$

for every $x \in E$. Then we get

$$Df(a)(u_j) = Df_1(a)(u_j)v_1 + \cdots + Df_i(a)(u_j)v_i + \cdots + Df_m(a)(u_j)v_m,$$

that is,

$$Df(a)(u_j) = \partial_j f_1(a)v_1 + \cdots + \partial_j f_i(a)v_i + \cdots + \partial_j f_m(a)v_m.$$

Since the jth column of the $m \times n$ matrix $J(f)(a)$ with respect to the bases $(u_1, \ldots, u_n)$ and $(v_1, \ldots, v_m)$ representing $Df(a)$ is equal to the components of the vector $Df(a)(u_j)$ over the basis $(v_1, \ldots, v_m)$, the linear map $Df(a)$ is determined by the $m \times n$ matrix $J(f)(a) = (\partial_j f_i(a))$:

$$J(f)(a) = \begin{pmatrix} \partial_1 f_1(a) & \partial_2 f_1(a) & \cdots & \partial_n f_1(a) \\ \partial_1 f_2(a) & \partial_2 f_2(a) & \cdots & \partial_n f_2(a) \\ \vdots & \vdots & \ddots & \vdots \\ \partial_1 f_m(a) & \partial_2 f_m(a) & \cdots & \partial_n f_m(a) \end{pmatrix},$$

or

$$J(f)(a) = \begin{pmatrix} \dfrac{\partial f_1}{\partial x_1}(a) & \dfrac{\partial f_1}{\partial x_2}(a) & \cdots & \dfrac{\partial f_1}{\partial x_n}(a) \\[2ex] \dfrac{\partial f_2}{\partial x_1}(a) & \dfrac{\partial f_2}{\partial x_2}(a) & \cdots & \dfrac{\partial f_2}{\partial x_n}(a) \\[2ex] \vdots & \vdots & \ddots & \vdots \\[2ex] \dfrac{\partial f_m}{\partial x_1}(a) & \dfrac{\partial f_m}{\partial x_2}(a) & \cdots & \dfrac{\partial f_m}{\partial x_n}(a) \end{pmatrix}.$$

This matrix is called the *Jacobian matrix* of Df at a. When $m = n$, the determinant $\det(J(f)(a))$ of $J(f)(a)$ is called the *Jacobian* (or *Jacobian determinant*) of $Df(a)$.

We know that this determinant depends only on $Df(a)$, and not on specific bases. However, partial derivatives give a means for computing it.

When $E = \mathbb{R}^n$ and $F = \mathbb{R}^m$, for any function $f \colon \mathbb{R}^n \to \mathbb{R}^m$, it is easy to compute the partial derivatives $\frac{\partial f_i}{\partial x_j}(a)$. We simply treat the function $f_i \colon \mathbb{R}^n \to \mathbb{R}$ as a function of its jth argument, leaving the others fixed, and compute the derivative as the usual derivative.

Example 18.1. For example, consider the function $f \colon \mathbb{R}^2 \to \mathbb{R}^2$, defined by

$$f(r, \theta) = (r\cos\theta, r\sin\theta).$$

Then we have

$$J(f)(r, \theta) = \begin{pmatrix} \cos\theta & -r\sin\theta \\ \sin\theta & r\cos\theta \end{pmatrix},$$

and the Jacobian (determinant) has value $\det(J(f)(r, \theta)) = r$.

In the case $E = \mathbb{R}$ (or $E = \mathbb{C}$), for any function $f \colon \mathbb{R} \to F$ (or $f \colon \mathbb{C} \to F$), the Jacobian matrix of $Df(a)$ is a column vector. In fact, this column vector is just $D_1 f(a)$. Then for every $\lambda \in \mathbb{R}$ (or $\lambda \in \mathbb{C}$), $Df(a)(\lambda) = \lambda D_1 f(a)$. This case is sufficiently important to warrant a definition.

Definition 18.8. Given a function $f \colon \mathbb{R} \to F$ (or $f \colon \mathbb{C} \to F$), where F is a normed space, the vector

$$Df(a)(1) = D_1 f(a)$$

is called the *vector derivative or velocity vector (in the real case)* at a. We usually identify $Df(a)$ with its Jacobian matrix $D_1 f(a)$, which is the column vector corresponding to $D_1 f(a)$. By abuse of notation, we also let $Df(a)$ denote the vector $Df(a)(1) = D_1 f(a)$.

When $E = \mathbb{R}$, the physical interpretation is that f defines a (parametric) curve that is the trajectory of some particle moving in $\mathbb{R}^m$ as a function of time, and the vector $D_1 f(a)$ is the *velocity* of the moving particle $f(t)$ at $t = a$.

Example 18.2.

1. When $A = (0,1)$ and $F = \mathbb{R}^3$, a function $f \colon (0,1) \to \mathbb{R}^3$ defines a (parametric) curve in $\mathbb{R}^3$. If $f = (f_1, f_2, f_3)$, its Jacobian matrix at $a \in \mathbb{R}$ is

$$J(f)(a) = \begin{pmatrix} \dfrac{\partial f_1}{\partial t}(a) \\[2ex] \dfrac{\partial f_2}{\partial t}(a) \\[2ex] \dfrac{\partial f_3}{\partial t}(a) \end{pmatrix}.$$

2. When $E = \mathbb{R}^2$ and $F = \mathbb{R}^3$, a function $\varphi \colon \mathbb{R}^2 \to \mathbb{R}^3$ defines a parametric surface. Letting $\varphi = (f, g, h)$, its Jacobian matrix at $a \in \mathbb{R}^2$ is

$$J(\varphi)(a) = \begin{pmatrix} \dfrac{\partial f}{\partial u}(a) & \dfrac{\partial f}{\partial v}(a) \\[2ex] \dfrac{\partial g}{\partial u}(a) & \dfrac{\partial g}{\partial v}(a) \\[2ex] \dfrac{\partial h}{\partial u}(a) & \dfrac{\partial h}{\partial v}(a) \end{pmatrix}.$$

3. When $E = \mathbb{R}^3$ and $F = \mathbb{R}$, for a function $f \colon \mathbb{R}^3 \to \mathbb{R}$, the Jacobian matrix at $a \in \mathbb{R}^3$ is

$$J(f)(a) = \begin{pmatrix} \dfrac{\partial f}{\partial x}(a) & \dfrac{\partial f}{\partial y}(a) & \dfrac{\partial f}{\partial z}(a) \end{pmatrix}.$$

More generally, when $f \colon \mathbb{R}^n \to \mathbb{R}$, the Jacobian matrix at $a \in \mathbb{R}^n$ is the row vector

$$J(f)(a) = \left(\frac{\partial f}{\partial x_1}(a) \; \cdots \; \frac{\partial f}{\partial x_n}(a) \right).$$

Its transpose is a column vector called the *gradient* of f at a, denoted by $\mathrm{grad} f(a)$ or $\nabla f(a)$. Then, given any $v \in \mathbb{R}^n$, note that

$$\mathrm{D}f(a)(v) = \frac{\partial f}{\partial x_1}(a) v_1 + \cdots + \frac{\partial f}{\partial x_n}(a) v_n = \mathrm{grad} f(a) \cdot v,$$

the scalar product of $\mathrm{grad} f(a)$ and v.

When E, F, and G have finite dimensions, $(u_1, \ldots, u_p)$ is a basis for E, $(v_1, \ldots, v_n)$ is a basis for F, and $(w_1, \ldots, w_m)$ is a basis for G, if A is an open subset of E, B is an open subset of F, for any functions $f\colon A \to F$ and $g\colon B \to G$ such that $f(A) \subseteq B$, for any $a \in A$, letting $b = f(a)$ and $h = g \circ f$, if $\mathrm{D}f(a)$ exists and $\mathrm{D}g(b)$ exists, then by Theorem 18.4, the Jacobian matrix $J(h)(a) = J(g \circ f)(a)$ with respect to the bases $(u_1, \ldots, u_p)$ and $(w_1, \ldots, w_m)$ is the product of the Jacobian matrices $J(g)(b)$ with respect to the bases $(v_1, \ldots, v_n)$ and $(w_1, \ldots, w_m)$, and $J(f)(a)$ with respect to the bases $(u_1, \ldots, u_p)$ and $(v_1, \ldots, v_n)$:

$$J(h)(a) = \begin{pmatrix} \frac{\partial g_1}{\partial y_1}(b) & \frac{\partial g_1}{\partial y_2}(b) & \cdots & \frac{\partial g_1}{\partial y_n}(b) \\[6pt] \frac{\partial g_2}{\partial y_1}(b) & \frac{\partial g_2}{\partial y_2}(b) & \cdots & \frac{\partial g_2}{\partial y_n}(b) \\[6pt] \vdots & \vdots & \ddots & \vdots \\[6pt] \frac{\partial g_m}{\partial y_1}(b) & \frac{\partial g_m}{\partial y_2}(b) & \cdots & \frac{\partial g_m}{\partial y_n}(b) \end{pmatrix} \begin{pmatrix} \frac{\partial f_1}{\partial x_1}(a) & \frac{\partial f_1}{\partial x_2}(a) & \cdots & \frac{\partial f_1}{\partial x_p}(a) \\[6pt] \frac{\partial f_2}{\partial x_1}(a) & \frac{\partial f_2}{\partial x_2}(a) & \cdots & \frac{\partial f_2}{\partial x_p}(a) \\[6pt] \vdots & \vdots & \ddots & \vdots \\[6pt] \frac{\partial f_n}{\partial x_1}(a) & \frac{\partial f_n}{\partial x_2}(a) & \cdots & \frac{\partial f_n}{\partial x_p}(a) \end{pmatrix}.$$

Thus, we have the familiar formula

$$\frac{\partial h_i}{\partial x_j}(a) = \sum_{k=1}^{k=n} \frac{\partial g_i}{\partial y_k}(b) \frac{\partial f_k}{\partial x_j}(a).$$

Given two normed spaces E and F of finite dimension, given an open subset A of E, if a function $f\colon A \to F$ is differentiable at $a \in A$, then its Jacobian matrix is well defined.

One should be warned that the converse is false. There are functions such that all the partial derivatives exist at some $a \in A$, yet the function is not differentiable at a, and not even continuous at a.

However, there are sufficient conditions on the partial derivatives for $\mathrm{D}f(a)$ to exist, namely, continuity of the partial derivatives. If f is differentiable on A, then f defines a function $\mathrm{D}f\colon A \to \mathscr{L}(E;F)$. It turns out that the continuity of the partial derivatives on A is a necessary and sufficient condition for $\mathrm{D}f$ to exist and to be continuous on A.

Theorem 18.5. *Given two normed affine spaces E and F, where E is of finite dimension n and where $(u_1, \ldots, u_n)$ is a basis of E, given any open subset A of E, given any function $f \colon A \to F$, the derivative $Df \colon A \to \mathscr{L}(E; F)$ is defined and continuous on A iff every partial derivative $\partial_j f$ (or $\frac{\partial f}{\partial x_j}$) is defined and continuous on A, for all j, $1 \le j \le n$. As a corollary, if F is of finite dimension m, and $(v_1, \ldots, v_m)$ is a basis of F, the derivative $Df \colon A \to \mathscr{L}(E; F)$ is defined and continuous on A iff every partial derivative $\partial_j f_i$ (or $\frac{\partial f_i}{\partial x_j}$) is defined and continuous on A, for all i, j, $1 \le i \le m$, $1 \le j \le n$.*

Definition 18.9. Given two normed affine spaces E and F and an open subset A of E, we say that a function $f \colon A \to F$ is a *C^0-function on A* if f is continuous on A. We say that $f \colon A \to F$ is a *C^1-function on A* if Df exists and is continuous on A.

Let E and F be two normed affine spaces, let $U \subseteq E$ be an open subset of E and let $f \colon E \to F$ be a function such that $Df(a)$ exists for all $a \in U$. If $Df(a)$ is injective for all $a \in U$, we say that f is an *immersion* (on U), and if $Df(a)$ is surjective for all $a \in U$, we say that f is a *submersion* (on U).

When E and F are finite-dimensional with $\dim(E) = n$ and $\dim(F) = m$, if $m \ge n$, then f is an immersion iff the Jacobian matrix $J(f)(a)$ has full rank (n) for all $a \in E$, and if $n \ge m$, then f is a submersion iff the Jacobian matrix $J(f)(a)$ has full rank (m) for all $a \in E$.

A very important theorem is the inverse function theorem. In order for this theorem to hold for infinite-dimensional spaces, it is necessary to assume that our normed spaces are complete.

Given a normed vector space E we say that a sequence $(u_n)_n$ with $u_n \in E$ is a *Cauchy sequence* if for every $\varepsilon > 0$, there is some $N > 0$ such that for all $m, n \ge N$,

$$\|u_n - u_m\| < \varepsilon.$$

A normed vector space E is *complete* iff every Cauchy sequence converges. A complete normed vector space is also called a *Banach space*, after Stefan Banach (1892–1945).

Fortunately, $\mathbb{R}, \mathbb{C}$, and every finite-dimensional (real or complex) normed vector space is complete. A real (resp. complex) vector space E is a real (resp. complex) *Hilbert space* if it is complete as a normed space with the norm $\|u\| = \sqrt{\langle u, u \rangle}$ induced by its Euclidean (resp. Hermitian) inner product (of course, positive definite).

Definition 18.10. Given two topological spaces E and F and an open subset A of E, we say that a function $f \colon A \to F$ is a *local homeomorphism from A to F* if for every $a \in A$, there are an open set $U \subseteq A$ containing a and an open set V containing $f(a)$ such that f is a homeomorphism from U to $V = f(U)$. If B is an open subset of F, we say that $f \colon A \to F$ is a *(global) homeomorphism from A to B* if f is a homeomorphism from A to $B = f(A)$.

If E and F are normed spaces, we say that $f \colon A \to F$ is a *local diffeomorphism from A to F* if for every $a \in A$, there are an open set $U \subseteq A$ containing a and an open set V containing $f(a)$ such that f is a bijection from U to V, f is a C^1-function on U, and f^{-1} is a C^1-function on $V = f(U)$. We say that $f \colon A \to F$ is a *(global) diffeomorphism from A to B* if f is a homeomorphism from A to $B = f(A)$, f is a C^1-function on A, and f^{-1} is a C^1-function on B.

Note that a local diffeomorphism is a local homeomorphism. Also, as a consequence of Proposition 18.5, if f is a diffeomorphism on A, then $\mathrm{D}f(a)$ is a bijection for every $a \in A$.

Theorem 18.6. *(Inverse function theorem) Let E and F be complete normed spaces, let A be an open subset of E, and let $f \colon A \to F$ be a C^1-function on A. The following properties hold:*

(1) For every $a \in A$, if $\mathrm{D}f(a)$ is invertible, then there exist some open subset $U \subseteq A$ containing a and some open subset V of F containing $f(a)$ such that f is a diffeomorphism from U to $V = f(U)$. Furthermore,

$$\mathrm{D}f^{-1}(f(a)) = (\mathrm{D}f(a))^{-1}.$$

For every neighborhood N of a, the image $f(N)$ of N is a neighborhood of $f(a)$, and for every open ball $U \subseteq A$ of center a, the image $f(U)$ of U contains some open ball of center $f(a)$.

(2) If $\mathrm{D}f(a)$ is invertible for every $a \in A$, then $B = f(A)$ is an open subset of F, and f is a local diffeomorphism from A to B. Furthermore, if f is injective, then f is a diffeomorphism from A to B.

Part (1) of Theorem 18.6 is often referred to as the "(local) inverse function theorem." It plays an important role in the study of manifolds and (ordinary) differential equations.

If E and F are both of finite dimension, the case in which $\mathrm{D}f(a)$ is just injective or just surjective is also important for defining manifolds, using implicit definitions.

We finally reach the best vantage point of our hike, the formal definition of (linear) Lie groups and Lie algebras.

18.8 Finale: Manifolds, Lie Groups, and Lie Algebras

In this section we attempt to define precisely Lie groups and Lie algebras. One of the reasons that Lie groups are nice is that they have a differential structure, which means that the notion of tangent space makes sense at any point of the group. Furthermore, the tangent space at the identity happens to have some algebraic structure, that of a Lie algebra. Roughly, the tangent space at the identity provides a "linearization" of the Lie group, and it turns out that many properties of a Lie group are reflected in its Lie algebra, and that the loss of information is not too severe. The challenge that we are facing is that unless our readers are already familiar with manifolds, the amount of basic differential geometry required to define Lie groups and Lie algebras in full generality is overwhelming.

Fortunately, all the Lie groups that we need to consider are subspaces of $\mathbb{R}^N$ for some sufficiently large N. In fact, they are all isomorphic to subgroups of $\mathbf{GL}(N, \mathbb{R})$ for some suitable N, even $\mathbf{SE}(n)$, which is isomorphic to a subgroup of $\mathbf{SL}(n+1)$. Such groups are called *linear Lie groups* (or *matrix groups*). Since the groups under consideration are subspaces of $\mathbb{R}^N$, we do not need the definition of an abstract manifold. We just have to define embedded submanifolds (also called submanifolds) of $\mathbb{R}^N$ (in the case of $\mathbf{GL}(n, \mathbb{R})$, $N = n^2$). This is the path that we will follow.

In general, the difficult part in proving that a subgroup of $\mathbf{GL}(n, \mathbb{R})$ is a Lie group is to prove that it is a manifold. Fortunately, there is a characterization of the linear groups that obviates much of the work. This characterization rests on two theorems. First, a Lie subgroup H of a Lie group G (where H is an embedded submanifold of G) is closed in G (see Warner [47], Chapter 3, Theorem 3.21, page 97). Second, a theorem of Von Neumann and Cartan asserts that a closed subgroup of $\mathbf{GL}(n, \mathbb{R})$ is an embedded submanifold, and thus, a Lie group (see Warner [47], Chapter 3, Theorem 3.42, page 110). Thus, a linear Lie group is a closed subgroup of $\mathbf{GL}(n, \mathbb{R})$.

Since our Lie groups are subgroups (or isomorphic to subgroups) of $\mathbf{GL}(n, \mathbb{R})$ for some suitable n, it is easy to define the Lie algebra of a Lie group using curves. This approach to define the Lie algebra of a matrix group is followed by a number of authors, such as Curtis [14]. However, Curtis is rather cavalier, since he does not explain why the required curves actually exist, and thus, according to his definition, Lie algebras could be the trivial vector space! Although we will not prove the theorem of Von Neumann and Cartan, we feel that it is important to make clear why the definitions make sense, i.e., why we are not dealing with trivial objects.

A small annoying technical problem will arise in our approach, the problem with discrete subgroups. If A is a subset of $\mathbb{R}^N$, recall that A inherits a topology from $\mathbb{R}^N$ called the *subspace topology*, and defined such that a subset V of A is open if

$$V = A \cap U$$

for some open subset U of $\mathbb{R}^N$. A point $a \in A$ is said to be *isolated* if there is there is some open subset U of $\mathbb{R}^N$ such that

$$\{a\} = A \cap U,$$

in other words, if $\{a\}$ is an open set in A.

The group $\mathbf{GL}(n, \mathbb{R})$ of real invertible $n \times n$ matrices can be viewed as a subset of $\mathbb{R}^{n^2}$, and as such, it is a topological space under the subspace topology (in fact, a dense open subset of $\mathbb{R}^{n^2}$). One can easily check that multiplication and the inverse operation are continuous, and in fact smooth (i.e., C^∞-continuously differentiable). This makes $\mathbf{GL}(n, \mathbb{R})$ a *topological group*. Any subgroup G of $\mathbf{GL}(n, \mathbb{R})$ is also a topological space under the subspace topology. A subgroup G is called a *discrete subgroup* if it has some isolated point. This turns out to be equivalent to the fact that every point of G is isolated, and thus, G has the discrete topology (every subset of G is open). Now, because $\mathbf{GL}(n, \mathbb{R})$ is Hausdorff, it can be shown that every discrete subgroup of $\mathbf{GL}(n, \mathbb{R})$ is closed (which means that its complement is open). Thus, discrete subgroups of $\mathbf{GL}(n,\mathbb{R})$ are Lie groups! But these are not very interesting Lie groups, and so we will consider only closed subgroups of $\mathbf{GL}(n,\mathbb{R})$ that are not discrete.

Let us now review the definition of an embedded submanifold. For simplicity, we restrict our attention to smooth manifolds. For detailed presentations, see DoCarmo [15, 16], Milnor [35], Lee [30], Tu [46], Marsden and Ratiu [33], Guillemin and Pollack [20], Berger and Gostiaux [8], or Warner [47]. For the sake of brevity, we use the terminology *manifold* (but other authors would say *embedded submanifolds*, or something like that).

The intuition behind the notion of a smooth manifold in $\mathbb{R}^N$ is that a subspace M is a manifold of dimension m if every point $p \in M$ is contained in some open subset set U of M (in the subspace topology) that can be parametrized by some function $\varphi \colon \Omega \to U$ from some open subset Ω of the origin in $\mathbb{R}^m$, and that φ has some nice properties that allow the definition of smooth functions on M and of the tangent space at p. For this, φ has to be at least a homeomorphism, but more is needed: φ must be smooth, and the derivative $\varphi'(0_m)$ at the origin must be injective (letting $0_m = \underbrace{(0,\ldots,0)}_{m}$).

Definition 18.11. Given any integers N, m, with $N \geq m \geq 1$, an *m-dimensional smooth manifold in $\mathbb{R}^N$, for short a manifold*, is a nonempty subset M of $\mathbb{R}^N$ such that for every point $p \in M$ there are two open subsets $\Omega \subseteq \mathbb{R}^m$ and $U \subseteq M$, with $p \in U$, and a smooth function $\varphi \colon \Omega \to \mathbb{R}^N$ such that φ is a homeomorphism between Ω and $U = \varphi(\Omega)$, and $\varphi'(t_0)$ is injective, where $t_0 = \varphi^{-1}(p)$. The function $\varphi \colon \Omega \to U$ is called a *(local) parametrization of M at p*. If $0_m \in \Omega$ and $\varphi(0_m) = p$, we say that $\varphi \colon \Omega \to U$ is *centered at p*.

Recall that $M \subseteq \mathbb{R}^N$ is a topological space under the subspace topology, and U is some open subset of M in the subspace topology, which means that $U = M \cap W$ for some open subset W of $\mathbb{R}^N$. Since $\varphi \colon \Omega \to U$ is a homeomorphism, it has an inverse $\varphi^{-1} \colon U \to \Omega$ that is also a homeomorphism, called a *(local) chart*. Since $\Omega \subseteq \mathbb{R}^m$, for every point $p \in M$ and every parametrization $\varphi \colon \Omega \to U$ of M at p, we have $\varphi^{-1}(p) = (z_1,\ldots,z_m)$ for some $z_i \in \mathbb{R}$, and we call $z_1,\ldots,z_m$ the *local coordinates of p (with respect to φ^{-1})*. We often refer to a manifold M without explicitly specifying its dimension (the integer m).

Intuitively, a chart provides a "flattened" local map of a region on a manifold. For instance, in the case of surfaces (2-dimensional manifolds), a chart is analogous to a planar map of a region on the surface. For a concrete example, consider a map giving a planar representation of a country, a region on the earth, a curved surface.

Remark: We could allow $m = 0$ in definition 18.11. If so, a manifold of dimension 0 is just a set of isolated points, and thus it has the discrete topology. In fact, it can be shown that a discrete subset of $\mathbb{R}^N$ is countable. Such manifolds are not very exciting, but they do correspond to discrete subgroups.

Example 18.3. The unit sphere S^2 in $\mathbb{R}^3$ defined such that

$$S^2 = \{(x,y,z) \in \mathbb{R}^3 \mid x^2 + y^2 + z^2 = 1\}$$

is a smooth 2-manifold, because it can be parametrized using the following two maps φ_1 and φ_2:

$$\varphi_1 : (u,v) \mapsto \left(\frac{2u}{u^2+v^2+1}, \frac{2v}{u^2+v^2+1}, \frac{u^2+v^2-1}{u^2+v^2+1} \right)$$

and

$$\varphi_2 : (u,v) \mapsto \left(\frac{2u}{u^2+v^2+1}, \frac{2v}{u^2+v^2+1}, \frac{1-u^2-v^2}{u^2+v^2+1} \right).$$

The map φ_1 corresponds to the inverse of the stereographic projection from the north pole $N = (0,0,1)$ onto the plane $z = 0$, and the map φ_2 corresponds to the inverse of the stereographic projection from the south pole $S = (0,0,-1)$ onto the plane $z = 0$, as illustrated in Figure 18.1. We leave as an exercise to check that the map φ_1 parametrizes $S^2 - \{N\}$ and that the map φ_2 parametrizes $S^2 - \{S\}$ (and that they are smooth, homeomorphisms, etc.). Using φ_1, the open lower hemisphere is parametrized by the open disk of center O and radius 1 contained in the plane $z = 0$.

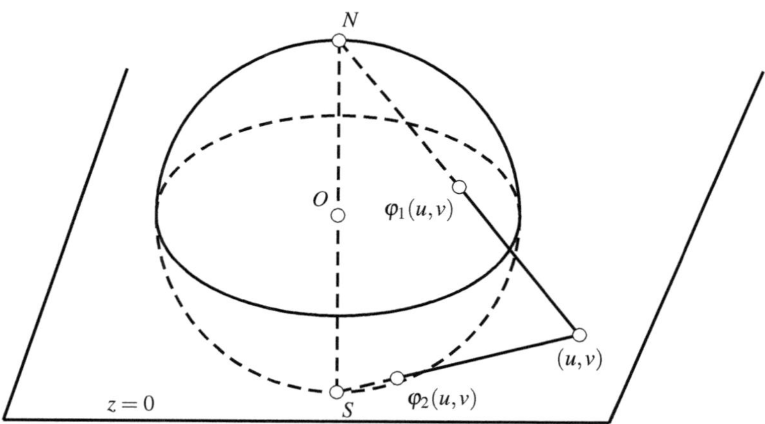

Fig. 18.1 Inverse stereographic projections.

The chart φ_1^{-1} assigns local coordinates to the points in the open lower hemisphere. If we draw a grid of coordinate lines parallel to the x and y axes inside the open unit disk and map these lines onto the lower hemisphere using φ_1, we get curved lines on the lower hemisphere. These "coordinate lines" on the lower hemisphere provide local coordinates for every point on the lower hemisphere. For this reason, older books often talk about *curvilinear coordinate systems* to mean the coordinate lines on a surface induced by a chart. We urge our readers to define a manifold structure on a torus. This can be done using four charts.

Every open subset of $\mathbb{R}^N$ is a manifold in a trivial way. Indeed, we can use the inclusion map as a parametrization. In particular, $\mathbf{GL}(n, \mathbb{R})$ is an open subset of $\mathbb{R}^{n^2}$, since its complement is closed (the set of invertible matrices is the inverse image of the determinant function, which is continuous). Thus, $\mathbf{GL}(n, \mathbb{R})$ is a manifold. We can view $\mathbf{GL}(n, \mathbb{C})$ as a subset of $\mathbb{R}^{(2n)^2}$ using the embedding defined as follows: For every complex $n \times n$ matrix A, construct the real $2n \times 2n$ matrix such that every entry $a + ib$ in A is replaced by the 2×2 block

$$\begin{pmatrix} a & -b \\ b & a \end{pmatrix},$$

where $a, b \in \mathbb{R}$. It is immediately verified that this map is in fact a group isomorphism. Thus, we can view $\mathbf{GL}(n, \mathbb{C})$ as a subgroup of $\mathbf{GL}(2n, \mathbb{R})$, and as a manifold in $\mathbb{R}^{(2n)^2}$.

A 1-manifold is called a *(smooth) curve*, and a 2-manifold is called a *(smooth) surface* (although some authors require that they also be connected).

The following two lemmas provide the link with the definition of an abstract manifold. The first lemma is easily proved using the inverse function theorem.

Lemma 18.10. *Given an m-dimensional manifold M in $\mathbb{R}^N$, for every $p \in M$ there are two open sets $O, W \subseteq \mathbb{R}^N$ with $0_N \in O$ and $p \in M \cap W$, and a smooth diffeomorphism $\varphi \colon O \to W$, such that $\varphi(0_N) = p$ and*

$$\varphi(O \cap (\mathbb{R}^m \times \{0_{N-m}\})) = M \cap W.$$

The next lemma is easily proved from Lemma 18.10 (see Berger and Gostiaux [8], Theorem 2.1.9, or DoCarmo [16], Chapter 0, Section 4). It is a key technical result used to show that interesting properties of maps between manifolds do not depend on parametrizations.

Lemma 18.11. *Given an m-dimensional manifold M in $\mathbb{R}^N$, for every $p \in M$ and any two parametrizations $\varphi_1 \colon \Omega_1 \to U_1$ and $\varphi_2 \colon \Omega_2 \to U_2$ of M at p, if $U_1 \cap U_2 \neq \emptyset$, the map $\varphi_2^{-1} \circ \varphi_1 \colon \varphi_1^{-1}(U_1 \cap U_2) \to \varphi_2^{-1}(U_1 \cap U_2)$ is a smooth diffeomorphism.*

The maps $\varphi_2^{-1} \circ \varphi_1 \colon \varphi_1^{-1}(U_1 \cap U_2) \to \varphi_2^{-1}(U_1 \cap U_2)$ are called *transition maps*. Lemma 18.11 is illustrated in Figure 18.2.

Using Definition 18.11, it may be quite hard to prove that a space is a manifold. Therefore, it is handy to have alternative characterizations such as those given in the next proposition:

Proposition 18.6. *A subset $M \subseteq \mathbb{R}^{m+k}$ is an m-dimensional manifold iff either*

(1) for every $p \in M$, there are some open subset $W \subseteq \mathbb{R}^{m+k}$ with $p \in W$ and a (smooth) submersion $f \colon W \to \mathbb{R}^k$ such that $W \cap M = f^{-1}(0)$,
or

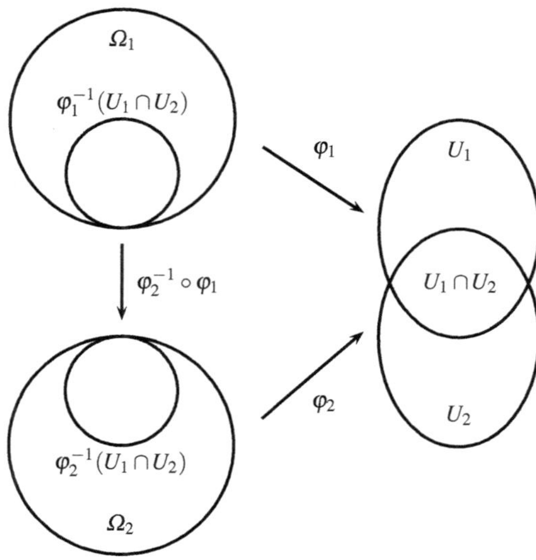

Fig. 18.2 Parametrizations and transition functions.

(2) for every $p \in M$, there are some open subset $W \subseteq \mathbb{R}^{m+k}$ with $p \in W$ and a (smooth) map $f: W \to \mathbb{R}^k$ such that $f'(p)$ is surjective and $W \cap M = f^{-1}(0)$.

Observe that condition (2), although apparently weaker than condition (1), is in fact equivalent to it, but more convenient in practice. This is because to say that $f'(p)$ is surjective means that the Jacobian matrix of $f'(p)$ has rank m, which means that some determinant is nonzero, and because the determinant function is continuous, this must hold in some open subset $W_1 \subseteq W$ containing p. Consequenly, the restriction f_1 of f to W_1 is indeed a submersion and $f_1^{-1}(0) = W_1 \cap f^{-1}(0) = W_1 \cap W \cap M = W_1 \cap M$.

A proof of Proposition 18.6 can be found in Lafontaine [27] or Berger and Gostiaux [8]. Lemma 18.10 and Proposition 18.6 are actually *equivalent* to Definition 18.11. This equivalence is also proved in Lafontaine [27] and Berger and Gostiaux [8].

The proof, which is somewhat illuminating, is based on two technical lemmas that are proved using the inverse function theorem (see Problem 18.24 and, for some help, Guillemin and Pollack [20], Chapter 1, Sections 3 and 4).

Lemma 18.12. *Let $U \subseteq \mathbb{R}^m$ be an open subset of $\mathbb{R}^m$ and pick some $a \in U$. If $f: U \to \mathbb{R}^n$ is a smooth immersion at a, i.e., df_a is injective (so $m \le n$), then there are an open set $V \subseteq \mathbb{R}^n$ with $f(a) \in V$, an open subset $U' \subseteq U$ with $a \in U'$ and $f(U') \subseteq V$, an open subset $O \subseteq \mathbb{R}^{n-m}$, and a diffeomorphism $\theta: V \to U' \times O$ such that*

$$\theta(f(x_1,\ldots,x_m)) = (x_1,\ldots,x_m,0,\ldots,0),$$

for all $(x_1,\ldots,x_m) \in U'$.

Lemma 18.13. *Let $W \subseteq \mathbb{R}^m$ be an open subset of $\mathbb{R}^m$ and pick some $a \in W$. If $f\colon W \to \mathbb{R}^n$ is a smooth submersion at a, i.e., df_a is surjective (so $m \geq n$), then there are an open set $V \subseteq W \subseteq \mathbb{R}^m$ with $a \in V$ and a diffeomorphism ψ with domain $O \subseteq \mathbb{R}^m$ such that $\psi(O) = V$ and*

$$f(\psi(x_1,\ldots,x_m)) = (x_1,\ldots,x_n),$$

for all $(x_1,\ldots,x_m) \in O$.

Using Lemmas 18.12 and 18.13, we can prove the following theorem, which confirms that all our characterizations of a manifold are equivalent.

Theorem 18.7. *A nonempty subset $M \subseteq \mathbb{R}^N$ is an m-manifold (with $1 \leq m \leq N$) iff any of the following conditions hold:*

(1) For every $p \in M$, there are two open subsets $\Omega \subseteq \mathbb{R}^m$ and $U \subseteq M$, with $p \in U$, and a smooth function $\varphi\colon \Omega \to \mathbb{R}^N$ such that φ is a homeomorphism between Ω and $U = \varphi(\Omega)$, and $\varphi'(0)$ is injective, where $p = \varphi(0)$.

(2) For every $p \in M$, there are two open sets $O, W \subseteq \mathbb{R}^N$ with $0_N \in O$ and $p \in M \cap W$, and a smooth diffeomorphism $\varphi\colon O \to W$, such that $\varphi(0_N) = p$ and

$$\varphi(O \cap (\mathbb{R}^m \times \{0_{N-m}\})) = M \cap W.$$

(3) For every $p \in M$, there are some open subset $W \subseteq \mathbb{R}^N$ with $p \in W$ and a smooth submersion $f\colon W \to \mathbb{R}^{N-m}$ such that $W \cap M = f^{-1}(0)$.

(4) For every $p \in M$, there are some open subset $W \subseteq \mathbb{R}^N$ and $N - m$ smooth functions $f_i\colon W \to \mathbb{R}$ such that the linear forms $df_1(p),\ldots,df_{N-m}(p)$ are linearly independent and
$$W \cap M = f_1^{-1}(0) \cap \cdots \cap f_{N-m}^{-1}(0).$$

Proof. If (1) holds, then by Lemma 18.12, replacing Ω by a smaller open subset $\Omega' \subseteq \Omega$ if necessary, there are some open subset $V \subseteq \mathbb{R}^N$ with $p \in V$ and $\varphi(\Omega') \subseteq V$, an open subset $O \subseteq \mathbb{R}^{N-m}$, and some diffeomorphism $\theta\colon V \to \Omega' \times O$ such that

$$(\theta \circ \varphi)(x_1,\ldots,x_m) = (x_1,\ldots,x_m,0,\ldots,0),$$

for all $(x_1,\ldots,x_m) \in \Omega'$. Observe that the above condition implies that

$$(\theta \circ \varphi)(\Omega') = \theta(V) \cap (\mathbb{R}^m \times \{(0,\ldots,0)\}).$$

Since φ is a homeomorphism between Ω and its image in M and since $\Omega' \subseteq \Omega$ is an open subset, $\varphi(\Omega') = M \cap W'$ for some open subset $W' \subseteq \mathbb{R}^N$, so if we let $W = V \cap W'$, because $\varphi(\Omega') \subseteq V$, it follows that $\varphi(\Omega') = M \cap W$ and

$$\theta(W \cap M) = \theta(\varphi(\Omega')) = \theta(V) \cap (\mathbb{R}^m \times \{(0,\ldots,0)\}).$$

However, θ is injective and $\theta(W \cap M) \subseteq \theta(W)$, so

$$\theta(W \cap M) = \theta(W) \cap \theta(V) \cap (\mathbb{R}^m \times \{(0,\ldots,0)\})$$
$$= \theta(W \cap V) \cap (\mathbb{R}^m \times \{(0,\ldots,0)\})$$
$$= \theta(W) \cap (\mathbb{R}^m \times \{(0,\ldots,0)\}).$$

If we let $O = \theta(W)$, we get

$$\theta^{-1}(O \cap (\mathbb{R}^m \times \{(0,\ldots,0)\})) = M \cap W,$$

which is (2).

If (2) holds, we can write $\varphi^{-1} = (f_1,\ldots,f_N)$, and because $\varphi^{-1}\colon W \to O$ is a diffeomorphism, $df_1(q),\ldots,df_N(q)$ are linearly independent for all $q \in W$, so the map

$$f = (f_{m+1},\ldots,f_N)$$

is a submersion $f\colon W \to \mathbb{R}^{N-m}$, and we have $f(x) = 0$ iff $f_{m+1}(x) = \cdots = f_N(x) = 0$ iff

$$\varphi^{-1}(x) = (f_1(x),\ldots,f_m(x),0,\ldots,0)$$

iff $\varphi^{-1}(x) \in O \cap (\mathbb{R}^m \times \{0_{N-m}\})$ iff $x \in \varphi(O \cap (\mathbb{R}^m \times \{0_{N-m}\}) = M \cap W$, because

$$\varphi(O \cap (\mathbb{R}^m \times \{0_{N-m}\})) = M \cap W.$$

Thus, $M \cap W = f^{-1}(0)$, which is (3).

The proof that (3) implies (2) uses Lemma 18.13 instead of Lemma 18.12. If $f\colon W \to \mathbb{R}^{N-m}$ is the submersion such that $M \cap W = f^{-1}(0)$ given by (3), then by Lemma 18.13, there are open subsets $V \subseteq W$, $O \subseteq \mathbb{R}^N$ and a diffeomorphism $\psi\colon O \to V$ such that

$$f(\psi(x_1,\ldots,x_N)) = (x_1,\ldots,x_{N-m})$$

for all $(x_1,\ldots,x_N) \in O$. If σ is the permutation of variables given by

$$\sigma(x_1,\ldots,x_m,x_{m+1},\ldots,x_N) = (x_{m+1},\ldots,x_N,x_1,\ldots,x_m),$$

then $\varphi = \psi \circ \sigma$ is a diffeomorphism such that

$$f(\varphi(x_1,\ldots,x_N)) = (x_{m+1},\ldots,x_N)$$

for all $(x_1,\ldots,x_N) \in O$. If we denote the restriction of f to V by g, it is clear that

$$M \cap V = g^{-1}(0),$$

and because $g(\varphi(x_1,\ldots,x_N)) = 0$ iff $(x_{m+1},\ldots,x_N) = 0_{N-m}$ and φ is a bijection,

$$M \cap V = \{(y_1,\ldots,y_N) \in V \mid g(y_1,\ldots,y_N) = 0\}$$
$$= \{\varphi(x_1,\ldots,x_N) \mid (\exists (x_1,\ldots,x_N) \in O)(g(\varphi(x_1,\ldots,x_N)) = 0)\}$$
$$= \varphi(O \cap (\mathbb{R}^m \times \{0_{N-m}\})),$$

which is (2).

If (2) holds, then $\varphi\colon O \to W$ is a diffeomorphism,

$$O \cap (\mathbb{R}^m \times \{0_{N-m}\}) = \Omega \times \{0_{N-m}\}$$

for some open subset $\Omega \subseteq \mathbb{R}^m$, and the map $\psi\colon \Omega \to \mathbb{R}^N$ given by

$$\psi(x) = \varphi(x, 0_{N-m})$$

is an immersion on Ω and a homeomorphism onto $U \cap M$, which implies (1).

If (3) holds, then if we write $f = (f_1, \ldots, f_{N-m})$ with $f_i\colon W \to \mathbb{R}$, then the fact that $df(p)$ is a submersion is equivalent to the fact that the linear forms $df_1(p), \ldots, df_{N-m}(p)$ are linearly independent and

$$M \cap W = f^{-1}(0) = f_1^{-1}(0) \cap \cdots \cap f_{N-m}^{-1}(0).$$

Finally, if (4) holds, then if we define $f\colon W \to \mathbb{R}^{N-m}$ by

$$f = (f_1, \ldots, f_{N-m}),$$

because $df_1(p), \ldots, df_{N-m}(p)$ are linearly independent we get a smooth map that is a submersion at p such that

$$M \cap W = f^{-1}(0).$$

Now, f is a submersion at p iff $df(p)$ is surjective, which means that a certain determinant is nonzero, and since the determinant function is continuous, this determinant is nonzero on some open subset $W' \subseteq W$ containing p, so if we restrict f to W', we get an immersion on W' such that $M \cap W' = f^{-1}(0)$. $\square$

Condition (4) says that locally (that is, in a small open set of M containing $p \in M$), M is "cut out" by $N - m$ smooth functions $f_i\colon W \to \mathbb{R}$ in the sense that the portion of the manifold $M \cap W$ is the intersection of the $N - m$ hypersurfaces $f_i^{-1}(0)$ (the zero-level sets of the f_i) and that this intersection is "clean," which means that the linear forms $df_1(p), \ldots, df_{N-m}(p)$ are linearly independent.

As an illustration of Theorem 18.7, we can show again that the sphere

$$S^n = \{x \in \mathbb{R}^{n+1} \mid \|x\|_2^2 - 1 = 0\}$$

is an n-dimensional manifold in $\mathbb{R}^{n+1}$. Indeed, the map $f\colon \mathbb{R}^{n+1} \to \mathbb{R}$ given by $f(x) = \|x\|_2^2 - 1$ is a submersion (for $x \neq 0$), since

$$df(x)(y) = 2 \sum_{k=1}^{n+1} x_k y_k.$$

We can also show that the rotation group $\mathbf{SO}(n)$ is an $\frac{n(n-1)}{2}$-dimensional manifold in $\mathbb{R}^{n^2}$. Indeed, $\mathbf{GL}^+(n)$ is an open subset of $\mathbb{R}^{n^2}$ (recall that $\mathbf{GL}^+(n) = \{A \in \mathbf{GL}(n) \mid \det(A) > 0\}$), and if f is defined by

$$f(A) = A^\top A - I,$$

where $A \in \mathbf{GL}^+(n)$, then $f(A)$ is symmetric, so $f(A) \in \mathbf{S}(n) = \mathbb{R}^{\frac{n(n+1)}{2}}$.

It is easy to show (using directional derivatives) that

$$df(A)(H) = A^\top H + H^\top A.$$

But then $df(A)$ is surjective for all $A \in \mathbf{SO}(n)$, because if S is any symmetric matrix, we see that

$$df(A)\left(\frac{AS}{2}\right) = S.$$

Since $\mathbf{SO}(n) = f^{-1}(0)$, we conclude that $\mathbf{SO}(n)$ is indeed a manifold.

A similar argument proves that $\mathbf{O}(n)$ is an $\frac{n(n-1)}{2}$-dimensional manifold. Using the map $f\colon \mathbf{GL}(n) \to \mathbb{R}$ given by $A \mapsto \det(A)$, we can prove that $\mathbf{SL}(n)$ is a manifold of dimension $n^2 - 1$.

Remark: We have $df(A)(B) = \det(A)\mathrm{tr}(A^{-1}B)$ for every $A \in \mathbf{GL}(n)$, where $f(A) = \det(A)$.

The third characterization of Theorem 18.7 suggests the following definition.

Definition 18.12. Let $f\colon \mathbb{R}^{m+k} \to \mathbb{R}^k$ be a smooth function. A point $p \in \mathbb{R}^{m+k}$ is called a *critical point (of f)* if df_p is *not* surjective, and a point $q \in \mathbb{R}^k$ is called a *critical value (of f)* if $q = f(p)$, for some critical point $p \in \mathbb{R}^{m+k}$. A point $p \in \mathbb{R}^{m+k}$ is a *regular point (of f)* if p is not critical, i.e., df_p is surjective, and a point $q \in \mathbb{R}^k$ is a *regular value (of f)* if it is not a critical value. In particular, any $q \in \mathbb{R}^k - f(\mathbb{R}^{m+k})$ is a regular value and $q \in f(\mathbb{R}^{m+k})$ is a regular value iff *every $p \in f^{-1}(q)$ is a regular point* (but in contrast, q is a critical value iff *some $p \in f^{-1}(q)$ is critical*).

Part (3) of Theorem 18.7 implies the following useful proposition:

Proposition 18.7. *Given any smooth function $f\colon \mathbb{R}^{m+k} \to \mathbb{R}^k$, for every regular value $q \in f(\mathbb{R}^{m+k})$, the preimage $Z = f^{-1}(q)$ is a manifold of dimension m.*

Definition 18.12 and Proposition 18.7 can be generalized to manifolds (see Problem 18.24). Regular and critical values of smooth maps play an important role in differential topology. Firstly, given a smooth map $f\colon \mathbb{R}^{m+k} \to \mathbb{R}^k$, almost every point of $\mathbb{R}^k$ is a regular value of f. To make this statement precise, one needs the notion of a *set of measure zero*. Then *Sard's theorem* says that the set of critical values of a smooth map has measure zero. Secondly, if we consider smooth functions $f\colon \mathbb{R}^{m+1} \to \mathbb{R}$, a point $p \in \mathbb{R}^{m+1}$ is critical iff $df_p = 0$. Then we can use second-order derivatives to further classify critical points. The *Hessian matrix* of f (at p) is the matrix of second-order partials

$$H_f(p) = \left(\frac{\partial^2 f}{\partial x_i \partial x_j}(p)\right),$$

and a critical point p is a *nondegenerate critical point* if $H_f(p)$ is a nonsingular matrix. The remarkable fact is that at a nondegenerate critical point p, the local behavior of f is completely determined, in the sense that after a suitable change of coordinates (given by a smooth diffeomorphism),

$$f(x) = f(p) - x_1^2 - \cdots - x_\lambda^2 + x_{\lambda+1}^2 + \cdots + x_{m+1}^2$$

near p, where λ, called the *index of f at p*, is an integer that depends only on p (in fact, λ is the number of negative eigenvalues of $H_f(p)$). This result is known as *Morse's lemma* (after Marston Morse, 1892-1977).

Smooth functions whose critical points are all nondegenerate are called *Morse functions*. It turns out that every smooth function $f\colon \mathbb{R}^{m+1} \to \mathbb{R}$ gives rise to a large supply of Morse functions by adding a linear function to it. More precisely, the set of $a \in \mathbb{R}^{m+1}$ for which the function f_a given by

$$f_a(x) = f(x) + a_1 x_1 + \cdots + a_{m+1} x_{m+1}$$

is not a Morse function has measure zero.

Morse functions can be used to study topological properties of manifolds. In a sense to be made precise and under certain technical conditions, a Morse function can be used to reconstruct a manifold by attaching cells, up to homotopy equivalence. However, these results are way beyond the scope of this book. A fairly elementary exposition of nondegenerate critical points and Morse functions can be found in Guillemin and Pollack [20] (Chapter 1, Section 7). Sard's theorem is proved in Appendix 1 of Guillemin and Pollack [20] and also in Chapter 2 of Milnor [35]. Morse theory (starting with Morse's lemma) and much more is discussed in Milnor [36], widely recognized as a mathematical masterpiece. An excellent and more leisurely introduction to Morse theory is given in Matsumoto [34], where a proof of Morse's lemma is also given.

Let us now review the definitions of a smooth curve in a manifold and the tangent vector at a point of a curve.

Definition 18.13. Let M be an m-dimensional manifold in $\mathbb{R}^N$. A *smooth curve γ in M* is any function $\gamma\colon I \to M$, where I is an open interval in $\mathbb{R}$, such that for every $t \in I$, letting $p = \gamma(t)$, there are some parametrization $\varphi\colon \Omega \to U$ of M at p and some open interval $]t - \varepsilon, t + \varepsilon[\subseteq I$ such that the curve $\varphi^{-1} \circ \gamma\colon]t - \varepsilon, t + \varepsilon[\to \mathbb{R}^m$ is smooth.

Using Lemma 18.11, it is easily shown that Definition 18.13 does not depend on the choice of the parametrization $\varphi\colon \Omega \to U$ at p.

Lemma 18.11 also implies that γ viewed as a curve $\gamma\colon I \to \mathbb{R}^N$ is smooth. Then the *tangent vector to the curve $\gamma\colon I \to \mathbb{R}^N$ at t*, denoted by $\gamma'(t)$, is the value of the derivative of γ at t (a vector in $\mathbb{R}^N$) computed as usual:

$$\gamma'(t) = \lim_{h \to 0} \frac{\gamma(t+h) - \gamma(t)}{h}.$$

Given any point $p \in M$, we will show that the set of tangent vectors to all smooth curves in M through p is a vector space isomorphic to the vector space $\mathbb{R}^m$. The tangent vector at p to a curve γ on a manifold M is illustrated in Figure 18.3.

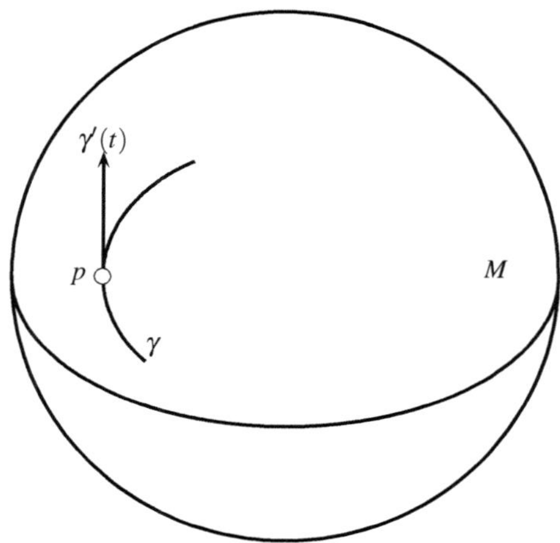

Fig. 18.3 Tangent vector to a curve on a manifold.

Given a smooth curve $\gamma \colon I \to M$, for any $t \in I$, letting $p = \gamma(t)$, since M is a manifold, there is a parametrization $\varphi \colon \Omega \to U$ such that $\varphi(0_m) = p \in U$ and some open interval $J \subseteq I$ with $t \in J$ and such that the function

$$\varphi^{-1} \circ \gamma \colon J \to \mathbb{R}^m$$

is a smooth curve, since γ is a smooth curve. Letting $\alpha = \varphi^{-1} \circ \gamma$, the derivative $\alpha'(t)$ is well-defined, and it is a vector in $\mathbb{R}^m$. But $\varphi \circ \alpha \colon J \to M$ is also a smooth curve, which agrees with γ on J, and by the chain rule,

$$\gamma'(t) = \varphi'(0_m)(\alpha'(t)),$$

since $\alpha(t) = 0_m$ (because $\varphi(0_m) = p$ and $\gamma(t) = p$). Observe that $\gamma'(t)$ is a vector in $\mathbb{R}^N$. Now, for every vector $v \in \mathbb{R}^m$, the curve $\alpha \colon J \to \mathbb{R}^m$ defined such that

$$\alpha(u) = (u - t)v$$

for all $u \in J$ is clearly smooth, and $\alpha'(t) = v$. This shows that the set of tangent vectors at t to all smooth curves (in $\mathbb{R}^m$) passing through 0_m is the entire vector space $\mathbb{R}^m$. Since every smooth curve $\gamma \colon I \to M$ agrees with a curve of the form $\varphi \circ \alpha \colon J \to M$ for some smooth curve $\alpha \colon J \to \mathbb{R}^m$ (with $J \subseteq I$) as explained above, and since it is assumed that $\varphi'(0_m)$ is injective, $\varphi'(0_m)$ maps the vector space $\mathbb{R}^m$

injectively to the set of tangent vectors to γ at p, as claimed. All this is summarized in the following definition.

Definition 18.14. Let M be an m-dimensional manifold in $\mathbb{R}^N$. For every point $p \in M$, the *tangent space* T_pM at p is the set of all vectors in $\mathbb{R}^N$ of the form $\gamma'(0)$, where $\gamma \colon I \to M$ is any smooth curve in M such that $p = \gamma(0)$. The set T_pM is a vector space isomorphic to $\mathbb{R}^m$. Every vector $v \in T_pM$ is called a *tangent vector to M at p*.

We can now define Lie groups (postponing defining smooth maps).

Definition 18.15. A *Lie group* is a nonempty subset G of $\mathbb{R}^N$ ($N \geq 1$) satisfying the following conditions:

(a) G is a group.
(b) G is a manifold in $\mathbb{R}^N$.
(c) The group operation $\cdot \colon G \times G \to G$ and the inverse map $^{-1} \colon G \to G$ are smooth.

(Smooth maps are defined in Definition 18.18). It is immediately verified that $\mathbf{GL}(n, \mathbb{R})$ is a Lie group. Since all the Lie groups that we are considering are subgroups of $\mathbf{GL}(n, \mathbb{R})$, the following definition is in order.

Definition 18.16. A *linear Lie group* is a subgroup G of $\mathbf{GL}(n, \mathbb{R})$ (for some $n \geq 1$) which is a smooth manifold in $\mathbb{R}^{n^2}$.

Let $\mathbf{M}(n, \mathbb{R})$ denote the set of all real $n \times n$ matrices (invertible or not). If we recall that the exponential map

$$\exp \colon A \mapsto e^A$$

is well defined on $\mathbf{M}(n, \mathbb{R})$, we have the following crucial theorem due to Von Neumann and Cartan.

Theorem 18.8. *A closed subgroup G of $\mathbf{GL}(n, \mathbb{R})$ is a linear Lie group. Furthermore, the set $\mathfrak{g}$ defined such that*

$$\mathfrak{g} = \{X \in \mathbf{M}(n, \mathbb{R}) \mid e^{tX} \in G \text{ for all } t \in \mathbb{R}\}$$

is a vector space equal to the tangent space T_IG at the identity I, and $\mathfrak{g}$ is closed under the Lie bracket $[-, -]$ defined such that $[A, B] = AB - BA$ for all $A, B \in \mathbf{M}(n, \mathbb{R})$.

Theorem 18.8 applies even when G is a discrete subgroup, but in this case, $\mathfrak{g}$ is trivial (i.e., $\mathfrak{g} = \{0\}$). For example, the set of nonnull reals $\mathbb{R}^* = \mathbb{R} - \{0\} = \mathbf{GL}(1, \mathbb{R})$ is a Lie group under multiplication, and the subgroup

$$H = \{2^n \mid n \in \mathbb{Z}\}$$

is a discrete subgroup of $\mathbb{R}^*$. Thus, H is a Lie group. On the other hand, the set $\mathbb{Q}^* = \mathbb{Q} - \{0\}$ of nonnull rational numbers is a multiplicative subgroup of $\mathbb{R}^*$, but it is not closed, since $\mathbb{Q}$ is dense in $\mathbb{R}$.

The proof of Theorem 18.8 involves proving that when G is not a discrete subgroup, there is an open subset $\Omega \subseteq \mathbf{M}(n,\mathbb{R})$ such that $0_{n,n} \in \Omega$, an open subset $W \subseteq \mathbf{M}(n,\mathbb{R})$ such that $I \in W$, and that exp: $\Omega \to W$ is a diffeomorphism such that

$$\exp(\Omega \cap \mathfrak{g}) = W \cap G.$$

If G is closed and not discrete, we must have $m \geq 1$, and $\mathfrak{g}$ has dimension m.

With the help of Theorem 18.8 it is now very easy to prove that $\mathbf{SL}(n)$, $\mathbf{O}(n)$, $\mathbf{SO}(n)$, $\mathbf{SL}(n,\mathbb{C})$, $\mathbf{U}(n)$, and $\mathbf{SU}(n)$ are Lie groups and to figure out what are their Lie algebras. (Of course, $\mathbf{GL}(n,\mathbb{R})$ is a Lie group, as we already know.)

For example, if $G = \mathbf{GL}(n,\mathbb{R})$, as e^{tA} is invertible for *every* matrix, $A \in \mathbf{M}(n,\mathbb{R})$, we deduce that the Lie algebra, $\mathfrak{gl}(n,\mathbb{R})$, of $\mathbf{GL}(n,\mathbb{R})$ is equal to $\mathbf{M}(n,\mathbb{R})$. We also claim that the Lie algebra, $\mathfrak{sl}(n,\mathbb{R})$, of $\mathbf{SL}(n,\mathbb{R})$ is the set of all matrices with zero trace. Indeed, $\mathfrak{sl}(n,\mathbb{R})$ is the subalgebra of $\mathfrak{gl}(n,\mathbb{R})$ consisting of all matrices $X \in \mathfrak{gl}(n,\mathbb{R})$ such that

$$\det(e^{tX}) = 1$$

for all $t \in \mathbb{R}$, and because $\det(e^{tX}) = e^{\mathrm{tr}(tX)}$, for $t = 1$, we get $\mathrm{tr}(X) = 0$, as claimed.

We can also prove that $\mathbf{SE}(n)$ is a Lie group as follows. Recall that we can view every element of $\mathbf{SE}(n)$ as a real $(n+1) \times (n+1)$ matrix

$$\begin{pmatrix} R & U \\ 0 & 1 \end{pmatrix}$$

where $R \in \mathbf{SO}(n)$ and $U \in \mathbb{R}^n$. In fact, such matrices belong to $\mathbf{SL}(n+1)$. This embedding of $\mathbf{SE}(n)$ into $\mathbf{SL}(n+1)$ is a group homomorphism, since the group operation on $\mathbf{SE}(n)$ corresponds to multiplication in $\mathbf{SL}(n+1)$:

$$\begin{pmatrix} RS & RV+U \\ 0 & 1 \end{pmatrix} = \begin{pmatrix} R & U \\ 0 & 1 \end{pmatrix} \begin{pmatrix} S & V \\ 0 & 1 \end{pmatrix}.$$

Note that the inverse is given by

$$\begin{pmatrix} R^{-1} & -R^{-1}U \\ 0 & 1 \end{pmatrix} = \begin{pmatrix} R^\top & -R^\top U \\ 0 & 1 \end{pmatrix}.$$

Also note that the embedding shows that as a manifold, $\mathbf{SE}(n)$ is diffeomorphic to $\mathbf{SO}(n) \times \mathbb{R}^n$ (given a manifold M_1 of dimension m_1 and a manifold M_2 of dimension m_2, the product $M_1 \times M_2$ can be given the structure of a manifold of dimension $m_1 + m_2$ in a natural way). Thus, $\mathbf{SE}(n)$ is a Lie group with underlying manifold $\mathbf{SO}(n) \times \mathbb{R}^n$, and in fact, a subgroup of $\mathbf{SL}(n+1)$.

Even though $\mathbf{SE}(n)$ is diffeomorphic to $\mathbf{SO}(n) \times \mathbb{R}^n$ as a manifold, it is *not* isomorphic to $\mathbf{SO}(n) \times \mathbb{R}^n$ as a group, because the group multiplication on $\mathbf{SE}(n)$ is not the multiplication on $\mathbf{SO}(n) \times \mathbb{R}^n$. Instead, $\mathbf{SE}(n)$ is a *semidirect product* of $\mathbf{SO}(n)$ and $\mathbb{R}^n$; see Chapter 2, Problem 2.19.

Returning to Theorem 18.8, the vector space $\mathfrak{g}$ is called the *Lie algebra* of the Lie group G. Lie algebras are defined as follows.

Definition 18.17. A *(real) Lie algebra* $\mathcal{A}$ is a real vector space together with a bilinear map $[\cdot, \cdot] \colon \mathcal{A} \times \mathcal{A} \to \mathcal{A}$ called the *Lie bracket* on $\mathcal{A}$ such that the following two identities hold for all $a, b, c \in \mathcal{A}$:

$$[a,\, a] = 0,$$

and the so-called *Jacobi identity*

$$[a,\, [b,\, c]] + [c,\, [a,\, b]] + [b,\, [c,\, a]] = 0.$$

It is immediately verified that $[b, a] = -[a, b]$.

In view of Theorem 18.8, the vector space $\mathfrak{g} = T_I G$ associated with a Lie group G is indeed a Lie algebra. Furthermore, the exponential map $\exp \colon \mathfrak{g} \to G$ is well-defined. In general, $\exp$ is neither injective nor surjective, as we observed earlier. Theorem 18.8 also provides a kind of recipe for "computing" the Lie algebra $\mathfrak{g} = T_I G$ of a Lie group G. Indeed, $\mathfrak{g}$ is the tangent space to G at I, and thus we can use curves to compute tangent vectors. Actually, for every $X \in T_I G$, the map

$$\gamma_X \colon t \mapsto e^{tX}$$

is a smooth curve in G, and it is easily shown that $\gamma_X'(0) = X$. Thus, we can use these curves. As an illustration, we show that the Lie algebras of $\mathbf{SL}(n)$ and $\mathbf{SO}(n)$ are the matrices with null trace and the skew-symmetric matrices.

Let $t \mapsto R(t)$ be a smooth curve in $\mathbf{SL}(n)$ such that $R(0) = I$. We have $\det(R(t)) = 1$ for all $t \in\,]-\varepsilon, \varepsilon[$. Using the chain rule, we can compute the derivative of the function

$$t \mapsto \det(R(t))$$

at $t = 0$, and we get

$$\det_I'(R'(0)) = 0.$$

It is an easy exercise to prove that

$$\det_I'(X) = \operatorname{tr}(X),$$

and thus $\operatorname{tr}(R'(0)) = 0$, which says that the tangent vector $X = R'(0)$ has null trace. Clearly, $\mathfrak{sl}(n, \mathbb{R})$ has dimension $n^2 - 1$.

Let $t \mapsto R(t)$ be a smooth curve in $\mathbf{SO}(n)$ such that $R(0) = I$. Since each $R(t)$ is orthogonal, we have

$$R(t)\, R(t)^\top = I$$

for all $t \in\,]-\varepsilon, \varepsilon[$. Taking the derivative at $t = 0$, we get

$$R'(0)\,R(0)^{\top} + R(0)\,R'(0)^{\top} = 0,$$

but since $R(0) = I = R(0)^{\top}$, we get

$$R'(0) + R'(0)^{\top} = 0,$$

which says that the tangent vector $X = R'(0)$ is skew-symmetric. Since the diagonal elements of a skew-symmetric matrix are null, the trace is automatically null, and the condition $\det(R) = 1$ yields nothing new. This shows that $\mathfrak{o}(n) = \mathfrak{so}(n)$. It is easily shown that $\mathfrak{so}(n)$ has dimension $n(n-1)/2$.

As a concrete example, the Lie algebra $\mathfrak{so}(3)$ of $\mathbf{SO}(3)$ is the real vector space consisting of all 3×3 real skew-symmetric matrices. Every such matrix is of the form

$$\begin{pmatrix} 0 & -d & c \\ d & 0 & -b \\ -c & b & 0 \end{pmatrix}$$

where $b, c, d \in \mathbb{R}$. The Lie bracket $[A, B]$ in $\mathfrak{so}(3)$ is also given by the usual commutator, $[A, B] = AB - BA$.

We can define an isomorphism of Lie algebras $\psi \colon (\mathbb{R}^3, \times) \to \mathfrak{so}(3)$ by the formula

$$\psi(b, c, d) = \begin{pmatrix} 0 & -d & c \\ d & 0 & -b \\ -c & b & 0 \end{pmatrix}.$$

It is indeed easy to verify that

$$\psi(u \times v) = [\psi(u),\ \psi(v)].$$

It is also easily verified that for any two vectors $u = (b, c, d)$ and $v = (b', c', d')$ in $\mathbb{R}^3$

$$\psi(u)(v) = u \times v.$$

The exponential map $\exp \colon \mathfrak{so}(3) \to \mathbf{SO}(3)$ is given by Rodrigues's formula (see Lemma 18.6):

$$e^A = \cos\theta\, I_3 + \frac{\sin\theta}{\theta} A + \frac{(1 - \cos\theta)}{\theta^2} B,$$

or equivalently by

$$e^A = I_3 + \frac{\sin\theta}{\theta} A + \frac{(1 - \cos\theta)}{\theta^2} A^2$$

if $\theta \neq 0$, where

$$A = \begin{pmatrix} 0 & -d & c \\ d & 0 & -b \\ -c & b & 0 \end{pmatrix},$$

$\theta = \sqrt{b^2 + c^2 + d^2}$, $B = A^2 + \theta^2 I_3$, and with $e^{0_3} = I_3$.

Using the above methods, it is easy to verify that the Lie algebras $\mathfrak{gl}(n, \mathbb{R})$, $\mathfrak{sl}(n, \mathbb{R})$, $\mathfrak{o}(n)$, and $\mathfrak{so}(n)$, are respectively $\mathbf{M}(n, \mathbb{R})$, the set of matrices with null trace, and the set of skew-symmetric matrices (in the last two cases). A similar computation can be done for $\mathfrak{gl}(n,\mathbb{C})$, $\mathfrak{sl}(n,\mathbb{C})$, $\mathfrak{u}(n)$, and $\mathfrak{su}(n)$, confirming the claims of Section 18.4. It is easy to show that $\mathfrak{gl}(n,\mathbb{C})$ has dimension $2n^2$, $\mathfrak{sl}(n,\mathbb{C})$ has dimension $2(n^2-1)$, $\mathfrak{u}(n)$ has dimension n^2, and $\mathfrak{su}(n)$ has dimension n^2-1.

For example, the Lie algebra $\mathfrak{su}(2)$ of $\mathbf{SU}(2)$ (or S^3) is the real vector space consisting of all 2×2 (complex) skew-Hermitian matrices of null trace. Every such matrix is of the form

$$i(d\sigma_1 + c\sigma_2 + b\sigma_3) = \begin{pmatrix} ib & c+id \\ -c+id & -ib \end{pmatrix},$$

where $b,c,d \in \mathbb{R}$, and $\sigma_1,\sigma_2,\sigma_3$ are the Pauli spin matrices (see Section 9.1), and thus the matrices $i\sigma_1, i\sigma_2, i\sigma_3$ form a basis of the Lie algebra $\mathfrak{su}(2)$. The Lie bracket $[A,B]$ in $\mathfrak{su}(2)$ is given by the usual commutator, $[A,B] = AB - BA$.

It is easily checked that the vector space $\mathbb{R}^3$ is a Lie algebra if we define the Lie bracket on $\mathbb{R}^3$ as the usual cross product $u \times v$ of vectors. Then we can define an isomorphism of Lie algebras $\varphi\colon (\mathbb{R}^3, \times) \to \mathfrak{su}(2)$ by the formula

$$\varphi(b,c,d) = \frac{i}{2}(d\sigma_1 + c\sigma_2 + b\sigma_3) = \frac{1}{2}\begin{pmatrix} ib & c+id \\ -c+id & -ib \end{pmatrix}.$$

It is indeed easy to verify that

$$\varphi(u \times v) = [\varphi(u),\, \varphi(v)].$$

Returning to $\mathfrak{su}(2)$, letting $\theta = \sqrt{b^2+c^2+d^2}$, we can write

$$d\sigma_1 + c\sigma_2 + b\sigma_3 = \begin{pmatrix} b & -ic+d \\ ic+d & -b \end{pmatrix} = \theta A,$$

where

$$A = \frac{1}{\theta}(d\sigma_1 + c\sigma_2 + b\sigma_3) = \frac{1}{\theta}\begin{pmatrix} b & -ic+d \\ ic+d & -b \end{pmatrix},$$

so that $A^2 = I$, and it can be shown that the exponential map $\exp\colon \mathfrak{su}(2) \to \mathbf{SU}(2)$ is given by

$$\exp(i\theta A) = \cos\theta\, \mathbf{1} + i\sin\theta\, A.$$

In view of the isomorphism $\varphi\colon (\mathbb{R}^3, \times) \to \mathfrak{su}(2)$, where

$$\varphi(b,c,d) = \frac{1}{2}\begin{pmatrix} ib & c+id \\ -c+id & -ib \end{pmatrix} = i\frac{\theta}{2}A,$$

the exponential map can be viewed as a map $\exp\colon (\mathbb{R}^3, \times) \to \mathbf{SU}(2)$ given by the formula

$$\exp(\theta v) = \left[\cos\frac{\theta}{2},\ \sin\frac{\theta}{2}v\right],$$

for every vector θv, where v is a unit vector in $\mathbb{R}^3$ and $\theta \in \mathbb{R}$. In this form, $\exp(\theta v)$ is a quaternion corresponding to a rotation of axis v and angle θ.

As we showed, $\mathbf{SE}(n)$ is a Lie group, and its lie algebra $\mathfrak{se}(n)$ described in Section 18.6 is easily determined as the subalgebra of $\mathfrak{sl}(n+1)$ consisting of all matrices of the form

$$\begin{pmatrix} B & U \\ 0 & 0 \end{pmatrix}$$

where $B \in \mathfrak{so}(n)$ and $U \in \mathbb{R}^n$. Thus, $\mathfrak{se}(n)$ has dimension $n(n+1)/2$. The Lie bracket is given by

$$\begin{pmatrix} B & U \\ 0 & 0 \end{pmatrix} \begin{pmatrix} C & V \\ 0 & 0 \end{pmatrix} - \begin{pmatrix} C & V \\ 0 & 0 \end{pmatrix} \begin{pmatrix} B & U \\ 0 & 0 \end{pmatrix} = \begin{pmatrix} BC - CB & BV - CU \\ 0 & 0 \end{pmatrix}.$$

We conclude by indicating the relationship between homomorphisms of Lie groups and homomorphisms of Lie algebras. First, we need to explain what is meant by a smooth map between manifolds.

Definition 18.18. Let M_1 (m_1-dimensional) and M_2 (m_2-dimensional) be manifolds in $\mathbb{R}^N$. A function $f\colon M_1 \to M_2$ is *smooth* if for every $p \in M_1$ there are parametrizations $\varphi\colon \Omega_1 \to U_1$ of M_1 at p and $\psi\colon \Omega_2 \to U_2$ of M_2 at $f(p)$ such that $f(U_1) \subseteq U_2$ and

$$\psi^{-1} \circ f \circ \varphi\colon \Omega_1 \to \mathbb{R}^{m_2}$$

is smooth.

Using Lemma 18.11, it is easily shown that Definition 18.18 does not depend on the choice of the parametrizations $\varphi\colon \Omega_1 \to U_1$ and $\psi\colon \Omega_2 \to U_2$. A smooth map f between manifolds is a *smooth diffeomorphism* if f is bijective and both f and f^{-1} are smooth maps.

We now define the derivative of a smooth map between manifolds.

Definition 18.19. Let M_1 (m_1-dimensional) and M_2 (m_2-dimensional) be manifolds in $\mathbb{R}^N$. For any smooth function $f\colon M_1 \to M_2$ and any $p \in M_1$, the function $f'_p\colon T_pM_1 \to T_{f(p)}M_2$, called the *tangent map of f at p, or derivative of f at p, or differential of f at p*, is defined as follows: For every $v \in T_pM_1$ and every smooth curve $\gamma\colon I \to M_1$ such that $\gamma(0) = p$ and $\gamma'(0) = v$,

$$f'_p(v) = (f \circ \gamma)'(0).$$

The map f'_p is also denoted by df_p or T_pf. Doing a few calculations involving the facts that

$$f \circ \gamma = (f \circ \varphi) \circ (\varphi^{-1} \circ \gamma) \quad \text{and} \quad \gamma = \varphi \circ (\varphi^{-1} \circ \gamma)$$

and using Lemma 18.11, it is not hard to show that $f'_p(v)$ does not depend on the choice of the curve γ. It is easily shown that f'_p is a linear map.

Finally, we define homomorphisms of Lie groups and Lie algebras and see how they are related.

Definition 18.20. Given two Lie groups G_1 and G_2, a *homomorphism (or map) of Lie groups* is a function $f\colon G_1 \to G_2$ that is a homomorphism of groups and a smooth map (between the manifolds G_1 and G_2). Given two Lie algebras $\mathscr{A}_1$ and $\mathscr{A}_2$, a *homomorphism (or map) of Lie algebras* is a function $f\colon \mathscr{A}_1 \to \mathscr{A}_2$ that is a linear map between the vector spaces $\mathscr{A}_1$ and $\mathscr{A}_2$ and that preserves Lie brackets, i.e.,

$$f([A,B]) = [f(A), f(B)]$$

for all $A, B \in \mathscr{A}_1$.

An *isomorphism of Lie groups* is a bijective function f such that both f and f^{-1} are maps of Lie groups, and an *isomorphism of Lie algebras* is a bijective function f such that both f and f^{-1} are maps of Lie algebras. It is immediately verified that if $f\colon G_1 \to G_2$ is a homomorphism of Lie groups, then $f'_I\colon \mathfrak{g}_1 \to \mathfrak{g}_2$ is a homomorphism of Lie algebras. If some additional assumptions are made about G_1 and G_2 (for example, connected, simply connected), it can be shown that f is pretty much determined by f'_I.

Alert readers must have noticed that we only defined the Lie algebra of a linear group. In the more general case, we can still define the Lie algebra $\mathfrak{g}$ of a Lie group G as the tangent space $T_I G$ at the identity I. The tangent space $\mathfrak{g} = T_I G$ is a vector space, but we need to define the Lie bracket. This can be done in several ways. We explain briefly how this can be done in terms of so-called adjoint representations. This has the advantage of not requiring the definition of left-invariant vector fields, but it is still a little bizarre!

Given a Lie group G, for every $a \in G$ we define *left translation* as the map $L_a\colon G \to G$ such that $L_a(b) = ab$ for all $b \in G$, and *right translation* as the map $R_a\colon G \to G$ such that $R_a(b) = ba$ for all $b \in G$. The maps L_a and R_a are diffeomorphisms, and their derivatives play an important role. The inner automorphisms $R_{a^{-1}} \circ L_a$ (also written as $R_{a^{-1}} L_a$) also play an important role. Note that

$$R_{a^{-1}} L_a(b) = aba^{-1}.$$

The derivative

$$(R_{a^{-1}} L_a)'_I\colon T_I G \to T_I G$$

of $R_{a^{-1}} L_a\colon G \to G$ at I is an isomorphism of Lie algebras, and since $T_I G = \mathfrak{g}$, we get a map denoted by $\mathrm{Ad}_a\colon \mathfrak{g} \to \mathfrak{g}$. The map $a \mapsto \mathrm{Ad}_a$ is a map of Lie groups

$$\mathrm{Ad}\colon G \to \mathbf{GL}(\mathfrak{g}),$$

called the *adjoint representation of G* (where $\mathbf{GL}(\mathfrak{g})$ denotes the Lie group of all bijective linear maps on $\mathfrak{g}$).

In the case of a linear group, one can verify that

$$\mathrm{Ad}(a)(X) = \mathrm{Ad}_a(X) = aXa^{-1}$$

for all $a \in G$ and all $X \in \mathfrak{g}$. The derivative

$$\mathrm{Ad}'_I \colon \mathfrak{g} \to \mathfrak{gl}(\mathfrak{g})$$

of $\mathrm{Ad}\colon G \to \mathbf{GL}(\mathfrak{g})$ at I is map of Lie algebras, denoted by $\mathrm{ad}\colon \mathfrak{g} \to \mathfrak{gl}(\mathfrak{g})$, called the *adjoint representation of* $\mathfrak{g}$. (Recall that Theorem 18.8 immediately implies that the Lie algebra $\mathfrak{gl}(\mathfrak{g})$ of $\mathbf{GL}(\mathfrak{g})$ is the vector space of all linear maps on $\mathfrak{g}$).

In the case of a linear group, it can be verified that

$$\mathrm{ad}(A)(B) = [A, B]$$

for all $A, B \in \mathfrak{g}$. One can also check that the Jacobi identity on $\mathfrak{g}$ is equivalent to the fact that ad preserves Lie brackets, i.e., ad is a map of Lie algebras:

$$\mathrm{ad}([A, B]) = [\mathrm{ad}(A), \mathrm{ad}(B)]$$

for all $A, B \in \mathfrak{g}$ (where on the right, the Lie bracket is the commutator of linear maps on $\mathfrak{g}$). Thus, we recover the Lie bracket from ad.

This is the key to the definition of the Lie bracket in the case of a general Lie group (not just a linear Lie group). We define the Lie bracket on $\mathfrak{g}$ as

$$[A, B] = \mathrm{ad}(A)(B).$$

To be complete, we would have to define the exponential map $\exp\colon \mathfrak{g} \to G$ for a general Lie group. For this we would need to introduce some left-invariant vector fields induced by the derivatives of the left translations, and integral curves associated with such vector fields.

This is not hard, but we feel that it is now time to stop our introduction to Lie groups and Lie algebras, even though we have not even touched many important topics, for instance vector fields and differential foms. Readers who wish to learn more about Lie groups and Lie algebras should consult (more or less listed in order of difficulty) Curtis [14], Sattinger and Weaver [42], Hall [21], and Marsden and Ratiu [33]. The excellent lecture notes by Carter, Segal, and Macdonald [11] constitute a very efficient (although somewhat terse) introduction to Lie algebras and Lie groups. Classics such as Weyl [48] and Chevalley [12] are definitely worth consulting, although the presentation and the terminology may seem a bit old fashioned. For more advanced texts, one may consult Abraham and Marsden [1], Warner [47], Sternberg [45], Bröcker and tom Dieck [9], and Knapp [26]. For those who read French, Mneimné and Testard [37] is very clear and quite thorough, and uses very little differential geometry, although it is more advanced than Curtis. Chapter 1, by Bryant, in Freed and Uhlenbeck [17] is also worth reading, but the pace is fast, and Chapters 7 and 8 of Fulton and Harris [18] are very good, but familiarity with manifolds is assumed.

18.9 Applications of Lie Groups and Lie Algebras

Some applications of Lie groups and Lie algebras to robotics and motion planning are discussed in Selig [44] and Murray, Li, and Sastry [39]. Applications to physics are discussed in Sattinger and Weaver [42] and Marsden and Ratiu [33].

The fact that the exponential maps $\exp\colon \mathfrak{so}(3) \to \mathbf{SO}(3)$ and $\exp\colon \mathfrak{se}(3) \to \mathbf{SE}(3)$ are surjective is important in robotics applications. Indeed, some matrices associated with joints arising in robot kinematics can be written as exponentials $e^{\theta s}$, where θ is a joint angle and $\mathbf{s} \in \mathfrak{se}(3)$ is the so-called *joint screw* (see Selig [44], Chapter 4). One should also observe that if a rigid motion (R, b) is used to define the position of a rigid body, then the velocity of a point p is given by $(R'p + b')$. In other words, the element (R', b') of the Lie algebra $\mathfrak{se}(3)$ is a sort of velocity vector.

The surjectivity of the exponential map $\exp\colon \mathfrak{se}(3) \to \mathbf{SE}(3)$ implies that there is a map $\log\colon \mathbf{SE}(3) \to \mathfrak{se}(3)$, although it is multivalued. Still, this log "function" can be used to perform motion interpolation. For instance, given two rigid motions $B_1, B_2 \in \mathbf{SE}(3)$ specifying the position of a rigid body B, we can compute $\log(B_1)$ and $\log(B_2)$, which are just elements of the Euclidean space $\mathfrak{se}(3)$, form the linear interpolant $(1 - t)\log(B_1) + t\log(B_2)$, and then apply the exponential map to get an interpolating rigid motion

$$e^{(1-t)\log(B_1) + t\log(B_2)}.$$

Of course, this can also be done for a sequence of rigid motions $B_1, \ldots, B_n$, where $n > 2$, and instead of using affine interpolation between two consecutive positions, a polynomial spline can be used to interpolate between the $\log(B_i)$'s in $\mathfrak{se}(3)$. This approach has been investigated by Kim, M.-J., Kim, M.-S. and Shin [24, 25], and Park and Ravani [40, 41].

R.S. Ball published a treatise on the theory of screws in 1900 [7]. Basically, Ball's screws are rigid motions, and his instantaneous screws correspond to elements of the Lie algebra $\mathfrak{se}(3)$ (they are rays in $\mathfrak{se}(3)$). A *screw system* is simply a subspace of $\mathfrak{se}(3)$. Such systems were first investigated by Ball [7]. The first heuristic classification of screw systems was given by Hunt [22]. Screw systems play an important role in kinematics, see McCarthy [31] and Selig [44], Chapter 8.

Lie groups and Lie algebras are also a key ingredient in the use of symmetries in motion, to reduce the number of parameters in the equations of motion, and in optimal control. Such applications are described in a very exciting paper by Marsden and Ostrowski [32] (see also the references in this paper).

18.10 Problems

18.1. Given a Hermitian space E, for every linear map $f\colon E \to E$, prove that there is an orthonormal basis $(u_1, \ldots, u_n)$ with respect to which the matrix of f is upper triangular. In terms of matrices, this means that there is a unitary matrix U and an upper triangular matrix T such that $A = UTU^*$.

Remark: This extension of Lemma 18.3 is usually known as *Schur's lemma.*

18.2. Prove that the torus obtained by rotating a circle of radius b contained in a plane containing the z-axis and whose center is on a circle of center O and radius b in the xy-plane is a manifold by giving four parametrizations. What are the conditions required on a, b?

Hint. What about

$$x = a\cos\theta + b\cos\theta\cos\varphi,$$
$$y = a\sin\theta + b\sin\theta\cos\varphi,$$
$$z = b\sin\varphi?$$

18.3. (a) Prove that the maps φ_1 and φ_2 parametrizing the sphere are indeed smooth and injective, that $\varphi_1'(u,v)$ and $\varphi_2'(u,v)$ are injective, and that φ_1 and φ_2 give the sphere the structure of a manifold.

(b) Prove that the map $\psi_1 : \Delta(1) \to S^2$ defined such that

$$\psi_1(x,y) = \left(x,\, y,\, \sqrt{1-x^2-y^2}\right),$$

where $\Delta(1)$ is the unit open disk, is a parametrization of the open upper hemisphere. Show that there are five other similar parametrizations, which, together with ψ_1, make S^2 into a manifold.

18.4. Use Lemma 18.11 to prove that Definition 18.13 does not depend on the choice of the parametrization $\varphi \colon \Omega \to U$ at p.

18.5. Given a linear Lie group G, for every $X \in T_I G$, letting γ be the smooth curve in G

$$\gamma_X : t \mapsto e^{tX},$$

prove that $\gamma_X'(0) = X$.

18.6. Prove that

$$\det_I'(X) = \operatorname{tr}(X).$$

Hint. Find the directional derivative

$$\lim_{t\to 0} \frac{\det(I+tX) - \det(I)}{t}.$$

18.7. Confirm that $\mathfrak{gl}(n,\mathbb{C})$, $\mathfrak{sl}(n,\mathbb{C})$, $\mathfrak{u}(n)$, and $\mathfrak{su}(n)$, are the vector spaces of matrices described in Section 18.4. Prove that $\mathfrak{gl}(n,\mathbb{C})$ has dimension $2n^2$, $\mathfrak{sl}(n,\mathbb{C})$ has dimension $2(n^2-1)$, $\mathfrak{u}(n)$ has dimension n^2, and $\mathfrak{su}(n)$ has dimension n^2-1.

18.8. Prove that the map $\varphi \colon (\mathbb{R}^3, \times) \to \mathfrak{su}(2)$ defined by the formula

$$\varphi(b,c,d) = \frac{i}{2}(d\sigma_1 + c\sigma_2 + b\sigma_3) = \frac{1}{2}\begin{pmatrix} ib & c+id \\ -c+id & -ib \end{pmatrix}$$

is an isomorphism of Lie algebras. If

$$A = \frac{1}{\theta}\begin{pmatrix} b & -ic+d \\ ic+d & -b \end{pmatrix},$$

where $\theta = \sqrt{b^2+c^2+d^2}$, prove that the exponential map $\exp\colon \mathfrak{su}(2) \to \mathbf{SU}(2)$ is given by

$$\exp(i\theta A) = \cos\theta\, \mathbf{1} + i\sin\theta\, A.$$

18.9. Prove that Definition 18.18 does not depend on the parametrizations $\varphi\colon \Omega_1 \to U_1$ and $\psi\colon \Omega_2 \to U_2$.

18.10. In Definition 18.19, prove that $f'_p(v)$ does not depend on the choice of the curve γ, and that f'_p is a linear map.

18.11. In the case of a linear group, prove that

$$\mathrm{Ad}(a)(X) = \mathrm{Ad}_a(X) = aXa^{-1}$$

for all $a \in G$ and all $X \in \mathfrak{g}$.

18.12. In the case of a linear group, prove that

$$\mathrm{ad}(A)(B) = [A, B]$$

for all $A, B \in \mathfrak{g}$.

Check that the Jacobi identity on $\mathfrak{g}$ is equivalent to the fact that ad preserves Lie brackets, i.e., ad is a map of Lie algebras:

$$\mathrm{ad}([A, B]) = [\mathrm{ad}(A), \mathrm{ad}(B)]$$

for all $A, B \in \mathfrak{g}$ (where on the right, the Lie bracket is the commutator of linear maps on $\mathfrak{g}$).

18.13. Consider the Lie algebra $\mathfrak{su}(2)$, whose basis is the Pauli spin matrices $\sigma_1, \sigma_2, \sigma_3$ (see Chapter 6, Section 9.1). The map $\mathrm{ad}(X)$ is a linear map for every $X \in \mathfrak{g}$, since $\mathrm{ad}\colon \mathfrak{g} \to \mathfrak{gl}(\mathfrak{g})$. Compute the matrices representing $\mathrm{ad}(\sigma_1)$, $\mathrm{ad}(\sigma_2)$, $\mathrm{ad}(\sigma_3)$.

18.14. (a) Consider the affine maps ρ of $\mathbb{A}^2$ defined such that

$$\rho\begin{pmatrix} x \\ y \end{pmatrix} = \begin{pmatrix} \cos\theta & -\sin\theta \\ \sin\theta & \cos\theta \end{pmatrix}\begin{pmatrix} x \\ y \end{pmatrix} + \begin{pmatrix} u \\ v \end{pmatrix},$$

where $\theta, u, v \in \mathbb{R}$.

Given any map ρ as above, letting

$$R = \begin{pmatrix} \cos\theta & -\sin\theta \\ \sin\theta & \cos\theta \end{pmatrix}, \quad X = \begin{pmatrix} x \\ y \end{pmatrix}, \quad \text{and} \quad U = \begin{pmatrix} u \\ v \end{pmatrix},$$

ρ can be represented by the 3×3 matrix

$$A = \begin{pmatrix} R & U \\ 0 & 1 \end{pmatrix} = \begin{pmatrix} \cos\theta & -\sin\theta & u \\ \sin\theta & \cos\theta & v \\ 0 & 0 & 1 \end{pmatrix}$$

in the sense that

$$\begin{pmatrix} \rho(X) \\ 1 \end{pmatrix} = \begin{pmatrix} R & U \\ 0 & 1 \end{pmatrix} \begin{pmatrix} X \\ 1 \end{pmatrix}$$

iff

$$\rho(X) = RX + U.$$

Prove that these maps are affine bijections and that they form a group, denoted by $\mathbf{SE}(2)$ (the *direct affine isometries, or rigid motions*, of $\mathbb{A}^2$). Prove that such maps preserve the inner product of $\mathbb{R}^2$, i.e., that for any four points $a, b, c, d \in \mathbb{A}^2$,

$$\rho(\overrightarrow{ac}) \cdot \rho(\overrightarrow{bd}) = \overrightarrow{ac} \cdot \overrightarrow{bd}.$$

If $\theta \neq k2\pi$ ($k \in \mathbb{Z}$), prove that ρ has a unique fixed point c_ρ, and that with respect to any frame with origin c_ρ, ρ is a rotation of angle θ and of center c_ρ.

(b) Let us now consider the set of matrices of the form

$$\begin{pmatrix} 0 & -\theta & u \\ \theta & 0 & v \\ 0 & 0 & 0 \end{pmatrix}$$

where $\theta, u, v \in \mathbb{R}$. Verify that this set of matrices is a vector space isomorphic to $(\mathbb{R}^3, +)$. This vector space is denoted by $\mathfrak{se}(2)$. Show that in general, $AB \neq BA$.

(c) Given a matrix

$$A = \begin{pmatrix} 0 & -\theta & u \\ \theta & 0 & v \\ 0 & 0 & 0 \end{pmatrix}$$

letting

$$\Omega = \begin{pmatrix} 0 & -\theta \\ \theta & 0 \end{pmatrix} \quad \text{and} \quad U = \begin{pmatrix} u \\ v \end{pmatrix}$$

we can write

$$A = \begin{pmatrix} \Omega & U \\ 0 & 0 \end{pmatrix}.$$

Prove that

$$A^n = \begin{pmatrix} \Omega^n & \Omega^{n-1}U \\ 0 & 0 \end{pmatrix}$$

where $\Omega^0 = I_2$. Prove that if $\theta = k2\pi$ ($k \in \mathbb{Z}$), then

$$e^A = \begin{pmatrix} I_2 & U \\ 0 & 1 \end{pmatrix},$$

and that if $\theta \neq k2\pi$ $(k \in \mathbb{Z})$, then

$$e^A = \begin{pmatrix} \cos\theta & -\sin\theta & \frac{u}{\theta}\sin\theta + \frac{v}{\theta}(\cos\theta - 1) \\ \sin\theta & \cos\theta & \frac{u}{\theta}(-\cos\theta + 1) + \frac{v}{\theta}\sin\theta \\ 0 & 0 & 1 \end{pmatrix}.$$

Hint. Letting $V = \Omega^{-1}(e^\Omega - I_2)$, prove that

$$V = I_2 + \sum_{k \geq 1} \frac{\Omega^k}{(k+1)!}$$

and that

$$e^A = \begin{pmatrix} e^\Omega & VU \\ 0 & 1 \end{pmatrix}.$$

Another proof consists in showing that

$$A^3 = -\theta^2 A,$$

and that

$$E^A = I_3 + \frac{\sin\theta}{\theta}A + \frac{1 - \cos\theta}{\theta^2}A^2.$$

(d) Prove that e^A is a direct affine isometry in $\mathbf{SE}(2)$. If $\theta \neq k2\pi$ $(k \in \mathbb{Z})$, prove that V is invertible, and thus prove that the exponential map $\exp: \mathfrak{se}(2) \to \mathbf{SE}(2)$ is surjective. How do you need to restrict θ to get an injective map?

Remark: Rigid motions can be used to describe the motion of rigid bodies in the plane. Given a fixed Euclidean frame $(O, (e_1, e_2))$, we can assume that some moving frame $(C, (u_1, u_2))$ is attached (say glued) to a rigid body B (for example, at the center of gravity of B) so that the position and orientation of B in the plane are completely (and uniquely) determined by some rigid motion

$$A = \begin{pmatrix} R & U \\ 0 & 1 \end{pmatrix},$$

where U specifies the position of C with respect to O, and R specifies the orientation (i.e., angle) of B with respect to the fixed frame $(O, (e_1, e_2))$. Then, a motion of B in the plane corresponds to a curve in the space $\mathbf{SE}(2)$. The space $\mathbf{SE}(2)$ is topologically quite complex (in particular, it is "curved"). The exponential map allows us to work in the simpler (noncurved) Euclidean space $\mathfrak{se}(2)$. Thus, given a sequence of "snapshots" of B, say $B_0, B_1, \ldots, B_m$, we can try to find an interpolating motion (a curve in $\mathbf{SE}(2)$) by finding a simpler curve in $\mathfrak{se}(2)$ (say, a B-spline) using the inverse of the exponential map. Of course, it is desirable that the interpolating motion be reasonably smooth and "natural." Computer animations of such motions can be easily implemented.

18.15. (a) Consider the set of affine maps ρ of $\mathbb{A}^3$ defined such that

$$\rho(X) = RX + U,$$

where R is a rotation matrix (an orthogonal matrix of determinant $+1$) and U is some vector in $\mathbb{R}^3$. Every such a map can be represented by the 4×4 matrix

$$\begin{pmatrix} R & U \\ 0 & 1 \end{pmatrix}$$

in the sense that

$$\begin{pmatrix} \rho(X) \\ 1 \end{pmatrix} = \begin{pmatrix} R & U \\ 0 & 1 \end{pmatrix} \begin{pmatrix} X \\ 1 \end{pmatrix}$$

iff

$$\rho(X) = RX + U.$$

Prove that these maps are affine bijections and that they form a group, denoted by **SE**(3) (the *direct affine isometries, or rigid motions*, of $\mathbb{A}^3$). Prove that such maps preserve the inner product of $\mathbb{R}^3$, i.e., that for any four points $a, b, c, d \in \mathbb{A}^3$,

$$\rho(\overrightarrow{ac}) \cdot \rho(\overrightarrow{bd}) = \overrightarrow{ac} \cdot \overrightarrow{bd}.$$

Prove that these maps do not always have a fixed point.

(b) Let us now consider the set of 4×4 matrices of the form

$$A = \begin{pmatrix} \Omega & U \\ 0 & 0 \end{pmatrix},$$

where Ω is a skew-symmetric matrix

$$\Omega = \begin{pmatrix} 0 & -c & b \\ c & 0 & -a \\ -b & a & 0 \end{pmatrix},$$

and U is a vector in $\mathbb{R}^3$.

Verify that this set of matrices is a vector space isomorphic to $(\mathbb{R}^6, +)$. This vector space is denoted by $\mathfrak{se}(3)$. Show that in general, $AB \neq BA$.

(c) Given a matrix

$$A = \begin{pmatrix} \Omega & U \\ 0 & 0 \end{pmatrix}$$

as in (b), prove that

$$A^n = \begin{pmatrix} \Omega^n & \Omega^{n-1}U \\ 0 & 0 \end{pmatrix}$$

where $\Omega^0 = I_3$. Given

$$\Omega = \begin{pmatrix} 0 & -c & b \\ c & 0 & -a \\ -b & a & 0 \end{pmatrix},$$

let $\theta = \sqrt{a^2 + b^2 + c^2}$. Prove that if $\theta = k2\pi$ $(k \in \mathbb{Z})$, then

$$e^A = \begin{pmatrix} I_3 & U \\ 0 & 1 \end{pmatrix},$$

and that if $\theta \neq k2\pi$ $(k \in \mathbb{Z})$, then

$$e^A = \begin{pmatrix} e^\Omega & VU \\ 0 & 1 \end{pmatrix},$$

where

$$V = I_3 + \sum_{k \geq 1} \frac{\Omega^k}{(k+1)!}.$$

(d) Prove that

$$e^\Omega = I_3 + \frac{\sin \theta}{\theta}\Omega + \frac{(1 - \cos \theta)}{\theta^2}\Omega^2$$

and

$$V = I_3 + \frac{(1 - \cos \theta)}{\theta^2}\Omega + \frac{(\theta - \sin \theta)}{\theta^3}\Omega^2.$$

Hint. Use the fact that $\Omega^3 = -\theta^2 \Omega$.

(e) Prove that e^A is a direct affine isometry in $\mathbf{SE}(3)$. Prove that V is invertible. *Hint.* Assume that the inverse of V is of the form

$$W = I_3 + a\Omega + b\Omega^2,$$

and show that a, b, are given by a system of linear equations that always has a unique solution.

Prove that the exponential map $\exp \colon \mathfrak{se}(3) \to \mathbf{SE}(3)$ is surjective. You may use the fact that $\exp \colon \mathfrak{so}(3) \to \mathbf{SO}(3)$ is surjective, where

$$\exp(\Omega) = e^\Omega = I_3 + \frac{\sin \theta}{\theta}\Omega + \frac{(1 - \cos \theta)}{\theta^2}\Omega^2.$$

Remark: Rigid motions can be used to describe the motion of rigid bodies in space. Given a fixed Euclidean frame $(O, (e_1, e_2, e_3))$, we can assume that some moving frame $(C, (u_1, u_2, u_3))$ is attached (say glued) to a rigid body B (for example, at the center of gravity of B) so that the position and orientation of B in space are completely (and uniquely) determined by some rigid motion

$$A = \begin{pmatrix} R & U \\ 0 & 1 \end{pmatrix},$$

where U specifies the position of C with respect to O, and R specifies the orientation of B with respect to the fixed frame $(O, (e_1, e_2, e_3))$. Then a motion of B in space corresponds to a curve in the space $\mathbf{SE}(3)$. The space $\mathbf{SE}(3)$ is topologically quite complex (in particular, it is "curved"). The exponential map allows us to work in the simpler (noncurved) Euclidean space $\mathfrak{se}(3)$. Thus, given a sequence of "snap-shots" of B, say $B_0, B_1, \ldots, B_m$, we can try to find an interpolating motion (a curve in $\mathbf{SE}(3)$) by finding a simpler curve in $\mathfrak{se}(3)$ (say, a B-spline) using the inverse of the exponential map. Of course, it is desirable that the interpolating motion be reasonably smooth and "natural." Computer animations of such motions can be easily implemented.

18.16. Let A and B be the 4×4 matrices

$$A = \begin{pmatrix} 0 & -\theta_1 & 0 & 0 \\ \theta_1 & 0 & 0 & 0 \\ 0 & 0 & 0 & -\theta_2 \\ 0 & 0 & \theta_2 & 0 \end{pmatrix}$$

and

$$B = \begin{pmatrix} \cos\theta_1 & -\sin\theta_1 & 0 & 0 \\ \sin\theta_1 & \cos\theta_1 & 0 & 0 \\ 0 & 0 & \cos\theta_2 & -\sin\theta_2 \\ 0 & 0 & \sin\theta_2 & \cos\theta_2 \end{pmatrix}$$

where $\theta_1, \theta_2 \geq 0$. (i) Compute A^2, and prove that

$$B = e^A,$$

where

$$e^A = I_n + \sum_{p \geq 1} \frac{A^p}{p!} = \sum_{p \geq 0} \frac{A^p}{p!},$$

letting $A^0 = I_n$. Use this to prove that for every orthogonal 4×4 matrix B there is a skew-symmetric matrix A such that

$$B = e^A.$$

(ii) Given a skew-symmetric 4×4 matrix A, prove that there are two skew-symmetric matrices A_1 and A_2 and some $\theta_1, \theta_2 \geq 0$ such that

$$A = A_1 + A_2,$$
$$A_1^3 = -\theta_1^2 A_1,$$
$$A_2^3 = -\theta_2^2 A_2,$$
$$A_1 A_2 = A_2 A_1 = 0,$$
$$\operatorname{tr}(A_1^2) = -2\theta_1^2,$$
$$\operatorname{tr}(A_2^2) = -2\theta_2^2,$$

and where $A_i = 0$ if $\theta_i = 0$ and $A_1^2 + A_2^2 = -\theta_1^2 I_4$ if $\theta_2 = \theta_1$.
    Using the above, prove that

$$e^A = I_4 + \frac{\sin\theta_1}{\theta_1}A_1 + \frac{\sin\theta_2}{\theta_2}A_2 + \frac{(1-\cos\theta_1)}{\theta_1^2}A_1^2 + \frac{(1-\cos\theta_2)}{\theta_2^2}A_2^2.$$

(iii) Given an orthogonal 4×4 matrix B, prove that there are two skew-symmetric matrices A_1 and A_2 and some $\theta_1, \theta_2 \geq 0$ such that

$$B = I_4 + \frac{\sin\theta_1}{\theta_1}A_1 + \frac{\sin\theta_2}{\theta_2}A_2 + \frac{(1-\cos\theta_1)}{\theta_1^2}A_1^2 + \frac{(1-\cos\theta_2)}{\theta_2^2}A_2^2,$$

where

$$A_1^3 = -\theta_1^2 A_1,$$
$$A_2^3 = -\theta_2^2 A_2,$$
$$A_1 A_2 = A_2 A_1 = 0,$$
$$\mathrm{tr}(A_1^2) = -2\theta_1^2,$$
$$\mathrm{tr}(A_2^2) = -2\theta_2^2,$$

and where $A_i = 0$ if $\theta_i = 0$ and $A_1^2 + A_2^2 = -\theta_1^2 I_4$ if $\theta_2 = \theta_1$. Prove that

$$\frac{1}{2}(B - B^\top) = \frac{\sin\theta_1}{\theta_1}A_1 + \frac{\sin\theta_2}{\theta_2}A_2,$$
$$\frac{1}{2}(B + B^\top) = I_4 + \frac{(1-\cos\theta_1)}{\theta_1^2}A_1^2 + \frac{(1-\cos\theta_2)}{\theta_2^2}A_2^2,$$
$$\mathrm{tr}(B) = 2\cos\theta_1 + 2\cos\theta_2.$$

(iv) Prove that if $\sin\theta_1 = 0$ or $\sin\theta_2 = 0$, then A_1, A_2, and the $\cos\theta_i$ can be computed from B. Prove that if $\theta_2 = \theta_1$, then

$$B = \cos\theta_1 I_4 + \frac{\sin\theta_1}{\theta_1}(A_1 + A_2),$$

and $\cos\theta_1$ and $A_1 + A_2$ can be computed from B.

(v) Prove that

$$\frac{1}{4}\mathrm{tr}\left(\left(B - B^\top\right)^2\right) = 2\cos^2\theta_1 + 2\cos^2\theta_2 - 4.$$

Prove that $\cos\theta_1$ and $\cos\theta_2$ are solutions of the equation

$$x^2 - sx + p = 0,$$

where

$$s = \frac{1}{2}\mathrm{tr}(B), \quad p = \frac{1}{8}(\mathrm{tr}(B))^2 - \frac{1}{16}\mathrm{tr}\left(\left(B - B^\top\right)^2\right) - 1.$$

Prove that we also have

$$\cos^2 \theta_1 \cos^2 \theta_2 = \det\left(\frac{1}{2}\left(B + B^\top\right)\right).$$

If $\sin \theta_i \neq 0$ for $i = 1, 2$ and $\cos \theta_2 \neq \cos \theta_1$, prove that the system

$$\frac{1}{2}\left(B - B^\top\right) = \frac{\sin \theta_1}{\theta_1} A_1 + \frac{\sin \theta_2}{\theta_2} A_2,$$

$$\frac{1}{4}\left(B + B^\top\right)\left(B - B^\top\right) = \frac{\sin \theta_1 \cos \theta_1}{\theta_1} A_1 + \frac{\sin \theta_2 \cos \theta_2}{\theta_2} A_2$$

has a unique solution for A_1 and A_2.

(vi) Prove that $A = A_1 + A_2$ has an orthonormal basis of eigenvectors such that the first two are a basis of the plane with respect to which B is a rotation of angle θ_1, and the last two are a basis of the plane with respect to which B is a rotation of angle θ_2.

Remark: I do not know a simple way to compute such an orthonormal basis of eigenvectors of $A = A_1 + A_2$, but it should be possible!

18.17. (a) Consider the map, $f \colon \mathbf{GL}^+(n) \to \mathbf{S}(n)$, given by

$$f(A) = A^\top A - I.$$

Check that

$$df(A)(H) = A^\top H + H^\top A,$$

for any matrix, H.

(b) Consider the map, $f \colon \mathbf{GL}(n) \to \mathbb{R}$, given by

$$f(A) = \det(A).$$

Prove that $df(I)(B) = \mathrm{tr}(B)$, the trace of B, for any matrix B (here, I is the identity matrix). Then, prove that

$$df(A)(B) = \det(A)\mathrm{tr}(A^{-1}B),$$

where $A \in \mathbf{GL}(n)$.

(c) Use the map $A \mapsto \det(A) - 1$ to prove that $\mathbf{SL}(n)$ is a manifold of dimension $n^2 - 1$.

(d) Let J be the $(n+1) \times (n+1)$ diagonal matrix

$$J = \begin{pmatrix} I_n & 0 \\ 0 & -1 \end{pmatrix}.$$

We denote by $\mathbf{SO}(n, 1)$ the group of real $(n+1) \times (n+1)$ matrices:

$$\mathbf{SO}(n, 1) = \{A \in \mathbf{GL}(n+1) \mid A^\top J A = J \quad \text{and} \quad \det(A) = 1\}.$$

Check that $\mathbf{SO}(n,1)$ is indeed a group with the inverse of A given by $A^{-1} = JA^\top J$ (this is the *special Lorentz group*). Consider the function $f\colon \mathbf{GL}^+(n+1) \to \mathbf{S}(n+1)$, given by

$$f(A) = A^\top JA - J,$$

where $\mathbf{S}(n+1)$ denotes the space of $(n+1) \times (n+1)$ symmetric matrices. Prove that

$$df(A)(H) = A^\top JH + H^\top JA$$

for any matrix H. Prove that $df(A)$ is surjective for all $A \in \mathbf{SO}(n,1)$ and that $\mathbf{SO}(n,1)$ is a manifold of dimension $\frac{n(n+1)}{2}$.

18.18. (a) Given any matrix

$$B = \begin{pmatrix} a & b \\ c & -a \end{pmatrix} \in \mathfrak{sl}(2,\mathbb{C}),$$

if $\omega^2 = a^2 + bc$ and ω is any of the two complex roots of $a^2 + bc$, prove that if $\omega \neq 0$, then

$$e^B = \cosh\omega\, I + \frac{\sinh\omega}{\omega} B,$$

and $e^B = I + B$ if $a^2 + bc = 0$. Observe that $\mathrm{tr}(e^B) = 2\cosh\omega$.

Prove that the exponential map $\exp\colon \mathfrak{sl}(2,\mathbb{C}) \to \mathbf{SL}(2,\mathbb{C})$ is *not* surjective. For instance, prove that

$$\begin{pmatrix} -1 & 1 \\ 0 & -1 \end{pmatrix}$$

is not the exponential of any matrix in $\mathfrak{sl}(2,\mathbb{C})$.

(b) Recall that a matrix N is *nilpotent* if there is some $m \geq 0$ such that $N^m = 0$. Let A be any $n \times n$ matrix of the form $A = I - N$, where N is nilpotent. Why is A invertible? prove that there is some B such that $e^B = I - N$ as follows: Recall that for any $y \in \mathbb{R}$ such that $|y - 1|$ is small enough, we have

$$\log(y) = -(1-y) - \frac{(1-y)^2}{2} - \cdots - \frac{(1-y)^k}{k} - \cdots.$$

Since N is nilpotent, we have $N^m = 0$, where m is the smallest integer with this propery. Then the expression

$$B = \log(I - N) = -N - \frac{N^2}{2} - \cdots - \frac{N^{m-1}}{m-1}$$

is well defined. Use a formal power series argument to show that $e^B = A$. We denote B by $\log(A)$.

(c) Let $A \in \mathbf{GL}(n,\mathbb{C})$. Prove that there is some matrix B so that $e^B = A$. Thus the exponential map $\exp\colon \mathfrak{gl}(n,\mathbb{C}) \to \mathbf{GL}(n,\mathbb{C})$ is surjective.

First, use the fact that A has a Jordan form PJP^{-1}. Then show that finding a log of A reduces to finding a log of every Jordan block of J. Since every Jordan block

J has a fixed nonzero constant λ on the diagonal, with 1's immediately above each diagonal entry, and zeros everywhere else, we can write J as $(\lambda I)(I - N)$, where N is nilpotent. Find B_1 and B_2 such that $\lambda I = e^{B_1}$, $I - N = e^{B_2}$, and $B_1 B_2 = B_2 B_1$. Conclude that $J = e^{B_1 + B_2}$.

18.19. Let $B_r = \{x = (x_1, \ldots, x_n) \in \mathbb{R}^n \mid x_1^2 + \cdots + x_n^2 < r\}$ be the open ball of radius r (centered at the origin) in $\mathbb{R}^n$ (where $r > 0$). Prove that the map

$$x \mapsto \frac{rx}{\sqrt{r^2 - (x_1^2 + \cdots + x_n^2)}}$$

is a diffeomorphism of B_r onto $\mathbb{R}^n$ (where $x = (x_1, \ldots, x_n)$).
Hint. Compute explicity the inverse of this map.

18.20. A smooth bijective map of manifolds need not be a diffeomorphism. For example, show that $f \colon \mathbb{R} \to \mathbb{R}$ given by $f(x) = x^3$ is not a diffeomorphism.

18.21. (a) Let $X \subseteq \mathbb{R}^M$ and $Y \subseteq \mathbb{R}^N$ be two smooth manifolds of dimension m and n respectively. We can make $X \times Y \subseteq \mathbb{R}^{M+N}$ into a smooth manifold of dimension $m + n$ as follows: for any $(p, q) \in X \times Y$, if $\varphi \colon \Omega_1 \to U$ and $\psi \colon \Omega_2 \to V$ are parametrizations at $p \in U \subseteq X$ and $q \in V \subseteq Y$ respectively, then show that $\varphi \times \psi \colon \Omega_1 \times \Omega_2 \to U \times V$ is indeed a parametrization at $(p, q) \in X \times Y$. Since the $U \times V$'s cover $X \times Y$, these parametrizations make $X \times Y$ into a manifold.

Check that $T_{(p,q)}(X \times Y) = T_p X \times T_q Y$.

(b) Given a set X, let $\Delta = \{(x, x) \mid x \in X\} \subseteq X \times X$, called the *diagonal of X*. If X is a manifold, then prove that Δ is a manifold diffeomorphic to X.

(c) The *graph* of a function $f \colon X \to Y$ is the subset of $X \times Y$ given by

$$\mathrm{graph}(f) = \{(x, f(x)) \mid x \in X\}.$$

Define $F \colon X \to \mathrm{graph}(f)$ by $F(x) = (x, f(x))$. Prove that if X and Y are smooth manifolds and if f is smooth, then F is a diffeomorphism and thus $\mathrm{graph}(f)$ is a manifold diffeomorphic to X.

(d) Given any (smooth) map $f \colon X \to X$, some $x \in X$ is a *fixed point* of f if $f(x) = x$. Prove that f has a fixed point iff $\mathrm{graph}(f) \cap \Delta \neq \emptyset$ (where Δ is the diagonal in $X \times X$).

18.22. Recall from Problem 12.6 the Cayley parametrization of rotation matrices in $\mathbf{SO}(n)$ given by
$$C(B) = (I - B)(I + B)^{-1},$$

where B is any $n \times n$ skew-symmetric matrix. In that problem, it was shown that $C(B)$ is a rotation matrix that does not admit -1 as an eigenvalue and that every such rotation matrix is of the form $C(B)$.

(a) If you have not already done so, prove that the map $B \mapsto C(B)$ is injective.

(b) Prove that

$$dC(B)(A) = D_A((I - B)(I + B)^{-1}) = -[I + (I - B)(I + B)^{-1}]A(I + B)^{-1}.$$

Hint. First, show that $D_A(B^{-1}) = -B^{-1}AB^{-1}$ (where B is invertible) and that

$$D_A(f(B)g(B)) = (D_Af(B))g(B) + f(B)(D_Ag(B)),$$

where f and g are differentiable matrix functions.

Deduce that $dC(B)$ is injective for every skew-symmetric matrix B. If we identify the space of $n \times n$ skew-symmetric matrices with $\mathbb{R}^{n(n-1)/2}$, show that the Cayley map $C \colon \mathbb{R}^{n(n-1)/2} \to \mathbf{SO}(n)$ is a parametrization of $\mathbf{SO}(n)$.

(c) Now consider $n = 3$, i.e., $\mathbf{SO}(3)$. Let E_1, E_2, and E_3 be the rotations about the x-axis, y-axis, and z-axis, respectively, by the angle π, i.e.,

$$E_1 = \begin{pmatrix} 1 & 0 & 0 \\ 0 & -1 & 0 \\ 0 & 0 & -1 \end{pmatrix}, \quad E_2 = \begin{pmatrix} -1 & 0 & 0 \\ 0 & 1 & 0 \\ 0 & 0 & -1 \end{pmatrix}, \quad E_3 = \begin{pmatrix} -1 & 0 & 0 \\ 0 & -1 & 0 \\ 0 & 0 & 1 \end{pmatrix}.$$

Prove that the four maps

$$B \mapsto C(B), \quad B \mapsto E_1C(B), \quad B \mapsto E_2C(B), \quad B \mapsto E_3C(B),$$

where B is skew-symmetric, are parametrizations of $\mathbf{SO}(3)$ and that the union of the images of C, E_1C, E_2C, and E_3C covers $\mathbf{SO}(3)$, so that $\mathbf{SO}(3)$ is a manifold.

(d) Let A be *any* square matrix (not necessarily invertible). Prove that there is some diagonal matrix E with entries $+1$ or -1 such that $EA + I$ is invertible.

(e) Prove that every rotation matrix $A \in \mathbf{SO}(n)$ is of the form

$$A = E(I - B)(I + B)^{-1},$$

for some skew-symmetric matrix B and some diagonal matrix E with entries $+1$ and -1, and where the number of -1 is even. Moreover, prove that every orthogonal matrix $A \in \mathbf{O}(n)$ is of the form

$$A = E(I - B)(I + B)^{-1},$$

for some skew-symmetric matrix B and some diagonal matrix E with entries $+1$ and -1. The above provide parametrizations for $\mathbf{SO}(n)$ (resp. $\mathbf{O}(n)$) that show that $\mathbf{SO}(n)$ and $\mathbf{O}(n)$ are manifolds. However, observe that the number of these charts grows exponentially with n.

18.23. Let J be the 2×2 matrix

$$J = \begin{pmatrix} 1 & 0 \\ 0 & -1 \end{pmatrix}$$

and let $\mathbf{SU}(1, 1)$ be the set of 2×2 complex matrices

$$\mathbf{SU}(1,1) = \{A \mid A^*JA = J, \det(A) = 1\},$$

where A^* is the conjugate transpose of A.

(a) Prove that $\mathbf{SU}(1,1)$ is the group of matrices of the form

$$A = \begin{pmatrix} a & b \\ \bar{b} & \bar{a} \end{pmatrix}, \quad \text{with} \quad a\bar{a} - b\bar{b} = 1.$$

If

$$g = \begin{pmatrix} 1 & -i \\ 1 & i \end{pmatrix},$$

prove that the map from $\mathbf{SL}(2,\mathbb{R})$ to $\mathbf{SU}(1,1)$ given by

$$A \mapsto gAg^{-1}$$

is a group isomorphism.

(b) Prove that the Möbius transformation

$$z \mapsto \frac{z-i}{z+i}$$

associated with g is a bijection between the upper half-plane H and the unit open disk $D = \{z \in \mathbb{C} \mid |z| < 1\}$. Prove that the map from $\mathbf{SU}(1,1)$ to $S^1 \times D$ given by

$$\begin{pmatrix} a & b \\ \bar{b} & \bar{a} \end{pmatrix} \mapsto (a/|a|, b/a),$$

is a continuous bijection (in fact, a homeomorphism). Conclude that $\mathbf{SU}(1,1)$ is topologically an open solid torus.

18.24. (a) Let $W \subseteq \mathbb{R}^m$ be an open subset of $\mathbb{R}^m$ and pick some $a \in W$. If $f \colon W \to \mathbb{R}^n$ is a smooth submersion at a, i.e., df_a is surjective (so $m \geq n$), prove that there are an open set $V \subseteq W \subseteq \mathbb{R}^m$ with $a \in V$ and a diffeomorphism ψ with domain $O \subseteq \mathbb{R}^m$ such that $\psi(O) = V$ and

$$f(\psi(x_1, \ldots, x_m)) = (x_1, \ldots, x_n),$$

for all $(x_1, \ldots, x_m) \in O$.

Hint. Since df_a is surjective, the rank of the Jacobian matrix $(\partial f_i/\partial x_j(a))$ $(1 \leq i \leq n, 1 \leq j \leq m)$ is n, and after some permutation of $\mathbb{R}^m$, we may assume that the square matrix $B = (\partial f_i/\partial x_j(a))$ $(1 \leq i, j \leq n)$ is invertible. Define the map $h \colon W \to \mathbb{R}^m$ by

$$h(x) = (f_1(x), \ldots, f_n(x), x_{n+1}, \ldots, x_m),$$

where $x = (x_1, \ldots, x_m)$. Check that the Jacobian matrix of h at a is invertible. Then apply the inverse function theorem and finish up.

(b) Let $f \colon M \to N$ be a map of smooth manifolds. A point $p \in M$ is called a *critical point (of f)* if df_p is *not* surjective, and a point $q \in N$ is called a *critical value (of f)* if $q = f(p)$, for some critical point $p \in M$. A point $p \in M$ is a *regular point (of f)* if p is not critical, i.e., df_p is surjective, and a point $q \in N$ is a *regular value (of f)* if it is not a critical value. In particular, any $q \in N - f(M)$ is a regular value and $q \in f(M)$ is a regular value if *every* $p \in f^{-1}(q)$ is a regular point (but in contrast, q is a critical value if *some* $p \in f^{-1}(q)$ is critical).

Prove that for every regular value $q \in f(M)$, the preimage $Z = f^{-1}(q)$ is a manifold of dimension $\dim(M) - \dim(N)$.

Hint. Pick any $p \in f^{-1}(q)$ and some parametrizations φ at p and ψ at q with $\varphi(0) = p$ and $\psi(0) = q$ and consider $h = \psi^{-1} \circ f \circ \varphi$. Prove that dh_0 is surjective and then apply (a).

(c) Under the same assumptions as (b), prove that for every point $p \in Z = f^{-1}(q)$, the tangent space $T_p Z$ is the kernel of $df_p \colon T_p M \to T_q N$.

(d) If $X, Z \subseteq \mathbb{R}^N$ are manifolds and $Z \subseteq X$, we say that Z *is a submanifold of* X. Assume that there is a smooth function $g \colon X \to \mathbb{R}^k$ and that $0 \in \mathbb{R}^k$ is a regular value of g. Then by (b), $Z = g^{-1}(0)$ is a submanifold of X of dimension $\dim(X) - k$. Let $g = (g_1, \ldots, g_k)$, with each g_i a function $g_i \colon X \to \mathbb{R}$. Prove that for any $p \in X$, dg_p is surjective iff the linear forms $(dg_i)_p \colon T_p X \to \mathbb{R}$ are linearly independent. In this case, we say that $g_1, \ldots, g_k$ are *independent at* p. We also say that Z is *cut out* by $g_1, \ldots, g_k$ when

$$Z = \{ p \in X \mid g_1(p) = 0, \ldots, g_k(p) = 0 \}$$

with $g_1, \ldots, g_k$ independent for all $p \in Z$.

Let $f \colon X \to Y$ be a smooth map of manifolds and let $q \in f(X)$ be a regular value. Prove that $Z = f^{-1}(q)$ is a submanifold of X cut out by $k = \dim(X) - \dim(Y)$ independent functions.

Hint. Pick some parametrization ψ at q such that $\psi(0) = q$ and check that 0 is a regular value of $g = \psi^{-1} \circ f$, so that $g_1, \ldots, g_k$ work.

(e) Let $U \subseteq \mathbb{R}^m$ be an open subset of $\mathbb{R}^m$ and pick some $a \in U$. If $f \colon U \to \mathbb{R}^n$ is a smooth immersion at a, i.e., df_a is injective (so $m \leq n$), prove that there are an open set $V \subseteq \mathbb{R}^n$ with $f(a) \in V$, an open subset $U' \subseteq U$ with $a \in U'$ and $f(U') \subseteq V$, an open subset $O \subseteq \mathbb{R}^{n-m}$, and a diffeomorphism $\varphi \colon V \to U' \times O$ such that

$$\varphi(f(x_1, \ldots, x_m)) = (x_1, \ldots, x_m, 0, \ldots, 0),$$

for all $(x_1, \ldots, x_m) \in U'$.

Hint. Since df_a is injective, the rank of the Jacobian matrix $(\partial f_i / \partial x_j(a))$ $(1 \leq i \leq n, 1 \leq j \leq m)$ is m, and after some permutation of $\mathbb{R}^n$, we may assume that the square matrix $B = (\partial f_i / \partial x_j(a))$ $(1 \leq i, j \leq m)$ is invertible. Define the map $g \colon U \times \mathbb{R}^{n-m} \to \mathbb{R}^n$ by

$$g(x, y) = (f_1(x), \ldots, f_m(x), y_1 + f_{m+1}(x), \ldots, y_{n-m} + f_n(x)),$$

where $x = (x_1, \ldots, x_m)$ and $y = (y_1, \ldots, y_{n-m})$. Check that the Jacobian matrix of g at $(a, 0)$ is invertible. Then apply the inverse function theorem and finish up.

Now assume that Z is a submanifold of X. Prove that locally, Z is cut out by independent functions. This means that if $k = \dim(X) - \dim(Z)$, the *codimension* of Z in X, then for every $z \in Z$, there are k independent functions $g_1, \ldots, g_k$ defined on some open subset $W \subseteq X$ with $z \in W$, such that $Z \cap W$ is the common zero set of the g_i's.

(f) We would like to generalize our result in (b) to the more general situation in which we have a smooth map $f\colon X \to Y$, but this time, we have a submanifold $Z \subseteq Y$ and we are investigating whether $f^{-1}(Z)$ is a submanifold of X. In particular, if X is also a submanifold of Y and f is the inclusion of X into Y, then $f^{-1}(Z) = X \cap Z$.

Convince yourself that in general, the intersection of two submanifolds is *not* a submanifold. Try examples involving curves and surfaces and you will see how bad the situation can be. What is needed is a notion generalizing that of a regular value, and this turns out to be the notion of transversality.

We say that f *is transversal to* Z if

$$df_p(T_pX) + T_{f(p)}Z = T_{f(p)}Y,$$

for all $p \in f^{-1}(Z)$. (Recall that if U and V are subspaces of a vector space E, then $U + V$ is the subspace $U + V = \{u + v \in E \mid u \in U, v \in V\}$). In particular, if f is the inclusion of X into Y, the transversality condition is

$$T_pX + T_pZ = T_pY,$$

for all $p \in X \cap Z$.

Draw several examples of transversal intersections to understand better this concept. Prove that if f is transversal to Z, then $f^{-1}(Z)$ is a submanifold of X of codimension equal to $\dim(Y) - \dim(Z)$.

Hint. The set $f^{-1}(Z)$ is a manifold if for every $p \in f^{-1}(Z)$, there is some open subset $U \subseteq X$ with $p \in U$ and $f^{-1}(Z) \cap U$ is a manifold. First, use (e) to assert that locally near $q = f(p)$, Z is cut out by $k = \dim(Y) - \dim(Z)$ independent functions $g_1, \ldots, g_k$, so that locally near p, the preimage $f^{-1}(Z)$ is cut out by $g_1 \circ f, \ldots, g_k \circ f$. If we let $g = (g_1, \ldots, g_k)$, it is an immersion, and the issue is to prove that 0 is a regular value of $g \circ f$ in order to apply (b). Show that transversality is just what is needed to show that 0 is a regular value of $g \circ f$.

(g) With the same assumptions as in (f) (f is transversal to Z), if $W = f^{-1}(Z)$, prove that for every $p \in W$,

$$T_pW = (df_p)^{-1}(T_{f(p)}Z),$$

the preimage of $T_{f(p)}Z$ by $df_p\colon T_pX \to T_{f(p)}Y$. In particular, if f is the inclusion of X into Y, then

$$T_p(X \cap Z) = T_pX \cap T_pZ.$$

(h) Let $X, Z \subseteq Y$ be two submanifolds of Y, with X compact, Z closed, $\dim(X) + \dim(Z) = \dim(Y)$, and X transversal to Z. Prove that $X \cap Z$ consists of a finite set of points.

References

1. Ralph Abraham and Jerrold E. Marsden. *Foundations of Mechanics.* Addison-Wesley, second edition, 1978.
2. Vincent Arsigny. *Processing Data in Lie Groups: An Algebraic Approach. Application to Non-Linear Registration and Diffusion Tensor MRI.* PhD thesis, École Polytechnique, Palaiseau, France, 2006. Thèse de Sciences.
3. Vincent Arsigny, Pierre Fillard, Xavier Pennec, and Nicholas Ayache. Log-Euclidean metrics for fast and simple calculus on diffusion tensors. *Magnetic Resonance in Medicine*, 56(2):411–421, 2006.
4. Vincent Arsigny, Pierre Fillard, Xavier Pennec, and Nicholas Ayache. Geometric means in a novel vector space structure on symmetric positive-definite matrices. *SIAM J. on Matrix Analysis and Applications*, 29(1):328–347, 2007.
5. Vincent Arsigny, Xavier Pennec, and Nicholas Ayache. Polyrigid and polyaffine transformations: a novel geometrical tool to deal with non-rigid deformations-application to the registration of histological slices. *Medical Image Analysis*, 9(6):507–523, 2005.
6. Michael Artin. *Algebra.* Prentice-Hall, first edition, 1991.
7. R.S. Ball. *The Theory of Screws.* Cambridge University Press, first edition, 1900.
8. Marcel Berger and Bernard Gostiaux. *Géométrie différentielle: variétés, courbes et surfaces.* Collection Mathématiques. Puf, second edition, 1992. English edition: Differential geometry, manifolds, curves, and surfaces, GTM No. 115, Springer-Verlag.
9. T. Bröcker and T. tom Dieck. *Representation of Compact Lie Groups.* GTM, Vol. 98. Springer-Verlag, first edition, 1985.
10. Henri Cartan. *Cours de Calcul Différentiel.* Collection Méthodes. Hermann, 1990.
11. Roger Carter, Graeme Segal, and Ian Macdonald. *Lectures on Lie Groups and Lie Algebras.* Cambridge University Press, first edition, 1995.
12. Claude Chevalley. *Theory of Lie Groups I.* Princeton Mathematical Series, No. 8. Princeton University Press, first edition, 1946.
13. Yvonne Choquet-Bruhat, Cécile DeWitt-Morette, and Margaret Dillard-Bleick. *Analysis, Manifolds, and Physics, Part I: Basics.* North-Holland, first edition, 1982.
14. Morton L. Curtis. *Matrix Groups.* Universitext. Springer-Verlag, second edition, 1984.
15. Manfredo P. do Carmo. *Differential Geometry of Curves and Surfaces.* Prentice-Hall, 1976.
16. Manfredo P. do Carmo. *Riemannian Geometry.* Birkhäuser, second edition, 1992.
17. R.L. Bryant. An introduction to Lie groups and symplectic geometry. In D.S. Freed and K.K. Uhlenbeck, editors, *Geometry and Quantum Field Theory*, pages 5–181. AMS, Providence, RI, 1995.
18. William Fulton and Joe Harris. *Representation Theory, A First Course.* GTM No. 129. Springer-Verlag, first edition, 1991.
19. S. Gallot, D. Hulin, and J. Lafontaine. *Riemannian Geometry.* Universitext. Springer-Verlag, second edition, 1993.
20. Victor Guillemin and Alan Pollack. *Differential Topology.* Prentice-Hall, first edition, 1974.
21. Brian Hall. *Lie Groups, Lie Algebras, and Representations. An Elementary Introduction.* GTM No. 222. Springer Verlag, first edition, 2003.
22. K.H. Hunt. *Kinematic Geometry of Mechanisms.* Clarendon Press, first edition, 1978.
23. Roger Howe. Very basic Lie theory. *American Mathematical Monthly*, 90:600–623, 1983.
24. M.J. Kim, M.S. Kim, and S.Y. Shin. A general construction scheme for unit quaternion curves with simple high-order derivatives. In *Computer Graphics Proceedings, Annual Conference Series*, pages 369–376. ACM, 1995.
25. M.J. Kim, M.S. Kim, and S.Y. Shin. A compact differential formula for the first derivative of a unit quaternion curve. *Journal of Visualization and Computer Animation*, 7:43–57, 1996.
26. Anthony W. Knapp. *Lie Groups Beyond an Introduction.* Progress in Mathematics, Vol. 140. Birkhäuser, first edition, 1996.
27. Jacques Lafontaine. *Introduction aux Variétés Différentielles.* PUG, first edition, 1996.
28. Serge Lang. *Real and Functional Analysis.* GTM 142. Springer-Verlag, third edition, 1996.

29. Serge Lang. *Undergraduate Analysis*. UTM. Springer-Verlag, second edition, 1997.
30. John M. Lee. *Introduction to Smooth Manifolds*. GTM No. 218. Springer Verlag, first edition, 2006.
31. J.M. McCarthy. *Introduction to Theoretical Kinematics*. MIT Press, first edition, 1990.
32. Jerrold E. Marsden and Jim Ostrowski. Symmetries in motion: Geometric foundations of motion control. *Nonlinear Science Today*, 1998.
33. Jerrold E. Marsden and T.S. Ratiu. *Introduction to Mechanics and Symmetry*. TAM, Vol. 17. Springer-Verlag, first edition, 1994.
34. Yukio Matsumoto. *An Introduction to Morse Theory*. Translations of Mathematical Monographs No 208. AMS, first edition, 2002.
35. John W. Milnor. *Topology from the Differentiable Viewpoint*. The University Press of Virginia, second edition, 1969.
36. John W. Milnor. *Morse Theory*. Annals of Math. Series, No. 51. Princeton University Press, third edition, 1969.
37. R. Mneimné and F. Testard. *Introduction à la Théorie des Groupes de Lie Classiques*. Hermann, first edition, 1997.
38. James R. Munkres. *Analysis on Manifolds*. Addison-Wesley, 1991.
39. R.M. Murray, Z.X. Li, and S.S. Sastry. *A Mathematical Introduction to Robotics Manipulation*. CRC Press, first edition, 1994.
40. F.C. Park and B. Ravani. Bézier curves on Riemannian manifolds and Lie groups with kinematic applications. *ASME J. Mech. Des.*, 117:36–40, 1995.
41. F.C. Park and B. Ravani. Smooth invariant interpolation of rotations. *ACM Transactions on Graphics*, 16:277–295, 1997.
42. D.H. Sattinger and O.L. Weaver. *Lie Groups and Algebras with Applications to Physics, Geometry, and Mechanics*. Applied Math. Science, Vol. 61. Springer-Verlag, first edition, 1986.
43. Laurent Schwartz. *Analyse II. Calcul Différentiel et Equations Différentielles*. Collection Enseignement des Sciences. Hermann, 1992.
44. J.M. Selig. *Geometrical Methods In Robotics*. Monographs In Computer Science. Springer-Verlag, first edition, 1996.
45. S. Sternberg. *Lectures On Differential Geometry*. AMS Chelsea, second edition, 1983.
46. Loring W. Tu. *An Introduction to Manifolds*. Universitext. Springer Verlag, first edition, 2008.
47. Frank Warner. *Foundations of Differentiable Manifolds and Lie Groups*. GTM No. 94. Springer-Verlag, first edition, 1983.
48. Hermann Weyl. *The Classical Groups. Their Invariants and Representations*. Princeton Mathematical Series, No. 1. Princeton University Press, second edition, 1946.

Chapter 19
Basics of the Differential Geometry of Curves

19.1 Introduction: Parametrized Curves

In this chapter we consider parametric curves, and we introduce two important invariants, curvature and torsion (in the case of a 3D curve).

Properties of curves can be classified into *local properties* and *global properties*. Local properties are the properties that hold in a small neighborhood of a point on a curve. Curvature is a local property. Local properties can be studied more conveniently by assuming that the curve is parametrized locally. Thus, it is important and useful to study parametrized curves. In order to study the global properties of a curve, such as the number of points where the curvature is extremal, the number of times that a curve wraps around a point, or convexity properties, topological tools are needed. A proper study of global properties of curves really requires the introduction of the notion of a manifold, a concept beyond the scope of this book. In this chapter we study only local properties of parametrized curves. Readers interested in learning about curves as manifolds and about global properties of curves are referred to do Carmo [7] and Berger and Gostiaux [2]. Kreyszig [15] is also an excellent source, which does a great job at tracing the origin of concepts. It turns out that it is easier to study the notions of curvature and torsion if a curve is parametrized by arc length, and thus we will discuss briefly the notion of arc length.

Let $\mathscr{E}$ be some normed affine space of finite dimension, for the sake of simplicity the Euclidean space $\mathbb{E}^2$ or $\mathbb{E}^3$. Recall that the Euclidean space $\mathbb{E}^m$ is obtained from the affine space $\mathbb{A}^m$ by defining on the vector space $\mathbb{R}^m$ the standard inner product

$$(x_1,\ldots,x_m)\cdot(y_1,\ldots,y_m) = x_1 y_1 + \cdots + x_m y_m.$$

The corresponding Euclidean norm is

$$\|(x_1,\ldots,x_m)\| = \sqrt{x_1^2 + \cdots + x_m^2}.$$

Inspired by a kinematic view, we can define a curve as a continuous map $f\colon\,]a,b[\,\to\mathscr{E}$ from an open interval $I=\,]a,b[$ of $\mathbb{R}$ to the affine space $\mathscr{E}$. From this point of view we can think of the parameter $t \in \,]a,b[$ as time, and the function f gives the position $f(t)$ at time t of a moving particle. The image $f(I) \subseteq \mathscr{E}$ of the interval I is the trajectory of the particle. In fact, asking only that f be continuous turns out to be too liberal, as rather strange curves turn out to be definable, such as "square-filling curves," due to Peano, Hilbert, Sierpiński, and others (see the problems).

Example 19.1. A very pretty square-filling curve due to Hilbert is defined by a sequence (h_n) of polygonal lines $h_n \colon [0,1] \to [0,1] \times [0,1]$ starting from the simple pattern h_0 (a "square cap" ⊓) shown on the left in Figure 19.1.

Fig. 19.1 A sequence of Hilbert curves h_0, h_1, h_2.

The curve h_{n+1} is obtained by scaling down h_n by a factor of $\frac{1}{2}$, and connecting the four copies of this scaled–down version of h_n obtained by rotating by $\pi/2$ (left lower part), rotating by $-\pi/2$ and translating right (right lower part), translating up (left upper part), and translating diagonally (right upper part), as illustrated in Figure 19.1.

It can be shown that the sequence (h_n) converges (pointwise) to a continuous curve $h \colon [0,1] \to [0,1] \times [0,1]$ whose trace is the entire square $[0,1] \times [0,1]$. The Hilbert curve h is nowhere differentiable. It also has infinite length! The curve h_5 is shown in Figure 19.2.

Actually, there are many fascinating curves that are only continuous, fractal curves being a major example (see Edgar [8]), but for our purposes we need the existence of the tangent at every point of the curve (except perhaps for finitely many

Fig. 19.2 The Hilbert curve h_5.

points). This leads us to require that $f\colon\,]a,b[\,\to\mathscr{E}$ be at least continuously differentiable. Recall that a function $f\colon\,]a,b[\,\to\mathbb{A}^n$ is *of class C^p, or is C^p-continuous,* if all the derivatives $f^{(k)}$ exist and are continuous for all k, $0\le k\le p$ (when $p=0$, $f^{(0)}=f$). Thus, we require f to be at least a C^1-function. However, asking that $f\colon\,]a,b[\,\to\mathscr{E}$ be a C^p-function for $p\ge 1$ still allows unwanted curves.

Example 19.2. The plane curve defined such that

$$f(t)=\begin{cases}(0,\,\mathrm{e}^{1/t}) & \text{if } t<0;\\[2pt] (0,\,0) & \text{if } t=0;\\[2pt] (\mathrm{e}^{-1/t},\,0) & \text{if } t>0;\end{cases}$$

is a C^∞-function, but $f'(0)=0$, and thus the tangent at the origin is undefined. What happens is that the curve has a sharp "corner" at the origin.

Example 19.3. Similarly, the plane curve defined such that

$$f(t)=\begin{cases}(-\mathrm{e}^{1/t},\,\mathrm{e}^{1/t}\sin(\mathrm{e}^{-1/t})) & \text{if } t<0;\\[2pt] (0,\,0) & \text{if } t=0;\\[2pt] (\mathrm{e}^{-1/t},\,\mathrm{e}^{-1/t}\sin(\mathrm{e}^{1/t})) & \text{if } t>0;\end{cases}$$

shown in Figure 19.3 is a C^∞-function, but $f'(0)=0$. In this case, the curve oscillates more and more rapidly as it approaches the origin.

The problem with the above examples is that the origin is a singular point for which $f'(0)=0$ (a stationary point).

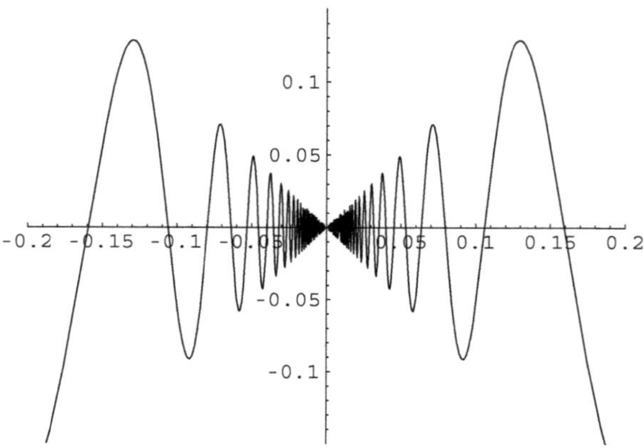

Fig. 19.3 Stationary point at the origin.

Although it is possible to define the tangent when f is sufficiently differentiable and when for every $t \in \,]a,b[\,$, $f^{(p)}(t) \neq 0$ for some $p \geq 1$ (where $f^{(p)}$ denotes the pth derivative of f), a systematic study is rather cumbersome. Thus, we will restrict our attention to curves having only regular points, that is, for which $f'(t) \neq 0$ for every $t \in \,]a,b[\,$. However, we will allow functions $f \colon \,]a,b[\, \to \mathscr{E}$ that are not necessarily injective, unless stated otherwise.

Definition 19.1. An *open curve (or open arc) of class C^p* is a map $f \colon \,]a,b[\, \to \mathscr{E}$ of class C^p, with $p \geq 1$, where $]a,b[$ is an open interval (allowing $a = -\infty$ or $b = +\infty$). The set of points $f(]a,b[)$ in $\mathscr{E}$ is called the *trace of the curve f*. A point $f(t)$ is *regular* at $t \in \,]a,b[$ if $f'(t)$ exists and $f'(t) \neq 0$, and *stationary* otherwise. A *regular open curve (or regular open arc) of class C^p* is an open curve of class C^p, with $p \geq 1$, such that every point is regular, i.e., $f'(t) \neq 0$ for every $t \in \,]a,b[\,$.

Note that Definition 19.1 is stated for an open interval $]a,b[$, and thus f may not be defined at a or b. If we want to include the boundary points at a and b in the curve (when $a \neq -\infty$ and $b \neq +\infty$), we use the following definition.

Definition 19.2. A *curve (or arc) of class C^p* is a map $f \colon [a,b] \to \mathscr{E}$, with $p \geq 1$, such that the restriction of f to $]a,b[$ is of class C^p, and where $f^{(i)}(a) = \lim_{t \to a, t > a} f^{(i)}(t)$ and $f^{(i)}(b) = \lim_{t \to b, t < b} f^{(i)}(t)$ exist, where $0 \leq i \leq p$. A *regular curve (or regular arc) of class C^p* is a curve of class C^p, with $p \geq 1$, such that every point is regular, i.e., $f'(t) \neq 0$ for every $t \in [a,b]$. The set of points $f([a,b])$ in $\mathscr{E}$ is called the *trace of the curve f*.

It should be noted that even if f is injective, the trace $f(I)$ of f may be self-intersecting.

Example 19.4. Consider the curve $f \colon \mathbb{R} \to \mathbb{E}^2$ defined such that

$$f_1(t) = \frac{t(1+t^2)}{1+t^4},$$
$$f_2(t) = \frac{t(1-t^2)}{1+t^4}.$$

The trace of this curve, shown in Figure 19.4, is called the "lemniscate of Bernoulli" and it has a self-intersection at the origin. The map f is continuous, and in fact bijective, but its inverse f^{-1} is not continuous. Self-intersection is due to the fact that

$$\lim_{t \to -\infty} f(t) = \lim_{t \to +\infty} f(t) = f(0).$$

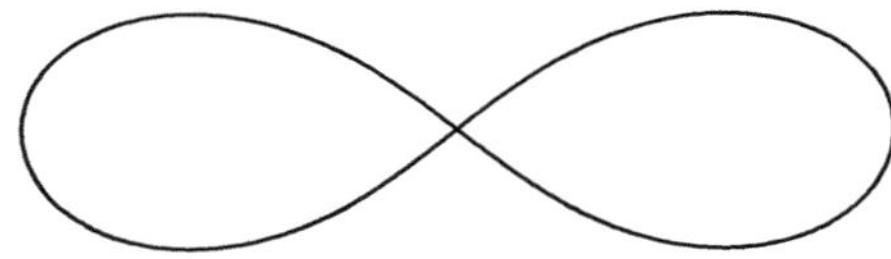

Fig. 19.4 Lemniscate of Bernoulli.

If we consider a curve $f \colon [a,b] \to \mathscr{E}$ and we assume that f is injective on the entire *closed* interval $[a,b]$, then the trace $f([a,b])$ of f has no self-intersection. Such curves are usually called *Jordan arcs*, or *simple arcs*. The theory of Jordan arcs $f \colon [a,b] \to \mathscr{E}$ where f is only required to be continuous is quite rich. Because $[a,b]$ is compact, f is in fact a homeomorphism between $[a,b]$ and $f([a,b])$. Many fractal curves are only continuous Jordan arcs that are not differentiable.

We can also define closed curves. A simple way to do so is to say that a closed curve is a curve $f \colon [a,b] \to \mathscr{E}$ such that $f(a) = f(b)$. However, this does not ensure that the derivatives at a and b agree, a situation that is quite undesirable. A better solution is to define a closed curve as an open curve $f \colon \mathbb{R} \to \mathscr{E}$, where f is periodic.

Definition 19.3. A *closed curve (or closed arc) of class C^p* is a map $f \colon \mathbb{R} \to \mathscr{E}$ such that f is of class C^p, with $p \geq 1$, and such that f is *periodic*, which means that there is some $T > 0$ such that $f(x+T) = f(x)$ for all $x \in \mathbb{R}$. A *regular closed curve (or regular closed arc) of class C^p* is a closed curve of class C^p, with $p \geq 1$, such that every point is regular, i.e., $f'(t) \neq 0$ for every $t \in \mathbb{R}$. The set of points $f([0,T])$ (or $f(\mathbb{R})$) in $\mathscr{E}$ is called the *trace of the curve f*.

A closed curve is a *Jordan curve (or a simple closed curve)* if f is injective on the interval $[0, T[$. A Jordan curve has no self-intersection. The ellipse defined by the map $t \mapsto (a\cos t, b\sin t)$ is an example of a closed curve of type C^∞ that is a Jordan curve. In this example, the period is $T = 2\pi$. Again, the theory of Jordan curves $f\colon [0, T] \to \mathscr{E}$ where f is only required to be continuous is quite rich.

An observant reader may have noticed that a curve has been defined as a map $f\colon\,]a, b[\, \to \mathscr{E}$ (or $f\colon [a, b] \to \mathscr{E}$), rather than as a certain set of points. In fact, it is possible for the trace of a curve to be defined by many parametrizations, as illustrated by the unit circle, which is the trace of the parametrized curves $f_k\colon\,]0, 2\pi[\, \to \mathscr{E}$ (or $f_k\colon [0, 2\pi] \to \mathscr{E}$), where $f_k(t) = (\cos kt, \sin kt)$, with $k \geq 1$. A clean way to handle this phenomenon is to define a notion of *geometric curve (or arc)*. Such a treatment is given in Berger and Gostiaux [2]. For our purposes it will be sufficient to define a notion of change of parameter that does not change the "geometric shape" of the trace. Recall that a *diffeomorphism* $g\colon\,]a, b[\, \to\,]c, d[\ of\ class$ C^p from an open interval $]a, b[$ to another open interval $]c, d[$ is a bijection such that both $g\colon\,]a, b[\, \to\,]c, d[$ and its inverse $g^{-1}\colon\,]c, d[\, \to\,]a, b[$ are C^p-functions. This implies that $f'(c) \neq 0$ for every $c \in\,]a, b[$.

Definition 19.4. Two regular curves $f\colon\,]a, b[\, \to \mathscr{E}$ and $g\colon\,]c, d[\, \to \mathscr{E}$ of class C^p, with $p \geq 1$, are C^p-*equivalent* if there is a diffeomorphism $\theta\colon\,]a, b[\, \to\,]c, d[$ of class C^p such that $f = g \circ \theta$.

It is immediately verified that Definition 19.4 yields an equivalence relation on open curves. Definition 19.4 is adapted to curves, by extending the notion of C^p-diffeomorphism to closed intervals in the obvious way.

Remark: Using Definition 19.4, we could define a *geometric curve (or arc) of class C^p* as an equivalence class of (parametrized) curves. This is done in Berger and Gostiaux [2].

From now on, in most cases we will drop the word "regular" when referring to regular curves, and simply say "curves." Also, when we refer to a point $f(t)$ on a curve, we mean that $t \in\,]a, b[$ for an open curve $f\colon\,]a, b[\, \to \mathscr{E}$, and $t \in [a, b]$ for a curve $f\colon [a, b] \to \mathscr{E}$. In the case of a closed curve $f\colon \mathbb{R} \to \mathscr{E}$, we can assume that $t \in [0, T]$, where T is the period of f, and thus closed curves will be treated simply as curves in the sequel. We now define tangent lines and osculating planes. According to Kreyszig [15], the term osculating plane was apparently first introduced by Tinseau in 1780.

19.2 Tangent Lines and Osculating Planes

We begin with the definition of a tangent line.

Definition 19.5. For any open curve $f\colon\,]a, b[\, \to \mathscr{E}$ of class C^p (or curve $f\colon [a, b] \to \mathscr{E}$ of class C^p), with $p \geq 1$, given any point $M_0 = f(t)$ on the curve, if f is locally injective at M_0 and if for any point $M_1 = f(t+h)$ near M_0 the line $T_{t,h}$ determined by the points M_0 and M_1 has a limit T_t when $h \neq 0$ approaches 0, we say that T_t is the *tangent line to f in $M_0 = f(t)$ at t.*

More precisely, if there is an open interval $]t - \eta, t + \eta[\subseteq]a,b[$ (with $\eta > 0$) such that $M_1 = f(t+h) \neq f(t) = M_0$ for all $h \neq 0$ with $h \in]-\eta, \eta[$ and the line $T_{t,h}$ determined by the points M_0 and M_1 has a limit T_t when $h \neq 0$ approaches 0 (with $h \in]-\eta, \eta[$), then T_t is the tangent line to f in M_0 at t.

For simplicity we will often say "tangent," instead of "tangent line." The definition is simpler when f is a simple curve (there is no danger that $M_1 = M_0$ when $h \neq 0$). In this chapter there will be situations where it is notationally more convenient to denote the vector $\overrightarrow{ab}$ by $b - a$. The following lemma shows why regular points are important.

Lemma 19.1. *For any open curve $f\colon\,]a,b[\to \mathcal{E}$ of class C^p (or curve $f\colon [a,b] \to \mathcal{E}$ of class C^p), with $p \geq 1$, given any point $M_0 = f(t)$ on the curve, if M_0 is a regular point at t, then the tangent line to f in M_0 at t exists and is determined by the derivative $f'(t)$ of f at t.*

Proof. Provided that $M_0 \neq M_1$, the line $T_{t,h}$ is determined by the point M_0 and the vector $M_1 - M_0 = f(t+h) - f(t)$. By the definition of $f'(t)$, we have

$$f(t+h) - f(t) = hf'(t) + h\varepsilon(h),$$

where $\lim_{h \to 0, h \neq 0} \varepsilon(h) = 0$. We claim that there must be an open interval $]t - \eta, t + \eta[\subseteq]a,b[$ (with $\eta > 0$) such that $f(t+h) \neq f(t)$ for all $h \neq 0$ with $-\eta < h < \eta$. Otherwise, since $f'(t)$ exists, for every $\alpha > 0$ there is some $\eta > 0$ such that

$$\left\| \frac{f(t+h) - f(t)}{h} - f'(t) \right\| \leq \alpha$$

for all h, with $-\eta < h < \eta$, and since $f(t+h) - f(t) = 0$ for some $h \neq 0$ with $h \in]-\eta, \eta[$, we would have $\|f'(t)\| \leq \alpha$. Since this holds for every $\alpha > 0$, we would have $f'(t) = 0$, a contradiction. Thus, the line $T_{t,h}$ is determined by the point M_0 and the vector $f'(t) + \varepsilon(h)$, which has the limit $f'(t)$ when $h \neq 0$ tends to 0, with $h \in]-\eta, +\eta[$. Thus, the line $T_{t,h}$ has for limit the line determined by M_0 and the derivative $f'(t)$ of f at t. $\square$

Remark: If $f'(t) = 0$, the above argument breaks down. However, if f is a C^p-function and $f^{(p)}(t) \neq 0$ for some $p \geq 2$, where p is the smallest integer with that property, we can show that the line $T_{t,h}$ has the limit determined by M_0 and the derivative $f^{(p)}(t)$. Thus, the tangent line may still exist at a stationary point. For example, the curve f defined by the map $t \mapsto (t^2, t^3)$ is a C^∞-function, but $f'(0) = 0$. Nevertheless, the tangent at the origin is defined for $t = 0$ (it is the x-axis). However, some strange things can happen at a stationary point. Assuming that a curve is of class C^p for p large enough, using Taylor's formula it is possible to study precisely the behavior of the curve at a stationary point.

Note that the tangent at a point can exist, even when the derivative f' is not continuous at this point.

Example 19.5. The C^0-curve f defined such that

$$f(t) = \begin{cases} (t, t^2 \sin(1/t)) & \text{if } t \neq 0; \\ (0,0) & \text{if } t = 0; \end{cases}$$

and shown in Figure 19.5 has a tangent at $t = 0$.

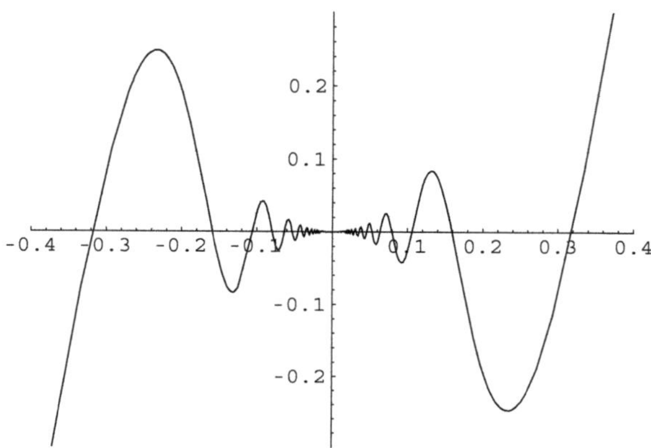

Fig. 19.5 Curve with tangent at O and yet f' discontinuous at O.

Indeed, $f(0) = (0, 0)$, and $\lim_{t \to 0} t \sin(1/t) = 0$, and the derivative at $t = 0$ is the vector $(1, 0)$. For $t \neq 0$,

$$f'(t) = (1, 2t \sin(1/t) - \cos(1/t)),$$

which has no limit as t tends to 0. Thus, f' is discontinuous at 0. What happens is that f oscillates more and more near the origin, but the amplitude of the oscillations decreases.

If $g = f \circ \theta$ is a curve C^p-equivalent to f, where θ is a C^p-diffeomorphism, the tangent at $\theta(t)$ to f exists iff the tangent at t to g exists, and the two tangents are identical. Indeed, $g'(t) = f'(u)\theta'(t)$, where $u = \theta(t)$, and since $\theta'(t) \neq 0$ because θ is a diffeomorphism, the result is clear. Thus, the notion of tangent is intrinsic to the geometric curve defined by f. We now consider osculating planes.

Definition 19.6. For any open curve $f \colon]a,b[\to \mathscr{E}$ of class C^p (or curve $f \colon [a,b] \to \mathscr{E}$ of class C^p), with $p \geq 2$, given any point $M_0 = f(t)$ on the curve, if the tangent T_t at M_0 exists, the point $M_1 = f(t+h)$ is not on T_t for $h \neq 0$ small enough, and the plane $P_{t,h}$ determined by the tangent T_t and the point M_1 has a limit P_t as $h \neq 0$ approaches 0, we say that P_t is the *osculating plane to f in $M_0 = f(t)$ at t.*

More precisely, if the tangent T_t at M_0 exists, there is an open interval $]t - \eta, t + \eta[$ (with $\eta > 0$) such that the point $M_1 = f(t+h)$ is not on T_t for every $h \neq 0$ with $h \in]-\eta, +\eta[$, and the plane $P_{t,h}$ determined by the tangent T_t and the point M_1 has a limit P_t when $h \neq 0$ approaches 0 (with $h \in]-\eta, +\eta[$), we say that P_t is the osculating plane to f in $M_0 = f(t)$ at t.

Again, the definition is simpler when f is a simple curve. The following lemma gives a simple condition for the existence of the osculating plane at a point.

Lemma 19.2. *For any open curve $f\colon]a,b[\to \mathscr{E}$ of class C^p (or curve $f\colon [a,b] \to \mathscr{E}$ of class C^p), with $p \geq 2$, given any point $M_0 = f(t)$ on the curve, if $f'(t)$ and $f''(t)$ are linearly independent (which implies that M_0 is a regular point at t), then the osculating plane to f in M_0 at t exists and is determined by the first and second derivatives $f'(t)$ and $f''(t)$ of f at t.*

Proof. The plane $P_{t,h}$ is determined by the point M_0, the vector $f'(t)$, and the vector $M_1 - M_0 = f(t+h) - f(t)$, provided that $M_1 - M_0$ and $f'(t)$ are linearly independent. By Taylor's formula, for $h > 0$ small enough we have

$$f(t+h) - f(t) = hf'(t) + \frac{h^2}{2} f''(t) + \frac{h^2}{2} \varepsilon(h),$$

where $\lim_{h \to 0, h \neq 0} \varepsilon(h) = 0$. By an argument similar to that used in Lemma 19.1, we can show that there is some open interval $]t - \eta, t + \eta[\subseteq]a,b[$ (with $\eta > 0$) such that for every $h \neq 0$ with $-\eta < h < \eta$, the point $M_1 = f(t+h)$ is not on the tangent T_t (otherwise, we could prove that $f''(t)$ is the limit of a sequence of vectors proportional to $f'(t)$, and thus that $f'(t)$ and $f''(t)$ are linearly dependent, a contradiction). Thus, for $h \neq 0$ with $h \in]-\eta, +\eta[$, the plane $P_{t,h}$ is determined by the point M_0, the vector $f'(t)$, and the vector $f''(t) + \varepsilon(h)$, which has the limit $f''(t)$ as $h \neq 0$ tends to 0, with $h \in]-\eta, +\eta[$. Thus, the plane $P_{t,h}$ has for limit the plane determined by M_0 and the derivatives $f'(t)$ and $f''(t)$ of f at t, since $f'(t)$ and $f''(t)$ are assumed to be linearly independent. $\square$

When $f'(t)$ and $f''(t)$ exist and are linearly independent, it is sometimes said that f is *biregular at t*, and that $f(t)$ *is a biregular point at t*. From the kinematic point of view, the osculating plane at time t is determined by the position of the moving particle $f(t)$, the velocity vector $f'(t)$, and the acceleration vector $f''(t)$.

Remark: If the curve f is a plane curve, then the osculating plane at every regular point is the plane containing the curve. Even when $f'(t)$ and $f''(t)$ are linearly dependent, the osculating plane may still exist, for instance, if there are two derivatives $f^{(p)}(t) \neq 0$ and $f^{(q)}(t) \neq 0$ that are linearly independent, with $p < q$, the smallest integers with that property.

In general, the curve crosses its osculating plane at the point of contact t.

If $g = f \circ \theta$ is a curve C^p-equivalent to f, where θ is a C^p-diffeomorphism, the osculating plane at $\theta(t)$ to f exists iff the osculating plane at t to g exists, and these two planes are identical. Indeed, $g'(t) = f'(u)\theta'(t)$, and

$$g''(t) = f''(u)\theta'(t)^2 + f'(u)\theta''(t),$$

where $u = \theta(t)$. Since $\theta'(t) \neq 0$ because θ is a diffeomorphism, the planes defined by $(f'(u), f''(u))$ and $(g'(t), g''(t))$ are identical. Thus, the notion of osculating plane is intrinsic to the geometric curve defined by f.

It should also be noted that the notions of tangent and osculating plane are affine notions, that is, preserved under affine bijections.

We now consider the notion of arc length. For this, we assume that the affine space $\mathscr{E}$ is a normed affine space of finite dimension with norm $\| \ \|$. For simplicity, we can assume that $\mathscr{E} = \mathbb{E}^n$.

19.3 Arc Length

Given an interval $[a,b]$ (where $a \neq -\infty$ and $b \neq +\infty$), a *subdivision* of $[a,b]$ is any finite increasing sequence $t_0, \ldots, t_n$ such that $t_0 = a$, $t_n = b$, and $t_i < t_{i+1}$, for all i, $0 \leq i \leq n-1$, where $n \geq 1$. Given any curve $f \colon [a,b] \to \mathscr{E}$ of class C^p, with $p \geq 0$, for any subdivision $\sigma = t_0, \ldots, t_n$ of $[a,b]$ we obtain a polygonal line $f(t_0), f(t_1), \ldots, f(t_n)$ with endpoints $f(a)$ and $f(b)$, and we define the length of this polygonal line as

$$l(\sigma) = \sum_{i=0}^{n-1} \|f(t_{i+1}) - f(t_i)\|.$$

Definition 19.7. For any curve $f \colon [a,b] \to \mathscr{E}$ of class C^p, with $p \geq 0$, if the set $\mathscr{L}(f)$ of the lengths $l(\sigma)$ of the polygonal lines induced by all subdivisions $\sigma = t_0, \ldots, t_n$ of $[a,b]$ is bounded, we say that f is *rectifiable*, and we call the least upper bound $l(f)$ of the set $\mathscr{L}(f)$ the *length of f*.

It is obvious that $\|f(b) - f(a)\| \leq l(f)$. If $g = f \circ \theta$ is a curve C^p-equivalent to f, where θ is a C^p-diffeomorphism, since $\theta'(t) \neq 0$, θ is a strictly increasing or decreasing function, and thus the set of sums of the form $l(\sigma)$ is the same for both f and g. Thus, the notion of length is intrinsic to the geometric curve defined by f. This is false if θ is not strictly increasing or decreasing. The following lemma can be shown.

Lemma 19.3. *For any curve $f \colon [a,b] \to \mathscr{E}$ of class C^p, with $p \geq 1$, f is rectifiable.*

Remark: In fact, Lemma 19.3 can be shown under the hypothesis that f is of class C^0, and that $f'(t)$ exists and $\|f'(t)\| \leq M$ for some $M \geq 0$, for all $t \in [a,b]$.

Definition 19.8. For any open curve $f \colon \ ]a,b[\ \to \mathscr{E}$ of class C^p (or curve $f \colon [a,b] \to \mathscr{E}$ of class C^p), with $p \geq 1$, for any closed interval $[t_0, t] \subseteq \ ]a,b[$ (or $[t_0, t] \subseteq [a,b]$, in the case of a curve), letting $f_{[t_0,t]}$ be the restriction of f to $[t_0,t]$, the length $l(f_{[t_0,t]})$ (which exists, by Lemma 19.3) is called the *arc length* of $f_{[t_0,t]}$. For any fixed $t_0 \in \ ]a,b[$ (or any fixed $t_0 \in [a,b]$, in the case of a curve), we define the function $s \colon \ ]a,b[\ \to \mathbb{R}$ (or $s \colon [a,b] \to \mathbb{R}$, in the case of a curve), called *algebraic arc* *length w.r.t.* t_0, as follows:

$$s(t) = \begin{cases} l(f_{[t_0,t]}) & \text{if } [t_0,t] \subseteq \,]a,b[\,; \\ -l(f_{[t_0,t]}) & \text{if } [t,t_0] \subseteq \,]a,b[\,; \end{cases}$$

(and similarly in the case of a curve, except that $[t_0,t] \subseteq [a,b]$ or $[t,t_0] \subseteq [a,b]$).

For the sake of brevity, we will often call s the arc length, rather than algebraic arc length w.r.t. t_0.

Lemma 19.4. *For any open curve $f: \,]a,b[\, \to \mathscr{E}$ of class C^p (or curve $f: [a,b] \to \mathscr{E}$ of class C^p), with $p \geq 1$, for any fixed $t_0 \in \,]a,b[$ (or $t_0 \in [a,b]$, in the case of a curve), the algebraic arc length $s(t)$ w.r.t. t_0 is of class C^p, and furthermore, $s'(t) = \|f'(t)\|$.*

Thus, the arc length is given by the integral

$$s(t) = \int_{t_0}^{t} \|f'(u)\| \, du.$$

In particular, when $\mathscr{E} = \mathbb{E}^n$ and the norm is the Euclidean norm, we have

$$s(t) = \int_{t_0}^{t} \sqrt{f_1'(u)^2 + \cdots + f_n'(u)^2} \, du.$$

where $f = (f_1, \ldots, f_n)$. The number $\|f'(t)\|$ is often called the *speed* of $f(t)$ at time t. For every regular point at t, the unit vector

$$\mathbf{t} = \frac{f'(t)}{\|f'(t)\|}$$

is called the *unit tangent (vector) at t.*

Now, if $f: \,]a,b[\, \to \mathscr{E}$ (or $f: [a,b] \to \mathscr{E}$) is a regular curve of class C^p, with $p \geq 1$, since $s'(t) = \|f'(t)\|$, and $f'(t) \neq 0$ for all $t \in \,]a,b[$ (or $t \in [a,b]$), we have $s'(t) > 0$ for all $t \in \,]a,b[$ (or $t \in [a,b]$). The mean value theorem implies that s is injective, and that $s: \,]a,b[\, \to \,]s(a),s(b)[$ (or $s: [a,b] \to [s(a),s(b)]$) is a diffeomorphism of class C^p. In particular, the curve $f \circ \varphi: \,]s(a),s(b)[\, \to \mathscr{E}$ (or $f \circ \varphi: [s(a),s(b)] \to \mathscr{E}$), with $\varphi = s^{-1}$, is C^p-equivalent to the original curve f, but it is parametrized by the arc length $s \in \,]s(a),s(b)[$ (or $s \in [s(a),s(b)]$). As a consequence, since $\varphi = s^{-1}$, we have

$$\varphi'(s(t)) = (s'(t))^{-1},$$

and letting $g = f \circ \varphi$, by the chain rule

$$g'(s(t)) = f'(\varphi(s(t)))\varphi'(s(t)) = f'(t)(s'(t))^{-1} = \frac{f'(t)}{\|f'(t)\|}.$$

This shows that $\|g'(s)\| = 1$, i.e., that when a regular curve is parametrized by arc length, its velocity vector has unit length. From a kinematic point of view, when a curve is parametrized by arc length, the moving particle travels at constant unit speed.

Remark: If a curve f (or a closed curve) is of class C^p, for $p \geq 1$, and it is a Jordan arc, then the algebraic arc length $s\colon [a,b] \to \mathbb{R}$ w.r.t. t_0 is strictly increasing, and thus injective. Thus, s^{-1} exists, and the curve can still be parametrized by arc length as $g = f \circ s^{-1}$. However, $g'(s)$ exists only when $s(t)$ corresponds to a regular point at t. Thus, it still seems necessary to restrict our attention to regular curves, in order to avoid complications.

We now consider the notion of curvature. In order to do so, we assume that the affine space $\mathscr{E}$ has a Euclidean structure (an inner product), and that the norm on $\mathscr{E}$ is the norm induced by this inner product. For simplicity, we assume that $\mathscr{E} = \mathbb{E}^n$.

19.4 Curvature and Osculating Circles (Plane Curves)

In a Euclidean space, orthogonality makes sense, and we can define normal lines and normal planes. We begin with plane curves, i.e., the case where $\mathscr{E} = \mathbb{E}^2$.

Definition 19.9. Given a regular plane curve $f\colon \,]a,b[\,\to \mathscr{E}$ (or $f\colon [a,b] \to \mathscr{E}$) of class C^p, with $p \geq 1$, the *normal line N_t to f at t* is the line through $f(t)$ and orthogonal to the tangent line T_t to f at t. Any nonnull vector defining the direction of the normal line N_t is called a *normal vector to f at t*.

From now on, we also assume that we are dealing with curves f that are biregular for all t. This means that $f'(t)$ and $f''(t)$ always exist and are linearly independent. A fairly intuitive way to introduce the notion of curvature is to study the variation of the normal line N_t to a curve f at t, in a small neighborhood of t. The intuition is that the normal N_t to f at t rotates around a certain point, and that the "speed" of rotation of the normal measures how much the curve bends around t. In other words, the rate at which the normal turns corresponds to the curvature of the curve at t. Another way to look at it is to focus on the point around which the normal turns, the center of curvature C at t, and to consider the radius $\mathscr{R}$ of the circle centered at C and tangent to the curve at $f(t)$ (i.e., tangent to the tangent line to f at t). Intuitively, the smaller $\mathscr{R}$ is, the faster the curve bends, and thus the curvature can be defined as $1/\mathscr{R}$.

Let us assume that some origin O is chosen in the affine plane, and to simplify the notation, for any curve f let us denote $f(t) - O$ by $\mathbf{M}(t)$ or $\mathbf{M}$, for any point P denote $P - O$ by $\mathbf{P}$, denote $P - M$ by $\overrightarrow{MP}$, and denote $f'(t)$ by $\mathbf{M}'(t)$ or $\mathbf{M}'$. The normal line N_t to f at t is the set of points P such that

$$\mathbf{M}' \cdot \overrightarrow{MP} = 0,$$

or equivalently

$$\mathbf{M}' \cdot \mathbf{P} = \mathbf{M}' \cdot \mathbf{M}.$$

Similarly, for any small $\delta \neq 0$ such that $f(t+\delta)$ is defined, the normal line $N_{t+\delta}$ to f at $t+\delta$ is the set of points Q such that

$$\mathbf{M}'(t+\delta) \cdot \mathbf{Q} = \mathbf{M}'(t+\delta) \cdot \mathbf{M}(t+\delta).$$

Thus, the intersection point P of N_t and $N_{t+\delta}$, if it exists, is given by the equations

$$\mathbf{M}' \cdot \mathbf{P} = \mathbf{M}' \cdot \mathbf{M},$$
$$\mathbf{M}'(t+\delta) \cdot \mathbf{P} = \mathbf{M}'(t+\delta) \cdot \mathbf{M}(t+\delta).$$

Thus, P would also satisfy the equation obtained by subtracting the first one from the second, that is,

$$(\mathbf{M}'(t+\delta) - \mathbf{M}') \cdot \mathbf{P} = \mathbf{M}'(t+\delta) \cdot \mathbf{M}(t+\delta) - \mathbf{M}' \cdot \mathbf{M}.$$

This equation can be written as

$$\left(\frac{\mathbf{M}'(t+\delta) - \mathbf{M}'}{\delta} \right) \cdot \mathbf{P} = \left(\frac{\mathbf{M}'(t+\delta) - \mathbf{M}'}{\delta} \right) \cdot \mathbf{M}(t+\delta)$$
$$+ \mathbf{M}' \cdot \left(\frac{\mathbf{M}(t+\delta) - \mathbf{M}}{\delta} \right),$$

and as $\delta \neq 0$ tends to 0, it has the following equation for limit:

$$\mathbf{M}'' \cdot \mathbf{P} = \mathbf{M}'' \cdot \mathbf{M} + \mathbf{M}' \cdot \mathbf{M}',$$

that is,

$$\mathbf{M}'' \cdot \overrightarrow{MP} = \|\mathbf{M}'\|^2.$$

Consequently, if it exists, P is the intersection of the two lines of equations

$$\mathbf{M}' \cdot \overrightarrow{MP} = 0$$
$$\mathbf{M}'' \cdot \overrightarrow{MP} = \|\mathbf{M}'\|^2.$$

Thus, if $\mathbf{M}'$ and $\mathbf{M}''$ are linearly independent, which is equivalent to saying that $f'(t)$ and $f''(t)$ are linearly independent, i.e., f is biregular at t, the above two equations have a unique solution P. Also, the above analysis shows that the intersection of the two normals N_t and $N_{t+\delta}$, for $\delta \neq 0$ small enough, has a limit C (really, $C(t)$). This limit is called the *center of curvature of f at t*. It is possible to compute the distance $\mathscr{R} = \|\overrightarrow{MC}\|$, the *radius of curvature at t*, and the coordinates of C, given any affine frame for the plane. It is worth noting that the equation

$$\mathbf{M}'' \cdot \mathbf{P} = \mathbf{M}'' \cdot \mathbf{M} + \mathbf{M}' \cdot \mathbf{M}'$$

is obtained by taking the derivative of the equation

$$\mathbf{M}' \cdot \mathbf{P} = \mathbf{M}' \cdot \mathbf{M}$$

with respect to t. This observation can be used to compute the coordinates of the center of curvature, but first we show that the radius of curvature has a very simple expression when the curve is parametrized by arc length. Indeed, in this case, $\|f'(s)\| = \|\mathbf{M}'\| = 1$, that is, $f'(s) \cdot f'(s) = 1$, and by taking the derivatives of both sides, we get

$$f''(s) \cdot f'(s) = 0,$$

which shows that $f''(s) = \mathbf{M}''$ and $f'(s) = \mathbf{M}'$ are orthogonal, and since the center of curvature C is determined by the equations

$$\mathbf{M}' \cdot \overrightarrow{MC} = 0,$$
$$\mathbf{M}'' \cdot \overrightarrow{MC} = \|\mathbf{M}'\|^2,$$

the vector $\overrightarrow{MC}$ must be collinear with $\mathbf{M}''$ (since it is orthogonal to $\mathbf{M}'$, which itself is orthogonal to $\mathbf{M}''$). Then, letting

$$\mathbf{n} = \frac{\mathbf{M}''}{\|\mathbf{M}''\|}$$

be the unit vector associated with the acceleration vector $\mathbf{M}''$, we have $\overrightarrow{MC} = \mathscr{R}\mathbf{n}$, and since $\|\mathbf{M}'\| = 1$, from $\mathbf{M}'' \cdot \overrightarrow{MC} = \|\mathbf{M}'\|^2$ we get

$$\mathbf{M}'' \cdot \overrightarrow{MC} = \mathbf{M}'' \cdot \mathscr{R}\,\frac{\mathbf{M}''}{\|\mathbf{M}''\|} = \mathscr{R}\,\frac{(\mathbf{M}'' \cdot \mathbf{M}'')}{\|\mathbf{M}''\|} = \mathscr{R}\,\frac{\|\mathbf{M}''\|^2}{\|\mathbf{M}''\|} = \mathscr{R}\,\|\mathbf{M}''\| = 1,$$

that is,

$$\mathscr{R} = \frac{1}{\|\mathbf{M}''\|} = \frac{1}{\|f''(s)\|}.$$

Thus, the radius of curvature is the inverse of the norm of the acceleration vector $f''(s)$. We define the *curvature* κ as the inverse of the radius of curvature $\mathscr{R}$, that is, as

$$\kappa = \|f''(s)\|.$$

In summary, when the curve f is parametrized by arc length, we found that the curvature κ and the radius of curvature $\mathscr{R}$ are defined by the equations

$$\boxed{\kappa = \|f''(s)\|, \quad \mathscr{R} = \frac{1}{\kappa}.}$$

We now come back to the general case. Assuming that $\mathbf{M}'$ and $\mathbf{M}''$ are linearly independent, we can write $\overrightarrow{MC} = \alpha\mathbf{M}' + \beta\mathbf{M}''$, for some unique α, β. Since C is determined by the equations

$$\mathbf{M}' \cdot \overrightarrow{MC} = 0,$$
$$\mathbf{M}'' \cdot \overrightarrow{MC} = \|\mathbf{M}'\|^2,$$

we get the system

$$(\mathbf{M}' \cdot \mathbf{M}')\,\alpha + (\mathbf{M}' \cdot \mathbf{M}'')\,\beta = 0,$$
$$(\mathbf{M}' \cdot \mathbf{M}'')\,\alpha + (\mathbf{M}'' \cdot \mathbf{M}'')\,\beta = \|\mathbf{M}'\|^2,$$

and we also note that

$$\mathscr{R}^2 = \overrightarrow{MC} \cdot \overrightarrow{MC} = \overrightarrow{MC} \cdot (\alpha\mathbf{M}' + \beta\mathbf{M}'') = \beta\|\mathbf{M}'\|^2.$$

The reader can verify that we obtain

$$\beta = \frac{\|\mathbf{M}'\|^4}{\|\mathbf{M}'\|^2\|\mathbf{M}''\|^2 - (\mathbf{M}' \cdot \mathbf{M}'')^2},$$

and thus

$$\mathscr{R}^2 = \frac{\|\mathbf{M}'\|^6}{\|\mathbf{M}'\|^2\|\mathbf{M}''\|^2 - (\mathbf{M}' \cdot \mathbf{M}'')^2}.$$

However, if we remember about the cross product of vectors and the Lagrange identity, we have

$$\|\mathbf{M}'\|^2\|\mathbf{M}''\|^2 - (\mathbf{M}' \cdot \mathbf{M}'')^2 = \|\mathbf{M}' \times \mathbf{M}''\|^2,$$

and thus

$$\mathscr{R} = \frac{\|\mathbf{M}'\|^3}{\|\mathbf{M}' \times \mathbf{M}''\|} = \frac{\|f'(t)\|^3}{\|f'(t) \times f''(t)\|},$$

and the curvature is given by

$$\kappa = \frac{\|\mathbf{M}' \times \mathbf{M}''\|}{\|\mathbf{M}'\|^3} = \frac{\|f'(t) \times f''(t)\|}{\|f'(t)\|^3}.$$

In summary, when the curve f is not necessarily parametrized by arc length, we found that the curvature κ and the radius of curvature $\mathscr{R}$ are defined by the equations

$$\boxed{\kappa = \frac{\|f'(t) \times f''(t)\|}{\|f'(t)\|^3}, \quad \mathscr{R} = \frac{1}{\kappa}.}$$

Note that from an analytical point of view, the curvature has the advantage of being defined at every regular point, since $\kappa = 0$ when either $f''(t) = 0$ or $f''(t)$ is collinear to $f'(t)$, whereas at such points, the radius of curvature goes to $+\infty$.

We leave as an exercise to show that if $g = f \circ \theta$ is a curve C^p-equivalent to f, where θ is a C^p-diffeomorphism, then

$$\kappa = \frac{\|f'(u) \times f''(u)\|}{\|f'(u)\|^3} = \frac{\|g'(t) \times g''(t)\|}{\|g'(t)\|^3},$$

where $u = \theta(t)$, i.e., κ has the same value for both f and g. Thus, the curvature is an *invariant* intrinsic to the geometric curve defined by f. In view of the above considerations, we give the following definition of the curvature, which is more intrinsic.

Definition 19.10. For any regular plane curve $f\colon \,]a,b[\, \to \mathscr{E}$ (or $f\colon [a,b] \to \mathscr{E}$) of class C^p parametrized by arc length, with $p \geq 2$, the *curvature* κ *at s* is defined as the nonnegative real number $\kappa = \|f''(s)\|$. For every s such that $f''(s) \neq 0$, letting $\mathbf{n} = f''(s)/\|f''(s)\|$ be the unit vector associated with $f''(s)$, we have $f''(s) = \kappa\mathbf{n}$, the point C defined such that $C - f(s) = \mathbf{n}/\kappa$ is the *center of curvature at s*, and $\mathscr{R} = 1/\kappa$ is the *radius of curvature at s*. The circle of center C and of radius $\mathscr{R}$ is called the *osculating circle to f at s*. When $f''(s) = 0$, by convention $\mathscr{R} = \infty$, and the center of curvature is undefined.

The locus of the center of curvature is a curve that is regular, except at points for which $\mathscr{R}' = 0$. Properties of this curve, called the *evolute*, will be given in Lemma 19.5.

Example 19.6. The evolute of an ellipse, the center of curvature corresponding to a specific point on the ellipse, and the osculating circle at that point are shown in Figure 19.6.

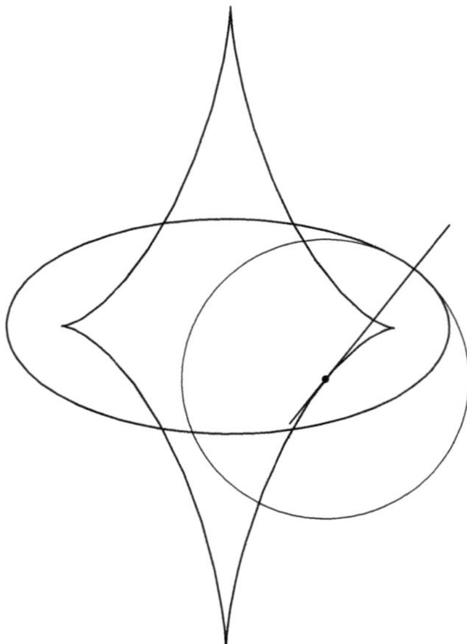

Fig. 19.6 The evolute of an ellipse, and an osculating circle.

It is also possible to define the notion of osculating circle more geometrically as a limit, in the spirit of the definition of a tangent.

Definition 19.11. Given any plane curve $f \colon\,]a,b[\to \mathcal{E}$ (or $f \colon [a,b] \to \mathcal{E}$) of class C^p, with $p \geq 1$, and given any point $M_0 = f(t)$ on the curve, if f is locally injective at M_0, the tangent T_t to f at t exists, and the circle $\Sigma_{t,h}$ tangent to T_t and passing through M_1 has a limit Σ_t as $h \neq 0$ approaches 0, we say that Σ_t is the *osculating circle to f in $M_0 = f(t)$ at t.*

More precisely, if there is an open interval $]t - \eta, t + \eta[\,\subseteq\,]a,b[$ (with $\eta > 0$) such that, $M_1 = f(t+h) \neq f(t) = M_0$ for all $h \neq 0$ with $h \in]-\eta, \eta[$, the tangent T_t to f at t exists, and the circle $\Sigma_{t,h}$ tangent to T_t and passing through M_1 has a limit Σ_t as $h \neq 0$ approaches 0 (with $h \in]-\eta, \eta[$), we say that Σ_t is the osculating circle to f in $M_0 = f(t)$ at t.

It is not hard to show that if the center of curvature C (and thus the radius of curvature $\mathscr{R}$) exists at t, then the osculating circle at t is indeed the circle of center C and radius $\mathscr{R}$ (also called *circle of curvature at t*).

Remark: It is possible that the osculating circle exists at a point t but that the center of curvature at t is undefined.

Example 19.7. Consider the curve defined such that

$$f(t) = \begin{cases} (t, t^2 + t^3 \sin(1/t)) & \text{if } t \neq 0; \\ (0,0) & \text{if } t = 0, \end{cases}$$

and shown in Figure 19.7.

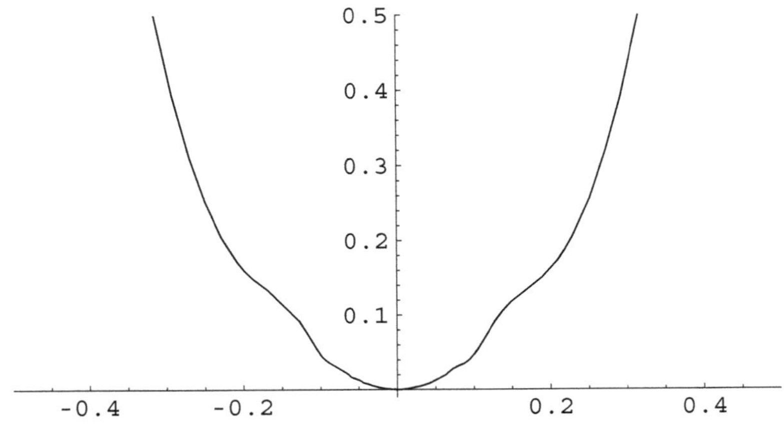

Fig. 19.7 Osculating circle at O exists and yet $f''(0)$ is undefined.

We leave as an exercise to show that the osculating circle for $t = 0$ is the circle of center $\left(0, \tfrac{1}{2}\right)$, but $f''(0)$ is undefined, so that the center of curvature is undefined at $t = 0$. This is because the intersection point of the normal line N_0 at $t = 0$ (the y-axis) and the normal N_δ for δ small oscillates forever as δ goes to zero.

In general, the osculating circle intersects the curve in another point besides the point of contact, which means that near the point of contact, one of the two branches of the curve is outside the osculating circle, and the other branch is inside. This property fails for points on an axis of symmetry for the curve, such as the points on the axes of an ellipse.

Osculating circles give a very good approximation of the curve around each (biregular) point. We will see in later examples that plotting enough osculating circles gives the illusion that the curve is plotted, when in fact it is not!

Recalling that we denoted the (unit) tangent vector $f'(s)$ at s by $\mathbf{t}$, and the unit normal vector $f''(s)/\|f''(s)\|$ by $\mathbf{n}$, since

$$\mathbf{t}' = f''(s) = \kappa\mathbf{n},$$

we have

$$\mathbf{t}' = \kappa\mathbf{n}.$$

Since $\mathbf{t}\cdot\mathbf{n} = 0$ and $\mathbf{n}\cdot\mathbf{n} = 1$, by taking derivatives of these equations we get $\mathbf{n}\cdot\mathbf{n}' = 0$ and $\mathbf{t}'\cdot\mathbf{n}+\mathbf{t}\cdot\mathbf{n}' = 0$. Since $\mathbf{n}'$ is orthognal to $\mathbf{n}$, it is collinear to $\mathbf{t}$, and from the second equation, since $\mathbf{t}' = \kappa\mathbf{n}$, we get

$$\kappa\mathbf{n}\cdot\mathbf{n}+\mathbf{t}\cdot\mathbf{n}' = \kappa+\mathbf{t}\cdot\mathbf{n}' = 0,$$

and thus

$$\mathbf{n}' = -\kappa\mathbf{t}.$$

Using the identity $\mathbf{n}' = -\kappa\mathbf{t}$, we can also show the following lemma, confirming the geometric characterization of the center of curvature.

Lemma 19.5. *For any regular plane curve $f\colon\,]a,b[\to \mathscr{E}$ (or $f\colon [a,b] \to \mathscr{E}$) of class C^p parametrized by arc length, with $p \geq 2$, assuming that $f''(s) \neq 0$, the center of curvature is on a curve c of class C^0 defined such that $c(s) - f(s) = \mathscr{R}\mathbf{n}$, where $\mathscr{R} = 1/\|f''(s)\|$ and $\mathbf{n} = f''(s)/\|f''(s)\|$, and whenever $\mathscr{R}'(s) \neq 0$, c is regular at s and $c'(s) = \mathscr{R}'\mathbf{n}$, which means that the normal to f at s is the tangent to c at s.*

Proof. Fixing any origin O in the plane, from $c(s) - f(s) = \mathscr{R}\mathbf{n}$ we have

$$c(s) - O = f(s) - O + \mathscr{R}\mathbf{n},$$

and thus

$$c'(s) = f'(s) + \mathscr{R}'\mathbf{n} + \mathscr{R}\mathbf{n}',$$

and since $\mathbf{n}' = -\kappa\mathbf{t}$, with $\kappa = 1/\mathscr{R}$, we get

$$c'(s) = \mathbf{t} + \mathscr{R}'\mathbf{n} - \mathbf{t} = \mathscr{R}'\mathbf{n}.$$

$\square$

In other words, for every s where κ'/κ^2 is defined and not equal to zero, the point $c(s)$ is regular. This is not the case for points for which the curvature is minimal or maximal. The example of an ellipse is typical (see below). The curve c defined in Lemma 19.5 is called the *evolute* of the curve f. Conversely, f is called the *involute* of c.

Summarizing the discussion before Definition 19.10, we also have the following lemma.

Lemma 19.6. *For any regular plane curve $f \colon\,]a,b[\to \mathcal{E}$ (or $f \colon [a,b] \to \mathcal{E}$) of class C^p, with $p \geq 2$, the curvature at t is given by the expression*

$$\kappa = \frac{\|f'(t) \times f''(t)\|}{\|f'(t)\|^3}.$$

Furthermore, whenever $f'(t) \times f''(t) \neq 0$, the center of curvature C defined such that $C - f(t) = \mathbf{n}/\kappa$ is the limit of the intersection of any normal $N_{t+\delta}$ and the normal N_t at t as $\delta \neq 0$ small enough approaches 0.

Lemma 19.6 gives us a way of calculating the curvature at any point, for any (regular) parametrization of a curve. Let us now determine the coordinates of the center of curvature (when defined). Let $(O,\mathbf{i},\mathbf{j})$ be an orthonormal frame for the plane, and let the curve be defined by the map $f(t) = O + u(t)\mathbf{i} + v(t)\mathbf{j}$. The equation of the normal to f at t is $(x-u)u' + (y-v)v' = 0$, or

$$u'x + v'y = uu' + vv'.$$

As we noted earlier, the center of curvature is obtained by intersecting this normal with the line whose equation is obtained by taking the derivative of the equation of the normal w.r.t. t. Thus, the center of curvature is the solution of the system

$$u'x + v'y = uu' + vv',$$
$$u''x + v''y = uu'' + vv'' + u'^2 + v'^2.$$

We leave as an exercise to verify that the solution is given by

$$x = u - \frac{v'(u'^2 + v'^2)}{u'v'' - v'u''},$$
$$y = v + \frac{u'(u'^2 + v'^2)}{u'v'' - v'u''},$$

provided that $u'v'' - v'u'' \neq 0$. One will also check that the radius of curvature is given by

$$\mathcal{R} = \frac{(u'^2 + v'^2)^{3/2}}{|u'v'' - v'u''|}.$$

This result can also be obtained from Lemma 19.6, by calculating the coordinates of the cross product $f'(t) \times f''(t)$.

We now give a few examples.

Example 19.8. If f is a straight line, then $f''(t) = 0$, and thus the curvature is null for every point of a line.

Example 19.9. A circle of radius a can be defined by

$$x = a\cos t,$$
$$y = a\sin t.$$

We have $u' = -a\sin t$, $v' = a\cos t$, $u'' = -a\cos t$, $v'' = -a\sin t$, and thus

$$u'v'' - v'u'' = (-a\sin t)(-a\sin t) - (a\cos t)(-a\cos t) = a^2$$

and

$$u'^2 + v'^2 = a^2(\sin^2 t + \cos^2 t) = a^2,$$

and thus

$$\mathscr{R} = \frac{(u'^2 + v'^2)^{3/2}}{|u'v'' - v'u''|} = a$$

and $\kappa = 1/a$. Thus, as expected, the circle has constant curvature $1/a$, where a is its radius, and the center of curvature is reduced to a single point, the center of the circle. Indeed, every normal to the circle goes through it!

Example 19.10. An ellipse is defined by

$$x = a\cos\theta,$$
$$y = b\sin\theta.$$

The equation of the normal to the ellipse at θ is

$$(x - a\cos\theta)(-a\sin\theta) + (y - b\sin\theta)(b\cos\theta) = 0,$$

or

$$a\sin\theta\, x - b\cos\theta\, y = \sin\theta\cos\theta(a^2 - b^2).$$

Assuming that $a \geq b$ (the other case being similar), and letting $c^2 = a^2 - b^2$, the above equation is

$$a\sin\theta\, x - b\cos\theta\, y = c^2\sin\theta\cos\theta.$$

We leave as an exercise to show that the radius of curvature is

$$\mathscr{R} = \frac{(a^2\sin^2\theta + b^2\cos^2\theta)^{3/2}}{ab},$$

and, that the center of curvature is on the curve defined by

$$x = \frac{c^2}{a}\cos^3\theta, \quad y = -\frac{c^2}{b}\sin^3\theta.$$

This curve has four cusps, corresponding to the two maxima and minima of the curvature. Letting

$$N = \left(\frac{c^2}{a} \cos \theta, 0 \right)$$

be the intersection of the normal to the point M on the ellipse with Ox, and $d = \|MN\|$ be the distance between M and N, we leave as an exercise to show that the radius of curvature is given by

$$\mathscr{R} = \frac{a^2}{b^4} d^3.$$

It is fun to plot the locus of the center of curvature and enough osculating circles to the ellipse. Figure 19.8 shows 64 osculating circles of the ellipse

$$\frac{x^2}{a^2} + \frac{y^2}{b^2} = 1$$

(with $a \geq b$), for $a = 4$, $b = 2$, and the locus of the center of curvature. Although the ellipse is not explicitly plotted, it seems to be present!

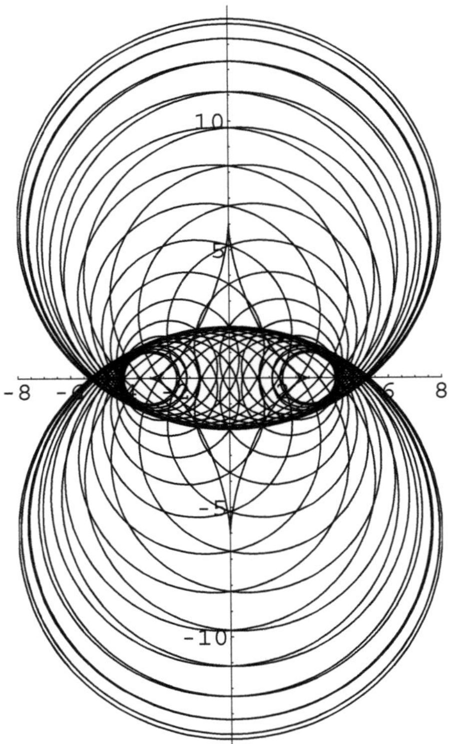

Fig. 19.8 Osculating circles of an ellipse.

Example 19.11. The logarithmic spiral given in polar coordinates by $r = a\,\mathrm{e}^{m\theta}$, or by

$$x = a\,\mathrm{e}^{m\theta}\cos\theta,$$
$$y = a\,\mathrm{e}^{m\theta}\sin\theta$$

(with $a > 0$), is particularly interesting. We leave as an exercise to show that the radius of curvature is

$$\mathscr{R} = a\sqrt{1+m^2}\,\mathrm{e}^{m\theta},$$

and that the center of curvature is on the spiral (in fact, equal to the original spiral) defined by

$$x = -ma\,\mathrm{e}^{m\theta}\sin\theta,$$
$$y = ma\,\mathrm{e}^{m\theta}\cos\theta.$$

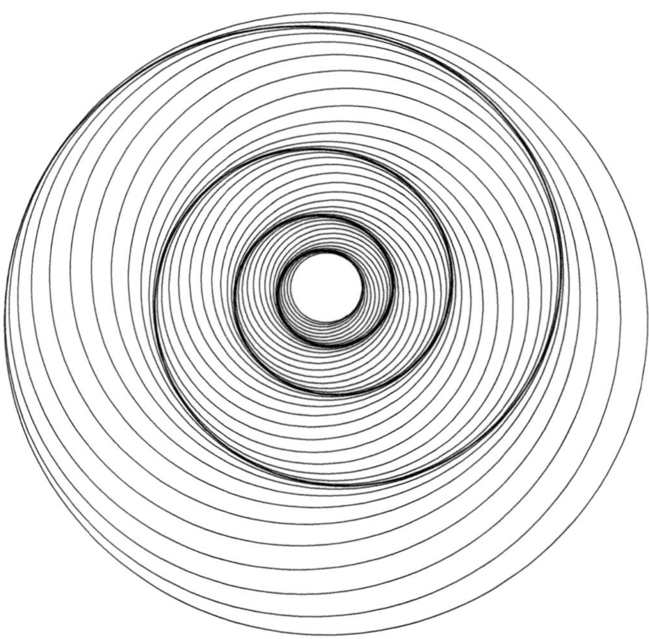

Fig. 19.9 Osculating circles of a logarithmic spiral.

Figure 19.9 shows 50 osculating circles of the logarithmic spiral given in polar coordinates by $r = a\,e^{m\theta}$, for $a = 0.6$ and $m = 0.1$. The spiral definitely shows up very clearly, even though it is not explicitly plotted. Also, note that since the radius of curvature is increasing, no two osculating circles intersect!

Example 19.12. The cardioid given in polar coordinates by $r = a(1 + \cos\theta)$, or by

$$x = a(1 + \cos\theta)\cos\theta,$$
$$y = a(1 + \cos\theta)\sin\theta,$$

is also a neat example. Figure 19.10 shows 50 osculating circles of the cardioid given in polar coordinates by $r = a(1 + \cos\theta)$, for $a = 2$, and the locus of the center of curvature.

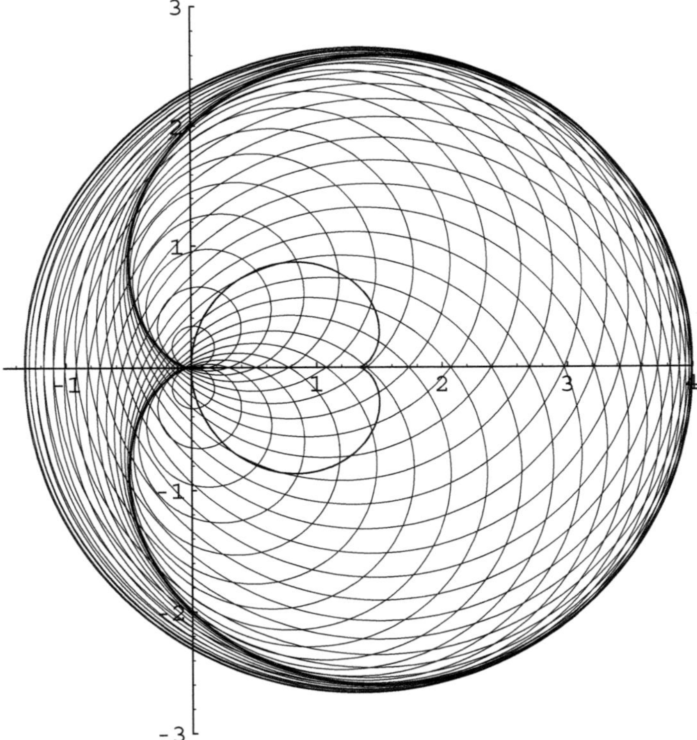

Fig. 19.10 Osculating circles of a cardioid.

We leave as an exercise to show that the radius of curvature is

$$\mathscr{R} = \left|\frac{2a}{3}\cos(\theta/2)\right|,$$

and that the center of curvature is on the cardioid defined by

$$x = \frac{2a}{3} + \frac{a}{3}(1 - \cos\theta)\cos\theta,$$

$$y = \frac{a}{3}(1 - \cos\theta)\sin\theta.$$

We conclude our discussion of the curvature of plane curves with a brief look at the algebraic curvature. Since a plane can be oriented, it is possible to give a sign to the curvature. Let us assume that the plane is oriented by an othonormal frame $(O, \mathbf{i}, \mathbf{j})$, assumed to have a positive orientation, and that the curve f is parametrized by arc length. Then, given any unit tangent vector $\mathbf{t}$ at s to a curve f, there exists a unit normal vector v such that $(O, \mathbf{t}, v)$ has positive orientation. In fact, if θ is the angle (mod 2π) between $\mathbf{i}$ and $\mathbf{t}$, so that

$$\mathbf{t} = \cos\theta\,\mathbf{i} + \sin\theta\,\mathbf{j},$$

we have

$$v = \cos(\theta + \pi/2)\,\mathbf{i} + \sin(\theta + \pi/2)\,\mathbf{j} = -\sin\theta\,\mathbf{i} + \cos\theta\,\mathbf{j}.$$

Note that this normal vector v is not necessarily equal to the unit normal vector $\mathbf{n} = f''(s)/\|f''(s)\|$: It can be of the opposite direction. Furthermore, v exists for every regular point, even when $f''(s) = 0$, which is not true of $\mathbf{n}$. We define the *algebraic curvature k at s* as the real number (negative, null, or positive) such that

$$f''(s) = kv.$$

We also define the *algebraic radius of curvature R* as $R = 1/k$. Clearly, $\kappa = |k|$ and $\mathscr{R} = |R|$. Thus, we also have

$$\mathbf{t}' = kv,$$

and it is immediately verified that the center of curvature is still given by $C - f(s) = Rv$, and that

$$v' = -k\mathbf{t}.$$

The algebraic curvature plays an important role in some global theorems of differential geometry. It is also possible to prove that if $c \colon \,]a, b[\, \to \mathbb{R}$ is a continuous function and $s_0 \in \,]a, b[$, then there is a unique curve $f \colon \,]a, b[\, \to \mathscr{E}$ such that $f(s_0)$ is any given point, $f'(s_0)$ is any given vector, and such that $c(s)$ is the algebraic curvature of f. Roughly speaking, the algebraic curvature k determines the curve completely, up to rigid motion.

One should be careful to note that this result fails if we consider the curvature κ instead of the algebraic curvature k. Indeed, it is possible that $k(s) = c(s) = 0$, and thus that $\kappa(s) = 0$. Such points may be inflection points, and counterexamples to the above result with κ instead of k are easily found. However, if we require $c(s) > 0$ for all s, the above result holds for the curvature κ.

We now consider curves in affine Euclidean 3D spaces (i.e. $\mathscr{E} = \mathbb{E}^3$).

19.5 Normal Planes and Curvature (3D Curves)

The first thing to do is to define the notion of a normal plane.

Definition 19.12. Given any regular 3D curve $f\colon\ ]a,b[\ \rightarrow \mathscr{E}$ (or $f\colon\ [a,b] \rightarrow \mathscr{E}$) of class C^p, with $p \geq 2$, the *normal plane N_t to f at t* is the plane through $f(t)$ and orthogonal to the tangent line T_t to f at t. The intersection of the normal plane and the osculating plane (if it exists) is called the *principal normal line to f at t*.

In order to get an intuitive idea of the notion of curvature, we need to look at the variation of the normal plane around t, since there are infinitely many normal lines to a given line in 3-space. This time, we will see that the normal plane rotates around a line perpendicular to the osculating plane (called the *polar axis at t*). The intersection of this line with the osculating plane is the center of curvature. But now, not only does the normal plane rotate around an axis, so do the osculating plane and the plane containing the tangent line and normal to the osculating plane, called the rectifying plane. Thus, a second quantity, called the torsion, will make its appearance. But let us go back to the intersection of normal planes around t.

Actually, the treatment that we gave for the plane extends immediately to space (in 3D). Indeed, the normal plane N_t to f at t is the set of points P such that

$$\mathbf{M}' \cdot \overrightarrow{MP} = 0,$$

or equivalently

$$\mathbf{M}' \cdot \mathbf{P} = \mathbf{M}' \cdot \mathbf{M}.$$

Similarly, for any small $\delta \neq 0$ such that $f(t+\delta)$ is defined, the normal plane $N_{t+\delta}$ to f at $t+\delta$ is the set of points Q such that

$$\mathbf{M}'(t+\delta) \cdot \mathbf{Q} = \mathbf{M}'(t+\delta) \cdot \mathbf{M}(t+\delta).$$

Thus, the intersection points P of N_t and $N_{t+\delta}$, if they exist, are given by the equations

$$\mathbf{M}' \cdot \mathbf{P} = \mathbf{M}' \cdot \mathbf{M},$$
$$\mathbf{M}'(t+\delta) \cdot \mathbf{P} = \mathbf{M}'(t+\delta) \cdot \mathbf{M}(t+\delta).$$

As in the planar case, for δ very small, the intersection of the two planes N_t and $N_{t+\delta}$ is given by the equations

$$\mathbf{M}' \cdot \overrightarrow{MP} = 0,$$
$$\mathbf{M}'' \cdot \overrightarrow{MP} = \|\mathbf{M}'\|^2.$$

Thus, if $\mathbf{M}'$ and $\mathbf{M}''$ are linearly independent, which is equivalent to saying that $f'(t)$ and $f''(t)$ are linearly independent, i.e., f is biregular at t, the above two equations define a unique line Δ orthogonal to the osculating plane. This line is called the *polar axis at t*. Also, the above analysis shows that the intersection of the two normal planes N_t and $N_{t+\delta}$, for $\delta \neq 0$ small enough, has the limit Δ. Since the line Δ is perpendicular to the osculating plane, it intersects the osculating plane in a single point C (really, $C(t)$), the *center of curvature of f at t*. The distance $\mathscr{R} = \|\overrightarrow{MC}\|$ is the *radius of curvature at t*, and its inverse $\kappa = 1/\mathscr{R}$ is the *curvature at t*. Note that C is on the normal line to the curve f at t contained in the osculating plane, i.e., on the principal normal at t.

19.6 The Frenet Frame (3D Curves)

When $f'(t)$ and $f''(t)$ are linearly independent, we can find a unit vector in the plane spanned by $f'(t)$ and $f''(t)$ and orthogonal to the unit tangent vector $\mathbf{t} = f'(t)/\|f'(t)\|$ at t, and equal to the unit vector $f''(t)/\|f''(t)\|$ when $f'(t)$ and $f''(t)$ are orthogonal, namely the unit vector

$$\mathbf{n} = \frac{-(f'(t) \cdot f''(t))\, f'(t) + \|f'(t)\|^2\, f''(t)}{\| - (f'(t) \cdot f''(t))\, f'(t) + \|f'(t)\|^2\, f''(t)\|}.$$

The unit vector $\mathbf{n}$ is called the *principal normal vector to f at t*. Note that $\mathbf{n}$ defines the direction of the principal normal at t. We define the *binormal vector* $\mathbf{b}$ at t as $\mathbf{b} = \mathbf{t} \times \mathbf{n}$. Thus, the triple $(\mathbf{t}, \mathbf{n}, \mathbf{b})$ is a basis of orthogonal unit vectors. It is usually called the *Frenet (or Frenet–Serret) frame at t* (this concept was introduced independently by Frenet, in 1847, and Serret, in 1850). This concept is sufficiently important to warrant the following definition.

Definition 19.13. Given a biregular 3D curve $f \colon \,]a,b[\,\rightarrow \mathscr{E}$ (or $f \colon [a,b] \rightarrow \mathscr{E}$) of class C^p, with $p \geq 2$, the *Frenet frame (or Frenet trihedron) at t* is the triple $(\mathbf{t}, \mathbf{n}, \mathbf{b})$ consisting of the three orthogonal unit vectors such that $\mathbf{t} = f'(t)/\|f'(t)\|$ is the unit tangent vector at t,

$$\mathbf{n} = \frac{-(f'(t) \cdot f''(t))\, f'(t) + \|f'(t)\|^2\, f''(t)}{\| - (f'(t) \cdot f''(t))\, f'(t) + \|f'(t)\|^2\, f''(t)\|}$$

is a unit vector orthogonal to $\mathbf{t}$ called the *principal normal vector to f at t*, and $\mathbf{b} = \mathbf{t} \times \mathbf{n}$ is the *binormal vector at t*. The plane containing $\mathbf{t}$ and $\mathbf{b}$ is called the *rectifying plane at t*.

As we will see shortly, the principal normal $\mathbf{n}$ has a simpler expression when the curve is parametrized by arc length. The calculations of $\mathscr{R}$ are still valid, since the cross product $\mathbf{M}' \times \mathbf{M}''$ makes sense in 3-space, and thus we have

$$\mathscr{R} = \frac{\|\mathbf{M}'\|^3}{\|\mathbf{M}' \times \mathbf{M}''\|} = \frac{\|f'(t)\|^3}{\|f'(t) \times f''(t)\|},$$

and the curvature is given by

$$\kappa = \frac{\|\mathbf{M}' \times \mathbf{M}''\|}{\|\mathbf{M}'\|^3} = \frac{\|f'(t) \times f''(t)\|}{\|f'(t)\|^3}.$$

Example 19.13. Consider the curve given by

$$f(t) = (t, t^2, t^3),$$

known as the *twisted cubic*. We have $f'(t) = (1, 2t, 3t^2)$ and $f''(t) = (0, 2, 6t)$, and thus at $t = 0$ (the origin), the vectors

$$f'(t) = (1, 0, 0) \quad \text{and} \quad f''(t) = (0, 2, 0)$$

are orthogonal, and the Frenet frame $(\mathbf{t}, \mathbf{n}, \mathbf{b})$ consists of the three unit vectors $i = (1, 0, 0)$, $j = (0, 1, 0)$, and $k = (0, 0, 1)$. Thus, the osculating plane is the *xy*-plane, the normal plane is the *yz*-plane, and the rectifying plane is the *xz*-plane. It is easily checked that

$$f' \times f'' = (6t^2, -6t, 2),$$

and the curvature at t is given by

$$\kappa(t) = \frac{2(9t^4 + 9t^2 + 1)^{1/2}}{(9t^4 + 4t^2 + 1)^{3/2}}.$$

In particular, $\kappa(0) = 2$, and the polar line is the vertical line in the *yz*-plane passing through the point $C = \left(0, \frac{1}{2}, 0\right)$, the center of curvature.

When the curve is parametrized by arc length, $\mathbf{t} = f'(s)$, and we obtain the same results as in the planar case, namely,

$$\mathscr{R} = \frac{1}{\|\mathbf{M}''\|} = \frac{1}{\|f''(s)\|}.$$

The radius of curvature is the inverse of the norm of the acceleration vector $f''(s)$, and the curvature κ is

$$\kappa = \|f''(s)\|.$$

Again, as in the planar case, the curvature is an invariant intrinsic to the geometric curve defined by f.

We now consider how the rectifying plane varies. This will uncover the torsion. According to Kreyszig [15], the term torsion was first used by de la Vallée in 1825. We leave as an easy exercise to show that the osculating plane rotates around the tangent line for points $t + \delta$ close enough to t.

19.7 Torsion (3D Curves)

Recall that the rectifying plane is the plane orthogonal to the principal normal at t passing through $f(t)$. Thus, its equation is

$$\mathbf{n} \cdot \overrightarrow{MP} = 0,$$

where $\mathbf{n}$ is the principal normal vector. However, things get a bit messy when we take the derivative of $\mathbf{n}$, because of the denominator, and it is easier to use the vector

$$\mathbf{N} = -(f'(t) \cdot f''(t)) f'(t) + \|f'(t)\|^2 f''(t),$$

which is collinear to $\mathbf{n}$, but not necessarily a unit vector. Still, we have $\mathbf{N} \cdot \mathbf{M}' = 0$, which is the important fact. Since the equation of the rectifying plane is $\mathbf{N} \cdot \overrightarrow{MP} = 0$ or

$$\mathbf{N} \cdot \mathbf{P} = \mathbf{N} \cdot \mathbf{M},$$

by familiar reasoning, the equation of a rectifying plane for $\delta \neq 0$ small enough is

$$\mathbf{N}(t + \delta) \cdot \mathbf{P} = \mathbf{N}(t + \delta) \cdot \mathbf{M}(t + \delta),$$

and we can easily prove that the intersection of these two planes is given by the equations

$$\mathbf{N} \cdot \overrightarrow{MP} = 0,$$
$$\mathbf{N}' \cdot \overrightarrow{MP} = \mathbf{N} \cdot \mathbf{M}' = 0,$$

since $\mathbf{N} \cdot \mathbf{M}' = 0$. Thus, if $\mathbf{N}$ and $\mathbf{N}'$ are linearly independent, the intersection of these two planes is a line in the rectifying plane, passing through the point $M = f(t)$. We now have to take a closer look at $\mathbf{N}'$. It is easily seen that

$$\mathbf{N}' = -(\|\mathbf{M}''\|^2 + \mathbf{M}' \cdot \mathbf{M}''')\mathbf{M}' + (\mathbf{M}' \cdot \mathbf{M}'')\mathbf{M}'' + \|\mathbf{M}'\|^2 \mathbf{M}'''.$$

Thus, $\mathbf{N}$ and $\mathbf{N}'$ are linearly independent iff $\mathbf{M}'$, $\mathbf{M}''$, and $\mathbf{M}'''$ are linearly independent. Now, since the line in question is in the rectifying plane, every point P on this line can be expressed as

$$\overrightarrow{MP} = \alpha \mathbf{b} + \beta \mathbf{t},$$

where α and β are related by the equation

$$(\mathbf{N}' \cdot \mathbf{b})\alpha + (\mathbf{N}' \cdot \mathbf{t})\beta = 0,$$

obtained from $\mathbf{N}' \cdot \overrightarrow{MP} = 0$. However, $\mathbf{t} = \mathbf{M}'/\|\mathbf{M}'\|$, and it is immediate that

$$\mathbf{b} = \frac{\mathbf{M}' \times \mathbf{M}''}{\|\mathbf{M}' \times \mathbf{M}''\|}.$$

Recalling that the radius of curvature is given by $\mathscr{R} = \|\mathbf{M}'\|^3/\|\mathbf{M}' \times \mathbf{M}''\|$, it is tempting to investigate the value of α when $\beta = \mathscr{R}$. Then the equation

$$(\mathbf{N}' \cdot \mathbf{b})\alpha + (\mathbf{N}' \cdot \mathbf{t})\beta = 0$$

becomes

$$(\mathbf{N}' \cdot (\mathbf{M}' \times \mathbf{M}''))\alpha + \|\mathbf{M}'\|^2(\mathbf{N}' \cdot \mathbf{M}') = 0.$$

Since

$$\mathbf{N}' = -(\|\mathbf{M}''\|^2 + \mathbf{M}' \cdot \mathbf{M}''')\mathbf{M}' + (\mathbf{M}' \cdot \mathbf{M}'')\mathbf{M}'' + \|\mathbf{M}'\|^2\mathbf{M}''',$$

we get

$$\mathbf{N}' \cdot (\mathbf{M}' \times \mathbf{M}'') = \|\mathbf{M}'\|^2(\mathbf{M}',\mathbf{M}'',\mathbf{M}'''),$$

where $(\mathbf{M}',\mathbf{M}'',\mathbf{M}''')$ is the mixed product of the three vectors, i.e., their determinant, and since $\mathbf{N} \cdot \mathbf{M}' = 0$, we get $\mathbf{N}' \cdot \mathbf{M}' + \mathbf{N} \cdot \mathbf{M}'' = 0$. Thus,

$$\mathbf{N}' \cdot \mathbf{M}' = -\mathbf{N} \cdot \mathbf{M}'' = (\mathbf{M}' \cdot \mathbf{M}'')^2 - \|\mathbf{M}'\|^2\|\mathbf{M}''\|^2 = -\|\mathbf{M}' \times \mathbf{M}''\|^2,$$

and finally, we get

$$\|\mathbf{M}'\|^2(\mathbf{M}',\mathbf{M}'',\mathbf{M}''')\alpha - \|\mathbf{M}'\|^2\|\mathbf{M}' \times \mathbf{M}''\|^2 = 0,$$

which yields

$$\alpha = \frac{\|\mathbf{M}' \times \mathbf{M}''\|^2}{(\mathbf{M}',\mathbf{M}'',\mathbf{M}''')}.$$

So finally, we have shown that the axis of rotation of the rectifying planes for $t + \delta$ close to t is determined by the vector

$$\overrightarrow{MP} = \alpha\mathbf{b} + \mathscr{R}\mathbf{t},$$

or equivalently, that

$$(\kappa\mathbf{t} + \tau\mathbf{b}) \cdot \overrightarrow{MP} = 0,$$

where κ is the curvature and $\tau = -1/\alpha$ is called the *torsion at t*, and is given by

$$\tau = -\frac{(\mathbf{M}',\mathbf{M}'',\mathbf{M}''')}{\|\mathbf{M}' \times \mathbf{M}''\|^2}.$$

Its inverse $\mathscr{T} = 1/\tau$ is called the *radius of torsion at t*. The vector $-\tau\mathbf{t} + \kappa\mathbf{b}$ giving the direction of the axis or rotation of the rectifying plane is called the *Darboux vector*. In summary, we have obtained the following formulae for the curvature and the torsion of a 3D-curve:

$$\boxed{\kappa = \frac{\|f'(t) \times f''(t)\|}{\|f'(t)\|^3}, \quad \tau = -\frac{(f'(t),f''(t),f'''(t))}{\|f'(t) \times f''(t)\|^2}.}$$

Example 19.14. Returning to the example of the twisted cubic

$$f(t) = (t, t^2, t^3),$$

since $f'(t) = (1, 2t, 3t^2)$, $f''(t) = (0, 2, 6t)$, and $f'''(t) = (0, 0, 6)$, we get

$$(f', f'', f''') = 12,$$

and since

$$f' \times f'' = (6t^2, -6t, 2),$$

the torsion at t is given by

$$\tau(t) = -\frac{3}{9t^4 + 9t^2 + 1}.$$

In particular, $\tau(0) = -3$, and the rectifying plane rotates around the line through the origin and of direction

$$-\tau \mathbf{t} + \kappa \mathbf{b} = (3, 2, 0).$$

The twisted cubic, the locus of the centers of curvature, the Frenet frame, the polar line (D), and the Darboux vector (Db) corresponding to $t = 0$ are shown in Figure 19.11.

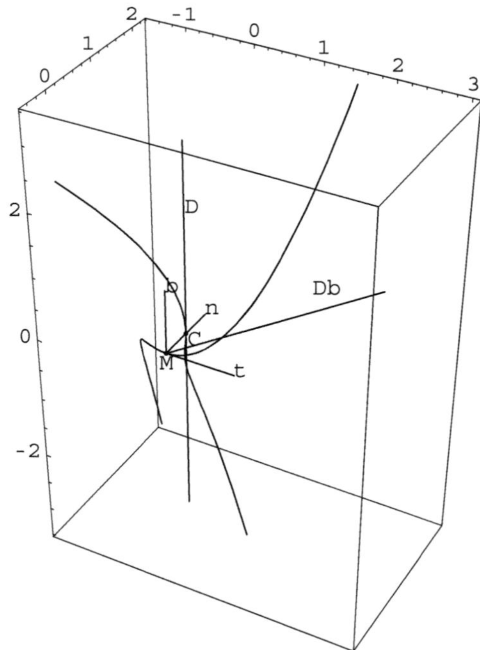

Fig. 19.11 The twisted cubic, the centers of curvature, a Frenet frame, a polar line, and a Darboux vector.

If $g = f \circ \theta$ is a curve C^p-equivalent to f, where θ is a C^p-diffeomorphism, we leave as an exercise to prove that

$$\tau = -\frac{(f'(u), f''(u), f'''(u))}{\|f'(u) \times f''(u)\|^2} = -\frac{(g'(t), g''(t), g'''(t))}{\|g'(t) \times g''(t)\|^2},$$

where $u = \theta(t)$, i.e., τ has the same value for both f and g. Thus, the torsion is an *invariant* intrinsic to the geometric curve defined by f.

19.8 The Frenet Equations (3D Curves)

Assuming that curves are parametrized by arc length, we are now going to see how κ and τ reappear naturally when we determine how the Frenet frame $(\mathbf{t}, \mathbf{n}, \mathbf{b})$ varies with s, and more specifically, in expressing $(\mathbf{t}', \mathbf{n}', \mathbf{b}')$ over the basis $(\mathbf{t}, \mathbf{n}, \mathbf{b})$. We claim that

$$\mathbf{t}' = \kappa \mathbf{n},$$
$$\mathbf{n}' = -\kappa \mathbf{t} - \tau \mathbf{b},$$
$$\mathbf{b}' = \tau \mathbf{n},$$

where κ is the curvature, and τ turns out to be the torsion.

We have $\mathbf{t}' = \kappa \mathbf{n}$ by definition of the curvature. Since $\|\mathbf{b}\| = \mathbf{b} \cdot \mathbf{b} = 1$ and $\mathbf{t} \cdot \mathbf{b} = 0$, by taking derivatives we get

$$\mathbf{b} \cdot \mathbf{b}' = 0$$

and

$$\mathbf{t}' \cdot \mathbf{b} = -\mathbf{t} \cdot \mathbf{b}',$$

and thus

$$\mathbf{t} \cdot \mathbf{b}' = -\mathbf{t}' \cdot \mathbf{b} = -\kappa \mathbf{n} \cdot \mathbf{b} = 0.$$

This shows that $\mathbf{b}'$ is collinear to $\mathbf{n}$, and thus that

$$\mathbf{b}' = \tau \mathbf{n},$$

for some real τ. From $\|\mathbf{n}\| = \mathbf{n} \cdot \mathbf{n} = 1$, $\mathbf{n} \cdot \mathbf{t} = 0$, and $\mathbf{n} \cdot \mathbf{b} = 0$, by taking derivatives we get

$$\mathbf{n} \cdot \mathbf{n}' = 0, \quad \mathbf{n}' \cdot \mathbf{t} = -\mathbf{n} \cdot \mathbf{t}', \quad \mathbf{n}' \cdot \mathbf{b} = -\mathbf{n} \cdot \mathbf{b}'.$$

Since $\mathbf{t}' = \kappa \mathbf{n}$ and $\mathbf{b}' = \tau \mathbf{n}$, we get

$$\mathbf{n}' \cdot \mathbf{t} = -\mathbf{n} \cdot \mathbf{t}' = -\mathbf{n} \cdot \kappa \mathbf{n} = -\kappa$$

and

$$\mathbf{n}' \cdot \mathbf{b} = -\mathbf{n} \cdot \mathbf{b}' = -\mathbf{n} \cdot \tau \mathbf{n} = -\tau.$$

But the components of $\mathbf{n}'$ over $(\mathbf{t}, \mathbf{n}, \mathbf{b})$ are indeed $\mathbf{n}' \cdot \mathbf{t}$, $\mathbf{n}' \cdot \mathbf{n}$, and $\mathbf{n}' \cdot \mathbf{b}$, and thus

$$\mathbf{n}' = -\kappa\mathbf{t} - \tau\mathbf{b}.$$

In matrix form we can write the equations know as the *Frenet (or Frenet–Serret) equations* as

$$\boxed{(\mathbf{t}',\mathbf{n}',\mathbf{b}') = (\mathbf{t},\mathbf{n},\mathbf{b}) \begin{pmatrix} 0 & -\kappa & 0 \\ \kappa & 0 & \tau \\ 0 & -\tau & 0 \end{pmatrix}.}$$

We can now verify that τ agrees with the geometric interpretation given before. The axis of rotation of the rectifying plane is the line given by the intersection of the two planes of equations

$$\mathbf{n} \cdot \overrightarrow{MP} = 0,$$
$$\mathbf{n}' \cdot \overrightarrow{MP} = 0,$$

and since

$$\mathbf{n}' = -\kappa\mathbf{t} - \tau\mathbf{b},$$

the second equation is equivalent to

$$(\kappa\mathbf{t} + \tau\mathbf{b}) \cdot \overrightarrow{MP} = 0.$$

This is exactly the equation that we found earlier with $\tau = -1/\alpha$, where

$$\alpha = \frac{\|\mathbf{M}' \times \mathbf{M}''\|^2}{(\mathbf{M}',\mathbf{M}'',\mathbf{M}''')}.$$

Remarks:

(1) Some authors, including Darboux ([6], Livre I, Chapter 1) and Élie Cartan ([5], Chapter VII, Section 2), define the torsion as $-\tau$, in which case

$$\tau = \frac{(\mathbf{M}',\mathbf{M}'',\mathbf{M}''')}{\|\mathbf{M}' \times \mathbf{M}''\|^2},$$

and the Frenet equations take the form

$$(\mathbf{t}',\mathbf{n}',\mathbf{b}') = (\mathbf{t},\mathbf{n},\mathbf{b}) \begin{pmatrix} 0 & -\kappa & 0 \\ \kappa & 0 & -\tau \\ 0 & \tau & 0 \end{pmatrix}.$$

A possible advantage of this choice is the elimination of the negative sign in the expression for τ above, and the fact that it may be slightly easier to remember the Frenet matrix, since signs on descending diagonals remain the same. Another possible advantage is that the Frenet matrix has a similar shape in higher dimension (≥ 4). Books on Computer-Aided Gemetric Design seem to prefer this choice. On the other hand, do Carmo [7] and Berger and Gostiaux [2] use the opposite convention (as we do).

(2) It should also be noted that if we let

$$\omega = \tau \mathbf{t} + \kappa \mathbf{b},$$

often called the *Darboux vector*, then (abbreviating three equations in one using a slight abuse of notation)

$$(\mathbf{t}', \mathbf{n}', \mathbf{b}') = \omega \times (\mathbf{t}, \mathbf{n}, \mathbf{b}),$$

which shows that the vectors $\mathbf{t}', \mathbf{n}', \mathbf{b}'$ are the velocities of the tips of the unit frame, and that the unit frame rotates around an instantaneous axis of rotation passing through the origin of the frame, whose direction is the vector $\omega = \tau \mathbf{t} + \kappa \mathbf{b}$.

We now summarize the above considerations in the following definition and lemma.

Definition 19.14. Given a biregular 3D curve $f \colon\,]a,b[\to \mathscr{E}$ (or $f \colon\, [a,b] \to \mathscr{E}$) of class C^p parametrized by arc length, with $p \geq 3$, given the Frenet frame $(\mathbf{t}, \mathbf{n}, \mathbf{b})$ at s, the *curvature* κ *at* s is the nonnegative real number such that $\mathbf{t}' = \kappa \mathbf{n}$, the *torsion* τ *at* s is the real number such that $\mathbf{b}' = \tau \mathbf{n}$, the *radius of curvature at* s is the nonnegative real number $\mathscr{R} = 1/\kappa$, the *radius of torsion at* s is the real number $\mathscr{T} = 1/\tau$, the *center of curvature at* s is the point C on the principal normal such that $C - f(s) = \mathscr{R}\mathbf{n}$, and the *polar axis at* s is the line orthogonal to the osculating plane passing through the center of curvature.

Again, we stress that the curvature κ and the torsion τ are intrinsic *invariants* of the geometric curve defined by f.

Lemma 19.7. *Given a biregular* 3D *curve* $f \colon\,]a,b[\to \mathscr{E}$ *(or* $f \colon\, [a,b] \to \mathscr{E}$*) of class* C^p *parametrized by arc length, with* $p \geq 3$*, given the Frenet frame* $(\mathbf{t}, \mathbf{n}, \mathbf{b})$ *at* s*, we have the Frenet (or Frenet–Serret) equations*

$$(\mathbf{t}', \mathbf{n}', \mathbf{b}') = (\mathbf{t}, \mathbf{n}, \mathbf{b}) \begin{pmatrix} 0 & -\kappa & 0 \\ \kappa & 0 & \tau \\ 0 & -\tau & 0 \end{pmatrix}.$$

Given any parametrization for f*, the curvature* κ *and the torsion* τ *are given by the expressions*

$$\kappa = \frac{\|f'(t) \times f''(t)\|}{\|f'(t)\|^3}$$

and

$$\tau = -\frac{(f'(t), f''(t), f'''(t))}{\|f'(t) \times f''(t)\|^2}.$$

Furthermore, for δ small enough, the normal plane at $t + \delta$ rotates around the polar axis, a line othogonal to the osculating plane and passing through the center of curvature, and the rectifying plane at $t + \delta$ rotates around the line defined by the point of contact at t and the vector $-\tau\mathbf{t} + \kappa\mathbf{b}$ (the Darboux vector).

The torsion measures how the osculating plane rotates around the tangent. Let us show that if f is a biregular curve and if $\tau = 0$ for all t, then f is a plane curve. We can assume that f is parametrized by arc length. Since $\mathbf{b}' = \tau\mathbf{n}$, and we are assuming that $\tau = 0$, we have $\mathbf{b}' = 0$, which means that $\mathbf{b}$ is a constant vector. Since f is biregular, $\mathbf{b} \neq 0$. But now, choosing any origin O and observing that

$$(Of(s) \cdot \mathbf{b})' = f'(s) \cdot \mathbf{b} + Of(s) \cdot \mathbf{b}' = \mathbf{t} \cdot \mathbf{b} + 0 = 0,$$

we conclude that $Of(s) \cdot \mathbf{b} = \lambda$ for some constant λ. Since $\mathbf{b} \neq 0$, we conclude that $f(s)$ is in a plane.

One should be careful to note that the above result is false if f has points that are not biregular, i.e., if $f''(s) = 0$ for some s. We leave as an exercise to find an example of a regular nonplanar curve such that $\tau = 0$.

As an example of the computation of the torsion, consider the circular helix defined by

$$f(t) = (a\cos t, a\sin t, kt).$$

It is easy to show that the curvature is given by

$$\kappa = \frac{a}{a^2 + k^2}$$

and that the torsion is given by

$$\tau = -\frac{k}{a^2 + k^2}.$$

Thus, both the curvature and the torsion are constant!

The intrinsic nature of the curvature and the torsion is illustrated by the following result. If $c\colon \,]a, b[\to \mathbb{R}_+$ is a continuous positive C^1 function, $d\colon \,]a, b[\to \mathbb{R}$ is a continuous function, and $s_0 \in \,]a, b[$, then there is a unique biregular 3D curve $f\colon \,]a, b[\to \mathscr{E}$ such that $f(s_0)$ is any given point, $f'(s_0)$ is any given vector, $f''(s_0)$ is any given vector, and such that $c(s)$ is the curvature of f, and $d(s)$ is the torsion of f. Roughly speaking, the curvature and the torsion determine a biregular curve completely, up to rigid motion.

The hypothesis that $c(s) > 0$ for all s is crucial, and the above result is false if this condition is not satisfied everywhere.

19.9 Osculating Spheres (3D Curves)

We conclude our discussion of curves in 3-space by discussing briefly the notion of osculating sphere. According to Kreyszig [15], osculating spheres were first considered by Fuss in 1806.

Definition 19.15. For any 3D curve $f: \,]a,b[\, \to \mathscr{E}$ (or $f: [a,b] \to \mathscr{E}$) of class C^p, with $p \geq 3$, and given any point $M_0 = f(t)$ on the curve, if the polar axis at t exists, f is locally injective at M_0, and the sphere $\Sigma_{t,h}$ centered on the polar axis and passing through the points M_0 and $M_1 = f(t+h)$ has a limit Σ_t as $h \neq 0$ approaches 0, we say that Σ_t is the *osculating sphere to f in $M_0 = f(t)$ at t*. More precisely, if the polar axis at t exists and if there is an open interval $]t-\eta, t+\eta[\, \subseteq \,]a,b[$ (with $\eta > 0$) such that the point $M_1 = f(t+h)$ is distinct from M_0 for every $h \neq 0$ with $h \in \,]-\eta, +\eta[$ and the sphere $\Sigma_{t,h}$ centered on the polar axis and passing through the points M_0 and M_1 has a limit Σ_t as $h \neq 0$ approaches 0 (with $h \in \,]-\eta, +\eta[$), we say that Σ_t is the osculating sphere to f in $M_0 = f(t)$ at t.

Again, the definition is simpler when f is a simple curve. The following lemma gives a simple condition for the existence of the osculating sphere at a point.

Lemma 19.8. *For any 3D curve $f: \,]a,b[\, \to \mathscr{E}$ (or $f: [a,b] \to \mathscr{E}$) of class C^p parametrized by arc length, with $p \geq 3$, given any point $M_0 = f(s)$ on the curve, if M_0 is a biregular point at s and if $\mathscr{R}'$ is defined, then the osculating sphere to f in M_0 at s exists and has its center Ω on the polar axis Δ, such that $\Omega - C = -\mathscr{T}\mathscr{R}'\mathbf{b}$, where $\mathscr{T}$ is the radius of torsion, $\mathscr{R}$ is the radius of curvature, C is the center of curvature, and $\mathbf{b}$ is the binormal, at s.*

According to Kreyszig [15], the formula

$$\Omega - M = \mathscr{R}\mathbf{n} - \mathscr{T}\mathscr{R}'\mathbf{b}$$

is due to de Saint Venant (1845). When s varies, the polar axis generates a surface, and the center Ω of the osculating sphere generates a curve on this surface. In general, this surface consists of the tangents to this curve (called *line of striction* or *edge of regression* of the ruled surface).

Figure 19.12 illustrates the Frenet frame, the polar axis, the center of curvature, and the osculating sphere. It also shows the osculating plane, the normal plane, and the rectifying plane.

The twisted cubic and the locus of the centers of osculating spheres are shown in Figure 19.13. The tangent surface, that is, the surface consisting of the tangent lines to the twisted cubic; the curve of centers of osculating spheres; and two great circles of osculating spheres corresponding to $t = \frac{1}{5}$, are shown in Figure 19.14. The tangent surface is the envelope of the osculating planes. The surface generated by the polar lines is shown in Figure 19.15. This surface is the envelope of the normal planes to the twisted cubic. The curve of the centers of osculating spheres is a *line of striction* (or *edge of regression*) on this surface.

Finally, we discuss the case of curves in Euclidean spaces of dimension $n \geq 4$.

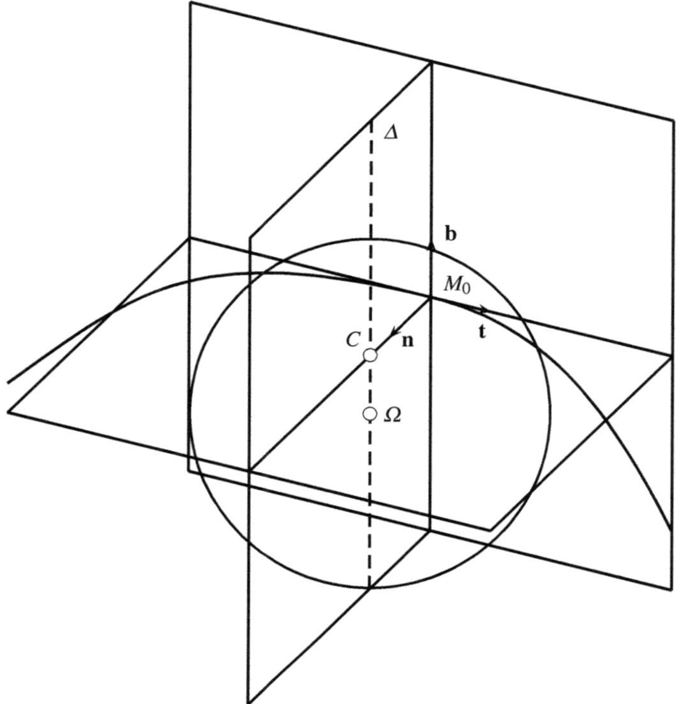

Fig. 19.12 The Frenet frame, polar axis, center of curvature, and osculating sphere.

19.10 The Frenet Frame for nD Curves ($n \geq 4$)

Given a curve $f \colon\ ]a,b[\ \to \mathbb{E}^n$ (or $f \colon [a,b] \to \mathbb{E}^n$) of class C^p, with $p \geq n$, it is interesting to consider families $(e_1(t),\ldots,e_n(t))$ of orthonormal frames. Moreover, if for every k, with $1 \leq k \leq n$, the kth derivative $f^{(k)}(t)$ of the curve $f(t)$ is a linear combination of $(e_1(t),\ldots,e_k(t))$ for every $t \in\ ]a,b[$, then such a frame plays the role of a generalized Frenet frame. This leads to the following definition:

Definition 19.16. Let $f \colon\ ]a,b[\ \to \mathbb{E}^n$ (or $f \colon [a,b] \to \mathbb{E}^n$) be a curve of class C^p, with $p \geq n$. A family $(e_1(t),\ldots,e_n(t))$ of orthonormal frames, where each $e_i \colon\ ]a,b[\ \to \mathbb{E}^n$ is C^{n-i}-continuous for $i = 1,\ldots,n-1$ and e_n is C^1-continuous, is called a *moving frame along f*. Furthermore, a moving frame $(e_1(t),\ldots,e_n(t))$ along f such that for every k, with $1 \leq k \leq n$, the kth derivative $f^{(k)}(t)$ of $f(t)$ is a linear combination of $(e_1(t),\ldots,e_k(t))$ for every $t \in\ ]a,b[$, is called a *Frenet n-frame* or *Frenet frame*.

If $(e_1(t),\ldots,e_n(t))$ is a moving frame, then

$$e_i(t) \cdot e_j(t) = \delta_{ij} \quad \text{for all } i,j,\ 1 \leq i,j \leq n.$$

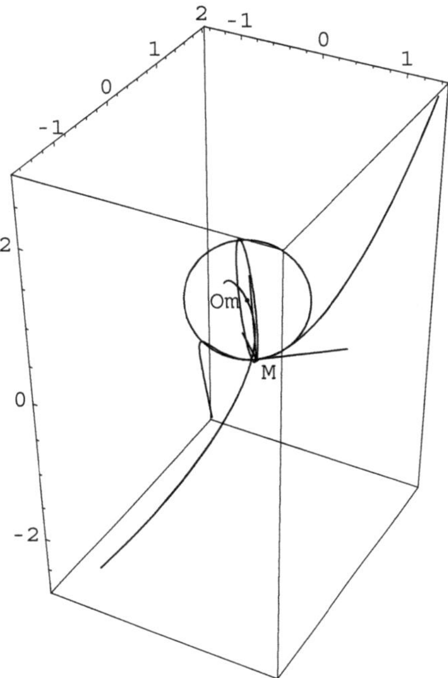

Fig. 19.13 The twisted cubic and the curve of centers of osculating spheres.

Lemma 19.9. *Let* $f:\ ]a,b[\to \mathbb{E}^n$ *(or* $f: [a,b] \to \mathbb{E}^n$*) be a curve of class* C^p*, with* $p \geq n$*, such that the derivatives* $f^{(1)}(t),\ldots,f^{(n-1)}(t)$ *of* $f(t)$ *are linearly independent for all* $t \in]a,b[$*. Then there is a unique Frenet n-frame* $(e_1(t),\ldots,e_n(t))$ *satisfying the following conditions:*

(1) The k-frames $(f^{(1)}(t),\ldots,f^{(k)}(t))$ *and* $(e_1(t),\ldots,e_k(t))$ *have the same orientation for all k, with* $1 \leq k \leq n-1$.
(2) The frame $(e_1(t),\ldots,e_n(t))$ *has positive orientation.*

Proof. Since $(f^{(1)}(t),\ldots,f^{(n-1)}(t))$ is linearly independent, we can use the Gram–Schmidt orthonormalization procedure (see Lemma 6.7) to construct $(e_1(t),\ldots,e_{n-1}(t))$ from $(f^{(1)}(t),\ldots,f^{(n-1)}(t))$. We use the generalized cross product to define e_n, where

$$e_n = e_1 \times \cdots \times e_{n-1}.$$

From the Gram–Schmidt procedure, it is easy to check that $e_k(t)$ is C^{n-k} for $1 \leq k \leq n-1$, and since the components of e_n are certain determinants involving the components of $(e_1,\ldots,e_{n-1})$, it is also clear that e_n is C^1. $\square$

The Frenet n-frame given by Lemma 19.9 is called the *distinguished Frenet n-frame*. We can now prove a generalization of the Frenet–Serret formula that gives an expression of the derivatives of a moving frame in terms of the moving frame itself.

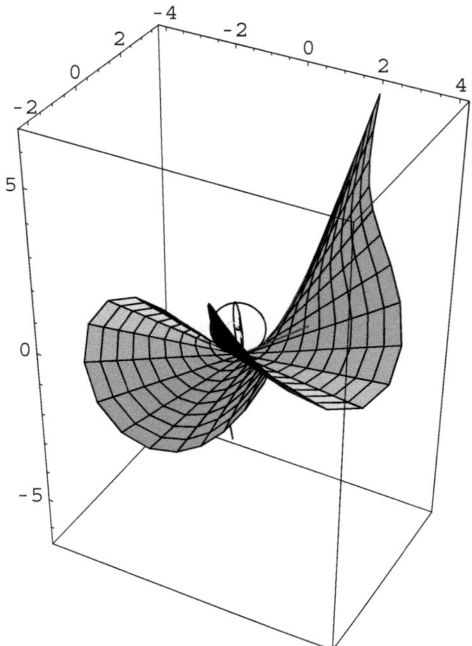

Fig. 19.14 The tangent surface and the centers of osculating spheres.

Lemma 19.10. *Let* $f\colon\ ]a,b[\ \to \mathbb{E}^n$ *(or* $f\colon\ [a,b] \to \mathbb{E}^n$*) be a curve of class* C^p, *with* $p \geq n$, *such that the derivatives* $f^{(1)}(t),\ldots,f^{(n-1)}(t)$ *of* $f(t)$ *are linearly independent for all* $t \in\ ]a,b[$. *Then for any moving frame* $(e_1(t),\ldots,e_n(t))$, *if we write* $\omega_{ij}(t) = e_i'(t)\cdot e_j(t)$, *we have*

$$e_i'(t) = \sum_{j=1}^{n} \omega_{ij}(t)e_j(t),$$

with

$$\omega_{ji}(t) = -\omega_{ij}(t),$$

and there are some functions $\alpha_i(t)$ *such that*

$$f'(t) = \sum_{i=1}^{n} \alpha_i(t)e_i(t).$$

Furthermore, if $(e_1(t),\ldots,e_n(t))$ *is the distinguished Frenet n-frame associated with* f, *then we also have*

$$\alpha_1(t) = \|f'(t)\|, \qquad \alpha_i(t) = 0 \quad for \quad i \geq 2,$$

and

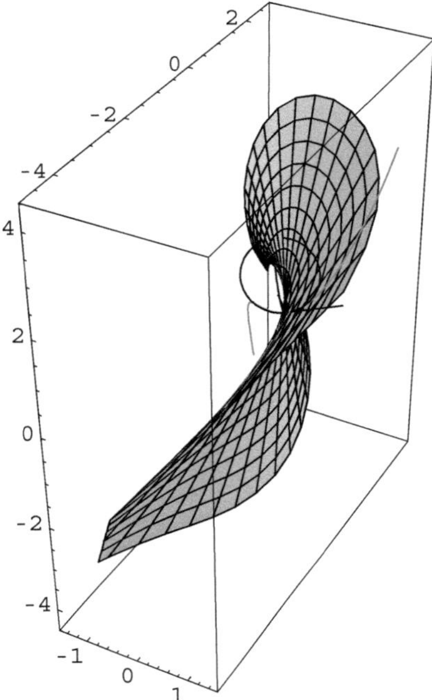

Fig. 19.15 The polar surface and the twisted cubic.

$$\omega_{ij}(t) = 0 \quad for \quad j > i+1.$$

Proof. Since $(e_1(t), \ldots, e_n(t))$ is a moving frame, it is an orthonormal basis, and thus $f'(t)$ and $e_i'(t)$ are linear combinations of $(e_1(t), \ldots, e_n(t))$. Also, we know that

$$e_i'(t) = \sum_{j=1}^{n} (e_i'(t) \cdot e_j(t)) e_j(t),$$

and since $e_i(t) \cdot e_j(t) = \delta_{ij}$, by differentiating, if we write $\omega_{ij}(t) = e_i'(t) \cdot e_j(t)$, we get

$$\omega_{ji}(t) = -\omega_{ij}(t).$$

Now if $(e_1(t), \ldots, e_n(t))$ is the distinguished Frenet frame, by construction, $e_i(t)$ is a linear combination of $f^{(1)}(t), \ldots, f^{(i)}(t)$, and so $e_i'(t)$ is a linear combination of $f^{(2)}(t), \ldots, f^{(i+1)}(t)$, hence of $(e_1(t), \ldots, e_{i+1}(t))$. $\square$

In matrix form, when $(e_1(t), \ldots, e_n(t))$ is the distinguished Frenet frame, the row vector $(e_1'(t), \ldots, e_n'(t))$ can be expressed in terms of the row vector $(e_1(t), \ldots, e_n(t))$ via a skew-symmetric matrix ω, as shown below:

$$(e_1'(t), \ldots, e_n'(t)) = -(e_1(t), \ldots, e_n(t))\omega(t),$$

where

$$\omega = \begin{pmatrix} 0 & \omega_{1\,2} & & & & \\ -\omega_{1\,2} & 0 & \omega_{23} & & & \\ & -\omega_{23} & 0 & \ddots & & \\ & & \ddots & & & \omega_{n-1\,n} \\ & & & & -\omega_{n-1\,n} & 0 \end{pmatrix}.$$

The next lemma shows the effect of a reparametrization and of a rigid motion.

Lemma 19.11. *Let $f\colon\,]a,b[\to \mathbb{E}^n$ (or $f\colon [a,b] \to \mathbb{E}^n$) be a curve of class C^p, with $p \ge n$, such that the derivatives $f^{(1)}(t),\ldots,f^{(n-1)}(t)$ of $f(t)$ are linearly independent for all $t \in\,]a,b[$. Let $h\colon \mathbb{E}^n \to \mathbb{E}^n$ be a rigid motion, and assume that the corresponding linear isometry is R. Let $\widetilde{f} = h \circ f$. The following properties hold:*

(1) For any moving frame $(e_1(t),\ldots,e_n(t))$, the n-tuple $(\widetilde{e_1}(t),\ldots,\widetilde{e_n}(t))$, where $\widetilde{e_i}(t) = R(e_i(t))$, is a moving frame along $\widetilde{f}$, and we have

$$\widetilde{\omega_{ij}}(t) = \omega_{ij}(t) \quad and \quad \|\widetilde{f}'(t)\| = \|f'(t)\|.$$

(2) For any orientation-preserving diffeormorphism $\rho\colon\,]c,d[\to\,]a,b[$ (i.e., $\rho'(t) > 0$ for all $t \in\,]c,d[$), if we write $\widetilde{f} = f \circ \rho$, then for any moving frame $(e_1(t),\ldots, e_n(t))$ on f, the n-tuple $(\widetilde{e_1}(t),\ldots,\widetilde{e_n}(t))$, where $\widetilde{e_i}(t) = e_i(\rho(t))$, is a moving frame on $\widetilde{f}$. Furthermore, if $\|\widetilde{f}'(t)\| \ne 0$, then

$$\frac{\widetilde{\omega_{ij}}(t)}{\|\widetilde{f}'(t)\|} = \frac{\omega_{ij}(\rho(t))}{\|f'(\rho(t))\|}.$$

The proof is straightforward and is omitted.

The above lemma suggests the definition of the curvatures $\kappa_1,\ldots,\kappa_{n-1}$.

Definition 19.17. Let $f\colon\,]a,b[\to \mathbb{E}^n$ (or $f\colon [a,b] \to \mathbb{E}^n$) be a curve of class C^p, with $p \ge n$, such that the derivatives $f^{(1)}(t),\ldots,f^{(n-1)}(t)$ of $f(t)$ are linearly independent for all $t \in\,]a,b[$. If $(e_1(t),\ldots,e_n(t))$ is the distinguished Frenet frame associated with f, we define the *ith curvature κ_i of f* by

$$\kappa_i(t) = \frac{\omega_{i\,i+1}(t)}{\|f'(t)\|},$$

with $1 \le i \le n-1$.

Observe that the matrix $\omega(t)$ can be written as

$$\omega(t) = \|f'(t)\|\,\kappa(t),$$

where

$$
\kappa = \begin{pmatrix} 0 & \kappa_{12} & & & & \\ -\kappa_{12} & 0 & \kappa_{23} & & & \\ & -\kappa_{23} & 0 & \ddots & & \\ & & \ddots & \ddots & \kappa_{n-1\,n} \\ & & & -\kappa_{n-1\,n} & 0 \end{pmatrix}.
$$

The matrix κ is sometimes called the *Cartan matrix*.

Lemma 19.12. *Let $f \colon\,]a,b[\,\to \mathbb{E}^n$ (or $f \colon [a,b] \to \mathbb{E}^n$) be a curve of class C^p, with $p \geq n$, such that the derivatives $f^{(1)}(t),\ldots,f^{(n-1)}(t)$ of $f(t)$ are linearly independent for all $t \in\,]a,b[$. Then for every i, with $1 \leq i \leq n-2$, we have $\kappa_i(t) > 0$.*

Proof. Lemma 19.9 shows that $e_1,\ldots,e_{n-1}$ are expressed in terms of $f^{(1)},\ldots,f^{(n-1)}$ by a triangular matrix (a_{ij}) whose diagonal entries a_{ii} are strictly positive, i.e., we have

$$
e_i = \sum_{j=1}^{i} a_{ij} f^{(j)},
$$

for $i = 1,\ldots,n-1$, and thus

$$
f^{(i)} = \sum_{j=1}^{i} b_{ij} e_j,
$$

for $i = 1,\ldots,n-1$, with $b_{ii} = a_{ii}^{-1} > 0$. Then, since $e_{i+1} \cdot f^{(j)} = 0$ for $j \leq i$, we get

$$
\|f'\| \kappa_i = \omega_{i\,i+1} = e_i' \cdot e_{i+1} = a_{ii} f^{(i+1)} \cdot e_{i+1} = a_{ii} b_{i+1\,i+1},
$$

and since $a_{ii} b_{i+1\,i+1} > 0$, we get $\kappa_i > 0$ $(i = 1,\ldots,n-2)$. $\square$

Our previous reasoning in the 3D case is immediately extended to show that the limit of the intersection of the normal hyperplane at $t + \delta$ with the normal hyperplane at t (for δ small) with the osculating plane is a point C such that $C - f(t) = (1/\kappa_1) e_1$. Thus, we obtain a geometric interpretation for the curvature κ_1, and it is also possible to obtain an interpretation for the other κ_i.

We conclude by exploring to what extent the curvatures $\kappa_1,\ldots,\kappa_{n-1}$ determine a curve satisfying the nondegeneracy conditions of Lemma 19.9. Basically, such curves are defined up to a rigid motion.

Lemma 19.13. *Let $f \colon\,]a,b[\,\to \mathbb{E}^n$ and $\widetilde{f} \colon\,]a,b[\,\to \mathbb{E}^n$ (or $f \colon [a,b] \to \mathbb{E}^n$ and $\widetilde{f} \colon [a,b] \to \mathbb{E}^n$) be two curves of class C^p, with $p \geq n$, and satisfying the nondegeneracy conditions of Lemma 19.9. Denote the distinguished Frenet frames associated with f and $\widetilde{f}$ by $(e_1(t),\ldots,e_n(t))$ and $(\widetilde{e}_1(t),\ldots,\widetilde{e}_n(t))$. If $\kappa_i(t) = \widetilde{\kappa}_i(t)$ for every i, with $1 \leq i \leq n-1$, and $\|f'(t)\| = \|\widetilde{f}'(t)\|$ for all $t \in\,]a,b[$, then there is a unique rigid motion h such that*

$$
\widetilde{f} = h \circ f.
$$

Proof. Fix $t_0 \in\,]a,b[$. First of all, there is a unique rigid motion h such that

$$h(f(t_0)) = \widetilde{f}(t_0) \quad \text{and} \quad R(e_i(t_0)) = \widetilde{e}_i(t_0),$$

for all i, with $1 \leq i \leq n$, where R is the linear isometry associated with h (in fact, a rotation). Consider the curve $\overline{f} = h \circ f$. The hypotheses of the lemma and Lemma 19.11 imply that

$$\overline{\omega_{ij}}(t) = \widetilde{\omega}_{ij}(t) = \omega_{ij}(t), \quad \|\overline{f}'(t)\| = \|\widetilde{f}'(t)\| = \|f'(t)\|,$$

and, by construction, $(\overline{e_1}(t_0), \ldots, \overline{e_n}(t_0)) = (\widetilde{e}_1(t_0), \ldots, \widetilde{e}_n(t_0))$ and $\overline{f}(t_0) = \widetilde{f}(t_0)$. Let

$$\delta(t) = \sum_{i=1}^{n} (\overline{e}_i(t) - \widetilde{e}_i(t)) \cdot (\overline{e}_i(t) - \widetilde{e}_i(t)).$$

Then we have

$$\delta'(t) = 2 \sum_{i=1}^{n} (\overline{e}_i(t) - \widetilde{e}_i(t)) \cdot (\overline{e}_i'(t) - \widetilde{e}_i'(t))$$

$$= -2 \sum_{i=1}^{n} (\overline{e}_i(t) \cdot \widetilde{e}_i'(t) + \widetilde{e}_i(t) \cdot \overline{e}_i'(t)).$$

Using the Frenet equations, we get

$$\delta'(t) = -2 \sum_{i=1}^{n} \sum_{j=1}^{n} \omega_{ij} \overline{e}_i \cdot \widetilde{e}_j - 2 \sum_{i=1}^{n} \sum_{j=1}^{n} \omega_{ij} \overline{e}_j \cdot \widetilde{e}_i$$

$$= -2 \sum_{i=1}^{n} \sum_{j=1}^{n} \omega_{ij} \overline{e}_i \cdot \widetilde{e}_j - 2 \sum_{j=1}^{n} \sum_{i=1}^{n} \omega_{ji} \overline{e}_i \cdot \widetilde{e}_j$$

$$= -2 \sum_{i=1}^{n} \sum_{j=1}^{n} \omega_{ij} \overline{e}_i \cdot \widetilde{e}_j + 2 \sum_{j=1}^{n} \sum_{i=1}^{n} \omega_{ij} \overline{e}_i \cdot \widetilde{e}_j$$

$$= 0,$$

since ω is skew-symmetric. Thus, $\delta(t)$ is constant, and since the Frenet frames at t_0 agree, we get $\delta(t) = 0$. Then $\overline{e}_i(t) = \widetilde{e}_i(t)$ for all i, and since $\|\overline{f}'(t)\| = \|\widetilde{f}'(t)\|$, we have

$$\overline{f}'(t) = \|\overline{f}'(t)\| \overline{e_1}(t) = \|\widetilde{f}'(t)\| \widetilde{e_1}(t) = \widetilde{f}'(t),$$

so that $\overline{f}(t) - \widetilde{f}(t)$ is constant. However, $\overline{f}(t_0) = \widetilde{f}(t_0)$, and so $\overline{f}(t) = \widetilde{f}(t)$ and $\widetilde{f} = \overline{f} = h \circ f$. $\square$

Finally, the lemma below settles the issue of the existence of a curve with prescribed curvature functions.

Lemma 19.14. *Let $\kappa_1, \ldots, \kappa_{n-1}$ be functions defined on some open $]a, b[$ containing 0 with κ_i C^{n-i-1}-continuous for $i = 1, \ldots, n-1$, and with $\kappa_i(t) > 0$ for $i = 1, \ldots, n-2$ and all $t \in]a, b[$. Then there is curve $f:]a, b[\rightarrow \mathbb{E}^n$ of class C^p, with $p \geq n$,* *satisfying the nondegeneracy conditions of Lemma 19.9 such that $\|f'(t)\| = 1$ and f has the $n-1$ curvatures $\kappa_1(t), \ldots, \kappa_{n-1}(t)$.*

Proof. Let $X(t)$ be the matrix whose columns consist of the vectors $e_1(t), \ldots, e_n(t)$ of the Frenet frame along f. Consider the system of ODEs,

$$X'(t) = -X(t)\kappa(t),$$

with initial conditions $X(0) = I$, where $\kappa(t)$ is the skew-symmetric matrix of curvatures. By a standard result in ODEs, there is a unique solution $X(t)$.

We claim that $X(t)$ is an orthogonal matrix. For this, note that

$$(XX^\top)' = X'X^\top + X(X^\top)' = -X\kappa X^\top - X\kappa^\top X^\top$$
$$= -X\kappa X^\top + X\kappa X^\top = 0.$$

Since $X(0) = I$, we get $XX^\top = I$. If $F(t)$ is the first column of $X(t)$, we define the curve f by

$$f(s) = \int_0^s F(t)\,dt,$$

with $s \in \,]a,b[\,$. It is easily checked that f is a curve parametrized by arc length, with Frenet frame $X(s)$, and with curvatures κ_is. $\square$

19.11 Applications

Many engineering problems can be reduced to finding curves having some desired properties. This is certainly true of mechanical engineering and robotics, where various trajectories must be computed, and of computer graphics and medical imaging, where contours of shapes, for instance organs, are modeled as curves. In most practical applications it is necessary to consider curves composed of various segments. The problem then arises to join these segments as smoothly as possible, without restricting too much the number of degrees of freedom required for the design. Various kinds of *splines* were invented to solve this problem. If the curve segments are defined parametrically in terms of polynomials, a simple way to achieve continuity is to enforce the agreement of enough derivatives at junction points. This leads to *parametric C^n-continuity* and to *B-splines*. The theory of *B*-splines is quite extensive. Among the many references, we recommend Farin [10, 9], Hoschek and Lasser [14], Bartels, Beatty, and Barsky [1], Fiorot and Jeannin [11, 12], Piegl and Tiller [17], or Gallier [13].

Because parametric continuity is easy to formulate, piecewise curves based on parametric continuity are popular. Additionally, there are occasions in which parametric continuity is required. For example, if a spline is used to represent the trajectory of an object, parametric continuity guarantees that the object moves smoothly at the junction between two curve segments. However, there are applications for which parametric continuity is too constraining, since it depends on details of the parametrization that are not relevant to the shape of the curve. For example, if a curve is used to represent the boundary of an object, then only the outline of the curve is important. Thus, more flexible continuity conditions (usually called *geometric continuity*) based only on the geometry of the curve have been investigated. For plane curves, one may consider tangent continuity, or curvature continuity. For space curves, one may consider tangent continuity, curvature continuity, or torsion continuity. One may also want to consider higher-order continuity of the curvature κ and of the torsion τ, which means considering the continuity of higher derivatives of κ and τ. Another notion is *geometric continuity*, or G^n-*continuity*. Roughly speaking, two curves join with G^n-continuity if there is a reparametrization (a diffeomorphism) after which the curves join with parametric C^n-continuity. As a consequence, geometric continuity may be defined using the chain rule, in terms of a certain *connection matrix*. Yet another notion is *Frenet frame continuity*. Again, there is a vast literature on these topics, and we refer the readers to Farin [10, 9], Hoschek and Lasser [14], Bartels, Beatty, and Barsky [1], and Piegl and Tiller [17].

Complex shapes are usually represented in a piecewise fashion, composed of primitive elements smoothly joined. Traditional methods focus on achieving a specific level of interelement continuity, but the resulting shapes often possess bulges and undulations, and thus are of poor quality. They lack *fairness*. Fairness refers to the quality of regularity of the curvature (and torsion, for a space curve) of a curve. For a curve to be fair, it is required that the curvature vary gradually and oscillate as little as possible. Furthermore, the maximum rate of change of curvature should be minimized. This suggests several approaches.

- Minimal energy curve (which bends as little as possible): Minimize

$$\int_C \kappa^2 ds$$

  where κ is the curvature.
- Minimal variation curve (which bends as smoothly as possible): Minimize

$$\int_C \left(\frac{d(\kappa\mathbf{n})}{ds} \right)^2 ds$$

  where κ is the curvature and $\mathbf{n}$ is the principal normal.

  Another possibility is to minimize

$$\int_C \left[\left(\frac{d\kappa}{ds} \right)^2 + \left(\frac{d\tau}{ds} \right)^2 \right] ds$$

where κ is the curvature and τ is the torsion.

These problems may be cast as constrained optimization problems. Interelement continuity is solved by incorporating a penalty function. Interested readers are referred to the Ph.D. dissertations of Moreton [16] and Welch [18] for more details.

It should also be mentioned that it is possible to define a notion of affine normal and a notion of affine curvature without appealing to the concept of an inner product. For some interesting applications, see Calabi, Olver, and Tannenbaum [4] and Calabi, Olver, Shakiban, Tannenbaum, and Haker [3].

19.12 Problems

19.1. Plot the curve f defined by

$$
f(t) = \begin{cases} (-e^{1/t}, e^{1/t}\sin(e^{-1/t})) & \text{if } t < 0; \\ (0,0) & \text{if } t = 0; \\ (e^{-1/t}, e^{-1/t}\sin(e^{1/t})) & \text{if } t > 0. \end{cases}
$$

Verify that $f'(0) = 0$ and that the curve oscillates around the origin.

19.2. Plot the curve f defined by

$$
f(t) = \begin{cases} (t, t^2\sin(1/t)) & \text{if } t \neq 0; \\ (0,0) & \text{if } t = 0. \end{cases}
$$

Show that $f'(0) = (1,0)$ and that $f'(t) = (1, 2t\sin(1/t) - \cos(1/t))$ for $t \neq 0$. Verify that f' is discontinuous at 0.

19.3. Let $f: \,]a,b[\to \mathcal{E}$ be and open curve of class C^∞. For some $t \in \,]a,b[$, assume that $f'(t) = 0$, but also that there exist some integers p, q with $1 \leq p < q$ such that $f^{(p)}(t)$ is the first derivative not equal to 0 and $f^{(q)}(t)$ is the first derivative not equal to 0 and not collinear to $f^{(p)}(t)$. Show that by Taylor's formula, for $h > 0$ small enough, we have

$$
f(t+h) - f(t) = \left(\frac{h^p}{p!} + \lambda_{p+1}\frac{h^{p+1}}{(p+1)!} + \cdots + \lambda_{q-1}\frac{h^{q-1}}{(q-1)!} \right) f^{(p)}(t)
$$
$$
+ \frac{h^q}{q!}f^{(q)}(t) + \frac{h^q}{q!}\varepsilon(h),
$$

where $\lim_{h \to 0, h \neq 0} \varepsilon(h) = 0$.

As a consequence, the curve is tangent to the line of direction $f^{(p)}(t)$ passing through $f(t)$. Show that the curve has the following appearance locally at t:

1. p is odd. The curve traverses every secant through $f(t)$.
1a. q is even. Locally, the curve is entirely on the same side of its tangent at $f(t)$. This looks like an ordinary point.
1b. q is odd. Locally, the curve has an *inflection point* at $f(t)$, i.e., the two arcs of the curve meeting at $f(t)$ are on different sides of the tangent.
2. p is even. The curve does not traverse any secant through $f(t)$. It has a *cusp*.

2a. q is even. In this case, the two arcs of the curve meeting at $f(t)$ are on the same side of the tangent. We say that we have a *cusp of the second kind*.

2b. q is odd. In this case, the two arcs of the curve meeting at $f(t)$ are on different sides of the tangent. We say that we have a *cusp of the first kind*.

Draw examples for $(p=1,q=2)$, $(p=1,q=3)$, $(p=2,q=3)$, and $(p=2,q=4)$.

19.4. Draw the curve defined such that

$$x(t) = \frac{2t^2}{1+t^2},$$

$$y(t) = \frac{2t^3}{1+t^2}.$$

Show that the point $(0,0)$ is a cusp and that the line of equation $x=2$ is an asymptote. This curve is called the *cissoid of Diocles*.

19.5. (a) Draw the curve defined such that

$$x(t) = \sin t,$$

$$y(t) = \cos t + \log \tan \frac{t}{2}.$$

Show that the point $(1,0)$ is a cusp and that the line of equation $x=0$ is an asymptote.

(b) Show that the length of the segment of the tangent of the curve between the point of contact and the y-axis is of constant length 1. For this reason, this curve is called a *tractrix*.

19.6. (a) Given a tractrix specified by

$$x(t) = a\sin t,$$

$$y(t) = a\cos t + a\log \tan \frac{t}{2},$$

show that the curvature is given by $\kappa = |\tan t|$.

(b) Show that the center of curvature is on the curve

$$x(t) = \frac{a}{\sin t},$$

$$y(t) = a\log \tan \frac{t}{2}.$$

Show that this curve has the implicit equation

$$x = a\cosh\left(\frac{y}{a}\right).$$

Draw this curve, called a *catenary*.

Note. Recall that the *hyperbolic functions* cosh and sinh are defined by

$$\cosh u = \frac{e^u + e^{-u}}{2} \quad \text{and} \quad \sinh u = \frac{e^u - e^{-u}}{2}.$$

19.7. (a) Draw the curve f defined such that

$$x(t) = a e^{-bt} \cos t,$$
$$y(t) = a e^{-bt} \sin t,$$

where $a, b > 0$.

Show that the curve approaches the origin $(0,0)$ as $t \to +\infty$, spiraling around it. This curve is called a *logarithmic spiral*.

(b) Show that $f'(t) \to (0,0)$ as $t \to +\infty$, and that

$$\lim_{t \to +\infty} \int_{t_0}^{t} \sqrt{x'(u)^2 + y'(u)^2}\, du$$

is finite. Conclude that f has finite arc length in $[t_0, \infty[$.

19.8. (A square-filling curve due to Hilbert) This version of the Hilbert curve is defined in terms of four maps f_1, f_2, f_3, f_4 defined by

$$x' = \frac{1}{2}x - \frac{1}{2}, \qquad\qquad y' = \frac{1}{2}y + 1,$$
$$x' = \frac{1}{2}x + \frac{1}{2}, \qquad\qquad y' = \frac{1}{2}y + 1,$$
$$x' = -\frac{1}{2}y + 1, \qquad\qquad y' = \frac{1}{2}x + \frac{1}{2},$$
$$x' = \frac{1}{2}y - 1, \qquad\qquad y' = -\frac{1}{2}x + \frac{1}{2}.$$

(a) Prove that these maps are affine. Can you describe geometrically what their action is (rotation, translation, scaling?)

(b) Given any polygonal line L, define the following sequence of poygonal lines:

$$S_0 = L,$$
$$S_{n+1} = f_1(S_n) \cup f_2(S_n) \cup f_3(S_n) \cup f_4(S_n).$$

Construct S_1 starting from the polygonal line $L = ((-1,0), (0,1)), ((0,1), (1,0))$.

Can you figure out what S_n looks like in general? (you may want to write a computer program, and iterate at least 6 times).

(c) Prove that S_n has a limit that is a continuous curve not C^1 anywhere and that is space–filling, in the sense that its image is the entire unit square.

19.9. Consider the curve f over $[0,1]$ defined such that

$$f(t) = \begin{cases} (t, t\sin(\pi/t)) & \text{if } t \neq 0, \\ (0,0) & \text{if } t = 0. \end{cases}$$

Show geometrically that the arc length of the portion of curve corresponding to the interval $[1/(n+1),\, 1/n]$ is at least $1/\left(n+\frac{1}{2}\right)$. Use this to show that the length of the curve in the interval $[1/N,\, 1]$ is greater than $2\sum_{n=1}^{N} 1/(n+1)$. Conclude that this curve is not rectifiable.

19.10. Consider a polynomial curve of degree m defined by the control points $(b_0,\dots,b_m)$ over $[0,1]$. Prove that the curvature at b_0 is

$$\kappa(0) = \frac{m-1}{m} \frac{\|\overrightarrow{b_0 b_1} \times \overrightarrow{b_1 b_2}\|}{\|\overrightarrow{b_0 b_1}\|^3},$$

and that the curvature at b_m is given by

$$\kappa(1) = \frac{m-1}{m} \frac{\|\overrightarrow{b_{m-1} b_m} \times \overrightarrow{b_{m-2} b_{m-1}}\|}{\|\overrightarrow{b_{m-1} b_m}\|^3}.$$

Show that the torsion at b_0 is given by

$$\tau(0) = -\frac{m-2}{m} \frac{(\overrightarrow{b_0 b_1}, \overrightarrow{b_0 b_2}, \overrightarrow{b_0 b_3})}{\|\overrightarrow{b_0 b_1} \times \overrightarrow{b_1 b_2}\|^2}.$$

If $a = \|\overrightarrow{b_0 b_1}\|$ and h is the distance from b_2 to the line (b_0, b_1), show that

$$\kappa(0) = \frac{m-1}{m} \frac{h}{a^2}.$$

If c is the distance from b_3 to the plane spanned by (b_0, b_1, b_2) (the osculating plane), show that

$$|\tau(0)| = \frac{m-2}{m} \frac{c}{ah}.$$

19.11. Consider the curve defined such that

$$f(t) = \begin{cases} (t, t^2 + t^3 \sin(1/t)) & \text{if } t \neq 0; \\ (0,0) & \text{if } t = 0. \end{cases}$$

Show that the osculating circle for $t = 0$ is the circle of center $\left(0, \frac{1}{2}\right)$ and that $f''(0)$ is undefined, so that the center of curvature is undefined at $t = 0$.

19.12. Show that the solution of the system

$$u'x + v'y = uu' + vv',$$
$$u''x + v''y = uu'' + vv'' + u'^2 + v'^2,$$

is given by

$$x = u - \frac{v'(u'^2 + v'^2)}{u'v'' - v'u''},$$

$$y = v + \frac{u'(u'^2 + v'^2)}{u'v'' - v'u''},$$

provided that $u'v'' - v'u'' \neq 0$. Show that the radius of curvature is given by

$$\mathscr{R} = \frac{(u'^2 + v'^2)^{3/2}}{|u'v'' - v'u''|}.$$

19.13. (a) Given an ellipse

$$x = a \cos \theta,$$
$$y = b \sin \theta,$$

show that the radius of curvature is given by

$$\mathscr{R} = \frac{(a^2 \sin^2 \theta + b^2 \cos^2 \theta)^{3/2}}{ab},$$

and that the center of curvature is on the curve defined by

$$x = \frac{c^2}{a} \cos^3 \theta,$$

$$y = -\frac{c^2}{b} \sin^3 \theta.$$

This curve is called an *astroid*.

(b) Letting $N = \left(\frac{c^2}{a} \cos \theta, 0 \right)$ be the intersection of the normal to the point M on the ellipse with Ox, and $d = \|MN\|$ be the distance between M and N, show that the radius of curvature is given by

$$\mathscr{R} = \frac{a^2}{b^4} d^3.$$

19.14. Given a parabola of equation $y^2 = 2px$, compute the radius of curvature and show that the center of curvature is on the curve of equation

$$y^2 = \frac{8}{27p} (x - p)^3.$$

Show that this is a cuspidal cubic with a cusp at $(p, 0)$.

19.15. Given a hyperbola

$$x = a \cosh \theta,$$
$$y = b \sinh \theta,$$

compute the radius of curvature and show that the center of curvature is on the curve defined by

$$x = \frac{c^2}{a}\cosh^3\theta,$$

$$y = -\frac{c^2}{b}\sinh^3\theta.$$

Note. The function cosh and sinh are defined in Problem 19.6.

19.16. Given a logarithmic spiral specified by

$$x = a\,e^{m\theta}\cos\theta,$$
$$y = a\,e^{m\theta}\sin\theta,$$

where $a > 0$, show that the radius of curvature is

$$\mathscr{R} = a\sqrt{1+m^2}\,e^{m\theta},$$

and that the center of curvature is on the spiral defined by

$$x = -ma\,e^{m\theta}\sin\theta,$$
$$y = ma\,e^{m\theta}\cos\theta.$$

Show that this is the original spiral

19.17. Given a cardioid

$$x = a(1+\cos\theta)\cos\theta,$$
$$y = a(1+\cos\theta)\sin\theta,$$

show that the radius of curvature is

$$\mathscr{R} = \left|\frac{2a}{3}\cos(\theta/2)\right|,$$

and that the center of curvature is on the cardioid defined by

$$x = \frac{2a}{3} + \frac{a}{3}(1-\cos\theta)\cos\theta,$$
$$y = \frac{a}{3}(1-\cos\theta)\sin\theta.$$

19.18. A plane curve is defined in *polar coordinates* if

$$x = \rho(\theta)\cos\theta,$$
$$y = \rho(\theta)\sin\theta,$$

for some function ρ of the polar angle θ.

(a) Prove that the element of arc length is given by

$$ds = \sqrt{\rho^2 + (\rho')^2}\, d\theta.$$

(b) Prove that the curvature is given by

$$\kappa = \frac{2(\rho')^2 - \rho\rho'' + \rho^2}{[(\rho')^2 + \rho^2]^{3/2}}.$$

19.19. Give an example of a regular nonplanar curve such that $\tau = 0$.

19.20. A circular helix is defined by

$$f(t) = (a\cos t,\, a\sin t,\, kt).$$

Show that the curvature is given by

$$\kappa = \frac{a}{a^2 + k^2}$$

and that the torsion is given by

$$\tau = -\frac{k}{a^2 + k^2}.$$

19.21. If C is a regular plane curve parametrized by arc length, let $C'(s) = \mathbf{t}$ be the tangent vector at s, and write

$$\mathbf{t} = \cos\varphi\mathbf{i} + \sin\varphi\mathbf{j},$$

where $(\mathbf{i}, \mathbf{j})$ is an orthonormal basis.

(a) Show that the algebraic curvature $k(s)$ is given by

$$k = \frac{d\varphi}{ds}.$$

(b) Letting

$$C(s) = x(s)\mathbf{i} + y(s)\mathbf{j},$$

we have $dx = \cos\varphi\, ds$ and $dy = \sin\varphi\, ds$. If $k(s) = f(s)$ for some C^0-function f, show that

$$\varphi = \int f(s)\, ds + \varphi_0$$

and thus that

$$x = \int \cos\varphi(s)\, ds + a,$$
$$y = \int \sin\varphi(s)\, ds + b,$$

for some constants φ_0, a, b.

Remark: Integrals of the above form are known as *Fresnel integrals*, and were first encountered by Fresnel (1788–1827) in the context of refraction problems.

(c) Study the curves defined such that $k = cs + d$, for some constants c, d (such curves are called *clothoids*, or *Cornu spirals*).

19.22. Write a computer program that takes as input the parametric equation (not necessarily arc length parametrized) of a curve. Your program will generate a graph of the curve and animate the Frenet frame, osculating circle, and osculating sphere, along the curve. Try your program on a C^2-continuous B-spline to observe discontinuities of the osculating sphere.

19.23. Given a circle C and a point O on C, consider the set of all lines Δ such that if $p \neq O$ is any point on C, the line Δ is the line passing through p and forming an angle with the normal N_p at p equal to the angle of N_p with pO (in other words, Δ is obtained by reflecting pO about the normal N_p at p). When $p = O$, the line Δ is the diameter through O. Prove that the lines Δ are tangent to a cardioid (see Problem 19.17).

Remark: The above problem can be viewed as a problem of optics. If a light source is placed at O, the reflections of the light rays emanating from O will have a cardioid as envelope. Such curves are also called *caustics*.

19.24. Using a recursion scheme in which $[0, 1]$ is initially subdivided into four equal intervals and the square $[0, 1] \times [0, 1]$ is initially subdivided into four equal subsquares, give an analytic definition for the functions $h_n \colon [0, 1] \to [0, 1] \times [0, 1]$ involved in defining the Hilbert curve (see Figure 19.1). Prove that the sequence h_n converges to a continuous function h. Prove that the h_n can be chosen to be injective but that h cannot be injective.

19.25. Two biregular curves f and g in $\mathbb{E}^3$ are called *Bertrand curves* if they have a common principal normal at any of their points.

(a) If f is a plane biregular curve, then prove that any involute of the locus of centers of curvatures of f is a Bertrand curve of f. Any two Bertrand curves are parallel, in the sense that the distance measured along the common principal normal, between corresponding points of the two Bertrand curves, is constant.

(b) If f^* and f are Bertrand curves, then f^* has an equation of the form

$$f^*(t) = f(t) + a(t)\mathbf{n},$$

where $\mathbf{n}$ is the principal normal to f at t. We will prove shortly that $a(t)$ must be a constant.

Assuming that f and f^* are Bertrand curves, using the fact that

$$f^*(t) = f(t) + a(t)\mathbf{n},$$

observe that

$$a^2(t) = (f^* - f) \cdot (f^* - f),$$

and prove that

$$\frac{d}{dt}(a^2) = 2(f^* - f) \cdot \left(\frac{d}{dt}(f^*) - \frac{d}{dt}(f) \right) = 0.$$

Conlude that $a(t)$ is constant.

Let $\mathbf{t}$ and $\mathbf{t}^*$ be the unit tangent vectors to f and f^*, respectively. Using the fact that

$$\frac{d}{dt}(\mathbf{t}^* \cdot \mathbf{t}) = \frac{d\mathbf{t}^*}{dt} \cdot \mathbf{t} + \mathbf{t}^* \cdot \frac{d\mathbf{t}}{dt},$$

prove that

$$\frac{d}{dt}(\mathbf{t}^* \cdot \mathbf{t}) = 0.$$

Let

$$\mathbf{t}^* \cdot \mathbf{t} = \cos \alpha,$$

a constant. Observe that α is the constant angle between the tangents at corresponding points of the Bertrand curves.

Now, assuming that f and f^* are both parametrized by arc lengths, s and s^*, respectively, we have

$$f^*(s) = f(s) + a(s)\mathbf{n}.$$

Prove that

$$\cos \alpha = \frac{ds}{ds^*}(1 - a\kappa).$$

Also prove that

$$\|\mathbf{t}^* \times \mathbf{t}\| = \left\| \frac{ds}{ds^*} a\tau \mathbf{n} \right\|.$$

Conclude that

$$a\tau \frac{ds}{ds^*} = \sin \alpha,$$

where the sign of α is suitably chosen. From

$$\frac{ds}{ds^*}(1 - a\kappa) = \cos \alpha \quad \text{and} \quad a\tau \frac{ds}{ds^*} = \sin \alpha,$$

prove that

$$\frac{1 - a\kappa}{a\tau} = \cot \alpha,$$

and thus, letting $c_1 = a$, $c_2 = a \cot \alpha$, that the linear equation

$$c_1 \kappa + c_2 \tau = 1$$

holds between κ and τ.

(c) Conversely, assume that that the linear equation

$$c_1 \kappa + c_2 \tau = 1$$

holds between κ and τ. We shall prove that f has the Bertrand curve

$$f^*(s) = f(s) + c_1 \mathbf{n}.$$

Prove that

$$\frac{df^*}{ds} = (1 - c_1 \kappa)\mathbf{t} + c_1 \tau \mathbf{b}.$$

In view of the equation

$$c_1 \kappa + c_2 \tau = 1,$$

letting $c = c_2/c_1$, prove that

$$\frac{df^*}{ds} = c_1 \tau (c\mathbf{t} + \mathbf{b}).$$

Conclude that the unit tangent vector to C^* is

$$\mathbf{t}^* = \frac{c\mathbf{t} + \mathbf{b}}{\sqrt{1 + c^2}},$$

that

$$\frac{d\mathbf{t}^*}{ds} = \frac{1}{\sqrt{1 + c^2}} (c\kappa - \tau)\mathbf{n},$$

and that C and C^* are Bertrand curves.

Thus, we have proved that a curve C has a Bertrand curve iff a linear equation

$$c_1 \kappa + c_2 \tau = 1$$

holds between κ and τ (Bertrand, 1850).

Extra Credit: Prove that a circular helix is the only nonplanar biregular curve having more than one Bertrand curve.

References

1. Richard H. Bartels, John C. Beatty, and Brian A. Barsky. *An Introduction to Splines for Use in Computer Graphics and Geometric Modelling*. Morgan Kaufmann, first edition, 1987.
2. Marcel Berger and Bernard Gostiaux. *Géométrie différentielle: variétés, courbes et surfaces*. Collection Mathématiques. Puf, second edition, 1992. English edition: Differential geometry, manifolds, curves, and surfaces, GTM No. 115, Springer-Verlag.
3. Eugenio Calabi, Peter J. Olver, C. Shakiban, Allen Tannenbaum, and Steven Haker. Differential and numerically invariant signature curves applied to object recognition. *International Journal of Computer Vision*, 26(2):107–135, 1998.

4. Eugenio Calabi, Peter J. Olver, and Allen Tannenbaum. Affine geometry, curve flows, and invariant numerical approximations. *Advances in Mathematics*, 124:154–196, 1996.

5. Élie Cartan. *Les systèmes différentiels extérieurs et leurs applications géométriques*. Hermann, first edition, 1945.

6. Gaston Darboux. *Leçons sur la théorie générale des surfaces, Première Partie*. Gauthier-Villars, second edition, 1914.

7. Manfredo P. do Carmo. *Differential Geometry of Curves and Surfaces*. Prentice-Hall, 1976.

8. Gerald A. Edgar. *Measure, Topology, and Fractal Geometry*. Undergraduate Texts in Mathematics. Springer-Verlag, first edition, 1992.

9. Gerald Farin. *NURB Curves and Surfaces, from Projective Geometry to Practical Use*. AK Peters, first edition, 1995.

10. Gerald Farin. *Curves and Surfaces for CAGD*. Academic Press, fourth edition, 1998.

11. J.-C. Fiorot and P. Jeannin. *Courbes et Surfaces Rationelles*. RMA 12. Masson, first edition, 1989.

12. J.-C. Fiorot and P. Jeannin. *Courbes Splines Rationelles*. RMA 24. Masson, first edition, 1992.

13. Jean H. Gallier. *Curves and Surfaces in Geometric Modeling: Theory and Algorithms*. Morgan Kaufmann, first edition, 1999.

14. J. Hoschek and D. Lasser. *Computer-Aided Geometric Design*. AK Peters, first edition, 1993.

15. Erwin Kreyszig. *Differential Geometry*. Dover, first edition, 1991.

16. Henry P. Moreton. *Minimum curvature variation curves, networks, and surfaces for fair free-form shape design*. PhD thesis, University of California, Berkeley, 1993.

17. Les Piegl and Wayne Tiller. *The NURBS Book*. Monograph in Visual Communications. Springer-Verlag, first edition, 1995.

18. William Welch. *Serious Putty: Topological Design for Variational Curves and Surfaces*. PhD thesis, Carnegie Mellon University, Pittsburgh, Pa., 1995.

Chapter 20
Basics of the Differential Geometry of Surfaces

20.1 Introduction

The purpose of this chapter is to introduce the reader to some elementary concepts of the differential geometry of surfaces. Our goal is rather modest: We simply want to introduce the concepts needed to understand the notion of Gaussian curvature, mean curvature, principal curvatures, and geodesic lines. Almost all of the material presented in this chapter is based on lectures given by Eugenio Calabi in an upper undergraduate differential geometry course offered in the fall of 1994. Most of the topics covered in this course have been included, except a presentation of the global Gauss–Bonnet–Hopf theorem, some material on special coordinate systems, and Hilbert's theorem on surfaces of constant negative curvature.

What is a surface? A precise answer cannot really be given without introducing the concept of a manifold. An informal answer is to say that a surface is a set of points in $\mathbb{R}^3$ such that for every point p on the surface there is a small (perhaps very small) neighborhood U of p that is continuously deformable into a little flat open disk. Thus, a surface should really have some topology. Also, locally, unless the point p is "singular," the surface looks like a plane.

Properties of surfaces can be classified into *local properties* and *global properties*. In the older literature, the study of local properties was called *geometry in the small*, and the study of global properties was called *geometry in the large*. Local properties are the properties that hold in a small neighborhood of a point on a surface. Curvature is a local property. Local properties can be studied more conveniently by assuming that the surface is parametrized locally. Thus, it is important and useful to study parametrized patches. In order to study the global properties of a surface, such as the number of its holes or boundaries, global topological tools are needed. For example, closed surfaces cannot really be studied rigorously using a single parametrized patch, as in the study of local properties. It is necessary to cover a closed surface with various patches, and these patches need to overlap in some clean fashion, which leads to the notion of a manifold.

Another more subtle distinction should be made between *intrinsic* and *extrinsic* properties of a surface. Roughly speaking, intrinsic properties are properties of a surface that do not depend on the way the surface is immersed in the ambient space, whereas extrinsic properties depend on properties of the ambient space. For example, we will see that the Gaussian curvature is an intrinsic concept, whereas the normal to a surface at a point is an extrinsic concept. The distinction between these two notions is clearer in the framework of Riemannian manifolds, since manifolds provide a way of defining an abstract space not immersed in some a priori given ambient space, but readers should have some awareness of the difference between intrinsic and extrinsic properties.

In this chapter we focus exclusively on the study of local properties, both intrinsic and extrinsic, and manifolds are completely left out. Readers eager to learn more differential geometry and about manifolds are refereed to do Carmo [12], Berger and Gostiaux [4], Lafontaine [29], and Gray [23]. A more complete list of references can be found in Section 20.11.

By studying the properties of the curvature of curves on a surface, we will be led to the first and second fundamental forms of a surface. The study of the normal and tangential components of the curvature will lead to the normal curvature and to the geodesic curvature. We will study the normal curvature, and this will lead us to principal curvatures, principal directions, the Gaussian curvature, and the mean curvature. In turn, the desire to express the geodesic curvature in terms of the first fundamental form alone will lead to the Christoffel symbols. The study of the variation of the normal at a point will lead to the Gauss map and its derivative, and to the Weingarten equations. We will also quote Bonnet's theorem about the existence of a surface patch with prescribed first and second fundamental forms. This will require a discussion of the *Theorema Egregium* and of the Codazzi–Mainardi compatibility equations. We will take a quick look at curvature lines, asymptotic lines, and geodesics, and conclude by quoting a special case of the Gauss–Bonnet theorem.

Since this chapter is just a brief introduction to the local theory of the differential geometry of surfaces, the following additional references are suggested. For an intuitive introduction to differential geometry there is no better source that the beautiful presentation given in Chapter IV of Hilbert and Cohn-Vossen [25]. The style is informal, and there are occasional mistakes, but there are amazingly powerful geometric insights. The reader will have a taste of the state of differential geometry in the 1920s. For a taste of the differential geometry of surfaces in the 1980s, we highly recommend Chapter 10 and Chapter 11 in Berger and Gostiaux [4]. These remarkable chapters are written as a guide, basically without proofs, and assume a certain familiarity with differential geometry, but we believe that most readers could easily read them after completing this chapter. For a comprehensive and yet fairly elementary treatment of the differential geometry of curves and surfaces we highly recommend do Carmo [12] and Kreyszig [28]. Another nice and modern presentation of differential geometry including many examples in *Mathematica* can be found in Gray [23]. The older texts by Stoker [42] and Hopf [26] are also recommended. For the (very) perseverant reader interested in the state of surface theory around the 1900s, nothing tops Darboux's four–volume treatise [9, 10, 7, 8]. Actually, Darboux is a real gold mine for all sorts of fascinating (often long forgotten) results. For a very interesting article on the history of differential geometry see Paulette Libermann's article in Dieudonné [11], Chapter IX. More references can be found in Section 20.11. Some interesting applications of the differential geometry of surfaces to geometric design can be found in the Ph.D. theses of Henry Moreton [38] and William Welch [44]; see Section 20.13 for a glimpse of these applications.

20.2 Parametrized Surfaces

In this chapter we consider exclusively surfaces immersed in the affine space $\mathbb{A}^3$. In order to be able to define the normal to a surface at a point, and the notion of curvature, we assume that some inner product is defined on $\mathbb{R}^3$. Unless specified otherwise, we assume that this inner product is the standard one, i.e.,

$$(x_1, x_2, x_3) \cdot (y_1, y_2, y_3) = x_1 y_1 + x_2 y_2 + x_3 y_3.$$

The Euclidean space obtained from $\mathbb{A}^3$ by defining the above inner product on $\mathbb{R}^3$ is denoted by $\mathbb{E}^3$ (and similarly, $\mathbb{E}^2$ is associated with $\mathbb{A}^2$).

Let Ω be some open subset of the plane $\mathbb{R}^2$. Recall that a map $X \colon \Omega \to \mathbb{E}^3$ is C^p-*continuous* if all the partial derivatives

$$\frac{\partial^{i+j} X}{\partial u^i \partial v^j}(u, v)$$

exist and are continuous for all i, j such that $0 \leq i + j \leq p$, and all $(u, v) \in \mathbb{R}^2$. A surface is a map $X \colon \Omega \to \mathbb{E}^3$, as above, where X is at least C^3-continuous. It turns out that in order to study surfaces, in particular the important notion of curvature, it is very useful to study the properties of curves on surfaces. Thus, we will begin by studying curves on surfaces. The curves arising as plane sections of a surface by planes containing the normal line at some point of the surface will play an important role. Indeed, we will study the variation of the "normal curvature" of such curves. We will see that in general, the normal curvature reaches a maximum value κ_1 and a minimum value κ_2. This will lead us to the notion of Gaussian curvature (it is the product $K = \kappa_1 \kappa_2$).

Actually, we will need to impose an extra condition on a surface X so that the tangent plane (and the normal) at any point is defined. Again, this leads us to consider curves on X.

A curve C on X is defined as a map $C \colon t \mapsto X(u(t), v(t))$, where u and v are continuous functions on some open interval I contained in Ω. We also assume that the plane curve $t \mapsto (u(t), v(t))$ is regular, that is, that

$$\left(\frac{du}{dt}(t), \frac{dv}{dt}(t) \right) \neq (0, 0) \text{ for all } t \in I.$$

For example, the curves $v \mapsto X(u_0, v)$ for some constant u_0 are called *u-curves*, and the curves $u \mapsto X(u, v_0)$ for some constant v_0 are called *v-curves*. Such curves are also called the *coordinate curves*.

We would like the curve $t \mapsto X(u(t), v(t))$ to be a regular curve for all regular curves $t \mapsto (u(t), v(t))$, i.e., to have a well-defined tangent vector for all $t \in I$. The tangent vector $dC(t)/dt$ to C at t can be computed using the chain rule:

$$\frac{dC}{dt}(t) = \frac{\partial X}{\partial u}(u(t), v(t))\frac{du}{dt}(t) + \frac{\partial X}{\partial v}(u(t), v(t))\frac{dv}{dt}(t).$$

Note that

$$\frac{dC}{dt}(t), \quad \frac{\partial X}{\partial u}(u(t), v(t)) \quad \text{and} \quad \frac{\partial X}{\partial v}(u(t), v(t))$$

are vectors, but for simplicity of notation, we omit the vector symbol in these expressions.[1]

It is customary to use the following abbreviations: The partial derivatives

$$\frac{\partial X}{\partial u}(u(t), v(t)) \quad \text{and} \quad \frac{\partial X}{\partial v}(u(t), v(t))$$

are denoted by $X_u(t)$ and $X_v(t)$, or even by X_u and X_v, and the derivatives

$$\frac{dC}{dt}(t), \quad \frac{du}{dt}(t) \quad \text{and} \quad \frac{dv}{dt}(t)$$

are denoted by $\dot{C}(t)$, $\dot{u}(t)$, and $\dot{v}(t)$, or even by $\dot{C}$, $\dot{u}$, and $\dot{v}$. When the curve C is parametrized by arc length s, we denote

$$\frac{dC}{ds}(s), \quad \frac{du}{ds}(s), \quad \text{and} \quad \frac{dv}{ds}(s)$$

by $C'(s)$, $u'(s)$, and $v'(s)$, or even by C', u', and v'. Thus, we reserve the prime notation to the case where the parametrization of C is by arc length.

 Note that it is the curve $C\colon t \mapsto X(u(t), v(t))$ that is parametrized by arc length, not the curve $t \mapsto (u(t), v(t))$.

Using this notation $\dot{C}(t)$ is expressed as follows:

$$\dot{C}(t) = X_u(t)\dot{u}(t) + X_v(t)\dot{v}(t),$$

or simply as

$$\dot{C} = X_u\dot{u} + X_v\dot{v}.$$

---

[1] Also, traditionally, the result of multiplying a vector u by a scalar λ is denoted by λu, with the scalar on the left. In the expressions above involving partial derivatives, the scalar is written on the right of the vector rather on the left. Although possibly confusing, this appears to be standard practice.

Now, if we want $\dot{C} \neq 0$ for all regular curves $t \mapsto (u(t), v(t))$, we must require that X_u and X_v be linearly independent. Equivalently, we must require that the cross product $X_u \times X_v$ be nonnull.

Definition 20.1. A *surface patch* X, for short a *surface* X, is a map $X \colon \Omega \to \mathbb{E}^3$ where Ω is some open subset of the plane $\mathbb{R}^2$ and where X is at least C^3-continuous. We say that the surface X is *regular at* $(u, v) \in \Omega$ if $X_u \times X_v \neq 0$, and we also say that $p = X(u, v)$ is *a regular point of* X. If $X_u \times X_v = 0$, we say that $p = X(u, v)$ is a *singular point of* X. The surface X is *regular on* Ω if $X_u \times X_v \neq 0$, for all $(u, v) \in \Omega$. The subset $X(\Omega)$ of $\mathbb{E}^3$ is called the *trace* of the surface X.

Remark: It often often desirable to define a (regular) surface patch $X \colon \Omega \to \mathbb{E}^3$ where Ω is a *closed* subset of $\mathbb{R}^2$. If Ω is a closed set, we assume that there is some open subset U containing Ω and such that X can be extended to a (regular) surface over U (i.e., that X is at least C^3-continuous).

   Given a regular point $p = X(u, v)$, since the tangent vectors to all the curves passing through a given point are of the form $X_u \dot{u} + X_v \dot{v}$, it is obvious that they form a vector space of dimension 2 isomorphic to $\mathbb{R}^2$ called the *tangent space at* p, and denoted by $T_p(X)$. Note that (X_u, X_v) is a basis of this vector space $T_p(X)$. The set of tangent lines passing through p and having some tangent vector in $T_p(X)$ as direction is an affine plane called the *affine tangent plane at* p. Geometrically, this is an object different from $T_p(X)$, and it should be denoted differently (perhaps as $AT_p(X)$?).[2] Nevertheless, we will use the notation $T_p(X)$ like everybody else, but by calling it *tangent plane* instead of *tangent space*, we hope that the potential confusion will be eliminated.

   The unit vector

$$\mathbf{N}_p = \frac{X_u \times X_v}{\|X_u \times X_v\|}$$

is called the *unit normal vector at* p, and the line through p of direction $\mathbf{N}_p$ is the *normal line to* X *at* p. This time, we can use the notation N_p for the line, to distinguish it from the vector $\mathbf{N}_p$.

Example 20.1. Let $\Omega = \,]-1, 1[\times \,]-1, 1[$, and let X be the surface patch defined by

$$x = \frac{2au}{u^2 + v^2 + 1}, \quad y = \frac{2bu}{u^2 + v^2 + 1}, \quad z = \frac{c(1 - u^2 - v^2)}{u^2 + v^2 + 1},$$

where $a, b, c > 0$. The surface X is a portion of an ellipsoid. Let

$$t \mapsto (t, t^2)$$

be the piece of parabola corresponding to $t \in \,]-1, 1[$. Then we obtain the curve $C(t) = X(t, t^2)$ on the surface X. It is easily verified that the unit normal to the

---

[2] It would probably be better to denote the tangent space by $\overrightarrow{T}_p(X)$ and the tangent plane by $T_p(X)$, but nobody else does!

surface is

$$\mathbf{N}_{(u,v)} = (2bcu/Z, 2acv/Z, ab(1-u^2-v^2)/Z),$$

where

$$Z^2 = 4b^2c^2u^2 + 4a^2c^2v^2 + a^2b^2(1-u^2-v^2)^2.$$

The portion of ellipsoid X, the curve C on X, some unit normals, and some tangent vectors (for $u = \frac{1}{3}, v = \frac{1}{9}$), are shown in Figure 20.1, for $a = 5, b = 4, c = 3$.

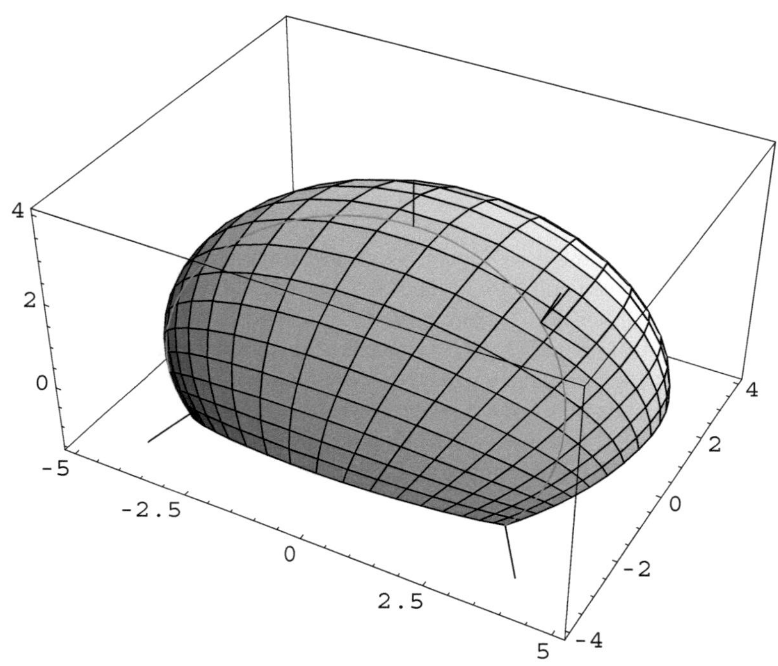

Fig. 20.1 A curve C on a surface X.

The fact that we are not requiring the map X defining a surface $X : \Omega \to \mathbb{E}^3$ to be injective may cause problems. Indeed, if X is not injective, it may happen that $p = X(u_0, v_0) = X(u_1, v_1)$ for some (u_0, v_0) and (u_1, v_1) such that $(u_0, v_0) \neq (u_1, v_1)$. In this case, the tangent plane $T_p(X)$ at p is not well-defined. Indeed, we really have two pairs of partial derivatives $(X_u(u_0, v_0), X_v(u_0, v_0))$ and $(X_u(u_1, v_1), X_v(u_1, v_1))$, and the planes spanned by these pairs could be distinct. In this case there are really two tangent planes $T_{(u_0, v_0)}(X)$ and $T_{(u_1, v_1)}(X)$ at the point p where X has a self-intersection. Similarly, the normal $\mathbf{N}_p$ is not well-defined, and we really have two normals $\mathbf{N}_{(u_0, v_0)}$ and $\mathbf{N}_{(u_1, v_1)}$ at p.

We could avoid the problem entirely by assuming that X is injective. This will rule out many surfaces that come up in practice. If necessary, we use the notation

$T_{(u,v)}(X)$ or $\mathbf{N}_{(u,v)}$, which removes possible ambiguities. However, it is a more cumbersome notation, and we will continue to write $T_p(X)$ and $\mathbf{N}_p$, being aware that this may be an ambiguous notation, and that some additional information is needed.

The tangent space may also be undefined when p is not a regular point.

Example 20.2. Considering the surface $X = (x(u,v), y(u,v), z(u,v))$ defined such that

$$x = u(u^2 + v^2),$$
$$y = v(u^2 + v^2),$$
$$z = u^2 v - v^3/3,$$

note that all the partial derivatives at the origin $(0,0)$ are zero. Thus, the origin is a singular point of the surface X. Indeed, one can check that the tangent lines at the origin do not lie in a plane.

It is interesting to see how the unit normal vector $\mathbf{N}_p$ changes under a change of parameters. Assume that $u = u(r,s)$ and $v = v(r,s)$, where $(r,s) \mapsto (u,v)$ is a diffeomorphism. By the chain rule,

$$X_r \times X_s = \left(X_u \frac{\partial u}{\partial r} + X_v \frac{\partial v}{\partial r} \right) \times \left(X_u \frac{\partial u}{\partial s} + X_v \frac{\partial v}{\partial s} \right)$$

$$= \left(\frac{\partial u}{\partial r} \frac{\partial v}{\partial s} - \frac{\partial u}{\partial s} \frac{\partial v}{\partial r} \right) X_u \times X_v$$

$$= \begin{vmatrix} \dfrac{\partial u}{\partial r} & \dfrac{\partial u}{\partial s} \\[2mm] \dfrac{\partial v}{\partial r} & \dfrac{\partial v}{\partial s} \end{vmatrix} X_u \times X_v$$

$$= \frac{\partial(u,v)}{\partial(r,s)} X_u \times X_v,$$

denoting the Jacobian determinant of the map $(r,s) \mapsto (u,v)$ by $\partial(u,v)/\partial(r,s)$. Then, the relationship between the unit vectors $\mathbf{N}_{(u,v)}$ and $\mathbf{N}_{(r,s)}$ is

$$\mathbf{N}_{(r,s)} = \mathbf{N}_{(u,v)} \operatorname{sign} \frac{\partial(u,v)}{\partial(r,s)}.$$

We will therefore restrict our attention to changes of variables such that the Jacobian determinant $\partial(u,v)/\partial(r,s)$ is positive.

One should also note that the condition $X_u \times X_v \neq 0$ is equivalent to the fact that the Jacobian matrix of the derivative of the map $X \colon \Omega \to \mathbb{E}^3$ has rank 2, i.e., that the derivative $DX(u,v)$ of X at (u,v) is injective. Indeed, the Jacobian matrix of the derivative of the map

$$(u,v) \mapsto X(u,v) = (x(u,v), y(u,v), z(u,v))$$

is

$$\begin{pmatrix} \dfrac{\partial x}{\partial u} & \dfrac{\partial x}{\partial v} \\[2mm] \dfrac{\partial y}{\partial u} & \dfrac{\partial y}{\partial v} \\[2mm] \dfrac{\partial z}{\partial u} & \dfrac{\partial z}{\partial v} \end{pmatrix},$$

and $X_u \times X_v \neq 0$ is equivalent to saying that one of the minors of order 2 is invertible. Thus, a regular surface is an *immersion* of an open set of $\mathbb{R}^2$ into $\mathbb{E}^3$.

To a great extent, the properties of a surface can be studied by studying the properties of curves on this surface. One of the most important properties of a surface is its curvature. A gentle way to introduce the curvature of a surface is to study the curvature of a curve on a surface. For this, we will need to compute the norm of the tangent vector to a curve on a surface. This will lead us to the first fundamental form.

20.3 The First Fundamental Form (Riemannian Metric)

Given a curve C on a surface X, we first compute the element of arc length of the curve C. For this, we need to compute the square norm of the tangent vector $\dot{C}(t)$. The square norm of the tangent vector $\dot{C}(t)$ to the curve C at p is

$$\|\dot{C}\|^2 = (X_u \dot{u} + X_v \dot{v}) \cdot (X_u \dot{u} + X_v \dot{v}),$$

where $\cdot$ is the inner product in $\mathbb{E}^3$, and thus,

$$\|\dot{C}\|^2 = (X_u \cdot X_u)\,\dot{u}^2 + 2(X_u \cdot X_v)\,\dot{u}\dot{v} + (X_v \cdot X_v)\,\dot{v}^2.$$

Following common usage, we let

$$E = X_u \cdot X_u, \quad F = X_u \cdot X_v, \quad G = X_v \cdot X_v,$$

and

$$\|\dot{C}\|^2 = E\,\dot{u}^2 + 2F\,\dot{u}\dot{v} + G\,\dot{v}^2.$$

Euler had already obtained this formula in 1760. Thus, the map $(x,y) \mapsto Ex^2 + 2Fxy + Gy^2$ is a quadratic form on $\mathbb{R}^2$, and since it is equal to $\|\dot{C}\|^2$, it is positive definite. This quadratic form plays a major role in the theory of surfaces, and deserves an official definition.

Definition 20.2. Given a surface X, for any point $p = X(u,v)$ on X, letting

$$E = X_u \cdot X_u, \quad F = X_u \cdot X_v, \quad G = X_v \cdot X_v,$$

the positive definite quadratic form $(x,y) \mapsto Ex^2 + 2Fxy + Gy^2$ is called the *first fundamental form of X at p*. It is often denoted by I_p, and in matrix form, we have

$$I_p(x,y) = (x,y) \begin{pmatrix} E & F \\ F & G \end{pmatrix} \begin{pmatrix} x \\ y \end{pmatrix}.$$

Since the map $(x,y) \mapsto Ex^2 + 2Fxy + Gy^2$ is a positive definite quadratic form, we must have $E \neq 0$ and $G \neq 0$. Then, we can write

$$Ex^2 + 2Fxy + Gy^2 = E\left(x + \frac{F}{E}y\right)^2 + \frac{EG - F^2}{E}y^2.$$

Since this quantity must be positive, we must have $E > 0$, $G > 0$, and also $EG - F^2 > 0$.

The symmetric bilinear form φ_I associated with I is an inner product on the tangent space at p, such that

$$\varphi_I((x_1,y_1),(x_2,y_2)) = (x_1,y_1) \begin{pmatrix} E & F \\ F & G \end{pmatrix} \begin{pmatrix} x_2 \\ y_2 \end{pmatrix}.$$

This inner product is also denoted by $\langle (x_1,y_1),(x_2,y_2)\rangle_p$. The inner product φ_I can be used to determine the angle of two curves passing through p, i.e., the angle θ of the tangent vectors to these two curves at p. We have

$$\cos \theta = \frac{\langle (\dot{u}_1,\dot{v}_1),(\dot{u}_2,\dot{v}_2)\rangle}{\sqrt{I(\dot{u}_1,\dot{v}_1)}\sqrt{I(\dot{u}_2,\dot{v}_2)}}.$$

For example, the angle between the u-curve and the v-curve passing through p (where u or v is constant) is given by

$$\cos \theta = \frac{F}{\sqrt{EG}}.$$

Thus, the u-curves and the v-curves are orthogonal iff $F(u,v) = 0$ on Ω.

Remarks:

(1) Since

$$\left(\frac{ds}{dt}\right)^2 = \|\dot{C}\|^2 = E\dot{u}^2 + 2F\dot{u}\dot{v} + G\dot{v}^2$$

represents the square of the "element of arc length" of the curve C on X, and since $du = \dot{u}dt$ and $dv = \dot{v}dt$, one often writes the first fundamental form as

$$ds^2 = E\,du^2 + 2F\,du\,dv + G\,dv^2.$$

Thus, the length $l(pq)$ of an arc of curve on the surface joining the points $p = X(u(t_0),v(t_0))$ and $q = X(u(t_1),v(t_1))$ is

$$l(p,q) = \int_{t_0}^{t_1} \sqrt{E\,\dot{u}^2 + 2F\,\dot{u}\dot{v} + G\,\dot{v}^2}\,dt.$$

One also refers to $ds^2 = E\,du^2 + 2F\,du\,dv + G\,dv^2$ as a *Riemannian metric*. The symmetric matrix associated with the first fundamental form is also denoted by

$$\begin{pmatrix} g_{11} & g_{12} \\ g_{21} & g_{22} \end{pmatrix},$$

where $g_{12} = g_{21}$.

(2) As in the previous section, if X is not injective, the first fundamental form I_p is not well-defined. What is well-defined is $I_{(u,v)}$. In some sense this is even worse, since one of the main themes of differential geometry is that the metric properties of a surface (or of a manifold) are captured by a Riemannian metric. Again, we will not worry too much about this, or we will assume X injective.

(3) It can be shown that the element of area dA on a surface X is given by

$$dA = \|X_u \times X_v\|\,du\,dv = \sqrt{EG - F^2}\,du\,dv.$$

We have just discovered that, in contrast to a flat surface, where the inner product is the same at every point, on a curved surface the inner product induced by the Riemannian metric on the tangent space at every point changes as the point moves on the surface. This fundamental idea is at the heart of the definition of an abstract Riemannian manifold. It is also important to observe that the first fundamental form of a surface does **not** characterize the surface.

Example 20.3. It is easy to see that the first fundamental form of a plane and the first fundamental form of a cylinder of revolution defined by

$$X(u,v) = (\cos u, \sin u, v)$$

are identical:

$$(E, F, G) = (1, 0, 1).$$

Thus $ds^2 = du^2 + dv^2$, which is not surprising.

A more striking example is that of the helicoid and the catenoid.

Example 20.4. The *helicoid* is the surface defined over $\mathbb{R} \times \mathbb{R}$ such that

$$x = u_1 \cos v_1,$$
$$y = u_1 \sin v_1,$$
$$z = v_1.$$

This is the surface generated by a line parallel to the xOy plane, touching the z-axis, and also touching a helix of axis Oz. It is easily verified that

$$(E,F,G) = (1,0,u_1^2+1).$$

Figure 20.2 shows a portion of helicoid corresponding to $0 \leq v_1 \leq 2\pi$.

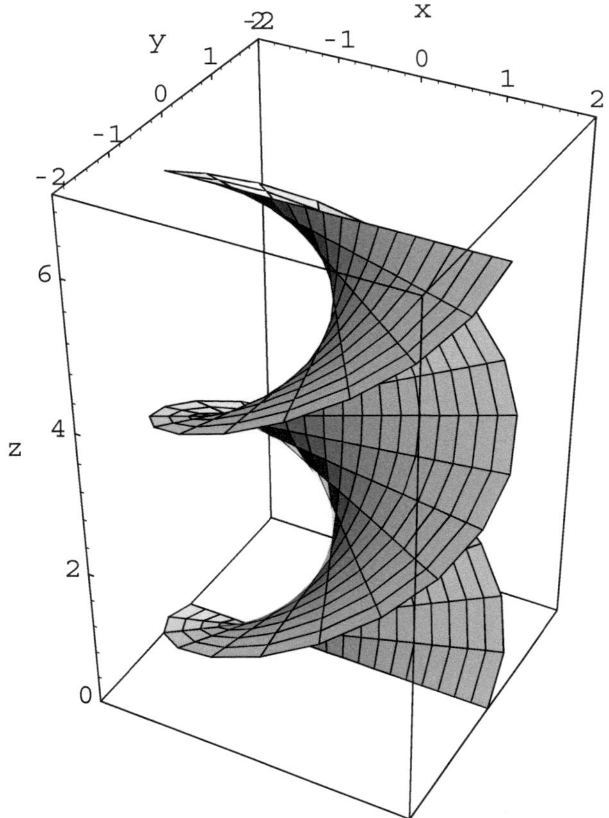

Fig. 20.2 A helicoid.

Example 20.5. The *catenoid* is the surface of revolution defined over $\mathbb{R} \times \mathbb{R}$ such that

$$x = \cosh u_2 \cos v_2,$$
$$y = \cosh u_2 \sin v_2,$$
$$z = u_2.$$

It is the surface obtained by rotating a *catenary* around the z-axis. (Recall that the *hyperbolic functions* cosh and sinh are defined by $\cosh u = (e^u + e^{-u})/2$ and

$\sinh u = (e^u - e^{-u})/2$. The catenary is the plane curve defined by $y = \cosh x$). It is easily verified that

$$(E, F, G) = (\cosh^2 u_2, 0, \cosh^2 u_2).$$

Figure 20.3 shows a portion of catenoid corresponding to $0 \leq v_2 \leq 2\pi$.

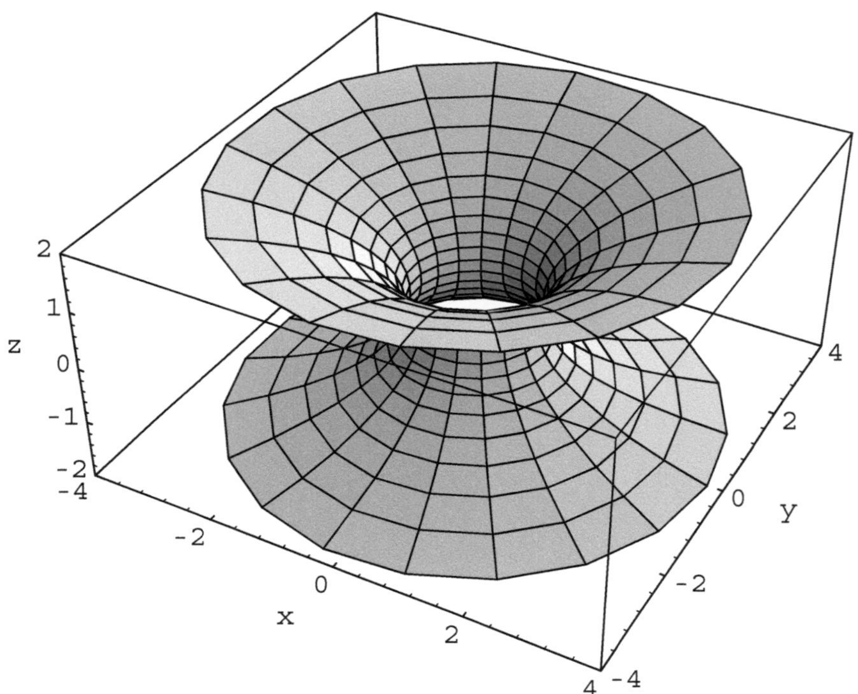

Fig. 20.3 A catenoid.

We can make the change of variables $u_1 = \sinh u_3$, $v_1 = v_3$, which is bijective and whose Jacobian determinant is $\cosh u_3$, which is always positive, obtaining the following parametrization of the helicoid:

$$x = \sinh u_3 \cos v_3,$$
$$y = \sinh u_3 \sin v_3,$$
$$z = v_3.$$

It is easily verified that

$$(E,F,G) = (\cosh^2 u_3, 0, \cosh^2 u_3),$$

showing that the helicoid and the catenoid have the same first fundamental form. What is happening is that the two surfaces are locally isometric (roughly, this means that there is a smooth map between the two surfaces that preserves distances locally). Indeed, if we consider the portions of the two surfaces corresponding to the domain $\mathbb{R} \times \,]0, 2\pi[$, it is possible to deform isometrically the portion of helicoid into the portion of catenoid (note that by excluding 0 and 2π, we have made a "slit" in the catenoid (a portion of meridian), and thus we can open up the catenoid and deform it into the helicoid). For more on this, we urge our readers to consult do Carmo [12], Chapter 4, Section 2, pages 218–227.

We will now see how the first fundamental form relates to the curvature of curves on a surface.

20.4 Normal Curvature and the Second Fundamental Form

In this section we take a closer look at the curvature at a point of a curve C on a surface X. Assuming that C is parametrized by arc length, we will see that the vector $X''(s)$ (which is equal to $\kappa\mathbf{n}$, where $\mathbf{n}$ is the principal normal to the curve C at p, and κ is the curvature) can be written as

$$\kappa\mathbf{n} = \kappa_N\mathbf{N} + \kappa_g\mathbf{n_g},$$

where $\mathbf{N}$ is the normal to the surface at p, and $\kappa_g\mathbf{n_g}$ is a tangential component normal to the curve. The component κ_N is called the normal curvature. Computing it will lead to the second fundamental form, another very important quadratic form associated with a surface. The component κ_g is called the geodesic curvature. It turns out that it depends only on the first fundamental form, but computing it is quite complicated, and this will lead to the Christoffel symbols.

Let $f \colon \,]a, b[\to \mathbb{E}^3$ be a curve, where f is at least C^3-continuous, and assume that the curve is parametrized by arc length. We saw in Section 19.6 that if $f'(s) \neq 0$ and $f''(s) \neq 0$ for all $s \in \,]a, b[$ (i.e., f is biregular), we can associate to the point $f(s)$ an orthonormal frame $(\mathbf{t}, \mathbf{n}, \mathbf{b})$ called the Frenet frame, where

$$\mathbf{t} = f'(s),$$
$$\mathbf{n} = \frac{f''(s)}{\|f''(s)\|},$$
$$\mathbf{b} = \mathbf{t} \times \mathbf{n}.$$

The vector $\mathbf{t}$ is the unit *tangent vector*, the vector $\mathbf{n}$ is called the *principal normal*, and the vector $\mathbf{b}$ is called the *binormal*. Furthermore, the curvature κ at s is $\kappa =$

$\|f''(s)\|$, and thus,

$$f''(s) = \kappa \mathbf{n}.$$

The principal normal $\mathbf{n}$ is contained in the osculating plane at s, which is just the plane spanned by $f'(s)$ and $f''(s)$. Recall that since f is parametrized by arc length, the vector $f'(s)$ is a unit vector, and thus $f'(s) \cdot f'(s) = 1$, and by taking derivatives, we get

$$f'(s) \cdot f''(s) = 0,$$

which shows that $f'(s)$ and $f''(s)$ are linearly independent and orthogonal, provided that $f'(s) \neq 0$ and $f''(s) \neq 0$.

Now, if $C\colon t \mapsto X(u(t), v(t))$ is a curve on a surface X, assuming that C is parametrized by arc length, which implies that

$$(s')^2 = E(u')^2 + 2Fu'v' + G(v')^2 = 1,$$

we have

$$X'(s) = X_u u' + X_v v',$$
$$X''(s) = \kappa \mathbf{n},$$

and $\mathbf{t} = X_u u' + X_v v'$ is indeed a unit tangent vector to the curve and to the surface, but $\mathbf{n}$ is the principal normal to the curve, and thus it is **not** necessarily orthogonal to the tangent plane $T_p(X)$ at $p = X(u(t), v(t))$.

Thus, if we intend to study how the curvature κ varies as the curve C passing through p changes, the Frenet frame $(\mathbf{t}, \mathbf{n}, \mathbf{b})$ associated with the curve C is not really adequate, since both $\mathbf{n}$ and $\mathbf{b}$ will vary with C (and $\mathbf{n}$ is undefined when $\kappa = 0$). Thus, it is better to pick a frame associated with the normal to the surface at p, and we pick the frame $(\mathbf{t}, \mathbf{n_g}, \mathbf{N})$ defined as follows.

Definition 20.3. Given a surface X, for any curve $C\colon t \mapsto X(u(t), v(t))$ on X and any point p on X, the orthonormal frame $(\mathbf{t}, \mathbf{n_g}, \mathbf{N})$ is defined such that

$$\mathbf{t} = X_u u' + X_v v',$$
$$\mathbf{N} = \frac{X_u \times X_v}{\|X_u \times X_v\|},$$
$$\mathbf{n_g} = \mathbf{N} \times \mathbf{t},$$

where $\mathbf{N}$ is the normal vector to the surface X at p. The vector $\mathbf{n_g}$ is called the *geodesic normal vector* (for reasons that will become clear later).

Observe that $\mathbf{n_g}$ is the unit normal vector to the curve C contained in the tangent space $T_p(X)$ at p.

If we use the frame $(\mathbf{t}, \mathbf{n_g}, \mathbf{N})$, we will see shortly that $X''(s) = \kappa \mathbf{n}$ can be written as

$$\kappa \mathbf{n} = \kappa_N \mathbf{N} + \kappa_g \mathbf{n_g}.$$

The component $\kappa_N\mathbf{N}$ is the orthogonal projection of $\kappa\mathbf{n}$ onto the normal direction $\mathbf{N}$, and for this reason κ_N is called the *normal curvature of C at p*. The component $\kappa_g\mathbf{n_g}$ is the orthogonal projection of $\kappa\mathbf{n}$ onto the tangent space $T_p(X)$ at p.

We now show how to compute the normal curvature. This will uncover the second fundamental form. Using the abbreviations

$$X_{uu} = \frac{\partial^2 X}{\partial u^2}, \quad X_{uv} = \frac{\partial^2 X}{\partial u \partial v}, \quad X_{vv} = \frac{\partial^2 X}{\partial v^2},$$

since $X' = X_u u' + X_v v'$, using the chain rule we get

$$X'' = X_{uu}(u')^2 + 2X_{uv}u'v' + X_{vv}(v')^2 + X_u u'' + X_v v''.$$

In order to decompose $X'' = \kappa\mathbf{n}$ into its normal component (along $\mathbf{N}$) and its tangential component, we use a neat trick suggested by Eugenio Calabi. Recall that

$$(u \times v) \times w = (u \cdot w)v - (w \cdot v)u.$$

Using this identity, we have

$$\begin{aligned}
(\mathbf{N} &\times (X_{uu}(u')^2 + 2X_{uv}u'v' + X_{vv}(v')^2) \times \mathbf{N} \\
&= (\mathbf{N} \cdot \mathbf{N})(X_{uu}(u')^2 + 2X_{uv}u'v' + X_{vv}(v')^2) \\
&\quad - (\mathbf{N} \cdot (X_{uu}(u')^2 + 2X_{uv}u'v' + X_{vv}(v')^2))\mathbf{N}.
\end{aligned}$$

Since $\mathbf{N}$ is a unit vector, we have $\mathbf{N} \cdot \mathbf{N} = 1$, and consequently, since

$$\kappa\mathbf{n} = X'' = X_{uu}(u')^2 + 2X_{uv}u'v' + X_{vv}(v')^2 + X_u u'' + X_v v'',$$

we can write

$$\begin{aligned}
\kappa\mathbf{n} = {} & (\mathbf{N} \cdot (X_{uu}(u')^2 + 2X_{uv}u'v' + X_{vv}(v')^2))\mathbf{N} \\
& + (\mathbf{N} \times (X_{uu}(u')^2 + 2X_{uv}u'v' + X_{vv}(v')^2)) \times \mathbf{N} + X_u u'' + X_v v''.
\end{aligned}$$

Thus, it is clear that the normal component is

$$\kappa_N\mathbf{N} = (\mathbf{N} \cdot (X_{uu}(u')^2 + 2X_{uv}u'v' + X_{vv}(v')^2))\mathbf{N},$$

and the normal curvature is given by

$$\kappa_N = \mathbf{N} \cdot (X_{uu}(u')^2 + 2X_{uv}u'v' + X_{vv}(v')^2).$$

Letting

$$L = \mathbf{N} \cdot X_{uu}, \quad M = \mathbf{N} \cdot X_{uv}, \quad N = \mathbf{N} \cdot X_{vv},$$

we have

$$\kappa_N = L(u')^2 + 2Mu'v' + N(v')^2.$$

It should be noted that some authors (such as do Carmo) use the notation

$$e = \mathbf{N} \cdot X_{uu}, \quad f = \mathbf{N} \cdot X_{uv}, \quad g = \mathbf{N} \cdot X_{vv}.$$

Recalling that

$$\mathbf{N} = \frac{X_u \times X_v}{\|X_u \times X_v\|},$$

using the Lagrange identity

$$(u \cdot v)^2 + \|u \times v\|^2 = \|u\|^2 \|v\|^2,$$

we see that

$$\|X_u \times X_v\| = \sqrt{EG - F^2},$$

and $L = \mathbf{N} \cdot X_{uu}$ can be written as

$$L = \frac{(X_u \times X_v) \cdot X_{uu}}{\sqrt{EG - F^2}} = \frac{(X_u, X_v, X_{uu})}{\sqrt{EG - F^2}},$$

where (X_u, X_v, X_{uu}) is the mixed product, i.e., the determinant of the three vectors (similar expressions are obtained for M and N). Some authors (including Gauss himself and Darboux) use the notation

$$D = (X_u, X_v, X_{uu}), \quad D' = (X_u, X_v, X_{uv}), \quad D'' = (X_u, X_v, X_{vv}),$$

and we also have

$$L = \frac{D}{\sqrt{EG - F^2}}, \quad M = \frac{D'}{\sqrt{EG - F^2}}, \quad N = \frac{D''}{\sqrt{EG - F^2}}.$$

These expressions were used by Gauss to prove his famous *Theorema Egregium*.

Since the quadratic form $(x, y) \mapsto Lx^2 + 2Mxy + Ny^2$ plays a very important role in the theory of surfaces, we introduce the following definition.

Definition 20.4. Given a surface X, for any point $p = X(u, v)$ on X, letting

$$L = \mathbf{N} \cdot X_{uu}, \quad M = \mathbf{N} \cdot X_{uv}, \quad N = \mathbf{N} \cdot X_{vv},$$

where $\mathbf{N}$ is the unit normal at p, the quadratic form $(x, y) \mapsto Lx^2 + 2Mxy + Ny^2$ is called the *second fundamental form of X at p*. It is often denoted by II_p. For a curve C on the surface X (parametrized by arc length), the quantity κ_N given by the formula

$$\kappa_N = L(u')^2 + 2Mu'v' + N(v')^2$$

is called the *normal curvature of C at p*.

The second fundamental form was introduced by Gauss in 1827. Unlike the first fundamental form, the second fundamental form is not necessarily positive or definite. Properties of the surface expressible in terms of the first fundamental form are called *intrinsic properties* of the surface X. Properties of the surface expressible in terms of the second fundamental form are called *extrinsic properties* of the surface X. They have to do with the way the surface is immersed in $\mathbb{E}^3$. As we shall see later, certain notions that appear to be extrinsic turn out to be intrinsic, such as the geodesic curvature and the Gaussian curvature. This is another testimony to the genius of Gauss (and Bonnet, Christoffel, et al.).

Remark: As in the previous section, if X is not injective, the second fundamental form II_p is not well-defined. Again, we will not worry too much about this, or we assume X injective.

It should also be mentioned that the fact that the normal curvature is expressed as

$$\kappa_N = L(u')^2 + 2Mu'v' + N(v')^2$$

has the following immediate corollary, known as *Meusnier's theorem* (1776).

Lemma 20.1. *All curves on a surface X and having the same tangent line at a given point $p \in X$ have the same normal curvature at p.*

In particular, if we consider the curves obtained by intersecting the surface with planes containing the normal at p, curves called *normal sections*, all curves tangent to a normal section at p have the same normal curvature as the normal section. Furthermore, the principal normal of a normal section is collinear with the normal to the surface, and thus $|\kappa| = |\kappa_N|$, where κ is the curvature of the normal section, and κ_N is the normal curvature of the normal section. We will see in a later section how the curvature of normal sections varies.

We obtained the value of the normal curvature κ_N assuming that the curve C is parametrized by arc length, but we can easily give an expression for κ_N for an arbitrary parametrization. Indeed, remember that

$$\left(\frac{ds}{dt}\right)^2 = \|\dot{C}\|^2 = E\dot{u}^2 + 2F\dot{u}\dot{v} + G\dot{v}^2,$$

and by the chain rule

$$u' = \frac{du}{ds} = \frac{du}{dt}\frac{dt}{ds},$$

and since a change of parameter is a diffeomorphism, we get

$$u' = \frac{\dot{u}}{\left(\frac{ds}{dt}\right)},$$

and from

$$\kappa_N = L(u')^2 + 2Mu'v' + N(v')^2,$$

we get

$$\boxed{\kappa_N = \frac{L\dot{u}^2 + 2M\dot{u}\dot{v} + N\dot{v}^2}{E\dot{u}^2 + 2F\dot{u}\dot{v} + G\dot{v}^2}.}$$

It is remarkable that this expression of the normal curvature uses both the first and the second fundamental forms!

We still need to compute the tangential part X_t'' of X''. We found that the tangential part of X'' is

$$X_t'' = (\mathbf{N} \times (X_{uu}(u')^2 + 2X_{uv}u'v' + X_{vv}(v')^2)) \times \mathbf{N} + X_u u'' + X_v v''.$$

This vector is clearly in the tangent space $T_p(X)$ (since the first part is orthogonal to $\mathbf{N}$, which is orthogonal to the tangent space). Furthermore, X'' is orthogonal to X' (since $X' \cdot X' = 1$), and by dotting $X'' = \kappa_N \mathbf{N} + X_t''$ with $\mathbf{t} = X'$, since the component $\kappa_N \mathbf{N} \cdot \mathbf{t}$ is zero, we have $X_t'' \cdot \mathbf{t} = 0$, and thus X_t'' is also orthogonal to $\mathbf{t}$, which means that it is collinear with $\mathbf{n_g} = \mathbf{N} \times \mathbf{t}$. Therefore, we have shown that

$$\kappa \mathbf{n} = \kappa_N \mathbf{N} + \kappa_g \mathbf{n_g},$$

where

$$\kappa_N = L(u')^2 + 2Mu'v' + N(v')^2$$

and

$$\kappa_g \mathbf{n_g} = (\mathbf{N} \times (X_{uu}(u')^2 + 2X_{uv}u'v' + X_{vv}(v')^2)) \times \mathbf{N} + X_u u'' + X_v v''.$$

The term $\kappa_g \mathbf{n_g}$ is worth an official definition.

Definition 20.5. Given a surface X, for any curve $C \colon t \mapsto X(u(t), v(t))$ on X and any point p on X, the quantity κ_g appearing in the expression

$$\kappa \mathbf{n} = \kappa_N \mathbf{N} + \kappa_g \mathbf{n_g}$$

giving the acceleration vector of X at p is called the *geodesic curvature of C at p.*

In the next section we give an expression for $\kappa_g \mathbf{n_g}$ in terms of the basis (X_u, X_v).

20.5 Geodesic Curvature and the Christoffel Symbols

We showed that the tangential part of the curvature of a curve C on a surface is of the form $\kappa_g \mathbf{n_g}$. We now show that κ_g can be computed only in terms of the first fundamental form of X, a result first proved by Ossian Bonnet circa 1848. The computation is a bit involved, and it will lead us to the Christoffel symbols, introduced in 1869.

Since $\mathbf{n_g}$ is in the tangent space $T_p(X)$, and since (X_u, X_v) is a basis of $T_p(X)$, we can write

$$\kappa_g \mathbf{n_g} = AX_u + BX_v,$$

for some $A, B \in \mathbb{R}$. However,

$$\kappa \mathbf{n} = \kappa_N \mathbf{N} + \kappa_g \mathbf{n_g},$$

and since $\mathbf{N}$ is normal to the tangent space, $\mathbf{N} \cdot X_u = \mathbf{N} \cdot X_v = 0$, and by dotting

$$\kappa_g \mathbf{n_g} = A X_u + B X_v$$

with X_u and X_v, since $E = X_u \cdot X_u$, $F = X_u \cdot X_v$, and $G = X_v \cdot X_v$, we get the equations

$$\kappa \mathbf{n} \cdot X_u = EA + FB,$$
$$\kappa \mathbf{n} \cdot X_v = FA + GB.$$

On the other hand,

$$\kappa \mathbf{n} = X'' = X_u u'' + X_v v'' + X_{uu}(u')^2 + 2 X_{uv} u' v' + X_{vv}(v')^2.$$

Dotting with X_u and X_v, we get

$$\kappa \mathbf{n} \cdot X_u = E u'' + F v'' + (X_{uu} \cdot X_u)(u')^2 + 2(X_{uv} \cdot X_u) u' v' + (X_{vv} \cdot X_u)(v')^2,$$
$$\kappa \mathbf{n} \cdot X_v = F u'' + G v'' + (X_{uu} \cdot X_v)(u')^2 + 2(X_{uv} \cdot X_v) u' v' + (X_{vv} \cdot X_v)(v')^2.$$

At this point it is useful to introduce the *Christoffel symbols (of the first kind)* $[\alpha \beta; \gamma]$, defined such that

$$[\alpha \beta; \gamma] = X_{\alpha \beta} \cdot X_\gamma,$$

where $\alpha, \beta, \gamma \in \{u, v\}$. It is also more convenient to let $u = u_1$ and $v = u_2$, and to denote $[u_\alpha v_\beta; u_\gamma]$ by $[\alpha \beta; \gamma]$. Doing so, and remembering that

$$\kappa \mathbf{n} \cdot X_u = EA + FB,$$
$$\kappa \mathbf{n} \cdot X_v = FA + GB,$$

we have the following equation:

$$\begin{pmatrix} E & F \\ F & G \end{pmatrix} \begin{pmatrix} A \\ B \end{pmatrix} = \begin{pmatrix} E & F \\ F & G \end{pmatrix} \begin{pmatrix} u_1'' \\ u_2'' \end{pmatrix} + \sum_{\substack{\alpha=1,2 \\ \beta=1,2}} \begin{pmatrix} [\alpha \beta; 1] \, u_\alpha' u_\beta' \\ [\alpha \beta; 2] \, u_\alpha' u_\beta' \end{pmatrix}.$$

However, since the first fundamental form is positive definite, $EG - F^2 > 0$, and we have

$$\begin{pmatrix} E & F \\ F & G \end{pmatrix}^{-1} = (EG - F^2)^{-1} \begin{pmatrix} G & -F \\ -F & E \end{pmatrix}.$$

Thus, we get

$$\begin{pmatrix} A \\ B \end{pmatrix} = \begin{pmatrix} u_1'' \\ u_2'' \end{pmatrix} + \sum_{\substack{\alpha=1,2 \\ \beta=1,2}} (EG - F^2)^{-1} \begin{pmatrix} G & -F \\ -F & E \end{pmatrix} \begin{pmatrix} [\alpha \beta; 1] \, u_\alpha' u_\beta' \\ [\alpha \beta; 2] \, u_\alpha' u_\beta' \end{pmatrix}.$$

It is natural to introduce the *Christoffel symbols (of the second kind)* Γ_{ij}^k, defined such that

$$\begin{pmatrix} \Gamma_{ij}^1 \\ \Gamma_{ij}^2 \end{pmatrix} = (EG - F^2)^{-1} \begin{pmatrix} G & -F \\ -F & E \end{pmatrix} \begin{pmatrix} [ij;1] \\ [ij;2] \end{pmatrix}.$$

Finally, we get

$$A = u_1'' + \sum_{\substack{i=1,2 \\ j=1,2}} \Gamma_{ij}^1 \, u_i' u_j',$$

$$B = u_2'' + \sum_{\substack{i=1,2 \\ j=1,2}} \Gamma_{ij}^2 \, u_i' u_j',$$

and

$$\kappa_g \mathbf{n_g} = \left(u_1'' + \sum_{\substack{i=1,2 \\ j=1,2}} \Gamma_{ij}^1 \, u_i' u_j' \right) X_u + \left(u_2'' + \sum_{\substack{i=1,2 \\ j=1,2}} \Gamma_{ij}^2 \, u_i' u_j' \right) X_v.$$

We summarize all the above in the following lemma.

Lemma 20.2. *Given a surface X and a curve C on X, for any point p on C, the tangential part of the curvature at p is given by*

$$\kappa_g \mathbf{n_g} = \left(u_1'' + \sum_{\substack{i=1,2 \\ j=1,2}} \Gamma_{ij}^1 \, u_i' u_j' \right) X_u + \left(u_2'' + \sum_{\substack{i=1,2 \\ j=1,2}} \Gamma_{ij}^2 \, u_i' u_j' \right) X_v,$$

where the Christoffel symbols Γ_{ij}^k are defined such that

$$\begin{pmatrix} \Gamma_{ij}^1 \\ \Gamma_{ij}^2 \end{pmatrix} = \begin{pmatrix} E & F \\ F & G \end{pmatrix}^{-1} \begin{pmatrix} [ij;1] \\ [ij;2] \end{pmatrix},$$

and the Christoffel symbols $[ij;k]$ are defined such that

$$[ij;k] = X_{ij} \cdot X_k.$$

Looking at the formulae

$$[\alpha\beta;\gamma] = X_{\alpha\beta} \cdot X_\gamma$$

for the Christoffel symbols $[\alpha\beta;\gamma]$, it does not seem that these symbols depend only on the first fundamental form, but in fact, they do! Firstly, note that

$$[\alpha\beta;\gamma] = [\beta\alpha;\gamma].$$

Next, observe that

$$X_{uu} \cdot X_u = \frac{1}{2}\frac{\partial(X_u \cdot X_u)}{\partial u} = \frac{1}{2}E_u,$$

$$X_{uv} \cdot X_u = \frac{1}{2}\frac{\partial(X_u \cdot X_u)}{\partial v} = \frac{1}{2}E_v,$$

$$X_{uv} \cdot X_v = \frac{1}{2}\frac{\partial(X_v \cdot X_v)}{\partial u} = \frac{1}{2}G_u,$$

$$X_{vv} \cdot X_v = \frac{1}{2}\frac{\partial(X_v \cdot X_v)}{\partial v} = \frac{1}{2}G_v,$$

and since

$$(X_u \cdot X_v)_v = X_{uv} \cdot X_v + X_u \cdot X_{vv}$$

and

$$X_{uv} \cdot X_v = \frac{1}{2}G_u,$$

we get

$$F_v = \frac{1}{2}G_u + X_u \cdot X_{vv},$$

and thus

$$X_{vv} \cdot X_u = F_v - \frac{1}{2}G_u.$$

Similarly, we get

$$X_{uu} \cdot X_v = F_u - \frac{1}{2}E_v.$$

In summary, we have the following formulae showing that the Christoffel symbols depend only on the first fundamental form:

$$[1\,1;\,1] = \frac{1}{2}E_u, \quad [1\,1;\,2] = F_u - \frac{1}{2}E_v,$$

$$[1\,2;\,1] = \frac{1}{2}E_v, \quad [1\,2;\,2] = \frac{1}{2}G_u,$$

$$[2\,1;\,1] = \frac{1}{2}E_v, \quad [2\,1;\,2] = \frac{1}{2}G_u,$$

$$[2\,2;\,1] = F_v - \frac{1}{2}G_u, \quad [2\,2;\,2] = \frac{1}{2}G_v.$$

Another way to compute the Christoffel symbols $[\alpha\beta;\,\gamma]$, is to proceed as follows. For this computation it is more convenient to assume that $u = u_1$ and $v = u_2$, and that the first fundamental form is expressed by the matrix

$$\begin{pmatrix} g_{11} & g_{12} \\ g_{21} & g_{22} \end{pmatrix} = \begin{pmatrix} E & F \\ F & G \end{pmatrix},$$

where $g_{\alpha\beta} = X_\alpha \cdot X_\beta$. Let

$$g_{\alpha\beta\,|\,\gamma} = \frac{\partial g_{\alpha\beta}}{\partial u_\gamma}.$$

Then, we have

$$g_{\alpha\beta\,|\,\gamma} = \frac{\partial g_{\alpha\beta}}{\partial u_\gamma} = X_{\alpha\gamma}\cdot X_\beta + X_\alpha\cdot X_{\beta\gamma} = [\alpha\,\gamma;\,\beta] + [\beta\,\gamma;\,\alpha].$$

From this, we also have

$$g_{\beta\gamma\,|\,\alpha} = [\alpha\,\beta;\,\gamma] + [\alpha\,\gamma;\,\beta]$$

and

$$g_{\alpha\gamma\,|\,\beta} = [\alpha\,\beta;\,\gamma] + [\beta\,\gamma;\,\alpha].$$

From all this we get

$$2[\alpha\,\beta;\,\gamma] = g_{\alpha\gamma\,|\,\beta} + g_{\beta\gamma\,|\,\alpha} - g_{\alpha\beta\,|\,\gamma}.$$

As before, the Christoffel symbols $[\alpha\,\beta;\,\gamma]$ and $\Gamma^\gamma_{\alpha\beta}$ are related via the Riemannian metric by the equations

$$\Gamma^\gamma_{\alpha\beta} = \begin{pmatrix} g_{11} & g_{12} \\ g_{21} & g_{22} \end{pmatrix}^{-1} [\alpha\,\beta;\,\gamma].$$

This seemingly bizarre approach has the advantage of generalizing to Riemannian manifolds. In the next section we study the variation of the normal curvature.

20.6 Principal Curvatures, Gaussian Curvature, Mean Curvature

We will now study how the normal curvature at a point varies when a unit tangent vector varies. In general, we will see that the normal curvature has a maximum value κ_1 and a minimum value κ_2, and that the corresponding directions are orthogonal. This was shown by Euler in 1760. The quantity $K = \kappa_1\kappa_2$, called the Gaussian curvature, and the quantity $H = (\kappa_1 + \kappa_2)/2$, called the mean curvature, play a very important role in the theory of surfaces. We will compute H and K in terms of the first and the second fundamental forms. We also classify points on a surface according to the value and sign of the Gaussian curvature.

Recall that given a surface X and some point p on X, the vectors X_u, X_v form a basis of the tangent space $T_p(X)$. Given a unit vector $\mathbf{t} = X_u x + X_v y$, the normal curvature is given by

$$\kappa_N(\mathbf{t}) = Lx^2 + 2Mxy + Ny^2,$$

since $Ex^2 + 2Fxy + Gy^2 = 1$. Usually, (X_u, X_v) is not an orthonormal frame, and it is useful to replace the frame (X_u, X_v) with an orthonormal frame. One verifies easily that the frame (e_1, e_2) defined such that

$$e_1 = \frac{X_u}{\sqrt{E}}, \quad e_2 = \frac{EX_v - FX_u}{\sqrt{E(EG-F^2)}}$$

is indeed an orthonormal frame. With respect to this frame, every unit vector can be written as $\mathbf{t} = \cos\theta\, e_1 + \sin\theta\, e_2$, and expressing (e_1, e_2) in terms of X_u and X_v, we have

$$\mathbf{t} = \left(\frac{w\cos\theta - F\sin\theta}{w\sqrt{E}}\right) X_u + \frac{\sqrt{E}\sin\theta}{w} X_v,$$

where $w = \sqrt{EG - F^2}$. We can now compute $\kappa_N(\mathbf{t})$, and we get

$$\kappa_N(\mathbf{t}) = L\left(\frac{w\cos\theta - F\sin\theta}{w\sqrt{E}}\right)^2 + 2M\left(\frac{(w\cos\theta - F\sin\theta)\sin\theta}{w^2}\right)$$
$$+ N\frac{E\sin^2\theta}{w^2}.$$

We leave as an exercise to show that the above expression can be written as

$$\kappa_N(\mathbf{t}) = H + A\cos 2\theta + B\sin 2\theta,$$

where

$$H = \frac{GL - 2FM + EN}{2(EG-F^2)},$$

$$A = \frac{L(EG - 2F^2) + 2EFM - E^2N}{2E(EG-F^2)},$$

$$B = \frac{EM - FL}{E\sqrt{EG-F^2}}.$$

Letting $C = \sqrt{A^2 + B^2}$, unless $A = B = 0$, the function

$$f(\theta) = H + A\cos 2\theta + B\sin 2\theta$$

has a maximum $\kappa_1 = H + C$ for the angles θ_0 and $\theta_0 + \pi$, and a minimum $\kappa_2 = H - C$ for the angles $\theta_0 + \pi/2$ and $\theta_0 + 3\pi/2$, where $\cos 2\theta_0 = A/C$ and $\sin 2\theta_0 = B/C$. The curvatures κ_1 and κ_2 play a major role in surface theory.

Definition 20.6. Given a surface X, for any point p on X, letting A, B, H be defined as above, and $C = \sqrt{A^2 + B^2}$, unless $A = B = 0$, the normal curvature κ_N at p takes a maximum value κ_1 and and a minimum value κ_2, called *principal curvatures at p*, where $\kappa_1 = H + C$ and $\kappa_2 = H - C$. The directions of the corresponding unit vectors are called the *principal directions at p*. The average $H = \kappa_1 + \kappa_2/2$ of the principal curvatures is called the *mean curvature*, and the product $K = \kappa_1\kappa_2$ of the principal curvatures is called the *total curvature*, or *Gaussian curvature*.

Observe that the principal directions θ_0 and $\theta_0 + \pi/2$ corresponding to κ_1 and κ_2 are orthogonal. Note that

$$K = \kappa_1 \kappa_2 = (H-C)(H+C) = H^2 - C^2 = H^2 - (A^2 + B^2).$$

We leave as an exercise to verify that

$$A^2 + B^2 = \frac{G^2 L^2 - 4FGLM + 4EGM^2 + 4F^2 LN - 2EGLN - 4EFMN + E^2 N^2}{4(EG - F^2)^2}$$

and

$$H^2 = \frac{G^2 L^2 - 4FGLM + 4F^2 M^2 + 2EGLN - 4EFMN + E^2 N^2}{4(EG - F^2)^2}.$$

From this we get

$$H^2 - A^2 - B^2 = \frac{LN - M^2}{EG - F^2}.$$

In summary, we have the following (famous) formulae for the mean curvature and the Gaussian curvature:

$$\boxed{\begin{aligned} H &= \frac{GL - 2FM + EN}{2(EG - F^2)}, \\ K &= \frac{LN - M^2}{EG - F^2}. \end{aligned}}$$

We have shown that the normal curvature κ_N can be expressed as

$$\kappa_N(\theta) = H + A\cos 2\theta + B\sin 2\theta$$

over the orthonormal frame (e_1, e_2). We also have shown that the angle θ_0 such that $\cos 2\theta_0 = A/C$ and $\sin 2\theta_0 = B/C$ plays a special role. Indeed, it determines one of the principal directions. If we rotate the basis (e_1, e_2) and pick a frame (f_1, f_2) corresponding to the principal directions, we obtain a particularly nice formula for κ_N. Indeed, since $A = C\cos 2\theta_0$ and $B = C\sin 2\theta_0$, letting $\varphi = \theta - \theta_0$, we can write

$$\begin{aligned} \kappa_N(\theta) &= H + A\cos 2\theta + B\sin 2\theta \\ &= H + C(\cos 2\theta_0 \cos 2\theta + \sin 2\theta_0 \sin 2\theta) \\ &= H + C(\cos 2(\theta - \theta_0)) \\ &= H + C(\cos^2(\theta - \theta_0) - \sin^2(\theta - \theta_0)) \\ &= H(\cos^2(\theta - \theta_0) + \sin^2(\theta - \theta_0)) + C(\cos^2(\theta - \theta_0) - \sin^2(\theta - \theta_0)) \\ &= (H+C)\cos^2(\theta - \theta_0) + (H-C)\sin^2(\theta - \theta_0) \\ &= \kappa_1 \cos^2 \varphi + \kappa_2 \sin^2 \varphi. \end{aligned}$$

Thus, for any unit vector $\mathbf{t}$ expressed as

$$\mathbf{t} = \cos\varphi f_1 + \sin\varphi f_2$$

with respect to an orthonormal frame corresponding to the principal directions, the normal curvature $\kappa_N(\varphi)$ is given by *Euler's formula* (1760)

$$\kappa_N(\varphi) = \kappa_1 \cos^2 \varphi + \kappa_2 \sin^2 \varphi.$$

Recalling that $EG - F^2$ is always strictly positive, we can classify the points on the surface depending on the value of the Gaussian curvature K and on the values of the principal curvatures κ_1 and κ_2 (or H).

Definition 20.7. Given a surface X, a point p on X belongs to one of the following categories:

(1) *Elliptic* if $LN - M^2 > 0$, or equivalently $K > 0$.
(2) *Hyperbolic* if $LN - M^2 < 0$, or equivalently $K < 0$.
(3) *Parabolic* if $LN - M^2 = 0$ and $L^2 + M^2 + N^2 > 0$, or equivalently $K = \kappa_1 \kappa_2 = 0$
    but either $\kappa_1 \neq 0$ or $\kappa_2 \neq 0$.
(4) *Planar* if $L = M = N = 0$, or equivalently $\kappa_1 = \kappa_2 = 0$.

Furthermore, a point p is an *umbilical point* (or *umbilic*) if $K > 0$ and $\kappa_1 = \kappa_2$.

Note that some authors allow a planar point to be an umbilical point, but we do not. At an elliptic point, both principal curvatures are nonnull and have the same sign. For example, most points on an ellipsoid are elliptic.

At a hyperbolic point, the principal curvatures have opposite signs. For example, all points on the catenoid are hyperbolic.

At a parabolic point, one of the two principal curvatures is zero, but not both. This is equivalent to $K = 0$ and $H \neq 0$. Points on a cylinder are parabolic.

At a planar point, $\kappa_1 = \kappa_2 = 0$. This is equivalent to $K = H = 0$. Points on a plane are all planar points!

Example 20.6. On a monkey saddle, there is a planar point, as shown in Figure 20.4. The principal directions at that point are undefined.

For an umbilical point we have $\kappa_1 = \kappa_2 \neq 0$. This can happen only when $H - C = H + C$, which implies that $C = 0$, and since $C = \sqrt{A^2 + B^2}$, we have $A = B = 0$. Thus, for an umbilical point, $K = H^2$. In this case the function κ_N is constant, and the principal directions are undefined. All points on a sphere are umbilics. A general ellipsoid (a, b, c pairwise distinct) has four umbilics.

It can be shown that a connected surface consisting only of umbilical points is contained in a sphere (see do Carmo [12], Section 3.2, or Gray [23], Section 28.2). It can also be shown that a connected surface consisting only of planar points is contained in a plane. A surface can contain at the same time elliptic points, parabolic points, and hyperbolic points. This is the case of a torus.

Example 20.7. The parabolic points are on two circles also contained in two tangent planes to the torus (the two horizontal planes touching the top and the bottom of the torus, as shown in Figure 20.5). The elliptic points are on the outside part of the torus (with normal facing outward), delimited by the two circles of parabolic

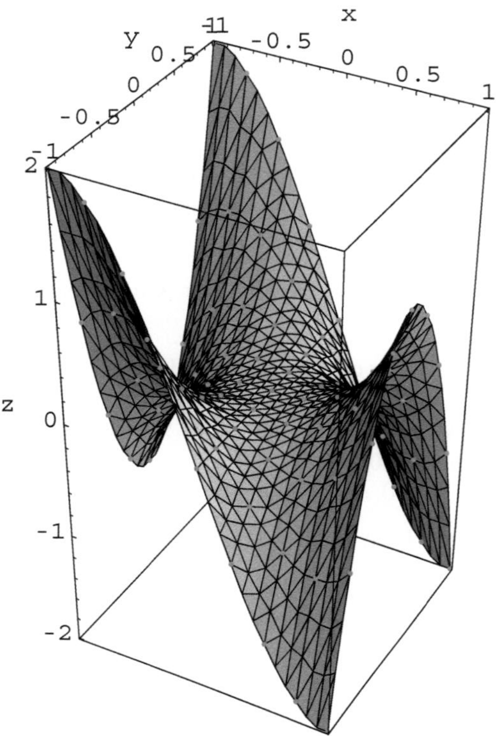

Fig. 20.4 A monkey saddle.

points. The hyperbolic points are on the inside part of the torus (with normal facing inward).

The normal curvature $\kappa_N(X_u x + X_v y) = Lx^2 + 2Mxy + Ny^2$ will vanish for some tangent vector $(x, y) \neq (0, 0)$ iff $M^2 - LN \geq 0$. Since

$$K = \frac{LN - M^2}{EG - F^2},$$

this can happen only if $K \leq 0$. If $L = N = 0$, then there are two directions corresponding to X_u and X_v for which the normal curvature is zero. If $L \neq 0$ or $N \neq 0$, say $L \neq 0$ (the other case being similar), then the equation

$$L\left(\frac{x}{y}\right)^2 + 2M\frac{x}{y} + N = 0$$

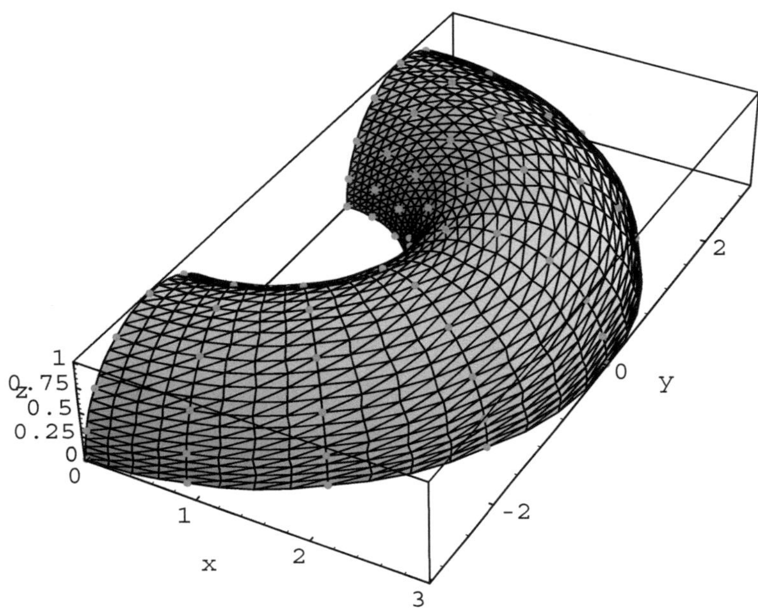

Fig. 20.5 Portion of torus.

has two distinct roots iff $K < 0$. The directions corresponding to the vectors $X_u x + X_v y$ associated with these roots are called the *asymptotic directions at p*. These are the directions for which the normal curvature is null at p.

There are surfaces of constant Gaussian curvature. For example, a cylinder or a cone is a surface of Gaussian curvature $K = 0$. A sphere of radius R has positive constant Gaussian curvature $K = 1/R^2$. Perhaps surprisingly, there are other surfaces of constant positive curvature besides the sphere. There are surfaces of constant negative curvature, say $K = -1$.

Example 20.8. A famous surfaces of constant negative curvature is the *pseudo-sphere*, also known as *Beltrami's pseudosphere*. This is the surface of revolution obtained by rotating a curve known as a *tractrix* around its asymptote. One possible parametrization is given by

$$x = \frac{2\cos v}{e^u + e^{-u}},$$

$$y = \frac{2\sin v}{e^u + e^{-u}},$$

$$z = u - \frac{e^u - e^{-u}}{e^u + e^{-u}},$$

over $]0, 2\pi[\times \mathbb{R}$. The pseudosphere has a circle of singular points (for $u = 0$). Figure 20.6 shows a portion of pseudosphere.

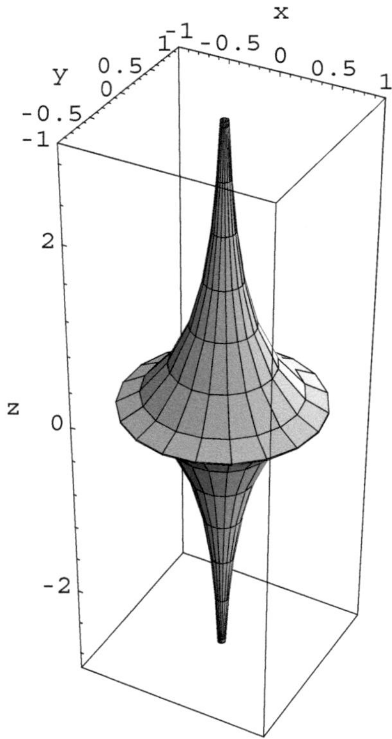

Fig. 20.6 A pseudosphere.

Again, perhaps surprisingly, there are other surfaces of constant negative curvature.
The Gaussian curvature at a point (x, y, x) of an ellipsoid of equation

$$\frac{x^2}{a^2} + \frac{y^2}{b^2} + \frac{z^2}{c^2} = 1$$

has the beautiful expression

$$K = \frac{p^4}{a^2 b^2 c^2},$$

where p is the distance from the origin $(0,0,0)$ to the tangent plane at the point (x,y,z).

There are also surfaces for which $H = 0$. Such surfaces are called *minimal surfaces*, and they show up in physics quite a bit. It can be verified that both the helicoid and the catenoid are minimal surfaces. The Enneper surface is also a minimal surface (see Example 20.9).

We will see shortly how the classification of points on a surface can be explained in terms of the Dupin indicatrix. The idea is to dip the surface in water, and to watch the shorelines formed in the water by the surface in a small region around a chosen point, as we move the surface up and down very gently. But first, we introduce the Gauss map, i.e., we study the variations of the normal $\mathbf{N}_p$ as the point p varies on the surface.

20.7 The Gauss Map and Its Derivative $d\mathbf{N}$

Given a surface $X\colon \Omega \to \mathbb{E}^3$ and any point $p = X(u,v)$ on X, we have defined the normal $\mathbf{N}_p$ at p (or really $\mathbf{N}_{(u,v)}$ at (u,v)) as the unit vector

$$\mathbf{N}_p = \frac{X_u \times X_v}{\|X_u \times X_v\|}.$$

Gauss realized that the assignment $p \mapsto \mathbf{N}_p$ of the unit normal $\mathbf{N}_p$ to the point p on the surface X could be viewed as a map from the trace of the surface X to the unit sphere S^2. If $\mathbf{N}_p$ is a unit vector of coordinates (x,y,z), we have $x^2 + y^2 + z^2 = 1$, and $\mathbf{N}_p$ corresponds to the point $N(p) = (x,y,z)$ on the unit sphere. This is the so-called *Gauss map of X*, denoted by $\mathbf{N}\colon X \to S^2$.

The derivative $d\mathbf{N}_p$ of the Gauss map at p measures the variation of the normal near p, i.e., how the surface "curves" near p. The Jacobian matrix of $d\mathbf{N}_p$ in the basis (X_u, X_v) can be expressed simply in terms of the matrices associated with the first and the second fundamental forms (which are quadratic forms). Furthermore, the eigenvalues of $d\mathbf{N}_p$ are precisely $-\kappa_1$ and $-\kappa_2$, where κ_1 and κ_2 are the principal curvatures at p, and the eigenvectors define the principal directions (when they are well-defined). In view of the negative sign in $-\kappa_1$ and $-\kappa_2$, it is desirable to consider the linear map $\mathscr{S}_p = -d\mathbf{N}_p$, often called the *shape operator*. Then it is easily shown that the second fundamental form $\mathrm{II}_p(\mathbf{t})$ can be expressed as

$$\mathrm{II}_p(\mathbf{t}) = \langle \mathscr{S}_p(\mathbf{t}), \mathbf{t} \rangle_p,$$

where $\langle -, - \rangle$ is the inner product associated with the first fundamental form. Thus, the Gaussian curvature is equal to the determinant of $\mathscr{S}_p$, and also to the determinant of $d\mathbf{N}_p$, since $(-\kappa_1)(-\kappa_2) = \kappa_1 \kappa_2$. We will see in a later section that the Gaussian curvature actually depends only of the first fundamental form, which is far from obvious right now!

Actually, if X is not injective, there are problems, because the assignment $p \mapsto \mathbf{N}_p$ could be multivalued, since there could be several different normals. We can either assume that X is injective, or consider the map from Ω to S^2 defined such that

$$(u,v) \mapsto \mathbf{N}_{(u,v)}.$$

Then we have a map from Ω to S^2, where (u,v) is mapped to the point $N(u,v)$ on S^2 associated with $\mathbf{N}_{(u,v)}$. This map is denoted by $\mathbf{N}\colon \Omega \to S^2$.

It is interesting to study the derivative $d\mathbf{N}$ of the Gauss map $\mathbf{N}\colon \Omega \to S^2$ (or $\mathbf{N}\colon X \to S^2$). As we shall see, the second fundamental form can be defined in terms of $d\mathbf{N}$. For every $(u,v) \in \Omega$, the map $d\mathbf{N}_{(u,v)}$ is a linear map $d\mathbf{N}_{(u,v)}\colon \mathbb{R}^2 \to \mathbb{R}^2$. It can be viewed as a linear map from the tangent space $T_{(u,v)}(X)$ at $X(u,v)$ (which is isomorphic to $\mathbb{R}^2$) to the tangent space to the sphere at $N(u,v)$ (also isomorphic to $\mathbb{R}^2$). Recall that $d\mathbf{N}_{(u,v)}$ is defined as follows: For every $(x,y) \in \mathbb{R}^2$,

$$d\mathbf{N}_{(u,v)}(x,y) = \mathbf{N}_u x + \mathbf{N}_v y.$$

Thus, we need to compute $\mathbf{N}_u$ and $\mathbf{N}_v$. Since $\mathbf{N}$ is a unit vector, $\mathbf{N} \cdot \mathbf{N} = 1$, and by taking derivatives, we have $\mathbf{N}_u \cdot \mathbf{N} = 0$ and $\mathbf{N}_v \cdot \mathbf{N} = 0$. Consequently, $\mathbf{N}_u$ and $\mathbf{N}_v$ are in the tangent space at (u,v), and we can write

$$\mathbf{N}_u = aX_u + cX_v,$$
$$\mathbf{N}_v = bX_u + dX_v.$$

The lemma below shows how to compute a,b,c,d in terms of the first and the second fundamental forms.

Lemma 20.3. *Given a surface X, for any point $p = X(u,v)$ on X, the derivative $d\mathbf{N}_{(u,v)}$ of the Gauss map expressed in the basis (X_u, X_v) is given by the equation*

$$d\mathbf{N}_{(u,v)} \begin{pmatrix} x \\ y \end{pmatrix} = \begin{pmatrix} a & b \\ c & d \end{pmatrix} \begin{pmatrix} x \\ y \end{pmatrix},$$

where the Jacobian matrix $J(d\mathbf{N}_{(u,v)})$ of $d\mathbf{N}_{(u,v)}$ is given by

$$\begin{pmatrix} a & b \\ c & d \end{pmatrix} = - \begin{pmatrix} E & F \\ F & G \end{pmatrix}^{-1} \begin{pmatrix} L & M \\ M & N \end{pmatrix},$$

that is,

$$\begin{pmatrix} a & b \\ c & d \end{pmatrix} = \frac{1}{EG - F^2} \begin{pmatrix} MF - LG & NF - MG \\ LF - ME & MF - NE \end{pmatrix}.$$

Proof. By dotting the equations

$$\mathbf{N}_u = aX_u + cX_v,$$
$$\mathbf{N}_v = bX_u + dX_v,$$

with X_u and X_v, we get

$$\mathbf{N}_u \cdot X_u = aE + cF,$$
$$\mathbf{N}_u \cdot X_v = aF + cG,$$
$$\mathbf{N}_v \cdot X_u = bE + dF,$$
$$\mathbf{N}_v \cdot X_v = bF + dG.$$

We can compute $\mathbf{N}_u \cdot X_u$, $\mathbf{N}_u \cdot X_v$, $\mathbf{N}_v \cdot X_u$, and $\mathbf{N}_v \cdot X_v$, using the fact that $\mathbf{N} \cdot X_u = \mathbf{N} \cdot X_v = 0$. By taking derivatives, we get

$$\mathbf{N} \cdot X_{uu} + \mathbf{N}_u \cdot X_u = 0,$$
$$\mathbf{N} \cdot X_{uv} + \mathbf{N}_v \cdot X_u = 0,$$
$$\mathbf{N} \cdot X_{vu} + \mathbf{N}_u \cdot X_v = 0,$$
$$\mathbf{N} \cdot X_{vv} + \mathbf{N}_v \cdot X_v = 0.$$

Thus, we have

$$\mathbf{N}_u \cdot X_u = -L,$$
$$\mathbf{N}_u \cdot X_v = -M,$$
$$\mathbf{N}_v \cdot X_u = -M,$$
$$\mathbf{N}_v \cdot X_v = -N,$$

and together with the previous equations, we get

$$-L = aE + cF,$$
$$-M = aF + cG,$$
$$-M = bE + dF,$$
$$-N = bF + dG.$$

This system can be written in matrix form as

$$-\begin{pmatrix} L & M \\ M & N \end{pmatrix} = \begin{pmatrix} a & c \\ b & d \end{pmatrix} \begin{pmatrix} E & F \\ F & G \end{pmatrix}.$$

Therefore, we have

$$\begin{pmatrix} a & c \\ b & d \end{pmatrix} = -\begin{pmatrix} L & M \\ M & N \end{pmatrix} \begin{pmatrix} E & F \\ F & G \end{pmatrix}^{-1},$$

which yields

$$\begin{pmatrix} \mathbf{N}_u \\ \mathbf{N}_v \end{pmatrix} = -\begin{pmatrix} L & M \\ M & N \end{pmatrix} \begin{pmatrix} E & F \\ F & G \end{pmatrix}^{-1} \begin{pmatrix} X_u \\ X_v \end{pmatrix}.$$

However, we have

$$\begin{pmatrix} E & F \\ F & G \end{pmatrix}^{-1} = \frac{1}{EG - F^2} \begin{pmatrix} G & -F \\ -F & E \end{pmatrix},$$

and thus

$$\begin{pmatrix} a & c \\ b & d \end{pmatrix} = \frac{-1}{EG - F^2} \begin{pmatrix} L & M \\ M & N \end{pmatrix} \begin{pmatrix} G & -F \\ -F & E \end{pmatrix},$$

that is,

$$\begin{pmatrix} a & c \\ b & d \end{pmatrix} = \frac{1}{EG - F^2} \begin{pmatrix} MF - LG & LF - ME \\ NF - MG & MF - NE \end{pmatrix}.$$

We shall now see that the Jacobian matrix $J(d\mathbf{N}_{(u,v)})$ of the linear map $d\mathbf{N}_{(u,v)}$ expressed in the basis (X_u, X_v) is the transpose of the above matrix. Indeed, as we saw earlier,

$$d\mathbf{N}_{(u,v)}(x,y) = \mathbf{N}_u x + \mathbf{N}_v y,$$

and using the expressions for $\mathbf{N}_u$ and $\mathbf{N}_v$, we get

$$d\mathbf{N}_{(u,v)}(x,y) = (aX_u + cX_v)x + (bX_u + dX_v)y = (ax + by)X_u + (cx + dy)X_v,$$

and thus

$$d\mathbf{N}_{(u,v)} \begin{pmatrix} x \\ y \end{pmatrix} = \begin{pmatrix} a & b \\ c & d \end{pmatrix} \begin{pmatrix} x \\ y \end{pmatrix},$$

and since $\begin{pmatrix} a & b \\ c & d \end{pmatrix}$ is the transpose of $\begin{pmatrix} a & c \\ b & d \end{pmatrix}$, we get

$$\begin{pmatrix} a & b \\ c & d \end{pmatrix} = -\begin{pmatrix} E & F \\ F & G \end{pmatrix}^{-1} \begin{pmatrix} L & M \\ M & N \end{pmatrix},$$

that is,

$$\begin{pmatrix} a & b \\ c & d \end{pmatrix} = \frac{1}{EG - F^2} \begin{pmatrix} MF - LG & NF - MG \\ LF - ME & MF - NE \end{pmatrix}.$$

This concludes the proof. $\square$

The equations

$$J(d\mathbf{N}_{(u,v)}) = \begin{pmatrix} a & b \\ c & d \end{pmatrix} = \frac{1}{EG - F^2} \begin{pmatrix} MF - LG & NF - MG \\ LF - ME & MF - NE \end{pmatrix}$$

are known as the *Weingarten equations* (in matrix form). If we recall from Section 20.6 the expressions for the Gaussian curvature and for the mean curvature

$$H = \frac{GL - 2FM + EN}{2(EG - F^2)},$$

$$K = \frac{LN - M^2}{EG - F^2},$$

we note that the trace $a+d$ of the Jacobian matrix $J(d\mathbf{N}_{(u,v)})$ of $d\mathbf{N}_{(u,v)}$ is $-2H$, and that its determinant is precisely K. This is recorded in the following lemma, which also shows that the eigenvectors of $J(d\mathbf{N}_{(u,v)})$ correspond to the principal directions.

Lemma 20.4. *Given a surface X, for any point $p = X(u,v)$ on X, the eigenvalues of the Jacobian matrix $J(d\mathbf{N}_{(u,v)})$ of the derivative $d\mathbf{N}_{(u,v)}$ of the Gauss map are $-\kappa_1, -\kappa_2$, where κ_1 and κ_2 are the principal curvatures at p, and the eigenvectors of $d\mathbf{N}_{(u,v)}$ correspond to the principal directions (when they are defined). The Gaussian curvature K is the determinant of the Jacobian matrix of $d\mathbf{N}_{(u,v)}$, and the mean curvature H is equal to $-\frac{1}{2}\mathrm{tr}(J(d\mathbf{N}_{(u,v)}))$.*

Proof. We have just observed that the trace $a+d$ of the matrix

$$J(d\mathbf{N}_{(u,v)}) = \begin{pmatrix} a & b \\ c & d \end{pmatrix} = \frac{1}{EG - F^2}\begin{pmatrix} MF - LG & NF - MG \\ LF - ME & MF - NE \end{pmatrix}$$

is $-2H$, and that its determinant is precisely K. However, the characteristic equation of the above matrix is

$$x^2 - \mathrm{tr}(J(d\mathbf{N}_{(u,v)}))x + \det(J(d\mathbf{N}_{(u,v)})) = 0,$$

which is just

$$x^2 + 2Hx + K = 0.$$

Since $\kappa_1 \kappa_2 = K$ and $\kappa_1 + \kappa_2 = 2H$, κ_1 and κ_2 are the roots of the equation

$$x^2 - 2Hx + K = 0.$$

This shows that the eigenvalues of $J(d\mathbf{N}_{(u,v)})$, which are the roots of the equation

$$x^2 + 2Hx + K = 0,$$

are indeed $-\kappa_1$ and $-\kappa_2$.

Recall that κ_1 and κ_2 are the maximum and minimum values of the normal curvature, which is given by the equation

$$\kappa_N(x,y) = \frac{Lx^2 + 2Mxy + Ny^2}{Ex^2 + 2Fxy + Gy^2}.$$

Thus, the partial derivatives $\partial \kappa_N(u',v')/\partial x$ and $\partial \kappa_N(u',v')/\partial y$ of the above function must be zero for the principal directions (u',v') associated with κ_1 and κ_2. It is easy to see that this yields the equations

$$(L - \kappa E)u' + (M - \kappa F)v' = 0,$$
$$(M - \kappa F)u' + (N - \kappa G)v' = 0,$$

where κ is either κ_1 or κ_2. On the other hand, the eigenvectors of $J(d\mathbf{N}_{(u,v)})$ also satisfy the equation

$$J(d\mathbf{N}_{(u,v)}) \begin{pmatrix} u' \\ v' \end{pmatrix} = -\kappa \begin{pmatrix} u' \\ v' \end{pmatrix},$$

that is

$$\frac{MF - LG}{EG - F^2}u' + \frac{NF - MG}{EG - F^2}v' = -\kappa u',$$
$$\frac{LF - ME}{EG - F^2}u' + \frac{MF - NE}{EG - F^2}v' = -\kappa v',$$

where $\kappa = \kappa_1$ or $\kappa = \kappa_2$. From the equations

$$(L - \kappa E)u' + (M - \kappa F)v' = 0,$$
$$(M - \kappa F)u' + (N - \kappa G)v' = 0,$$

we get

$$Lu' + Mv' = \kappa(Eu' + Fv'),$$
$$Mu' + Nv' = \kappa(Fu' + Gv'),$$

which reads in matrix form as

$$\begin{pmatrix} L & M \\ M & N \end{pmatrix} \begin{pmatrix} u' \\ v' \end{pmatrix} = \kappa \begin{pmatrix} E & F \\ F & G \end{pmatrix} \begin{pmatrix} u' \\ v' \end{pmatrix},$$

which yields

$$\begin{pmatrix} E & F \\ F & G \end{pmatrix}^{-1} \begin{pmatrix} L & M \\ M & N \end{pmatrix} \begin{pmatrix} u' \\ v' \end{pmatrix} = \kappa \begin{pmatrix} u' \\ v' \end{pmatrix},$$

that is,

$$\frac{1}{EG - F^2} \begin{pmatrix} G & -F \\ -F & E \end{pmatrix} \begin{pmatrix} L & M \\ M & N \end{pmatrix} \begin{pmatrix} u' \\ v' \end{pmatrix} = \kappa \begin{pmatrix} u' \\ v' \end{pmatrix},$$

which yields precisely

$$\frac{LG - MF}{EG - F^2}u' + \frac{MG - NF}{EG - F^2}v' = \kappa u',$$
$$\frac{ME - LF}{EG - F^2}u' + \frac{NE - MF}{EG - F^2}v' = \kappa v',$$

or equivalently

$$\frac{MF - LG}{EG - F^2}u' + \frac{NF - MG}{EG - F^2}v' = -\kappa u',$$
$$\frac{LF - ME}{EG - F^2}u' + \frac{MF - NE}{EG - F^2}v' = -\kappa v'.$$

Therefore, the eigenvectors of $J(d\mathbf{N}_{(u,v)})$ correspond to the principal directions at p. $\quad\square$

The fact that $\mathbf{N}_u = -\kappa X_u$ when κ is one of the principal curvatures and when X_u corresponds to the corresponding principal direction (and similarly $\mathbf{N}_v = -\kappa X_v$ for the other principal curvature) is known as the formula of Olinde Rodrigues (1815).

The somewhat irritating negative signs arising in the eigenvalues $-\kappa_1$ and $-\kappa_2$ of $d\mathbf{N}_{(u,v)}$ can be eliminated if we consider the linear map $\mathscr{S}_{(u,v)} = -d\mathbf{N}_{(u,v)}$ instead of $d\mathbf{N}_{(u,v)}$. The map $\mathscr{S}_{(u,v)}$ is called the *shape operator at* p, and the map $d\mathbf{N}_{(u,v)}$ is sometimes called the *Weingarten operator*. The following lemma shows that the second fundamental form arises from the shape operator, and that the shape operator is self-adjoint with respect to the inner product $\langle -, - \rangle$ associated with the first fundamental form.

Lemma 20.5. *Given a surface X, for any point $p = X(u,v)$ on X, the second fundamental form of X at p is given by the formula*

$$II_{(u,v)}(\mathbf{t}) = \langle \mathscr{S}_{(u,v)}(\mathbf{t}), \mathbf{t} \rangle,$$

for every $\mathbf{t} \in \mathbb{R}^2$. The map $\mathscr{S}_{(u,v)} = -d\mathbf{N}_{(u,v)}$ is self-adjoint, that is,

$$\langle \mathscr{S}_{(u,v)}(x), y \rangle = \langle x, \mathscr{S}_{(u,v)}(y) \rangle,$$

for all $x, y \in \mathbb{R}^2$.

Proof. For any tangent vector $\mathbf{t} = X_u x + Y_v y$, since

$$\mathscr{S}_{(u,v)}(X_u x + X_v y) = -d\mathbf{N}_{(u,v)}(X_u x + X_v y) = -\mathbf{N}_u x - \mathbf{N}_v y,$$

we have

$$\langle \mathscr{S}_{(u,v)}(X_u x + X_v y), (X_u x + X_v y) \rangle = \langle (-\mathbf{N}_u x - \mathbf{N}_v y), (X_u x + X_v y) \rangle$$
$$= -(\mathbf{N}_u \cdot X_u)x^2 - (\mathbf{N}_u \cdot X_v + \mathbf{N}_v \cdot X_u)xy$$
$$- (\mathbf{N}_v \cdot X_v)y^2.$$

However, we already showed in the proof of Lemma 20.3 that

$$L = \mathbf{N} \cdot X_{uu} = -\mathbf{N}_u \cdot X_u,$$
$$M = \mathbf{N} \cdot X_{uv} = -\mathbf{N}_v \cdot X_u,$$
$$M = \mathbf{N} \cdot X_{vu} = -\mathbf{N}_u \cdot X_v$$
$$N = \mathbf{N} \cdot X_{vv} = -\mathbf{N}_v \cdot X_v,$$

and thus that

$$\langle \mathscr{S}_{(u,v)}(X_u x + X_v y),\ (X_u x + X_v y)\rangle = Lx^2 + 2Mxy + Ny^2,$$

the second fundamental form. To prove that $\mathscr{S}_{(u,v)}$ is self-adjoint, it is sufficient to prove it for the basis (X_u, X_v). This amounts to proving that

$$\langle \mathbf{N}_u,\ X_v\rangle = \langle X_u,\ \mathbf{N}_v\rangle.$$

However, we just proved that $\mathbf{N}_v \cdot X_u = \mathbf{N}_u \cdot X_v = -M$, and the proof is complete.
$\square$

Thus, in some sense, the shape operator contains all the information about curvature.

Remark: The fact that the first fundamental form I is positive definite and that $\mathscr{S}_{(u,v)}$ is self-adjoint with respect to I can be used to give a fancier proof of the fact that $\mathscr{S}_{(u,v)}$ has two real eigenvalues, that the eigenvectors are orthonormal, and that the eigenvalues correspond to the maximum and the minimum of I on the unit circle. For such a proof, see do Carmo [12]. Our proof is more basic and from first principles.

20.8 The Dupin Indicatrix

The second fundamental form shows up again when we study the deviation of a surface from its tangent plane in a neighborhood of the point of tangency. A way to study this deviation is to imagine that we dip the surface in water, and watch the shorelines formed in the water by the surface in a small region around a chosen point, as we move the surface up and down very gently. The resulting curve is known as the Dupin indicatrix (1813). Formally, consider the tangent plane $T_{(u_0,v_0)}(X)$ at some point $p = X(u_0, v_0)$, and consider the perpendicular distance $\rho(u,v)$ from the tangent plane to a point on the surface defined by (u,v). This perpendicular distance can be expressed as

$$\rho(u,v) = (X(u,v) - X(u_0,v_0)) \cdot \mathbf{N}_{(u_0,v_0)}.$$

However, since X is at least C^3-continuous, by Taylor's formula, in a neighborhood of (u_0, v_0) we can write

$$
\begin{aligned}
X(u,v) = {} & X(u_0,v_0) + X_u(u - u_0) + X_v(v - v_0) \\
& + \frac{1}{2}\left(X_{uu}(u-u_0)^2 + 2X_{uv}(u-u_0)(v-v_0) + X_{vv}(v-v_0)^2\right) \\
& + ((u-u_0)^2 + (v-v_0)^2)h_1(u,v),
\end{aligned}
$$

where $\lim_{(u,v)\to(u_0,v_0)} h_1(u,v) = 0$. However, recall that X_u and X_v are really evaluated at (u_0, v_0) (and so are X_{uu}, $X_{u,v}$, and X_{vv}), and so they are orthogonal to $\mathbf{N}_{(u_0,v_0)}$. From this, dotting with $\mathbf{N}_{(u_0,v_0)}$, we get

$$\rho(u,v) = \frac{1}{2}\left(L(u-u_0)^2 + 2M(u-u_0)(v-v_0) + N(v-v_0)^2\right)$$
$$+ \left((u-u_0)^2 + (v-v_0)^2\right)h(u,v),$$

where $\lim_{(u,v)\to(u_0,v_0)} h(u,v) = 0$. Therefore, we get another interpretation of the second fundamental form as a way of measuring the deviation from the tangent plane.

For ε small enough, and in a neighborhood of (u_0, v_0) small enough, the set of points $X(u,v)$ on the surface such that $\rho(u,v) = \pm\frac{1}{2}\varepsilon^2$ will look like portions of the curves of equation

$$\frac{1}{2}\left(L(u-u_0)^2 + 2M(u-u_0)(v-v_0) + N(v-v_0)^2\right) = \pm\frac{1}{2}\varepsilon^2.$$

Letting $u-u_0 = \varepsilon x$ and $v-v_0 = \varepsilon y$, these curves are defined by the equations

$$Lx^2 + 2Mxy + Ny^2 = \pm 1.$$

These curves are called the *Dupin indicatrix*. It is more convenient to switch to an orthonormal basis where e_1 and e_2 are eigenvectors of the Gauss map $d\mathbf{N}_{(u_0,v_0)}$. If so, it is immediately seen that

$$Lx^2 + 2Mxy + Ny^2 = \kappa_1 x^2 + \kappa_2 y^2,$$

where κ_1 and κ_2 are the principal curvatures. Thus, the equation of the Dupin indicatrix is
$$\kappa_1 x^2 + \kappa_2 y^2 = \pm 1.$$

There are several cases, depending on the sign of $\kappa_1\kappa_2 = K$, i.e., depending on the sign of $LN - M^2$.

(1) If $LN - M^2 > 0$, then κ_1 and κ_2 have the same sign. This is the case of an *elliptic point*. If $\kappa_1 \neq \kappa_2$, and $\kappa_1 > 0$ and $\kappa_2 > 0$, we get the ellipse of equation

$$\frac{x^2}{\sqrt{\frac{1}{\kappa_1}}} + \frac{y^2}{\sqrt{\frac{1}{\kappa_2}}} = 1,$$

and if $\kappa_1 < 0$ and $\kappa_2 < 0$, we get the ellipse of equation

$$\frac{x^2}{\sqrt{-\frac{1}{\kappa_1}}} + \frac{y^2}{\sqrt{-\frac{1}{\kappa_2}}} = 1.$$

When $\kappa_1 = \kappa_2$, i.e., an *umbilical point*, the Dupin indicatrix is a circle.

(2) If $LN - M^2 = 0$ and $L^2 + M^2 + N^2 > 0$, then $\kappa_1 = 0$ or $\kappa_2 = 0$, but not both. This is the case of a *parabolic point*. In this case, the Dupin indicatrix degenerates to two parallel lines, since the equation is either

$$\kappa_1 x^2 = \pm 1$$

or

$$\kappa_2 y^2 = \pm 1.$$

(3) If $LN - M^2 < 0$, then κ_1 and κ_2 have different signs. This is the case of a *hyperbolic point*. In this case, the Dupin indicatrix consists of the two hyperbolae of equations

$$\frac{x^2}{\sqrt{\frac{1}{\kappa_1}}} - \frac{y^2}{\sqrt{-\frac{1}{\kappa_2}}} = 1$$

if $\kappa_1 > 0$ and $\kappa_2 < 0$, or of equation

$$-\frac{x^2}{\sqrt{-\frac{1}{\kappa_1}}} + \frac{y^2}{\sqrt{\frac{1}{\kappa_2}}} = 1$$

if $\kappa_1 < 0$ and $\kappa_2 > 0$. These hyperbolae share the same asymptotes, which are the asymptotic directions as defined in Section 20.6, and are given by the equation

$$Lx^2 + 2Mxy + Ny^2 = 0.$$

(4) If $L = M = N$, we have a *planar point*, and in this case, the Dupin indicatrix is undefined.

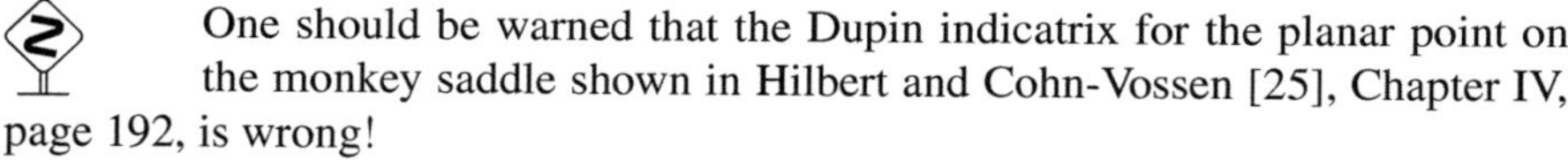

One should be warned that the Dupin indicatrix for the planar point on the monkey saddle shown in Hilbert and Cohn-Vossen [25], Chapter IV, page 192, is wrong!

Therefore, analyzing the shape of the Dupin indicatrix leads us to rediscover the classification of points on a surface in terms of the principal curvatures. It also lends some intuition to the meaning of the words elliptic, hyperbolic, and parabolic (the last one being a bit misleading). The analysis of $\rho(u,v)$ also shows that in the elliptic case, in a small neighborhood of $X(u,v)$, all points of X are on the same side of the tangent plane. This is like being on the top of a round hill. In the hyperbolic case, in a small neighborhood of $X(u,v)$ there are points of X on both sides of the tangent plane. This is a saddle point or a valley (or mountain pass).

20.9 The *Theorema Egregium* of Gauss, the Equations of Codazzi–Mainardi, and Bonnet's Theorem

In Section 20.5 we expressed the geodesic curvature in terms of the Christoffel symbols, and we also showed that these symbols depend only on E, F, G, i.e., on the first fundamental form. In Section 20.7, we expressed $\mathbf{N}_u$ and $\mathbf{N}_v$ in terms of the coefficients of the first and the second fundamental forms. At first glance, given any six functions E, F, G, L, M, N that are at least C^3-continuous on some open subset U of $\mathbb{R}^2$, and where $E, F > 0$ and $EG - F^2 > 0$, it is plausible that there is a surface X defined on some open subset Ω of U and having $Ex^2 + 2Fxy + Gy^2$ as its first fundamental form and $Lx^2 + 2Mxy + Ny^2$ as its second fundamental form. However, this is false. The problem is that for a surface X, the functions E, F, G, L, M, N are not independent.

In this section we investigate the relations that exist among these functions. We will see that there are three compatibility equations. The first one gives the Gaussian curvature in terms of the first fundamental form only. This is the famous *Theorema Egregium* of Gauss (1827). The other two equations express $M_u - L_v$ and $N_u - M_v$ in terms of L, M, N and the Christoffel symbols. These equations are due to Codazzi (1867) and Mainardi (1856). They were discovered independently by Peterson in 1852 (see Gamkrelidze [20]). Remarkably, these compatibility equations are just what it takes to ensure the existence of a surface (at least locally) with $Ex^2 + 2Fxy + Gy^2$ as its first fundamental form and $Lx^2 + 2Mxy + Ny^2$ as its second fundamental form, an important theorem shown by Ossian Bonnet (1867).

Recall that

$$
\begin{aligned}
X'' &= X_u u_1'' + X_v u_2'' + X_{uu}(u_1')^2 + 2X_{uv}u_1'u_2' + X_{vv}(u_2')^2, \\
&= (L(u_1')^2 + 2Mu_1'u_2' + N(u_2')^2)\mathbf{N} + \kappa_g \mathbf{n_g},
\end{aligned}
$$

and since

$$
\kappa_g \mathbf{n_g} = \left(u_1'' + \sum_{\substack{i=1,2 \\ j=1,2}} \Gamma_{ij}^1 u_i'u_j' \right) X_u + \left(u_2'' + \sum_{\substack{i=1,2 \\ j=1,2}} \Gamma_{ij}^2 u_i'u_j' \right) X_v,
$$

we get the equations (due to Gauss)

$$
\begin{aligned}
X_{uu} &= \Gamma_{11}^1 X_u + \Gamma_{11}^2 X_v + L\mathbf{N}, \\
X_{uv} &= \Gamma_{12}^1 X_u + \Gamma_{12}^2 X_v + M\mathbf{N}, \\
X_{vu} &= \Gamma_{21}^1 X_u + \Gamma_{21}^2 X_v + M\mathbf{N}, \\
X_{vv} &= \Gamma_{22}^1 X_u + \Gamma_{22}^2 X_v + N\mathbf{N},
\end{aligned}
$$

where the Christoffel symbols Γ_{ij}^k are defined such that

$$\begin{pmatrix} \Gamma_{ij}^1 \\ \Gamma_{ij}^2 \end{pmatrix} = \begin{pmatrix} E & F \\ F & G \end{pmatrix}^{-1} \begin{pmatrix} [i\,j;\,1] \\ [i\,j;\,2] \end{pmatrix},$$

and where

$$[1\,1;\,1] = \frac{1}{2}E_u, \quad [1\,1;\,2] = F_u - \frac{1}{2}E_v,$$

$$[1\,2;\,1] = \frac{1}{2}E_v, \quad [1\,2;\,2] = \frac{1}{2}G_u,$$

$$[2\,1;\,1] = \frac{1}{2}E_v, \quad [2\,1;\,2] = \frac{1}{2}G_u,$$

$$[2\,2;\,1] = F_v - \frac{1}{2}G_u, \quad [2\,2;\,2] = \frac{1}{2}G_v.$$

Also, recall from Section 20.7 that we have the Weingarten equations

$$\begin{pmatrix} \mathbf{N}_u \\ \mathbf{N}_v \end{pmatrix} = \begin{pmatrix} a & c \\ b & d \end{pmatrix} \begin{pmatrix} X_u \\ X_v \end{pmatrix} = - \begin{pmatrix} L & M \\ M & N \end{pmatrix} \begin{pmatrix} E & F \\ F & G \end{pmatrix}^{-1} \begin{pmatrix} X_u \\ X_v \end{pmatrix}.$$

From the Gauss equations and the Weingarten equations

$$X_{uu} = \Gamma_{11}^1 X_u + \Gamma_{11}^2 X_v + L\mathbf{N},$$

$$X_{uv} = \Gamma_{12}^1 X_u + \Gamma_{12}^2 X_v + M\mathbf{N},$$

$$X_{vu} = \Gamma_{21}^1 X_u + \Gamma_{21}^2 X_v + M\mathbf{N},$$

$$X_{vv} = \Gamma_{22}^1 X_u + \Gamma_{22}^2 X_v + N\mathbf{N},$$

$$\mathbf{N}_u = aX_u + cX_v,$$

$$\mathbf{N}_v = bX_u + dX_v,$$

we see that the partial derivatives of X_u, X_v and $\mathbf{N}$ can be expressed in terms of the coefficients E, F, G, L, M, N and their partial derivatives. Thus, a way to obtain relations among these coefficients is to write the equations expressing the commutation of partials, i.e.,

$$(X_{uu})_v - (X_{uv})_u = 0,$$

$$(X_{vv})_u - (X_{vu})_v = 0,$$

$$\mathbf{N}_{uv} - \mathbf{N}_{vu} = 0.$$

Using the Gauss equations and the Weingarten equations, we obtain relations of the form

$$A_1 X_u + B_1 X_v + C_1 \mathbf{N} = 0,$$

$$A_2 X_u + B_2 X_v + C_2 \mathbf{N} = 0,$$

$$A_3 X_u + B_3 X_v + C_3 \mathbf{N} = 0,$$

where $A_i, B_i,$ and C_i are functions of E, F, G, L, M, N and their partial derivatives, for $i = 1, 2, 3$. However, since the vectors $X_u, X_v,$ and $\mathbf{N}$ are linearly independent, we obtain the nine equations

$$A_i = 0, \quad B_i = 0, \quad C_i = 0, \quad \text{for } i = 1, 2, 3.$$

Although this is very tedious, it can be shown that these equations are equivalent to just three equations. Due to its importance, we state the *Theorema Egregium* of Gauss.

Theorem 20.1. *Given a surface X and a point $p = X(u, v)$ on X, the Gaussian curvature K at (u, v) can be expressed as a function of E, F, G, and their partial derivatives. In fact,*

$$(EG - F^2)^2 K = \begin{vmatrix} C & F_v - \frac{1}{2} G_u & \frac{1}{2} G_v \\ \frac{1}{2} E_u & E & F \\ F_u - \frac{1}{2} E_v & F & G \end{vmatrix} - \begin{vmatrix} 0 & \frac{1}{2} E_v & \frac{1}{2} G_u \\ \frac{1}{2} E_v & E & F \\ \frac{1}{2} G_u & F & G \end{vmatrix},$$

where

$$C = \frac{1}{2}(-E_{vv} + 2F_{uv} - G_{uu}).$$

Proof. Following Darboux [7] (Volume III, page 246), a way of proving Theorem 20.1 is to start from the formula

$$K = \frac{LN - M^2}{EG - F^2}$$

and to go back to the expressions of L, M, N using D, D', D'' as determinants:

$$L = \frac{D}{\sqrt{EG - F^2}}, \quad M = \frac{D'}{\sqrt{EG - F^2}}, \quad N = \frac{D''}{\sqrt{EG - F^2}},$$

where

$$D = (X_u, X_v, X_{uu}), \quad D' = (X_u, X_v, X_{uv}), \quad D'' = (X_u, X_v, X_{vv}).$$

Then we can write

$$(EG - F^2)^2 K = (X_u, X_v, X_{uu})(X_u, X_v, X_{vv}) - (X_u, X_v, X_{uv})^2,$$

and compute these determinants by multiplying them out. One will eventually get the expression given in the theorem! $\square$

It can be shown that the other two equations, known as the *Codazzi–Mainardi equations*, are the equations

$$M_u - L_v = \Gamma_{11}^2 N - (\Gamma_{12}^2 - \Gamma_{11}^1)M - \Gamma_{12}^1 L,$$
$$N_u - M_v = \Gamma_{12}^2 N - (\Gamma_{22}^2 - \Gamma_{12}^1)M - \Gamma_{22}^1 L.$$

We conclude this section with an important theorem of Ossian Bonnet. First, we show that the first and the second fundamental forms determine a surface up to rigid motion. More precisely, we have the following lemma.

Lemma 20.6. *Let $X \colon \Omega \to \mathbb{E}^3$ and $Y \colon \Omega \to \mathbb{E}^3$ be two surfaces over a connected open set Ω. If X and Y have the same coefficients E, F, G, L, M, N over Ω, then there is a rigid motion mapping $X(\Omega)$ onto $Y(\Omega)$.*

The above lemma can be shown using a standard theorem about ordinary differential equations (see do Carmo, [12] Appendix to Chapter 4, pp. 309–314). Finally, we state Bonnet's theorem.

Theorem 20.2. *Let E, F, G, L, M, N be C^3-continuous functions on some open set $U \subset \mathbb{R}^2$, and such that $E > 0$, $G > 0$, and $EG - F^2 > 0$. If these functions satisfy the Gauss formula (of the Theorema Egregium) and the Codazzi–Mainardi equations, then for every $(u,v) \in U$ there is an open set $\Omega \subseteq U$ such that $(u,v) \in \Omega$, and a surface $X \colon \Omega \to \mathbb{E}^3$ such that X is a diffeomorphism, and E, F, G are the coefficients of the first fundamental form of X, and L, M, N are the coefficients of the second fundamental form of X. Furthermore, if Ω is connected, then $X(\Omega)$ is unique up to a rigid motion.*

20.10 Lines of Curvature, Geodesic Torsion, Asymptotic Lines

Given a surface X, certain curves on the surface play a special role, for example, the curves corresponding to the directions in which the curvature is maximum or minimum. More precisely, we have the following definition.

Definition 20.8. Given a surface X, a *line of curvature* is a curve $C \colon t \mapsto X(u(t), v(t))$ on X defined on some open interval I and having the property that for every $t \in I$, the tangent vector $C'(t)$ is collinear with one of the principal directions at $X(u(t), v(t))$.

Note that we are assuming that no point on a line of curvature is either a planar point or an umbilical point, since principal directions are undefined as such points. The differential equation defining lines of curvature can be found as follows. Remember from Lemma 20.4 of Section 20.7 that the principal directions are the eigenvectors of $d\mathbf{N}_{(u,v)}$. Therefore, we can find the differential equation defining lines of curvature by eliminating κ from the two equations from the proof of Lemma 20.4:

$$\frac{MF - LG}{EG - F^2}u' + \frac{NF - MG}{EG - F^2}v' = -\kappa u',$$

$$\frac{LF - ME}{EG - F^2}u' + \frac{MF - NE}{EG - F^2}v' = -\kappa v'.$$

It is not hard to show that the resulting equation can be written as

$$\begin{vmatrix} (v')^2 & -u'v' & (u')^2 \\ E & F & G \\ L & M & N \end{vmatrix} = 0.$$

From the above equation we see that the u-lines and the v-lines are the lines of curvature iff $F = M = 0$. Generally, this differential equation does not have closed-form solutions.

There is another notion that is useful in understanding lines of curvature, the geodesic torsion. Let $C\colon s \mapsto X(u(s),v(s))$ be a curve on X assumed to be parametrized by arc length, and let $X(u(0),v(0))$ be a point on the surface X, and assume that this point is neither a planar point nor an umbilic, so that the principal directions are defined. We can define the orthonormal frame $(e_1, e_2, \mathbf{N})$, known as the *Darboux frame*, where e_1 and e_2 are unit vectors corresponding to the principal directions, $\mathbf{N}$ is the normal to the surface at $X(u(0),v(0))$, and $\mathbf{N} = e_1 \times e_2$.

It is interesting to study the quantity $d\mathbf{N}_{(u,v)}(0)/ds$. If $\mathbf{t} = C'(0)$ is the unit tangent vector at $X(u(0),v(0))$, we have another orthonormal frame considered in Section 20.4, namely $(\mathbf{t}, \mathbf{n_g}, \mathbf{N})$, where $\mathbf{n_g} = \mathbf{N} \times \mathbf{t}$, and if φ is the angle between e_1 and $\mathbf{t}$, we have

$$\mathbf{t} = \cos\varphi\, e_1 + \sin\varphi\, e_2,$$
$$\mathbf{n_g} = -\sin\varphi\, e_1 + \cos\varphi\, e_2.$$

In the following lemma we show that

$$\frac{d\mathbf{N}_{(u,v)}}{ds}(0) = -\kappa_N \mathbf{t} + \tau_g \mathbf{n_g},$$

where κ_N is the normal curvature and where τ_g is a quantity called the *geodesic torsion*.

Lemma 20.7. *Given a curve* $C\colon s \mapsto X(u(s),v(s))$ *parametrized by arc length on a surface* X, *we have*

$$\frac{d\mathbf{N}_{(u,v)}}{ds}(0) = -\kappa_N \mathbf{t} + \tau_g \mathbf{n_g},$$

where κ_N *is the normal curvature, and where the geodesic torsion* τ_g *is given by*

$$\tau_g = (\kappa_1 - \kappa_2)\sin\varphi\cos\varphi.$$

Proof. Since $-\kappa_1$ and $-\kappa_2$ are the eigenvalues of $d\mathbf{N}_{(u(0),v(0))}$ associated with the eigenvectors e_1 and e_2 (where κ_1 and κ_2 are the principal curvatures), it is immediate that

$$\frac{d\mathbf{N}_{(u,v)}}{ds}(0) = d\mathbf{N}_{(u(0),v(0))}(\mathbf{t}) = -\kappa_1 \cos\varphi\, e_1 - \kappa_2 \sin\varphi\, e_2,$$

which shows that this vector is a linear combination of $\mathbf{t}$ and $\mathbf{n_g}$. By projection onto $\mathbf{n_g}$ we get that the geodesic torsion τ_g given by

$$\begin{aligned}
\tau_g &= d\mathbf{N}_{(u(0),v(0))}(\mathbf{t}) \cdot \mathbf{n_g}, \\
&= (-\kappa_1 \cos\varphi\, e_1 - \kappa_2 \sin\varphi\, e_2) \cdot (-\sin\varphi\, e_1 + \cos\varphi\, e_2), \\
&= (\kappa_1 - \kappa_2) \sin\varphi \cos\varphi.
\end{aligned}$$

Using Euler's formula (see Section 20.6)

$$\kappa_N = \kappa_1 \cos^2\varphi + \kappa_2 \sin^2\varphi,$$

it is immediately verified that

$$d\mathbf{N}_{(u(0),v(0))}(\mathbf{t}) \cdot \mathbf{t} = -\kappa_N,$$

which proves the lemma. $\square$

From the formula

$$\tau_g = (\kappa_1 - \kappa_2) \sin\varphi \cos\varphi,$$

since φ is the angle between the tangent vector to the curve C and a principal direction, it is clear that the lines of curvature are characterized by the fact that $\tau_g = 0$. One will also observe that orthogonal curves have opposite geodesic torsions (same absolute value and opposite signs).

If $\mathbf{N}$ is the principal normal, τ is the torsion of C at $X(u(0),v(0))$, and θ is the angle between $\mathbf{N}$ and $\mathbf{n}$, so that $\cos\theta = \mathbf{N} \cdot \mathbf{n}$, we claim that

$$\tau_g = \tau - \frac{d\theta}{ds},$$

which is often known as *Bonnet's formula*.

Lemma 20.8. *Given a curve $C\colon s \mapsto X(u(s),v(s))$ parametrized by arc length on a surface X, the geodesic torsion τ_g is given by*

$$\tau_g = \tau - \frac{d\theta}{ds} = (\kappa_1 - \kappa_2) \sin\varphi \cos\varphi,$$

where τ is the torsion of C at $X(u(0),v(0))$, and θ is the angle between $\mathbf{N}$ and the principal normal $\mathbf{n}$ to C at $s = 0$.

Proof. We differentiate

$$\cos\theta = \mathbf{N} \cdot \mathbf{n}.$$

This yields

$$-\sin\theta\, \frac{d\theta}{ds} = \frac{d\mathbf{N}}{ds} \cdot \mathbf{n} + \mathbf{N} \cdot \frac{d\mathbf{n}}{ds},$$

and since by the Frenet–Serret formulae

$$\frac{d\mathbf{n}}{ds} = -\kappa\mathbf{t} - \tau\mathbf{b},$$

and by Lemma 20.7

$$\frac{d\mathbf{N}}{ds} = -\kappa_N \mathbf{t} + \tau_g \mathbf{n_g},$$

we get

$$-\sin\theta \frac{d\theta}{ds} = (-\kappa_N \mathbf{t} + \tau_g \mathbf{n_g}) \cdot \mathbf{n} + \mathbf{N} \cdot (-\kappa \mathbf{t} - \tau \mathbf{b})$$
$$= \tau_g (\mathbf{n_g} \cdot \mathbf{n}) - \tau (\mathbf{N} \cdot \mathbf{b})$$
$$= \tau_g \sin\theta - \tau \sin\theta,$$

since

$$\mathbf{n_g} \cdot \mathbf{n} = \mathbf{N} \cdot \mathbf{b} = \sin\theta.$$

Therefore, when $\theta \neq 0$, we get

$$-\frac{d\theta}{ds} = \tau_g - \tau,$$

and by continuity, when $\theta = 0$,

$$0 = \tau_g - \tau.$$

Therefore, in all cases we obtain the formula

$$\tau_g = \tau - \frac{d\theta}{ds},$$

which proves the lemma. $\square$

Note that the geodesic torsion depends only on the tangent of curves C. Also, for a curve for which $\theta = 0$, we have $\tau_g = \tau$. Such a curve is also characterized by the fact that the geodesic curvature κ_g is null. As we will see shortly, such curves are called geodesics, which explains the name geodesic torsion for τ_g.

Lemma 20.8 can be used to give a quick proof of a beautiful theorem of Dupin (1813). Dupin's theorem has to do with families of surfaces forming a triply orthogonal system. Given some open subset U of $\mathbb{E}^3$, three families $\mathscr{F}_1$, $\mathscr{F}_2$, $\mathscr{F}_3$ of surfaces form a *triply orthogonal system* for U if for every point $p \in U$ there is a unique surface from each family $\mathscr{F}_i$ passing through p, where $i = 1, 2, 3$, and any two of these surfaces intersect orthogonally along their curve of intersection. Then Dupin's theorem is as follows.

Theorem 20.3. *The surfaces of a triply orthogonal system intersect each other along lines of curvature.*

Proof. Here is a sketch of the proof. First, we note that if two surfaces X_1 and X_2 intersect along a curve C, and if they form a constant angle along C, then the geodesic torsion τ_g^1 of C on X_1 is equal to the geodesic torsion τ_g^2 of C on X_2. Indeed, if θ_1 is the angle between $\mathbf{N_1}$ and $\mathbf{n}$, and θ_2 is the angle between $\mathbf{N_2}$ and $\mathbf{n}$, where $\mathbf{N_1}$ is the normal to X_1, $\mathbf{N_2}$ is the normal to X_2, and $\mathbf{n}$ is the principal normal to C, then

$$\theta_1 - \theta_2 = \lambda,$$

where λ is some constant, and thus

$$\frac{d\theta_1}{ds} = \frac{d\theta_2}{ds},$$

which shows that

$$\tau_g^1 = \tau - \frac{d\theta_1}{ds} = \tau - \frac{d\theta_2}{ds} = \tau_g^2.$$

Now, if the system of surfaces is triply orthogonal, letting τ_{ij} be the geodesic curvature of the curve of intersection C_{ij} between $X_i \in \mathscr{F}_i$ and $X_j \in \mathscr{F}_j$ (where $1 \leq i < j \leq 3$), which is well defined, since X_i and X_j intersect orthogonally, from a previous observation the geodesic torsions of orthogonal curves are opposite, and thus

$$\tau_{12} = -\tau_{13}, \quad \tau_{23} = -\tau_{12}, \quad \tau_{13} = -\tau_{23},$$

from which we get that

$$\tau_{12} = \tau_{23} = \tau_{13} = 0.$$

However, this means that the curves of intersection are lines of curvature. $\square$

A nice application of Theorem 20.3 is that it is possible to find the lines of curvature on an ellipsoid. Indeed, a system of confocal quadrics is triply orthogonal! (see Berger and Gostiaux [4], Chapter 10, Sections 10.2.2.3, 10.4.9.5, and 10.6.8.3, and Hilbert and Cohn-Vossen [25], Chapter 4, Section 28).

We now turn briefly to asymptotic lines. Recall that asymptotic directions are defined only at points where $K < 0$, and at such points they correspond to the directions for which the normal curvature κ_N is null.

Definition 20.9. Given a surface X, an *asymptotic line* is a curve $C: t \mapsto X(u(t), 4v(t))$ on X defined on some open interval I where $K < 0$, and having the property that for every $t \in I$, the tangent vector $C'(t)$ is collinear with one of the asymptotic directions at $X(u(t), v(t))$.

The differential equation defining asymptotic lines is easily found, since it expresses the fact that the normal curvature is null:

$$L(u')^2 + 2M(u'v') + N(v')^2 = 0.$$

Such an equation generally does not have closed-form solutions. Note that the u-lines and the v-lines are asymptotic lines iff $L = N = 0$ (and $F \neq 0$).

Example 20.9. Perseverant readers are welcome to compute E, F, G, L, M, N for the *Enneper surface*

$$x = u - \frac{u^3}{3} + uv^2,$$

$$y = v - \frac{v^3}{3} + u^2v,$$

$$z = u^2 - v^2.$$

Then they will be able to find closed-form solutions for the lines of curvature and the asymptotic lines.

Parabolic lines are defined by the equation

$$LN - M^2 = 0,$$

where $L^2 + M^2 + N^2 > 0$. In general, the locus of parabolic points consists of several curves and points. For fun, the reader should look at Klein's experiment as described in Hilbert and Cohn-Vossen [25], Chapter IV, Section 29, page 197. We now turn briefly to geodesics.

20.11 Geodesic Lines, Local Gauss–Bonnet Theorem

Geodesics play a very important role in surface theory and in dynamics. One of the main reasons why geodesics are so important is that they generalize to curved surfaces the notion of "shortest path" between two points in the plane (**warning**: As we shall see, this is true only *locally, not globally*). More precisely, given a surface X and any two points $p = X(u_0, v_0)$ and $q = X(u_1, v_1)$ on X, let us look at all the regular curves C on X defined on some open interval I such that $p = C(t_0)$ and $q = C(t_1)$ for some $t_0, t_1 \in I$. It can be shown that in order for such a curve C to minimize the length $l_C(pq)$ of the curve segment from p to q, we must have $\kappa_g(t) = 0$ along $[t_0, t_1]$, where $\kappa_g(t)$ is the geodesic curvature at $X(u(t), v(t))$. In other words, the principal normal $\mathbf{n}$ must be parallel to the normal $\mathbf{N}$ to the surface along the curve segment from p to q. If C is parametrized by arc length, this means that the acceleration must be normal to the surface.

It it then natural to define geodesics as those curves such that $\kappa_g = 0$ everywhere on their domain of definition. Actually, there is another way of defining geodesics in terms of vector fields and covariant derivatives (see do Carmo [12] or Berger and Gostiaux [4]), but for simplicity, we stick to the definition in terms of the geodesic curvature (however, see Section 20.12).

Definition 20.10. Given a surface $X \colon \Omega \to \mathbb{E}^3$, a *geodesic line, or geodesic,* is a regular curve $C \colon I \to \mathbb{E}^3$ on X such that $\kappa_g(t) = 0$ for all $t \in I$.

Note that by regular curve we mean that $\dot{C}(t) \neq 0$ for all $t \in I$, i.e., C is really a curve, and not a single point. Physically, a particle constrained to stay on the surface and not acted on by any force, once set in motion with some nonnull initial velocity (tangent to the surface), will follow a geodesic (assuming no friction).

Since $\kappa_g = 0$ iff the principal normal $\mathbf{n}$ to C at t is parallel to the normal $\mathbf{N}$ to the surface at $X(u(t), v(t))$, and since the principal normal $\mathbf{n}$ is a linear combination of the tangent vector $\dot{C}(t)$ and the acceleration vector $\ddot{C}(t)$, the normal $\mathbf{N}$ to the surface at t belongs to the osculating plane.

The differential equations for geodesics are obtained from Lemma 20.2. Since the tangential part of the curvature at a point is given by

$$\kappa_g \mathbf{n_g} = \left(u_1'' + \sum_{\substack{i=1,2 \\ j=1,2}} \Gamma_{ij}^1 u_i' u_j' \right) X_u + \left(u_2'' + \sum_{\substack{i=1,2 \\ j=1,2}} \Gamma_{ij}^2 u_i' u_j' \right) X_v,$$

the differential equations for geodesics are

$$u_1'' + \sum_{\substack{i=1,2 \\ j=1,2}} \Gamma_{ij}^1 u_i' u_j' = 0,$$

$$u_2'' + \sum_{\substack{i=1,2 \\ j=1,2}} \Gamma_{ij}^2 u_i' u_j' = 0,$$

or more explicitly (letting $u = u_1$ and $v = u_2$),

$$u'' + \Gamma_{11}^1 (u')^2 + 2\Gamma_{12}^1 u'v' + \Gamma_{22}^1 (v')^2 = 0,$$
$$v'' + \Gamma_{11}^2 (u')^2 + 2\Gamma_{12}^2 u'v' + \Gamma_{22}^2 (v')^2 = 0.$$

In general, it is impossible to find closed-form solutions for these equations. Nevertheless, from the theory of ordinary differential equations, the following lemma showing the local existence of geodesics can be shown (see do Carmo [12], Chapter 4, Section 4.7).

Lemma 20.9. *Given a surface X, for every point $p = X(u, v)$ on X and every nonnull tangent vector $v \in T_{(u,v)}(X)$, there is some $\varepsilon > 0$ and a unique curve $\gamma\colon\,]-\varepsilon,\, \varepsilon[\to \mathbb{E}^3$ on the surface X such that γ is a geodesic, $\gamma(0) = p$, and $\gamma'(0) = v$.*

To emphasize that the geodesic γ depends on the initial direction v, we often write $\gamma(t, v)$ instead of $\gamma(t)$. The geodesics on a sphere are the great circles (the plane sections by planes containing the center of the sphere). More generally, in the case of a surface of revolution (a surface generated by a plane curve rotating around an axis in the plane containing the curve and not meeting the curve), the differential equations for geodesics can be used to study the geodesics.

Example 20.10. For example, the meridians are geodesics (meridians are the plane sections by planes through the axis of rotation: They are obtained by rotating the original curve generating the surface). Also, the parallel circles such that at every point p the tangent to the meridian through p is parallel to the axis of rotation is a geodesic. In general, there are other geodesics. For more on geodesics on surfaces of revolution, see do Carmo [12], Chapter 4, Section 4, and the problems.

The geodesics on an ellipsoid are also fascinating; see Berger and Gostiaux [4], Section 10.4.9.5, and Hilbert and Cohn-Vossen [25], Chapter 4, Section 32.

It should be noted that geodesics can be self-intersecting or closed. A deeper study of geodesics requires a study of vector fields on surfaces and would lead us too far. Technically, what is needed is the exponential map, which we now discuss briefly.

The idea behind the exponential map is to parametrize locally the surface X in terms of a map from the tangent space to the surface, this map being defined in terms of short geodesics. More precisely, for every point $p = X(u,v)$ on the surface, there is some open disk B_ε of center $(0,0)$ in $\mathbb{R}^2$ (recall that the tangent plane $T_p(X)$ at p is isomorphic to $\mathbb{R}^2$) and an injective map

$$\exp_p \colon B_\varepsilon \to X(\Omega)$$

such that for every $v \in B_\varepsilon$ with $v \neq 0$,

$$\exp_p(v) = \gamma(1,v),$$

where $\gamma(t,v)$ is the unique geodesic segment such that $\gamma(0,v) = p$ and $\gamma'(0,v) = v$. Furthermore, for B_ε small enough, $\exp_p$ is a diffeomorphism. It turns out that $\exp_p(v)$ is the point q obtained by "laying off" a length equal to $\|v\|$ along the unique geodesic that passes through p in the direction v. Of course, to make sure that all this is well-defined, it is necessary to prove a number of facts. We state the following lemmas, whose proofs can be found in do Carmo [12].

Lemma 20.10. *Given a surface* $X \colon \Omega \to \mathbb{E}^3$, *for every* $v \neq 0$ *in* $\mathbb{R}^2$, *if*

$$\gamma(-,v) \colon \,]-\varepsilon, \varepsilon[\to \mathbb{E}^3$$

is a geodesic on the surface X, then for every $\lambda > 0$, *the curve*

$$\gamma(-,\lambda v) \colon \,]-\varepsilon/\lambda, \varepsilon/\lambda[\to \mathbb{E}^3$$

is also a geodesic, and

$$\gamma(t,\lambda v) = \gamma(\lambda t,v).$$

From Lemma 20.10, for $v \neq 0$, if $\gamma(1,v)$ is defined, then

$$\gamma\left(\|v\|, \frac{v}{\|v\|} \right) = \gamma(1,v).$$

This leads to the definition of the exponential map.

Definition 20.11. Given a surface $X \colon \Omega \to \mathbb{E}^3$ and a point $p = X(u,v)$ on X, the *exponential map* $\exp_p$ is the map

$$\exp_p \colon U \to X(\Omega)$$

defined such that

$$\exp_p(v) = \gamma\left(\|v\|, \frac{v}{\|v\|}\right) = \gamma(1, v),$$

where $\gamma(0, v) = p$ and U is the open subset of $\mathbb{R}^2 (= T_p(X))$ such that for every $v \neq 0$, $\gamma(\|v\|, v/\|v\|)$ is defined. We let $\exp_p(0) = p$.

It is immediately seen that U is star-like. One should realize that in general, U is a proper subset of Ω. For example, in the case of a sphere, the exponential map is defined everywhere. However, given a point p on a sphere, if we remove its antipodal point $-p$, then $\exp_p(v)$ is undefined for points on the circle of radius π. Nevertheless, $\exp_p$ is always well-defined in a small open disk.

Lemma 20.11. *Given a surface* $X\colon \Omega \to \mathbb{E}^3$, *for every point* $p = X(u, v)$ *on X there is some* $\varepsilon > 0$, *some open disk* B_ε *of center* $(0, 0)$, *and some open subset V of* $X(\Omega)$ *with* $p \in V$ *such that the exponential map* $\exp_p\colon B_\varepsilon \to V$ *is well-defined and is a diffeomorphism.*

A neighborhood of p on X of the form $\exp_p(B_\varepsilon)$ is called a *normal neighborhood of p*. The exponential map can be used to define special local coordinate systems on normal neighborhoods, by picking special coordinate systems on the tangent plane. In particular, we can use polar coordinates (ρ, θ) on $\mathbb{R}^2$. In this case, $0 < \theta < 2\pi$. Thus, the closed half-line corresponding to $\theta = 0$ is omitted, and so is its image under $\exp_p$. It is easily seen that in such a coordinate system $E = 1$ and $F = 0$, and ds^2 is of the form

$$ds^2 = dr^2 + G d\theta^2.$$

The image under $\exp_p$ of a line through the origin in $\mathbb{R}^2$ is called a *geodesic line*, and the image of a circle centered at the origin is called a *geodesic circle*. Since $F = 0$, these lines are orthogonal. It can also be shown that the Gaussian curvature is expressed as follows:

$$K = -\frac{1}{\sqrt{G}}\frac{\partial^2(\sqrt{G})}{\partial \rho^2}.$$

Polar coordinates can be used to prove the following lemma showing that geodesics locally minimize arc length.

However, globally, geodesics generally do not minimize arc length. For instance, on a sphere, given any two nonantipodal points p, q, since there is a unique great circle passing through p and q, there are two geodesic arcs joining p and q, but only one of them has minimal length.

Lemma 20.12. *Given a surface* $X\colon \Omega \to \mathbb{E}^3$, *for every point* $p = X(u, v)$ *on X there is some* $\varepsilon > 0$ *and some open disk* B_ε *of center* $(0, 0)$ *such that for every* $q \in \exp_p(B_\varepsilon)$ *and geodesic* $\gamma\colon \,]-\eta, \eta[\to \mathbb{E}^3$ *in* $\exp_p(B_\varepsilon)$ *such that* $\gamma(0) = p$ *and* $\gamma(t_1) = q$, *and for every regular curve* $\alpha\colon [0, t_1] \to \mathbb{E}^3$ *on X such that* $\alpha(0) = p$ *and* $\alpha(t_1) = q$, *we have*

$$l_\gamma(pq) \leq l_\alpha(pq),$$

where $l_\alpha(pq)$ denotes the length of the curve segment α from p to q (and similarly for γ). Furthermore, $l_\gamma(pq) = l_\alpha(pq)$ iff the trace of γ is equal to the trace of α between p and q.

As we already noted, Lemma 20.12 is false globally, since a geodesic, if extended too much, may not be the shortest path between two points (example of the sphere). However, the following lemma shows that a shortest path must be a geodesic segment.

Lemma 20.13. *Given a surface $X \colon \Omega \to \mathbb{E}^3$, let $\alpha \colon I \to \mathbb{E}^3$ be a regular curve on X parametrized by arc length. For any two points $p = \alpha(t_0)$ and $q = \alpha(t_1)$ on α, assume that the length $l_\alpha(pq)$ of the curve segment from p to q is minimal among all regular curves on X passing through p and q. Then α is a geodesic.*

At this point, in order to go further into the theory of surfaces, in particular closed surfaces, it is necessary to introduce differentiable manifolds and more topological tools. However, this is beyond the scope of this book, and we simply refer the interested readers to the following sources. For the foundations of differentiable manifolds, see Berger and Gostiaux [4], do Carmo [12, 13, 14], Guillemin and Pollack [24], Warner [43], Sternberg [41], Boothby [5], Lafontaine [29], Lehmann and Sacré [31], Gray [23], Stoker [42], Gallot, Hulin, and Lafontaine [19], Milnor [36], Lang [30], Malliavin [33], and Godbillon [21]. Abraham and Marsden [1] contains a compact and yet remarkably clear and complete presentation of differentiable manifolds and Riemannian geometry (and a lot of Lagrangian and Hamiltonian mechanics!). For the differential topology of surfaces, see Guillemin and Pollack [24], Milnor [36, 37], Hopf [26], Gramain [22], Lehmann and Sacré [31], and for the algebraic topology of surfaces, see Chapter 1 of Massey [35, 34] and Chapter 1 of Ahlfors and Sario [2], which is remarkable. A lively and remarkably clear introduction to algebraic topology, including the classification theorem for surfaces, can be found in Fulton [17]. For a detailed presentation of differential geometry and Riemannian geometry, see do Carmo [14], Gallot, Hulin, and Lafontaine [19], Sternberg [41], Gray [23], Sharpe [40], Lang [30], Lehmann and Sacré [31], and Malliavin [33]. Choquet-Bruhat [6] also covers a lot of geometric analysis, differential geometry, and topology, and stresses applications to physics. Volume 28 of the *Encyclopaedia of Mathematical Sciences* edited by Gamkrelidze [20] contains a very interesting survey of the field of differential geometry, understood in a broad sense.

Nevertheless, we cannot resist to state one of the "gems" of the differential geometry of surfaces, the local Gauss–Bonnet theorem.

The local Gauss–Bonnet theorem deals with regions on a surface homeomorphic to a closed disk whose boundary is a closed piecewise regular curve α without self-intersection. Such a curve has a finite number of points where the tangent has a discontinuity. If there are n such discontinuities $p_1, \ldots, p_n$, let θ_i be the exterior angle between the two tangents at p_i. More precisely, if $\alpha(t_i) = p_i$, and the two tangents at p_i are defined by the vectors

$$\lim_{t \to t_i, t < t_i} \alpha'(t) = \alpha'_-(t_i) \neq 0,$$

and

$$\lim_{t \to t_i, t > t_i} \alpha'(t) = \alpha'_+(t_i) \neq 0,$$

the angle θ_i is defined as follows. Let θ_i be the angle between $\alpha'_-(t_i)$ and $\alpha'_+(t_i)$ such that $0 < |\theta_i| \leq \pi$, its sign being determined as follows. If p_i is not a cusp, which means that $|\theta_i| \neq \pi$, we give θ_i the sign of the determinant

$$(\alpha'_-(t_i), \alpha'_+(t_i), \mathbf{N}_{\mathbf{p_i}}).$$

If p_i is a cusp, which means that $|\theta_i| = \pi$, it is easy to see that there is some $\varepsilon > 0$ such that the determinant

$$(\alpha'(t_i - \eta), \alpha'(t_i + \eta), \mathbf{N}_{\mathbf{p_i}})$$

does not change sign for $\eta \in]-\varepsilon, \varepsilon[$, and we give θ_i the sign of this determinant. Let us call a region defined as above a *simple region*. In order to state a simpler version of the theorem, let us also assume that the curve segments between consecutive points p_i are geodesic lines. We will call such a curve a *geodesic polygon*. Then the *local Gauss–Bonnet theorem* can be stated as follows.

Theorem 20.4. *Given a surface $X \colon \Omega \to \mathbb{E}^3$, assuming that X is injective, $F = 0$, and that Ω is an open disk, for every simple region R of $X(\Omega)$ bounded by a geodesic polygon with n vertices $p_1, \ldots, p_n$, letting $\theta_1, \ldots, \theta_n$ be the exterior angles of the geodesic polygon, we have*

$$\iint_R K dA + \sum_{i=1}^{n} \theta_i = 2\pi.$$

Remark: The assumption that $F = 0$ is not essential, it simply makes the proof easier.

Some clarification regarding the meaning of the integral $\iint_R K dA$ is in order. Firstly, it can be shown that the element of area dA on a surface X is given by

$$dA = \|X_u \times X_v\| du\, dv = \sqrt{EG - F^2}\, du\, dv.$$

Secondly, if we recall from Lemma 20.3 that

$$\begin{pmatrix} \mathbf{N}_u \\ \mathbf{N}_v \end{pmatrix} = - \begin{pmatrix} L & M \\ M & N \end{pmatrix} \begin{pmatrix} E & F \\ F & G \end{pmatrix}^{-1} \begin{pmatrix} X_u \\ X_v \end{pmatrix},$$

it is easily verified that

$$\mathbf{N}_u \times \mathbf{N}_v = \frac{LN - M^2}{EG - F^2} X_u \times X_v = K(X_u \times X_v).$$

Thus,

$$\iint_R KdA = \iint_R K\|X_u \times X_v\|du\,dv = \iint_R \|\mathbf{N}_u \times \mathbf{N}_v\|du\,dv,$$

the latter integral representing the area of the spherical image of R under the Gauss map. This is the interpretation of the integral $\iint_R KdA$ that Gauss himself gave.

If the geodesic polygon is a triangle, and if A, B, C are the interior angles, so that $A = \pi - \theta_1$, $B = \pi - \theta_2$, $C = \pi - \theta_3$, the Gauss–Bonnet theorem reduces to what is known as the *Gauss formula*:

$$\iint_R KdA = A + B + C - \pi.$$

The above formula shows that if $K > 0$ on R, then $\iint_R KdA$ is the excess of the sum of the angles of the geodesic triangle over π. If $K < 0$ on R, then $\iint_R KdA$ is the defficiency of the sum of the angles of the geodesic triangle over π. And finally, if $K = 0$, then $A + B + C = \pi$, which we know from the plane!

For the global version of the Gauss–Bonnet theorem, we need the topological notion of the Euler–Poincaré characteristic. If S is an orientable compact surface with g holes, the *Euler–Poincaré characteristic* $\chi(S)$ of S is defined by

$$\chi(S) = 2(1 - g).$$

Then the Gauss–Bonnet theorem states that

$$\iint_S KdA = 2\pi\chi(S).$$

What is remarkable about the above formula is that it relates the topology of the surface (its *genus g*, the number of holes) and the geometry of S, i.e., how it curves. However, all this is beyond the scope of this book. For more information the interested reader is referred to Berger and Gostiaux [4], do Carmo [12, 13, 14], Hopf [26], Milnor [36], Lehmann and Sacré [31], Chapter 1 of Massey [35, 34], Chapter 1 of Ahlfors and Sario [2], and Fulton [17].

20.12 Covariant Derivative, Parallel Transport, Geodesics Revisited

Another way to approach geodesics is in terms of covariant derivatives. The notion of covariant derivative is a key concept of Riemannian geometry, and this section provides a down-to-earth presentation of this notion in the case of a surface.

Let $X : \Omega \to \mathbb{E}^3$ be a surface. Given any open subset U of X, a *vector field on U* is a function w that assigns to every point $p \in U$ some tangent vector $w(p) \in T_pX$ to X at p. A vector field w on U is *differentiable at p* if when expressed as $w = aX_u + bX_v$

in the basis (X_u, X_v) (of T_pX), the functions a and b are differentiable at p. A vector field w is *differentiable on U* when it is differentiable at every point $p \in U$.

Definition 20.12. Let w be a differentiable vector field on some open subset U of a surface X. For every $y \in T_pX$, consider a curve $\alpha: \,]-\varepsilon, \varepsilon[\to U$ on X with $\alpha(0) = p$ and $\alpha'(0) = y$, and let $w(t) = (w \circ \alpha)(t)$ be the restriction of the vector field w to the curve α. The normal projection of $dw/dt(0)$ onto the plane T_pX, denoted by

$$\frac{Dw}{dt}(0), \quad \text{or} \quad D_{\alpha'}w(p), \quad \text{or} \quad D_y w(p),$$

is called the *covariant derivative of w at p relative to y*.

The definition of $Dw/dt(0)$ seems to depend on the curve α, but in fact, it depends only on y and the first fundamental form of X. Indeed, if $\alpha(t) = X(u(t), v(t))$, from

$$w(t) = a(u(t), v(t))X_u + b(u(t), v(t))X_v,$$

we get

$$\frac{dw}{dt} = a(X_{uu}\dot{u} + X_{uv}\dot{v}) + b(X_{vu}\dot{u} + X_{vv}\dot{v}) + \dot{a}X_u + \dot{b}X_v.$$

However, we obtained earlier the following formula (due to Gauss) for X_{uu}, X_{uv}, X_{vu}, and X_{vv}:

$$X_{uu} = \Gamma_{11}^1 X_u + \Gamma_{11}^2 X_v + L\mathbf{N},$$
$$X_{uv} = \Gamma_{12}^1 X_u + \Gamma_{12}^2 X_v + M\mathbf{N},$$
$$X_{vu} = \Gamma_{21}^1 X_u + \Gamma_{21}^2 X_v + M\mathbf{N},$$
$$X_{vv} = \Gamma_{22}^1 X_u + \Gamma_{22}^2 X_v + N\mathbf{N}.$$

Now Dw/dt is the tangential component of dw/dt. Thus by dropping the normal components, we get

$$\frac{Dw}{dt} = (\dot{a} + \Gamma_{11}^1 a\dot{u} + \Gamma_{12}^1 a\dot{v} + \Gamma_{21}^1 b\dot{u} + \Gamma_{22}^1 b\dot{v})X_u$$
$$+ (\dot{b} + \Gamma_{11}^2 a\dot{u} + \Gamma_{12}^2 a\dot{v} + \Gamma_{21}^2 b\dot{u} + \Gamma_{22}^2 b\dot{v})X_v.$$

Thus, the covariant derivative depends only on $y = (\dot{u}, \dot{v})$ and the Christoffel symbols, but we know that those depend only on the first fundamental form of X.

Definition 20.13. Let $\alpha: I \to X$ be a regular curve on a surface X. A *vector field along α* is a map w that assigns to every $t \in I$ a vector $w(t) \in T_{\alpha(t)}X$ in the tangent plane to X at $\alpha(t)$. Such a vector field is differentiable if the components a, b of $w = aX_u + bX_v$ over the basis (X_u, X_v) are differentiable. The expression $Dw/dt(t)$ defined in the above equation is called the *covariant derivative of w at t*.

Definition 20.13 extends immediately to piecewise regular curves on a surface.

Definition 20.14. Let $\alpha\colon I \to X$ be a regular curve on a surface X. A vector field along α is *parallel* if $Dw/dt = 0$ for all $t \in I$.

Thus, a vector field along a curve on a surface is parallel iff its derivative is normal to the surface. For example, if C is a great circle on the sphere S^2 parametrized by arc length, the vector field of tangent vectors $C'(s)$ along C is a parallel vector field. We get the following alternative definition of a geodesic.

Definition 20.15. Let $\alpha\colon I \to X$ be a nonconstant regular curve on a surface X. Then α is a *geodesic* if the field of its tangent vectors $\dot{\alpha}(t)$ is parallel along α, that is,

$$\frac{D\dot{\alpha}}{dt}(t) = 0$$

for all $t \in I$.

If we let $\alpha(t) = X(u(t), v(t))$, from the equation

$$\frac{Dw}{dt} = (\dot{a} + \Gamma_{11}^1 a\dot{u} + \Gamma_{12}^1 a\dot{v} + \Gamma_{21}^1 b\dot{u} + \Gamma_{22}^1 b\dot{v})X_u$$
$$+ (\dot{b} + \Gamma_{11}^2 a\dot{u} + \Gamma_{12}^2 a\dot{v} + \Gamma_{21}^2 b\dot{u} + \Gamma_{22}^2 b\dot{v})X_v,$$

with $a = \dot{u}$ and $b = \dot{v}$, we get the equations

$$\ddot{u} + \Gamma_{11}^1 (\dot{u})^2 + \Gamma_{12}^1 \dot{u}\dot{v} + \Gamma_{21}^1 \dot{u}\dot{v} + \Gamma_{22}^1 (\dot{v})^2 = 0,$$
$$\ddot{v} + \Gamma_{11}^2 (\dot{u})^2 + \Gamma_{12}^2 \dot{u}\dot{v} + \Gamma_{21}^2 \dot{u}\dot{v} + \Gamma_{22}^2 (\dot{v})^2 = 0,$$

which are indeed the equations of geodesics found earlier, since $\Gamma_{12}^1 = \Gamma_{21}^1$ and $\Gamma_{12}^2 = \Gamma_{21}^2$ (except that α is not necessarily parametrized by arc length).

Lemma 20.14. *Let $\alpha\colon I \to X$ be a regular curve on a surface X, and let v and w be two parallel vector fields along α. Then the inner product $\langle v(t), w(t)\rangle$ is constant along α (where $\langle -, -\rangle$ is the inner product associated with the first fundamental form, i.e., the Riemannian metric). In particular, $\|v\|$ and $\|w\|$ are constant and the angle between $v(t)$ and $w(t)$ is also constant.*

Proof. The vector field $v(t)$ is parallel iff dv/dt is normal to the tangent plane to the surface X at $\alpha(t)$, and so

$$\langle v'(t), w(t)\rangle = 0$$

for all $t \in I$. Similarly, since $w(t)$ is parallel, we have

$$\langle v(t), w'(t)\rangle = 0$$

for all $t \in I$. Then

$$\langle v(t), w(t)\rangle' = \langle v'(t), w(t)\rangle + \langle v(t), w'(t)\rangle = 0$$

for all $t \in I$, which means that $\langle v(t), w(t)\rangle$ is constant along α. $\square$

As a consequence of Corollary 20.14, if $\alpha \colon I \to X$ is a nonconstant geodesic on X, then $\|\dot{\alpha}\| = c$ for some constant $c > 0$. Thus, we may reparametrize α with respect to the arc length $s = ct$, and we note that the parameter t of a geodesic is proportional to the arc length of α.

Lemma 20.15. *Let $\alpha \colon I \to X$ be a regular curve on a surface X, and for any $t_0 \in I$, let $w_0 \in T_{\alpha(t_0)}X$. Then there is a unique parallel vector field $w(t)$ along α such that $w(t_0) = w_0$.*

Lemma 20.15 is an immediate consequence of standard results on ODEs. This lemma yields the notion of parallel transport.

Definition 20.16. Let $\alpha \colon I \to X$ be a regular curve on a surface X, and for any $t_0 \in I$, let $w_0 \in T_{\alpha(t_0)}X$. Let w be the parallel vector field along α, so that $w(t_0) = w_0$, given by Lemma 20.15. Then for any $t \in I$, the vector $w(t)$ is called the *parallel transport of w_0 along α at t*.

It is easily checked that the parallel transport does not depend on the parametrization of α. If X is an open subset of the plane, then the parallel transport of w_0 at t is indeed a vector $w(t)$ parallel to w_0 (in fact, equal to w_0). However, on a curved surface, the parallel transport may be somewhat counterintuitive.

If two surfaces X and Y are tangent along a curve $\alpha \colon I \to X$, and if $w_0 \in T_{\alpha(t_0)}X = T_{\alpha(t_0)}Y$ is a tangent vector to both X and Y at t_0, then the parallel transport of w_0 along α is the same whether it is relative to X or relative to Y. This is because Dw/dt is the same for both surfaces, and by uniqueness of the parallel transport, the assertion follows. This property can be used to figure out the parallel transport of a vector w_0 when Y is locally isometric to the plane.

In order to generalize the notion of covariant derivative, geodesic, and curvature to manifolds more general than surfaces, the notion of *connection* is needed.

If M is a manifold, we can consider the space $\mathscr{X}(M)$ of smooth vector fields X on M. They are smooth maps that assign to every point $p \in M$ some vector $X(p)$ in the tangent space T_pM to M at p. We can also consider the set $\mathscr{C}^\infty(M)$ of smooth functions $f \colon M \to \mathbb{R}$ on M. Then an *affine connection D on M* is a differentiable map

$$D \colon \mathscr{X}(M) \times \mathscr{X}(M) \to \mathscr{X}(M),$$

denoted by $D_X Y$ (or $\nabla_X Y$), satisfying the following properties:

(1) $D_{fX+gY}Z = fD_X Z + gD_Y Z$;
(2) $D_X(\lambda Y + \mu Z) = \lambda D_X Y + \mu D_X Z$;
(3) $D_X(fY) = fD_X Y + X(f)Y$,

for all $\lambda, \mu \in \mathbb{R}$, all $X, Y, Z \in \mathscr{X}(M)$, and all $f, g \in \mathscr{C}^\infty(M)$, where $X(f)$ denotes the directional derivative of f in the direction X.

Thus, an affine connection is $\mathscr{C}^\infty(M)$-linear in X, $\mathbb{R}$-linear in Y, and satisfies a "Leibniz"-type law in Y. For any chart $\varphi \colon U \to \mathbb{R}^m$, denoting the coordinate functions by $x_1, \ldots, x_m$, if X is given locally by

$$X(p) = \sum_{i=1}^{m} a_i(p) \frac{\partial}{\partial x_i},$$

then

$$X(f)(p) = \sum_{i=1}^{m} a_i(p) \frac{\partial (f \circ \varphi^{-1})}{\partial x_i}.$$

It can be checked that $X(f)$ does not depend on the choice of chart.

The intuition behind a connection is that $D_X Y$ is the directional derivative of Y in the direction X. The notion of covariant derivative can be introduced via the following lemma:

Lemma 20.16. *Let M be a smooth manifold and assume that D is an affine connection on M. Then there is a unique map D associating with every vector field V along a curve $\alpha\colon I \to M$ on M another vector field DV/dt along c (the covariant derivative of V along c), such that:*

(1)

$$\frac{D}{dt}(\lambda V + \mu W) = \lambda \frac{DV}{dt} + \mu \frac{DW}{dt},$$

(2)

$$\frac{D}{dt}(fV) = \frac{df}{dt} V + f \frac{DV}{dt},$$

(3) if V is induced by a vector field $Y \in \mathscr{X}(M)$, in the sense that $V(t) = Y(\alpha(t))$, then

$$\frac{DV}{dt} = D_{\alpha'(t)} Y.$$

Then in local coordinates, DV/dt can be expressed in terms of the Chistoffel symbols, pretty much as in the case of surfaces. Parallel vector fields, parallel transport, geodesics, are defined as before.

Affine connections are uniquely induced by Riemannian metrics, a fundamental result of Levi-Civita. In fact, such connections are *compatible with the metric*, which means that for any smooth curve α on M and any two parallel vector fields X, Y along α, the inner product $\langle X, Y \rangle$ is constant. Such connections are also *symmetric*, which means that

$$D_X Y - D_Y X = [X, Y],$$

where $[X, Y]$ is the Lie bracket of vector fields.

For more on all this, consult do Carmo [12, 13], Gallot, Hulin, and Lafontaine [19], or any other text on Riemannian geometry.

20.13 Applications

We saw in Section 19.11 that many engineering problems can be reduced to finding curves having some desired properties. Surfaces also play an important role in engineering problems where modeling 3D shapes is required. Again, this is true of computer graphics and medical imaging, where 3D contours of shapes, for instance organs, are modeled as surfaces. As in the case of curves, in most practical applications it is necessary to consider surfaces composed of various patches, and the problem then arises to join these patches as smoothly as possible, without restricting too much the number of degrees of freedom required for the design. Various kinds of *spline surfaces* were invented to solve this problem. But this time, the situation is more complex than in the case of curves, because there are two kinds of surface patches, rectangular and triangular. Roughly speaking, since rectangular patches are basically products of curves, their spline theory is rather well understood. This is not the case for triangular patches, for which the theory of splines is very sparse (Loop [32] being a noteworthy exception). Thus, we will restrict our brief discussion to rectangular patches. As for curves, there is a notion of *parametric C^n-continuity* and of *B-spline*. The theory of *B*-splines is quite extensive. Among the many references, we recommend Farin [16, 15], Hoschek and Lasser [27], Bartels, Beatty, and Barsky [3], Piegl and Tiller [39], or Gallier [18]. However, since parametric continuity is sometimes too constraining, more flexible continuity conditions have been investigated. There are various notions of *geometric continuity*, or *G^n-continuity*. Roughly speaking, two surface patches join with G^n-continuity if there is a reparametrization (a diffeomorphism) after which the patches join with parametric C^n-continuity along the common boundary curve. As a consequence, geometric continuity may be defined using the chain rule, in terms of a certain *connection matrix*.

One of the most important applications of geometric continuity occurs when two or more rectangular patches are stitched together. In such cases polygonal holes can occur between patches. It is often impossible to fill these holes with patches that join with parametric continuity, and a geometrically continuous solution must be used instead. There are also variations on the theme of geometric continuity, which seems to be a topic of current interest. Again, we refer the readers to Farin [16, 15], Hoschek and Lasser [27], Bartels, Beatty, and Barsky [3], Piegl and Tiller [39], and Loop [32].

As in the case of curves, traditional methods for surface design focus on achieving a specific level of interelement continuity, but the resulting shapes often possess bulges and undulations, and thus are of poor quality. They lack *fairness*. Fairness refers to the quality of regularity of the curvature of a surface. The maximum rate of change of curvature should be minimized. This suggests several approaches.

- Minimal energy surface (which bends as little as possible): Minimize

$$\int_S \left(\kappa_1^2 + \kappa_2^2 \right) dA$$

   where κ_1 and κ_2 are the principal curvatures.
- Minimal variation surface (which bends as smoothly as possible): Minimize

$$\int_S \left[(D_{e_1} \kappa_1)^2 + (D_{e_2} \kappa_2)^2 \right] dA$$

where κ_1 and κ_2 are the principal curvatures and e_1 and e_2 are unit vectors giving the principal directions.

As in the case of curves, these problems can be cast as constrained optimization problems. More details on this approach, called *variational surface design*, can be found in the Ph.D. theses of Henry Moreton [38] and William Welch [44].

20.14 Problems

20.1. Consider the surface X defined by

$$x = v\cos u,$$
$$y = v\sin u,$$
$$z = v.$$

(i) Show that F is regular at every point (u,v), except when $v = 0$.
(ii) Show that X is the set of points such that

$$x^2 + y^2 = z^2.$$

What does this surface look like?

20.2. Let $\alpha\colon I \to \mathbb{E}^3$ be a regular curve whose curvature is nonzero for all $t \in I$, where $I = \,]a,b[$. Let X be the surface defined over $I \times \mathbb{R}$ such that

$$X(u,v) = \alpha(u) + v\alpha'(u).$$

Show that X is regular for all (u,v) where $v \neq 0$.

Remark: The surface X is called the *tangent surface of* α. The curve α is a *line of striction* on X.

20.3. Let $\alpha\colon I \to \mathbb{E}^3$ be a regular curve whose curvature is nonzero for all $t \in I$, where $I = \,]a,b[$, and assume that α is parametrized by arc length. For any $r > 0$, let X be the surface defined over $I \times \mathbb{R}$ such that

$$X(u,v) = \alpha(u) + r(\cos v\,\mathbf{n}(u) + \sin v\,\mathbf{b}(u)),$$

where $(\mathbf{t},\mathbf{n},\mathbf{b})$ is the Frenet frame of α at u.

Show that for every (u,v) such that $X(u,v)$ is regular, the unit normal vector $\mathbf{N}_{(u,v)}$ to X at (u,v) is given by

$$\mathbf{N}_{(u,v)} = -(\cos v\,\mathbf{n}(u) + \sin v\,\mathbf{b}(u)).$$

Remark: The surface X is called the *tube of radius r around* α.

20.4. (i) Show that the normals to a regular surface defined by

$$x = f(v)\cos u,$$
$$y = f(v)\sin u,$$
$$z = g(v),$$

all pass through the z-axis.

Remark: Such a surface is called a *surface of revolution.*

(ii) If S is a connected regular surface and all its normals meet the z axis, show that S has a parametrization as in (i).

20.5. Show that the first fundamental form of a plane and the first fundamental form of a cylinder of revolution defined by

$$X(u,v) = (\cos u, \sin u, v)$$

are both $(E,F,G) = (1,0,1)$.

20.6. Given a helicoid defined such that

$$x = u_1\cos v_1,$$
$$y = u_1\sin v_1,$$
$$z = v_1,$$

show that $(E,F,G) = (1,0,u_1^2+1)$.

20.7. Given a catenoid defined such that

$$x = \cosh u_2\cos v_2,$$
$$y = \cosh u_2\sin v_2,$$
$$z = u_2,$$

show that $(E,F,G) = \left(\cosh^2 u_2, 0, \cosh^2 u_2\right)$.

20.8. Recall that the Enneper surface is given by

$$x = u - \frac{u^3}{3} + uv^2$$
$$y = v - \frac{v^3}{3} + u^2 v$$
$$z = u^2 - v^2.$$

(i) Show that the first fundamental form is given by

$$E = G = (1 + u^2 + v^2)^2, \quad F = 0.$$

(ii) Show that the second fundamental form is given by

$$L = 2, \quad M = 0, \quad N = -2.$$

(iii) Show that the principal curvatures are

$$\kappa_1 = \frac{2}{(1+u^2+v^2)^2}, \quad \kappa_2 = -\frac{2}{(1+u^2+v^2)^2}.$$

(iv) Show that the lines of curvature are the coordinate curves.
(v) Show that the asymptotic curves are the curves of the form $u+v=C$, $u-v=C$, for some constant C.

20.9. Show that at a hyperbolic point, the principal directions bisect the asymptotic directions.

20.10. Given a pseudosphere defined such that

$$x = \frac{2\cos v}{e^u + e^{-u}},$$

$$y = \frac{2\sin v}{e^u + e^{-u}},$$

$$z = u - \frac{e^u - e^{-u}}{e^u + e^{-u}},$$

show that $K = -1$.

20.11. Prove that a general ellipsoid of equation

$$\frac{x^2}{a^2} + \frac{y^2}{b^2} + \frac{z^2}{c^2} = 1$$

(a, b, c pairwise distinct) has four umbilics.

20.12. Prove that the Gaussian curvature at a point (x, y, x) of an ellipsoid of equation

$$\frac{x^2}{a^2} + \frac{y^2}{b^2} + \frac{z^2}{c^2} = 1$$

has the expression

$$K = \frac{p^4}{a^2 b^2 c^2},$$

where p is the distance from the origin $(0,0,0)$ to the tangent plane at the point (x, y, z).

20.13. Show that the helicoid, the catenoid, and the Enneper surface are minimal surfaces, i.e., $H = 0$.

20.14. Consider two parabolas P_1 and P_2 in two orthogonal planes and such that each one passes through the focal point of the other. Given any two points $q_1 \in P_1$ and $q_2 \in P_2$, let H_{q_1,q_2} be the bisector plane of (q_1, q_2), i.e., the plane orthogonal to (q_1, q_2) and passing through the midpoint of (q_1, q_2). Prove that the envelope of the planes H_{q_1,q_2} when q_1 and q_2 vary on the parabolas P_1 and P_2 is the Enneper surface (i.e., the Enneper surface is the surface to which each H_{q_1,q_2} is tangential).

20.15. Show that if a curve on a surface S is both a line of curvature and a geodesic, then it is a planar curve.

20.16. Given a regular surface X, a *parallel surface to X* is a surface Y defined such that

$$Y(u,v) = X(u,v) + a\mathbf{N}_{(u,v)},$$

where $a \in \mathbb{R}$ is a given constant.

(i) Prove that

$$Y_u \times Y_v = (1 - 2Ha + Ka^2)(X_u \times X_v),$$

  where K is the Gaussian curvature of X and H is the mean curvature of X.
(ii) Prove that the Gaussian curvature of Y is

$$\frac{K}{1 - 2Ha + Ka^2}$$

  and the mean curvature of Y is

$$\frac{H - Ka}{1 - 2Ha + Ka^2},$$

  where K is the Gaussian curvature of X and H is the mean curvature of X.
(iii) Assume that X has constant mean curvature $c \neq 0$. If $K \neq 0$, prove that the parallel surface Y corresponding to $a = 1/(2c)$ has constant Gaussian curvature equal to $4c^2$. Prove that the parallel surface Y corresponding to $a = 1/(2c)$ is regular except at points where $K = 0$.
(iv) Again, assume that X has constant mean curvature $c \neq 0$ and is not contained in a sphere. Show that there is a unique value of a such that the parallel surface Y has constant mean curvature $-c$. Furthermore, this parallel surface is regular at (u,v) iff $X(u,v)$ is not an umbilical point, and the Gaussian curvature of Y at (u,v) has the opposite sign to that of X.

Remark: The above results are due to Ossian Bonnet.

20.17. Given a torus of revolution defined such that

$$x = (a + b\cos\varphi)\cos\theta,$$
$$y = (a + b\cos\varphi)\sin\theta,$$
$$z = b\sin\varphi,$$

prove that the Gaussian curvature is given by

$$K = \frac{\cos\varphi}{b(a+b\cos\varphi)}.$$

Show that the mean curvature is given by

$$H = \frac{a+2b\cos\varphi}{2b(a+b\cos\varphi)}.$$

20.18. (i) Given a surface of revolution defined such that

$$x = f(v)\cos u,$$
$$y = f(v)\sin u,$$
$$z = g(v),$$

show that the first fundamental form is given by

$$E = f(v)^2, \quad F = 0, \quad G = f'(v)^2 + g'(v)^2.$$

(ii) Show that the Christoffel symbols are given by

$$\Gamma^1_{11} = 0, \ \Gamma^2_{11} = -\frac{ff'}{(f')^2+(g')^2}, \ \Gamma^1_{12} = \frac{ff'}{f^2},$$
$$\Gamma^2_{12} = 0, \ \Gamma^1_{22} = 0, \qquad\qquad \Gamma^2_{22} = \frac{f'f''+g'g''}{(f')^2+(g')^2}.$$

(iii) Show that the equations of the geodesics are

$$u'' + \frac{2ff'}{f^2}u'v' = 0,$$
$$v'' - \frac{ff'}{(f')^2+(g')^2}(u')^2 + \frac{f'f''+g'g''}{(f')^2+(g')^2}(v')^2 = 0.$$

Show that the meridians parametrized by arc length are geodesics. Show that a
parallel is a geodesic iff it is generated by the rotation of a point of the generating
curve where the tangent is parallel to the axis of rotation.

(iv) Show that the first equation of geodesics is equivalent to

$$f^2 u' = c,$$

for some constant c. Since the angle θ, $0 \le \theta \le \pi/2$, of a geodesic with a parallel
that intersects it is given by

$$\cos\theta = \frac{|X_u \cdot (X_u u' + X_v v')|}{\|X_u\|} = |fu'|,$$

and since $f = r$ is the radius of the parallel at the intersection, show that

$$r\cos\theta = c$$

for some constant $c > 0$. The equation $r\cos\theta = c$ is known as *Clairaut's relation*.

20.19. (i) Given a surface of revolution defined such that

$$\begin{aligned}
x &= f(v)\cos u,\\
y &= f(v)\sin u,\\
z &= g(v),
\end{aligned}$$

show that the second fundamental form is given by

$$L = -fg', \quad L = 0, \quad M = g'f'' - g''f'.$$

Conclude that the parallels and the meridians are lines of curvature.

(ii) Recall from Problem 20.18 that the first fundamental form is given by

$$E = f(v)^2, \quad F = 0, \quad G = f'(v)^2 + g'(v)^2.$$

Show that the Gaussian curvature is given by

$$K = -\frac{g'(g'f'' - g''f')}{f}.$$

Show that the parabolic points are the points where the tangent to the generating curve is perpendicular to the axis of rotation, or the points of the generating curve where the curvature is zero.

If we assume that $G = 1$, which is the case if the generating curve is parametrized by arc length, show that

$$K = -\frac{f''}{f}.$$

(iii) Show that the principal curvatures are given by

$$\kappa_1 = \frac{L}{E} = \frac{-g'}{f}, \quad \kappa_2 = \frac{N}{G} = g'f'' - g''f'.$$

20.20. (i) Is it true that if a principal curve is a plane curve, then it is a geodesic?

(ii) Is it true that if a geodesic is a plane curve, then it is a principal curve?

20.21. If X is a surface with negative Gaussian curvature, show that the asymptotic curves have the property that the torsion at (u,v) is equal to $\pm\sqrt{-K_{(u,v)}}$.

20.22. *Show that if all the geodesics of a connected surface are planar curves, then this surface is contained in a plane or a sphere.*

20.23. A *ruled surface* $X(t,v)$ is defined by a pair $(\alpha(t), w(t))$, where $\alpha(t)$ is some regular curve and $w(t)$ is some nonnull C^1-continuous vector in $\mathbb{R}^3$, both defined over some open interval I, with

$$X(t,v) = \alpha(t) + vw(t).$$

In other words, X consists of the one-parameter family of lines $\langle \alpha(t), w(t) \rangle$, called *rulings*. Without loss of generality, we can assume that $\|w(t)\| = 1$. In this problem we will also assume that $w'(t) \neq 0$ for all $t \in I$, in which case we say that X is *noncylindrical*.

(i) Consider the ruled surface defined such that α is the unit circle in the xy-plane, and

$$w(t) = \alpha'(t) + e_3,$$

where $e_3 = (0,0,1)$. Show that X can be parametrized as

$$X(t,v) = (\cos t - v\sin t, \ \sin t + v\cos t, \ v).$$

Show that X is the quadric of equation

$$x^2 + y^2 - z^2 = 1.$$

What happens if we choose $w(t) = -\alpha'(t) + e_3$?

(ii) Prove that there is a curve $\beta(t)$ on X (called the *line of striction of X*) such that

$$\beta(t) = \alpha(t) + u(t)w(t) \quad \text{and} \quad \beta'(t) \cdot w'(t) = 0$$

for all $t \in I$, for some function $u(t)$.

Hint. Show that $u(t)$ is uniquely defined by

$$u = -\frac{\alpha' \cdot w'}{w' \cdot w'}.$$

Prove that β depends only on the surface X in the following sense: If α_1 and α_2 are two curves such that

$$\alpha_2(t) + vw(t) = \alpha_1(t) + \delta(v)w(t)$$

for all $t \in I$ and all $v \in \mathbb{R}$ for some C^3-function δ, and β_1, β_2 are the corresponding lines of striction, then $\beta_1 = \beta_2$.

(iii) Writing $X(t,v)$ as

$$X(t,v) = \beta(t) + vw(t),$$

show that there is some function $\lambda(t)$ such that $\beta' \times w = \lambda w'$ and

$$\|X_t \times X_v\|^2 = (\lambda^2 + v^2)\|w'\|^2.$$

Furthermore, show that

$$\lambda = \frac{(\beta', w, w')}{\|w'\|^2}.$$

Show that the singular points (if any) occur along the line of striction $v = 0$, and that they occur iff $\lambda(t) = 0$.

(iv) Show that

$$M = \frac{(\beta', w, w')}{\|X_t \times X_v\|}, \quad N = 0,$$

and

$$K = -\frac{\lambda^2}{(\lambda^2 + v^2)^2}.$$

Conclude that the Gaussian curvature of a ruled surface satisfies $K \leq 0$, and that $K = 0$ only along those rulings that meet the line of striction at a singular point.

20.24. As in Problem 20.23, let X be a *ruled surface* $X(t,v)$ where

$$X(t,v) = \alpha(t) + vw(t)$$

and with $\|w(t)\| = 1$. In this problem we will assume that

$$(w, w', \alpha') = 0,$$

and we call such a ruled surface *developable*.

(i) Show that

$$M = \frac{(\alpha', w, w')}{\|X_t \times X_v\|}, \quad N = 0.$$

Conclude that $M = 0$, and thus that $K = 0$.

(ii) If $w(t) \times w'(t) = 0$ for all $t \in I$, show that $w(t)$ is constant and that the surface is a cylinder over a plane curve obtained by intersecting the cylinder with a plane normal to w.

If $w(t) \times w'(t) \neq 0$ for all $t \in I$, then $w'(t) \neq 0$ for all $t \in I$. Using Problem 20.23, there is a line of striction β and a function $\lambda(t)$. Check that $\lambda = 0$. If $\beta'(t) \neq 0$ for all $t \in I$, then show that the ruled surface is the tangent surface of β. If $\beta'(t) = 0$ for all $t \in I$, then show that the ruled surface is a cone.

20.25. (i) Let $\alpha : I \to \mathbb{R}^3$ be a curve on a regular surface S, and consider the ruled surface X defined such that

$$X(u,v) = \alpha(u) + v\mathbf{N}_{(u(t),v(t))},$$

where $\mathbf{N}_{(u(t),v(t))}$ is the unit normal vector to S at $\alpha(t)$. Prove that α is a line of curvature on S iff X is developable.

(ii) Let X be a regular surface without parabolic, planar, or umbilical points. Consider the two surfaces Y and Z (called *focal surfaces of X*, or *caustics of X*) defined such that

$$Y(u,v) = X(u,v) + \frac{1}{\kappa_1} \mathbf{N}_{(u,v)},$$

$$Z(u,v) = X(u,v) + \frac{1}{\kappa_2} \mathbf{N}_{(u,v)},$$

where κ_1 and κ_2 are the principal curvatures at (u,v).

Prove that if $(\kappa_1)_u$ and $(\kappa_2)_v$ are nowhere zero, then Y and Z are regular surfaces.

(iii) Show that the focal surfaces Y and Z are generated by the lines of strictions of the developable surfaces generated by the normals to the lines of curvatures on X. This means that if we consider the two orthogonal families $\mathscr{F}_1$ and $\mathscr{F}_2$ of lines of curvatures on X, for any line of curvature C in $\mathscr{F}_1$ (or in $\mathscr{F}_2$), the line of striction of the developable surface generated by the normals to the points of C lies on Y (or Z), and when $C \in \mathscr{F}_1$ varies on X, the corresponding line of striction sweeps Y (or Z).

What are the positions of the focal surfaces with respect to X, depending on the sign of the Gaussian curvature K? Is it possible for Y and Z to be reduced to a single point? Is it possible for Y and Z to be reduced to a curve? If Y reduces to a curve, show that X is the envelope of a one-parameter family of spheres.

20.26. Given a nonplanar regular curve f in $\mathbb{E}^3$, the surface F generated by the tangents lines to f is called the *tangent surface* of f. The tangent surface F of f may be defined by the equation

$$F(t,v) = f(t) + v\mathbf{t},$$

where $\mathbf{t} = f'(t)$. Assume that f is biregular. An *involute* of f is a curve g contained in the tangent surface of f and such that g intersects orthogonally every tangent of f. Assuming that f is parametrized by arc length, this means that every involute g of f is defined by an equation of the form

$$g(s) = f(s) + v(s)\mathbf{t}(s),$$

where $v(s)$ is a C^1-function, $\mathbf{t} = f'(s)$, and where $g'(s) \cdot \mathbf{t}(s) = 0$.

(a) Prove that
$$g'(s) = \mathbf{t} + v(s)\kappa\mathbf{n} + v'(s)\mathbf{t},$$

where $\mathbf{n}$ is the principal normal vector to f at s. Prove that the equation

$$1 + v'(s) = 0$$

must hold. Conclude that
$$v(s) = C - s,$$

where C is some constant, and thus that every involute has an equation of the form

$$g(s) = f(s) + (C - s)\mathbf{t}(s).$$

Remark: There is a physical interpretation of involutes. If a thread lying on the curve is unwound so that the unwound portion of it is always held taut in the direction of the tangent to the curve while the rest of it lies on the curve, then every point of the thread generates an involute of the curve during this motion.

(b) Consider the twisted cubic defined by

$$f(t) = \left(t, \frac{t^2}{2}, \frac{t^3}{6} \right).$$

Prove that the element of arc length is

$$ds = \left(1 + \frac{t^2}{2} \right) dt.$$

Give the equation of any involute of the twisted cubic.

Extra Credit: Plot the twisted cubic in some suitable interval and some of its involutes.

20.27. Let $\Omega : X \to \mathbb{E}^3$ be a surface.

(a) Assume that every point of X is an umbilic. Prove that X is contained in a sphere.

Hint. If $\kappa_1 = \kappa_2 = \kappa$ for every $(u,v) \in \Omega$, then $d\mathbf{N}(w) = -\kappa w$ for all tangent vectors $w = X_u x + X_v y$, which implies that $\mathbf{N}_u = -\kappa X_u$ and $\mathbf{N}_v = -\kappa X_v$. Prove that κ does not depend on $(u,v) \in \Omega$, i.e., it is a nonnull constant. Then prove that $X + \mathbf{N}/\kappa$ is a constant vector.

(b) Assume that every point of X is a planar point. Prove that X is contained in a plane.

Hint. This time, $\kappa_1 = \kappa_2 = 0$ for every $(u,v) \in \Omega$. Prove that $\mathbf{N}$ does not depend on $(u,v) \in \Omega$, i.e., it is a constant vector, and compute $(X \cdot \mathbf{N})_u$ and $(X \cdot \mathbf{N})_v$.

References

1. Ralph Abraham and Jerrold E. Marsden. *Foundations of Mechanics*. Addison-Wesley, second edition, 1978.
2. Lars V. Ahlfors and Leo Sario. *Riemann Surfaces*. Princeton Math. Series, No. 2. Princeton University Press, 1960.
3. Richard H. Bartels, John C. Beatty, and Brian A. Barsky. *An Introduction to Splines for Use in Computer Graphics and Geometric Modelling*. Morgan Kaufmann, first edition, 1987.
4. Marcel Berger and Bernard Gostiaux. *Géométrie différentielle: variétés, courbes et surfaces*. Collection Mathématiques. Puf, second edition, 1992. English edition: Differential geometry, manifolds, curves, and surfaces, GTM No. 115, Springer-Verlag.
5. William M. Boothby. *An Introduction to Differentiable Manifolds and Riemannian Geometry*. Academic Press, second edition, 1986.
6. Yvonne Choquet-Bruhat, Cécile DeWitt-Morette, and Margaret Dillard-Bleick. *Analysis, Manifolds, and Physics, Part I: Basics*. North-Holland, first edition, 1982.
7. Gaston Darboux. *Leçons sur la théorie générale des surfaces, Troisième Partie*. Gauthier-Villars, first edition, 1894.
8. Gaston Darboux. *Leçons sur la théorie générale des surfaces, Quatrième Partie*. Gauthier-Villars, first edition, 1896.
9. Gaston Darboux. *Leçons sur la théorie générale des surfaces, Première Partie*. Gauthier-Villars, second edition, 1914.
10. Gaston Darboux. *Leçons sur la théorie générale des surfaces, Deuxième Partie*. Gauthier-Villars, second edition, 1915.

11. Jean Dieudonné. *Abrégé d'Histoire des Mathématiques, 1700–1900*. Hermann, first edition, 1986.

12. Manfredo P. do Carmo. *Differential Geometry of Curves and Surfaces*. Prentice-Hall, 1976.

13. Manfredo P. do Carmo. *Riemannian Geometry*. Birkhäuser, second edition, 1992.

14. Manfredo P. do Carmo. *Differential Forms and Applications*. Universitext. Springer-Verlag, first edition, 1994.

15. Gerald Farin. *NURB Curves and Surfaces, from Projective Geometry to Practical Use*. AK Peters, first edition, 1995.

16. Gerald Farin. *Curves and Surfaces for CAGD*. Academic Press, fourth edition, 1998.

17. William Fulton. *Algebraic Topology, A First Course*. GTM No. 153. Springer-Verlag, first edition, 1995.

18. Jean H. Gallier. *Curves and Surfaces in Geometric Modeling: Theory and Algorithms*. Morgan Kaufmann, first edition, 1999.

19. S. Gallot, D. Hulin, and J. Lafontaine. *Riemannian Geometry*. Universitext. Springer-Verlag, second edition, 1993.

20. R.V. Gamkrelidze (Ed.). *Geometry I*. Encyclopaedia of Mathematical Sciences, Vol. 28. Springer-Verlag, first edition, 1991.

21. Claude Godbillon. *Géométrie Différentielle et Mécanique Analytique*. Collection Méthodes. Hermann, first edition, 1969.

22. André Gramain. *Topologie des Surfaces*. Collection Sup. Puf, first edition, 1971.

23. A. Gray. *Modern Differential Geometry of Curves and Surfaces*. CRC Press, second edition, 1997.

24. Victor Guillemin and Alan Pollack. *Differential Topology*. Prentice-Hall, first edition, 1974.

25. D. Hilbert and S. Cohn-Vossen. *Geometry and the Imagination*. Chelsea Publishing Co., 1952.

26. Heinz Hopf. *Differential Geometry in the Large*. LNCS, Vol. 1000. Springer-Verlag, second edition, 1989.

27. J. Hoschek and D. Lasser. *Computer-Aided Geometric Design*. AK Peters, first edition, 1993.

28. Erwin Kreyszig. *Differential Geometry*. Dover, first edition, 1991.

29. Jacques Lafontaine. *Introduction aux Variétés Différentielles*. PUG, first edition, 1996.

30. Serge Lang. *Differential and Riemannian Manifolds*. GTM No. 160. Springer-Verlag, third edition, 1995.

31. Daniel Lehmann and Carlos Sacré. *Géométrie et Topologie des Surfaces*. Puf, first edition, 1982.

32. Charles Loop. A G^1 triangular spline surface of arbitrary topological type. *Computer-Aided Geometric Design*, 11:303–330, 1994.

33. Paul Malliavin. *Géométrie Différentielle Intrinsèque*. Enseignement des Sciences, No. 14. Hermann, first edition, 1972.

34. William S. Massey. *Algebraic Topology: An Introduction*. GTM No. 56. Springer-Verlag, second edition, 1987.

35. William S. Massey. *A Basic Course in Algebraic Topology*. GTM No. 127. Springer-Verlag, first edition, 1991.

36. John W. Milnor. *Topology from the Differentiable Viewpoint*. The University Press of Virginia, second edition, 1969.

37. John W. Milnor. *Morse Theory*. Annals of Math. Series, No. 51. Princeton University Press, third edition, 1969.

38. Henry P. Moreton. *Minimum curvature variation curves, networks, and surfaces for fair free-form shape design*. PhD thesis, University of California, Berkeley, 1993.

39. Les Piegl and Wayne Tiller. *The NURBS Book*. Monograph in Visual Communications. Springer-Verlag, first edition, 1995.

40. Richard W. Sharpe. *Differential Geometry. Cartan's Generalization of Klein's Erlangen Program*. GTM No. 166. Springer-Verlag, first edition, 1997.

41. S. Sternberg. *Lectures On Differential Geometry*. AMS Chelsea, second edition, 1983.

42. J.J. Stoker. *Differential Geometry*. Wiley Classics. Wiley-Interscience, first edition, 1989.

43. Frank Warner. *Foundations of Differentiable Manifolds and Lie Groups*. GTM No. 94. Springer-Verlag, first edition, 1983.

44. William Welch. *Serious Putty: Topological Design for Variational Curves and Surfaces*. PhD thesis, Carnegie Mellon University, Pittsburgh, Pa., 1995.

Chapter 21
Appendix

21.1 Hyperplanes and Linear Forms

This appendix covers two topics. First, we prove that hyperplanes are precisely the kernels of nonzero linear forms. Second, we review the definitions of metric spaces and normed vector spaces.

Given a vector space E over a field K, a linear map $f : E \to K$ is called a *linear form*. The set of all linear forms $f : E \to K$ is a vector space called the *dual space of E* and denoted by E^*. We now prove that hyperplanes are precisely the Kernels of nonzero linear forms.

Lemma 21.1. *Let E be a vector space. The following properties hold:*

(a) Given any nonzero linear form $f \in E^*$, its kernel $H = \operatorname{Ker} f$ is a hyperplane.

(b) For any hyperplane H in E, there is a (nonzero) linear form $f \in E^*$ such that $H = \operatorname{Ker} f$.

(c) Given any hyperplane H in E and any (nonzero) linear form $f \in E^*$ such that $H = \operatorname{Ker} f$, for every linear form $g \in E^*$, $H = \operatorname{Ker} g$ iff $g = \lambda f$ for some $\lambda \neq 0$ in K.

Proof. (a) If $f \in E^*$ is nonzero, there is some vector $v_0 \in E$ such that $f(v_0) \neq 0$. Let $H = \operatorname{Ker} f$. For every $v \in E$, we have

$$f\left(v - \frac{f(v)}{f(v_0)} v_0\right) = f(v) - \frac{f(v)}{f(v_0)} f(v_0) = f(v) - f(v) = 0.$$

Thus,

$$v - \frac{f(v)}{f(v_0)} v_0 = h \in H$$

and $v = h + (f(v)/f(v_0))v_0$, that is, $E = H + Kv_0$. Also, since $f(v_0) \neq 0$, we have $v_0 \notin H$, that is, $H \cap Kv_0 = 0$. Thus, $E = H \oplus Kv_0$, and H is a hyperplane.

(b) If H is a hyperplane, $E = H \oplus Kv_0$ for some $v_0 \notin H$. Then every $v \in E$ can be written in a unique way as $v = h + \lambda v_0$. Thus there is a well–defined function

$f\colon E \to K$ such that $f(v) = \lambda$ for every $v = h + \lambda v_0$. We leave as a simple exercise the verification that f is a linear form. Since $f(v_0) = 1$, the linear form f is nonzero. Also, by definition it is clear that $\lambda = 0$ iff $v \in H$, that is, $\mathrm{Ker}\, f = H$.

(c) Let H be a hyperplane in E, and let $f \in E^*$ be any (nonzero) linear form such that $H = \mathrm{Ker}\, f$. Clearly, if $g = \lambda f$ for some $\lambda \neq 0$, then $H = \mathrm{Ker}\, g$. Conversely, assume that $H = \mathrm{Ker}\, g$ for some nonzero linear form g. From (a) we have $E = H \oplus K v_0$, for some v_0 such that $f(v_0) \neq 0$ and $g(v_0) \neq 0$. Then observe that

$$g - \frac{g(v_0)}{f(v_0)}\, f$$

is a linear form that vanishes on H, since both f and g vanish on H, but also vanishes on $K v_0$. Thus, $g = \lambda f$, with $\lambda = g(v_0)/f(v_0)$. $\square$

If E is a vector space of finite dimension n and $(u_1, \ldots, u_n)$ is a basis of E, for any linear form $f \in E^*$ and every $x = x_1 u_1 + \cdots + x_n u_n \in E$, we have

$$f(x) = \lambda_1 x_1 + \cdots + \lambda_n x_n,$$

where $\lambda_i = f(u_i) \in K$, for every i, $1 \leq i \leq n$. Thus, with respect to the basis $(u_1, \ldots, u_n)$, $f(x)$ is a linear combination of the coordinates of x, as expected.

21.2 Metric Spaces and Normed Vector Spaces

Thorough expositions of the material of this section can be found in Lang [2, 3] and Dixmier [1]. We begin with metric spaces. Recall that $\mathbb{R}_+ = \{x \in \mathbb{R} \mid x \geq 0\}$.

Definition 21.1. A *metric space* is a set E together with a function $d\colon E \times E \to \mathbb{R}_+$, called a *metric, or distance*, assigning a nonnegative real number $d(x, y)$ to any two points $x, y \in E$ and satisfying the following conditions for all $x, y, z \in E$:

(D1) $d(x, y) = d(y, x)$.   (symmetry)
(D2) $d(x, y) \geq 0$, and $d(x, y) = 0$ iff $x = y$.   (positivity)
(D3) $d(x, z) \leq d(x, y) + d(y, z)$.   (triangle inequality)

Geometrically, condition (D3) expresses the fact that in a triangle with vertices x, y, z, the length of any side is bounded by the sum of the lengths of the other two sides. From (D3), we immediately get

$$|d(x, y) - d(y, z)| \leq d(x, z).$$

Let us give some examples of metric spaces. Recall that the *absolute value* $|x|$ of a real number $x \in \mathbb{R}$ is defined such that $|x| = x$ if $x \geq 0$, $|x| = -x$ if $x < 0$, and for a complex number $x = a + ib$, as $|x| = \sqrt{a^2 + b^2}$.

Example 21.1. Let $E = \mathbb{R}$ and $d(x, y) = |x - y|$, the absolute value of $x - y$. This is the so-called natural metric on $\mathbb{R}$.

Example 21.2. Let $E = \mathbb{R}^n$ (or $E = \mathbb{C}^n$). We have the Euclidean metric

$$d_2(x, y) = \left(|x_1 - y_1|^2 + \cdots + |x_n - y_n|^2\right)^{\frac{1}{2}},$$

the distance between the points $(x_1, \ldots, x_n)$ and $(y_1, \ldots, y_n)$.

Example 21.3. For every set E we can define the *discrete metric*, defined such that $d(x, y) = 1$ iff $x \neq y$, and $d(x, x) = 0$.

Example 21.4. For any $a, b \in \mathbb{R}$ such that $a < b$, we define the following sets:

1. $[a,b] = \{x \in \mathbb{R} \mid a \leq x \leq b\}$,   (closed interval)
2. $[a,b[= \{x \in \mathbb{R} \mid a \leq x < b\}$,   (interval closed on the left, open on the right)
3. $]a,b] = \{x \in \mathbb{R} \mid a < x \leq b\}$,   (interval open on the left, closed on the right)
4. $]a,b[= \{x \in \mathbb{R} \mid a < x < b\}$,   (open interval)

Let $E = [a,b]$, and $d(x, y) = |x - y|$. Then $([a,b], d)$ is a metric space.

We now consider a very important special case of metric spaces: Normed vector spaces.

Definition 21.2. Let E be a vector space over a field K, where K is either the field $\mathbb{R}$ of reals or the field $\mathbb{C}$ of complex numbers. A *norm on E* is a function $\| \ \| : E \to \mathbb{R}_+$ assigning a nonnegative real number $\|u\|$ to any vector $u \in E$ and satisfying the following conditions for all $x, y, z \in E$:

(N1) $\|x\| \geq 0$, and $\|x\| = 0$ iff $x = 0$.   (positivity)
(N2) $\|\lambda x\| = |\lambda| \, \|x\|$.   (scaling)
(N3) $\|x + y\| \leq \|x\| + \|y\|$.   (convexity inequality)

A vector space E together with a norm $\| \ \|$ is called a *normed vector space*.

Condition (N3) is also called the *triangle inequality*, and it is illustrated in Figure 21.1. From (N3), we easily get

$$\big| \|x\| - \|y\| \big| \leq \|x - y\|.$$

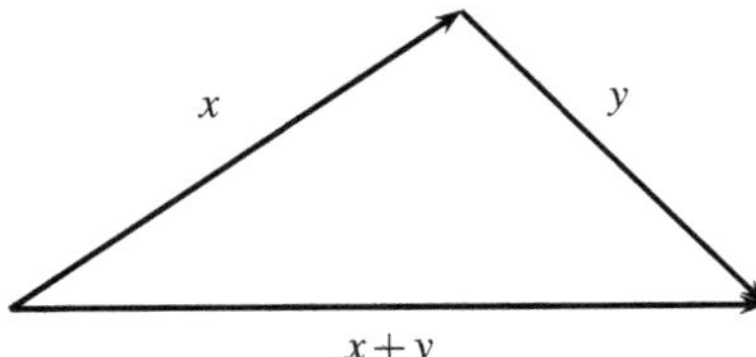

Fig. 21.1 The triangle inequality.

Given a normed vector space E, if we define d such that

$$d(x, y) = \|x - y\|,$$

it is easily seen that d is a metric. Thus, every normed vector space is immediately a metric space. Note that the metric associated with a norm is invariant under translation, that is,

$$d(x + u, y + u) = d(x, y).$$

Let us give some examples of normed vector spaces.

Example 21.5. Let $E = \mathbb{R}$ and $\|x\| = |x|$, the absolute value of x. The associated metric is $|x - y|$, as in Example 21.1.

Example 21.6. Let $E = \mathbb{R}^n$ (or $E = \mathbb{C}^n$). There are three standard norms. For every $(x_1, \ldots, x_n) \in E$, we have the norm $\|x\|_1$, defined such that

$$\|x\|_1 = |x_1| + \cdots + |x_n|,$$

we have the Euclidean norm $\|x\|_2$, defined such that

$$\|x\|_2 = \left(|x_1|^2 + \cdots + |x_n|^2\right)^{\frac{1}{2}},$$

and the *sup*-norm $\|x\|_\infty$, defined such that

$$\|x\|_\infty = \max\{|x_i| \mid 1 \leq i \leq n\}.$$

For geometric applications, we will need to consider affine spaces (E, E) where the associated space of translations E is a vector space equipped with a norm.

Definition 21.3. Given an affine space (E, E), where the space of translations E is a vector space over $\mathbb{R}$ or $\mathbb{C}$, we say that (E, E) is a *normed affine space* if E is a normed vector space with norm $\|\ \|$.

Given a normed affine space, there is a natural metric on E itself, defined such that

$$d(a, b) = \|\overrightarrow{ab}\|.$$

Observe that this metric is invariant under translation, that is,

$$d(a + u, b + u) = d(a, b).$$

References

1. Jacques Dixmier. *General Topology*. UTM. Springer-Verlag, first edition, 1984.
2. Serge Lang. *Real and Functional Analysis*. GTM 142. Springer-Verlag, third edition, 1996.
3. Serge Lang. *Undergraduate Analysis*. UTM. Springer-Verlag, second edition, 1997.

Symbol Index

Index